AIR-SEA INTERACTION
Instruments and Methods

AIR-SEA INTERACTION
Instruments and Methods

Edited by

F. Dobson
Bedford Institute of Oceanography
Dartmouth, Nova Scotia, Canada

L. Hasse
Meteorologisches Institut
Universität Hamburg
Hamburg, German Federal Republic

and

R. Davis
Scripps Institution of Oceanography
La Jolla, California

PLENUM PRESS · NEW YORK AND LONDON

Library of Congress Cataloging in Publication Data

Main entry under title:

Air-sea interaction.

Includes bibliographical references and index.
1. Ocean-atmosphere interaction–Instruments. 2. Ocean-atmosphere interaction–Methodology. I. Dobson, F. II. Hasse, L. III. Davis, Russ.
GC190.5.A37 551.47 80-17895
ISBN: 978-1-4615-9184-9 e-ISBN-13: 978-1-4615-9182-5
DOI: 10.1007/978-1-4615-9182-5

Softcover reprint of the hardcover 1st edition 1980
A Division of Plenum Publishing Corporation
227 West 17th Street, New York, N.Y. 10011

Preface

During the past decade, man's centuries-old interest in marine meteorology and oceanography has broadened. Ocean and atmosphere are now treated as coupled parts of one system; the resulting interest in air-sea interaction problems has led to a rapid growth in the sophistication of instruments and measurement techniques. This book has been designed as a reference text which describes, along with the instruments themselves, the accumulated practical experience of experts engaged in field observations of air-sea interactions. It is meant to supplement rather than replace manuals on standard routine observations or instrumentation handbooks.

At the inception a textbook was planned, which would contain only well tested methods and instruments. It was quickly discovered that for the book to be useful many devices and techniques would have to be included which are still evolving rapidly. The reader is therefore cautioned to take nothing in these pages for granted. Certainly, every contributor is an expert, but while some are backed up by generations of published work, others are pioneers. The choice of topics, of course, is debatable. The types of observations included are not exhaustive and topics such as marine aerosols and radio-tracers are omitted, as was the general subject of remote sensing, which was felt to be too broad and evolving too rapidly. The guideline adopted in limiting size was maximum usefulness to 'a trained experimentalist new to the field'.

Each contribution has been subjected to peer review, and, following revision, to open-forum discussion during a two-week meeting held at Ustaoset, Norway, in May of 1978. The chapters are arranged by parameter to be measured (with an additional section on platforms) so, for instance, articles on temperature measurements in air and water are to be found side by side. The wise reader will browse.

Bibliographies have been a problem. Many authors have been forced to refer to 'grey' literature such as unpublished technical reports, since no other source of technical information was available. We have done our best to identify sources for all references and the authors themselves have agreed to supply copies of the more obscure ones, so if you can't locate something, write them!

We have tried to goad all the contributors into admitting faults and pointing out pitfalls, hoping thus to engender a healthy disrespect for the optimistic pronouncements of land-based instrument designers, and a healthy respect for the elements.

The Editors

Acknowledgments

We gratefully acknowledge the support - moral and monetary - of the NATO Science Committee Special Programme Panel on Air-Sea Interaction and the U.S. Office of Naval Research. Dr. J.A. Strømme of the Christian Michelsen Institute in Bergen organized a memorable contributors' meeting at Ustaoset. Dr. S.D. Smith gave freely of his time to help the manuscript through the final production phase. To all the people who worked overtime for us when they didn't really have to, we offer our sincere thanks.

Contents

VI. THE SEA SURFACE

VII. RADIATION

VIII. PRECIPITATION

IX. THE ATMOSPHERIC BOUNDARY LAYER

X. PLATFORMS

Introduction

The marine environment places many obstacles in the path of the would-be experimentalist; this has led to the adoption of specialized instruments and methods which, in the absence of experience, may seem unsophisticated or overly conservative. In the first place, instrument design is influenced by environmental constraints. The motion of the ocean surface imposes mechanical problems for sensors (inundation of 'air' sensors cannot always be excluded) and requires supporting structures which have their own effects on the measurements. Salt water and spray are corrosive, hygroscopic, and electrically conducting; protective measures are required which further degrade performance.

In the second place, the very nature of the work influences the methods employed. Using a ship is extremely costly, so experiments cannot readily be repeated and techniques limited to calm sea states are inappropriate. Unattended operation is necessary if long records are required and all instruments must be rugged and reliable. If the data are stored internally, a high priority must be placed on the survival of the recording package. In attended operation some method of visualizing the data as it is being stored is invaluable as it provides an opportunity to monitor instrument performance and check that the results are safely recorded. In addition to the essential pre- and post-calibrations, in situ calibration is highly desirable. Similarly, sources of noise and calibration instability must be documented both in the laboratory and under field conditions.

The investigation of air-sea interaction processes demands the recording of various types of independent environmental information, such as, routine hourly meteorological and oceanographic data, cloud photographs, and incoming solar radiation. Such independent data are useful in the field for experimental planning and are indispensable for the subsequent analysis, both as a description of the environment and as a source of information for debugging the data.

Because this book concerns observational methods, little is included about theory or dynamics. For the reader new to the field, a brief explanation of some terms and concepts is included here; more complete treatments are to be found, for example, in Lumley and Panofsky (1964), Roll (1965), Kraus (1972), Kitaigorodsky (1973), Phillips (1977), Hinze (1959), and Tennekes and Lumley (1972).

The structure of the air above the sea surface is commonly described as consisting of a 'Planetary' or 'Ekman' boundary layer a few hundreds of metres thick, in which horizontal pressure gradients are balanced by Coriolis and friction forces and in which buoyancy forces play an important role in the dynamics; a 'constant flux' or 'friction' or 'Prandtl' layer a few tens of metres thick, in which the flow is dominated by the presence of the boundary; and a 'diffusive sublayer' a few millimetres thick next to the surface itself, where molecular transport and viscous forces dominate (a similar 'molecular' layer exists in the water which limits the rate at which most gases, except the highly soluble ones, are transferred between air and sea). The effective sublayer thickness varies with the molecular diffusivity of the transferred quantity: gas, heat, momentum, etc.

Meteorological air-sea interaction studies have concentrated mainly on vertical fluxes of momentum, heat, water vapour and gases. While in principle fluxes through the interface are required, in most cases, for practical reasons, the fluxes are measured in the 'friction' layer at a few metres height. This practice is based on the assumption that the divergence of the fluxes between the interface and the instrument level is negligible - a reasonable assumption except for weak sensible heat flux, where radiational divergence may become important. It is customary to divide the fluxes into transports by the 'mean' state and by turbulent fluctuations; this assumes that a meaningful separation of scales can be made. In the atmospheric boundary layer such a separation is useful but in the oceanic mixed layer, dominated by surface and internal waves, it is less clear that this is a useful notion for the experimentalist. A few of the methods of determining fluxes are summarized at the end of this introduction; for more complete descriptions see Pond (1972) and Kraus (1977, Ch. 6).

The principal difference between the marine and terrestrial boundary layers is the fact that the ocean has a free surface. This leads to the generation of surface waves, which interact with and modify the air and water boundary layers in ways that are still far from understood (see Favre and Hasselmann, 1978). Whereas in the surface layer over land the dominant scales of the problem are height above the surface and roughness length, over water another length scale, the wavelength λ, appears; but, perhaps of even greater importance, there is an additional velocity scale - the wave speed c. These last two scales are related by the 'dispersion relation' $c^2 = g/k \tanh [kD]$ where $k = 2\pi/\lambda$ is the wavenumber and D the water depth. The presence of the waves causes wave-coherent flows in the air and the water. In the air, which has a strong vertical shear in horizontal speed, the wave-coherent flows are very difficult to observe except very close to the interface. They are strongly dependent on the relative direction of the wind and wave vectors and on the ratio of wind speed to wave phase speed.

Pressure fluctuations above waves are of direct interest in the study of wave generation by the wind. Unfortunately, pressure fluctuations caused by upwind-travelling wave components are much larger than those produced by growing waves, whose velocities approximate that of the wind. Since upwind-travelling wave components are ubiquitous in the ocean, great care must be taken to account for their presence when making observations of wave-induced pressures. In fact, the only reliable way is to make the observations with horizontal arrays of wave and pressure sensors, which can distinguish between upwind- and downwind-travelling wave components.

The mean state of the upper ocean may frequently be described as a strongly density-stratified layer (the seasonal pycnocline with a buoyancy frequency of several cycles per hour) which separates a reasonably isopycnal layer above (the 'mixed' layer) from the less strongly stratified permanent pycnocline below. To some extent, the mixed layer can be regarded as a mirror image of the atmospheric boundary layer and divided into a diffusive sublayer blending into a fully turbulent layer which is bounded below by a stratified entrainment layer. Observations show, however, that the two boundary layers have quite different structures and this suggests they may have fundamentally different dynamics (see Phillips, 1977).

The high static stability of the air-sea interface and the ocean's density stratification support, respectively, surface and internal waves (e.g. Kinsman, 1965; LeBlond and Mysak, 1978). In the ocean, waves dominate the velocity spectrum. Near the surface, rms velocities of 0.5 to 2 m s^{-1} are typical, associated with surface waves having frequencies from 0.3 to 0.05 Hz. Internal wave frequencies are bounded by the inertial frequency (the sine of the latitude divided by 12 hours) and the maximum buoyancy frequency, usually found in the seasonal pycnocline. In mid-latitudes the velocity spectrum in this frequency range is dominated by near-inertial frequency motions, particularly in the mixed layer; typical rms velocities are 0.05 to 0.3 m s^{-1}, generally greatest in thin mixed layers. The pycnocline displacement spectrum has a maximum somewhat above the inertial frequency and also at tidal frequencies; a few metres rms displacement is typical in the upper layers, decreasing to zero at the surface. Although internal waves require stratification somewhere in the water column to exist, motions at internal wave frequencies appear to dominate the mixed layer spectrum from the inertial frequency to several cycles per hour.

The domination of the velocity spectrum by wave motion seriously impedes the observation of 'turbulent' fluxes. The downwards momentum fluxes on both sides of the air-water interface are the same, but because of the ratio of air and water densities the oceanic turbulence velocities are weak with friction velocities of the order the wind speed divided by 1000. Thus determination of fluxes requires velocity resolution to a few millimetres per second in the presence of surface and internal wave signals of order 1 m s^{-1}. It may also be speculated that in the presence of vigorous wave motions the oceanic transport processes differ significantly from those in the atmosphere. Observational evidence (Pollard and Millard, 1970) of downward transport of energy by inertial frequency motions supports this conjecture.

Mixing in the oceanic upper layer is caused by mechanically-induced turbulent stirring, convection (buoyancy-driven processes) and molecular diffusion of heat and salt. All three processes may be active at a given point in time and space but the resulting structures and the most useful investigative tools may differ according to which process dominates [see Kraus (1977) for a review]. Water may be mechanically mixed horizontally or vertically by large scale horizontal eddies or inertial motions, medium scale oceanic 'fronts' or horizontal vortical motions aligned with the wind ('Langmuir' cells), and small scale turbulence caused by shear-flow instabilities or the breaking of surface or internal waves. Convection results from surface cooling by evaporation, radiation, and direct heat exchange. Molecular diffusion works most effectively with large gradients over small distances. It is important everywhere but particularly at the boundaries between layers, where the different diffusion rates of heat and salt produce a variety of layering effects (Turner, 1973).

The strong stratification in the main and seasonal pycnoclines distinctly affects the mixing processes, introducing high anisotropy and 'intrusions'. The structure of temperature and salinity variability changes markedly from regions of active mixing to regions where property variation results from lateral spreading following some past mixing event. The old patches are highly anisotropic with temperature and salinity highly correlated; this suggests that they are intrusions of anomalous properties moving laterally along the level of their own density. This is particularly marked at the base of the mixed layer and precludes estimating static stability from temperature alone. The temporal and spatial changes in degree and nature of the anisotropy of the patches greatly complicates the interpretation of observations made along lines rather than over volumes (see Garrett and Munk, 1975).

Much of the challenge of upper ocean observation of air-sea interaction processes results from the small size of the signals of interest compared with wave motion, advection of gradients past the sensor, or instrument noise levels. Improvements must be made in sensor linearity and noise level, and the platforms which support the instruments must be understood and matched to sensor capabilities. Platform motion and flow interference must be made predictable and as far as possible decoupled from the signal; otherwise noise will result, including harmonics generated by nonlinearities in the system. The fact that in a typical wave field the difference between mean velocity measurements made with Eulerian (fixed) and Lagrangian (drifting) sensors is comparable with the expected mean current indicates that platform motion effects may cause the greater difficulty. Yet only after such problems are overcome can real progress be made in understanding the dynamics of the upper ocean.

FLUX MEASUREMENT TECHNIQUES

We append here a summary of widely used flux measurement techniques. We only introduce the terms, leaving precise definitions to the references; instead we concentrate on the practical questions of how to make useful, reliable flux measurements.

Eddy Correlation Method

The upwards turbulent flux of a specific property s is $F_s = \overline{\rho s'w'}$, where w is the vertical component of flow, and the overbar denotes averaging so $s' = s - \bar{s}$ and $w' = w - \bar{w}$ are fluctuations. For most purposes the density, ρ, can be taken as constant and removed from the average.

Observation of s' and w' is the only way of making a direct measurement of turbulent fluxes. The method involves no assumptions aside from the homogeneous steady state condition in which the fluctuations can be separated from the mean flow, and the existence of a constant flux layer between the surface and the measuring heights. As a guideline, contributions to the flux may be expected at frequencies $0.005 < f < 10$, where $f = nz/u$ is a dimensionless frequency, obtained from the natural frequency n, height z, and mean wind speed u. Since the fluxes are obtained from second-order moments, a sufficiently large sample (30-60 minutes) is required to ensure stable statistics.

Accurate instrument levelling is desired, especially in the case of momentum flux measurements, to avoid contamination of the vertical wind component by the relatively large horizontal wind, and the resulting spurious high correlation (typically the rms vertical wind fluctuation $\sigma_w \simeq 0.05\,\bar{u}$). Usually any remaining bias is corrected in the course of the data reduction by assuming $\bar{w} = 0$. From the continuity equation with the assumption of stationarity, horizontal homogeneity and no mass flux through the interface, $\overline{\rho w} = 0$, making $\bar{w} = 0$ a reasonable assumption in most cases provided that a clear separation can be made between 'mean' and 'fluctuating' quantities. It is necessary to eliminate any drift or linear trend from the records of s and w, since these would also give spuriously high correlations $\overline{s'w'}$. At sea the use of the method is much more difficult, since platform movements can introduce highly correlated errors in the wind components and flow distortion by towers, buoys, and ships can produce a nonzero $\overline{w}$.

The Profile Method

The 'profile' is the vertical distribution of a mean property near the boundary. The turbulent fluxes are almost always down the gradient and functionally related to it. A simple flux-gradient relationship is that for the time-averaged wind speed $\overline{u}(z)$ in conditions of neutral stability (zero vertical density gradient),

$$\overline{u}(z) = \frac{u_*}{\kappa} \ln \left(\frac{z}{z_0}\right)$$

where $u_* = (\tau/\rho)^{\frac{1}{2}}$

and τ is the vertical flux of horizontal momentum. u_* is called the 'friction velocity', κ is the von Karman constant $\simeq 0.4$, and the 'roughness length' z_0 is related to the size of the roughness which typifies the surface underlying the boundary layer. This apparently simple method is in fact indirect and limited in its reliability by the empirical relations required. Further, it is difficult to achieve the necessary accuracy . The largest changes in velocity with height occur very close to the sea surface, but measurements are usually not possible below the crests of the largest waves; in addition, the waves have a significant influence on the profile at low levels. At higher levels typical velocity differences are smaller and measurement errors more important: the difference of wind speed between, say, 3 and 10 m height, is of the order of 10% of the total difference between the 10 m level wind and the surface current. This makes questions of good sensor exposure and calibration stability more important than is commonly perceived.

The flux-gradient relationship is a function of density stratification. Only in conditions of neutral stability are the profiles logarithmic. Dyer (1974) reviews a number of empirical stability-dependent profile formulae. The change of the profile slope is much more pronounced than the change in curvature, which is difficult to detect in the usual height range of profile measurements. It is therefore a popular, but inexcusable error to take 'logarithmic-looking' profiles as being near-neutral. In every air-sea interaction experiment, stability must be measured independently or estimated from the mean wind speed and the difference of temperature and humidity between air and water.

Bulk Aerodynamic 'Method'

A bulk transfer coefficient, C_s, may be defined so that the flux, F_s, is

$$F_s = \rho C_s \bar{u} \overline{\Delta s}$$

where $\bar{u}$ is mean wind speed, $\overline{\Delta s} = \bar{s}_a - \bar{s}_w$ the mean difference of the property s between air and water, and C_s^w is a bulk transfer coefficient. Usually, transfer coefficients are referred to a 'standard' height or converted to that height with the aid of the flux-gradient relationship or the logarithmic profile. The formula can be conceived as an integration of the flux-gradient relationship from the roughness height z_0 to the reference height. The stability dependence of the transfer coefficients is smaller than that of the gradients. While $\bar{s}_w$ is defined as the value of the property s at the surface, it is usually obtained by sampling the water at some distance of the order of a metre below the surface. Thus, rather complicated transfer processes are lumped together in the transfer coefficient C_s. The bulk aerodynamic 'method' therefore is a method of parameterization rather than of measurement, but there is no doubt that it is useful for large scale applications and for interpretation of more refined measurements.

Dissipation Method and Others

It is found that over a range of conditions the conservation equations for second order moments (like $\overline{u'w'}$) are dominated by a balance between 'production' terms, which involve fluxes and mean gradients explicitly, and a 'dissipation' term. One or two terms in the 'dissipation' are measured directly and the rest computed assuming small-scale isotropy. Coupled with observation of the mean gradient or assumption of a profile form this allows determination of the fluxes. In a variant of this method, more properly called the 'inertial' technique, the spectral density is measured in the inertial subrange, and from the theory of the inertial subrange [see, for example, Tennekes and Lumley (1972)] an estimate of the dissipation is obtained. Although well suited for use on ships or buoys, these methods are open to criticism because the underlying assumptions are often not fulfilled. A check on the validity of these assumptions requires much greater experimental effort.

Aside from the aforementioned techniques, a variety of other, less direct, methods for the determination of fluxes are in use, each of which poses different demands on instrumentation and requires assumptions which must be examined for relevance in any particular application. For example, surface fluxes can be obtained by budget calculation of a closed volume which requires only observation of mean quantities. This is an especially useful technique in wind tunnel work, while in larger scale atmospheric applications it is a method of analysis rather than of measurement. The 'ageostrophic' method is also in use; see Roll (1965), Chapter 4. Various measurements of fluxes, expressed as bulk aerodynamic coefficients, are summarized by Garratt (1977) for momentum, and Friehe and Schmitt (1976) for heat and water vapour.

Stability parameters

The fluxes at the interface are dependent on the energy of the turbulence in the surface layer. It is therefore necessary to take the effects of stability into account. As a measure of static stability in the surface layer above the water, the 'Monin-Obukhov length' L is used, which together with the height of observation z gives a dimensionless stability parameter

$$\frac{z}{L} = -\frac{g}{T}\frac{\frac{H_v}{c_p \rho}}{u_*^3}\kappa z$$

where g is gravitational acceleration, T is temperature, H_v is the virtual sensible heat flux $\rho c_p \overline{\theta'_v w'}$ (θ_v is defined below), u_* the friction velocity, and c_p is the specific heat of air at constant pressure. Since the density stratification is the numerator in all the stability parameters, the effect of humidity on density must be considered. The effect is most apparent in near-neutral conditions, and is normally accounted for by correcting the heat fluxes and temperature gradient with terms proportional to the latent heat flux and the buoyant effect of moisture in the air.

The Monin-Obukhov length is formed from the turbulent fluxes u_*^2 and H_v. Since these fluxes in general are unknown or are determined experimentally, the Richardson number Ri may be used instead, if the profiles are measured:

$$Ri = +\frac{g}{T}\frac{\partial\overline{\theta}_v/\partial z}{(\partial\overline{u}/\partial z)^2}$$

where θ_v is virtual potential temperature (the temperature of dry air of equivalent density). It is even more advantageous to use

$$Ri_{bulk} = +\frac{g}{T}\frac{\Delta\overline{\theta}_v}{(\overline{u})^2}$$

where $\Delta\overline{\theta}_v$ is the mean air-sea virtual temperature difference, since the parameters $\overline{u}$, $\Delta\overline{\theta}_v$ can be accurately obtained from simple measurements. The bulk transfer coefficients which are used to convert Ri_{bulk} to z/L are less stability-dependent than the flux-gradient relationship used to convert Ri to z/L.

REFERENCES

DYER, A.J. 1974. A review of flux-profile relationships. *Boundary-Layer Meteorology,* 7: 363-372.

FAVRE, A. and K. HASSELMANN. 1978. *Turbulent fluxes through the sea surface, wave dynamics, and prediction.* Plenum Press, New York and London, 677 pp.

FRIEHE, C.A. and K.F. SCHMITT. 1976. Parameterization of air-sea interface fluxes of sensible heat and moisture by the bulk aerodynamic method. *Journal of Physical Oceanography,* 6: 801-809.

GARRATT, J.R. 1977. Review of drag coefficients over oceans and continents. *Monthly Weather Review,* 105: 915-929.

GARRETT, C.J.R. and W.H. MUNK. 1975. Space-time scales of internal waves: a progress report. *Journal of Geophsyical Research,* 80: 291-297.

HINZE, J.O. 1959. *Turbulence.* McGraw Hill, New York, Toronto and London, 586 pp.

HAUGEN, D.A. (editor). 1973. *Workshop on Micrometeorology.* American Meteorological Society of Boston, 392 pp.

JENKINS, G.M. and D.G. WATTS. 1968. *Spectral Analysis and Its Applications.* Holden-Day, San Francisco, 525 pp.

KINSMAN, B. 1965. *Wind Waves.* Prentice-Hall Inc., Englewood Cliffs, N.J., U.S.A., 676 pp.

KITAIGORODSKY, S.A. 1973. The physics of air-sea interaction. (English ed.) U.S. Department of Commerce, National Technical Information Service, Springfield, VA 22151, 237 pp.

KOOPMANS, L.H. 1974. *The Spectral Analysis of Time Series.* Academic Press, New York and London, 366 pp.

KRAUS, E.B. 1972. *Atmosphere-Ocean Interaction.* Clarendon Press, Oxford, 275 pp.

LE BLOND, P.H. and L.A. MYSAK. 1978. *Waves in the ocean.* Elsevier Scientific Publishing Co., Amsterdam, Oxford and New York, 602 pp.

LUMLEY, J.L. and H.A. PANOFSKY. 1964. *The Structure of Atmospheric Turbulence.* Interscience Publishers, New York, 230 pp.

PHILLIPS, O.M. 1977. *The Dynamics of the Upper Ocean.* 2nd ed. Cambridge University Press, 336 pp.

POLLARD, R.T. and R.C. MILLARD, JR. 1970. Comparison between observed and simulated wind-generated inertial oscillations. *Deep-Sea Research,* 17: 813-821.

POND, S. 1975. The exchanges of momentum, heat and moisture at the ocean-atmosphere interface. In: *Numerical models of ocean circulation, Proceedings of the Symposium.* National Academy of Sciences, Washington, D.C., U.S.A.: 26-38.

ROLL, H.U. 1965. *Physics of the Marine Atmosphere.* Academic Press, New York, 426 pp.

TENNEKES, H. and J.L. LUMLEY. 1972. *A First Course in Turbulence.* MIT Press, Cambridge, MA, 300 pp.

TURNER, J.S. 1973. *Buoyancy Effects in Fluids.* Cambridge University Press, London and New York, 367 pp.

1

Cups, Vanes, Propellers, and Laser Anemometers

N.E. Busch, O. Christensen, L. Kristensen, L. Lading, and S.E. Larsen

1. INTRODUCTION

This chapter deals with instrumentation for measurements of wind speed and direction. We shall concern ourselves with such 'classical' instruments as cup anemometers, wind vanes, and propeller anemometers. A brief introduction is given to laser-based wind speed sensors, a new and promising technology.

Cup anemometers, vanes, and propeller anenometers have been employed by meteorologists for a hundred years or more. The literature is replete with technical descriptions and theoretical treatments of their design and properties. We shall not in any way attempt to give a review, nor shall we dwell on technical details and specific applications. Since most users purchase such instruments and do not develop them themselves, we shall concentrate on the basic properties of the instruments and on the simplified theory behind these properties. In this way we hope to enable users to evaluate the available instruments, their potentials and limitations.

2. CUP ANEMOMETERS

Cup anemometers are widely used because they are simple, sturdy, and reliable instruments that generally require only a minimum of maintenance. A major operational advantage is that alignment into the wind direction is unnecessary; hence cup anemometers are ideal for continuous measurements.

Problems of overspeeding and angular response notwithstanding, well designed and calibrated cup anemometers are typically accurate to within ±1% of the actual reading above 5 m s^{-1} and to ±5 cm s^{-1} below. Cup anemometers are easily made geometrically and mechanically alike, so calibrations are the same within limits largely comparable to this accuracy.

In this section we shall discuss the kinematics of cup anemometers (i.e., steady state calibration), the dynamics (i.e., the frequency response), overspeeding, and angular response.

2.1 Equation of Motion - Steady State Calibration

For cup anemometers rotating around a vertical axis in a horizontal mean wind a rather general equation of motion may be established

$$J \frac{\partial \Omega}{\partial t} = T(R\Omega, U, W) - f_d \qquad (1)$$

where J denotes the moment of inertia, R the cup-arm radius, Ω the angular velocity, T the torque exerted on the instrument by the wind field with the total horizontal component U and the vertical component W, and f_d the frictional torque stemming from the bearings. The angular position of the cup wheel does not appear in the equation because Ω and T represent quantities that are smoothed over the period of time that it takes the cup wheel to turn the angle $2\pi/N$, where N is the number of cups.

In a steady horizontal wind and neglecting friction we obtain

$$T(R\Omega_0, U_0, 0) = 0, \qquad (2)$$

which yields the steady state calibration relation between U_0 and Ω_0 for the frictionless anemometer.

Simple geometrical considerations led Busch (1965) to suggest that the torque may be expressed

$$T = \tfrac{1}{2}\rho A R U^2 \, g\left(\frac{\Omega R}{U}, \frac{W}{U}\right), \qquad (3)$$

where ρ is the air density and $A = \pi r^2$ the cup area with r being the radius of the cup; for geometrically similar cup systems the 'drag coefficient' g is a function only of the ratio between the cup speed and the wind speed and angle of attack.

It follows from Equation 3 that the frictionless, steady-state calibration expression is linear:

$$\frac{R\Omega_0}{U_0} = x_0 \,. \qquad (4)$$

For a given anemometer the calibration factor x_0 is a characteristic constant with values between 0.18 and 0.41 depending on the geometry of the cup anemometer (Busch, 1965; Kondo et al., 1971; Lindley, 1975).

If we assume that the friction term f_d in Equation 1 is so small that g(x) in Equation 3 may be linearized, i.e.

$$g\left(\frac{R\Omega}{U}\right) = \alpha\left(x_0 - \frac{R\Omega}{U}\right) , \tag{5}$$

and if we assume that the friction term may be approximated by a sum of two terms

$$f_d = C_1U + C_2U^2, \tag{6}$$

then we obtain the usual linear calibration function

$$U = U_s + CR\Omega \tag{7}$$

with a calibration factor

$$C = \left(x_0 - \frac{C_2}{\alpha\rho AR/2}\right)^{-1} \tag{8}$$

and an apparent starting speed

$$U_s = \frac{C_1}{\alpha x_0\rho AR/2 - C_2} \tag{9}$$

which is called 'apparent' since the actual starting speed in general will be somewhat greater. The coefficient α depends on the cup wheel geometry but will normally be around 3. This simple model is very crude; the friction term is modelled ad hoc and complex aerodynamic effects such as cup interactions have not been discussed. The model, however, does provide some background for the linear calibration expression, Equation 7, and offers a means by which the relative importance of cup wheel size versus friction can be estimated.

Instead of using dimensional arguments, an anemometer equation can be obtained by integrating the moments exerted by the relative flow around the cups over one revolution, using known drag coefficients for hemispherical cups under the appropriate angle of attack. This leads to a quadratic equation in wind speed and speed of revolution. Since the wake effect from the preceding cup cannot be taken into account in a general way, the coefficients of such an equation have to be determined empirically. Therefore, a series expansion such as in Equation 10 (below) is appropriate.

2.2 Cup Anemometer Dynamics

Wyngaard et al. (1974) suggest more generally that the torque $T(R\Omega,U,W)$ can be expressed as a three-variable Taylor series expansion about the equilibrium $T(R\Omega_0,U_0,0) = 0$. To second order they obtain

$$T = \frac{J\Omega_0}{\tau_0} \{a_1 \frac{U-U_0}{U_0} - \frac{\Omega-\Omega_0}{\Omega_0} + a_3\left(\frac{\Omega-\Omega_0}{\Omega_0}\right)^2 + a_4\left(\frac{U-U_0}{U_0}\right)^2$$

$$+ a_6\left(\frac{U-U_0}{U_0}\right)\left(\frac{\Omega-\Omega_0}{\Omega_0}\right) + a_5\left(\frac{W}{U_0}\right)^2\} \quad (10)$$

for a cup anemometer which has a symmetric W-response such that its torque characteristics are independent of the sign of W. The coefficients a_i depend solely on the mean aerodynamic properties of the cup arrangement.

If we require that Equation 3 be identical to Equation 10 to first order in small deviations from the equilibrium, then $a_1 = 1$ and the time constant

$$\tau_0 = 2J/(\alpha U \rho A R^2) \quad (11)$$

For later convenience we introduce the so-called distance constant ℓ_0 by

$$\ell_0 = \tau_0 U = 2J/(\alpha \rho A R^2). \quad (12)$$

Neglecting the influence of vertical velocity and of friction and using Equation 10 to the first order limits consideration to small deviations from equilibrium; i.e., either small amplitudes or slow fluctuations in wind speed around the stationary mean. The equation of motion then becomes

$$\frac{\partial \Omega'/\Omega_0}{\partial t} = \frac{1}{\tau_0}\left[\frac{U'}{U_0} - \frac{\Omega'}{\Omega_0}\right], \quad (13)$$

which is the equation for a simple linear first-order system with the power transfer function

$$H(\omega) = \frac{1}{1 + (\tau_0\omega)^2}, \quad (14)$$

where $\Omega' = \Omega-\Omega_0$ and $U' = U-U_0$ denote the fluctuating parts of Ω and U, and ω is the frequency in rad s^{-1}.

Cup anemometers can only approximately be considered as linear systems, but Equation 14 indicates that the energy in oscillations with periods less than 2 to 5 times τ_0 (or wavelengths less than 2 to 5 times the distance constant) will be severely damped (see Fig. 1).

The distance constant of an extremely lightweight cup anemometer (made from expanded polystyrene, say) may be about 0.5 m. Typical micrometeorological cup anemometers with plastic cups will have distance constants in the range 1 to 2 m, whereas some heavy old-fashioned cup anemometers may have distance constants as large as 15 to 20 m. It is illuminating to notice that Equation 12 leads to

$$\ell_0 = 6 \frac{\rho_c}{\rho} \frac{t}{\alpha} \left[1 + \left(\frac{r}{R}\right)^2\right] , \qquad (15)$$

where t is the thickness of the cup material and ρ_c its density; the moment of inertia was calculated for a three-cup anemometer with hemispherical cups neglecting the inertia of everything but the cups. Expression 15 shows why it is difficult to shorten the distance constant: it requires low density material and thin-walled cups, which lead to frail instruments.

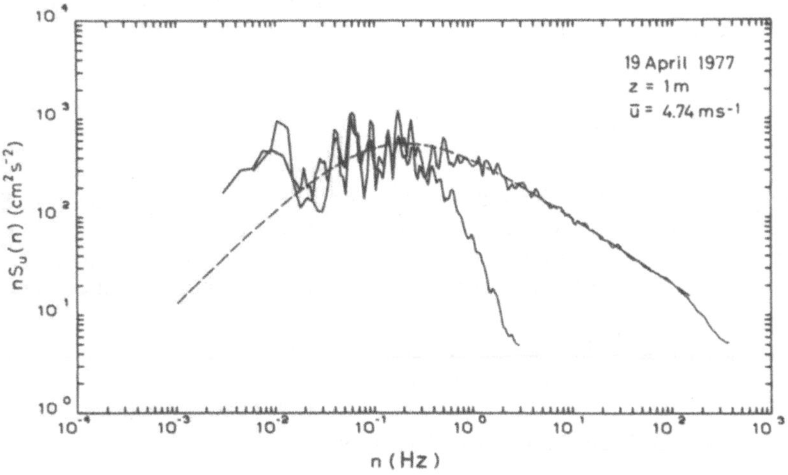

Fig. 1 Spectra of horizontal velocity fluctuations measured by a Risø 70 cup anemometer and a vertical hot wire. The two measuring heights were within 1 cm of each other and the horizontal distance between the two instruments was approximately 25 cm. The 'cup spectrum' was computed from a 655 s long time series, whereas the 'hot wire spectrum' was computed from the first half of the same period, hence the slight difference in the low frequency parts of the spectra. The distance constant for the Risø 70 cup anemometer is about 1.5 m (from Peterson et al., 1979).

2.3 Overspeeding

A disadvantage which is inherent in all cup anemometers is 'overspeeding' caused by nonlinear response to fluctuating winds. Cup anemometers respond more quickly to an increase in the wind speed than to a decrease of the same magnitude. Consequently, in a turbulent flow the mean wind speed will be overestimated if the instrument has been calibrated in a laminar flow. The problem has recently been discussed by Kaganov and Yaglom (1975) and by Busch and Kristensen (1976). Here we shall mention only the most basic results.

Using Equations 1 and 10, it is possible to show that the relative overspeeding

$$\frac{\Delta U}{U_0} = a_4 \frac{\sigma_U^2}{U_0^2} \int_{-\infty}^{+\infty} \frac{(\ell_0 \kappa)^2}{1 + (\ell_0 \kappa)^2} \Phi_U(\kappa) d\kappa + a_5 \frac{\sigma_W^2}{U_0^2}, \tag{16}$$

where σ_U^2 and σ_W^2 are the variances of the horizontal and the vertical wind component, U_0 the mean wind speed, κ the wavenumber, and $\Phi_U(\kappa)$ the normalized velocity spectrum for the fluctuating part of the horizontal wind velocity:

$$\int_{-\infty}^{+\infty} \Phi_U(\kappa) d\kappa = 1. \tag{17}$$

A number of spectral shapes were used in an evaluation of Equation 16 in terms of $\ell_0 \kappa_0$, where κ_0 is the wavenumber at which the logarithmic spectrum $\kappa \Phi_U(\kappa)$ has its maximum. Most realistic atmospheric spectra cannot be integrated analytically. The spectrum

$$\Phi_U(\kappa) = \frac{1}{\pi} \frac{1}{\kappa_0} \frac{1}{1 + (\kappa/\kappa_0)^2} \tag{18}$$

can be integrated in Equation 16 to yield

$$\frac{\Delta U}{U_0} = a_4 \frac{\sigma_U^2}{U_0^2} J_U + a_5 \frac{\sigma_W^2}{U_0^2} \tag{19}$$

with

$$J_U = \ell_0 \kappa_0 / (1 + \ell_0 \kappa_0). \tag{20}$$

The resulting J_U-function obtained for various spectral shapes is displayed in Figure 2, which shows, as expected, that a fast responding anemometer (small ℓ_0) exhibits less overspeeding than a slow anemometer, if the two are exposed to the same wind field. The coefficient a_4 is generally in the vicinity of unity (Wyngaard et al., 1974). The coefficient a_5 is considerably more uncertain, but believed always to be between zero and one. The general conclusion that the overspeeding is equal to or less than the turbulence intensity squared σ_U^2/U_0^2 is in good agreement with much earlier results (McCready, 1965; Busch, 1965).

2.4 Angular Response

The ideal angular response of an anemometer is the so-called 'cosine response'. For a cup anemometer with cosine response the angular velocity is - by definition - independent of the wind velocity component parallel to the anemometer axis.

It is possible to derive an expression based on Equation 10 for the angular response for small angles of attack. The angle ϕ is defined as the angle between the wind velocity and the plane of rotation. By demanding that T = 0, setting

$$U = U_0 \cos\phi \tag{21}$$

and

$$W = U_0 \sin\phi, \tag{22}$$

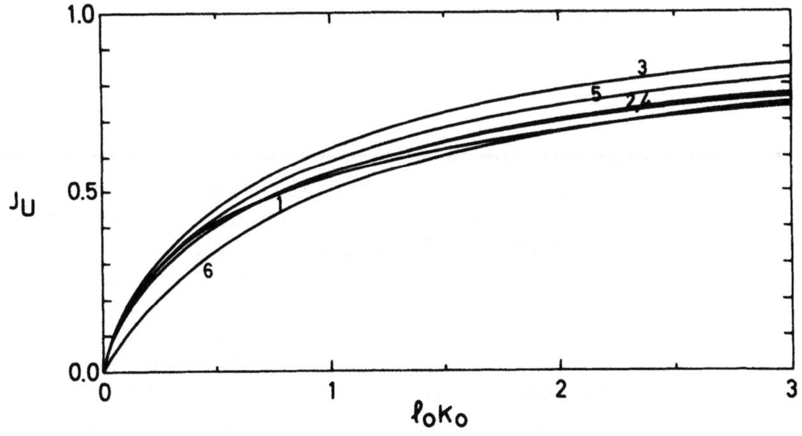

Fig. 2 The relative overspeeding divided by $a_4(\simeq 1)$ and the square of the turbulence intensity σ_U/U in a strictly horizontal wind (see Eq. 19). Curve number 6 shows Equation 20 and the other five numbered curves correspond to various choices for the spectral shape of $\Phi_U(\kappa)$ (see Busch and Kristensen, 1976).

Equation 10 becomes a quadratic equation in the angular response quantity $\Omega(\phi)/\Omega(0)$, where $\Omega(\phi)$ is the angular velocity for the angle of attack ϕ and the wind speed equal to U_0. A simplified analysis assuming Equation 4 shows the significance of the coefficient a_5 in Equation 10. In this case it can be shown (Wyngaard et al., 1974) that

$$a_1 = 1 \tag{23}$$

and

$$a_3 + a_4 + a_6 = 0 \tag{24}$$

These equations in turn imply that $\Omega(\phi)/\Omega(0)$ is greater than or equal to unity for $a_3 \geqslant a_5 \geqslant 0.5$. Qualitatively this means that

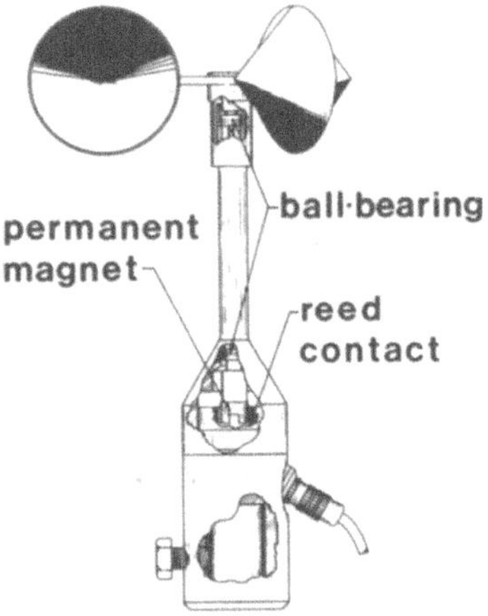

Fig. 3 The Risø 70 cup anemometer in a slightly modified form, in which the electrical pulses are created by means of a stationary reed contact and a permanent magnet mounted on the rotating shaft. The anemometer height, measured along the axis, is 260 mm. Cup diameter and radius, defined as the distance from the instrument axis to the centre of a cup, are 70 mm and 58 mm, respectively. The cups are made of carbon reinforced plastic and the weight of the cup wheel, including the arms, is approximately 40 g. The calibration of this instrument is given by $U = U_s + L\Omega$, where $U_s \simeq 0.2$ m s^{-1}, $L \simeq 0.2$ m, and Ω is the angular velocity in radians per second. The apparent starting speed U_s and L vary from one anemometer to the next of the same type.

only for weak torque dependence of W, i.e. a_5 smaller than 0.5, is it possible to obtain an angular response approaching the cosine response for ϕ approaching 0°.

The angular responses of the Risø 70 cup anemometer (Fig. 3) and of that discussed by Wyngaard et al. (1974) are shown in Fig. 4. Whereas the first is a decreasing function for numerically small

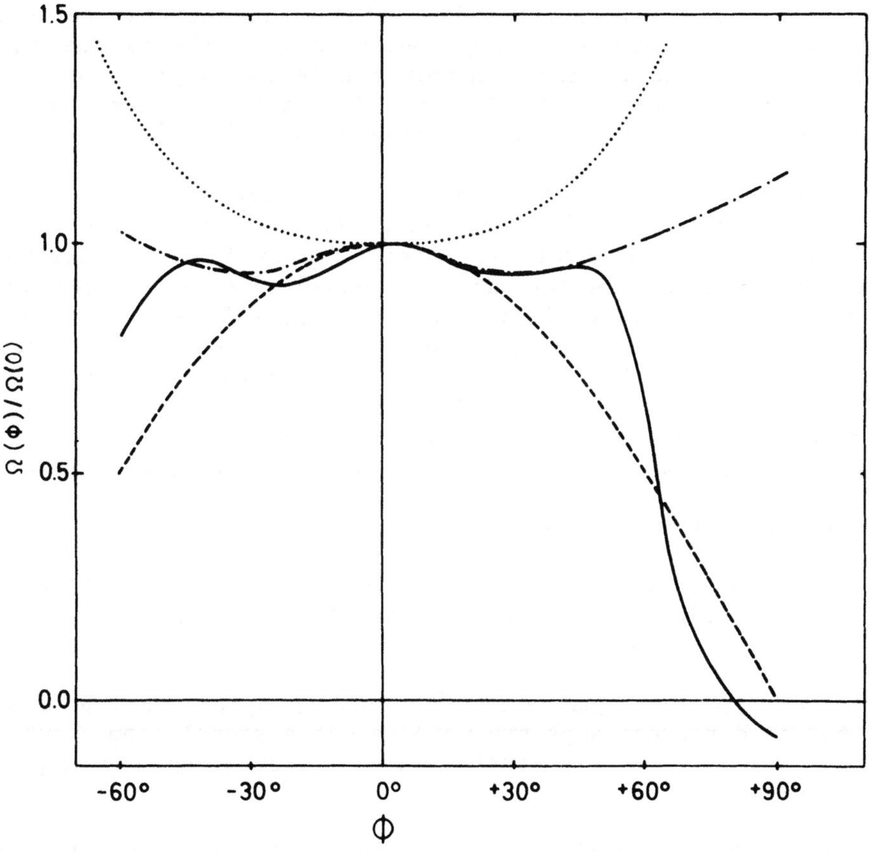

Fig. 4 Angular responses of cup anemometers. A positive angle indicates that the wind comes from above (negative W). The full line is the directly measured angular response of the Risø 70 model. The dot-dashed line is the best fit to this response on the basis of the theory outlined in section 2.4. The same theory yields the dotted line for the angular response of the anemometer described by Wyngaard et al. (1974). The dashed line is the ideal or cosine response.

angles of attack, the second is an increasing function based on the values $a_3 = -0.23$, $a_4 = 0.96$, and $a_5 = 0.67$. The directly measured angular response of the Risø 70 anemometer represents an interval in U_0 from 7 to 18 m s^{-1}, and the individual responses agree with the mean response shown here within 0.001 for angles numerically less than 45°. It is worth noticing the lack of symmetry with respect to $\phi = 0°$. Qualitatively, the two sides of the curves have similar behaviour, both having a relative minimum and a relative maximum. On the positive side $\Omega(\phi)/\Omega(0)$ follows $\cos\phi$ quite closely up to 20°. On the negative side it falls considerably below down to about -20°. This asymmetry, which undoubtedly is due to the effect of the wake of the anemometer body for negative angles of attack, cannot be explained without adding a term proportional to W in an odd power to the right-hand side of Equation 10.

MacCready (1966) found angular responses similar to those shown in Figure 4. For $|\phi|$ in the interval 0° to 20° his 'sensitive small cups' behaved qualitatively similarly to the Risø 70 anemometer, whereas his 'standard small cups' responded in a way similar to the cup anemometer discussed by Wyngaard et al. (1976). For some cup anemometers $\Omega(\phi)/\Omega(0)$ increases with $|\phi|$ in the neighbourhood of 0°. It seems that the long, thin cup arms on these anemometers are responsible for this non-ideal angular response; the ratio of the cup diameter to the radius is equal to 1.21 for the Risø 70 model and 0.73 for the anemometer discussed by Wyngaard et al. (1974).

2.5 Operation of Cup Anemometers

When it comes to the actual use of cup anemometers in air-sea interaction experiments it seems worthwhile to consider a few practical considerations.

First, the type of signal and the recording system must be chosen. A number of different ways are used to get a signal from a cup anemometer. Here we shall mention two, illustrating two different principles.

For remote areas with limited access to electrical power it is convenient to let the rotation generate electrical power by means of an ac generator, say, installed in the anemometer body. The rms voltage thus generated is proportional to the angular velocity of the cup wheel.

Another method, which is widely used at places where the availability of power is not of primary concern, is to let a circular disc or cylinder with equally spaced holes, mounted coaxially on the anemometer axis, chop a light beam in order to produce electrical pulses in a photodiode. The pulse rate is proportional to the angular velocity, and counting of the pulses over a period provides a number, which is proportional to the mean angular velocity over that period.

A second item for consideration is the environment in which the anemometer is installed. Wind, salt spray, icing, and rain can influence the instrument's performance in several ways.

Strong winds can easily break a light cup wheel, and it will in many cases be necessary, at the expense of fast response, to use rather thick-walled cups to secure more than a few hours of operation in these situations.

Salt spray can easily get into the interior of the anemometer, causing changes in calibration due to corrosion. Lubrication of ball bearings with a water-repelling, strongly adhesive grease seems to diminish the problem.

Icing, caused, in particular, by supercooled rain or droplets, occurs frequently at coastal sites and at sea during the winter. If the ice does not stop the cup wheel completely, the ice coating, which can be several millimetres thick, will change the calibration by changing the geometry (see section 2.1). The only remedy for icing seems to be electrical heating of the moving parts.

Dentler (1978) analyzed the problem of how rainfall, by momentum transfer to the cups, gives rise to rain-induced errors in cup anemometer readings. By assuming a general drop-size distribution he concluded that the magnitude of these errors, even for heavy rainfall, amounts to no more than a few per cent.

For low wind speeds the cup wheel angular velocity is not even close to being a linear function of the wind speed, and this presents a special problem when mean wind speeds and wind profiles are to be measured. In fact, the nonlinearity for small wind speeds is a discontinuity in the calibration. In order to overcome the friction in the bearings the wind speed has to be greater than a threshold value. In order to determine the correct mean wind speed it is necessary to identify the periods when the cup wheel did not rotate and make suitable corrections. In marine environments the rapid deterioration of bearings enhances the problem, and it is often advisable to use extra cup anemometers to provide redundant information.

Finally, a cup anemometer mounted on a buoy can give readings which are strongly influenced by the buoy's motion. Both the change in the angle of attack and the anemometer's own velocity with respect to a hypothetical fixed platform will give rise to misreadings of the instantaneous wind. If the data are to be used later for spectral analysis, then corrections seem unavoidable. The wave spectrum and the anemometer angular response are necessary ingredients in the calculation of these corrections. The mean wind speed is only weakly influenced by the buoy's motion, and a correction can easily be calculated if the anemometer's angular response is known.

3. WIND VANES

Wind vanes are probably the oldest instruments used in meteorology. We have detailed knowledge about simple wind vanes used to measure the horizontal wind direction from as early as the first century B.C.

More complicated types (bivanes) that can rotate around two axes have been in use since the start of this century, occasionally mounted with propellers (trivanes) to yield information about all three wind velocity components (MacCready and Jex, 1964).

3.1 Linear Response Theory of a Wind Vane

To derive the linear response of a wind vane we use Figure 5, where the vane is placed in a homogeneous velocity field, $\vec{U}$, the angle of which, relative to an arbitrary reference direction, θ, varies with time. We shall keep α, the angle between $\vec{U}$ and the vane, small.

The governing equation is

$$\frac{d^2\Theta}{dt^2} = M/I - D/I \tag{25}$$

where $\Theta = \theta + \alpha$ is the angle from the reference direction to the vane, I the moment of inertia, M the forcing torque due to the wind, and D the damping torque due to mechanical friction.

The effective forcing wind velocity, Δv_r, is the vector difference

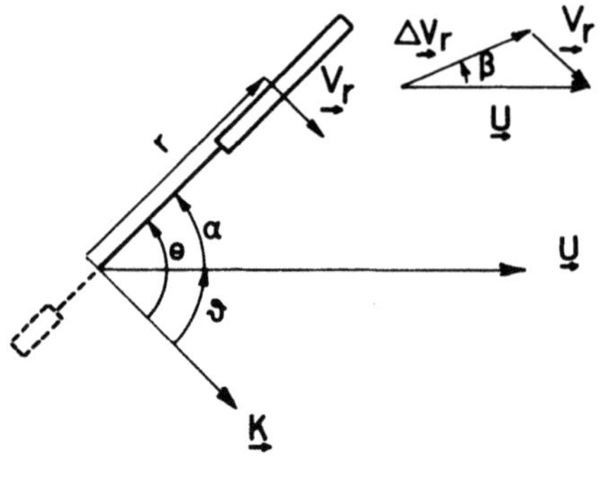

Fig. 5 Schematic drawing of a wind vane seen from above. $\vec{U}$ is the velocity field. $\vec{K}$ is an arbitrary reference direction. The figure shows how the effective attack velocity $\Delta \vec{v}_r$ differs from $\vec{U}$, due to the motion of the vane. Also shown is the counterweight used on many wind vanes.

between the wind velocity, $\underset{\rightarrow}{U}$, and the vane velocity, $\underset{\rightarrow}{v}_r$, the subscript r indicating the approximate distance from the axis of rotation to the middle of the vane. The magnitude of the vane velocity can be written

$$v_r = -r\frac{d\Theta}{dt}. \tag{26}$$

If we assume that the effective velocity attacks the vane at a fixed distance r from the axis we can find Δv_r from Figure 5. Application of the additional assumption that $U >> v_r$ yields the following result

$$\Delta v_r \simeq U$$

$$\beta \simeq v_r/U = -\frac{r}{U}\frac{d\Theta}{dt} \tag{27}$$

Hence we see that, under the assumptions made, the role of the vane's velocity is to shift the angle of attack from α to $\alpha-\beta$, while the effective forcing speed remains equal to the wind speed.

In order to calculate the total wind force, F, perpendicular to the vane, we apply the semi-empirical theories developed for thin air foils. The result for small angles is (Larsen and Busch, 1974)

$$F = \tfrac{1}{2}\rho U^2 AK(\alpha - \beta), \tag{28}$$

where A is the surface area of the vane, ρ is the density of air, and K is a constant dependent only on the physical characteristics of the vane.

Since the D/I term in Equation 25 is due to mechanical friction in the bearings, it depends on the speed of revolution. In MacCready (1965) the following expression is suggested

$$D/I \simeq (C_1 + C_2U^2)\frac{d\Theta}{dt} \tag{29}$$

where C_1 takes into account the weight on the bearings and C_2U^2 the wind load.

Combining these equations we can write Equation 25 as

$$\frac{1}{\omega_e^2}\frac{d^2\Theta}{dt^2} + 2\zeta\frac{1}{\omega_e}\frac{d\Theta}{dt} + \Theta = \theta \tag{30}$$

where we have used $\alpha = \Theta - \theta$ and $M = Fr$ and introduced the characteristic frequency, ω_e, and the damping ratio, ζ, by

$$\omega_e = 2\pi \frac{U}{\lambda_e} \tag{31}$$

and

$$\zeta = \frac{C_1}{4\pi} \frac{\lambda_e}{U} + \pi \frac{r}{\lambda_e} + \frac{C_2}{4\pi} U\lambda_e, \tag{32}$$

where

$$\lambda_e = 2\pi \sqrt{\frac{2I}{\rho r\ AK}} \tag{33}$$

In Equation 32 the first and last terms are due to mechanical damping, while the middle term describes the aerodynamic damping.

Equation 30 relates the wind vane deflection, $\Theta(t)$, to the true wind direction, $\theta(t)$. It is seen that the wind vane acts as a second order filter with a transfer function

$$H(\omega) = \left[1 - \left(\frac{\omega}{\omega_e}\right)^2 - 2i\ \zeta \left(\frac{\omega}{\omega_e}\right)\right]^{-1} \tag{34}$$

The behaviour of the amplitude transfer function, $|H(\omega)|$, is shown in Figure 6, and the vane response to a step change in θ in Figure 7.

It should be noted that the basic characteristic scale associated with a wind vane is a length scale, λ_e, not a time scale. Indeed, if we transform Equation 34 using Taylor's hypothesis, $\lambda = 2\pi\ U/\omega$, and Equation 31, we see that $\omega/\omega_e \rightarrow \lambda_e/\lambda$, and hence that the vane in 'λ-variables' had a wind speed-independent response, notwithstanding the velocity dependence of ζ.

Before proceeding to the more practical considerations concerning wind vanes, it is appropriate to point out a number of limitations for the validity of the derived expressions.

One basic limitation is that the angles α and α-β should be small. This will usually present no problem in the measuring situation, but needs to be considered when testing a wind vane in a wind tunnel (see Section 3.2).

Another set of limitations is associated with the variability of the real wind velocity, $U(t)$. In practice we will usually interpret the velocity scale in Equation 31 as the mean velocity. However, if the turbulence intensity is too large this is no longer a good approximation and the system becomes highly nonlinear.

Finally, scales of variability of U, which are not much larger than the dimensions of the wind vane, will be subjected to spatial averaging over the surface of the vane. The consequences of this can be illustrated by taking λ_e as the characteristic physical scale of the wind vane and applying Taylor's hypothesis. In this way we find the following upper frequency ω_{upper} for the application of Equation 34

$$\omega_{upper} \simeq 2\pi \frac{U}{\lambda_e} = \omega_e \qquad (35)$$

Hence we should be careful not to try to correct measured data for the influence of the transfer function for frequencies larger than ω_e. The choice of λ_e as the characteristic scale for the spatial averaging of the vane is of course somewhat arbitrary, but in most cases is quite realistic.

3.2 Design and Testing of Wind Vanes

In the design of wind vanes due attention must be given to strength and stiffness of the system, but aside from that, general criteria can be derived from considerations of the parameters λ_e and ζ. We usually wish a large ω_e; this calls for a small λ_e. From Figures 6 and 7 it appears that a ζ-value between 0.5 and 0.8 seems advantageous; this in turn gives a lower bound for λ_e according to Equation 32. Traditionally a ζ-value around 0.6 is recommended, but it should be mentioned that the best choice from a theoretical point of view is $\zeta = 1/\sqrt{2}$. Obviously, small values of the friction parameters C_1 and C_2 will minimize the velocity dependence of ζ.

These points suggest that the vane should be designed with minimum weight and friction (this is an argument against the use of a counterweight often seen on commercial wind direction sensors). For the vane itself, experience indicates that a simple, rough flat plate is superior to more elaborate structures (see e.g. Wieringa, 1967). The significance of the dimensions of the vane is discussed by Wieringa (1967) and Larsen and Busch (1974). A lightweight plate made of expanded polystyrene is well suited for vane material. If no counterweight is used and the vane geometry and material have been chosen it is seen from Equations 32 and 33 that λ_e and ζ are both roughly proportional to $\sqrt{r}$ and wind tunnel tests can be used to obtain the proper ζ value.

Results of such a test are shown in Figure 8. The curves illustrate the difficulties in keeping the frictional terms small, especially the C_1-term in Equation 32, even for very light wind vanes.

A simple way of performing such a wind tunnel test is to move the vane a small angle, $\Delta\Theta$, away from its equilibrium position in the air stream and release it. The vane response will be as shown in

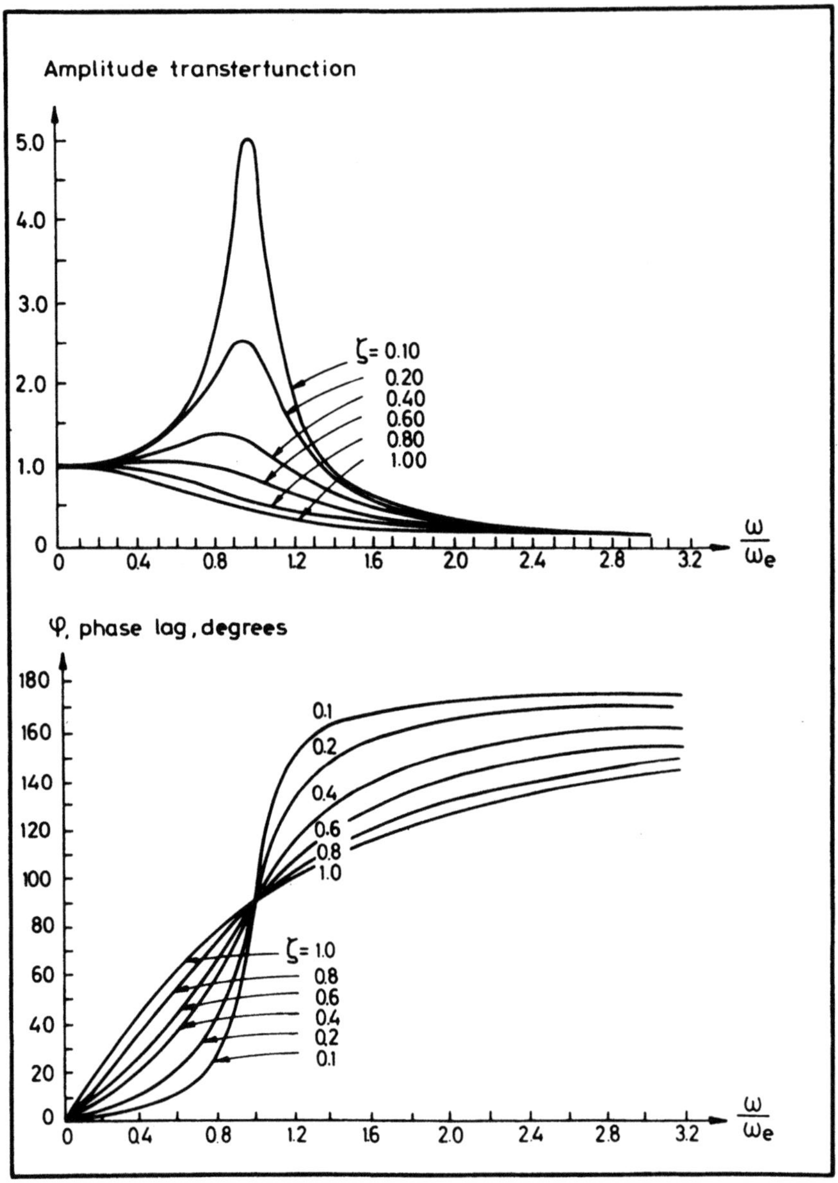

Fig. 6 Behaviour of the wind vane amplitude transfer function and the phase lag for different values of the damping ratio (from MacCready, 1965).

Figure 7 and is given by

$$\Theta(t) = \Delta\Theta \left[1 - \frac{\exp\{-\omega_e \zeta t\}}{\sqrt{1-\zeta^2}} \cos\left(\omega_\varepsilon \sqrt{1-\zeta^2}\, t + \cos^{-1}(\sqrt{1-\zeta^2})\right)\right] \quad (36)$$

Therefore from records of the vane response ω_e and ζ can be estimated (MacCready, 1965).

Equation 36 suggests another characteristic time scale for the wind vane system, $\tau = (\omega_e \zeta)^{-1}$, which, using Taylor's hypothesis and Equation 37, is converted into a length scale

$$L = \frac{1}{2\pi} \frac{\lambda_e}{\zeta} \quad (37)$$

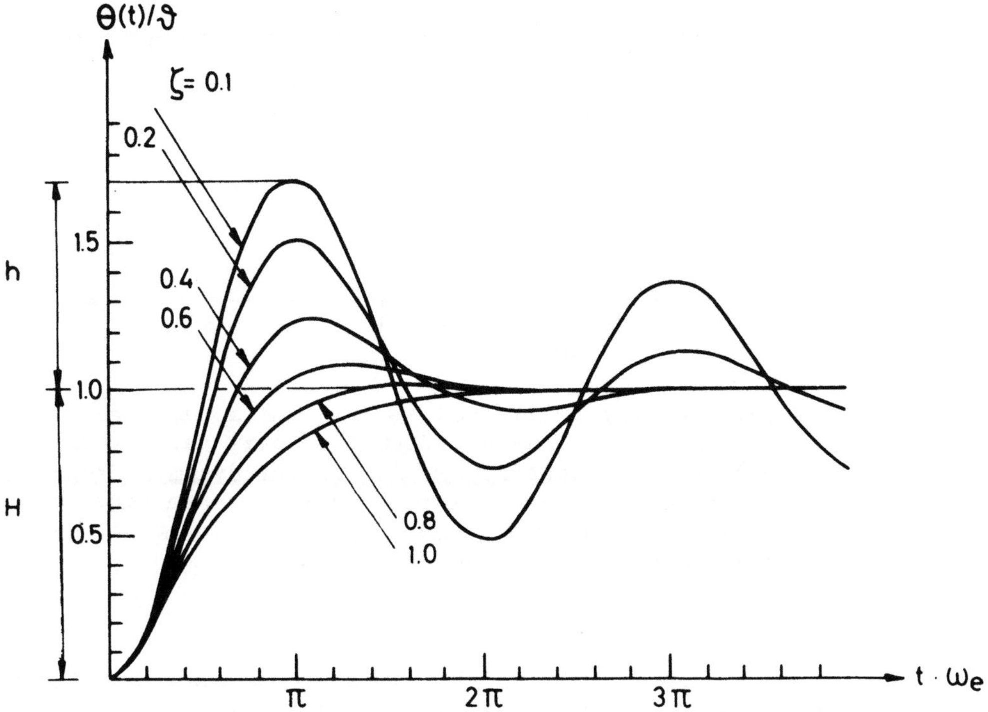

Fig. 7 Response of a wind vane to a step function input (from MacCready, 1965). Here the arbitrary zero point of Θ and θ is taken to be the direction before the step change. H and h are the normalized new angle and overshoot, respectively.

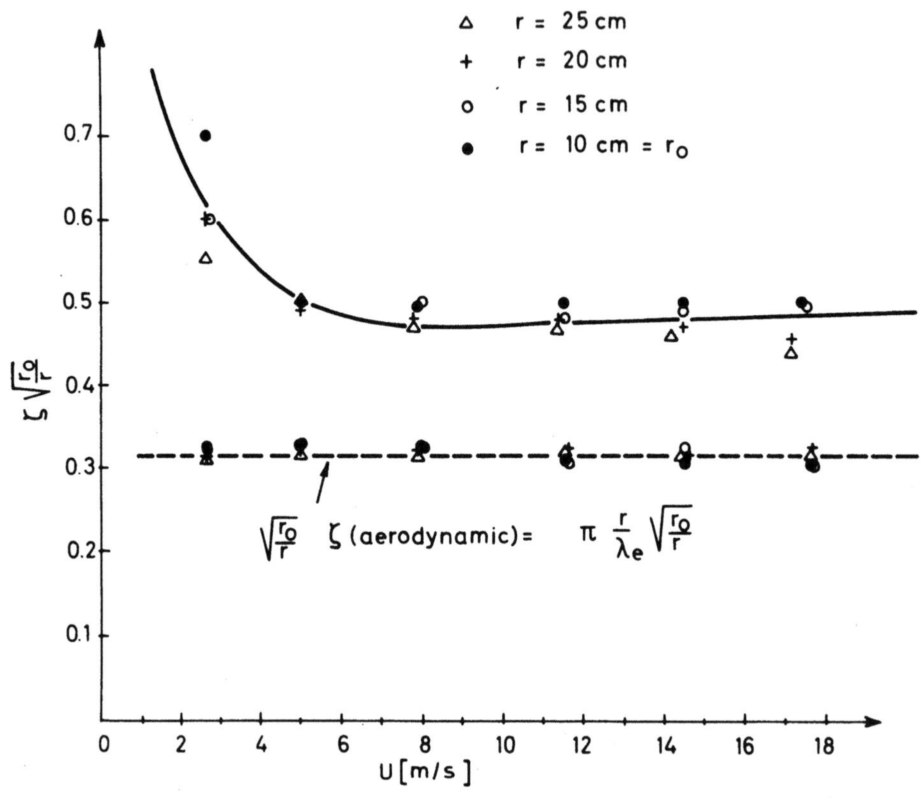

Fig. 8 Experimentally determined values for a lightweight experimental wind vane. Both the total value (full line) and the contribution due to aerodynamic damping (dashed line) are shown. All values are normalized to the same arm length r_0. The curves illustrate the analytical expression (32).

This length scale corresponds to the distance constant for first order systems such as cup and propeller anemometers.

3.3 Output and Signal Processing

The basic problem in handling output data from most wind vanes is the discontinuity between 0° and 360°, which is of little importance in an experimental situation, but inconvenient in climatological use of the wind vane. In Table 1 a number of different methods are summarized.

Table 1

Wind Vane Data Recovery Techniques

Type	Signal	Advantages	Disadvantages
1. Potentiometer	Analogue	Simple, 1 signal	Discontinuous signal; finite range without signal; subject to wear
2. Shaft digitizer	Digital	No friction, 1 signal	Additional moment of inertia; discontinuity
3. Synchrogenerator	Strip chart	No discontinuity	Heavy, strong damping
4. Sine/cosine Generator	Digital or Analogue	No discontinuity, $\overline{\Theta}$ and σ_θ from mean values	2 signals

Also:

(a) Potentiometer wear is often not detected before it causes serious errors;

(b) Possibilities 1 and 2 can be expanded by using two detectors, which will remove the discontinuity problem, but introduce two output signals;

(c) A synchrogenerator electrically couples the motion of the vane with the motion of the recorder and demands use of ac driving; and

(d) For the cosine/sine-generator, determination of wind-direction variance σ_Θ^2 from the mean values $\overline{\cos\Theta}$ and $\overline{\sin\Theta}$ first discussed by Camuffo (1976), is given simply by Busch et al. (1976):

for small Θ'

$$\overline{\cos\Theta} = \overline{\cos(\overline{\Theta}+\Theta')} \simeq (1-\tfrac{1}{2}\overline{\Theta'^2})\cos\overline{\Theta}$$

and

$$\overline{\sin\Theta} \approx (1-\tfrac{1}{2}\overline{\Theta'^2})\,\sin\overline{\Theta}$$

and hence

$$\sigma_\Theta^2 \approx 1 - \overline{\cos\Theta}^2 - \overline{\sin\Theta}^2. \tag{38}$$

3.4 Special Considerations

So far bivane systems without or with propellers (trivanes) have only been mentioned briefly. With respect to vane response a simple extension of the above described relations will suffice. However, these systems all have a serious practical problem, namely that vertical motion has to be balanced and that the balance is easily disturbed by rain, ice, and dust. Some of the vane types in use are even made of hygroscopic material, which certainly aggravates the problem. Discussion of such systems and the associated data handling procedures are given by Wieringa (1972) and Busch et al. (1970).

A special problem with propeller vanes is the fact that propeller rotation induces a gyroscopic stability in the vane system; hence, its response characteristics are probably different for strong and weak winds. An experimental investigation of this gyro effect has not yet been undertaken, and it is strongly recommended that potential users of propeller vanes look into the problem.

Finally, it should be mentioned that the motion of a vane, especially if it is white, seems to trigger aggressive behaviour in birds such as sea gulls, so that if the vane is made of an easily-destroyed material it is advisable to make provisions for easy replacement.

4. PROPELLER ANEMOMETER

For measurements of vertical velocities or the three-dimensional characteristics of atmospheric flows in general, the propeller anemometer is an attractive alternative to a cup anemometer-bivane system (Holmes et al., 1964).

The speed of revolution of a propeller is linearly related to the magnitude of the wind component directed along the propeller axis, i.e. perpendicular to the plane in which the propeller rotates. For an arbitrary but constant angle of attack, a doubling of the wind speed will lead to a doubling of the rate of rotation of the propeller. However, the speed of rotation is smaller than would have been obtained with the axial component of the wind acting alone. In other words, the propeller sensor does not have a perfect cosine response.

Several researchers have attempted to design a propeller with improved angular response (Gill, 1975; Davis and Weller, see Chapter 8). Both experience and unpublished computations indicate that a simple propeller cannot be expected to have a perfect cosine response. This argument is based on the simple aerodynamic model detailed below; the main reason is that stalling at the propeller blades will occur for certain blade positions at small flow angles with respect to the axle. An improvement of the off-axis response can be achieved by more complicated systems, e.g. through ducting. Since one has to interfere with the flow in order to obtain a better cosine response, one expects to pay for it. That is, it may be possible to improve the angular response at certain angles of attack, by sacrificing the cosine response at other angles and/or increasing the distance constant.

A propeller of helicoidal shape which is exposed to a steady wind along the propeller axis will assume a rate of revolution at which the torque exerted by the lift forces on the blades balances the torque stemming from the very small drag forces. If the drag forces could be eliminated, then the lift forces would be zero everywhere on the blades. A nonhelicoidal shape would produce positive and negative lift forces in order to maintain zero turning torque. In the case of the cup anemometer, a sudden increase in the wind speed will produce a positive torque from unbalanced drag forces. A propeller anemometer will experience a torque stemming from unbalanced lift forces, and since lift forces can exceed drag forces considerably in magnitude, it may be possible to design a propeller anemometer with a faster response than that of a cup anemometer with the same moment of inertia. However, a major problem must be recognized: it is considerably easier to construct a low-weight cup wheel which is capable of resisting gale force winds than it is to make a robust low-weight propeller with this property.

A commercially available, widely-used propeller anemometer is the so-called 'Gill' propeller anemometer (Holmes et al., 1964; Gill, 1975; Teunissen, 1976), which was developed at the University of Michigan in the early 1960s (Fig. 9). The propeller is made of polystyrene and the two or four blades are of helicoidal shape. The blade chord increases and the blade thickness decreases with distance from the propeller axis. The propeller has a very low starting speed and a distance constant of about 1 m. Due to the low strength of polystyrene, the propeller cannot sustain wind speeds of more than 25 m s^{-1}, or 20 m s^{-1} if the turbulence intensity is high.

In the following we shall discuss propeller response to axial and nonaxial flow. We shall limit ourselves to the behaviour of a

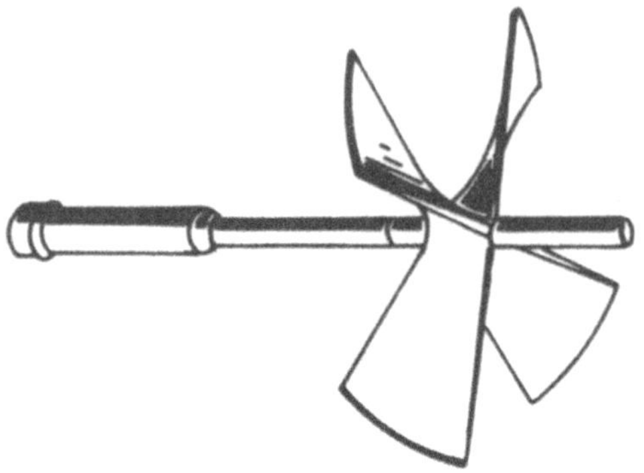

Fig. 9 Helicoidal four-bladed 'Gill' propeller.

Gill-propeller, i.e. to a propeller with helicoidal shaped blades; the conclusions, however, are applicable to other propeller types.

4.1 Propeller Response to Axial Flow

Aerodynamically a propeller may be regarded as a number of airfoils attached perpendicularly to a central axis. An airfoil which is exposed to a relative wind u_r at an angle of attack ε will experience a lift force F_L perpendicular to the wind vector and a drag force F_D along the wind vector which may be written

$$F_L = \tfrac{1}{2}\,\rho|u_r|^2\,C_L A \qquad \text{and} \qquad F_D = \tfrac{1}{2}\,\rho|u_r|^2\,C_D A \qquad (39)$$

where A is the surface area of the airfoil and ρ the air density; C_L and C_D are the lift and drag coefficients, respectively. These coefficients depend on the cross-sectional shape of the airfoil and on the angle of attack. For small values of ε, C_L increases linearly with ε. The cross-section determines both the limit $\varepsilon = \varepsilon_s$ (the angle of 'stall') beyond which linearity is no longer a valid approximation, and how C_L behaves for still larger ε.

For a flat plate ε_s is about 8° and C_L slowly decreases with larger ε. Cross-sections which are aerodynamically shaped to yield larger lift forces may have $\varepsilon_s \simeq$ 15 to 20°, but C_L drops abruptly for slightly larger values of ε. The drag coefficient C_D depends on the angle of attack, Reynolds number, aspect ratio (i.e. length to width), and the roughness of the foil surface. It is desirable to make C_D as small as possible in order to diminish the skewness of the dynamic response of the propeller.

When a propeller is rotating in a steady and axial flow, then the angle which the relative wind forms with the propeller plane will be

$$\alpha(r) = \tan^{-1} \left(\frac{u}{r\ \omega}\right) , \tag{40}$$

where u is the wind speed, r is the radial distance from the axis, and ω is the angular velocity of revolution.

If the propeller blades are twisted such that the angle between the blade chords and the propeller plane is

$$\alpha_b(r) = \tan^{-1} \left(\gamma_R \frac{R}{r}\right) , \tag{41}$$

where γ_R is the so-called pitch factor and R the length of the blades, the propeller is helicoidally shaped. In the absence of drag and friction forces, a steadily rotating propeller in a steady, axial wind, i.e. no lift forces, will rotate such that $\alpha = \alpha_b$ or - in other terms - such that

$$u = \gamma_R\ R\ \omega \tag{42}$$

It may be shown that in the presence of drag forces one should expect the angular velocity to be reduced by a factor k, say, which is rather insensitive to changes in wind speed. The bearings, which are needed to support the propeller shaft, provide friction that to a first approximation may be assumed to be proportional to the speed of revolution, in which case the calibration of a helicoidally shaped propeller takes the form

$$u = \gamma_R kR\omega + \Delta u_f \tag{43}$$

where Δu_f is a constant which is never negative.

The Gill propellers are manufactured in four versions: two- or four-bladed propellers with diameters of either 18 or 23 cm. All versions have the same pitch factor (γ_R = 0.43). Calibration experiments with several propellers support Equation 43.

A few comments on the dynamic response of propeller anemometers may be in order. A perfectly linear first-order wind speed sensor has a distance constant which is independent of the magnitude and sign of imposed step changes of wind speed. The distance constant may be defined as the length of the air column which must pass the sensor in order to provide 63% [i.e. 1 - exp(-1)] of the change towards the new equilibrium.

Using quasi-steady theory for airfoils together with the so-called 'strip method' for calculation of the total torque from all propeller blade segments, it is possible to estimate and compare distance constants for different velocity steps and various values of the lift-to-drag ratio C_L/C_D. Unpublished calculations at Risø National Laboratory show that the distance constant for deceleration becomes smaller as the magnitude of the velocity step is increased. In contrast, the distance constant increases in an accelerating flow as the velocity step is made larger. The effect of drag forces appears to be quite pronounced. The smaller the drag coefficient - everything being equal - the less is the nonlinearity of the dynamic response. The skewness of response to accelerating and decelerating flow may or may not be significant from a practical point of view, but it is interesting to note that whereas a cup anemometer 'overspeeds' in a fluctuating wind, a propeller anemometer 'underspeeds.'

Distance constants are usually measured in wind tunnels by releasing a propeller from rest in a steady wind and observing the time necessary for 63% of the change towards steady rotation to occur. Calculations show that the time constants thus obtained may be a factor 2 to 4 larger than those which pertain to a 10% increase in wind speed. The distance constant for a 23 cm diameter Gill propeller has been determined to be approximately 1 m in wind tunnel experiments.

4.2 Propeller Response to Nonaxial Flow

When the wind blows along the propeller axis, the turning torque is independent of blade position; all blades contribute equally to the forcing. If, however, the wind vector has a component perpendicular to the axis, then the blades do not contribute equally, and the torque becomes a function of blade position. Basically the propeller motion is governed by a nonlinear, second-order differential equation, which we will ignore in the present context.

We introduce an angle of attack

$$\theta = \tan^{-1} \frac{v}{u} \tag{44}$$

where v is the wind component in the propeller plane and u the axial component perpendicular to that plane. For large lift-to-drag ratios the nonlinearity of the system is weak enough to ensure validity of Equation 43 for small angles of attack, i.e. to yield a cosine response. For larger angles of attack stalling occurs on the blades and nonlinear effects force the average speed of rotation to become smaller than predicted by Equation 43.

The airfoil theory mentioned earlier leads us to define an angular steady-state response function

$$S(\theta) = k\gamma_R \frac{R\omega}{u - \Delta u_f}, \tag{45}$$

where ω is the average angular velocity. The theory predicts that $S(\theta)$ is a function of γ_R and C_L/C_D, but also that it is independent of u if $u \gg \Delta u_f$.

The response function $S(\theta)$ has been evaluated experimentally for a Gill propeller with a diameter of 23 cm and a pitch factor γ_R = 0.43. The results, shown in Figure 10, may be approximated to within a few per cent by a polynomial expansion in $\cos(\theta)$:

$$S(\theta) = \cos\theta \sum_{m=0}^{3} b_{2m} \cos^{2m}\theta \tag{46}$$

with b_0 = 0.656, b_2 = 0.408, b_4 = 0.064, and b_6 = -0.128. The lack of cosine response implies that the angle of attack must be known in order to permit the axial wind component to be determined. This of course means that three propeller sensors with nonparallel axes are needed unless information about the angle of attack is furnished either by other instruments or by constraints on the flow field.

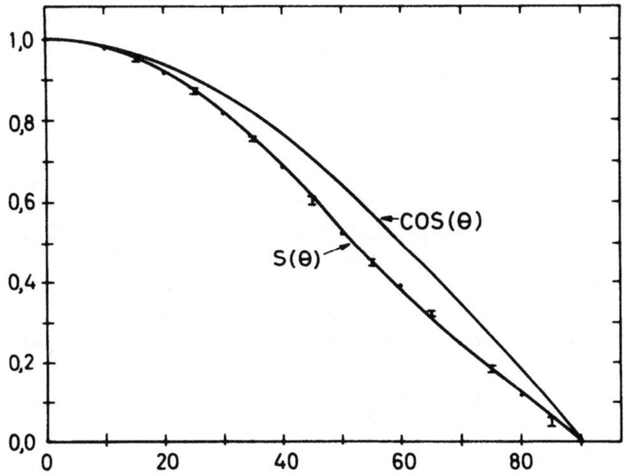

Fig. 10 Steady-state angular response versus offwind angle in degrees for the 23 cm Gill propeller. Dots represent experimental values at 17 m s^{-1}; vertical bars indicate the range of measurements for wind speeds varying from 1 to 26 m s^{-1}.

The distance constant for different angles of attack may still be defined in terms of step changes in the axial component of the flow as in the preceding section. Theoretical computations as well as wind tunnel measurements indicate that the distance constant is largely independent of the angle of attack, if this angle does not exceed 15 to 25°. The extent to which this statement is true depends on the aerodynamic properties of the propeller. An increase in C_L/C_D or a decrease in γ_R improves the approximation.

Measurements of distance constants for accelerating flow have been reported in the literature for the 23 cm Gill propeller. The results are fairly well approximated by

$$L(\theta) = L(0)\cos^{\frac{1}{2}}\theta \tag{47}$$

where $L(\theta)$ is the distance constant for the angle of attack θ (Hicks, 1972; Garratt, 1975; Brook, 1977). For propellers with γ_R different from 0.43 one should expect a different relation. From Equation 47 as θ approaches 90° the time constant $T = L(\theta)/U(\cos\theta)$ becomes large. (See also Fichtl and Kumar, 1974, for examples from atmospheric turbulent flow.)

4.3 Wind Measurements by Means of Propellers

In the natural wind one propeller alone can only yield approximate information about the axial component. In this section we shall briefly discuss how measurements of vertical wind components are obtained by means of a propeller anemometer and a cup anemometer, and how complete three-dimensional wind measurements can be performed with three propellers.

A cup anemometer and a propeller with a vertical axis may be combined to yield the vertical and horizontal wind components independent of wind direction. By simultaneously measuring the speeds of revolution of the two devices, the magnitude of the wind vector and its angle of elevation can be computed. The basis of the computations is the expressions pertaining to the steady state calibration of both instruments. Thus an important question arises: what are the high-frequency cutoff characteristics of the system?

The cutoff characteristics can only be properly assessed through comparison with fast-responding instruments such as a sonic anemometer or hot-wire anemometers. Some idea of the value of the cutoff frequencies may be obtained by evaluating the time constants for the actual mean wind speed on the basis of the distance constants discussed in preceding sections.

In a horizontal mean wind a propeller with a vertical axis must have a very large time constant. The only way of reducing this time constant is by tilting the propeller into the mean wind, in which case the combined system loses its independence of wind direction.

Instead of using a cup anemometer as a speed sensor, Gill suggested the use of two propellers with perpendicular horizontal axes. The system can provide complete three-dimensional wind measurements, but the lack of cosine response necessitates the use of a computer, if accuracy is to be obtained. An iterative scheme may be devised by means of which the three wind components can be derived (Christensen, 1971; Horst, 1973).

The three propeller axes form a Cartesian coordinate system in which we have

$$u_i = \frac{\cos\theta_i}{S(\theta_i)}\,[\gamma_R R\omega_i + \Delta u_f S(\theta_i)]; \qquad i = 1,2,3$$

$$\cos\theta_i = u_i/U \tag{48}$$

$$U^2 = u_1{}^2 + u_2{}^2 + u_3{}^2$$

These seven equations with seven unknowns are to be solved for each set of instantaneously recorded ω_i. Subscript i = 1, 2, or 3 denotes the number of the propeller. In the normal mode of operation the orthogonal propeller system is kept fixed for all wind directions. This may cause disturbances of one propeller by another for certain wind directions. Furthermore, the structure supporting the propeller prevents complete symmetry in angular response for positive and negative angles of attack (a shaft extension may be used to improve symmetry). Also, vertically oriented propellers will stop with changing direction of the vertical component, producing a nonlinear response.

The characteristics of the system may be improved by tilting it into the mean wind. Such a system will have time constants, which are close to being equal for all three propellers; furthermore, each propeller will experience positive angles of attack except in cases of the most extreme turbulence intensities.

The three propellers of a tilted orthogonal system which is positioned in an optimal way will on the average be approached at angles of attack of 55°. The range of wind directions yielding positive angles of attack for all three propellers is about 100°, if the wind inclination is less than 25°. In order to fully exploit the possible accuracy of such a propeller system, it is necessary to keep the array adjusted into the mean wind and to rely on computer processing to iteratively correct for the non-ideal angular response.

Having once accepted these complications one may further improve the dynamic response and range in terms of changes in wind direction by abandoning the orthogonality of the array. The cost is a relatively small increase in computer time. If the propeller axes are arranged equi-angularly on a conical surface with a solid angle of 0.2 ster (half opening angle 15°) and with the axis of the cone pointing into the mean wind, then most of the time the angle of attack will be about 15°; hence there will be a reduction in the time constant as compared to the orthogonal system. The acceptable wind direction range within which the angle of attack is positive for all three propellers will be about 150°, if the inclination of the wind is less than 45°.

4.4 Operation of Propeller Anemometers

Experience with Gill propellers indicates that the calibrations for speed and angular response are very stable in time. Further, the variability from one propeller to another is very small. The main problem, from an operational point of view, is their mechanical frailty. A gale force wind will nearly always break such a propeller, especially if the intensity of turbulence is large.

For climatological use the Gill system has a very useful option which prolongs the lifetime of the microbearings and hence controls the bearing friction. By means of a special blower, the inside of the pipes constituting a Gill orthogonal UVW system are maintained at a slightly higher pressure than the pressure outside. The air drawn through the pipes is sucked through an oily filter and blown through the bearings. By means of this system the bearings are well oiled at all times, and foreign objects such as dust or salt spray are kept out of the bearings.

The scarcity of experience with Gill propellers (or any other propeller system) in air-sea interaction studies (Francey and Garratt, 1978) is probably due to their fragile nature. The problems concerning resistance of bearings in hostile marine environments (salty water spray) are not greater than the problems encountered through the usage of cup anemometers, and, if the above mentioned option on the Gill system is adopted, the problem may well have been eliminated.

5. LASER ANEMOMETRY

The following is a short introduction to laser anemometry. Laser anemometers are now available commercially and have proven useful in wind/water tunnel work; so far, there have been no published results from use at sea. The requirements of a solid support and a large power supply seem prohibitive. However, they still may be useful on fixed platforms.

Remote measurement of velocities in fluids using light may be based either on scattering from small particles suspended in the fluid (e.g. aerosols) or on disturbances caused by fluctuations in the refractive index of the fluid due to turbulence.

5.1 Devices Based on Scattering by Aerosols

The Doppler effect has been applied to a range of velocity measuring devices using some kind of coherent radiation.

Yeh and Cummings (1964) showed that laser light could be used for localized fluid flow measurements. A basic ingredient in their method was the use of the photodetector as a mixer: scattered light was mixed with a reference beam on a photodetector, which

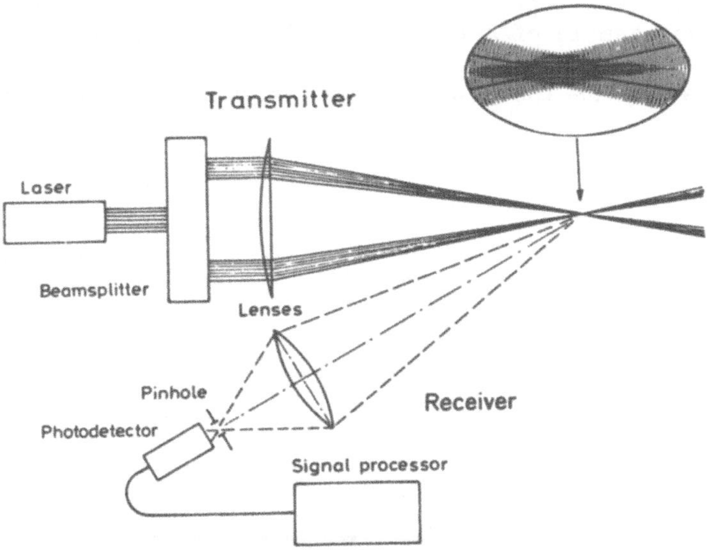

Fig. 11 Schematic diagram of a dual-beam laser Doppler anemometer. The two spatially separated laser beams emerging from the beam-splitter are refracted and focused by the transmitter lens, thus generating an interference pattern. Scattered light from particles passing this interference pattern is collected and detected in the receiver. The transmitter and receiver may have a common optical axis and be placed either on the same side of the measuring volume (back-scattering) or on opposite sides (forward-scattering). An essential feature of such a setup is that it is calibrated by the laser wavelength and the geometry of the transmitter.

provided a photocurrent oscillating with a frequency given by the difference between the scattered light and the reference beam frequencies. This principle of 'optical heterodyning' has been the basis for most methods evolved since then. Figure 11 shows the principles of one of the most used configurations. Only measurements performed on very high speed flows allow for a direct optical determination of the frequency shift.

An alternative system is the 'time-of-flight' laser anemometer based on the time of flight between two small volumes in space of either single particles or particle patterns. It has been shown by Lading (1976) that this principle is closely related to the Doppler method, and that the same very good space-time resolution can be obtained - in some cases even better. Such anemometers have so far been most successful for nonintrusive measurements on fairly short ranges (i.e. less than a few metres) in liquids and gases. Figure 12 shows an example of a time-of-flight anemometer.

The processing of the photodetector signals depends very much on

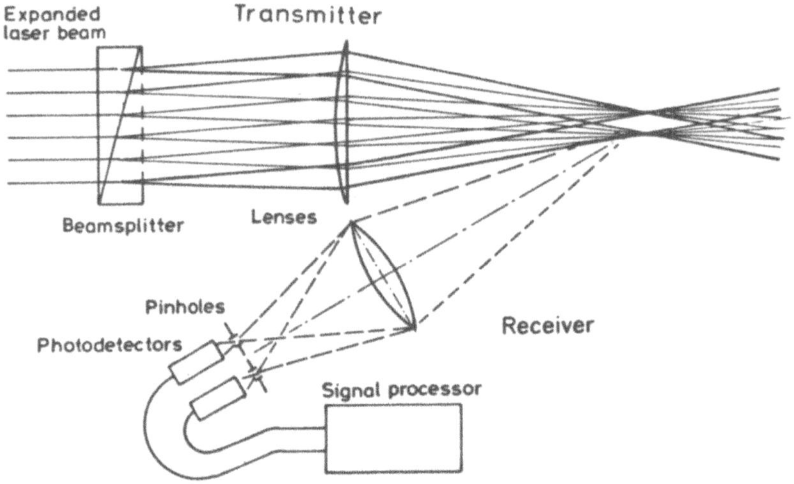

Fig. 12 Schematic diagram of a time-of-flight anemometer. The beam-splitter divides the incoming beam into two angularly (but not necessarily spatially) separated beams. The transmitter lens refracts and focuses these two beams into two spatially separated spots. Scattered light from particles passing the two focal volumes is collected and detected in the receiver. As in Figure 11, the transmitter and receiver may have a common optical axis either in a forward- or back-scattering configuration.

the specific circumstances, that is, on particle concentration, scattered light power, layout of the optical system and, of course, on what kind of velocity information is required. From a fluid mechanical point of view, the processors can be categorized into three groups:

(1) The first group essentially give a continuous output of which the expected value is proportional to the instantaneous velocity (George and Lumley, 1973); the instrumental averaging time is much shorter than the integral time scale.

(2) In the second group the instrumental averaging time is much longer than the integral time scale. The output is typically a correlation function or a power spectrum of the photocurrent. From this correlation function or power spectrum it is in principle possible to calculate all the moments of the velocity probability distribution; in practice, often only the mean and variance. The spectrum of the velocity fluctuations cannot be determined.

(3) For the third group the velocity is (conditionally) sampled at random, e.g. the particle concentration can be so low that the mean number of particles in the measuring volume is much less than one. It is - again in principle - possible to determine the correlation function or power spectrum for the turbulence if all three components are measured. However, in practice, the estimates of the mean and variance will often be biased.

For long-range measurements, two Doppler systems have appeared. One uses visible light from an argon-ion laser having essentially a backscattering layout as shown in Figure 11. This system will

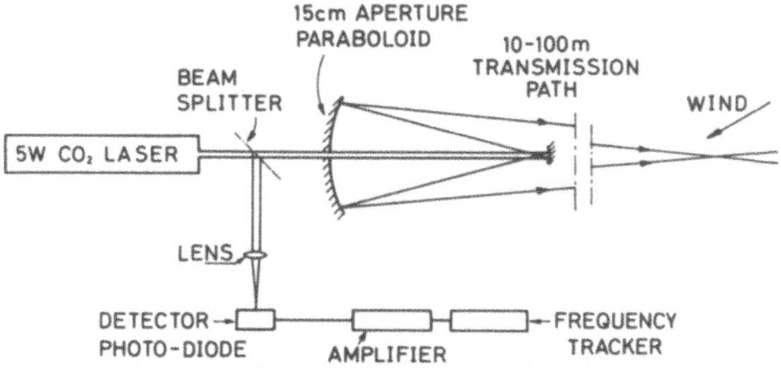

Fig. 13 Schematic diagram of a CO_2 laser anemometer.

measure the velocity component perpendicular to the optical axis. The second, used in most successful long-range systems, uses a CO_2-laser with a wavelength of 10.6 μ (infrared). The layout of such a system can be as shown in Figure 13. Measurements have been made at ranges up to 100 m with a 5 W laser (Hughes et al., 1972).

A pulsed CO_2-laser Doppler system has been developed at the NASA Marshall Space Flight Center (Huffaker, 1975) for the detection of clear air turbulence (CAT). The range is here determined by the time of flight of a laser pulse, as opposed to the continuous wave systems where the measuring volume is confined by the spatial filtering properties of the setup.

Recent experiments have proved that the time-of-flight system can be used for measurements at longer range (Lading et al., 1978). The velocity component perpendicular to the optical axis is measured. Ranges up to 70 m have been obtained with 0.5 W laser power from an argon-ion laser.

5.2 Devices Based on Turbulent Fluctuations in the Refractive Index

Turbulence induces fluctuations in the refractive index of the fluid in which it occurs. The movement of 'frozen turbulence' patterns (Taylor's hypothesis) can be utilized for mean velocity measurements. The smallest scale of turbulence in the atmosphere which causes detectable disturbances is generally of the order of 1 mm. The cone within which light is scattered (diffracted) is then limited to roughly 1 mrad (this is a simple result of diffraction theory), which means that only forward-scattering configurations can be used. This is in contrast to systems using longer wavelength radiation, e.g. microwave systems or acoustic sounders. Since the velocity is sampled by the turbulence pattern, it is naturally

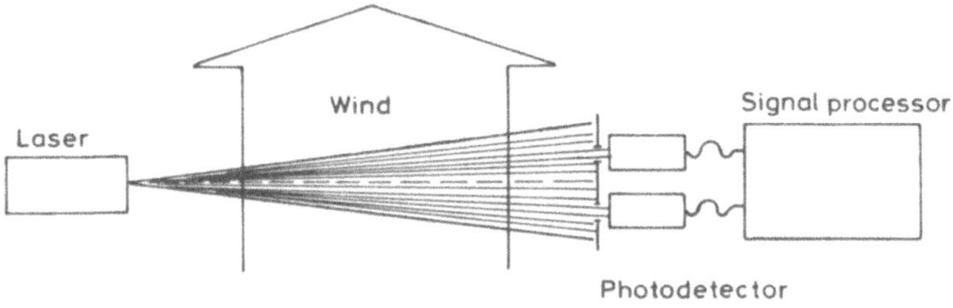

Fig. 14 Layout of a cross-wind anemometer used for measuring the average component of the wind across a laser beam, parallel to the spacing of the two detectors.

impossible either spatially or temporally to resolve that part of the turbulence which generates the detected scintillation pattern. The output must necessarily represent an average over many 'eddies'. Practical systems have used a time-of-flight principle, or a principle derived from it.

Figure 14 shows an example of a 'cross-wind anemometer', which measures the wind averaged along the path of the laser beam. The output is the derivative of the cross-correlation function at zero delay (Fig. 15). It has been shown that this quantity is proportional to the mean wind velocity and that the calibration is independent of the velocity distribution along the path. This is not the case if the peak of the cross-correlation function is assumed to be proportional to the average wind velocity (Lawrence et al., 1972). The system can measure average cross-winds over paths from 300 m to 10 km under normal weather conditions. A more recent version utilizes a spatial filter in the form of an array of photodiodes at the receiver end. This also allows for measurement of the refractive-index structure parameter (C_n^2). In order to get path-resolved measurements, configurations with crossed beams and sets of detector arrays have been proposed and at test model built. A resolution of roughly one-sixth of the total path length has been obtained (Clifford et al., 1974).

In the case of very strong turbulence, saturation may occur in the scintillation, leading to erroneous results. A system has been

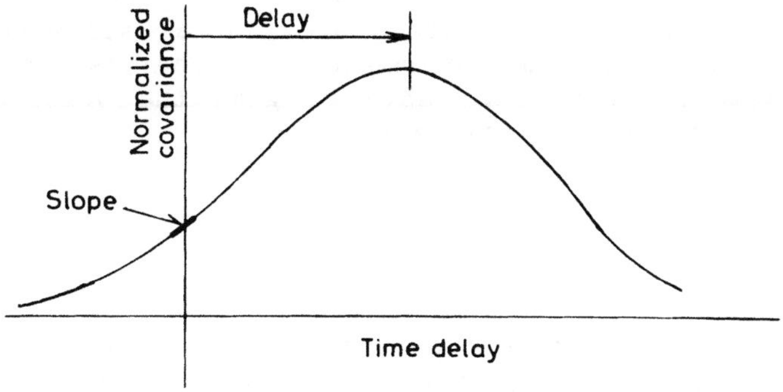

Fig. 15 A normalized covariance function of the detector signals in Figure 14. In the system of Lawrence et al. (1972) the slope at zero time lag, rather than the delay to the peak, is used as a measure of velocity.

developed to overcome this saturation problem. It uses an incandescent lamp and a larger receiver aperture, and operates partly in what could be called a 'shadowgraph' mode (a shadowgraph may be observed on the bottom of shallow waters on a sunny day, caused by the focusing and defocusing effects produced by water surface irregularities). A disadvantage of using an incandescent lamp is that stray light cannot be rejected as well as in the case of a laser source (Ochs et al., 1976).

The passive remote cross-wind system uses light from a natural (distant) source and therefore only requires access to one end of the path. This system can, of course, only be used in daylight and has to be 'calibrated' for the specific scene from which light is collected. A system using a pulsed high-power laser for illumination is being investigated. The laser is co-located with the receiver optics.

5.3 Use of Laser Anemometers

The following remarks may be appropriate for practical applications of laser anemometers. A stable platform will in general be necessary for any setup, but especially for setups where the transmitter and receiver are situated at different locations.

Devices based on turbulence fluctuations in refractive index will give a path-averaged output. They can operate with a low power laser, or even an incandescent lamp.

Aerosol-based systems give a very good spatial and temporal resolution - the best for the visible laser light systems. The light sources have to be high-power lasers. Input power requirements for long range visible laser systems are several kilowatts and water cooling is necessary. At the moment, only the visible and infrared laser anemometers have been able to give the velocity component perpendicular to the optical axis.

REFERENCES

BROOK, R.R. 1977. Effective dynamic response of paired Gill anemometers. *Boundary-Layer Meteorology,* 11: 33-37.

BUSCH, N.E. 1965. A micrometeorological data-handling system and some preliminary results. Risø Report No. 99, 92 pp., Risø National Laboratory, DK4000, Roskilde.

BUSCH, N.E., R.M. BROWN and J.A. FRIZZOLA. 1970. Vertical velocity variances and Reynolds stresses at Brookhaven. *Journal of Applied Meteorology,* 9: 583-587.

BUSCH, N.E., P. DORPH-PETERSEN, C.J. CHRISTENSEN and L. KRISTENSEN. 1976. Wind-direction measurements at the Risø tower. Risø Report No. 352: 67-71. Risø National Laboratory, DK4000 Roskilde.

BUSCH, N.E. and L. KRISTENSEN. 1976. Cup anemometer overspeeding. *Journal of Applied Meteorology*, 15: 1328-1332.

CAMUFFO, D. 1976. How to obtain mean value and variance of wind direction by using a sine-cosine transducer. *Atmospheric Environment*, 10: 167-168.

CLIFFORD, S.F., G.R. OCHS and TING-I WANG. 1974. Theoretical analysis and experimental evaluation of a prototype passive sensor to measure crosswind. Report NOAA TR ERL 312-WPL 35, NOAA Wave Propagation Laboratory, Boulder, Colorado, 28 pp.

CHRISTENSEN, O. 1971. Wind velocity sensing by means of four-bladed helicoid propellers. Report of University of Michigan, Department of Meteorology and Oceanography, 100 pp. (can be obtained by writing to: Dr. O. Christensen, Risø National Laboratory, DK 4000 Roskilde).

DENTLER, F.-U. 1978. The effect of rainfall on measurements of mean wind speed with cup anemometers in the surface layer at sea. *Boundary-Layer Meteorology*, 14: 134-140.

FICHTL, G.H. and P. KUMAR. 1974. The response of a propeller anemometer to turbulent flow with the mean wind vector perpendicular to the axis of rotation. *Boundary-Layer Meteorology*, 6: 363-379.

FRANCEY, R.J. and J.R. GARRATT. 1978. Eddy flux measurements over the ocean and related transfer coefficients. *Boundary-Layer Meteorology*, 14: 153-166.

GARRATT, J.R. 1975. Limitations of the Eddy-Correlation technique for the determination of turbulent fluxes near the surface. *Boundary-Layer Meteorology*, 8: 255-259.

GEORGE, W.K. and J.L. LUMLEY. 1973. The laser-Doppler velocimeter and its application to the measurement of turbulence. *Journal of Fluid Mechanics*, 60: 321-362.

GILL, G.C. 1975. Development and use of the Gill UVW anemometer. *Boundary-Layer Meteorology*, 8: 475-495.

HICKS, B.B. 1972. Propeller anemometers as sensors of atmospheric turbulence. *Boundary-Layer Meteorology*, 3: 214-228.

HOLMES, R.M., G.C. GILL and H.W. CARSON. 1964. A propeller-type vertical anemometer. *Journal of Applied Meteorology*, 3: 802-804.

HORST, T.W. 1973. Corrections for response errors in a three-component propeller anemometer. *Journal of Applied Meteorology*, 12: 716-725.

HUFFAKER, R.M. 1975. CO_2 laser Doppler systems for the measurement of atmospheric winds and turbulence. *Atmospheric Technology, NCAR*, 6: 71-76.

HUGHES, A.J., J. O'SHAUGHNESSY, E.R. PIKE, A. McPHERSON, C.SPAVIUS and T.H. CLIFTON. 1972. Long range anemometry using a CO_2 laser. *Opto-electronics*, 4: 379-384.

KAGANOV, E.I. and A.M. YAGLOM. 1975. Errors in wind speed measurements by rotation anemometers. *Boundary-Layer Meteorology*, 4: 289-309.

KONDO, J., G. NAITO and Y. FUJINAWA. 1971. Response of cup anemometer in turbulence. *Journal of the Meteorological Society of Japan,* 49: 63-74.

LADING, L. 1976. Comparing a laser Doppler anemometer with a laser correlation anemometer. *The Engineering Uses of Coherent Optics,* edited by E.R. Robertson, Cambridge University Press: 493-510.

LADING, L., A.S. JENSEN, C. FOG and H. ANDERSEN. 1978. Time-of-flight laser anemometer for velocity measurements in the atmosphere. *Applied Optics,* 17: 1486-1488.

LARSEN, S.E. and N.E. BUSCH. 1974. Hot-wire measurements in the atmosphere. Part I: calibration and response characteristics. DISA Information No. 16: 15-36.

LAWRENCE, R.S., G.R. OCHS and S.F. CLIFFORD. 1972. Use of scintillation to measure average wind across a light beam. *Applied Optics,* 11: 239-243.

LINDLEY, D. 1975. The design and performance of a 6-cup anemometer. *Journal of Applied Meteorology,* 14: 1135-1145.

MacREADY, P.B., JR. 1965. Dynamic response characteristics of meteorological sensors. *Bulletin of American Meteorological Society,* 46: 553-558.

MacREADY, P.B., JR. 1966. Mean wind speed measurements in turbulence. *Journal of Applied Meteorology,* 5: 219-225.

MacCREADY, P.B., JR. and H.R. JEX. 1964. Response characteristics and meteorological utilization of propeller and vane wind sensors. *Journal of Applied Meteorology,* 3: 182-193.

OCHS, G.R., S.F. CLIFFORD and TING-I WANG. 1976. Laser wind sensing: the effects of saturation of scintillation. *Applied Optics,* 15: 403-408.

PETERSON, E.W., N.O. JENSEN and J. HØJSTRUP. 1979. Observations of downwind development of wind speed and variance profiles at Bognaes and comparison with theory. *Quarterly Journal of the Royal Meteorological Society,* 105 (in press).

TEUNISSEN, H.W. 1976. Comments on development and use of the Gill UVW anemometer. *Boundary-Layer Meteorology,* 10: 515-516.

WIERINGA, J. 1967. Evaluation and design of wind vanes. *Journal of Applied Meteorology,* 6: 1114-1122.

WIERINGA, J. 1972. Tilt errors and precipitation effects in trivane measurements of turbulent fluxes over open water. *Boundary-Layer Meteorology,* 2: 406-426.

WYNGAARD, J.C., J.T. BAUMAN and R.A. LYNCH. 1974. Cup anemometer dynamics. *Proceedings of the Instrument Society of America,* Pittsburgh, Pennsylvania, 10-14 May, 1971, 1: 701-708.

YEH, Y. and H.Z. CUMMINS. 1964. Localized flow with a He-Ne laser spectrometer. *Applied Physics Letters,* 4: 176-178.

2

Hot Wire and Hot Film Anemometers

L. Hasse and M. Dunckel

1. INTRODUCTION

Hot wire and hot film sensors are used for measurements of fluctuations in the wind. Such anemometers are in competition with both mechanical and sonic anemometers. At least some of the mechanical sensors are more stable in calibration, and less expensive. The high frequency resolution of the hot wire/hot film sensors, however, cannot be matched by any other anemometers. Also, they can be made small enough and fast enough to measure within the dissipation range. Hot wire sensors have been used in atmospheric turbulence measurements since 1936; there are review articles by Bradshaw (1963), Comte Bèllot (1976), Corrsin (1963), and Sandborn (1972).

2. SENSOR TYPES

In wind measurements the cooling of a heated wire or film is dependent on the velocity and density of the flow past the wire. Hot wire sensors can be made by welding thin wires (of order 10 μm) to supports, or by using Wollaston wire (a platinum wire core embedded in a silver mantle). The Wollaston wire is soldered to the support and the desired sensor length is obtained by etching the silver mantle away. In a similar way, the sensitive length of welded wires is determined by plating the ends with conducting material. Since the resistance of the core alone is much higher than of the mantle plus core, only the piece of bare core is heated and is wind sensitive. Typical dimensions are: 5 μm wire diameter, 1.25 mm sensitive length, 3 mm total length of wire. Commercially available sensors are generally used, since accurate production requires some skill. This is even truer for hot film sensors. These usually consist of a cylindrical quartz or glass core, covered with a nickel or platinum film which is in turn electrically insulated with a very thin quartz or ceramic coating. Typical dimensions are: 1.25 mm active length, 50 μm to 70 μm diameter of core rod, less than 0.1 μm film thickness, 2 μm thickness of coating. The actual length of the rod is typically 3 mm; the length of the active film is determined by gold plating the ends. Due to the greater length of rod compared to film, any influence from the prongs on the flow around the wire is probably small except when the flow is nearly parallel to the sensor. The typical maximum frequency resolution for a single sensor is 30 kHz with higher frequencies attainable with wire sensors (when used in the constant temperature mode and assuming that the feedback loop is fast enough not to limit the frequency response). In most meteorological applications the maximum frequency is not determined by the frequency limit of the sensor but rather by the spacing between the different sensors used in a system.

The hot wire has the advantage that, by making the ratio of length (L) to diameter (D) large, the temperature distribution along the wire is more homogeneous and thus the effect of flow not perpendicular to the wire is more predictable. Typically, a ratio L/D = 200 is used. In order to keep L short, D must be made small; this decreases mechanical stability. Also, thinner wires are more effective samplers of atmospheric contamination. Sea salt particles accumulating on the wires affect the heat exchange process, giving rise to rapid changes in calibration.

With a film sensor, mechanical strength and film thickness are independent. The main advantage, compared to a traditional wire, lies in the much greater mechanical strength of the film sensor for similar electrical and thermal properties. Hot film sensors have a more stable calibration than wires; this is probably a result of their larger diameter. Smaller diameter film sensors are less stable in the marine atmosphere; however, 150 μm diameter sensors could be used for time spans of the order of 6 hr. Calibration stability is better since the larger diameter film sensors are less efficient as collectors and, with the same amount of salt deposit, less change occurs in flow and thermal properties. Also, film sensors may be cleaned and then regain their original calibration.

Both wire and film sensors have been used in the marine atmosphere. Mechanical sturdiness and stability of calibration are usually decisive arguments in favour of the hot film sensor rather than the hot wire, with its higher frequency response.

Recently, split film sensors have been introduced; the film on a cylindrical rod is in two parts. The ordinary hot wire/film sensor cannot distinguish between flow components in a plane perpendicular to the wire. Even with a three-dimensional system an ambiguity of 180 degrees exists, and the split film was invented to eliminate this. Such sensors would be advantageous for fixed installations which have a wide range of wind directions. For a three-dimensional split film system, however, the useful range of wind direction will be severely limited because analysis becomes impracticable for winds nearly parallel to the sensor. Carroll (1979) found the useful range restricted to horizontal angles of $\pm$ 18 degrees from the symmetry axis. The advantages of the split film system are paid for by increased complexities of data evaluation and calibration (which are feasible only for an experienced group with ample resources). To the best of our knowledge the split film has not yet been used at sea.

Hot film sensors are available in other shapes as well: conical, wedge-shaped, hemispheric, and flush surface. These shapes, as far as we know, have not yet been used at sea, although they offer some advantage in bad weather since they are more rugged. Such sensors have been used successfully on airplanes during rain (Merceret, 1976).

Cylindrical hot film sensors, which are geometrically similar to hot wire sensors, are used to obtain fluctuating velocity components in the surface boundary layer over the water. In the following we will use the term 'hot film' sensor for cylindrical elements, if not stated otherwise.

The theory originally developed for hot wire anemometry is also applicable to hot film elements. There are two modes of operation: constant current, or constant temperature. Constant current systems are electronically less demanding, but in this mode the temperature of the element depends very much on the wind speed perpendicular to the wire, and in low wind speeds overheating occurs. This was especially detrimental with the old-fashioned platinum wire sensors, which aged from excessive heating. Also, the calibration curve of a constant current system is more nonlinear. Consequently, constant current systems are no longer used in air-sea interaction studies, and will not be treated further here.

The constant temperature mode is electrically more demanding. An electronic servo loop is used to maintain the temperature constant; the power consumption is a measure of the cooling of the element. The main advantage of the constant temperature mode is its high frequency resolution. Even the comparatively large thermal mass of the rod of a hot film element does not limit its frequency response, since its temperature remains essentially constant. The power is used mainly to replace the energy loss to the air. The frequency response of a hot film sensor is more complicated than that of a hot wire due to the different time constants of the film and the substrata.

Another advantage of constant temperature operation is that one can use higher overheat without risk of burning out the probe; the higher overheating reduces the temperature sensitivity of the hot sensor relative to its wind sensitivity. Typical overheat for a constant current system is 60°C, while for constant temperature it is 200°C at 5 m s^{-1} wind speed. It may be noted that the term 'constant temperature' is correct only in the mean, since the temperature along a wire or over a film is not truly constant (see Champagne et al., 1967).

3. VELOCITY RESPONSE

A concise introduction into the theory of hot wire anemometry is given by Hinze (1959). A general expression for the heat flux from a long wire (large length to diameter ratio) in terms of Prandtl and Reynolds numbers is determined experimentally. Since there are always side effects (for example, by thermal conduction at the end of the active length of the wire) the general expression is used as a guideline only. The calibration curve is written (for flow perpendicular to the wire)

$$\frac{I^2 R_w}{R_w - R_a} = A' + B'U^n \qquad (1)$$

where I is the electrical current, U is the wind velocity, and R_w and R_a are the resistances at the operating temperature of the wire and at the air temperature. Often $n = \frac{1}{2}$ is used, but values of $n \sim 0.45$, decreasing with increasing wind, are quoted (this may be important for use with aircraft). Since the actual sensors are not infinitely long, A' and B' are determined experimentally and verified over the desired range of speeds. For constant temperature operation $R_w/(R_w - R_a)$ is taken as constant and instead of Equation 1 a simpler form is used:

$$I^2 = A + B\sqrt{U} \qquad (2)$$

Equations 1 and 2 are called King's law, although King's theoretical derivation has been found unacceptable by Hinze (1959). Figure 1 shows an actual calibration curve for a hot film. At very low speeds (say, below 1 m s^{-1}) the flow around the wire is dominated by buoyancy due to overheating rather than by the mean flow; this leads to a systematic error.

4. DEPENDENCE ON TEMPERATURE AND HUMIDITY

Using the constant temperature mode, the sensor temperature is held constant, but the properties of the surrounding air may vary. Thus the heat loss from the wire is affected not only by the velocity but also by the temperature difference between sensor and air, and by the density of the air. Since for calibration Equation 2 is used instead of Equation 1, a small error is introduced. Automatic air temperature compensation could be built into the circuit by using an extra wire/film sensor (Dunckel et al., 1974). This increases the complexity of the sensor head and the circuitry, and thus the chance of failure. Since the influence of air temperature on the temperature difference becomes smaller with larger overheat, simpler methods can be used. In principle, Equation 1 could be used with a knowledge of mean air temperature and manufacturer's specifications. In practice, one can check the calibration by calibrating at different mean temperatures.

Air temperature also enters through the density, since the heat loss is dependent on ρU rather than on U alone (that is, in Equations 1 and 2, B' and B should be multiplied by $(\rho/\rho_c)^n$, where n is the same power as that used on U, and ρ_c is the density during calibration). The water vapour content of the air also affects the calibration through the density because the coefficient for heat exchange into water vapour is different from that into dry air. A detailed discussion is given by Larsen and Busch (1974). If only terms of first order are kept, A and A' contain a factor (1 + 1.34 e/p) and B and B' a factor (1 + 0.63 e/p), where e is water vapour pressure and p air pressure. Here it is assumed that the density is for dry air and the virtual water vapour correction to the

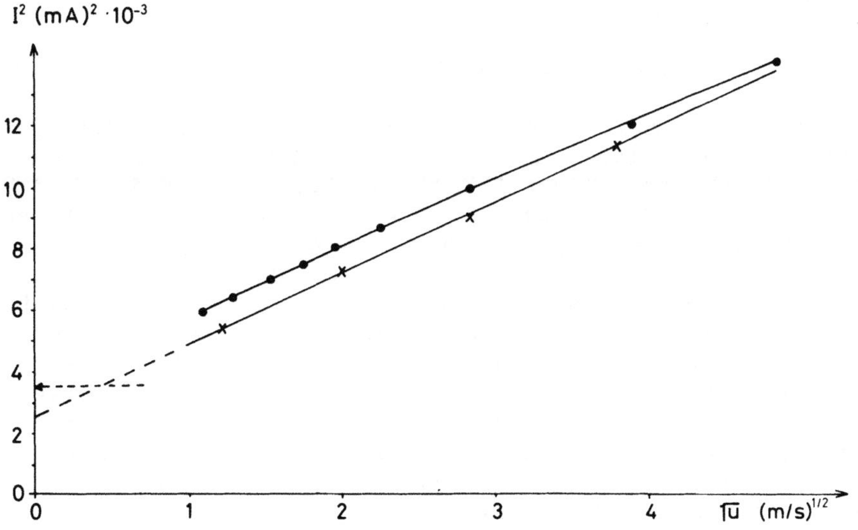

Fig. 1 Actual calibration curves for two cylindrical hot film sensors. The slightly curved line would be more linear if U^n with n = 0.45 instead of n = 0.50 were used.

density is included in the coefficients B and B'. The numerical values are for large L/D wires, but should be sufficient for other cylindrical elements. For large humidity differences between calibration and measurement the correction factors may vary by 3% (the vapour pressure over a tropical ocean can be 3 kPa). It is useful to note air temperature, air pressure, and humidity separately with each calibration and data run (however, the above corrections may be too large, see Larsen and Busch, 1979).

5. DIRECTIONAL RESPONSE

Calibration Equations 1 and 2 are for flows perpendicular to the wire. However, the cooling of the wire depends on the components of flow perpendicular to the wire and, to a lesser degree, on the flow component parallel to the wire. Equations 3 and 4 (below) show that the hot wire/film sensors have a non-cosine response, necessitating a more elaborate evaluation (see Section 6). The effective velocity sensed by the wire is called the 'cooling velocity', U. Let v_1, v_2, and v_3 be the velocity components in a co-ordinate system fixed to the wire: v_1 perpendicular to the wire and parallel to the mean flow, v_2 perpendicular to v_1 and to the wire, and v_3 parallel to the wire. Then

$$U^2 = v_1^2 + k_1^2 v_2^2 + k^2 v_3^2 \tag{3}$$

For flow in one plane (V is wind speed, β the yaw angle between the wind vector and the sensor normal) and taking $k_1 = 1$, this reduces to the more familiar form

$$U^2 = V^2(\cos^2\beta + k^2\sin^2\beta) \tag{4}$$

(e.g. Champagne et al., 1967). k and k_1 vary with velocity angle of attack, sensor type, and especially with the L/D ratio. The larger L/D, the more 'ideal' k and k_1 become; that is, $k \to 0$ and $k_1 \to 1$ for $L/D \to \infty$. k_1 was introduced by Jørgensen (1971) and is discussed extensively by Larsen and Busch (1974); it allows for asymmetry of the wire and flow properties around the axis parallel to the wire. An asymmetry could be induced, for example, by the influence of prongs and supports. In most commercially available hot film sensors (except split film sensors, of course) k_1 is nearly 1. Larsen and Busch quote values for k_1 of 1.02 to 1.06, Freytag (1976) gives 1.01, while Jørgensen gives values up to 1.12 depending on sensor configuration and wind speed (the higher values being applicable to compact sensors and higher speeds). It is common practice to use $k_1 = 1$.

k is more troublesome. Values for k of 0.1 to 0.2 are quoted. Kjellstrøm and Hedberg (1970) found a linear dependence of k^2 on wind speed for hot wires of L/D = 220. k is presumed to approach zero for large L/D. A definite advantage of wire over film sensors is large L/D, and hence a better known and smaller k. For film sensors L/D is typically 25, if based on the total diameter (core plus film) of the probe. But since the core is of poorly conducting material, a direct comparison of film sensors with wires is not possible. Jørgensen (1971) gives k values of 0.5 to 0.2 for a hot film probe. Since at small yaw angles the influence of the $k^2\sin^2\beta$ term is negligible, it appears sufficient to use an approximate value for k for yaw angles up to 70 degrees. On the other hand, the uncertainty of k makes it impractical to resolve the wind components if the flow is parallel or nearly parallel to the wire.

6. DETERMINATION OF FLUCTUATING COMPONENTS

Calibration is usually done in a wind tunnel with low-intensity turbulence and could be called a static calibration; the same calibration is generally used to obtain the fluctuating velocities. Equations for the fluctuating quantities are given by Hinze (1959), Champagne and Sleicher (1967), and Larsen and Busch (1974). Typical sensor configurations are: a single sensor (perpendicular to the flow or slanting), an X-configuration, and a three-dimensional system (see Fig. 2).

6.1 Single Sensor

For a sensor parallel to the z axis and flow in the x,y plane with x the downwind coordinate,

$$U = \sqrt{(\bar{u}_1 + u_1')^2 + u_2'^2 + ku_3'^2} \tag{5}$$

where u_1, u_2, u_3 are the wind components in the natural coordinates with u_1 downwind and u_3 vertical; the mean and fluctuating quantities are denoted by overbars and dashes respectively. Such a wire is to first order sensitive only to the mean flow and to fluctuations of the downwind component; the lateral component enters only to second order. For a single, slanting wire explicit expressions are given by Larsen and Busch (1974). Since single slanting sensors are rarely used, these expressions are not repeated here.

6.2 X-Configuration

For the X- or V-configuration consider the wind in a natural coordinate system. With an X-sensor system in the x,z-plane, pointing in the downwind direction, the downwind and vertical fluctuating components are to be measured, but the sensors are also sensitive to the lateral component. From the cooling velocities U_1 and U_2 obtained by the two sensors (1 and 2), apparent horizontal velocities $V_1 = (U_1 + U_2)/\sqrt{2}$ and $V_3 = (U_1 - U_2)/\sqrt{2}$ are defined. It can be shown (e.g. Larsen and Busch, 1974) that, neglecting some smaller terms,

$$V_1 = \bar{u}_1 + u_1' + \frac{1}{1 + k^2} \frac{u_2'^2}{\bar{u}_1}$$

$$V_3 = \frac{1 - k^2}{1 + k^2} u_3' \left(1 - \frac{u_2'^2}{\bar{u}_1^2}\right) \tag{6}$$

Unless the lateral component is known an error, of the order of 10% for typical atmospheric conditions, occurs in the measurement of the downwind component. The effect on the vertical component is smaller (order 1% or less), but it remains advisable to keep the system pointing into the mean wind direction.

6.3 Three-dimensional Configuration

Due to the contamination of the X-configuration by the lateral component it is better to use a three-dimensional system, which

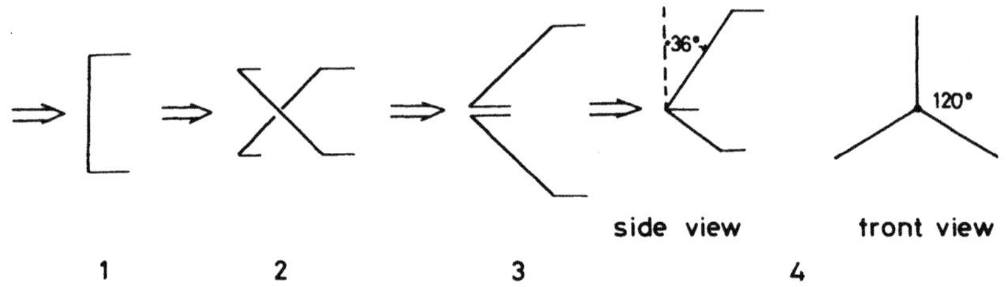

Fig. 2 Typical sensor configuration:
(1) Single hot film, e.g. vertical to measure horizontal wind component.
(2) X-configuration: two hot film elements, 45° and -45° from vertical in one plane, to measure the two flow components in this plane system.
(3) Two sensor elements like a standing V is less susceptible to crosstalk than the X-configuration. Two standing Vs in diverging directions have been used in parallel to obtain inclination of the wind vector, with little sensitivity to varying azimuthal angles of attack.
(4) 3D system, where three hot film sensors are mutually perpendicular, forming three adjacent edges of a cube, the diagonal of which is along the probe axis.

usually consists of three mutually perpendicular wires. These wires define a cartesian coordinate system, say a,b,c. The velocity components in the a,b,c system are obtained from the cooling velocities, using Equation 3. From this one proceeds via a matrix operation to the x,y,z (natural) coordinate system.

For the special three-dimensional system where the sensors form the adjacent edges of a cube, the equations are given below, assuming ideal geometry (negligible deviation from right angles) and equal coefficients, k, for the three sensors, as well as $k_1 = 1.00$. Let U_i, $i = 1,2,3$ denote the cooling velocities of the three sensors, and v_i, $i = 1,2,3$ the wind components along the sensors with the same index; then

$$\begin{pmatrix} U_1^2 \\ U_2^2 \\ U_3^2 \end{pmatrix} = \begin{pmatrix} k^2 & 1 & 1 \\ 1 & k^2 & 1 \\ 1 & 1 & k^2 \end{pmatrix} \begin{pmatrix} v_1^2 \\ v_2^2 \\ v_3^2 \end{pmatrix} \tag{7}$$

Hence $$V^2 = \sum_i v_i^2 = \frac{1}{2 + k^2} \sum_i U_i^2 \tag{8}$$

$$v_i = \sqrt{\frac{V^2 - U_i^2}{1 - k^2}} \tag{9}$$

We may orient the probe so that the x axis (i.e. horizontal downwind) is in the principal diagonal of the cube and the No. 3 wire is in the x,z plane. Then the (true) wind components are

$$\begin{aligned} u_1 &= (v_1 + v_2 + v_3)/\sqrt{3} \\ u_2 &= (-v_1 + v_2)/\sqrt{2} \\ u_3 &= (2v_3 - v_1 - v_2)/\sqrt{6} \end{aligned} \tag{10}$$

With proper alignment, these are the desired instantaneous wind components from which the mean and fluctuating parts may be obtained. It may be noted that for this system the wires form an angle of 35.26 degrees with a plane perpendicular to the x axis; this is an average yaw angle (see Fig. 2).

6.4 Influence of Temperature Fluctuations

Temperature fluctuations can appear as errors in the fluctuations of the wind components through temperature dependence of A and B.

The error is of the same order as neglecting density fluctuations compared to velocity fluctuations, which is common practice for the atmospheric boundary layer. The influence of temperature fluctuations may be more important in the determination of the sensible heat flux $\overline{\theta' u_3'}$, where they produce spurious correlations. In the latter case, over the sea, other limitations may be more severe (see Larsen et al., Chapter 15). Corrections of wind fluctuations for the effects of temperature fluctuations are given by Larsen and Busch (1976). For hot film measurements of velocity fluctuations in water, the influence of temperature fluctuations is definitely important (see Gibson and Deaton, Chapter 18).

7. CALIBRATION CONSIDERATIONS

Evaluation of wind components from two- and three-dimensional probes using the equations in the preceding section depends on the sensor configuration having very close to the assumed geometry. With the tiny probes in common use today, this cannot be expected to hold true. Deviations of a few degrees are likely, which produce about the same percentage error in the indicated wind. It is recommended that the sensor system be checked during calibration, for non-ideal geometry.

Another common error for field use is the misalignment of the probe with its support (e.g. wind vane), assuming that the orientation of the support is known accurately. Mounting of probes at sea (with a system of sensors fixed relative to each other) may for example produce a small deviation from the vertical. Geometry and misalignment errors are similar in effect for a given wind direction, but the effects will behave differently for larger changes of the wind direction. Calibration changes will produce similar errors in the data, except that the angles describing probe geometry errors and mounting alignment errors should be constant for one probe, while changes in calibration may be either sudden or drift-like.

It is not very practical to do a full calibration at sea. But it has been found useful to have a small wind tunnel and to check at least one calibration point for each sensor prior to and after each exposure. It is also recommended that a 'running calibration' be obtained by comparing the mean wind speed and direction of the fast response system with the averaged output of a cup anemometer and a wind vane. In the surface layer at sea the vertical component can be checked by assuming $\overline{\rho u_3} = 0$ over a long run. Another possible assumption, $\overline{u_2' u_3'} = 0$ near the ground, is more questionable, since the equations of motion show that the wind stress vector changes its direction more rapidly with height than the wind direction; this assumption is, however, still a useful tool as a consistency check. Additionally, if corrections are obtained from these assumptions, their variation with time and changing wind speed and direction can be checked, making it possible to assign errors to changes of calibration, misalignment and non-ideal geometry. A running calibration has been used with X- or V-shaped probes, and is the approach we are working on now for a three-dimensional rectangular system.

8. METHODS OF EXPOSURE

The hot wire/film cannot discriminate directions in the plane perpendicular to the wire. X- or V-configurations are often used at sea to determine the horizontal downwind and the vertical velocity. For these, it is important to keep the horizontal crosswind component small. This is achieved by mounting the X-configuration on a rapid-response wind vane. For determination of stress $\tau = \overline{\rho u_1' u_3'}$, the contamination of u_1' by u_2', as mentioned above, is not detrimental since $\overline{u_2' u_3'}$ is small compared to $\overline{u_1' u_3'}$, except perhaps under special sea state conditions.

Mounting X or V probes on a wind vane has the somewhat disconcerting effect that the horizontal component measured along the vane direction does not correspond fully to the true downwind component. The wind vane follows those low frequency turbulent fluctuations that are not filtered out by the vane's response (Busch et al., see Chapter 1). In order to calculate the stress in the mean downwind direction corresponding to the Reynolds averaging of the Navier-Stokes equations, an averaging time is required such that the periods of low frequency fluctuations, which still contribute significantly to the stress, are short compared to the averaging time. For this averaging time (typically 10 minutes) the downwind direction is taken as constant. In order to properly resolve the wind components it is also necessary to record the vane direction.

The above effect can be conceived as adding a term $\frac{1}{2}u_2'^2/\overline{u_1}$ to u_1' (where u_1' is the fluctuating component in the downwind direction, and u_2' perpendicular to it). Thus the effect is of order $\frac{1}{2}u_2'/\overline{u_1}$ compared to one, which amounts to an error of a few percent. The resulting influence on stress determinations will usually be less than 1% except under extremely unstable conditions, where it may reach the order of 5%. Such conditions occur at sea in calm weather, which in practice is not very suitable for hot wire/film anemometry anyway.

For a three-dimensional probe, the requirement that the probe be oriented in the mean wind direction is not as stringent, since the lateral component is measured; it suffices to have the wind vector within the appropriate solid angle. Since for wind speeds of, say, $5\ m\ s^{-1}$ and higher, the wind is usually quite steady at sea, it often suffices to put the sensor on a remotely controlled shaft, which is either activated by hand or automatically by a wind vane if a certain range of wind direction is exceeded.

9. RESOLUTION AND LINEARIZATION

Since the calibration curve of heated wires and films is strongly nonlinear, their sensitivity is dependent on the mean wind speed. To increase accuracy at higher wind speeds and to make optimum use of recording dynamic range, electronic linearizers can be used; this is especially desirable for analogue recording.

Instead of in situ linearization, an improvement in analogue recording accuracy can be obtained by automatic range switching. In this case, an extra channel is necessary to record the range switching. It is well known that velocity spectra fall off with the minus five-thirds power of frequency. In order to go to higher frequencies when analogue recording, it may be advisable to filter the output (for example, of the linearizer) and record low and high frequency parts separately. An alternate approach would be high-resolution digital recording.

As a design guideline for measurements in the turbulence production range, a typical record to be used for the eddy correlation technique in the surface boundary layer would have a record length of 30 to 60 minutes, a Nyquist frequency of 10 Hz, and a 12-bit resolution (including range switching if applicable) for wind speeds of up to 20 m s^{-1}. For airborne measurements, with typical aircraft speeds of 100 m s^{-1}, the frequencies and duration of the record are shifted correspondingly. If direct measurements of turbulent dissipation are desired, a wave number resolution up to $1/\eta$, where $\eta = (\nu^3/\varepsilon)^{\frac{1}{4}}$, the Kolmogorov length scale, is necessary. η is of the order of centimetres to millimetres, decreasing with increasing wind speed.

10. SPECIAL CONSIDERATIONS AT SEA

10.1 Platform Motions

As with other instruments that measure wind components, it is essential to relate the axes of the sensor system to the true vertical and horizontal. Since ships, airplanes, tethered balloons, and buoys do move, it is necessary to record the platform motions and correct for them. This increases the amount of data handling (and hence the chance of introducing errors) by a factor of 9 compared to mounting on a fixed support. Even when the sensor motions are known, movement through the turbulent field above the wavy sea surface considerably increases the complexity of the analysis problem. It may therefore be advisable to use a servo mechanism for tilt stabilization if moving platforms are to be used (see also Kaimal, Chapter 4, and Bean and Emmanuel, Chapter 30). The misalignment error has been mentioned already in Section 7; a short review of the desired accuracy is given by Kaimal in Chapter 4.

To avoid contamination of measurements by platform motions, Deacon (1959) has recommended using Kolmogorov's inertial subrange hypothesis for determining the wind stress. This is an appealing idea, since it involves spectral energy at frequencies higher than the frequency range of platform motions. Also, since local isotropy is assumed at higher frequencies, it should suffice to measure only one, say the downwind, component. From the Kolmogorov hypothesis one can determine the energy dissipation and, via the energy balance equation of the turbulence, the stress. The method is not without difficulties. Most measurements in the inertial subrange have been at frequencies between 1 and 20 Hz, which obviously is not high enough (Wucknitz, 1979). Since the spectra fall off as $f^{-5/3}$, the average spectral energy in a given frequency band is biased towards the lower frequencies, which may not yet be in the inertial subrange. Plotting the data as $f^{5/3}E(f)$ versus frequency [where E(f) is the spectral energy] will show any deviation from the -5/3 behavior. In addition, horizontal and vertical velocity components should be measured if the inertial dissipation method is to be used.

10.2 Salt Spray

Aside from questions of good exposure and platform motions, the main difficulty in hot wire/film anemometry at sea is contamination by salt spray deposits. This was a severe problem with hot wire sensors and is still a matter which requires due attention with hot film sensors. Some wire material (e.g. tungsten) is susceptible to corrosion and will thereby change resistance, with the subsequent danger of local overheating. It seems that the most important effect, applicable to both wire and film sensors, is that sensors of smaller diameter are more efficient collectors of aerosols. Friehe (pers. comm.) has reported rapid changes in the calibration of a 60 μm hot film sensor over the sea, which we have not experienced with 150 mm diameter hot film sensors. The diameter should, however, be kept small enough so that its Reynolds number is below the critical value in order to avoid von Karman vortex shedding. Contamination also increases the thermal time constant, and this effect is often used to detect such contamination. Many commercial systems have a built-in response time test mode.

10.3 Hot Film Anemometer Use in Rain

From an aircraft, hot film sensors of the more sturdy wedge-shaped variety have been used in rain in the subcloud layer, while within raining clouds, data acquisition was not possible (Merceret, 1976). Cylindrical hot film sensors of 150 μm diameter have also been used in tropical rains, an example is shown in Figure 3: the impact of drops produces sharp spikes. The initial recovery of the film sensor is very fast, but total recovery takes about one-tenth of a second (see Fig. 4). After longer exposure sensors may be damaged in rain. Splitting of the hot film coating experienced during GATE has been supposed to be due to sudden cooling by rain. Fortunately, raindrop spikes can easily be detected in the records and removed in the course of data reduction. If not removed, spikes would be most detrimental in measurements with inclined sensors: a spike at one sensor would show up in two or three velocity components and produce a spuriously high correlation. The error introduced in cross-correlation cancels somewhat, if the average error contribution of all three wires is considered, depending on the angle of attack to the symmetry axis of the system. An error of the order of 2% in the stress was estimated (Dentler, 1976) if records were used without spike elimination, which is not recommended.

10.4 Submerging Hot Film Probe

Hot wires have been used for velocity measurements just above the waves in a wind water tank by allowing the wire to be submerged by the wave crests. With a large overheat the wire dries rapidly enough to allow velocity masurements between the wave crests. Precautions are necessary to avoid local overheat when irregular waves only partially submerge the wire. The rapid accumulation of contaminants on the wire results in calibration changes. For further details see Wills (1976).

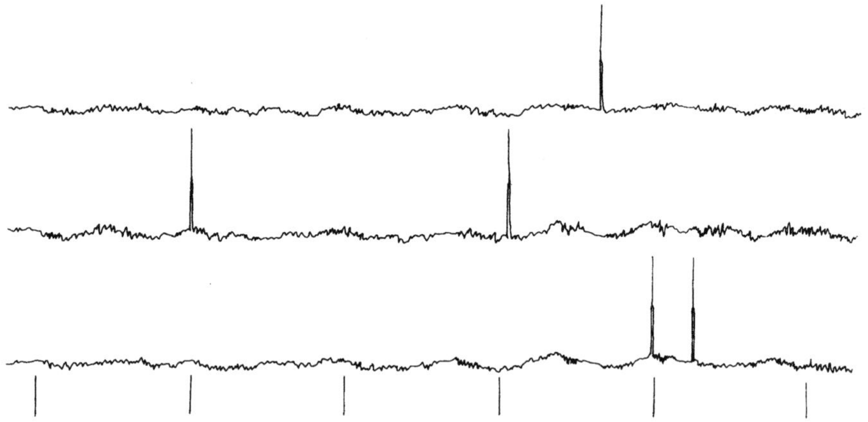

Fig. 3 Time trace of the cooling velocities of a three-dimensional system during heavy rain. The record shown is 50 seconds long, 10 seconds between time marks (from Dentler, 1976).

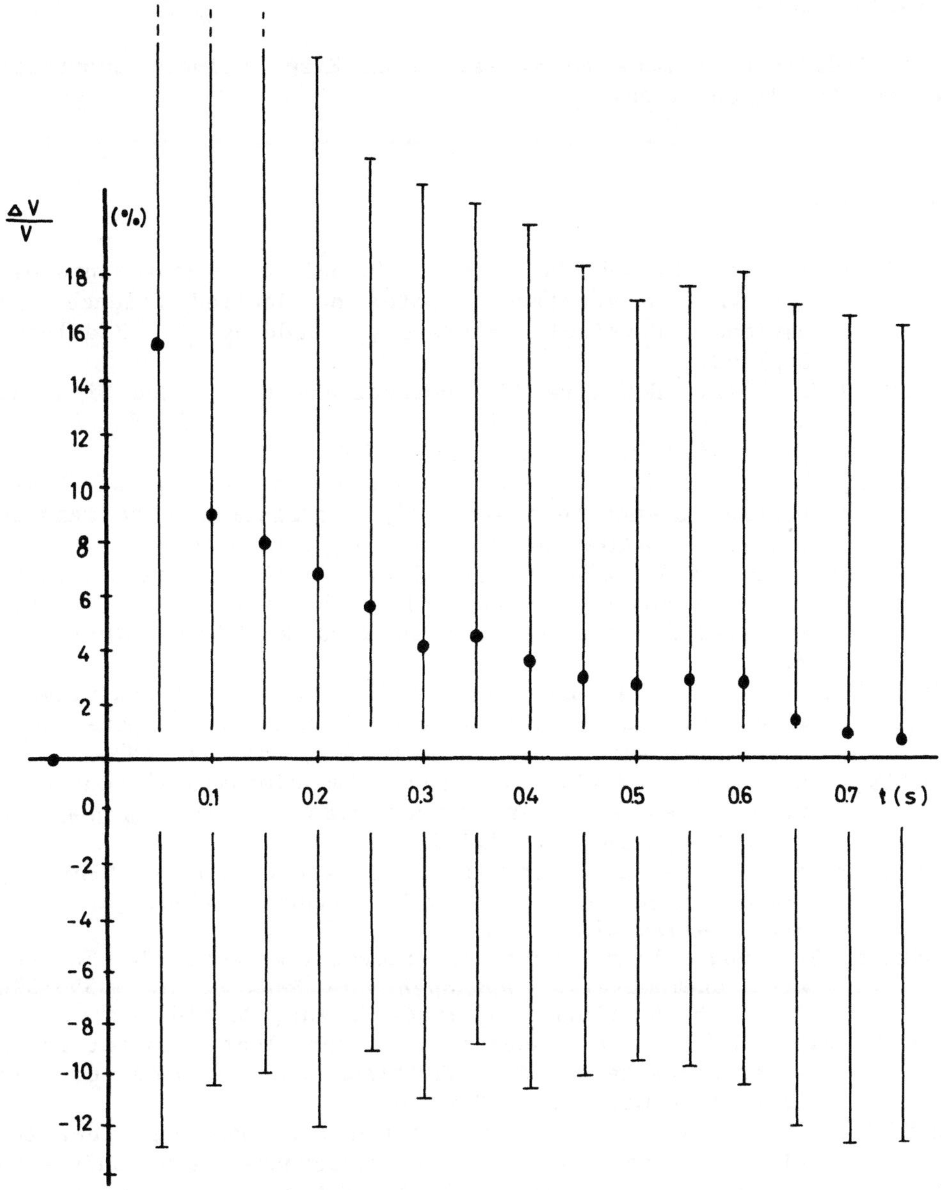

Fig. 4 Effect of rain on the cooling velocity, from actual data obtained during GATE. $\Delta V/V$ is the cooling velocity at a given time as a percentage deviation from the cooling velocity immediately prior to the rain-induced spike. Average of about 100 drops. The vertical bars indicate standard deviations, including real velocity fluctuations (from Dentler, 1976).

ACKNOWLEDGEMENT

Acknowledgement is made to S. Larsen of Risø National Laboratory for helpful discussions.

REFERENCES

BRADSHAW, P. and R.F. JOHNSON. 1963. Turbulence measurements with hot wire anemometers. Notes on Applied Science, 33, National Physical Laboratory, Teddington, Middlesex, England.

BRUUN, H.H. 1972. Hot wire data corrections in low and high turbulence intensity flows. *Journal of Physics E, Scientific Instruments*, 5: 812-818.

CARROLL, J.J. 1979. A note of caution on the use of 3 axis thermal anemometers for eddy correlation flux measurements. *Boundary-Layer Meteorology*, in press.

CHAMPAGNE, F.H. C.A. SLEICHER and O.H. WEHRMANN. 1967. Turbulence measurements with inclined hot-wires. Part 1. Heat Transfer Experiments with Inclined Hot Wires. *Journal of Fluid Mechanics*, 28: 153-175.

CHAMPAGNE, F.H and C.A. SLEICHER. 1967. Turbulence measurements with inclined hot wires. Part 2. Hot Wire Response Equations. *Journal of Fluid Mechanics*, 28: 175-183.

COLLIS, D.C. and M.J. WILLIAMS. 1959. Two dimensional convection from heated wires at low Reynolds numbers. *Journal of Fluid Mechanics*, 6: 357-384.

COMTE BELLOT, G. 1976. Hot-wire anemometry. *Annual Review of Fluid Mechanics*, 8: 208-229. Annual Review, Inc., Palo Alto, California.

CORRSIN, S. 1963. Turbulence: experimental methods, B. The hot-wire anemometer. *Handbuch der Physik*, 8: 555-590. Edited by S. Flügge, Springer-Verlag, Berlin.

DEACON, E.L. 1959. The measurement of turbulent transfer in the lower atmosphere. In: *Advances in Geophysics*, 6: 211-228. Academic Press, New York.

DENTLER, F.U. 1976. Niederschlagseinfluss auf Messungen des turblenten Windfledes in der wassernahen atmosphärischen Grenzschicht. Diplomarbeit, Meteorologisches Institut Universität Hamburg, Bundesstr. 55, D 2000 Hamburg, 96 pp.

DRUBKA, R.E., J. TANATICHAT and H.M. NAGIB. 1977. Analysis of temperature compensating circuits for hot wires and hot films. *DISA Information*, 22: 5-14.

DUNCKEL, M., L. HASSE, L. KRUGERMEYER, D. SCHRIEVER and J. WUCKNITZ. 1974. Turbulent fluxes of momentum, heat and water vapor in the atmospheric surface layer at sea during ATEX. *Boundary-Layer Meteorology*, 6: 81-106.

FREYTAG, C. 1976. Statistische Eigenschaften der Energiedissipation. Wissenschaftliche Mitteilungen 27, Universität München, Meteorologisches Institut, Theresienstr. 37, D 8000 München.

FRIEHE, C. and W.H. SCHWARZ. 1968. Deviations from the cosine law for yawed cylindrical anemometer sensors. *Journal of Applied Mechanics*: 655-662.

HINZE, J.O. 1959. *Turbulence*. McGraw-Hill, New York, 586 pp.

JØRGENSEN, F.E. 1971. Directional sensitivity of wire and fiber-film probes. *DISA Information*, 11: 31-37.

KJELLSTROM, B. and S. HEDBERG. 1970. Calibration of a DISA hot-wire anemometer and measurements in a circular channel for confirmation of the calibration. *DISA Information*, 9: 8-21.

LARSEN, S.E. and N.E. BUSCH. 1974. Hot wire measurements in the atmosphere. Part I. Calibration and response characteristics. *DISA Information*, 16: 15-34.

LARSEN, S.E. and N.E. BUSCH. 1976. Hot-wire measurements in the atmospehre. Part II. A field experiment in the surface boundary layer. *DISA Information*, 20: 5-22.

LARSEN, S.E. and N. BUSCH. 1979. On the humidity sensitivity of hot wire measurements. *DISA Information*, 25.

MERCERET, F.J. 1976. Measuring atmospheric turbulence with airborne hot fim anemometers. *Journal of Applied Meteorology*, 15: 482-490.

SANDBORN, V.A. 1972. *Resistance Temperature Transducers*. Metrology Press, Fort Collins, Colorado.

WILLS, J.A.B. 1976. A submerging hot wire for flow measurements over waves. *DISA Information*, 20: 31-34.

WUCKNITZ, J. 1979. The influence of anisotropy on stress estimates by the indirect dissipation method. Accepted for publication in *Boundary-Layer Meteorology*.

WYNGAARD, J.C. 1968. Measurement of small-scale turbulence structure with hot wires. *Journal of Scientific Instruments (Journal of Physics E)*, 2 (1): 1105-1108.

3

Dynamic Anemometers

S.D. Smith

1. RESPONSE OF DYNAMIC ANEMOMETERS

In eddy-flux measurements of wind stress, heat exchange, and evaporation over the sea a knowledge of the instantaneous wind vector is essential. Dynamic anemometers measure wind velocity by sensing either pressure or drag force on an object placed in the flow. Three-axis thrust anemometers sense the wind thrust, or drag, on a spherical object. Pressure-sensing dynamic anemometers are seen to have similar characteristics. These anemometers, designed for measurements of turbulence and eddy fluxes, are normally mounted on stable platforms. Dynamic anemometers which measure horizontal wind speed and direction and require minimal maintenance may be suitable for general use on buoys, ships, etc., where rotating anemometers (see Busch, Chapter 1) are more often used. In many cases the anemometer is used to determine the wind stress on the sea surface,

$$\tau = \overline{-\rho u_1{}'u_3{}'} \simeq -\rho\overline{u_1{}'u_3{}'} \qquad (1)$$

where u'_1 and u'_3 are wind velocity fluctuations along the mean wind direction and in the vertical direction, respectively. Since the rms downwind and vertical fluctuations over the sea are typically 10% and 5%, respectively, of the mean wind speed, the anemometer must have good resolution and low noise in both wind speed and tilt angle, relative to these figures.

2. THREE-AXIS THRUST ANEMOMETERS

The wind thrust on a highly-perforated spherical shell follows a quadratic law, with the thrust (F) and the velocity (V) aligned in the same direction,

$$F_i = \tfrac{1}{2}\rho A C_D |V| V_i \qquad (2)$$

where ρ is the density of air, A is the cross-sectional area, C_D is the drag coefficient of the sphere and i = 1, 2, and 3 indicate right-handed orthogonal coordinates, usually with the first from north to south, the second from west to east, and the third vertically upwards. Because precise vertical alignment is often not possible at sea we use a separate symbol V_i to denote a velocity component in anemometer coordinates. The quadratic response causes determination of each velocity component to depend on all three thrust components. If the anemometer is to be used to detect small variations in light winds, it must be capable of resolving a very small fraction of its full-scale range since on differentiating Equation 2 the sensitivity to gusts is seen to be proportional to the wind velocity.

The air density enters into the calibration, with the sensitivity to velocity being proportional to the square root of density. The mean air density at sea level varies by a few per cent with time and location, and correcting the results for these variations is easily accomplished. However, the air density is inversely proportional to the absolute temperature and so the indicated velocity fluctuations are subject to errors coupled to the temperature fluctuations, and in a similar way, but to a lesser extent, to humidity fluctuations also. Fortunately, these errors are small enough to be neglected in most cases. Sonic anemometers (see Kaimal, Chapter 4) have a velocity response which is proportional to the absolute temperature, while dynamic anemometer response is proportional to the square root of absolute temperature. Errors due to temperature fluctuations are therefore only half as large in dynamic anemometers as in sonic anemometers.

The advantages of thrust anemometers for air-sea interaction studies are that they can be left unattended for months at a time because they can be made weatherproof, have non-wearing parts, need no day-to-day adjustments, and consume very little electrical power. The main disadvantage of thrust anemometers is that they are basically spring balances, and so it is not possible to entirely eliminate drift in the signals at zero thrust associated with temperature changes and with aging of the springs. Furthermore, there are bound to be imperfections in the directional response associated with the springs, the sphere, and deflection of the air flow by the case of the anemometer. If no corrections are made to Equation 2 for anemometer irregularities, directional errors in the BIO (Bedford Institute of Oceanography) anemometers amount to about ±2° in azimuth and in wind tilt.

The problem of drift can be solved by covering the anemometer and recording zero-wind signal levels immediately before and after each data run. This makes it necessary to build a remote-controlled cover if, as is most often the case, the anemometer is operated in an inaccessible location. The cover should have a highly reflective surface to prevent heating the anemometer on sunny days when it is closed, which could cause problems with thermal drift.

Other disadvantages of thrust anemometers are that they do not give an absolute measurement but must rely on calibration, that vibration or acceleration of their mounting can cause errors, and that the nonlinear response makes the calibration and data processing more complex.

2.1 Bedford Institute of Oceanography

The application of thrust anemometers to air-sea interaction studies was pioneered by L.A.E. Doe at New York University, Woods Hole Oceanographic Institution, and Bedford Institute of Oceanography.

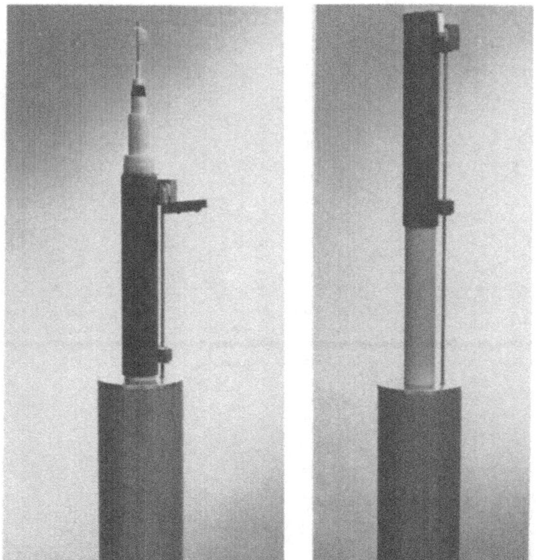

Fig. 1 (a) The Mark 8 Thrust Anemometer with cover open.
(b) With cover closed.

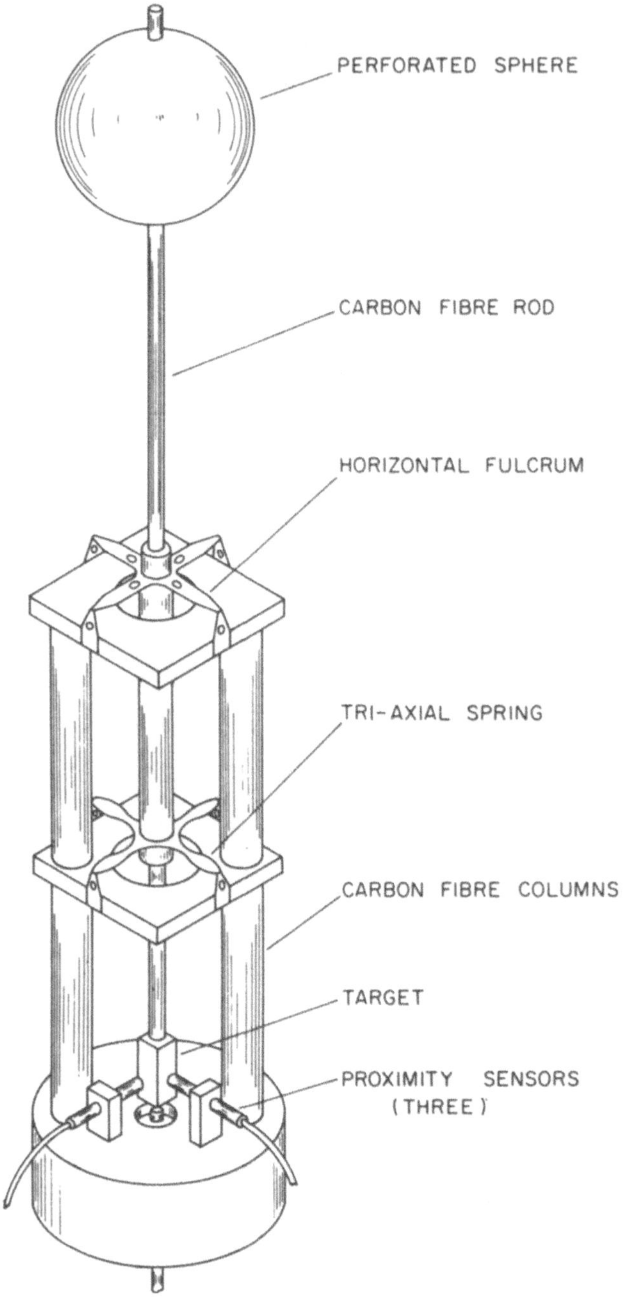

Fig. 2 Schematic mechanical design of Mark 8 Thrust Anemometer.

Doe (1963) tested anemometers with various smooth and rough spheres, finding (as have others independently) that a highly-perforated spherical shell gave low noise levels in a steady wind, and a nearly constant drag coefficient of 0.7 over a wide range of wind speeds (Reynolds numbers).

Subsequent work at New York University used Mark 4 thrust anemometers (Doe, 1967; Kirwan et al., 1975), but the NYU group is no longer measuring wind turbulence.

The Mark 8 thrust anemometer (Fig. 1, 2) is the most recent model, designed to (1) be suitable for eddy flux measurements (McBean, 1972), (2) operate for several months at sea without maintenance, (3) measure winds from all directions, and (4) draw modest power from a battery supply. The sphere is a ping pong ball (3.8 cm diameter), drilled with closely-spaced holes, attached to a carbon fibre rod supported by springs. The upper spring is a 'fulcrum' which is nearly rigid for horizontal deflections but allows vertical movement. The rod and sphere assembly is balanced at the fulcrum, making the anemometer insensitive to horizontal accelerations. The lower spring provides restoring force for horizontal wind thrust, while both springs act together for vertical deflections. The bottom of the rod carries targets for three proximity sensors. This design may prove suitable for limited production should a demand arise. Earlier thrust anemometers, such as Mark 6 used by Smith (1970, 1974) and Smith and Banke (1975), have a similar sphere and overall shape and size, but use differential transformers and have a more complicated spring arrangement (Smith, 1969).

To evaluate and where necessary correct for irregularities in thrust anemometer response, Smith (1969) used a series of wind-tunnel tests at a number of constant wind speeds, azimuths, and tilt angles to determine by least-square fits a set of coefficients for equations to derive velocity components from the thrust anemometer signals.

The BIO thrust anemometers are calibrated at wind speeds from 6 to 30 m s^{-1}. They are not used to measure turbulence at wind speeds below about 6 m s^{-1} because signal levels become too low in relation to noise and drift. The upper limit is imposed by the wind tunnel used, and by extrapolation the full-scale range is approximately 40 m s^{-1}.

Results of a calibration of the Mark 8 thrust anemometer are shown in Figures 3 to 5. The quadratic response to wind velocity (Fig. 3) is a very close approximation, and the fitted lines generally have correlation coefficients greater than 0.9999. The indicated horizontal angle $\tan^{-1}(F_2/F_1)$ (Fig. 4) lies within ±2° of the true wind direction, which is considered sufficiently accurate.

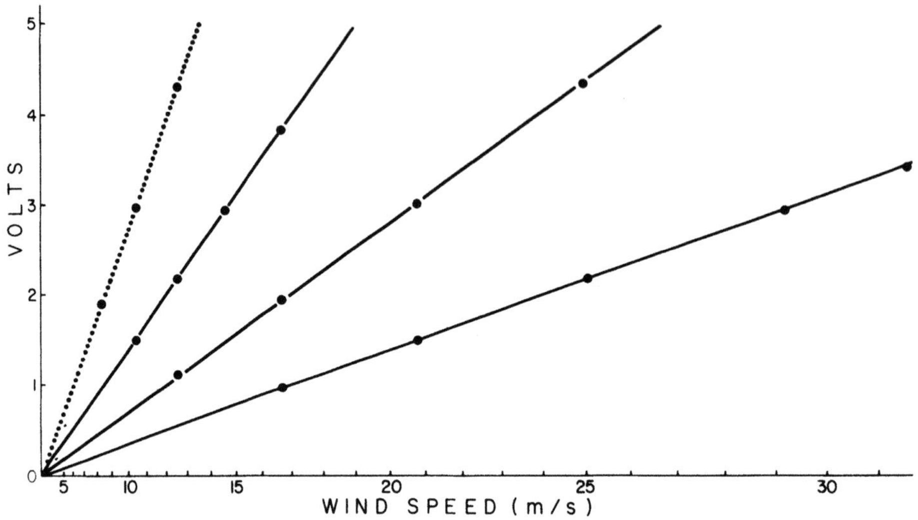

Fig. 3 Wind speed calibration of Mark 8 thrust anemometer for four amplifier ranges.

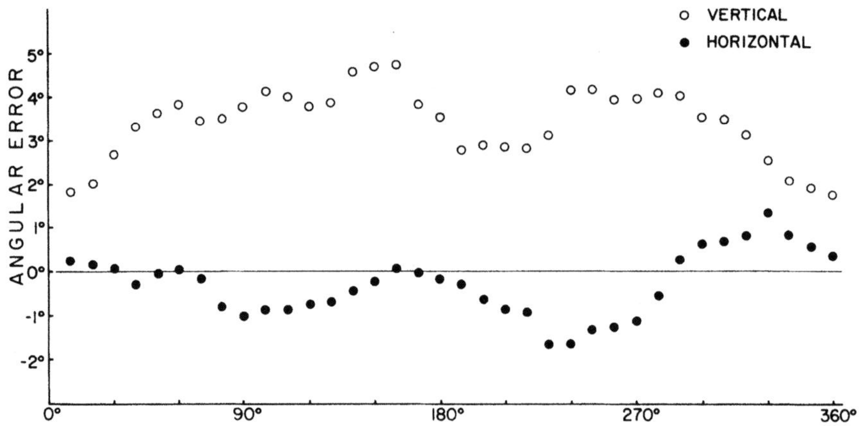

Fig. 4 Vertical and horizontal angular errors of Mark 8 thrust anemometer for horizontal winds of 21 m s^{-1}.

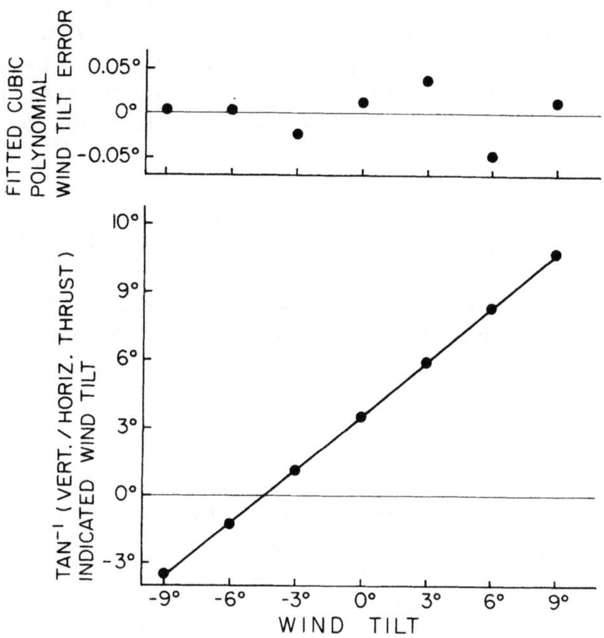

Fig. 5 Indicated wind tilt and deviation from fitted cubic polynomial for Mark 8 thrust anemometer in 20 m s^{-1} south wind.

The indicated vertical angle, however, shows a mean updraft of 3° which is associated with deflection of the flow by the case of the anemometer. There is also a distressing variation of ±2° of indicated tilt with wind azimuth, which has not been greatly improved by use of a jig to ensure uniform spacing of holes in the sphere nor by redesigning the spring linkage. The variation of indicated tilt with actual wind tilt (Fig. 5) is regular but differs greatly from Equation 1. Fortunately these patterns are reproducible for a particular anemometer.

Field data, recorded on magnetic tape in frequency-modulated form, are played back through an analogue-to-digital converter and a digital computer is used to derive velocity components from the thrust signals. The total (vector sum) thrust determines the total velocity using the appropriate calibration line from Figure 3. The wind azimuth counterclockwise is $\tan^{-1}(F_2/F_1)$ where F_1 and F_2 are thrust components from the north and from the west, respectively. The wind tilt angle is calculated from a cubic polynomial

$$\theta = A_0 + A_1R + A_2R^2 + A_3R^3$$

in the ratio $R = F_3/(F_1^2 + F_2^2)^{\frac{1}{2}}$ of vertical to horizontal thrust. The zero-order term is selected according to wind direction from Figure 4 while the first, second, and third order terms are obtained by fitting the data in Figure 5. The polynomial fits the calibration points within ±0.05° (Fig. 5), but may not be quite as good at other wind speeds and directions.

The Mark 8 thrust anemometer has constant thrust sensitivity within ±5% over a temperature range from -20 to 25°C. Lower viscosity damping fluid would have to be used below -10°C. Some improvements in thermal drift of the sensors may be achieved by their manufacturer by the time additional copies of the Mark 8 anemometer are built.

Thrust anemometers have been tested in the rain simultaneously with other turbulence anemometers and have been found to operate quite satisfactorily. Problems with weight of water on the sphere or impact forces of raindrops might be anticipated in heavy rain and light winds. Light, dry snow has been shown on one occasion not to cause detectable errors. No tests have been carried out in wet snow or freezing rain, which would stick to the sphere and change its weight and its aerodynamic drag. Ice might build up in these circumstances so that operation of the cover mechanism would damage the anemometer.

The frequency response of thrust anemometers is limited by the mechanical characteristics of the springs and by the size of the sphere. If the drag responds to wavelengths of the order of 10 sphere diameters (Laferty and Schmidt, 1969) then the BIO anemometer should respond to wavelengths of 38 cm or more, corresponding to a frequency of 26 Hz in a wind speed of 10 m s^{-1}. There is a design trade-off between higher sensitivity (softer springs) and higher frequency response (firmer springs). A mechanical resonance at 69 Hz for the Mark 8 anemometer (40 Hz for Mark 6), combined with viscous damping (underdamped) achieved by filling the anemometer case with silicone fluid, introduces amplitude errors at frequencies above 30 Hz and phase errors at even lower frequencies. For eddy flux measurements a few metres above the surface, frequency response to 10 Hz is quite adequate. Once the anemometer case has been filled, the anemometer cannot be tipped much more than 45°. The fluid can be drained for shipping.

The Mark 6 thrust anemometer has been compared with a sonic anemometer (see chapter 4) at a tower in Lake Ontario by Smith (1974). For seven simultaneous runs the rms difference in drag coefficients was 0.10×10^{-3}, the mean and standard deviation for the thrust anemometer were $(1.20 \pm 0.13) \times 10^{-3}$. The thrust anemometer indicated 11% lower rms vertical turbulence than the sonic anemometer but otherwise the two anemometers agreed well in wind speed and in the shape of velocity spectra and cospectra from 0.006 Hz to 10 Hz.

Spectral analysis of field data shows the characteristic -5/3 power law behaviour of boundary layer turbulence, while the more-or-less white spectrum characteristic of noise does not intrude at frequencies below 10 Hz.

2.2 Central Electricity Research Laboratories, Great Britain

Thrust anemometers with a 10 cm perforated sphere surrounding a vertical tube containing a three-axis spring assembly are described by Hopley and Tunstall (1971). A remote-controlled tubular cover is used to obtain zero-wind signals. These anemometers are presently used for studies of wind loading on towers and chimneys, but would appear to be suitable also for air-sea interaction studies.

2.3 Pennsylvania State University

A thrust anemometer has recently been developed (Norman et al., 1976; Perry et al., 1978) using cylindrical drag elements 1.5 cm in diameter and 4.5 cm long, supported on a rod equipped with two-axis strain gauges. Two of these are combined to derive three components of wind (Fig.6), with a vertical cylinder sensing two horizontal wind thrust components and a horizontal cylinder sensing vertical thrust. An analogue linearizing circuit produces voltages proportional to u_i and to V. Motorized tubular covers are again used to obtain zero-wind signals. These low-cost anemometers were developed for studies of turbulence structure in light winds over plowed or snow-covered fields, but could probably be adapted to air-sea interaction studies by weatherproofing and by increasing

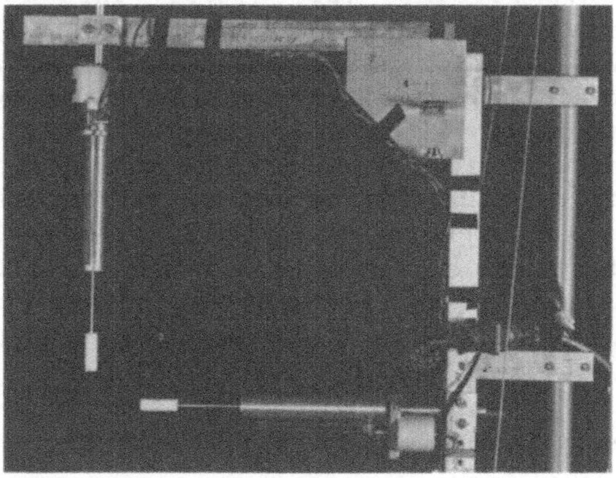

Fig. 6 The Pennsylvania State University Thrust Anemometer with tubular covers retracted (from Norman et al., 1976).

their wind speed range from the present 0.4 to 11 m s^{-1}. They have a resonant frequency of 20 Hz with viscous damping, and have temperature compensation from -12 to 35°C. Orthogonality of vector components is claimed to be within ±0.1°, which is far better than the BIO thrust anemometers have achieved.

2.4 University of Uppsala, Sweden

Högström (1974) developed a thrust anemometer for micrometeorological studies which used a rectangular drag plate to sense downwind thrust and fins to sense lateral and vertical components. He has discontinued use of this instrument because of irregular directional response; in some cases the same output represented more than one wind direction.

3. PRESSURE ANEMOMETERS

3.1 Pitot Tube

A familiar example of the pressure anemometer is the Pitot tube (Prandtl and Tietjens, 1934, pp. 229-231), which consists of two concentric tubes pointed in the air flow. The inner tube has a small dynamic pressure port at the end, while the outer tube is closed at the end with a rounded contour and has a ring of small static pressure ports a short distance back. Pressure p and velocity V are related by Bernoulli's equation

$$p = \tfrac{1}{2}a\rho V^2 \qquad (3)$$

The shape of the tube and the location of the static pressure ports have been designed to make 'a' very close to 1. The Prandtl type Pitot tube need not be aimed very closely into the wind; it can tolerate misalignment up to 17° with negligible (<1%) error so that in the relatively low turbulence levels found in the atmospheric boundary layer over the sea the Pitot tube could be used to sense the total instantaneous velocity. It would have to be aimed into the mean wind direction. Frequency response, determined by the size of the Pitot tube and the amount of tubing used to attach it to a pressure sensor, should be up to 20 Hz with appropriate design. Because they are extremely simple and have no moving parts they reliably maintain their calibration, and are commonly used for reference in wind tunnels. In field use, care would be required to prevent blocking of the ports by moisture or dust.

3.2 The IMFL Anemoclinometer

Three-axis pressure-sphere anemometers (Martinot-Lagarde et al., 1952) have been developed at Institut de Mécanique des Fluides de Lille (IMFL), France. The model used by Thurtell et al. (1970) consists of a spherical probe 3 cm in diameter (Fig. 7). A Pitot tube is centered in a venturi bored in the sphere. Eight of the twelve ports on the surface of the sphere, located on a circle at an angle of 47.5° to the axis of the venturi, are connected to a pressure-averaging cavity and serve as reference ports for the Pitot tube. The other four holes lie on a circle 45° from the axis, two lying in a horizontal and two in a vertical plane. Tubes run downwind from the sphere to three differential pressure transducers. The pitot tube and reference pressure difference are used to obtain the total wind velocity with a = 1.015 in Equation 3. The pressure differences between the horizontal and vertical pairs of ports are

$$p_i = b\rho V_1 V_i,\ i = 2,3 \qquad (4)$$

where b ≃ 2 is a function of probe geometry and of velocity (Reynolds number). The velocity components can be obtained by solving Equations 3 and 4. Equations 3 and 4 do not differ greatly from Equation 2 since $V \simeq V_1$ for low turbulence levels; V_1 is the horizontal downwind component.

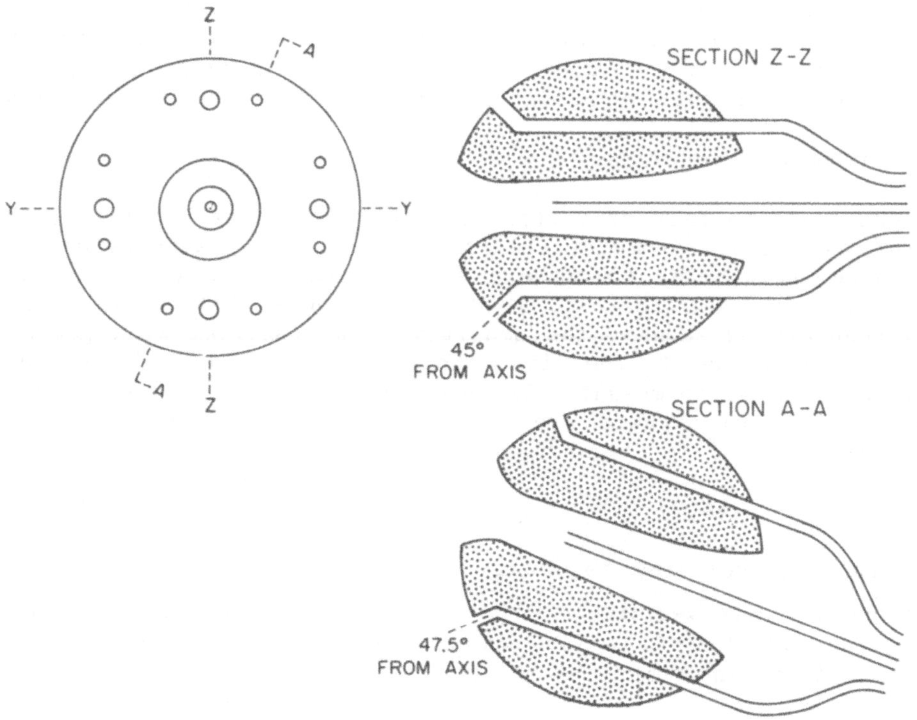

Fig. 7 (a) Front view and (b) Cross section view of Anemoclinometer (from Thurtell et al., 1970).

The Anemoclinometer must be used in a flow which does not deviate more than 20 to 30° from the probe axis, and so it would be necessary to aim the probe into the mean wind by means of a vane or rotor. The pressure sensors selected (see Dobson, Chapter 13) determine the sensitivity, range, stability, and power requirements. As with thrust anemometers, the quadratic response makes the sensitivity to gusts proportional to wind velocity so that the sensor must operate over a very wide range of pressures.

The frequency response of the Anemoclinometer is limited by the tubing from the sensing head to the transducers. Thurtell et al. (1970) found that the amplitude response in pressure was flat within 5% from dc to 15 Hz, but that there was a phase lag increasing with frequency which amounted to 48° at 10 Hz. This response is quite adequate for eddy flux measurements at a height of 3 m or more from the sea surface. They have compared wind stress from an Anemoclinometer with direct drag plate measurements and found agreement generally within 10%. Simultaneous wind profile measurements indicated about 20% higher stress.

The Anemoclinometer is presently used in studies over land and in wind tunnels. Blocking of the ports by dust, spray, condensation, rain, or corrosion would be a problem in air-sea interaction studies. It would be necessary to find some way to prevent blocking, perhaps by heating the probe to keep it dry, and by providing periodically for an outflow of dry pressurized air to clear the ports.

3.3 Yaw Sphere Anemometer

A sphere with a pair of vertically-separated ports, in principle a simplified anemoclinometer, has been used with a fast-response resistance thermometer to measure sensible heat flux over land surfaces (Tanner and Thurtell, 1970; Yap and Oke, 1974). Because of the quadratic response, the sensitivity to vertical wind gusts is again proportional to the wind velocity (Equation 4) and a cup anemometer is used to measure the mean wind speed. The pressure and temperature (t) signals are combined so that the output of the instrument is

$$e = \rho(VV_3t) \tag{5}$$

If the mean velocity is horizontal and the instrument carefully aligned with the vertical so that $V_3 \simeq u_3'$, and if $\overline{Vu_3't} \simeq \overline{V}\,(\overline{u_3't})$ then the turbulent heat flux may be obtained by dividing the mean output of the instrument by the mean wind speed,

$$H \simeq c_p\overline{e}/\overline{V} \tag{6}$$

where c_p is the specific heat of air.

3.4 Royal Netherlands Meteorological Institute Tubular Outflow Anemometer

An anemometer is being developed specially for turbulence and eddy flux studies in the marine boundary layer. Compressed air flows through a constriction and out to the atmosphere through two open-ended tubular arms pointing in opposite directions. Either pressure or velocity differencees between the opposing tubes are sensed. Three pairs of tubes are used to make a three-dimensional anemometer, and the wind vector components are derived from the three signals. A possible drawback in field applications is the need for a pump or other source of compressed air, and the associated power requirements and weight and bulk of equipment. A digital radio telemetry and remote control sytem has been developed for operation of this anemometer on unmanned towers. Data are analyzed by digital computer, using stored wind-tunnel calibration data for a large number of wind directions and speeds. A field evaluation of the performance of this anemometer is planned in the near future. Oost (1974) describes earlier versions of this instrument; he may be contacted for information on current developments.

4. DYNAMIC ANEMOMETERS TO MEASURE HORIZONTAL WIND

4.1 Vortex Anemometers

The frequency of shedding of vortices in the wake of a cylindrical object is nearly proportional to Reynolds number and so to the flow velocity. An anemometer may be built to measure the frequency of vortices behind a vertical cylinder and calibrated in terms of horizontal wind. The wake may be sensed acoustically by a microphone, or by a hot wire or hot film probe; in either case the sensing element does not require precise calibration. The only moving part required is a vane to keep the anemometer aimed into the wind and to sense wind direction. Such a device should be suitable for long-term use at sea without change of calibration. This type of anemometer may have suitable frequency response for estimation of wind stress by the 'dissipation' method. There appears to be nothing in the published literature relating to its reliability at sea.

A commercially manufactured vortex anemometer is described by Colton (1976). The user may wish to adjust the calibration for variation of Reynolds number with ambient air density, and to allow for high-frequency noise associated with natural variability in the vortex frequency even in a steady flow.

4.2 Aerowatt Thrust Anemometer

This two-axis anemometer measures horizontal wind thrust on a vertical cylinder, which is a corrugated stub of pipe 30 cm long and 7 cm in diameter. It has a range of 60 m s^{-1} and operates at temperatures from -20 to 50°C. The anemometer is counterbalanced to reduce sensitivity to acceleration, and is recommended by the manufacturer for use on buoys and ships. Optional outputs are thrust in rectangular coordinates or velocity in polar coordinates, and in the latter case an accuracy of 2% in speed and ±2.5° in direction is claimed. A similar two-axis anemometer (Norwood et al., 1966) was produced for a time in the United States.

5. SUMMARY

Several types of dynamic anemometers are in use for atmospheric turbulence and eddy flux measurements. At the present time only the Bedford Institute of Oceanography thrust anemometer is being used for air-sea interaction studies, but the performance of the various anemometers is quite similar and most of the anemometers described could be adapted by suitable weatherproofing.

Dynamic anemometers offer the advantages of (1) faster response and freedom from bearing wear when compared with rotating anemometers, (2) better calibration stability than hot-wire and hot-film anemometers, and (3) less power consumption, better operation in rain, probably less need for adjustment, and lower cost than sonic anemometers.

Engineering problems in design of dynamic anemometers are: (1) regularity of angular response, (2) thermal or other drift of zero points and sensitivity, (3) resistance to rain, salt spray, snow and ice, and (4) minimizing response to acceleration and vibration. In the case of BIO thrust anemometers we still wish to make improvements in the first two areas. To adapt other dynamic anemometers for marine use, special attention to the last two areas would be required.

These anemometers are research instruments, custom-made at various laboratories. Rotating (cup and propeller), hot-wire, sonic, and vortex anemometers are manufactured commercially with good success, but no thrust anemometer directly suitable for eddy-flux measurements has, to the author's knowledge, been marketed.

A problem common for the use at sea of all three-component fast-response anemometers is proper alignment; see the corresponding remarks by Kaimal, Chapter 4, and Hasse and Dunckel, Chapter 2.

REFERENCES

COLTON, R.F. 1976. Vortex anemometers - second generation. *Instrument Society of America Transactions*, 15: 343-353.

DOE, L.A.E. 1963. A three component thrust anemometer for studies of vertical transports above the sea surface. Report 63-1, Bedford Institute of Oceanography, Dartmouth, N.S., Canada, 87 pp.

DOE, L.A.E. 1967. A series of three-component thrust anemometers. *Proceedings of the First Canadian Conference on Micrometeorology*: 105-114. Atmospheric Environment Service, ... Downsview, Ont., Canada.

HOGSTROM, U. 1974. A critical examination of the turbulence instrument used in the Marsta micro-meteorological field project. *Tellus*, 26: 672-681.

HOPLEY, C.E. and M.J. TUNSTALL. 1971. A fast-response anemometer for measuring the turbulence characteristics of the natural wind. *Journal of Physics E, Scientific Instruments*, 4: 489-494.

KIRWAN, A.D., JR., G.J. McNALLY and E. MEHR. 1975. Response characteristics of a three-dimensional thrust anemometer. *Boundary-Layer Meteorology*, 8: 365-381.

LAFERTY, J.D. and L.V. SCHMIDT. 1969. Aerodynamic frequency response measurements of a wind anemometer. *Journal of Spacecraft and Rockets*, 6: 1080-1081.

MARTINOT-LAGARDE, A., A. FAUQUET and F.N. FRENKIEL. 1952. The IMFL Anemoclinometer - an instrument for the investigation of a fluctuating velocity vector. *Review of Scientific Instruments*, 23: 661-666.

McBEAN, G.A. 1972. Instrument requirements for eddy correlation measurements. *Journal of Applied Meteorology*, 11: 1078-1084.

NORMAN, J.M., S.G. PERRY and H.A. PANOFSKY. 1976. Measurement and theory of horizontal coherence at a two-meter height. Reprints of Third Symposium on Atmospheric Turbulence, Diffusion, and Air Quality, Raleigh, N.C., U.S.A., October 19-22, 1976, American Meteorological Society: 26-31.

NORWOOD, M.H., A.E. CARIFFE and V.E. OLSZEWSKI. 1966. Drag force solid state anemometer and vane. *Journal of Applied Meteorology*, 5: 887-892.

OOST, W.A. 1974. A fast spray-proof anemometer. Report WR 74-5, Royal Netherlands Meteorological Institute, deBilt, Netherlands, 7 pp.

PERRY, S.G., J.M. NORMAN, H.A. PANOFSKY, and J.D. MARTSOLF. 1978. Horiziontal coherence decay near large mesoscale variations in topography. *Journal of the Atmospheric Sciences*, 35: 1884-1887.

PRANDTL, L. and O.G. TIETJENS. 1934. *Applied Aero- and Hydro-Mechanics*. Dover edition, New York, 1957, 311 pp.

SMITH, S.D. 1969. A sensor system for wind stress measurement. Report 1969-4, Bedford Institute of Oceanography, Dartmouth, N.S., Canada, 64 pp.

SMITH, S.D. 1970. Thrust-anemometer measurements of wind turbulence, Reynolds stress and drag coefficient over the sea. *Journal of Geophysical Research,* 75: 6758-6770.

SMITH, S.D. and E.G. BANKE. 1975. Variation of the sea surface drag coefficient with wind speed. *Quarterly Journal of the Royal Meteorological Society,* 101: 665-673.

TANNER, C.B. and G.W. THURTELL. 1970. Sensible heat flux measurements with a yaw sphere and thermometer. *Boundary-Layer Meteorology,* 1: 195-200.

THURTELL, G.W., C.B. TANNER and M.L. WESLEY. 1970. Three-dimensional pressure-sphere anemometer system. *Journal of Applied Meteorology,* 9: 379-385.

YAP, D. and T.R. OKE. 1974. Eddy correlation measurements of sensible heat flux over a grass surface. *Boundary-Layer Meteorology,* 7: 151-163.

4

Sonic Anemometers

J.C. Kaimal

1 INTRODUCTION

The sonic anemometer measures wind velocity components from arrival times (or phase) of acoustic signals transmitted across a fixed path. Since there are no moving parts to come into dynamic equilibrium with the flow, it responds rapidly to velocity fluctuations. Its frequency response is limited only through the attenuation in spatial response imposed by line averaging along the path. It responds linearly to wind velocity and, with proper design, is relatively free of contamination from other velocity components or temperature. As an absolute instrument, its calibration is established by its design parameters Because of these advantages, the sonic anemometer has become a prime research instrument for measuring turbulent velocity fluctuations in the atmosphere.

The use of sonic anemometers in micrometeorological studies was pioneered by Suomi (1957). In his instrument, acoustic pulses were propagated in opposite directions along a one-metre path to determine wind velocity along the path. He was able to demonstrate the feasibility of sonic anemometry for turbulence measurement but encountered substantial difficulties in detecting the leading edge of the received pulses. This problem led later workers in the United States (Kaimal and Businger, 1963) and the USSR (Gurvich, 1960; Bovsheverov and Voronov, 1960) to turn to continuous wave techniques which avoided some of the difficulties of the pulse technique. Measuring transit time differences with phase comparators proved more dependable, so the early field instruments were the continuous-wave types. However, inherent drifts in the zero-wind calibration due to thermal drifts in the transducers, and the need for a multiplicity of frequencies in any three-axis configuration, led investigators to re-examine the pulse approach and overcome some of the problems encountered in the earlier versions. The instruments described by Mitsuta (1974) and Kaimal et al. (1974) use the pulse approach to advantage. In the USSR, continuous-wave systems still remain in use (Koprov and Sokolov, 1973) and a three-dimensional version utilizing one transmitter and four receivers is described by Bovsheverov et al. (1973). The author is aware of two recent developments in sonic anemometry. A three-axis anemometer using only one transducer at the end of each path, switching alternately between transmitter and receiver modes, is now being offered commercially. The benefits include reduced interference to airflow and smaller temperature drifts in the measurements, but sampling rates can be expected to be lower than in conventional arrays where transducer settling times are not so critical. The other development is a continuous wave phase-locked loop anemometer at the Department of Atmospheric Sciences of the University of Washington. A one-axis prototype has been tested successfully.

2. PRINCIPLE OF OPERATION

A schematic drawing of a single-axis, dual-path sonic anemometer is shown in Figure 1. The paths are parallel and closely spaced to minimize errors arising from velocity and temperature differences between the two paths. Assuming a uniform wind and temperature field within the array the transit times for two opposing pulses can be approximated by

$$t_1 = \frac{d}{c\,\cos\alpha + V_d} \quad \text{and} \quad t_2 = \frac{d}{c\,\cos\alpha - V_d} \tag{1}$$

where t_1 and t_2 are transit times for pulses travelling along paths 1 and 2, d is the path length, V_d is the velocity component along the path, c is the velocity of sound in air, and $\alpha = \sin^{-1}(V_n/c)$, with V_n being the velocity component normal to the path. The transit time difference is therefore

$$\Delta t = (t_2 - t_1) = \frac{2V_d d}{c^2 - V^2}, \tag{2}$$

where $V^2 = V_d^{\,2} + V_n^{\,2}$. Assuming, $V^2 << c^2$,

$$\Delta t = \left(\frac{2d}{c^2}\right)V_d\,. \tag{3}$$

The speed of sound is a function of temperature and water vapour content and can be expressed in the form (Kaimal and Businger, 1963):

$$c^2 = 403T(1 + 0.32e/P), \tag{4}$$

where T is the absolute temperature, e is the vapour pressure of water, and P is the atmospheric pressure with all terms expressed in SI units. The vapour pressure term is small and usually negligible, as are fluctuations in T compared with the mean absolute temperatures in the lower atmosphere. From Equations 3 and 4 the velocity component along the path becomes

$$V_d = \left[\frac{201.5\ \overline{T}}{d}\right]\Delta t. \tag{5}$$

Thus, with T known, the measurement of velocity simply reduces to a measurement of transit time difference while the sign of Δt indicates the direction of the velocity component. Therefore, the accuracy of the measured velocity depends critically on the stability, resolution, and accuracy of the Δt measurement. In the pulse sonic anemometer, a trigger pulse generated at a specific point on the received wave form marks the arrival time of the acoustic pulse

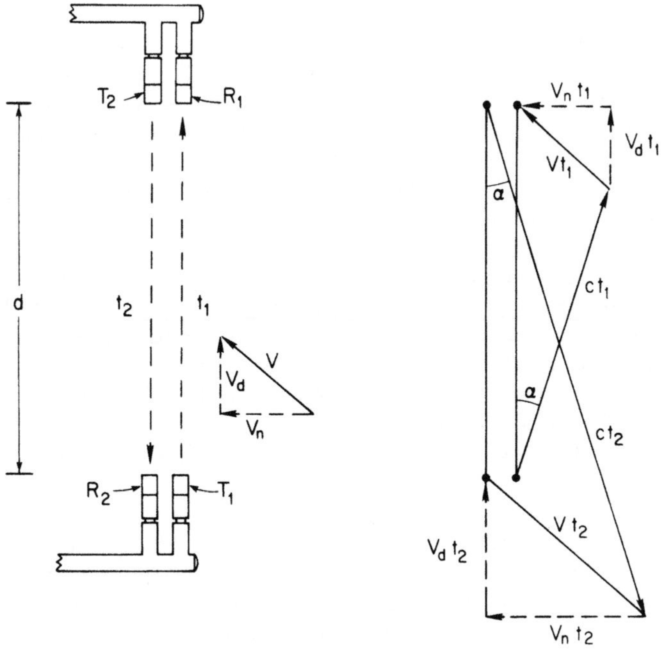

Fig. 1 Acoustic paths for one axis of the sonic anemometer and ray vectors showing principle of operation. Letters T and R denote transmitter and receiver respectively.

at that receiver. The time interval between trigger pulses in the two channels can be measured with analogue circuitry (Mitsuta, 1974) or with up-down counters and a 10 MHz clock (Kaimal et al., 1974). In continuous-wave systems, heterodyning and phase-comparison techniques are employed to determine Δt (Gurvich, 1959; Kaimal and Businger, 1963).

The error introduced by neglecting the sound speed fluctuations in the assumption $c^2 = 403\overline{T}$ has been analyzed in detail by Friehe (1976). Separating the terms in Equation 4 into their mean and fluctuating parts (denoted by overbars and primes, respectively), and neglecting the higher order terms,

$$\frac{c'}{\overline{c}} \simeq \left(\frac{T'}{2\overline{T}}\right) + 0.16\left(\frac{e'}{\overline{P}}\right) . \qquad (6)$$

The humidity fluctuation term is usually negligible. The effect of the temperature fluctuations on the velocity measurement can then be expressed as

$$(V_d)_m \simeq V_d\left(1 - \frac{T'}{\overline{T}}\right) , \qquad (7)$$

where the subscript m indicates the measured value. The error is usually less than 1%, but could be large if a significant temperature trend occurs during the observation period. Hourly updating of T is essential especially during transition periods. Friehe has shown that the error introduced by the temperature fluctuation term in Equation 7 is negligibly small in the variance and heat flux calculations but can be as large as -3.6% in the momentum flux.

However, there is an alternate approach that avoids the velocity error discussed above. If, instead of $(t_2 - t_1)$, the difference of their reciprocals $(1/t_1 - 1/t_2)$ is computed, the resulting term, $2V_d/d$, is independent of c. What diminishes the attractiveness of this approach to the instrument designer is the difficulty of measuring the difference between the reciprocals accurately. This approach is now offered in commercially available digital sonic anemometers.

Finally, the effect of variations in the path length should be considered since the velocity sensitivity depends directly on d. Thermal expansion has a negligible effect on accuracy. The coefficient of expansion for aluminum is about 20 ppm $°C^{-1}$ and is five times smaller for stainless steel, the material used in many of the probes. However, distortions of the frame, which introduce relative changes in the two paths, can show up as drifts in the zero calibration.

In addition to wind measurement, some sonic anemometers (Mitsuta, 1974) provide temperature measurements by computing the sum of the transit times. The temperature fluctuations thus obtained are a satisfactory approximation of the true temperature fluctuations under unstable conditions, when the fluctuations are large, but they are subject to serious error under near-neutral and stable conditions. Humidity and wind components normal to the path affect both t_1 and t_2 the same way, so they appear prominently in the measured temperature signal when the actual temperature fluctations are small. From the relationship given by Kaimal and Businger (1963), the measured temperature fluctuation T_m' can be expressed as

$$T_m' = T' + 0.32 \frac{\overline{T}}{\overline{P}} e' - 2 \frac{\overline{T}\,\overline{V}_n}{\overline{c^2}} V_n' . \tag{8}$$

The error due to water vapour turns out to be less important than is generally believed (~10%) because of the high correlation between e' and T' in the surface layer. V_n, on the other hand, can be large, since in most sonic anemometers that measure temperature, V_n would be the horizontal component of the wind. Kaimal (1969) measured errors of +7%, +19%, and +33% under unstable, stable, and neutral conditions respectively, in heat fluxes computed with sonic temperature fluctuations. A bias results in the estimation of the time of transition through neutral stability (because of the contribution from $\overline{V_n'w'}$) which shifts the apparent transition further into the stable regime. These basic uncertainties should be recognized when using sonic temperature data in any covariance computations. Not surprisingly, the error in the horizontal heat flux is even larger.

3. MEASURING THE THREE-DIMENSIONAL WIND FIELD

The velocity components of particular interest in studies of atmospheric turbulence are the longitudinal (streamwise), lateral, and vertical components. They are denoted by u, v, and w, respectively, along coordinates x, y, and z. Independent velocity measurements along three separate directions are needed to construct the three-dimensional wind field. At first glance the choices appear infinite, but, when proper consideration is given to transducer placement for minimum obstruction to airflow along the acoustic paths and to avoiding the possibility of pulses from other axes interfering with reception from the correct transmitter, not many options remain.

One successful array geometry which has evolved in recent years for surface layer work (Kaimal et al., 1974; Mitsuta, 1974) is the non-orthogonal system shown in Figure 2. One of the three axes is vertical; the other two are horizontal, and form a 120° angle. A wider range of wind directions can be detected without interference from the transducers in this array, as compared with an orthogonal

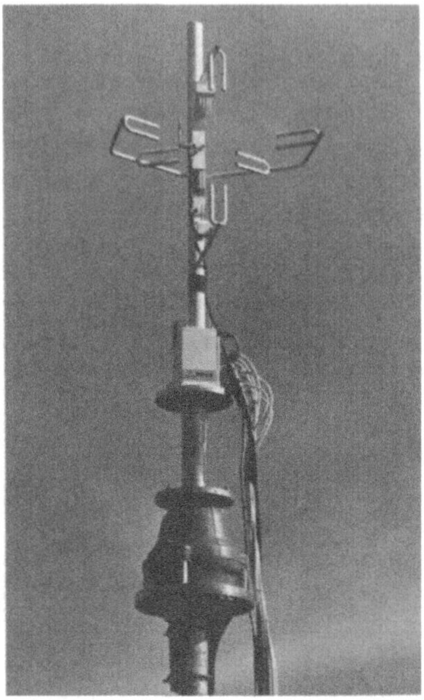

Fig. 2 A three-axis sonic anemometer array (EG&G Model 198-3) with fast-response platinum-wire thermometer mounted inside the array.

array. The only drawback is the need to re-orient the array into the mean wind as the wind direction changes. The w component is obtained directly from the vertical axis, but u and v must be computed from velocities V_A and V_B, which are measured along the horizontal paths A and B (see Figure 3). If α is the angle between the vector-mean wind coordinates (x and y) and the axis of symmetry (X and Y) for the array,

$$u = V_A(\cos\alpha - \frac{1}{\sqrt{3}}\sin\alpha) + V_B(\cos\alpha + \frac{1}{\sqrt{3}}\sin\alpha); \qquad (9a)$$

$$v = -V_A\left(\frac{1}{\sqrt{3}}\cos\alpha + \sin\alpha\right) + V_B\left(\frac{1}{\sqrt{3}}\cos\alpha - \sin\alpha\right); \qquad (9b)$$

$$\alpha = \tan^{-1}\left[\frac{\overline{V}_B - \overline{V}_A}{\sqrt{3}(\overline{V}_B + \overline{V}_A)}\right]. \qquad (9c)$$

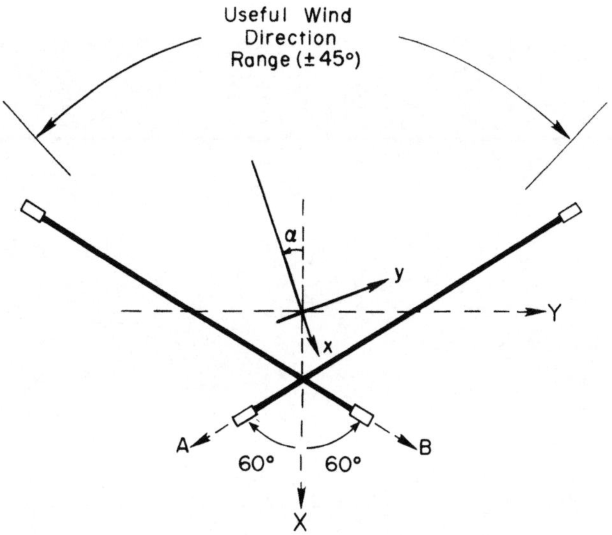

Fig. 3 Relationship between the horizontal axes of the sonic anemometer and the vector-mean axes of the wind. Coordinate transformation for converting wind measurements to components along the vector-mean axes are given in Equation 1.

The overbar denotes time average over a suitably long interval. In the case of $\alpha = 0$ we have $u = V_A + V_B$ and $v = (1/\sqrt{3})\ (V_B - V_A)$.

Although this arrangement is ideal for surface layer measurements where observational periods can be conveniently broken up into hourly segments for re-orienting the array, it is not particularly suited for continuous unattended operation on tall towers. A different configuration of the axes is needed. One such array, designed for use on the Boulder Atmospheric Observatory 300 m tower, is shown in Figure 4. (Similar arrays are also available commercially.) The axes are orthogonal, with the vertical axis kept upwind of the other axes to optimize the vertical velocity measurement. The acceptable azimuth range for the vertical axis is clearly much larger in this array than in the array in Figure 2; the unacceptable segment would be useless anyway because of shadowing caused by the tower. However, the horizontal wind measurements fare less well in this arrangement. An underestimation of about 10% in the wind (Mitsuta, 1974) can be expected when the mean wind direction is along either axis. This error can be corrected in the data processing routine by a first order approximation of the wind direction. Interference from supports for the crosswind axis is also minimized by vertically separating the horizontal paths. The

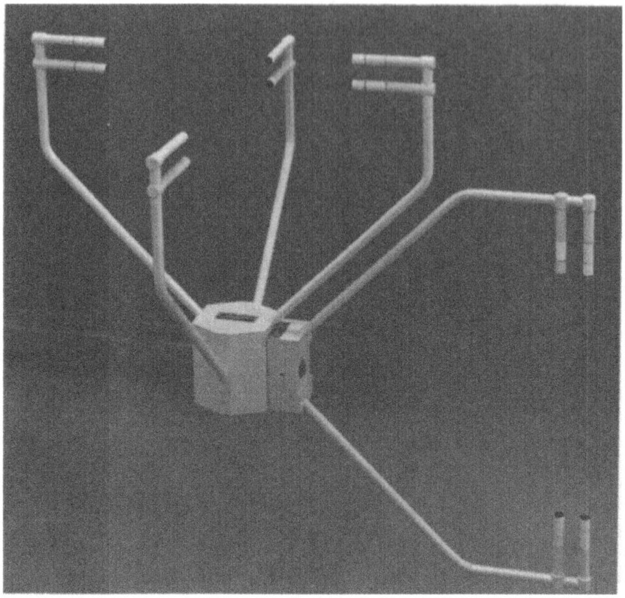

Fig. 4 A three-axis sonic anemometer (Ball Bros. Model 125-197, 198) for use on tall towers. The array is optimized for best azimuth coverage in the vertical component measurements.

main disadvantage for surface layer measurements is the spatial separation between the horizontal axes and the vertical axis. The effect of such separation is discussed in the following section.

4. SOME CONSEQUENCES OF ARRAY GEOMETRY

The dimensions of the array and the configuration of the acoustic path have pronounced effects on spectral response to fluctuations with length scales comparable to that of the array. Line-averaging along the acoustic path attenuates wavelengths smaller than $2\pi d$. The shape of the transfer function varies with the orientation of the path to the mean wind direction (Kaimal et al., 1968). In the cross-wind direction, the function resembles that of a single-pole low-pass filter with its half-power point at wavelength d; in the streamwise direction it has the $\sin^2 x/x^2$ response of a moving average filter, which rolls off more sharply.

When measurements along different paths are combined to resolve the wind fluctuations along the x and y directions (Eq. 9a and 9b), the response is more complex because of the spatial separation between the axes. A detailed analysis of this effect for the nonorthogonal array is given by Kaimal et al. (1968). The transfer functions for the most common configuration (d = 20 cm; spatial separation between mid-points of paths A and B, s = 12 cm; and angle α = 0) are given in Figure 5. (The term transfer function is used loosely to indicate the ratio of measured to true power spectrum; strictly speaking it is not a transfer function since it assumes a spectral form for the velocity field.) Horst (1973) has extended the analysis of Kaimal et al. to accommodate some aspects of a commercial array. The new geometry provided for two parallel paths on each axis and a vertical separation between the two horizontal axes. He found the v and w transfer functions relatively unaffected by the changes, but the u transfer functions showed 12% more energy at one-metre wavelength than does curve T_u in Figure 5.

The effects of spatial separation merit further discussion. Distortions in the spectral response of u and v arise because wavelengths smaller than $2\pi s$ are not properly transformed by Equation 9. As the wavelength decreases, the Fourier components of velocity observed along A and B become increasingly uncorrelated. In the limit when no correlation exists, the u spectrum is underestimated by a factor of 2.5 and the v spectrum by a factor of 0.625 (Kaimal et al., 1974).

The spatial separation between sensors (e.g., the w path with respect to the fast-response temperature and humidity sensors)

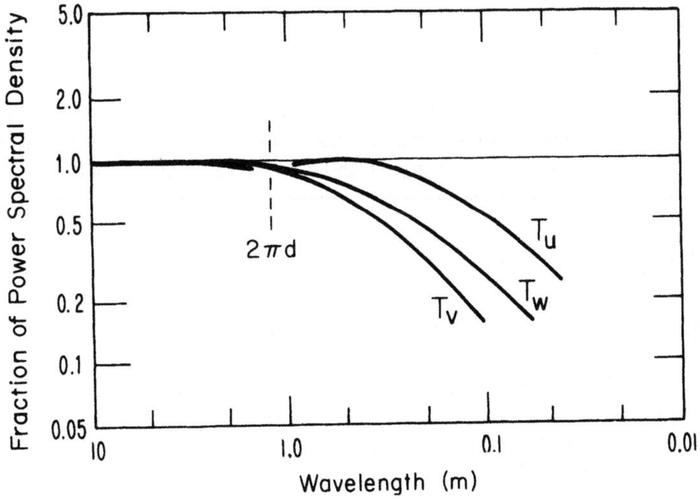

Fig. 5 Transfer functions for u, v, and w components for a nonorthogonal three-axis array with d = 0.2 m, s = 0.6 d, and α = 0. Distortions in the spectral response become apparent at wavelengths smaller than $2\pi d$ (1.26 m for this array).

introduces a limitation on the high-frequency response in flux cospectra. A diminishing correlation between the sensors for wavelengths smaller than 2π times the separation distance can be expected, which leads to the simple requirement that the separation distance be made no larger than the vertical path length in the sonic anemometer.

Minimum criteria for operating sonic anemometers can be obtained from considerations of line averaging in the array and known spectral and cospectral behavior in the surface layer (Kaimal et al., 1972). As mentioned earlier in this section, the effects of line averaging become apparent at wavelengths $\lambda < 2\pi d$. This limiting wavelength can be related to bandwidth requirements for spectral and flux calculations.

In unstable and near-neutral lapse rates the spectra of all three velocity components follow a -5/3 power law at nondimensional frequencies $nz/V > 1.5$, where z is the height above ground. If undistorted spectral information is available up to that point the high-frequency end can be filled in through simple extrapolation. Allowing another octave for possible filtering and aliasing effects, the desired upper frequency limit becomes $nz/V = 3$ or $\lambda = z/3$ (the cospectral contributions to the fluxes become negligible beyond this frequency). Equating the two limiting conditions for λ, we have

$$(z)_{min} = 6\pi d, \tag{10}$$

or roughly 4 m for d = 0.2 m.

In strongly stable lapse rates, the criterion for -5/3 spectral behavior and negligible cospectra is given by $\lambda = L/10$, where L is the Oboukhov length. A comparable expression for stable layers would be

$$(L)_{min} = 20\pi d, \tag{11}$$

which turns out to be 12.6 m for the same d.

Another type of error can result from small offsets in the array geometry and in the levelling of the array. The parameter most affected is the covariance of u and w, which represents the vertical momentum flux. Kaimal and Haugen (1969, 1971) have shown that the error can be very large in unstable air (typically 25% for 1° tilt) and recommend a 0.1° accuracy in the mechanical alignment of the probe and in the levelling procedure. Analyzing the tilt-error through its effect on an empirical formulation of the cospectrum, Rayment and Readings (1971) found the error to be much smaller, 15.5% for a 5° tilt (see also Deacon, 1968). In the author's view, this figure may be realistic for near-neutral and stable air but underestimates the error one is likely to find under light wind convective conditions. It is a common practice to correct measurements of the vertical velocity component in the surface layer for tilt errors with aid of the assumption $\overline{w} = 0$. Heat flux and variances, on the other hand, are not as sensitive to alignment and tilt errors under unstable conditions, but the error can be half as large as the momentum flux error under stable conditions (Wieringa, 1972).

5. DATA SAMPLING

The upper frequency limit dictated by the array dimensions also determines the maximum sampling rate for sonic anemometer data. Allowing an additional octave at the high-frequency end for aliasing effects, the required sampling frequency n_s may be specified as

$$n_s = \frac{2V}{\pi d}, \tag{12}$$

which can be rounded off to 20 Hz for most atmospheric surface layer work with a 0.2 m path sonic anemometer. Depending on its output characteristics, the wind information may be digitized at discrete time intervals (0.05 s) with an analogue-to-digital converter or, if it is available in parallel digital form, may be simply transferred from holding registers to computer memory.

The pulse technique lends itself easily to digital time interval processing and parallel data transfer. If the two transmitters are pulsed simultaneously, the transit time difference can be measured directly with a 10 MHz up-down counter. In practice (Kaimal et al., 1974) the transmitters are pulsed at a frequency 20 times n_s and the successive counts are accumulated in the counters before transfer to the holding registers, so that each reading is a block average over the preceding sampling interval. This prefiltering reduces distortions at the high-frequency end of spectra due to aliasing. In a sonic anemometer with analogue outputs, the prefiltering can be done prior to sampling and digital conversion.

If the data are needed solely for computations of time-averaged turbulence properties such as variances and covariances, the sampling requirement can be relaxed to 1.0 Hz or even 0.1 Hz, provided the full 10 Hz bandwidth is maintained in the sensors and in the data handling equipment, and provided the averaging is done over a large enough sample size to yield stable estimates. Preliminary tests conducted at the Wave Propagation Laboratory with various sampling rates indicate that the error in estimates computed from instantaneous samples obtained once every 10 s are of the same order as the deviations from the ensemble average determined by the properties of the turbulent flow. The departures from estimates computed with the entire time series over a one-hour averaging period is roughly ±5% in the variances and the vertical heat flux and about ±15% in the Reynolds stress.

6. OTHER CONSIDERATIONS

The two most critical considerations in the design of any sonic anemometer are the transducer design and the acoustic isolation of transducers from the frame. An ideal transducer must be small in size, unaffected by changes in ambient temperature, highly directional, and weatherproof. Not surprisingly, such transducers are rarely, if ever, available as off-the-shelf items. Sonic anemometer manufacturers, therefore, fabricate their own transducers. Earlier designs often fell short of expectations as they tended to break down at low temperatures or after a few years of use. Acoustic isolation of the transducers can now be achieved by relatively simple means (Fig. 6).

Transducers operate most efficiently at their resonant frequencies, but, in continuous-wave systems, the advantage of signal strength is offset by drifts due to spurious phase shifts introduced by temperature induced changes in the resonant frequency. Hourly zero-wind checks in anechoic boxes are needed to correct the data for such drifts. An obvious solution is the use of transducers with flat frequency response in the range of interest. High quality condenser microphones meet this requirement, but they are seldom very small, they require large bias voltages, and they are not designed for all-weather operation.

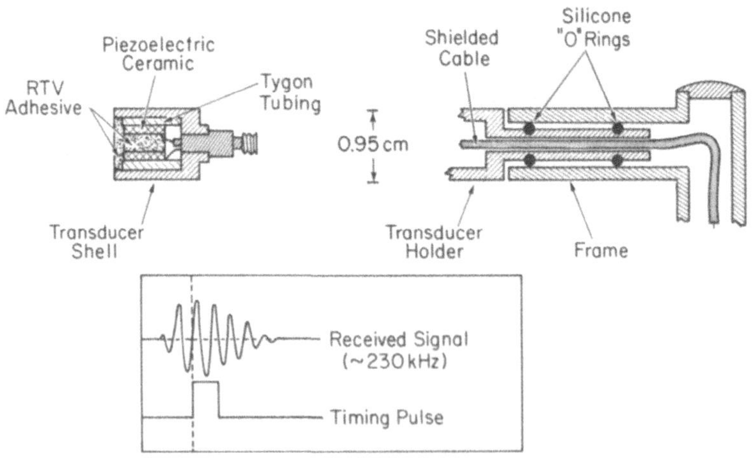

Fig. 6 Details of transducer design and acoustic isolation used in the EG&G anemometer.

The temperature stability requirement is much less serious in pulse systems where resonant frequencies are at least an order of magnitude higher than in continuous wave systems, but damping becomes an important consideration. Signal amplitude in the transmitters and the receivers must build up and decay rapidly for reliable triggering and for avoiding interference with adjacent channels. A transducer design that has proved successful is shown in simplified form in Figure 6. A cylindrical piezoelectric ceramic element filled with silicone rubber adhesive fits in a short PVC tube inside a thin aluminum shell. The transmitting end is capped off with a thin layer of silicone rubber, while the rear is fitted with a microdot connector with leads attached to the inner and outer surfaces of the cylindrical element. The transducer is small, weatherproof, and easily replaced. It resonates at approximately 230 kHz and, when properly coupled to the preamplifier input, produces a waveform which reaches its maximum amplitude after its third positive zero crossing. The timing pulse marking the arrival of the received signal corresponds in time to this zero-crossing (see inset in Fig. 6). The simple acoustic isolator used with the above transducer is also shown in Figure 6. Two silicone 'O' rings separate the transducer holder from the frame, providing weather protection as well as acoustic isolation.

To minimize noise pickup and cross-talk, triaxial cables are used to transmit the receiver signals to preamplifier circuits 5 m away. This permits effective isolation of the transducer leads from the array frame. Aside from occasional adjustments of the triggering level in the receiver logic circuits and zero-wind checks with small cylindrical anechoic tubes enclosing the acoustic paths, the sonic anemometer requires very little maintenance. Some transducer-preamplifier combinations tend to be more stable than others. A trial and error period is often needed to match up the components. In two months of continuous operation at Boulder the author has found that sonic anemometers at six of the eight levels on the tower require no adjustment or supervision, while the other two have needed attention from time to time and will continue to do so until the weak links in the receiver channels are identified and corrected. Temporary loss of signals from all levels accompanied periods of wet snow accumulation and heavy rain, but seldom was the operation affected by the mere wetting of the transducers.

7. APPLICATIONS TO AIR-SEA INTERACTION

Although sonic anemometers have been used more widely over land, the need for detailed turbulence measurements over water is just as great. However, measurements over water are complicated by such factors as typically low turbulence levels, contamination of signals from wave-induced motions of the instrument, and a more hostile environment for the instrument and the operator. A number of measurements over water (Miyake et al., 1970; Pond et al., 1971;

Mitsuta and Fujitani, 1974; Smith, 1974) and over ice (Banke et al., 1976) have been reported. Recent sonic anemometer data collected by Friehe (private communication) off the coast of San Diego, California, suggest that the high-frequency spectral behavior in the first few metres over water departs from behavior observed with the same instrument over land.

Its sensitivity to small velocity fluctuations, and the absence of exposed elements and moving parts that could corrode in a marine environment, make the sonic anemometer an ideal tool for air-sea interaction studies. Weatherproof connectors and weathertight boxes should provide adequate protection for the various components associated with the acoustic array. However, the performance of the transducers in some early sonic anemometers was affected adversely by salt accumulation, spray and rain. Sonic anemometers experience loss of signal when large water drops collect at the ends of downward pointing transducers, a common occurrence in heavy fog.

Another problem common to all wind observations over water is contamination from wave-induced motions of the instrument. The large spikes that inevitably show up at the wave frequencies in velocity spectra can be identified and removed, but their effects on the variances and covariances cannot be easily isolated. Use of gyrostabilized instrument platforms (e.g., Brocks and Hasse, 1969) can greatly reduce the tilt error. The motions of the instrument platform are often complex, involving both translation velocity and tilt. Friehe et al. (1975) have used an airline inertial navigation unit to measure the three components of both angular rotation and translation in order to correct the data during digital data processing (most commercial inertial navigation systems contain three accelerometers mounted orthogonally on a gyroscopically stabilized platform). This correction introduces an added level of complexity which seems unavoidable if high-quality turbulence data is the desired objective.

REFERENCES

BANKE, E.G., S.D. SMITH and R.J. ANDERSON. 1976. Recent measurements of wind stress on Arctic Sea Ice. *Journal of the Fisheries Research Board of Canada,* 33: 2307-2317.

BOVSHEVEROV, V.M. and V.P. VORONOV. 1960. Acoustic anemometer. *Academy of Sciences of the USSR Bulletin, Geophysics Series,* 6: 882-885.

BOVSHEVEROV, V.M., B.M. KOPROV and M.I. MORDUKHOVICH. 1973. A three-component acoustic anemometer. *Atmospheric and Oceanic Physics,* 9: 240-241.

BROCKS, K. and L. HASSE. 1969. Eine neigungsstabilisierte Boje zur Messung der turbulenten Vertikalflüsse über dem Meer. *Archiv fur Merteorologie, Geophysik und Bioklimatologie, Series A,* 18: 331-344.

DEACON, E.L. 1968. The leveling error in Reynolds stress measurement. *Bulletin of the American Meteorological Society,* 49: 836.

FRIEHE, C.A. 1976. Effects of sound speed fluctuations on sonic anemometer measurements. *Journal of Applied Meteorology,* 15: 607-610.

FRIEHE, C.A., C.H. GIBSON, F.H. CHAMPAGNE and J.C. LaRUE. 1975. Turbulence measurement in the marine boundary layer. *Atmospheric Technology,* 7 (National Center for Atmospheric Research, Boulder, Colorado): 15-23.

GURVICH, A.S. 1960. Frequency spectra and functions of distribution of probabilities of vertical velocity components. Izvestia Geophysical Series 7: 1042-1055 (AGU English translation: 695-703).

HORST, T.W. 1973. Spectral transfer functions for a three-component sonic anemometer. *Journal of Applied Meteorology,* 12: 1072-1075.

KAIMAL, J.C. 1969. Measurement of momentum and heat flux variations in the surface boundary layer. *Radio Science,* 4: 1147-1153.

KAIMAL, J.C. and J.A. BUSINGER. 1963. A continuous wave sonic anemometer-thermometer. *Journal of Applied Meteorology,* 2: 156-164.

KAIMAL, J.C. and D.A. HAUGEN. 1969. Some errors in the measurement of Reynolds stress. *Journal of Applied Meteorology,* 8: 460-462.

KAIMAL, J.C. and D.A. HAUGEN. 1971. Comments on "Minimizing the levelling error in Reynolds stress measurement by filtering". *Journal of Applied Meteorology,* 10: 337-339.

KAIMAL, J.C., J.T. NEWMAN, A. BISBERG and K. COLE. 1974. An improved three-component sonic anemometer for investigation of atmospheric turbulence. In *Flow - Its Measurement and Control in Science and Industry,* 1 (Instrument Society of America): 349-359.

KAIMAL, J.C., J.C. WYNGAARD and D.A. HAUGEN. 1968. Deriving power spectra from a three-component sonic anemometer. *Journal of Applied Meteorology,* 7: 827-837.

KAIMAL, J.C., J.C. WYNGAARD, Y. IZUMI and O.R. COTE. 1972. Spectral characteristics of surface layer turbulence. *Quarterly Journal of the Royal Meteorological Society,* 98: 563-589.

KOPROV, B.M. and D. Y. SOKOLOV. 1973. Spatial correlation functions of velocity and temperature components in the surface layer of the atmosphere. *Atmospheric and Oceanic Physics,* 9: 95-98.

MITSUTA, Y. 1966. Sonic anemometer thermometer for general use. *Journal of the Meteorological Society of Japan,* 44: 12-24.

MITSUTA, Y. 1974. Sonic anemometer-thermometer for atmospheric turbulence measurements. In *Flow - Its Measurement and Control in Science and Industry,* 1 (Instrument Society of America): 341-347.

MITSUTA, Y. and T. FUJITANI. 1974. Direct measurement of turbulent fluxes on a cruising ship. *Boundary-Layer Meteorology,* 6: 203-217.

MIYAKE, M., M. DONELAN, G. McBEAN, C. PAULSON, F. BADGLEY and E. LEAVITT. 1970. Comparison of turbulent fluxes over water determined by profile and eddy correlation techniques. *Quarterly Journal of the Royal Meteorological Society,* 96: 132-137.

POND, S., G.T. PHELPS, J.E. PAQUIN, G. McBEAN and R.W. STEWART. 1971. Measurement of turbulent fluxes of momentum, moisture and sensible heat over the ocean. *Journal of Atmospheric Sciences,* 28: 901-917.

RAYMENT, R. and C.J. READINGS. 1971. The importance of instrument tilt on measurements of atmospheric turbulence. *Quarterly Journal of the Royal Meteorological Society,* 97: 124-130.

SMITH, S.D. 1974. Eddy flux measurements over Lake Ontario. *Boundary-Layer Meteorology,* 6: 235-255.

SUOMI, V.E. 1957. Sonic anemometer. In *Exploring the Atmosphere's First Mile,* Pergamon, New York, 1: 256-266.

WIERINGA, J. 1972. Tilt errors and precipitation effects in trivane measurements of turbulent fluxes over open water. *Boundary-Layer Meteorology,* 2: 406-426.

5

Pilot Balloon Techniques

L. Hasse and D. Schriever

1. INTRODUCTION

Within the planetary boundary layer, the difference between the actual wind and the geostrophic wind is called the ageostrophic wind. This is the result of an imbalance of forces, which generates the momentum (wind stress) that is transferred from the atmosphere to the ocean. The actual wind profile within the planetary boundary layer can be obtained from the tracking of pilot balloons (pibals). Of the various methods of tracking, the use of a theodolite is the least demanding on instrumentation and trained observers. Determination of the wind stress from such pilot balloon observations was one of the first methods to give reasonable stress estimates. The analysis of the data, however, depends very much on the assumptions used. The best examples of determining stress from pilot balloon observations at sea are the Scilly and Heligoland wind profiles (Lettau, 1957; Lettau and Hoeber, 1964). Pilot balloon observations may be used for other applications, e.g. in studies of flow over coastal bluffs or land/sea breeze flows.

2. INSTRUMENTATION

2.1 Balloon Theodolites

In order to track balloons, almost any theodolite capable of measuring azimuth and elevation to at least 0.1° may be used. In practice it is advantageous to have an unlimited vernier drive on both axes. With simpler instruments the scales must be read by a second observer or photographed. Instruments are available that print the scale readings on waxed paper. Some older theodolites of this type, however, shake so badly while printing that the observer loses track of the balloon. More expensive instruments (such as those used for aircraft approach surveillance) have electrical readouts, using either potentiometers or digital encoders fastened to the axes of the vernier drives.

Theodolites for tracking balloons released over the ocean, have been used both from islands and from ships. On land, levelling is easily obtained with a spirit level; at sea, special precautions are necessary to eliminate the ship's motion. In early years, shipborne balloon theodolites were mounted in gimbals and held level by a heavy, damped counterweight (see Fig. 1). Alternatively, the theodolite can be mounted rigidly to the ship with a recording instrument monitoring ship motion (ideally, the ship's pitch and roll angles). Levelling at sea must be referred to the horizon. The observer follows the balloon continuously, as well as possible, and the ship motions are later removed from the data by reference to the ship's motion record in the case of the recording

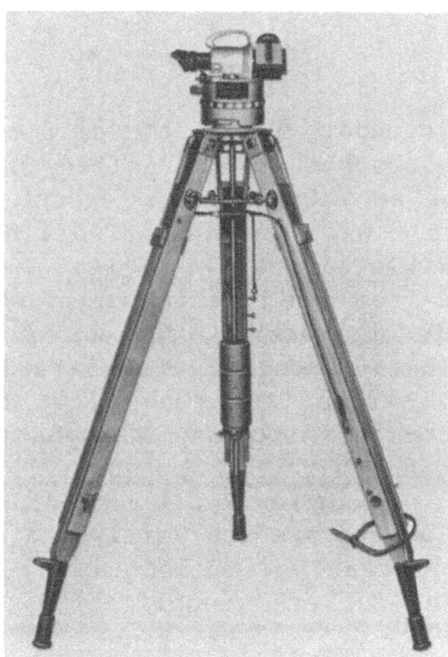

Fig. 1 Balloon theodolite on tripod. A typical feature is the 90° bend between eyepiece and objective. This allows for tracking balloons overhead, which is a desirable feature if the balloon is released near one observing station (e.g. on the same ship), especially under low wind convective conditions. Note the counterweight in the middle of the tripod and the rubber cords for damping for shipboard use (photo courtesy Labor Dr. J. Rosenhagen, Hamburg).

theodolite, or by filtering (which is less effective) when no such record exists. Although this technique has been used, e.g. during GATE, it is very difficult when there is a lot of ship motion. To avoid the difficulties of levelling, one might use a sextant and a compass. This was found to be inconvenient but led to the development of the mirror theodolite, which has been used extensively at sea by the Deutsche Seewarte and other marine meteorological services (Kuhlbrodt and Reger, 1933). This instrument combines the features of a sextant and a balloon theodolite; that is, it uses the horizon as a reference for elevation.

2.2 Double versus Single Theodolite Ascent

In double theodolite ascents, two theodolites are installed at the end points of a baseline which is a few kilometres long. The length of the baseline plus the four angles provide one surplus observation. In order to use the information, Thyer (1962) suggested that the best position of the balloon is given by the point which divides the nearest distance between the rays originating from the two theodolites (see Fig. 2). The double theodolite method gives the three-dimensional trajectory of the balloon without other assumptions.

With the single theodolite technique, the height is often assumed to be known from the time of ascent and a predetermined ascent rate. Since pilot balloons are not perfect and perform erratic motions, their ascent rates will vary considerably. Thus the horizontal velocity will be uncertain. Single theodolite observations

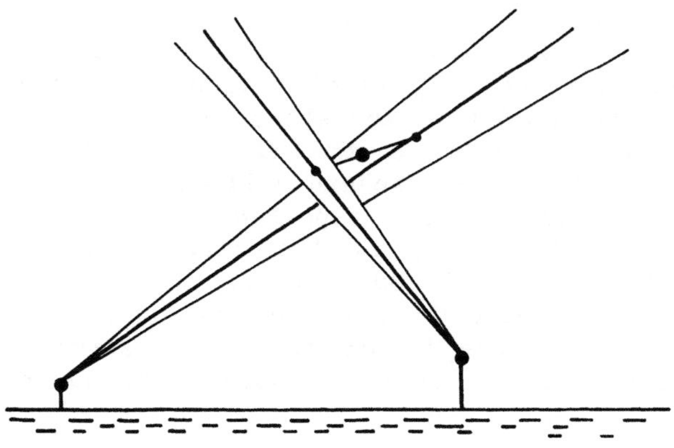

Fig. 2 Schematic diagram of Thyer's (1962) method for obtaining balloon position as the half point of the shortest distance between two rays.

are therefore recommended only if height or slant range can be obtained by independent means. Both optical distance meters for slant range and pressure from simple radiosondes for height have been used.

2.3 Radar Tracking

Radar tracking is more expensive than ordinary pilot balloon theodolite observation and usually gives a less detailed and less accurate wind profile. It is difficult to track a balloon within 100 m of the radar, which is obviously a disadvantage when launching balloons from a ship, especially since the important ageostrophic components of the wind are found near the surface. For this reason it is advisable to use the 'up and down' balloon technique, which adds to the cost because of the extra parachute and the time- or pressure-actuated release device. The drawback of the lesser accuracy of the radar tracking can be compensated by averaging a series of profiles; this is done in any case with pibal observations to get rid of turbulent fluctuations.

When using radar, it might also be useful to track the balloon by theodolite, thus utilizing the better direction from the theodolite and the reasonably accurate slant range from the radar.

2.4 Tracking by Photogrammetry (Loeser Technique)

Over land, pibal swarms have been tracked by photogrammetry at night (Chamberlain et al., 1957). Tracking by photogrammetry should also be useful from a ship. To allow for ship motions, the horizon could be used as a reference using illuminated marker buoys.

2.5 Balloon Material

Standard weather service 35 g pilot balloons are ordinarily used. White ones are preferred for blue skies, red ones otherwise. During the night, balloons carry a small dry cell with built-on socket and a flashlight bulb. The standard balloons perform irregular motions. Roughened balloons are reported to perform better (Scoggins, 1964). The roughening of the balloons is done by gluing paper cups to the balloon in an irregular distribution. It seems questionable if this effort is really worthwhile; increasing the number of ascents per time is probably more useful. Averaging is necessary at any rate, in order to obtain the desired profile of mean velocity.

3. AGEOSTROPHIC METHOD

Pilot balloon observations are often used to infer the wind stress from the ageostrophic wind components; this is commonly called the ageostrophic method. It should be noted that the results depend very much on the assumptions used - there is a large variety of ageostrophic methods. For this reason a short review of the theory is given here.

The ageostrophic method starts from the equations of mean motion, usually written in the form:

$$
\begin{aligned}
f(v - v_g) + \frac{\partial}{\partial z}\left(\frac{\tau_x}{\rho}\right) &= 0 \\
-f(u - u_g) + \frac{\partial}{\partial z}\left(\frac{\tau_y}{\rho}\right) &= 0
\end{aligned}
\qquad (1)
$$

From these, the stress components (τ_x, τ_y) at the surface can be obtained if the vertical profiles of the mean actual wind u,v and the geostrophic wind u_g,v_g are known and the x and y components of stress for at least one height each are known. In Equation 1 u(z) and v(z) are profiles of the mean wind components, free from turbulent motions. In practice, this means averaging 10 to 20 balloon ascents to obtain a mean profile. In the more general case, when conditions are nonstationary and/or advective,

$$
\begin{aligned}
v_g - v &= \frac{1}{f} \cdot \frac{\partial}{\partial z}\left(\frac{\tau_x}{\rho}\right) - \frac{1}{f} \cdot \left(\frac{\partial u}{\partial t} + u\,\frac{\partial u}{\partial x} + v\,\frac{\partial u}{\partial y} + w\,\frac{\partial u}{\partial z}\right) \\
-u_g + u &= \frac{1}{f} \cdot \frac{\partial}{\partial z}\left(\frac{\tau_y}{\rho}\right) - \frac{1}{f} \cdot \left(\frac{\partial v}{\partial t} + u\,\frac{\partial v}{\partial x} + v\,\frac{\partial v}{\partial y} + w\,\frac{\partial v}{\partial z}\right)
\end{aligned}
\qquad (2)
$$

These equations can be reduced to the form (1) again by using those parts of the inertial terms that are known from observation as a correction of the observed wind profile and by combining the other parts with the geostrophic wind to form a generalized frictionless wind (Lettau, 1957; Hasse, 1976).

In addition to the mean wind profile, certain assumptions are required in order to estimate the surface stress. Some that have been tried are listed below:

1. Obtain u_g,v_g from the weather map. This is not a good assumption; this geostrophic wind is unreliable (Lettau, 1957).

2. Obtain u_g,v_g from a precision mesoscale pressure recording network. Since the local pressure is sensitive to local circulations and disturbances, it is essential to

have a considerable number of redundant stations to reduce the rms pressure variability. This would only be applicable if the inertial terms were known.

3. Obtain the thermal wind from weather map or climatological information. This is not accurate enough.

4. Assume a vanishing thermal wind. This is a dangerous assumption, which is not valid in general.

5. Assume that the actual wind approaches the geostrophic wind at higher levels. This is not too good, either, since pibal observations become more unreliable at greater heights. As the ageostrophic components are integrated, errors are added up. Note that this assumption cannot be made for a generalized frictionless wind.

6. Assume that the stress vanishes at greater heights. This may not be valid, since the thermal wind and the mixing may not vanish at the top of the boundary layer.

7. Assume that the surface stress has the same direction as the surface wind.

8. Assume an eddy diffusivity description. $\tau_x = \rho K (\partial u)/(\partial z)$ $\tau_y = \rho K (\partial v)/(\partial z)$. It follows that τ_x or τ_y is zero where the respective component of the wind profile has an extremum.

9. Assume that the vertical eddy diffusivity for momentum is equal for downwind and crosswind components.

10. Assume the thermal wind is constant with height.

Only assumptions 7 to 10 seem to be applicable, with some restrictions, while use of assumption 6 depends on the given situation. The validity of assumption 10 should be checked if there is a low-lying inversion.

Using combinations of assumption 6 through 10, Lettau (1957) has been able to obtain stress estimates from pibal observations. Lettau, following Sutcliffe (1936), did not use observed geostrophic and thermal wind, but rather obtained these from the wind profiles with the aid of the above assumptions and Equation 1. The advantage of his method is that he, in fact, incorporated the additional terms of Equation 2 into the generalized frictionless wind, which were determined indirectly from the profiles. Thus, the full equations of motion were used, and this is probably the reason for his success. Even so, the eddy diffusivities at greater heights sometimes became negative, which makes the validity of the assumptions used or the accuracy of observations dubious at greater heights.

Lettau used a kind of graphic trial and error method. Schriever (1966) followed Lettau's idea to obtain a generalized frictionless wind from the observed wind profile by the method of least squares using assumptions 8, 9, and 10. Note that since assumption 7 is not used, his system could also be used for elevated layers and thus for verification of assumption 10. A difficulty encountered with Schriever's approach is that if too many levels are used (in order to obtain redundant observations for the method of least squares) the equations at the different levels become less independent and the determinant of the system of equations becomes very small.

Obviously, the weather type will govern which analysis can be used. Since assumptions 8 and 9 imply an eddy diffusivity type of approach, these methods will probably fail in a layer of free convection, i.e. vanishing wind shear. In strong convection it might be necessary to measure $w\ (\partial u/\partial z)$ and $w\ (\partial v/\partial z)$.

For design purposes, it may be helpful to note that the height of the friction layer above the sea under average wind conditions, say 8 m s^{-1} surface wind, is of the order of 0.5 km in the temperate zones (compared to the often quoted 1 km over land). The height of the planetary boundary layer itself is dependent both on the friction at the surface (say, proportional to u_*/f) and on the static stability (e.g. height of the inversion).

4. CONCLUSION

Pilot balloon observations seem a useful way to achieve meaningful wind profile data for the planetary boundary layer with low instrument cost and relatively untrained observers. It is a drawback of optical tracking that the balloon is lost sight of in low clouds and low visibility; thus there is a bias in the data towards fine weather situations. Due precaution must be taken in the analysis to obtain reasonable stress estimates from the ageostrophic method. Since the stress is obtained by integration, some scatter in the data will average out, such as the turbulence induced balloon motions and any residual ship motions left over after filtering.

REFERENCES

CARSON, D.J., and F.B. SMITH. 1973. The Leipzig wind profile and the boundary layer wind-stress relationship. *Quarterly Journal of the Royal Meteorological Society,* 99: 171-177.

CHAMBERLIN, L.C., M.L. BARAD, R. ELY, and H.H. LETTAU. 1957. Loeser technique of wind profile determination - daytime smoke-puff and night-time pibal swarm measurements. In: *Exploring the Atmosphere's First Mile,* Edited by Heinz H. Lettau and Ben Davidson, Pergamon Press: 276-292.

CHARNOCK, H., J.R.D. FRANCIS, and P.A. SHEPPARD. 1957. An investigation of wind structure in the trades: Anegada 1953. *Philosophical Transactions of the Royal Society of London, Series A,* 249: 179-234.

DUNCKEL, M., L. HASSE, and D. SCHRIEVER. 1971. A refined system for pibal theodolite observations. *IEEE Transactions on Geoscience Electronics,* GE-9: 208-211.

HASSE, L. 1976. A resistance-law hypothesis for the nonstationary advective planetary boundary layer. *Boundary Layer Meteorology,* 10: 393-407.

KUHLBRODT, E., and J. REGER. 1933. Die aerologischen Methoden und das aerologische Beobachtungsmaterial. *Wissenschaftliche Ergebnisse Deutsche Atlantische Expedition "Meteor" 1925-1927,* 15: 1-34.

LETTAU, H. 1957. Windprofil, innere Reibung und Energieumsatz in den unteren 500 m über dem Meer. *Beiträge zur Physik der Atmosphäre,* 30: 78-96.

LETTAU, H.H., and H. HOEBER. 1964. Über die Bestimmung der Höhenverteilung von Schubspannung und Austauschkoeffizient in der atmosphärischen Reibungsschicht. *Beiträge zur Physik der Atmosphäre,* 37: 105-118.

RIDER, L.J., and M. ARMENDARIZ. 1966. A comparison of tower and pibal wind measurements. *Journal of Applied Meteorology,* 5: 43-48.

ROSSBY, C.G., and R. MONTGOMERY. 1935. The layer of frictional influence in wind and ocean currents. *Papers in Physical Oceanography and Meteorology,* 3: 1-101.

SCHRIEVER, D. 1966. Zur Berechnung der Schubspannung aus der vertikalen Windverteilung. *Naturwissenschaften,* 53: 404-405.

SCOGGINS, J.R. 1964. Aerodynamics of spherical balloon wind sensors. *Journal of Geophysical Research,* 69: 591-598.

SUTCLIFFE, R.C. 1936. Surface resistance in atmospheric flow. *Quarterly Journal of the Royal Meteorological Society* 62: 3-14.

THYER, N. 1962. Double theodolite pibal evaluation by computer. *Journal of Applied Meteorology,* 1: 66-68.

6

Survey of Techniques for Measuring Currents near the Ocean Surface

J.R. McCullough

1. INTRODUCTION

Air-sea interaction studies of momentum, heat, and salt fluxes in the upper ocean require detailed knowledge of the mean and fluctuating components of near-surface flow. Wind waves, swell, and mooring motions, however, make the slowly varying mean current component difficult to measure. This unfavourable 'signal-to-noise' ratio for mean currents is further complicated by the broadband aspect of the noise in both time and space. To accommodate the varied oceanic flow scales, many current measuring devices (perhaps 10^4 published papers) have been proposed. The slow and expensive evolution of suitable equipment attests to the difficulty of making what at first appears to be a rather simple measurement. Observational scatter is frequently a large part of the measured signal, and identification of systematic errors remains difficult.

Basic technical problems for long-term, near-surface moored operation at sea now include: mooring motion, sensor response, endurance, fouling, and single instruments versus profiling operation.

At present the preferred sensor types are: propellers, acoustic travel time (ATT) difference probes, electromagnetic (EM) spheres and open coils, and range-gated acoustic Doppler backscatter sensors. In this text Berteaux, Davis/Weller, Gibson/Deaton, Gytre, Halpern, Jones, Osborn/Crawford, Pinkel, Van Leer, and others discuss these technologies in detail. This paper presents an overview of moored current meter techniques, alludes to some other techniques, and comments on the general problem of calibration.

2. WAVE ZONE CURRENT MEASUREMENTS

2.1 Background

In a way, it seems curious that both time and distance can be measured with extraordinary precision while only crude estimates of the vector flow fields can be made near the sea surface. This seeming paradox arises naturally from the large continuous scales of motion in the water. There is no single velocity of the water, but many, which are characterized by their temporal and spatial spectra. Implicit then in the concept of a fluid 'velocity' is knowledge of the temporal and spatial averaging processes used in measuring it. Imprecise, or worse, inappropriate modes of averaging in time and/or space now represent the most prominent source of error in near-surface flow measurements.

2.2 Sensor Attributes

Ideally, sensor arrays would consist of individual elements which:

(a) measure only one bidirectional component of the flow and are insensitive to all other off-axis components,

(b) have sufficient frequency response to follow variations of interest while limiting the bandwidth of input variations to the linear range of the sensor,

(c) average over clearly defined temporal and spatial scales without disturbing the flow, and

(d) are sensitive, stable, rugged, small, inexpensive, easily calibrated, easily deployed, and require little power.

The importance of other current meter considerations such as compass, resolver, multiplexer, control, averager, size, packaging, reliability, cost, etc., are well recognized technical problems but are not emphasized in this review. The considerable importance of rectification due to sensor nonlinearities and flow interference by the sensor is discussed.

For optimum performance, conflicting design goals should be quantified and the necessary compromises made. In practice, however, most measurements are made with available instruments which often are not well matched to the scientific requirement. This has been particularly true for studies of oscillating flows associated with surface gravity waves.

Flow sensors must respond fast enough to follow the signals of interest, and must be small enough to resolve their spatial variations. Transducer outputs must be sampled rapidly enough in time and close enough in space to avoid aliasing the sampled data. Where knowledge of the highest frequency and wavenumber processes present is not required, the time and spatial passbands of the sensors can be reduced by low-pass filtering.

The spatial scale and sampling frequency of the sensors should, where possible, match the process being measured. Clearly defined, space and time averaging of limited bandwidth is preferable; if the passband of the sensor extends beyond the sensor's linear range, nonlinear distortion levels and consequent systematic error bounds will remain unknown. To avoid such harmonic and parametric distortion (ringing, rectifying, smearing, etc.) the sensor should be bandwidth limited to its linear range, and for reversing flows have a zero (or known) bias with linear response through zero.

Limiting the system bandwidth at the transducer may seem contradictory to the sound common sense practice of recording as much as possible and reducing the data later, but unless the sensor is linear over its own passband and insensitive outside it, irrecoverable contamination of the data may result. Worse yet, as is often the case, the data may 'look good' but contain large transient errors caused by nonlinear transducer response.

Off-axis sensitivity, or 'cross talk,' greatly complicates the signal processing needed to reconstruct the desired orthogonal flow components. Also, it introduces the additional requirement that each axis have matched frequency and spatial response or that all the flow sensors be linear over the entire input signal bandwidth. If no cross talk exists, each sensing element measures a single bidirectional scalar quantity and need only be linear over its own passband, a much less demanding requirement than that of accommodating the entire unknown bandwidth of the applied signal. Evaluation of a particular sensor for such basic specifications as percent harmonic distortion, flow distortion, dynamic response, cross talk, linearity, stability, and the like are difficult, time consuming, and expensive tasks. The recommendation is: in the wave zone, use a linear, single-axis sensor if at all possible.

2.3 Platforms

In broad kinematic terms, flow sensors are either located on fixed Eulerian platforms or drift with the particle motions as Lagrangian tracers. The Eulerian and Lagrangian techniques do not measure the same phenomena, so care must be taken when comparing results. Further, most practical platforms such as moorings and drogues provide only approximations to true Eulerian and Lagrangian reference frames. Platform motion not only can change the signal-to-noise ratio at the current meter, but in waves or other shears it can change the real value of the apparent current. Two ideal current sensors (ones with no errors), in identical waves, one below a surface-following float and the other below a spar at the same mean depth will not measure the same mean current even if both meters are working perfectly! The difference arises from the meters occupying different parts of the wave shear at synchronized phases of the wave cycle.

In a similar way, horizontal oscillatory motion of a perfect current meter in waves introduces errors. For example, if the meter moves horizontally with the wave orbital motion at its depth, it spends slightly longer under the wave crests (it moves along with the crest) than under the wave troughs. Figure 1 illustrates the magnitude of such effects as a function of depth.

Profiling platforms, reviewed in this text by Van Leer (see Chapter 36), provide a means for better depth coverage at the cost of reduced temporal sampling and platform stability. Since the profiling sensors move, the designer must somehow resolve problems of space and time aliasing, motion compensation, duty cycles, etc. The use of a single package, however, greatly simplifies calibration and allows a variety of high quality sensors in a single, convenient unit. Since both amplitude and phase information are needed in mean-flow wave zone measurements, current sensors must average at constant depth for several wave cycles. Where simultaneous measurements at several depths are needed to compute shears, fluxes, etc., chains of current sensors on stable platforms may be required.

In much the same way, tow fish (see Chapter 38 by Nasmyth) increase horizontal spatial coverage. Ships provide natural moving platforms for tow fish and remote sensors. Several systems for ships-of-opportunity are now being tested at sea.

The next generation satellite navigation system, called Global Positioning System (GPS) or NAVSTAR, is intended to give surface positions to ±10 m and velocities to ±10 cm s^{-1} in all three dimensions at each fix. Ships carrying GPS receivers and two-axis speed logs could provide extensive high precision ship drift data for the world's oceans.

By virtue of their long life and remote positioning capabilities, drogues, Swallow floats, and drifting moorings (see Chapter 11 by Vachon) can traverse tens of thousands of kilometres. Just as the profiling moored instrument can trade high frequency resolution for depth coverage, the drifting mooring can extend horizontal coverage at the expense of low frequency resolution at any particular location.

During the last few years, huge deep-water oil platforms and semi-submersibles have been under construction; several hundred now exist. Their unprecedented stability and the long reach of their

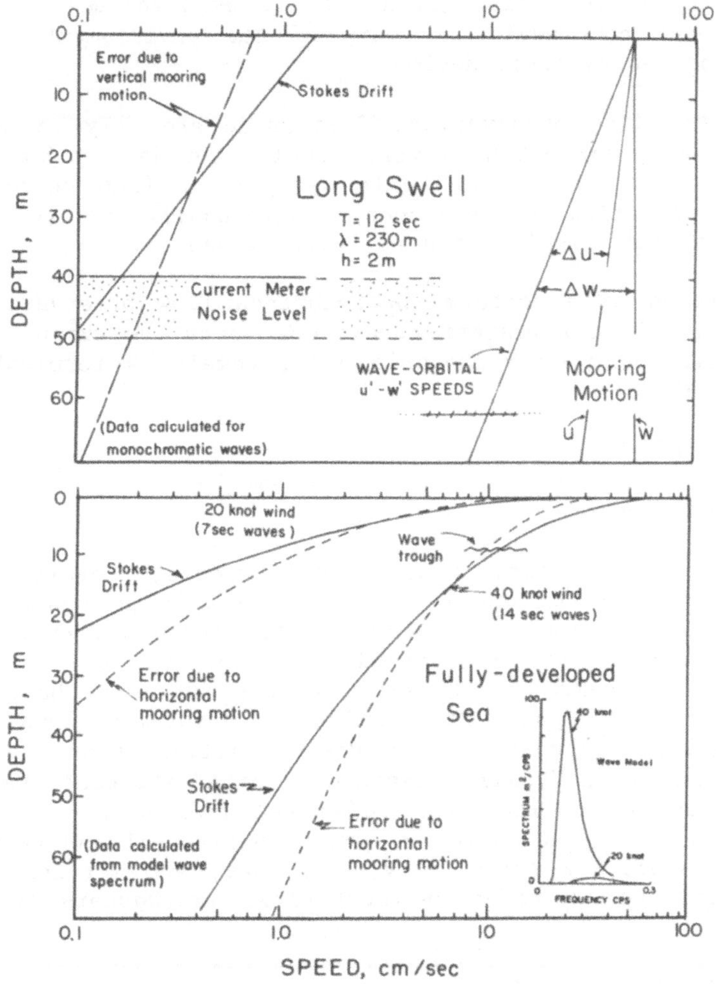

Fig. 1 Eulerian speed errors versus depth. Error contributions by vertical mooring motion calculated for long swell (upper frame) and by horizontal mooring motions in 20 and 40 knot fully developed seas (lower frame). Upper frame also shows wave orbital speeds u' and w' for the long swell conditions tabulated. Relative wave speeds Δu and Δw seen by instruments on a surface following mooring increases with depth (after McCullough, 1978).

cranes give them unique capabilities as flow observatories. Lower performance platforms will continue to be used for most purposes, but users must continually be wary of the environment-modulated errors introduced by their motions.

Techniques for flow observations from space are only in an early stage of development but will very likely provide a major tool in near-surface flow research. World synoptic surface currents may find wide application in areas such as ship navigation, recreation, search and rescue, fishing, and pollution research.

The acquisition of long-term, near-surface flow observations, by any technique, in all weather conditions, over the expanse of the world oceans, and at a reasonable cost, remains a formidable engineering challenge.

3. SENSOR TYPES

3.1 Mechanical Sensors

Mechanical sensors (rotors, vanes, paddle wheels, propellers, drogues, dye, ship drift, etc.), have collectively provided most of our direct information about near-surface ocean currents. When properly calibrated and maintained they can give quite reliable measurements for years in steady flow, but typically have rather complex nonlinear response in turbulent or reversing flows. Their general limitations include: low speed stalling, poor high speed endurance, nonlinear dynamic response, limited frequency response, susceptibility to fouling, and mechanical frailness. Attractive features, such as good cosine response (horizontal and vertical), calibration stability, simplicity, low cost, and servicing ease, however, continue to recommend mechanical transducers for ocean use.

(a) Large Vanes and Rotors

A variety of in situ experiments (see Halpern, Chapter 7) have provided compelling evidence that current meters of the large-vane and rotor type are not suitable for near-surface moored observations in waves. When wave-zone currents reverse with each wave cycle, the distance scale of the vane does not provide adequate directional information for proper velocity averaging. The large vane makes the observed speed too high. For the nonreversing coplanar case, the meter again reads high due to rotor rectification or 'overspeeding'. When the wave orbital velocities and mean velocity are at an angle to each other (the usual case) small flow scales again appear as erroneous contributions to the mean. Saunders (1976) observes in moderate seas, that while the mean current directions are generally correct, the vane response in waves causes speeds to be too large by a factor of between 1 and 10, with a typical value of about 2.

One possible exception is represented by the tethered spar-Aanderaa current meter combination described by Beardsley et al. (1977) in the Current Meter Intercomparison Experiment 1976 (CMICE 76). A small spar was designed to move the meter vertically (via damping plate drag) and horizontally (via the tethered spar motion) with the wave orbital motion at the 3.5 m depth of the single current meter. The spar motion apparently reduced the oscillating component of relative flow sufficiently to allow the meter to work quite well. It is unlikely, however, that the intended match between wave speed and meter can be made adequate in all sea states.

Another large-vane method that shows promise is a tethered neutrally buoyant meter with a reversing ducted-impeller as described by Brainard and Lukens (1975). For measurement at the air-sea interface, this neutrally-buoyant tethered meter can be ballasted slightly light and attached to a slack mooring. Preliminary experiments suggest that this mode of operation may be suitable for mean flow measurements in the upper 0.5 m, in waves. Additional tests and improved response modelling are needed, however, to support these conjectures.

(b) Small Vanes and Rotors

The AMF vector averaging current meter (VACM) uses a Savonious rotor and vane with similar response length scales (McCullough, 1975). The rotor-vane response lengths cannot be accurately matched, however, since the rotor length scale changes (it accelerates about three times faster than it decelerates). In situ intercomparisons (see Halpern, Chapter 7) indicate that despite errors introduced by the three nonlinear, polar coordinate sensors of the VACM (rotor, vane, and magnetically-coupled vane follower), useful measurements of mean near-surface currents can be made with VACMs in waves and from surface-following moorings. The approximate match of the rotor-vane response scales and relatively high sampling rate (eight samples per rotor revolution) appear to average wave flow to a near-zero mean. In the CMICE 76 intercomparison, for example, electromagnetic and VACM hourly means agree to within about ±3 cm s^{-1} and ±5°, in 1 to 4 m waves, at 7.4 m depth, for 6 weeks, in varying wind conditions.

It should not be concluded, however, that VACMs will necessarily give correct mean currents in waves. Laboratory measurements by McCullough (1974) show that for 7 s period orbital motion and colinear mean speeds of equal magnitude (nonreversing flow), the rotor overspeeds by approximately the rms value of the oscillatory component. Gould and Sambuco (1975) also show that the VACM is not suitable for mean flow observations at mid-depths on surface following deep-sea moorings.

Allowing current meters to move about in some complex, unspecified manner, as on a tether or surface-following mooorring, in order to reduce the effects of poor sensor response, is an expedient compromise at best. The preferred Eulerian method uses stable platforms and fast linear component current sensors. Various single-axis orthogonal sensors (discussed next) allow such measurements.

(c) Component Propellers

Propeller history and design for oceanographic applications are treated by Davis and Weller in this text (see Chapter 8). The advantages of component propellers for wave zone studies were recognized a decade ago by Shonting (1967) and others, but the need for general purpose deep water current meters with low speed thresholds favoured rotor sensors. Well designed propellers offer attractive advantages for both air (see Chapter 1 by Busch et al.) and sea measurements. They provide simple construction, single-axis sensitivity (i.e., good horizontal and vertical cosine response), stable calibrations, adequate frequency response, and the ability to be easily and economically multiplexed in arrays sharing common signal-processing and data-logging. Steady-state response can be made very nearly linear by making the blade lift forces large compared with bearing friction. Overall construction, calibration, and servicing costs are similar to or less than those of the nonmoving part techniques. The outstanding advantage of propellers for wave zone applications is good static and dynamic cosine response.

Propeller disadvantages include nonlinear response through zero speed and susceptibility to fouling. The extent to which such limitations influence practical measurements is presently being studied.

3.2 Acoustic Travel-Time (ATT) Sensors

Acoustic travel-time difference (ATT) sensors are reviewed in this text by Gytre (see Chapter 9). They measure the travel-time difference of sound sent simultaneously in opposite directions along the same acoustic path. Two or more such paths are used to establish the flow vector. ATT sensors have the interesting property that the narrow acoustic path over which the flow is averaged can be varied in length from centimetres to tens of kilometres and perhaps to scales as large as ocean passages and basins. Further, they can be used to directly measure shear and vorticity (Rossby, 1975). The spatial-averaging scale is clearly defined: the response is fast and inherently linear. Individual transducer elements are relatively inexpensive and easily multiplexed. Sensitivity is fixed by flow-independent parameters.

Disadvantages include flow interference by the transducer supports and electronic complexity.

Three detection techniques for acoustic travel-time difference are in general use. (a) Pulse edge detection: Acoustic pulses are transmitted simultaneously from two transducers; when received (preferably by the transmitters acting as receivers), the arrival time difference of a zero crossing of the received pulses is measured. (b) Continuous wave (CW): Continuous acoustic signals (or pulsed CW signals) are transmitted in opposite directions and the component of the total phase shift caused by the fluid motion is measured. (c) Sing-around: One pulse is transmitted; when it is received by the opposite transducer a new pulse is sent in the same direction and so on. The pulse repetition frequency is a measure of the total travel time in one direction; reversing transmitter and receiver functions gives the travel time in the opposite direction; the difference gives the desired speed component along the acoustic path. These three electronic techniques and numerous others have been used for roughly four decades in sonic anemometers, sound-speed meters, and current meters.

In the pulse technique, the travel-time difference signal, Δt, is related to the flow speed V, at an angle θ to the acoustic path of length L, by

$$\Delta t = \frac{2\ LV \cos\theta}{C^2}$$

where C is the speed of sound in the water. The measured speed, V, is the average component along the acoustic path or device aperture L. For a device with a 15 cm aperture, the on-axis sensitivity, $\Delta t/V$, is only 1.3 ns per cm s^{-1}, but present electronic circuits allow noise levels in Δt as low as an equivalent flow of 0.01 cm s^{-1}.

Increasing L increases the sensitivity ($\Delta t/V$) without changing either the electronic or flow noise appreciably. Signal attenuation, however, increases with L and the square of the frequency, F. The bandwidth is limited by the pulse travel time L/C and echo decay times. In the 15 cm example a bandwidth of at least 1000 Hz is practical. Other design problems involve electrical and mechanical cross talk between axes and within each axis. Transducer damping, for example, must be high enough (100 μs scale) to separate transmit/receive functions within each transducer. Design optimizations for particular applications are given by Gytre in this text (see Chapter 9).

In the continuous wave (CW) technique, the time difference, Δt, measured as a phase change, $\Delta\phi$, at frequency F is

$$\Delta\phi = \frac{4\pi FLV \cos\theta}{C^2} .$$

Phase detection is technically more difficult to achieve but uses less power (lower frequency detection with the desired phase shift is accomplished by beating the transmitted frequency with one which is only slightly different than the carrier; the heterodyned beat-frequency retains the phase characteristics of the transmitted signal). Potential disadvantages of the method include a maximum speed range set by phase ambiguity and less rapid (or more complex) multiplexed operation.

Meter operation using a sing-around technique is described by Hardies (1975).

Unfortunately, both the maximum sensitivity and maximum flow disturbance by the ATT probes lie along the path separating the sensors. The wake behind the sensors has two adverse effects; it reduces the mean flow and causes it to oscillate (shadowing and vortex shedding effects). Both are Reynolds number dependent and introduce speed and time-dependent nonlinearities. The net effect can be reduced, however, by increasing the separation ratio, L/D, where D is the characteristic obstruction diameter.

For cylindrical probes in the Reynolds number range of interest for moored meters, the maximum speed attentuation due to the probe wakes is about 35% for L/D = 10. The integrated attenuation increases only slightly beyond L ~ 10D. Thus beyond L/D = 10, doubling the separation decreases the relative wake effects by approximately a factor of two. At L/D = 80 the maximum speed attenuation will be about 35%/8, or less than 5%.

Large separations and small transducers lead to fabrication and handling problems, so additional strategies have evolved to reduce wake effects. First, the acoustic path can be folded with a mirror. Although this adds the wake of the mirror and its support to the flow interference, it moves the primary interference off axis. Second, since the principal velocity defect is concentrated in a zone about 60° wide behind each support, redundant paths can be used to allow selection of those least contaminated by wakes. Finally, support structures can be streamlined, faired or extended to reduce wake effects along preferred axes.

Since the sensitivity changes with the square of the sound speed, C, some measure of C is needed. Some designs measure C directly from the sound travel-time and apply an electronic correction; others rely on measured temperature and assumed salinity and pressure to compute C in the calibration. If the instrument is calibrated at room temperature in fresh water and used at depth in the ocean the sound speed correction is about 12%, while the maximum surface correction is about 7%.

Aeration of the water limits the minimum depth at which ATT sensors can operate (Medwin, 1977). Because bubbles at a given depth will be larger under wave troughs than under crests, bubble reflection and absorption will be synchronized with the waves. Unless electronically identified in some manner, systematic biases due to bubbles may prove troublesome in near surface in situ averages.

Fouling, structural rigidity, flow disturbance, and electronic techniques are the main ATT design issues. Considerable progress has been made in recent years on the electronic power, size and stability problems. Appropriate structural design for near-surface operation and multiplexed sensors has begun (see McCullough, 1978; Williams and Tochko, 1977). Other oceanographic ATT meters have been described by Middleton (1955), Hardies (1975), Lawson et al. (1976), Lowell (1977), and others. Commercial versions are now available from Japan, France, Norway, and the United States.

In principle, ATT probes with known calibration, frequency-bandwidth, and spatial aperture can be constructed. There is adequate bandwidth to directly measure the near-surface wave fluctuations, u', v', and w'. When combined with the temperature fluctuations T' such observed quantities give direct measures of the momentum and heat fluxes. Since ATT sensors can operate in both air and water, air-sea interaction studies could use ATT velocity spars to measure air and sea motions simultaneously from a central instrument package. Conceptually, an acoustically instrumented cubic metre (or kilometre, Munk and Wunsch, 1979) could measure u, v, w, u', v', w', shear, and vorticity within the cube with high sensitivity and accuracy. Techniques for making such measurements, however, are only in an early stage of development.

3.3 Electromagnetic (EM) Sensors

Oceanographic EM sensors fall into two broad categories: those using the earth's magnetic field and those using local magnets. Sanford et al. (1978) discuss the earth's field technique in some detail. In this text, Jones (see Chapter 12) reviews the theory and practice of the electro-magnet type. Both operate by measuring the electric field induced in the water as it moves through a magnetic field (Faraday's principle). The electro-chemical voltages (battery type) generated at the measuring electrodes are typically 10 to 1000 times larger than the motion-induced voltages (generator types), so some sort of frequency separation of the low frequency electro-chemical voltages is made. To do this the earth's field devices (GEKs) are towed on changing courses while free-drop versions are made to spin. With the local field type, the electric polarity is changed by reversing the electro-magnet current or by rotating a permanent magnet. Electrical leads in the electromagnet type act as transformer windings and must be carefully placed to avoid unwanted voltages in quadrature with the signals. Practical designs typically evolve from sinusoidal to chopped-DC magnetic excitation.

Various strategies have been tried to reduce the flow disturbance while maintaining a practical, rugged sensor shape, the most popular shapes now being the cylinder, disc, sphere, and open coil. Concentric poles, isolated poles, tubes, etc., are also used. Protruding electrodes of various shapes give additional control of the flow disturbance effects. Each of the magnet-electrode shapes represents some compromise between rugged, small probes and good, three-dimensional flow response, a situation not unlike that for mechanical and ATT sensors.

Accurate vertical cosine response in the wave zone is critical (see McCullough, 1978). The vertical cosine response of the EM disc is inadequate, that for the cylinder marginal, the sphere with protruding electrodes (Marsh McBirney Inc., 1977) is better, while the open coil (Olson, 1972) appears best for wave zone measurements. The open coil design, however, has received relatively little attention. One practical problem is that as the open coil(s) grows larger and consequently more fragile, control of electrical leakage becomes more difficult (see Collar and Griffiths, 1977). For all EM sensor configurations, the new low noise, operational amplifiers make the voltage detection part of the problem quite tractable. An inexpensive, expendable, 750 m depth, profiling, earth's field EM current meter, looking much like an XBT, is now being developed (Sanford and Drever, personal communication, 1977).

The parameter sensed by EM current meters is the electrical potential, E, given by

$$E = \int (\vec{v} \times \vec{B} - \vec{j}/\sigma) \cdot d\vec{\ell}$$

where $\vec{v} \times \vec{B}$ is the voltage generated by the motion $\vec{v}$ of the water through the magnetic field $\vec{B}$, and $\vec{j}/\sigma$ is the voltage caused by currents, $\vec{j}$, flowing in the water of conductivity σ when curl $(\vec{B} \times \vec{v}) \neq 0$. If the magnetic and flow fields were uniform each axis of the EM meter would be linear and only thermally noise-limited in sensitivity - a nearly perfect current sensor. In practice, however, the probe distorts the flow around it in the region where the local magnetic field is strongest. The resulting electric field is influenced by flow variations in the flow boundary layer, and the voltage integral above becomes a function of the flow dynamics.

Inadequate knowledge of EM sensor nonlinearities requires empirical laboratory calibration in steady flow. The assumption is then made that the response to time-varying flow is approximately equal to that for steady flow. Testing the error introduced by this assumption becomes a major cost factor in EM sensor development. Expected ocean turbulence levels are poorly defined and expensive to simulate. Adequate numeric models are not available to guide the tests or help interpret the results. Limited surface roughness, vibration, and turbulence sensitivity studies made so far suggest due caution on the part of critical users (see Bivins, 1975).

Without adequate predictive models, it has been difficult to assess the relative importance of the various structural parameters and overall probe scaling in the dynamic response range of interest for near-surface work. It is, after all, the packaging that limits EM sensors in ocean research, not the electronic, electro-chemical, power or signal-processing problems. Progress is hampered by the expensive and time-consuming process of 'cut and try' full-scale probe construction and testing. Proposed improvements often take years to develop, test, and gain general acceptance.

On the positive side, EM sensors can be made small, rugged, and very sensitive. Numerous commercial versions are marketed and have been used extensively in mixed-layer work. Single EM probes of several designs can measure all three components of flow.

3.4 Acoustic Backscatter Sensors

When sound passes through water some of it is scattered back toward the sound source. This echo can be analyzed for Doppler shift to find the relative velocity component of the scatters in the radial direction along the beam. Using an array of sensors, the backscattered sound can also be processed by correlation techniques to find the other two velocity components in the plane normal to the transmitted-received beams. In principle, all three components of velocity can be measured remotely at great distance in the sea.

The Doppler shift Δf is related to the transmitted frequency f, the flow speed V, and the speed of sound C at the transducer by

$$\Delta f = \frac{2fV\cos\theta}{C}$$

where θ is the angle between the sound beam axis and V. In a more general sense that takes into account the time rate of change of sound velocity along the acoustic path, the Doppler frequency is equal to the rate at which the travel time of sound between source (scatterers) and receiver is changing; i.e.,

$$\Delta f \propto \frac{dT}{dt}$$

where T is the travel time.

In the correlation sonar technique under development by Dickey and Edwards (1978) and Jungner Instrument Division (1977), a short line of receivers is used to measure the spatial variation of backscattered sound. When the temporal correlation of the scatterers is sufficient, the return from two pulses (CW can also be used) can be analyzed by correlation methods to detect the relative motion of the line array past the scatterers. The time lag Δt producing maximum correlation is related to the length L of the receiving line array and the relative velocity V by

$$\Delta t = \frac{L}{2V \cos\theta}$$

where V and L lie in parallel planes and θ is the angle between the line array and the projection of the flow vector V in the plane of the array.

As discussed by Pinkel in this text (see Chapter 10), Doppler backscatter techniques have been used successfully to measure horizontal and vertical shears over distances of the order of 1 km. Methods for making vertical profiles of horizontal currents from ships using backscatter techniques are also being developed. In the future, similar devices might be used on subsurface moorings or Swallow floats to look up at violent surface motions, or placed flush with the bottom in shelf applications to avoid both trawlers and storms.

For short-range moored applications, the Doppler transducer is smaller, more rugged and less susceptible to fouling than are other sensors of comparable sensitivity. It allows profiling of undisturbed flow in all directions from a single location. Signal dropout has not been a problem in the upper 200 m for towed Doppler sensors. Power and signal processing complexity are the present limiting parameters but appear to be electronically tractable (see Frescura and Meindl, 1976).

3.5 Laser Doppler Velocimetry (LDV)

The laser Doppler velocimeter (see Chapter 1 by Busch et al.) is particularly well suited for small scale remote observations in the laboratory and atmosphere. With very high-powered pulsed lasers, the range is only limited by the transparency of the fluid. The technique has been tested experimentally in the ocean but at present is rather expensive, cumbersome, and energy inefficient. Recent oceanographic applications are described by Wannamaker (1976), Terry and Williams (1977), etc.

In the usual LDV configuration, a small oval-shaped test volume about 1 mm or so long is illuminated with two crossed, spatially coherent beams from the same laser. The beams set up closely spaced, parallel planes of interference, normal to the plane of the laser beams and parallel to their bisector. Small particles moving through these planes scatter light of varying intensity as they pass through the alternately dark and light interference zones. The observed modulation frequency F is

$$\Delta F = \frac{2Vn\sin\phi}{\lambda}$$

where V is the component of the particle velocity normal to the planes of interference, n is the index of refraction of the fluid, λ is the laser wavelength in free space, and ϕ is the half angle between the laser beams.

Alternately, we can understand ΔF as follows: randomly scattered light from particles moving in the test volume is Doppler shifted in the opposite sense for each of the two beams. The composite scattered light, now at two slightly different frequencies, is optically heterodyned at the receiver, producing a beat frequency, ΔF, which is directly proportional to the speed component of the scatterers in the plane of the beams and normal to their bisector. Thus it is not the colour (Doppler shift) of the light that is directly observed, but a beat frequency caused by two slightly different Doppler shifts (the absolute frequency change

$$\frac{\Delta F}{F} = \frac{2Vn\sin\phi}{C}$$

is exceedingly small, being typically about 10^{-11} per cm s^{-1} of flow).

The sensitivity of typical LDVs is high, about 1500 Hz per cm s^{-1} of flow. Since the sensitivity axis is normal to the beams, the instrument need not interfere with the flow being measured. The forward scattered light, however, may typically be 3 orders of magnitude more intense than that backscattered, so an additional instrument housing or a mirror is often added opposite the laser housing. Sufficient scatterers for continuous operation are found even in clear ocean surface water (Wannamaker, 1976).

To find the sign of the Doppler shift, a Doppler offset can be established by frequency shifting one of the laser beams with a Bragg cell (an acoustic-optic device using a moving acoustic grating principle). The second orthogonal component of flow requires optical multiplexing or a third beam. The third component can also be detected at reduced sensitivity in a two-beam system.

Note that the acoustic Doppler and radar Doppler (discussed below) sense the radial speed (one component), while the laser Doppler and correlation sonar methods sense the velocity (two components) normal to the transmitted beams.

The laser Doppler technique provides an excellent, reproducible standard for small scale flow calibration: (a) the calibration depends only on n, λ, and ϕ, which are all stable, accurately reproducible, and readily measured; (b) it is intrinsically linear, has high sensitivity, and has excellent temporal and spatial resolution (1 to 1000 Hz and 1 to 0.1 mm); (c) it senses at a clearly defined, illuminated, remote location which is essentially undisturbed by the measurement; (d) it senses motion of particles whose size can be made very small, allowing them to accurately track the local flow accelerations (flow-induced particle accelerations at low Reynolds numbers vary inversely with the square of their size).

Laser penetration of the sea from the air has been demonstrated to a depth of order 100 m in clear water. In principle, very high-powered lasers, now being developed for other purposes, could be used to remotely profile the three components of velocity, species densities, and temperature (Leonard et al., 1977) below a ship, aircraft, or satellite. Perhaps one day satellites will map the mixed layer flow and temperature fields of the world's oceans by means of high-powered, solar-energized, pulsed lasers.

3.6 Hot-Film and Hot-Wire Anemometry

Hot films and hot wires are widely used for small scale, laboratory turbulence measurements. Ocean applications of heated elements are discussed in this text by Gibson and Deaton (see Chapter 18). Typically these sensors require careful handling and frequent cleaning. They have been used successfully in short term (several day) ocean turbulence studies at sea but have not demonstrated very adequate long term stability. Techniques of sonic, chemical, radioactive and mechanical cleaning have been studied as have the nonlinear signal processing requirements of these sensors. Hot-films have been suggested for air-deployable miniature current sensors. The details of laboratory hot-wire techniques are well developed and presented in the literature. Disa Electronics (1974, 1977), Thermo-Systems, Inc. (1977), and others describe hot-films, hot-wires, and laser Doppler systems.

3.7 Radar Backscatter and Aerial Surveys

Radar pulses (1-30 MHz, HF, dekameter radio waves) transmitted near the ocean surface are selectively backscattered (Bragg or diffraction grating type scattering) by surface waves of wavelength equal to one-half the radar wavelength. Since the phase speed of deep water waves can be calculated from the known wavelength, the mean radial velocity component of the water wave speed due to advection can be extracted from the total Doppler shift of the backscattered radar pulses. The travel time gives the range. Phased line arrays of dipole antennas are used for reception directivity. By measuring at several frequencies, waves of different wavelength and hence effective water depth can be observed to give estimates of vertical shear. By observing at two or more locations the two horizontal velocity components in each sector can be resolved (see Chapter 24 by Stewart).

Barrick et al. (1977) describe a shore-based Gulf Stream test of a two-station system with a range of 70 km, a peak pulse power of 2.5 kW, and an average power of 50 W. The tests averaged horizontal surface velocities in 3x3 km squares over an area of 2000 km^2 in about one-half hour, giving an estimated precision in each square of better than ± 15 cm s^{-1}. Stewart and Joy (1974) describe an earlier system. R.H. Stewart of Scripps and C. Teague and H.T. Howard of Stanford University tested a single ship version in the JASIN 78 experiment. The recent revival of interest in the radar technique for Gulf Stream ring tracking, search and rescue, shelf studies, etc., may make it more widely used in the next few years. Over-the-horizon radar can be used in a similar way, but ionospheric Doppler effects complicate the interpretation.

In a similar way, optical processing of photographic sequences can be used to map flow from above (Butterfield and Andrews, 1972). Saunders (1973) describes an isotherm advection technique for measuring surface currents. Colour, runoff, oil slick, and other signatures have also been used. A number of airborne and satellite remote current-measuring systems seem feasible but have received little attention.

3.8 Near-Surface Tracers

Dye, drogues, ship drift, bottles, temperature structures, oil slicks, radioactive materials, paper, wood chips, ice, trees, flora, flauna, etc., have all been used to study the surface flow of the oceans. The drift technique is simple: tag the water and observe the tag's displacement as a function of time. The basic problems are locating the tracer (navigation), keeping it in the area of interest (re-seeding), determining how well or poorly it follows the water (slip), and finding how much it and/or the water it represents has spread (dispersion). Various systems, with varying degrees of cost versus accuracy-coverage-convenience, are discussed by Vachon in this text (see Chapter 11).

Near-surface drogue slip speeds are typically of the order of ½ to 1% of the average wind speed (~10 cm s^{-1} in a 20-knot wind). Low profile drogues and constant pressure, subsurface floats (Wells, 1973) can be used where lower slip values are required. Dye, on the other hand, follows the water almost exactly but disperses so rapidly that it is difficult to detect after more than a day or two. The fact that dye and floats disperse (or come together) with time complicates the mean flow analysis but provides additional information about dispersion, eddy diffusivities, vorticity, divergence and deformation rates (Okubo and Ebbesmeyer, 1976). For surface mixing studies dye, camera, and fluorometer are natural tools (Kullenberg, 1976). The dye techniques give a valuable overview, nicely demonstrated in the motion pictures of underwater waves filmed by Woods (1977). Near-surface, acoustically-tracked floats are being used in JASIN 1978 experiments to measure near-surface flow. Some nearshore dye tracing techniques are detailed in EG&G (1976).

4. CALIBRATION

Tow tanks, flumes, wave tanks, submerged jets, wind tunnels, 'sloshers', turbulence screens, and in situ intercomparisons are used to calibrate and evaluate near-surface current sensors. Since most current sensors work well in steady flow, the major calibration difficulties for wave zone sensors only become apparent in dynamic tests. The practical calibration problem is complicated by the general lack of dynamic flow facilities large enough for full scale sensor testing. Unfortunately, testing costs and/or delays have been prohibitive in many programs.

The response of most flow sensors is so complex or unpredictable that it must be determined directly in a tow tank or flume. Some devices, such as continuous wave acoustic Doppler, should not be calibrated in a tow tank. Mechanical sensors can be designed in wind tunnels and calibrated in water as a check of the scaling procedure. Submerged water jets are useful in conjunction with tow tanks to reveal unwanted turbulence effects (Bivins, 1975). Wave tanks should be useful for testing near-surface flow sensors but have not been widely used. Lakes, with low mean currents, are particularly useful for long range acoustic profilers.

A typical sequence for evaluating a moored current sensor involves finding its reading in still water (zero bias and stability), its response to steady flow as a function of speed (sensitivity, linearity, and gain stability), and its immunity to off-axis flow components (horizontal and vertical cosine response). If these steady tests look promising, some sort of dynamic tests involving oscillating tows (sloshers) and turbulence should (but often don't) follow. The next step is to check the sensor and meter as a system in the ocean (moored intercomparisons) to evaluate overall performance and identify unanticipated problems. Intercomparisons at sea provide a useful but generally insufficient test since only differences, not absolute errors, are found. A new dye and drogue technique intended to help establish absolute performance at sea is described by McCullough (1977). Acoustic techniques also look promising for this purpose.

Since full-scale dynamic calibrations are expensive and since the wave zone forcing and turbulence levels are not well known, sensors with independently determined response functions are desirable. The ability to predict dynamic response is excellent for acoustic and laser Doppler sensors; it is not yet possible for EM and rotor sensors.

As discussed by Vachon in this text (see Chapter 11), drogue slip values are difficult to measure since they may vary considerably with even minor component alterations. In present practice the slip function of a particular drogue design is rarely determined. The necessary in situ wave, shear, and wind vectors are usually unknown, so slip corrections can only be estimated. A simple direct means of measuring slip would provide a valuable operational tool for this complex calibration problem. Techniques for estimating non-Lagrangian tracking due to vertical drogue motions in wave shear are also needed.

The need for flow standards for laboratory and particularly for open ocean testing is widely recognized. The continuing difficulty and cost of in situ intercomparisons emphasizes this need. A moored test range(s) with fixed acoustic and laser standards may be a possible, although expensive, solution. On the other hand, prolonged misapplication of current meters and numerous in situ intercomparisons are not inexpensive alternatives.

A concerted, international program for near-surface current meter calibration and evaluation is much needed. Such a program would be particularly timely now, with the new propeller, ATT, EM, and Doppler sensor developments.

5. CONCLUSION

Waves, turbulence, and mooring motion make mean current measurement difficult near the sea surface. Propeller, acoustic travel-time difference, and electromagnetic sensors, however, have developed rapidly in the last few years and make good near-surface observations technically feasible. Remote radar, laser Doppler, and acoustic Doppler hold promise for undisturbed flow profiling in both large and small samples. We now lack only the necessary tests to select optimum designs and establish their error bounds. Methods of temporal and spatial averaging need special consideration. Errors due to platform motion also need further study. Better standards for calibration and in situ testing are needed. There is much work to be done; considerable progress is anticipated during the next decade.

ACKNOWLEDGEMENTS

It is a pleasure to acknowledge the assistance of J. Dean, R. Koehler, W. Terry, and A. Williams in the preparation of the manuscript, and R. Davis for editorial encouragement. This work was supported by the Office of Naval Research under Contract No. N00014-76-0197 NR 083-400.

REFERENCES

BARRICK, D.E., M.W. EVANS and B.L. WEBER. 1977. Ocean surface currents mapped by radar. *Science,* 198: 138-198.

BEARDSLEY, R.C., W. BOICOURT, L.C. HUFF and J.SCOTT. 1977. CMICE 76: A current meter intercomparison experiment conducted off Long Island in February-March 1976. *Woods Hole Oceanographic Institution* Ref. 77-62.

BIVINS, L.E. 1975. *Turbulence effects on current measurements.* Master's thesis, University of Miami, 104 pp.

BRAINARD, E.C., II, and R.J. LUKENS. 1975. A comparison of the accuracies of various continuous recording current meters for offshore use. *Proceedings of the Offshore Technology Conference 1975,* Houston, Texas, Paper 2295: 485-490.

BUTTERFIELD, R. and K.Z. ANDREWS. 1972. Current measurements from aerial photographs. *Proceedings of the Society of Underwater Technology,* 2: 48-52.

COLLAR, P. and G. GRIFFITHS. 1977. Laboratory evaluation of current sensors. *POLYMODE News, 29,* Woods Hole Oceanographic Institution (unpublished manuscript).

DICKEY, F.R. and J.A. EDWARDS. 1978. Velocity measurements using correlation sonar. IEEE, *Plans 1978, Position, Location and Navigation Symposium*: 255-264.

DISA ELECTRONICS. 1974. *Bibliography of Laser Doppler Anemometry Literature.* Disa Electronics, 779 Susquehanna Ave., Franklin Lakes, N.J. 07417, 55 pp.

DISA ELECTRONICS. 1977. *Laser and Hot-Wire, Hot-Film Anemometry Systems.* Disa Information, 779 Susquehanna Ave., Franklin Lakes, N.J. 07417. Technical Series No. 1-23.

DURST, F., A. MELLING and J.H. WHITELAW. 1976. *Principles and practice of laser-Doppler anemometry.* Academic Press, New York, 405 pp.

EG&G. 1976. *Forecasting power plant effects on the coastal zone.* Final Report No. B-4441, EG&G, Bear Hill Rd., Waltham, Mass. 02154, 335 pp.

FRESCURA, B.L. and J.D. MEINDL. 1976. Micropower integrated circuits for an implantable bidirectional blood flowmeter. *IEEE Journal of Solid State Circuits,* 11: 817-825.

GOULD, W.J. and E. SAMBUCO. 1975. The effect of mooring type on measured values of ocean currents. *Deep-Sea Research,* 22: 55-62.

HARDIES, C.E. 1975. An advanced two-axis acoustic current meter. *Proceedings of the Offshore Technology Conference 1975,* Houston, Texas, Paper 2293: 465-476.

JUNGNER INSTRUMENT DIVISION. 1977. SAL-ACCOR Marine Log, Svetsarvagen 15, Fac S-17120 Solna, Stockholm, Sweden, 31 pp.

KULLENBERG, G.E.B. 1976. On vertical mixing and the energy transfer from the wind to the water. *Tellus,* 28: 159-165.

LAWSON, K.D., N.L. BROWN, D.H. JOHNSON and R.A. MATTEY. 1976. A three-axis acoustic current meter for small scale turbulence. *Instrument Society of America, ASI No. 76269:* 501-508.

LEONARD, D.A., B. CAPUTO, R.L. JOHNSON and F.E. HOGE. 1977. Experimental remote sensing of subsurface temperature in natural ocean water. *Geophysical Research Letters,* 4: 279-281.

LOWELL, F.C., JR. 1977. Designing open channel acoustic flow meters for accuracy: parts 1 and 2. *Water & Sewage Works,* July & August 1977, 11 pp.

MARSH McBIRNEY, INC. 1977. *Adaptive Recording Current Meter, ARC Model 585.* Marsh McBirney, Inc., Gaithersburg, Md. 20760, 5 pp.

McCULLOUGH, J.R. 1974. In search of moored current sensors. *Marine Technology Society,* 10th Annual Conference, Washington, D.C., Woods Hole Oceanographic Institution Contribution No. 3385: 31-54.

McCULLOUGH, J.R. 1975. Vector averaging current meter speed calibration and recording technique. *Woods Hole Oceanographic Institution* Ref. 75-44, 35 pp.

McCULLOUGH, J.R. 1977. Problems in measuring currents near the ocean surface. Marine Technology Society IEEE *Oceans '77* Conference, 2: 46A1-7.

McCULLOUGH, J.R. 1978. Near-surface ocean current sensors: Problems and performance. *Proceedings of a Working Conference on Current Measurements,* edited by W. Woodward, C.N.K. Mooers, and K. Jensen. Technical Report DEL-SG-3-78, University of Delaware, Newark, DE, 19711: 9-33.

MEDWIN, H. 1977. In situ acoustic measurements of microbubbles at sea. *Journal of Geophysical Research,* 82: 971-976.

MIDDLETON, F.H. 1955. An ultrasonic current meter for estuarine research. *Journal of Marine Research,* 14: 176-186.

MUNK, W. and C. WUNSCH. 1979. Ocean acoustic tomography: A scheme for large scale monitoring. *Deep-Sea Research,* 26A: 123-161.

OKUBO, A. and C.C. EBBESMEYER. 1976. Determination of vorticity divergence and deformation rates from analysis of drogue observations. *Deep-Sea Research,* 23: 349-352.

OLSON, J.R. 1972. Two component electromagnetic flow meter. *Marine Technology Society Journal,* 6: 19-24.

POLLARD, R.T. 1973. Interpretation of near-surface meter observations. *Deep-Sea Research,* 20: 261-268.

ROSSBY, T. 1975. An oceanic vorticity meter. *Journal of Marine Research,* 33: 213-222.

SANFORD, T.B., R.G. DREVER and J.H. DUNLAP. 1978. A velocity profiler based on the principles of geomagnetic induction. *Deep-Sea Research,* 25: 183-209.

SAUNDERS, P.M. 1973. Tracing surface flow with surface isotherms. *Societe Royale des Sciences de Liege Memoires. Collection,* 6: 99-108.

SAUNDERS, P.M. 1976. Near-surface current measurements. *Deep-Sea Research,* 23: 249-257.

SHONTING, D.H. 1967. Observations of particle motions in ocean waves. Naval Underwater Weapons Research and Engineering Station, Newport, R.I., *Technical Memo* No. 377, 1, 176 pp.

STEWART, R.H. and J.W. JOY. 1974. HF radio measurements of surface currents. *Deep-Sea Research,* 21: 1039-1049.

TERRY, W.E. and A.J. WILLIAMS, III. 1977. Ocean applications of laser Doppler velocimetry. *POLYMODE News, 28,* Woods Hole Oceanographic Institution (unpublished manuscript), 13 pp.

THERMO-SYSTEMS, INC. 1977. *Laser and Hot-Wire, Hot-Film Anemometer Systems.* Thermo-Systems, Inc., 2500 Cleveland Ave. N., St. Paul, Minn. 55113, 100 pp.

WANNAMAKER, B.W. 1976. *The STARESO current meter intercomparison experiment: a preliminary analysis.* Master's thesis, University of Southampton, U.K., 43 pp.

WELLS, R.K. 1973. The design and construction of a constant depth current tracking buoy. Massachusetts Institute of Technology, Ocean Engineering, *Technical Report* No. 73-14, 17 pp.

WILLIAMS, A.J., III, and J.S. TOCHKO. 1977. An acoustic sensor of velocity for benthic boundary layer studies. In: *Bottom Turbulence, Proceedings 8th International Liege Colloquium on Ocean Hydrodynamics,* Elsevier Oceanographic Series, edited by J.C.J. Nihoul, Elsevier: 83-97.

WOODS, J.D. 1977. Parameterization of unresolved motions. In: *Modelling and Prediction of the Upper Layers of the Ocean,* edited by E.B. Kraus, Pergamon Press Ltd.: 118-140.

7

Moored Current Measurements in the Upper Ocean

D. Halpern

1. INTRODUCTION

A goal of air-sea interaction studies is to parameterize the mixing processes in the atmospheric and oceanic boundary layers in a form suitable for inclusion in coupled ocean-atmosphere models. The parameterizations of upper ocean processes which are characterized by short time and small space scales are too crude at present to allow for their proper representation. The objective of particular studies is to compare the observed response of the upper ocean (i.e., the changes in velocity, temperature, and salinity distributions) with theoretical responses calculated on the basis of physical measurements and different upper ocean models. Though changes in the vertical distribution of temperature in the upper ocean produced by a storm have been documented (e.g., Tabata et al., 1965; Leipper, 1967), the vertical profile of wind-generated currents in the mixed layer and thermocline regions awaits adequate description. Wind-generated shears in the mixed layer greater than 0.01 s^{-1}, which will contribute significantly to the turbulent energy budget (Pollard, 1977), have been observed on a few occasions (e.g., Gonella, 1971; Halpern, 1976; Perkins and Van Leer, 1977).

Accurate horizontal velocity measurements recorded frequently by instruments which remain at a fixed position for extended periods of time (i.e., moored current measurements) are extremely difficult to make in the harsh environment of the upper ocean. Because measurements in the upper ocean require instruments be placed near the surface, surface moorings must be used and the reliability of surface moorings is about one-half compared to subsurface moorings (Walden and Panicker, 1973). Descriptions of the design, deployment and recovery of moored current measurement systems, such as the one shown in Figure 1, are given by Heinmiller (1976), Berteaux (1975), Halpern (1972), Pillsbury et al. (1969), and others. Current meters are attached to the mooring either in-line with the mooring cable or adjacent to the mooring cable; both cases are shown in Figure 1. When a current meter is suspended in-line in a deep-sea surface mooring, the instrument must be sturdy enough to withstand tensions of nearly 2500 kg.

2. NEAR-SURFACE CURRENT MEASUREMENTS

The technology for recording accurate current measurements within the uppermost few hundred metres of the deep ocean for extended intervals of time needs to surmount two major problems: error-free current measuring sensors (see McCullough, Chapter 6; McCullough, 1978, 1977), and motionless measurement platforms (see Berteaux, Chapter 35; Chhabra, 1977). The presence of surface waves affects the quality of moored current measurements in several ways. First, the large amplitude high-frequency orbital wave motions are not easily recorded and are aliased into the low-frequency current variations required for mixed layer studies. The effect of this type of contamination of the data decreases with depth from the surface if the current meter is mounted rigidly in space. Second, if the current meter is placed near the surface beneath a buoy which moves horizontally (e.g., a spar-buoy), the relative motion recorded by the instrument will not be equal to the absolute water velocity (e.g., see Davis et al., 1978). Third, if the surface float is displaced vertically by surface waves, the current meter is moved through a vertical shear occurring in the water and, in addition, axial and transverse vibrations of the mooring line produce spurious motions of the velocity sensors. Even if a perfect current measuring sensor which recorded error-free data were developed, Pollard (1973) has indicated that near-surface measurements made beneath a surface-following buoy would contain spurious currents.

There are available today a wide variety of current meters designed to measure water flow. Each current meter is capable of recording accurately only a segment of the spectrum of water motions because of the influence of the mooring arrangement, the kind of velocity sensor used, and the sampling and recording scheme of the instrument. On occasions moored current meters are placed in regions of the ocean where the variability of the motions and the response characteristics of the instrument are poorly matched.

In the early 1960s several electromechanical internal-recording current meters were developed for unattended use on moorings. One such instrument was the Richardson current meter (Richardson et al., 1963), which contained a Savonius rotor (Savonius, 1931) and a small vane with a time constant of about ½ s which was capable of burst recording (Webster, 1967) to reduce the contribution of large amplitude surface wave generated high-frequency motions. Although

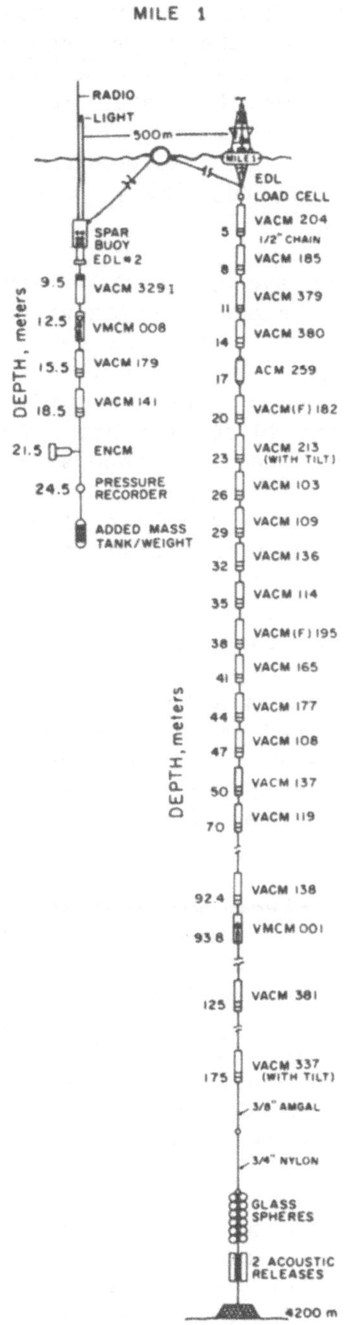

Fig. 1 Schematic diagram of the surface mooring system used in the Mixed Layer Experiment (MILE) during August and September 1977 in the northeast Pacific. The VACM and VMCM were described in the text, the ENCM was an Endeco Model 174 current meter, and the ACM was an acoustic current meter provided by James R. McCullough of Woods Hole Institution of Oceanography.

the Richardson current meter recorded 'noisy' data near the surface, the noise level seemed to be less than that recorded by other current meters available during the sixties (SCOR Working Group 21, 1975). Vector averaging the burst-sampled Richardson current measurements apparently reduced the amplitudes of the high-frequency motions to levels less than the values of the low-frequency wind-generated currents [e.g., inertial oscillations (Sverdrup et al., 1942)] because analyses of near-surface Richardson current measurements made beneath surface buoys were in good agreement with theoretical results (e.g., Pollard and Millard, 1970; Halpern, 1974; Garwood, 1977; Marchuk et al., 1977).

The development in 1971 of in situ vector averaging of numerous measurements represented a significant advancement of oceanographic technology. The first vector-averaging current meter was designed and developed at the Woods Hole Oceanographic Institution (McCullough, 1975), and is usually denoted VACM. The VACM contains a magnetic compass and at the lower end of the instrument are a Savonius rotor of 10 cm diameter with four S-shaped rotors and a small (17x9 cm) vane, both turning on a vertical axis. The instrument measures speed continuously and direction in a nearly continuous manner. Every eighth of a revolution of the rotor, vane and compass orientations are measured and internally combined into a discrete current direction which is converted internally to Cartesian components of velocity. For the duration of the sampling interval the east and north speed values are summed internally. The number of data samples used in each vector average is a function of the current speed; e.g., at 10 cm s^{-1} each recorded value is the sum of 2400 individual direction determinations over a 15-min recording cycle. Such a sampling scheme has the effect of reducing contamination of the current spectrum by surface wave 'noise' (Halpern and Pillsbury, 1976b; Saunders, 1976a).

The VACM represents a step in development between previous instruments using integrated rotor revolutions as a measure of current speed plus a single observation of vane direction to indicate current direction (e.g., Aanderaa, 1964) and the development of vector-averaging current meters which measure the current by acoustic (Gytre, 1976; Lawson et al., 1976), propeller (Davis and Weller, 1977), electromagnetic (Capart, 1969), and laser (Kullenberg and Woods, 1975) sensors, to name a few. Halpern and Pillsbury (1976a) found that Aanderaa measurements were severely contaminated by mooring motions.

The unequalled response times of the rotor for accelerating and decelerating flows (Fofonoff and Ercan, 1967; Gaul et al., 1963), biological fouling, the nonlinearity of the rotor calibration curves for unsteady flow (Karweit, 1974), and spurious rotor rotations or rotor pumping produced by high frequency cable and sensor displacements (Gould and Sambuco, 1975) limit the usefulness of rotor instruments for current measurements in the upper ocean. However, even with its deficiencies the VACM will be widely used until a new generation of reliable vector-averaging current meters becomes available.

3. IN SITU INTERCOMPARISON TESTS OF CURRENT METERS

Current meters are calibrated generally under flow conditions nonexistent in the ocean; e.g., fastened to a rigid mount in a steady-state flow regime rather than on a flexible mooring line in flow containing a myriad of temporal and spatial scales. In situ intercomparison of current meters represents a technique to determine the least favourable operating environment of different types of current meters. Because of our general lack of knowledge of the quality of moored current measurements in the oceanic environment, another purpose of intercomparison tests is to provide a quality-control or consistency-check of a composite data set formed from measurements made from different kinds of current meters. In contrast to the widespread use of current meters, only a limited number of intercomparisons of current meters has been made. An overview of 14 intercomparison tests was presented elsewhere (Halpern, 1977). Appendix 1 lists these intercomparison tests and others which have been performed more recently.

Several in situ intercomparison tests (Halpern, 1978; Saunders, 1976a; Pollard, 1974; Halpern et al., 1974) indicated that reliable VACM measurements could be made with ~2 cm s^{-1} resolution for frequencies less than about 2 cph within the uppermost ~50 m beneath surface-following buoys moored in deep and shallow waters. Differences of 10% to 20% were found among measurements recorded by a VACM, an electromagnetic current meter, and an acoustic current meter (Beardsley et al., 1977; Saunders, 1976b). The good agreement Halpern (1976) reported between the near-surface Ekman transport computed from VACM measurements and from the wind stress provided further evidence of the reliability of near-surface VACM data. However, the VACM did not yield reliable data when placed at depths greater than a few hundred metres beneath surface-following buoys (Gould et al., 1974) because the data were affected by surface wave energy transmitted to instruments at depth by motion of the mooring cable. Though laboratory experiments have demonstrated the amplification of rotor speeds by high frequency axial motions (Gaul, 1963), the quantitative relationship between cable motion and the rotation of a rotor is unknown; i.e., it is not known what surface wave height will produce a spurious 1 cm s^{-1} current by high frequency displacements of the mooring line and instruments.

Soon after the completion of the 1972 SCOR Working Group 21 (1975) Intercomparison Test it was widely conjectured that erroneous measurements made from all current meters available at that time would be obtained at all depths beneath single-point moorings using a surface-following buoy. Though surface wave and mooring motions are large near the surface, the low-frequency wind-generated currents are also large there. At depths below a few hundred metres atmospheric-forced currents have small amplitude relative to the motions of the mooring. In summary, the signal-to-noise ratio becomes the limiting factor for recording reliable data. Answers to the question "to what depth can reliable current records be made beneath a surface-following buoy?" are being sought from data obtained during an August 1977 severe storm in the northeast Pacific and from data recorded during the 1978 JASIN Project. Prior to 1977 the intercomparison tests were made under conditions of relatively low wind speeds, small sea heights, and large current speeds, and thus were not entirely suitable for studying the in situ characteristics of vector-averaging current meters during conditions typical of storm-generated mixed layer deepening.

Initial trials of the vector-measuring dual orthogonal unshrouded fast-response fan propeller current meter (called a VMCM) designed by Davis and Weller (Weller, 1978; Winant et al., 1978; Davis and Weller, Chapter 8) indicate that this instrument represents a further advancement in near-surface current meter technology. During a 5-day test in August 1977 at Ocean Weather Station P in the Gulf of Alaska when wind speeds were greater than 18 m s^{-1} for 36 hours and significant wave heights reached 5 m, a VMCM averaged the high frequency motions better than a VACM. Figure 2 shows that for frequencies between 2 and 10 cph the kinetic energy density levels of the 2-min averaged VMCM data at 12.5 m depth were 40% to 50% smaller than the corresponding energy levels of 1.875-min averaged VACM data at 9.5 m and 15.5 m. The lower energy levels recorded at high frequencies by the VMCM indicate that the VMCM was the more effective averager of the two instruments. The VACM measured speed and direction with relatively slow-response sensors from which Cartesian speeds were computed nonlinearly, whereas the VMCM measured Cartesian speeds directly with relatively fast-response sensors. The larger VACM values at low frequencies were perhaps the result of the natural current shear, the limited response of the VACM vane, the over-response of the VACM rotor to accelerating than decelerating flows, and the under-response of the VMCM propellers. These instruments were suspended beneath a spar-buoy tethered to a surface mooring shown in Figure 1. The 5-day average root-mean-square difference between the east (or north) component series was about 4 cm s^{-1}. The vector-mean speeds differed by less than 1 cm s^{-1}.

In conclusion, the small number of intercomparison tests conducted with current meters placed near the surface beneath surface-following buoys indicated that spurious currents produced by surface wave motions and mooring line vibrations and displacements were greatly reduced by fast-responding velocity sensors and an internal vector-averaging recording mechanism. The use of faired cables reduces

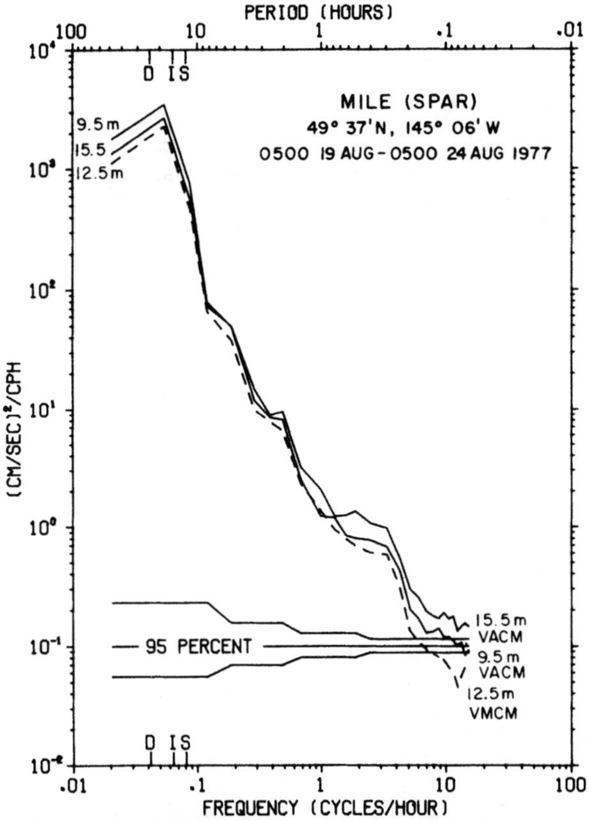

Fig. 2 Kinetic energy density spectral estimates of 1.875-min averaged VACM and 2-min averaged VMCM (see text for definition of current meters) measurements made beneath a spar-buoy. I, D and S indicate the inertial, diurnal, and semidiurnal frequencies. The "95 percent" represents the 95% confidence levels determined from the chi-square distribution and applicable to each curve.

the amplitude of mooring line motions (Bell, 1977; Pattison, 1977; Softley et al., 1977).

Appendix 1

A list of intercomparison tests of moored current measurements.
Z = current meter depth (m), H = water depth (m),
SPEC = specialized mooring such as a rigid tower or spar-buoy,
SUR = single-point taut-line surface mooring,
SUBSUR = subsurface mooring

REFERENCE	CURRENT METER	Z	H	SPEC	SUR	SUBSUR
Beardsley et al., 1977	Aanderaa RCM4, EG&G CT/3, Endeco 105, Geodyne 850, Marsh McBirney 711, VACM	7-25	28	x	x	x
Boicourt, 1973	Braincon 1381	3-13	15	x		x
Briscoe and Hayes, 1973	Geodyne 850, VACM	50-1000	2600		x	x
Cheng, 1978	Aanderaa RCM4, Endeco 174, Oceanics 6011, Marsh McBirney 711	3	8	x		
Chiocchio, 1976	Geodyne 850, Plessey M021	15-68	28-161			x
Gonella and Lamy, 1974	Aanderaa	4	85	x	x	
Gould et al., 1974	Geodyne 850, VACM	1500	2600		x	x
Gould and Sambuco, 1975	Geodyne	100-2500	2600		x	x
Halpern et al., 1974	Aanderaa RCM4, Geodyne 850, VACM	18-20	100		x	x
Halpern and Pillsbury, 1967a	Aanderaa RCM4	43	50			x

Halpern and Pillsbury, 1976b	Aanderaa RCM4, VACM	8	50			x
Halpern, 1978	VACM	9	4200	x		x
Kullenberg and Woods, 1975	Laser Doppler, acoustic, hot-film, electromagnetic	2-9	20	x		
Loeng, 1976	Aanderaa RCM4, Ekman, propeller	4-22	23	x	x	
Magnell, 1977	Endeco 105, EG&G CT/3, Geodyne 102, VACM	4-5,10	13			x
Paramonov and Kushnir, 1977	Alexaev, Geodyne	575-4000	4300		x	x
Petrie, 1977	Braincon 381, Endeco 174	17-20	170			x
Pollard, 1974	VACM	10,32	3000	x	x	
Ramster and Howarth, 1975	Aanderaa RCM4, Plessey M021	16-80	38-86			x
Saunders, 1976a	Aanderaa RCM4, VACM	10,12	3000	x	x	
Saunders, 1976b	Acoustic VACM	35	4000	x		
SCOR WG 21, 1969	Bergen 42-66, Geodyne A100 and 850, Plessey M021, Tiefen, Hydrowerkstatten	500	2600			x
SCOR WG 21, 1974	Alexaev BPV, Bergen RCM4, Braincon 381, Geodyne 850, LSK 801, Plessey M021	50-100	5100		x	
SCOR WG 21, 1975	Alexaev, Geodyne 850, LSK, VACM	50-1000	2600		x	x
Winant et al., 1978	SIO VMCM, electromagnetic SIO VMCM, VACM	6 10.5	20 4500	 x		x

ACKNOWLEDGEMENTS

In Figure 2 the VMCM data were kindly provided to me by Dr. Russ Davis, Scripps Institution of Oceanography, and the VACM data were obtained with support from NORDA's Office of Naval Research (NR-083-421) and NOAA's Environmental Research Laboratories. Contribution Number 380 from the NOAA/ERL Pacific Marine Environmental Laboratory.

REFERENCES

AANDERAA, I.R. 1964. A recording and telemetering instrument. Technical Report 16, Fixed Buoy Project, NATO Subcommittee for Oceanographic Research, 53 pp.

BEARDSLEY, R.C., J. SCOTT and W. BOICOURT. 1977. CMICE 76: A current meter intercomparison experiment conducted off Long Island in February-March 1976. Technical Report 77-62, Woods Hole Oceanographic Institution, Woods Hole, Massachusetts, 123 pp.

BELL, W.H. 1977. The use of extruded plastic fairing for a subsurface mooring. Report 77-8, Institute of Ocean Sciences, Patricia Bay, Victoria, British Columbia, 13 pp.

BERTEAUX, H.O. 1975. *Buoy Engineering.* John Wiley, New York, 319 pp.

BOICOURT, W. 1973. Current measurements from a subsurface mooring and a rigid tower. *Mode Hot-Line News,* No. 44. (Unpublished manuscript available from Woods Hole Oceanographic Institution, Woods Hole, Massachusetts.)

BRISCOE, M. and S. HAYES. 1973. Current measurements from surface and subsurface moorings. *Mode Hot-Line News,* No. 40. (Unpublished manuscript available from Woods Hole Oceanographic Institution, Woods Hole, Massachusetts.)

CAPART, G. 1969. Description of an electromagnetic oceanographic current meter. Technical Report 50, NATO Subcommittee on Oceanographic Research, Brussels, 60 pp.

CHENG, R.T. 1978. Comparison of a few recording current meters in San Francisco Bay, California. *Proceedings of a Working Conference on Current Measurement,* Technical Report DEL-SG-3-78, College of Marine Studies, University of Delaware, Newark: 293-304.

CHHABRA, N.K. 1977. Correction of vector-averaging current meter records from the MODE-1 central mooring for the effects of low-frequency mooring line motion. *Deep-Sea Research,* 24: 279-288.

CHIOCCHIO, F. 1976. Evaluation of the use of moored current meters in the Great Lakes, 154 pp. (Unpublished manuscript available from Applied Research Division, Canada Centre for Inland Waters, Burlington, Ontario.)

DAVIS, R.E. and R. WELLER. 1977. SIO propeller current meters. *Polymode News*, No. 40. (Unpublished manuscript available from Woods Hole Oceanographic Institution, Woods Hole, Massachusetts.)

DAVIS, R.E., T.P. BARNETT and C.S. COX. 1978. Variability of near-surface currents observed during the POLE Experiment. *Journal of Physical Oceanography*, 8: 290-301.

FOFONOFF, N.P. and Y. ERCAN. 1967. Response characteristics of a Savonius rotor current meter. Technical Report 67-33, Woods Hole Oceanographic Institution, Woods Hole, 36 pp.

GARWOOD, R.W. 1977. An oceanic mixed layer model capable of simulating cyclic states. *Journal of Physical Oceanography*, 7: 455-468.

GAUL, R.D. 1963. Influence of vertical motion on the Savonius rotor current meter. Technical Report 63-4T, Texas A&M University, College Station, Texas, 29 pp.

GAUL, R.D., J.M. SNODGRASS and D.J. CRETZLER. 1963. Some dynamical properties of the Savonius rotor current meter. *Marine Science Instrumentation*, 2: 115-125.

GONELLA, J. 1971. The drift current from observations made on the Bouée Laboratoire. *Cahiers Oceanographique*, 23: 19-33.

GONELLA, J. and A. LAMY. 1974. Comparison de courantometres Aanderaa - Influence de la nature des supports. Interne Report, Laboratoire d'Oceanographie Physique, Paris, 11 pp.

GOULD, W.J. and E. SAMBUCO. 1975. The effect of mooring type on measured values of ocean currents. *Deep-Sea Research*, 22: 55-62.

GOULD, W.J., W.M. SCHMITZ and C. WUNSCH. 1974. Preliminary field results for a Mid Ocean Dynamics Experiment (MODE-0). *Deep-Sea Research*, 21: 911-932.

GYTRE, T. 1976. The use of a high sensitivity ultrasonic current meter in an oceanographic data acquisition system. *The Radio and Electronic Engineer*, 46: 617-623.

HALPERN, D. 1972. Description of an experimental investigation on the response of the upper ocean to variable winds. NOAA Technical Report ERL 231-POL 9, U.S. Government Printing Office, 51 pp.

HALPERN, D. 1974. Observations of the deepening of the wind-mixed layer in the northeast Pacific Ocean. *Journal of Physical Oceanography*, 4: 454-466.

HALPERN, D. 1976. Structure of a coastal upwelling event observed off Oregon during July 1973. *Deep-Sea Research*, 23: 495-508.

HALPERN, D. 1977. Review of intercomparisons of moored current measurements. Oceans '77 Conference Record, Marine Technology Society, Washington, D.C.: 46D1-46D6.

HALPERN, D. 1978. Mooring motion influences on current measurements. *Proceedings of a Working Conference on Current Measurement*, Technical Report DEL-SG-3-78, College of Marine Studies, University of Delaware, Newark: 69-76.

HALPERN, D., R.D. PILLSBURY and R.L. SMITH. 1974. An intercomparison of three current meters operating in shallow water. *Deep-Sea Research,* 21: 489-497.

HALPERN, D. and R.D. PILLSBURY. 1976a. Influence of surface waves upon subsurface current measurements in shallow water. *Limnology and Oceanography,* 21: 611-616.

HALPERN, D. and R.D. PILLSBURY. 1976b. Near-surface moored current meter measurements. *Marine Technology Society Journal,* 10: 32-38.

HEINMILLER, R.H. 1976. Mooring operations and techniques of the Buoy Project of the Woods Hole Oceanographic Institution. Technical Report 76-69, Woods Hole Oceanographic Institution, Woods Hole, Massachusetts, 93 pp.

KARWEIT, M. 1974. Response of a Savonius rotor to unsteady flow. *Journal of Marine Research,* 32: 359-364.

KULLENBERG, G. and J.D. WOODS. 1975. Preliminary report on current meter intercomparisons at Stareso, August 1975. Institut due Fysisk Oceanografic, Kobenhauns Universitat, Copenhagen, 37 pp.

LAWSON, K.D., Jr., N.B. BROWN, D.H. JOHNSON and R.A. MATTEY. 1976. A three-axis acoustic current meter for small scale turbulence. *Instrument Society of America,* ASI, No. 76269: 501-508.

LEIPPER, D.F. 1967. Observed ocean conditions and hurricane Hilda, 1964. *Journal of Atmospheric Sciences,* 24: 182-196.

LOENG, H. 1976. A comparison of current meters and mooring methods in the Kattegat. Technical Report 43, Geophysical Institute, University of Bergen, Bergen, 17 pp.

MAGNELL, B. 1977. Intercomparision of the EG&G Model 102 current meter with EG&G Model CT/3, Endeco Model 105, and AMF VACM current meters in shallow coastal waters. Environmental Consultants Monograph B-4518, EG&G, Environmental Consultants, Waltham, Massachusetts.

MARCHUK, I.G., V.P. KOCHERGIN, V.I. KLIMOK and V.A. SUKHORUKOV. 1977. On the dynamics of the ocean surface mixed layer. *Journal of Physical Oceanography,* 7: 865-875.

McCULLOUGH, J.R. 1975. Vector-averaging current meter speed calibration and recording technique. Technical Report 75-44, Woods Hole Oceanographic Institution, Woods Hole, Massachusetts, 35 pp.

McCULLOUGH, J.R. 1977. Problems in measuring currents near the ocean surface. Oceans '77 Conference Record, Marine Technology Society, Washington, D.C.: 46A1-46A7.

McCULLOUGH, J.R. 1978. Near-surface ocean current sensors: problems and performance. *Proceedings of a Working Conference on Current Measurement,* Technical Report DEL-SG-3-78. College of Marine Studies, University of Delaware, Newark: 9-33.

PARAMONOV, A. and V. KUSHNIR. 1977. An intercomparison of ocean current meters. *Polymode News,* No. 38. (Unpublished manuscript available from Woods Hole Oceanographic Institution, Woods Hole, Massachusetts.)

PATTISON, J.H. 1977. Measurement technique to obtain strumming characteristics of model mooring cables in uniform currents. Report SPD 766-01, David W. Taylor Naval Ship Research and Development Center, Bethesda, MD 20084.

PERKINS, H. and J. VAN LEER. 1977. Simultaneous current-temperature profiles in the equatorial countercurrent. *Journal of Physical Oceanography,* 7: 264-271.

PETRIE, B. 1977. An intercomparison of data from Endeco 174 and Braincon 381 current meters. Technical Report BI-R-77-10, Bedford Institute of Oceanography, Dartmouth, N.S., Canada, 18 pp.

PILLSBURY, R.D., R.L. SMITH and R.C. TIPPER. 1969. A reliable low-cost mooring system for oceanographic instrumentation. *Limnology and Oceanography,* 14: 307-311.

POLLARD, R.T. 1973. Interpretation of near-surface current meter observations. *Deep-Sea Research,* 20: 261-268.

POLLARD, R.T. 1974. The joint air-sea interaction trial, JASIN 1972. *Memoires de la Societe Royale des Sciences de Liege,* 6 Serie, Tome VI: 17-34.

POLLARD, R.T. 1977. Observations and models of the structure of the upper ocean. In *Modelling and Prediction of the Upper Layers of the Ocean,* edited by E.B. Kraus, Pergamon Press: 102-117.

POLLARD, R.T. and R.C. MILLARD. 1970. Comparisons between observed and simulated wind-generated inertial oscillations. *Deep-Sea Research,* 17: 813-821.

RAMSTER, J.W. and M.J. HOWARTH. 1975. A detailed comparison of the data recorded by Aanderaa Model 4 and Plessey MO21 recording current meters moored in two shelf-sea locations, each with strong tidal currents. *Sonderdruck aus der Deutschen Hydrographischen Zeitschrift,* Band 28: 1-25.

RICHARDSON, W.S., P.B. STIMSON and C.H. WILKINS. 1963. Current measurements from moored buoys. *Deep-Sea Research,* 10: 369-388.

SAUNDERS, P.M. 1976a. Near-surface current measurements. *Deep-Sea Research,* 23: 249-258.

SAUNDERS, P.M. 1976b. Intercomparison of a VACM and acoustic current meter. *Polymode News,* No. 12. (Unpublished manuscript available from Woods Hole Oceanographic Institution, Woods Hole, Massachusetts.)

SAVONIUS, S.J. 1931. The s-rotor and its applications. *Mechanical Engineering,* 53: 333-338.

SCOR WORKING GROUP 21. 1969. An intercomparison of some current meters, I. UNESCO Technical Papers in Marine Sciences, No. 11, UNESCO, Paris, 70 pp.

SCOR WORKING GROUP 21. 1974. An intercomparison of some current meters, II. UNESCO Technical Papers in Marine Sciences, No. 17, UNESCO, Paris, 116 pp.

SCOR WORKING GROUP 21. 1975. An intercomparison of some current meters, III. UNESCO Technical Papers in Marine Sciences, No. 23, UNESCO, Paris, 42 pp.

SOFTLEY, E.J., J.F. DILLEY and D.A. ROGERS. 1977. An experiment to correlate strumming and fishbite events on deep ocean moorings. Report No. 77SDR2181, General Electric Re-entry and Environmental Systems Division, Philadelphia, Pennsylvania, 49 pp. (NTIS No. ADA040617).

SVERDRUP, H.U., M.W. JOHNSON and R.H. FLEMING. 1942. *The Oceans*. Prentice-Hall, Englewood Cliffs, New Jersey, 1087 pp.

TABATA, S., N.E.J. BOSTON and F.M. BOYCE. 1965. The relation between wind speed and summer isothermal surface layer of water at Ocean Station "P" in the eastern subarctic Pacific Ocean. *Journal of Geophysical Research*, 70: 3867-3878.

WALDEN, R.G. and N.N. PANICKER. 1973. Performance analysis of Woods Hole taut moorings. Technical Report WHOI 76-31, Woods Hole Oceanographic Institution, Woods Hole, Massachusetts, 47 pp.

WEBSTER, F. 1967. A scheme for sampling deep-sea currents from moored buoys. *2nd International Buoy Technology Symposium*, Marine Technology Society, Washington, D.C.: 419-431.

WELLER, R. 1978. Observations of horizontal velocity in the upper ocean made with a new vector measuring current meter. Ph.D. Dissertation, Scripps Institution of Oceanography, La Jolla, California, 169 pp.

WINANT, C.D., R.E. DAVIS and R. WELLER. 1978. Shallow current measurements. *Proceedings of a Working Conference on Current Measurement*, Technical Report DEL-SG-3-78, College of Marine Studies, University of Delaware, Newark: 129-136.

8

Propeller Current Sensors

R.E. Davis and R.A. Weller

1. INTRODUCTION

The primary problem in measurement of upper ocean currents is separation of the weak low frequency 'signal' from the energetic high frequency 'noise' of surface waves and the mooring motion they produce. Overcoming this problem requires either perfectly responsive, although possibly nonlinear, sensors or completely linear sensors, possibly with limited frequency response. If neither requirement is met, a rectification error must result. A well known example of such rectification errors is the result of computing mean velocity from a speed sensor and a vane sensor which respond slowly; the mean velocity cannot be computed from the mean speed and mean direction because the relation between speed and direction and the desired velocity components is nonlinear. A slow response velocity <u>component</u> sensor, on the other hand, yields the mean velocity.

Propeller sensors are approximately velocity component sensors, being primarily sensitive to the flow component along their axle, and are therefore promising for use in current meters. Propeller sensors are not novel in current meters; Ekman's ingenious current meter employed a propeller and vane together with a simple form of vector averaging. More recently, propellers have found use as the speed sensor in several commercially available propeller/vane current meters and as velocity component sensors in two- and three-axis arrays (Shonting, 1967; Cannon and Pritchard, 1971; Gill and Michelena, 1971; Smith, 1974; Dahlen and Shillingford, 1976).

The intent here is to present a brief outline of the theory of propeller sensor dynamics in order to underscore the elements of propeller design which influence performance and then to give some examples of the performance of propellers which have been tested. The literature on propeller anemometer theory and performance is both pertinent and extensive (see Busch et al., Chapter 1).

2. ELEMENTARY DESIGN CONSIDERATIONS

The essentials of propeller dynamics can best be understood by considering an element of propeller blade as depicted in Figure 1. The angle of attack of the relative flow past the blade is

$$\beta = \arctan\left(\frac{u}{\omega r - v \sin\phi}\right) - \alpha. \quad (1)$$

The hydrodynamic forces on the blade element depend on the relative flow squared through lift and drag coefficients, C_L and C_D, as depicted in Figure 1b. The contribution of these forces to the torque around the axle is

$$r\rho[\cos(\alpha+\beta)\, C_L - \sin(\alpha+\beta)\, C_D]\, \frac{1}{2}\, [u^2 + (\omega r - v \sin\phi)^2]dA, \quad (2)$$

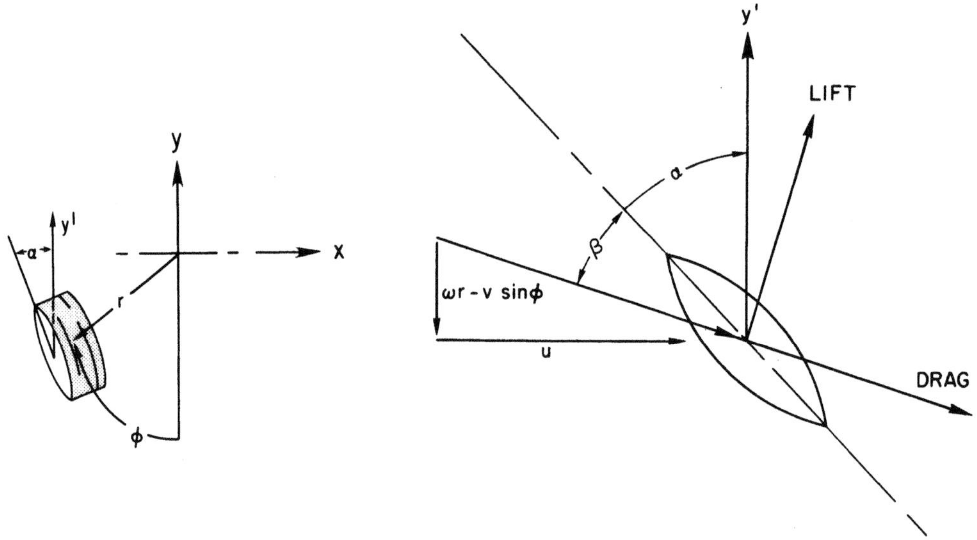

Fig. 1 An element of propeller blade at radius r from the axle (the x axis) with pitch α. The propeller is taken to have angular position ϕ and angular velocity $\omega = d\phi/dt$. The fluid velocity is taken to have components u and v along the x and y axes, respectively.

where dA is the area of the element of blade and ρ is the fluid density.

The dependence of C_L and C_D on angle of attack β, Reynolds number, aspect ratio, and background turbulence level is discussed in standard texts such as Hoerner (1975). The features common to all thin airfoils are (a) C_L increases nearly linearly with β up to the angle β_s, where separation occurs, and the slope of C_L versus β depends little on airfoil shape, (b) β_s, which determines the maximum lift, depends on Reynolds number, airfoil shape, and turbulence level but generally falls between 10° and 16°, and (c) the drag is approximately quadratic in β for $\beta << \beta_s$ and jumps markedly as separation is approached. These features are represented in Figure 2.

With v = 0 and u steady the dynamics of a propeller driven by the torque in Equation 2 is simple. The rotation rate adjusts until the applied torque balances the torque from bearings on the axle. If this friction is small the equilibrium is determined by the attack angle such that $C_L \cos(\alpha+\beta) = C_D \sin(\alpha+\beta)$. If C_D is small the equilibrium is very near $\beta = C_D = 0$ giving the linear calibration

$$\omega = \frac{u}{r \tan \alpha} .$$

In this ideal case the propeller operates as a mechanical nulling servo-feedback loop which accelerates the propeller until it ceases to be subject to hydrodynamic forces. Thus despite the quadratic

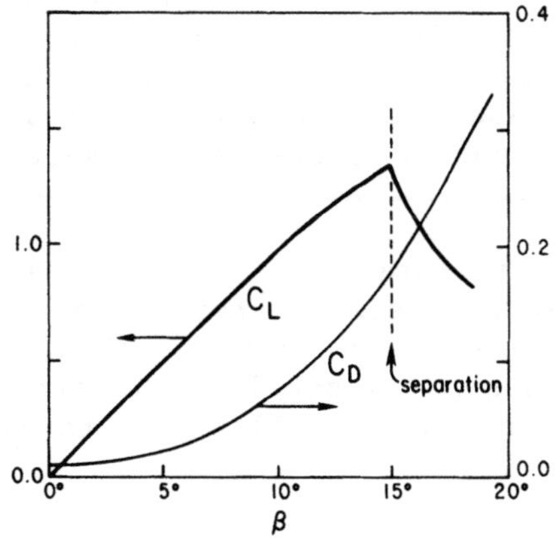

Fig. 2 A schematic representation of lift and drag coefficients (C_L and C_D respectively) for a typical airfoil as a function of angle of attack, β. Note that as β exceeds the separation value, C_L/C_D decreases markedly.

drag law which provides the restoring force, the equiliibrium is a linear u versus ω calibration which depends only on geometry.

Fortunately, elements of this ideal design are achievable. Bearing friction can generally be made negligible (although this may require using high pitch propellers to limit axial loading of the bearings) and the Reynolds number can be made large enough that changes in $C_L(\beta)/C_D(\beta)$ with u and v are negligible. These features ensure that the equilibrium is geometric and ω varies linearly with u,v. Unfortunately, other elements of the ideal propeller are not achievable. When $v \neq 0$, the average over ϕ of the torque in Equation 2 will generally depend on v and the propeller will be an imperfect component sensor. Only for axial flow and a helicoid propeller (where $\tan \alpha$ varies as $1/r$) is the angle of attack β constant along the blade. In any other case equilibrium involves balancing opposite torques at different radii, the sensor is not a nulling feedback device, and the quadratic lift and drag laws affect the dynamics. Similarly when the flow is unsteady the angle of attack must increase, the quadratic laws become important and the potential for nonlinear response increases.

It is clear from the form of β in Equation 1 that the influence of v on the equilibrium is minimized when ωr is large. This favours low pitch blades. In opposition are the increased bearing loads associated with small pitch. More importantly the response in unsteady currents is degraded by low pitch blades which require a greater angular acceleration to adjust to a change in u. The reader is directed to Walrod's (1970) discussion of the factors influencing the response of propellers to the off-axis component of flow. In practice propeller sensors in steady flow generally show high linearity between ω and u for fixed angle of incidence $\theta = \arctan v/u$. Most propellers show some dependence on v, but this can be minimized by tuning propeller geometry as will be shown later.

The major limitation of propellers, as other mechanical sensors, is that in unsteady flow the relation between average rotation rate ω and the average flow u depends on the unsteady component of flow, that is to say, the sensor is not linear. Some causes of unsatisfactory performance in unsteady flow can be seen from the model equation

$$M\, r^2 \frac{d\omega}{dt} = \frac{\rho r}{2} [\cos(\alpha+\beta)\, C_L - \sin(\alpha+\beta)\, C_D][u^2 + (\omega r)^2], \quad (3)$$

describing response to axial flow of a propeller as in Figure 1 with mass per projected area M and an applied hydrodynamic torque given by Equation 2. In unsteady flow the propeller inertia prevents the maintenance of the ideal equilibrium $\beta = C_L = 0$, and the nonlinear dependence of $\cos(\alpha+\beta)\, C_L$ and $\sin(\alpha+\beta)\, C_D$ on β comes into play leading to possible rectification of the oscillatory portion of u. This effect, often producing 'overspeeding', is well known in anemometry. McMichael and Klebanoff (1975) present a clear theory for the effect but neglect flow separation and the hydrodynamic drag described by C_D. While appropriate to the usual applications in anemometry, these simplifications are inappropriate for the case when oscillatory motion is comparable to mean flow, the usual oceanographic situation.

The simplest unsteady flow is a small change, u', of the mean flow U. The response of Equation 3 then exhibits an exponential decay towards the new mean flow of the form

$$\exp(-Ut/L)$$

where L is the distance constant describing unsteady response. In Equation 3

$$L = \frac{2M}{\rho \sin \alpha} \Big/ \left(\frac{dC_L}{d\beta}\right) \beta = 0 \; .$$

Since $(dC_L/d\beta)_0 \simeq 5$ for most reasonable airfoils, the propeller response would appear to be determined primarily by its mass/area and sin α. This demonstrates the trade-off between minimizing the influence of nonaxial flow, which favours small pitch, and improving resonse to unsteady flow, which favours large pitch.

The solution to Equation 3 when the imposed flow is of the form $u' \cos \sigma t + U$ has been computed for propellers described by various realistic forms of $C_L(\beta)$ and $C_D(\beta)$. The response is a function of the two parameters

$$\mu = U/u', \qquad \lambda = \sigma L/u',$$

where λ is also the ratio of L to the amplitude of water particle oscillations in the unsteady flow. In general, it is found that the average flow inferred from the average propeller rotation differs from U by an error $\varepsilon u'$. A typical plot of the dimensionless error ε is given in Figure 3. General features of the rectification error and its relation to propeller design are (a) the error becomes small as $\mu \to \infty$ or $\lambda \to 0$, (b) for $\mu > 1$ and moderate or small λ the flow generally does not separate and the error is 'overspeeding', (c) for $\mu < 1$ separation cannot be avoided because the flow reverses direction, and (d) when flow separation occurs the error is typically 'underspeeding' which results from the marked increase of C_D/C_L when separation occurs.

These general results can be useful in designing propellers with good response characteristics. They are, however, derived from a highly oversimplified model which neglects interaction of the blades with each other and neglects the fact that when $\mu < 1$ the propeller is generally operating in its own wake. Laboratory tests described below indicate that wake interference is the major cause of rectification when $\mu < 1$. This greatly limits the utility of models like Equation 3 but it still seems safest to make the propeller distance constant as small as possible so that in all cases $\lambda << 1$; otherwise rectification errors, ε, of order (0.1 to 0.5) λ appear unavoidable. Unfortunately, laboratory tests also indicate that the distance constant is not simply proportional to M and suggest that only small radius propellers can have small L when the fluid and propeller material have similar densities.

3. STEADY FLOW RESPONSE

Steady flow response of propellers is determined by two characteristics, namely, propeller rotation versus flow speed for fixed angle of flow incidence (measured from the propeller axis) and rotation versus angle of incidence at fixed flow speed. Most calibrations are accomplished by assuming these to be separable so that

$$\omega = S(q)\ H(\phi)$$

where q is the flow speed and ϕ the angle of incidence so that the axial component of flow is $q\cos\phi$. Ideally $S(q) = Kq$ and $H(\phi) = \cos\phi$.

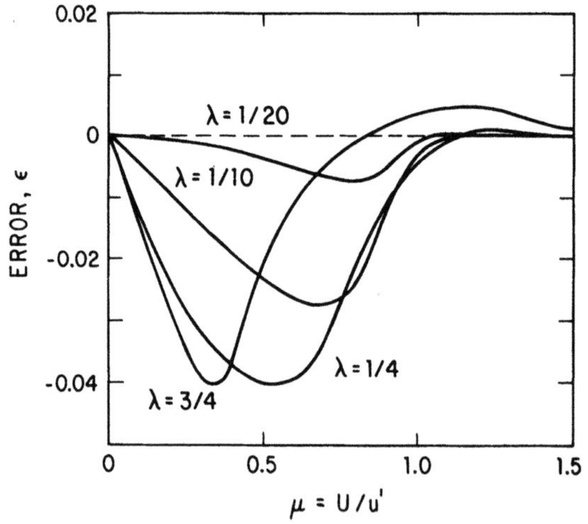

Fig. 3 A typical model result from Equation 3 for the rectification error of a propeller in mean plus oscillatory flow. The symbols are defined in the text. Note that for weak mean flows the error is 'underspeeding'.

Review of published tests of various propeller current sensors shows them generally to have a linear response once flow speed is sufficient to overcome bearing friction. Cannon and Pritchard (1971), using a 38 cm diameter propeller with six blades, report a threshold of 0.6 cm s^{-1} and a response linear to 0.1 cm s^{-1} for $q >$ 1 cm s^{-1}. Gill and Michelena (1971), using four slender blades of length about 15 cm, obtain less than 0.6 cm s^{-1} deviation from linearity above 6 cm s^{-1}. Most other designs of comparable dimension appear to fall somewhere between these. The threshold and low speed response are apparently determined primarily by a combination of bearing quality and the projected area of the blades. For near surface work most propellers appear to have satisfactory linearity of speed response although Smith (1974) finds it desirable to apply a nonlinear response curve in his work on oceanic turbulence.

Angular response is highly variable depending on various design parameters such as whether or not the propeller is 'ducted' (surrounded by a cylindrical tube parallel to the axle), the shape and number of blades, the pitch angle, and the number of propeller stages on the axle. Weller (1978) has performed exhaustive studies determining the effect of these factors. His conclusions, supported by the calibration data presented in the references, are as follows:

(1) If $H(0) = 1$ then ducting tends to increase H around 30° but leads to $H < \cos\theta$ for large angles of incidence. Frequently, ducted propellers do not respond consistently for $\theta > 70°$, probably because of flow separation from the shroud itself. If the ducting tube is sufficiently long, H is almost independent of θ for small θ.

(2) Unducted propellers generally under-respond for θ around 45°. In fact, for $\theta < 60°$ response is frequently well fitted by $H = \cos^n\theta$ where n is around 2. Unducted propellers do not, however, suffer serious degradation at large θ and generally respond at high angles of incidence until the axial flow component falls below the threshold.

(3) Low angles of pitch tend to produce responses most nearly fitting $\cos\theta$, as was anticipated in the previous section. The effects of blade aspect ratio and the number of blades seem to depend primarily on the total frontal area covered by blades. As the covered area increases $H(\theta)$ generally more nearly approximates $\cos\theta$.

(4) By placing two propellers on a single shaft and tuning blade geometry, it is possible to achieve an $H(\theta)$ which matches $\cos\theta$ to within 1%. This is demonstrated in Figure 4 which shows the function $H(\theta)$ for a propeller sensor developed by the authors (see Weller, 1978) which is sketched in Figure 5.

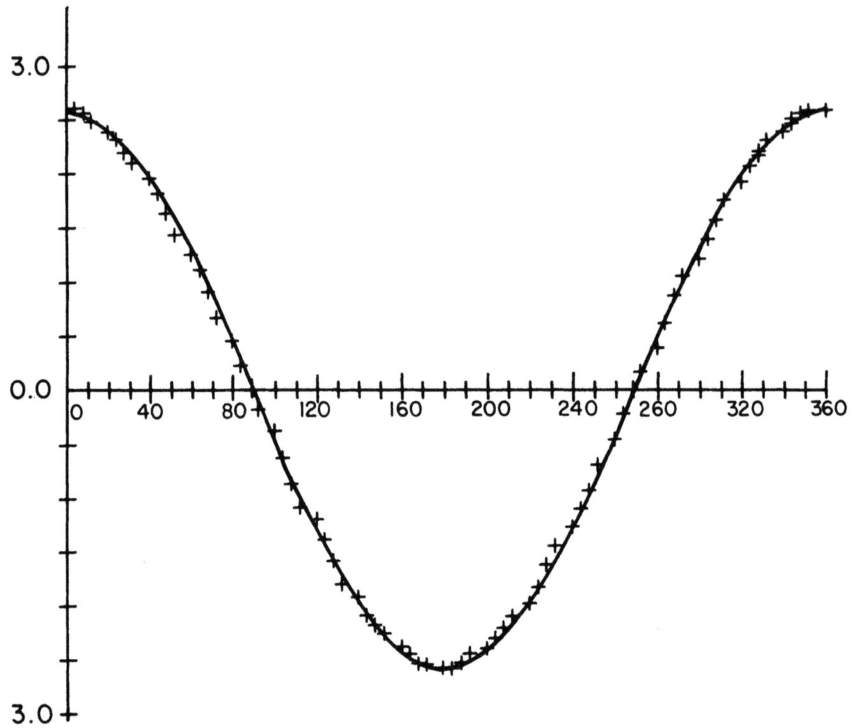

Fig. 4 The directional response of the SIO propeller current meter. The rms error from a cos θ fit is 1.5% of full scale.

If sensor response is adequately fast it is not necessary to achieve a good fit of directional response to cos θ. If data are recorded at a sufficient rate that flow at every time is known, it is quite practical to find θ from the known H(θ). For this, unducted propellers which respond for all θ appear optimum. However, if averaging is to be feasibly accomplished within an unattended instrument a true cos θ response is highly desirable and if sensor response is limited it is, as discussed in the introduction, essential.

It is important to note that in most applications instrument supports and pressure vessels provide large disturbances to the flow and can dramatically alter the directional response of any type of sensor. It is, therefore, desirable to minimize disturbances near the sensor and essential to calibrate the entire instrument, including mooring attachments and pressure cases. The importance of making the complete current meter a linear component sensor is demonstrated in Chapter 6 by McCullough, where rectification errors produced by wave motion are discussed. He shows that acoustic travel time current meters, with sensors which are in principle perfect components except for hydrodynamic flow distortion, can be subject to significant rectification errors.

4. UNSTEADY RESPONSE

The distance constant, L, is a readily measured characteristic of transient response. Cannon and Pritchard (1971) report a distance constant of approximtely 10 cm. Smith (1974) reports his much smaller propellers to have unit response to 5 Hz but does not report the mean flow speed so L cannot be computed. The 22 cm diameter S10 propeller has a measured L = 9 cm. Surprisingly, when the mass and moment of inertia of this propeller are reduced by a factor of 2, leaving the size and geometry unchanged, no detectable reduction in L is noted. This suggests that the dynamics of the Elementary Design Consideration section, which do not include the

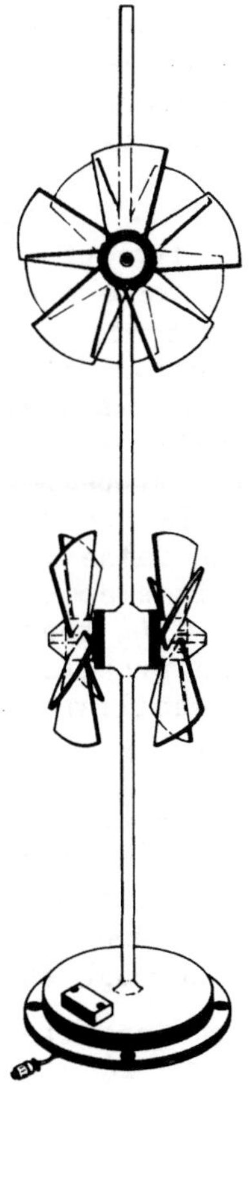

Fig. 5 Sketch of a biaxial propeller sensor. Propeller diameters are 22 cm; the distance between the propellers is 40 cm. Pick-ups to sense propeller revolution are mounted internally, and the wires from the pick-ups lead down inside the tube between the propeller sensors and out the base of the sensor.

'virtual mass' of entrained flow, are inadequate to allow prediction of the transient response. This casts doubt on Dahlen and Shillingford's (1976) quoted design values of L = 2 cm, which are based on a similar theory.

The authors have carried out extensive laboratory tests to determine the extent to which propellers placed in mean plus oscillatory flow can correctly measure the mean flow. For observation of upper ocean currents in the presence of wave-induced water velocities and mooring motions, this is the critical performance characteristic. From geometric considerations alone, it can be seen that, in contrast to rotor/vane sensors, a symmetric propeller will yield no spurious mean flow in a flow consisting of a symmetric oscillatory component and zero mean flow along the propeller axle.

Our test procedure was to oscillate a propeller while towing at uniform velocity, U. The oscillation was approximately sinusoidal with various peak-to-peak excursions, A, and various oscillation amplitudes $u' = A\pi/T$. The mean velocity error, U_E, is defined as the mean velocity inferred from the sensor minus the tow velocity, U, and is usually negative, implying a tendency to under-respond. The dimensionless error $\varepsilon = U_E/u'$ depends on the relative orientation of mean flow, oscillating flow, and propeller axis as well as U/u' and the frequency of oscillation as measured by the dimensionless parameter A/L, where L is the response distance constant.

Two fundamentally different causes of error are suspected. Most serious is the fact that when the total flow reverses direction, the propeller senses the velocity of water parcels which were previously in its wake. This phenomenon is clearly evidenced in traces of propeller rotation during an oscillation cycle. Flow interference of this type is common to all in situ current meters, but may be more serious for mechanical sensors which rely on drag to sense velocity. A less serious error occurs when the propeller axle and mean flow are parallel and both are perpendicular to the oscillation. In this case, large oscillations appear to raise the effective threshold speed causing the propeller to remain stationary for extended periods. This error is not significant if there is a significant component of oscillatory flow along the propeller axle but could be important if mooring motions cause oscillatory vertical motion much larger than the horizontal flow velocity.

The largest errors occur when propeller axle, mean flow, and oscillation are parallel. Here wake interference is most serious, particularly when u' and U are comparable and the sensor spends protracted periods in its own wake. The errors for this case are plotted in Figure 6. The dependence of U_E on frequency, through the factor $(A/L)^{\frac{1}{2}}$, was inferred empirically from observations spanning A/L from 1 to 22. Experimental scatter is somewhat less for the

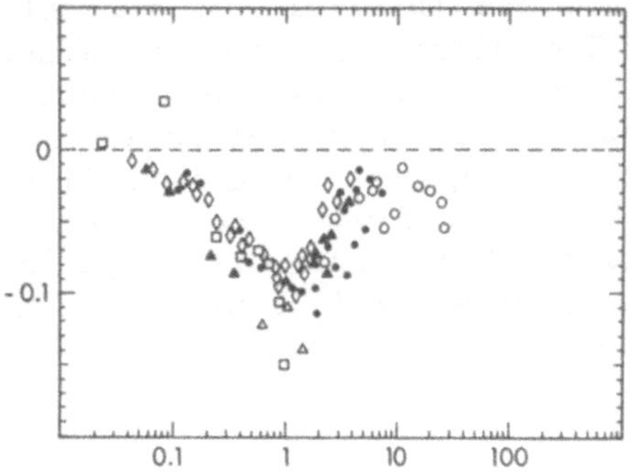

Fig. 6 Error in unsteady flow when propeller axle, mean flow, and oscillation are parallel. Data from tests with A = 32 cm, and T = 1.7 s (□), 2.3 s (▲), 2.8 s (●), 5.4 s (o), and with T = 50 cm, A = from 8 to 177 cm (◇).

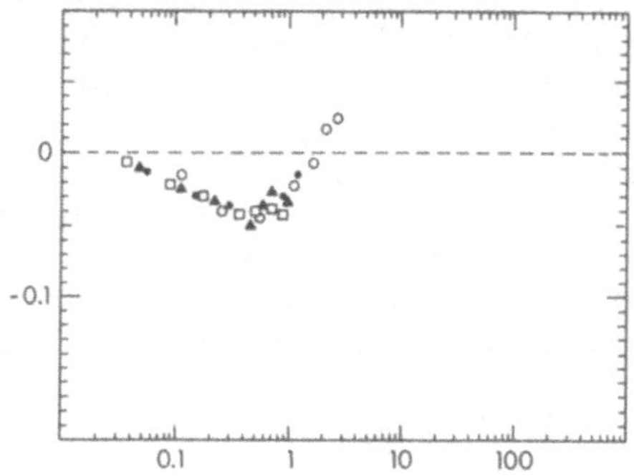

Fig. 7 Error in unsteady flow when mean flow and oscillation are parallel and propeller axle is at 45° to mean flow and oscillation. All tests run with A = 32 cm. Different periods of oscillation are represented by the different symbols: □, T = 1.7 s; ▲, T = 2.3 s; ●, T = 2.8 s; and o, T = 5.4 s.

exponent 1/2 than for 1/3 or 2/3 but there may be some additional dependence on a combination of A/L and U/u'. It will be noted that U_E/u' is greatest around U = u' when wake interference is greatest. For fixed A/L, the fractional error U_E/U is approximately constant for all U < u', so long as U exceeds the propeller threshold. As U/u' increases beyond unity the fractional error U_E/U decreases quite rapidly. For U/u' > 5 the observations of U_E/u' become less reliable but U_E/U shows no sign of increasing. On theoretical grounds one anticipates that for sufficiently large U/u' the error may become 'overspeeding'. Figure 7 shows that for the same flow geometry the errors are lessened considerably when the propeller is at 45° to both mean and oscillatory flow, presumably in some measure a result of the reduction of frontal area presented to the flow. When the propeller is at 90° to both oscillation and mean flow, it always registers no flow as anticipated.

Similar tests were carried out for the case of mean flow perpendicular to oscillation. The frequency dependence of the error was not determined, but the results are presented using the same dependence as in Figures 6 and 7 to simplify comparison. Figure 8 shows the error when the propeller and mean flow are parallel and are both perpendicular to the oscillation. The under-response is caused primarily by an increase of the effective threshold and reduces rapidly as U/u' approaches unity. Many of hte points at small U/u' represent the propeller being stalled during much of the oscillation cycle. This error is dramatically reduced when the propeller is turned 45° so that the oscillating flow keeps the propeller turning, as is shown in Figure 9. Again no error is found when the propeller is perpendicular to the mean flow.

5. RECOMMENDATIONS

It appears that propellers offer an improvement as upper ocean current meters when compared with rotor/vane meters, which can significantly over-respond to a mean flow in the presence of oscillatory motions. Propellers, too, suffer from rectification errors, primarily when the flow reverses direction carrying the sensor wake

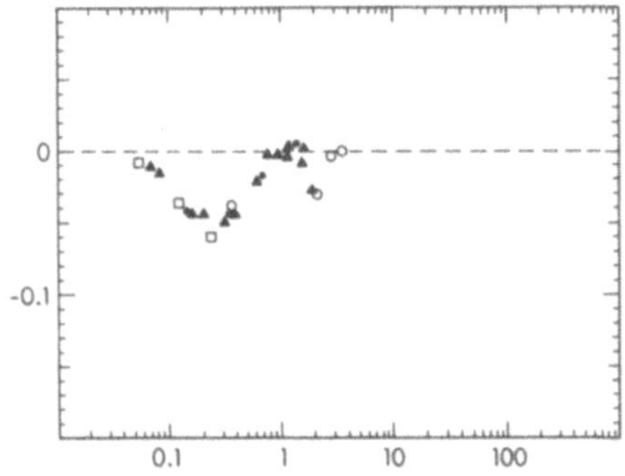

Fig. 8 Error in unsteady flow when propeller axle and mean flow are parallel and are both perpendicular to oscillation. All tests run with A = 32 cm. Different periods of oscillation are represented by the different symbols: □, T = 1.7 s; ▲, T = 2.3 s; ●, T = 2.8 s; and o, T = 5.4 s.

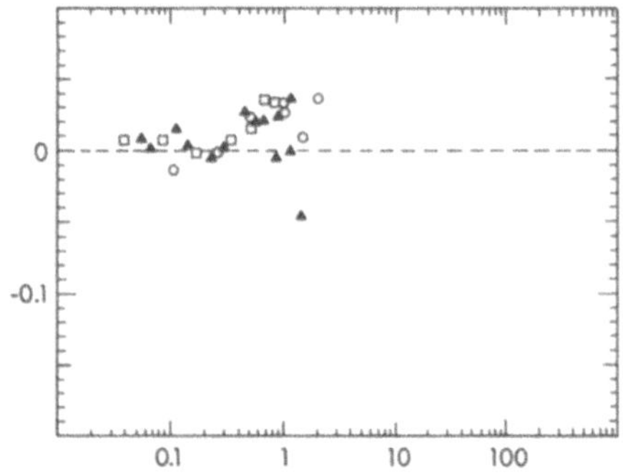

Fig. 9 Error in unsteady flow when mean flow and oscillation are perpendicular and the axle is at 45° to both. All tests run with A = 32 cm. Different periods of oscillation are represented by the different symbols: □, T = 1.7 s; ▲, T = 2.3 s; ●, T = 2.8 s; and o, T = 5.4 s.

back to the propeller. Reduced response lengths and more efficient propeller designs can improve this situation but field realities will limit such improvements. Hydrodynamic interference errors probably cannot be avoided by any instrument which measures the velocity near the sensor itself, and acoustic sensors, with the ability to sample large volumes of water from small sensors, may be the only way to solve this problem.

Laboratory tests provide the opportunity to unambiguously compare various current sensors. Unfortunately they are tedious and few sensors have been exhaustively tested. This makes meaningful intercomparison difficult and can lead to false optimism about poorly tested but promising sensors. However, laboratory error measurements are not easily used to estimate the errors encountered in the field. Barring a laboratory proof of an 'absolute standard' sensor, field intercomparisons will serve only as a guide, and only when mooring motions are well understood will it be possible to predict the field performance of anything but a perfect current sensor.

REFERENCES

CANNON, G.A. and D.W. PRITCHARD. 1971. Biaxial propeller current-meter system for fixed-mount applications. *Journal of Marine Research,* 29: 181-190.

DAHLEN, J. and J. SHILLINGFORD. 1976. The microscale sensing array. Report of the Charles Stark Draper Laboratory, Cambridge, Mass., 02139.

GILL, C.G. and E. MICHELENA. 1971. An improved biaxial water meter. *Proceedings of the 14th Conference on Great Lakes Research,* Publication 205, Department of Meteorology and Oceanography, University of Michigan, Ann Arbor, Michigan 48109: 681-689.

HOERNER, A.F. and H.V. BORST. 1975. Fluid dynamic lift. *Hoerner Fluid Dynamics,* Brick Town, N.J., U.S.A.

McMICHAEL, J.M. and P.S. KLEBANOFF. 1975. The dynamic response of helicoid anemometers. Report NBSIR 75-772, National Bureau of Standards, Washington, D.C., 20234.

SHONTING, D.H. 1967. Measurement of particle motions in surface waves. *Journal of Marine Research,* 25: 162-181.

SMITH, J.D. 1974. Turbulent structure of the surface boundary layer in an ice-covered ocean, pp. 53-65 in *International Council for the Exploration of the Sea, 1974-1975,* G. Kullenberg and J.W. Talbot, eds.

WALROD, R.A. 1970. Propeller current sensor response to off-axis flow. M.Sc. thesis, Department of Naval Architecture and Marine Engineering, M.I.T., Cambridge, Mass., 02139.

WELLER, R.A. 1978. Ph.D. thesis, University of California San Diego, La Jolla, CA, 92093.

9

Acoustic Travel Time Current Meters

T. Gytre

1. INTRODUCTION

All acoustic current meters are based on the principle that the resultant velocity of an acoustic wave propagating in a moving fluid is the vectorial sum of the fluid velocity and the sound velocity in the fluid at rest.

Since the 1930s various attempts have been made to obtain current measurements by means of such a non-moving current sensor. Only since the 1960s, however, have electronic components reached the degree of perfection needed to produce reliable and stable ultrasonic current measurements.

Figure 1 illustrates how an ultrasonic current measurement is made. Two piezoelectric transducers, denoted as A and B, are placed in a fluid of velocity $\vec{U}$ shown in the figure as a unidirectional field $\vec{V}(y)$ directed along the x-axis. Three different direct signal processing systems are now in general use: the 'travel time' or 'leading edge' system, the 'sing around' system, and the 'phase difference' system.

In the travel time or leading edge system A and B are simultaneously excited by voltage steps from 100 to 400 V at a repetition rate f_r. For each excitation a burst of exponentially damped acoustic oscillations is generated at each transducer surface. The resulting acoustic wavetrains travel towards the opposite transducer with a velocity $C \pm \vec{U} \cdot \vec{S}$ where C is the sound velocity in the resting fluid, and $\vec{S}$ is a unit vector parallel to the sound propagation path $d\vec{l}$. The travel time of the leading edge of the signal from A to B is

$$T_{AB} = \int_A^B \frac{d\vec{l}}{C + \vec{U} \cdot \vec{S}} \,. \tag{1}$$

As shown by Kaimal (see Chapter 4) the travel time difference is

$$T_{BA} - T_{AB} = \Delta t = \frac{2\,L}{C^2}\,|\vec{U}_m|\,\cos\,\theta \tag{2}$$

where L is the straight line distance between A and B,
$\vec{U}_m$ is the mean velocity along the path between A and B,
and θ is the angle between $\vec{U}_m$ and the line between A and B.

To compensate for changes in the sound velocity, the average travel time $T_A \simeq T_B = T$ is simultaneously measured and a computation of

$$\frac{\dot{\Delta\tau}}{T^2} \approx \frac{\Delta t}{[L/C]^2} = k_1|\vec{U}_m|\cos\,\theta, \tag{3}$$

where k_1 is a constant, is carried out for each ultrasonic burst. To get a continuous output, $\Delta T/T^2$ is converted to an analogue or digital signal which is updated in an output-storing circuit for each new ultrasonic burst. The repetition frequency is chosen to correspond to the needed bandwidth.

In the sing around system bursts of ultrasonic signals are first sent from A to B. The reception of a leading edge at B causes the

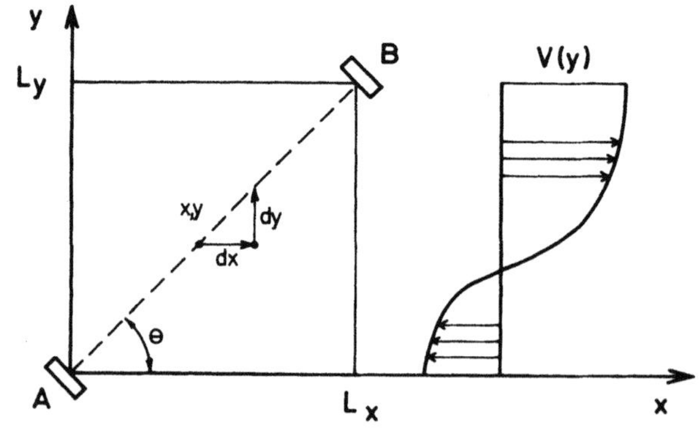

Fig. 1 Principle of operation for acoustic current meters. Two transducers in a one-dimensional flow field V(y) emit acoustic signals to each other at an angle θ with the current direction.

generation of a new burst of signals in the same direction. Thus an 'upstream' frequency $f_{AB} = 1/T_{AB}$ can be defined. Similarly a corresponding 'downstream' frequency $f_{BA} = 1/T_{BA}$ is determined by sending signals in the opposite direction, from B to A. Finally

$$f_{AB} - f_{BA} = k_2 \, |\vec{U}_m| \cos \theta \qquad (4)$$

is calculated. Thus in the sing around system, C cancels from the equation with no further calculations.

In the phase difference system continuous ultrasonic signals of frequency f_A are transmitted from A to B while similar signals of frequency f_B are simultaneously sent from B to A. By means of several possible signal processing methods (e.g. as described by Lawson 1976) the phase difference $\phi_B - \phi_A$ can be transformed into a signal proportional to

$$\frac{k_3}{C^2} \, |\vec{U}_m| \cos \theta$$

where k_3 is a constant.

2. CURRENT METER PROPERTIES

When selecting a current meter a number of properties are desirable. The most important of them are:

- stable, low noise output signals
- high sensitivity
- wide dynamic range
- good linearity
- good long term zero point stability
- low power consumption
- small dimensions
- ease of operation
- mechanical strength

Well designed current meters based on Equations 1 to 5 can now offer a noise level corresponding to less than 1 mm s^{-1}, sensitivity better than 1 mm s^{-1}, dynamic range from 0 to ±3 m s^{-1}, linearity better than 1%, a zero point long term stability better than 5 mm s^{-1} per month, a continuous power consumption from 50 to 500 mW, a bandwidth in excess of 50 Hz, and the possibility of measuring more than one current component at the same time.

All the listed properties cannot be combined in one instrument. When making (or buying) an instrument, compromises must be made to favour the specific problem.

3. INSTRUMENT DESIGN

An acoustic current meter consists of a current sensor (probes) in combination with electronic circuits, batteries, and mechanical protection for the parts that cannot be directly exposed to the sea.

3.1 Acoustic Design

Piezoelectric transducer elements are available in a range of frequencies and dimensions. A typical transducer could be resonant at 3.5 MHz and have a diameter of 10 mm. Typical thickness is 0.5 to 1 mm.

The designer usually tries to generate an acoustic beam with as high intensity as possible at the receiving end. In the leading edge type current meter non-perfect impedance matching between transducers and the surroundings causes a ringing which may modulate into the signal baseline. This ringing must be completely damped out before the received signal appears. Sufficient damping is normally obtained by backing the transducer with an absorbent material of characteristic impedance Z_B as close to the transducer impedance as possible. The transducers must always be protected by acoustic windows in the probe. The ideal protecting material must be strong enough to withstand the expected maximum ambient water pressure without noticeable deformation and have a characteristic impedance as close to

$$Z_P = (Z_1 Z_W)^{\frac{1}{2}} \qquad (6)$$

as possible (Kinsler and Frey, 1962). Z_1 is the characteristic impedance of the transducer material, and Z_W that of the surrounding fluid. The thickness of the protecting material should ideally be chosen according to the formula

$$t_p = (2n - 1)\,\frac{\lambda}{4} \qquad (7)$$

where n is an integer and λ is the wavelength in the protecting material. Good results have been obtained by using protecting discs of ceramic materials which are both strong, non-absorbing, and have a characteristic impedance close to the optimum value of around 6.6 x 10^6 m^{-2} s^{-1}. The rules given for the design of pulsing type transducers also apply to continuous signal instruments.

3.2 Ultrasonic Field Distribution

According to Equation 2, the current signal is proportional to the distance between the probes. This distance willl depend on what is being measured: small-scale microstructure or turbulence will require a small distance; small average currents, a large distance. Possible ranges in L vary from about 0.01 to 2 m. In any case, care must be taken to obtain a sufficient ultrasonic intensity at the receiving transducers.

For continuous signals the intensity along the acoustic axis is described by the general equation

$$\frac{I_z}{I_0} = \sin^2 \frac{\pi}{\lambda} [(a^2 + z^2)^{\frac{1}{2}} - z] \tag{8}$$

where I_z is the sound intensity at a distance z from the transducer, I_0 is the intensity at the transducer face, a is the transducer radius, and λ is the wavelength in the propagating medium.

For short bursts of ultrasonic signals, as used in leading edge type instruments, the axial field distribution is much more difficult to describe. A well written treatment of both the axial and off-axis ultrasonic field intensity distribution for pulsed ultrasound has been given by Foster and Hunt (1978).

Figure 2 shows both computed and measured on-axis values of the field intensity for pulsed and continuous ultrasonic signals. In

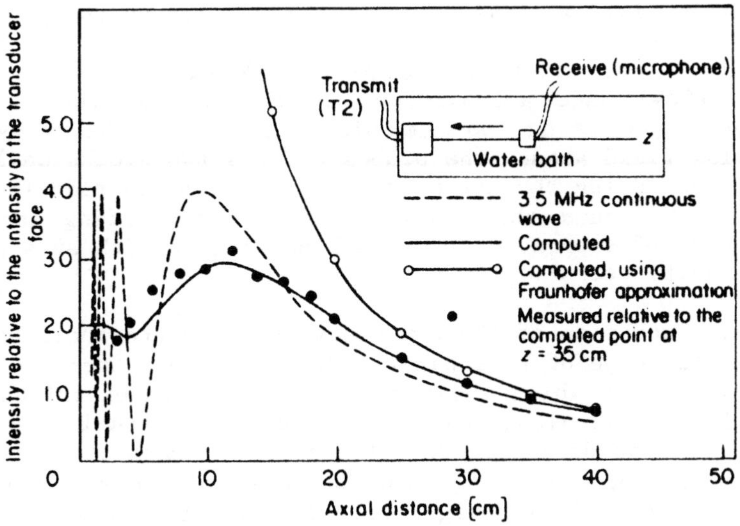

Fig. 2 On axis acoustic field intensity for both continuous and pulsed ultrasonic signals (Foster and Hunt, 1978).

order to make the received signal intensity non-critical to slight misalignments or bending during use, the main acoustic lobe must not be made too narrow.

3.3 Hydrodynamic Probe Design

The probes must emit acoustic signals in the desired direction, protect the piezoelectric transducers, and have a physical shape that combines mechanical strength with a minimum of interference to the flow field. Good mechanical probe design is the most difficult part of making good current meters.

A basic probe design, shown in Figure 3a, is to mount the piezoelectric transducers in cylindrical prongs which are encapsulated in an acoustic window made from glass, ceramic or similar material. The wake behind the leading transducer may affect the velocity field and hence contribute to errors.

Both the designer and user want to know:

- how different prong configurations will modulate the flow field along the acoustic path;

- where the zones of separation will occur and how they will fluctuate with varying velocity, water viscosity, changing geometry, etc.;

- how the instrument pressure case to which the probes are usually fixed will disturb the current around the probes.

Exact answers are difficult to give. Working with the author, Laukholm (1974) made approximate calculations of the conditions around each probe using two-dimensional potential theory by assuming the flow field around the probes to be a two-dimensional potential flow and adding the effect of separation by introducing circulation (lift) around the prongs, and by representing each prong with its zone of separation by a Rankine ellipse with axes parallel to the velocity vector.

A detailed discussion of the equations involved and their solutions is beyond the scope of this paper. One practical conclusion of the theoretical work - which seems to be verified by experiments - is that the flow field along the acoustic path is virtually undisturbed if the current direction deviates from the acoustic path direction by more than about 25° (for the case of 15 mm thick prongs with $L \simeq 15$ cm). This design criterion may be satisfied in four ways:

1. by using vertically-mounted prongs combined with a vane that makes the sound paths cross the current direction at an angle of approximately 45° (Fig. 3a);

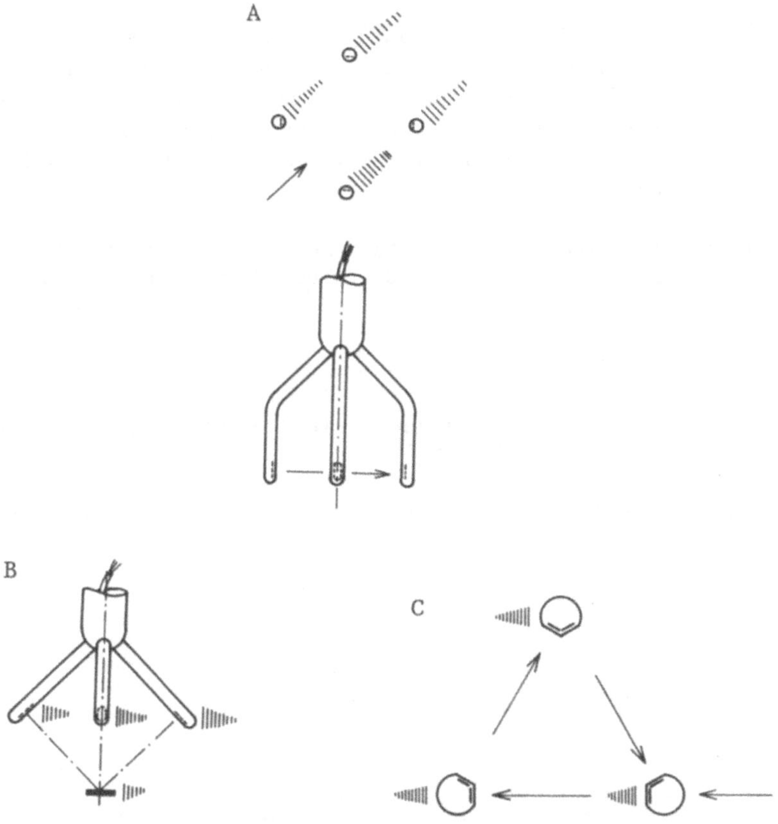

Fig. 3 Basic two-axis probe designs: a. direct transmission; b. reflected signal; c. three direct signal paths.

2. by using a reflector to obtain a 'V'-shaped acoustic path (Fig. 3b);

3. by using three pairs of transducers with acoustic paths crossing each other at an angle of 60°, sorting out the signals from the wake-disturbed path (Fig. 3c) during subsequent signal processing; or

4. by reducing the d/L ratio, either by decreasing the probe diameter d or by increasing L until the disturbed part of the path length can be neglected.

Figure 4 shows a selection of probe designs developed at the Christian Michelsen Institute, Bergen, Norway.

3.4 Electronic Design

The electronic circuits in acoustic current meters must generate the acoustic signals and convert the resulting travel time, frequency, or phase difference into an equivalent electrical signal in a suitable format for further processing or recording. Since most current meters operate on batteries, it is also important to keep power consumption low.

The basic problem with acoustic measurement is the very small time differences that must be detected. From Figure 1, assuming L = 10 cm, C = 1500 m s^{-1}, and $|\vec{U}_m|$ = 1 mm s^{-1}, equation 2 gives

$$\Delta t \approx 10^{-10} \text{ s.}$$

To get stable and low noise signals in the order of 1 mm s^{-1}, two basic problems must be solved:

1. precise detection of the moment when a specified part of the received signals arrive,

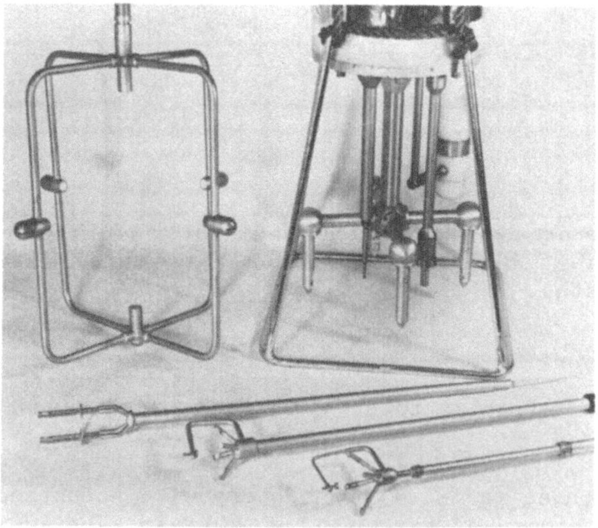

Fig. 4 Examples of different probe designs. Upper left: three-axis direct signal probe; Upper right: fork-shaped direct signal probe (head section of a profiling instrument); Below: miniature probes for turbulence measurements.

2. elimination of apparent changes in Δt due to drift in the electronic components.

Figure 5 illustrates the first problem. The figure shows the situation when the received signals arrive at A and B at the same time (the current is zero). The time-measuring circuits are adjusted to start when the signal's leading edges exceed V_0. V_A and V_B are the peak amplitudes of the received signals. Assuming sinusoidal signals, the apparent travel time difference is:

$$\Delta t = \frac{1}{2\pi f}\left(\sin^{-1}\frac{V_0}{V_A} - \sin^{-1}\frac{V_0}{V_B}\right) \qquad (9)$$

With a typical ultrasonic frequency of 4 MHz, a resolution of 1 mm s^{-1} requires a stability in the phase angle detection of at least one-tenth of a degree. In travel time difference current meters this is solved by setting V_0 as close to zero as possible, and by making both V_A and V_B equal and as large as possible. In continuous type current meters the best way is to beat the received signal with a stable oscillator and to detect the phase changes in the difference frequency (Lawson et al., 1976).

Detecting time differences in the order of 10^{-10} s with electronic components that may have rise times in the order of several nanoseconds and in addition may vary by several nanoseconds with

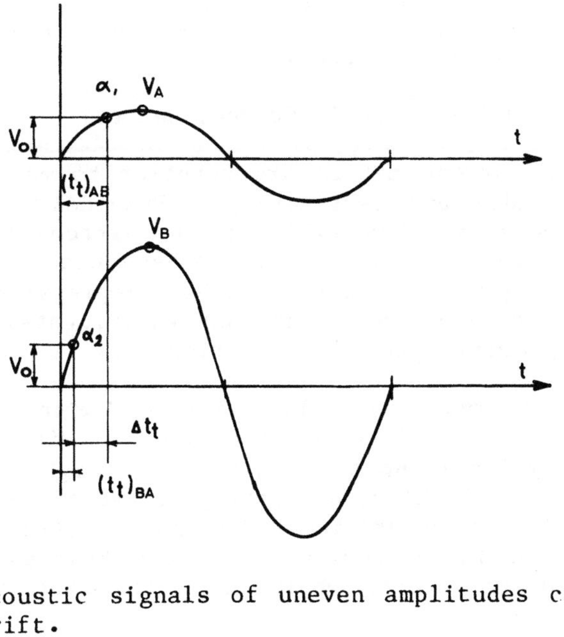

Fig. 5 Acoustic signals of uneven amplitudes causing zero drift.

changes in temperature represents another basic problem. It can be partially solved by accumulating the Δt values from a succession of measurements (Hardies, 1975), and satisfactorily solved by comparing the measured Δt with a known Δt_0 reference (Gytre, 1976).

The power needed to run an acoustic current meter is a function of sampling rate, number of channels, and its integration into a recording system. The choice of components and ingenuity in design are also very important factors. Presently the continuous type of current meter has the lowest power consumption (around 50 mW), but this is not necessarily a permanent situation.

4. ULTRASONIC CURRENT METER DESIGNS

Prototypes of ultrasonic current meters, both for special as well as for general applications, have been in use for several years.

Figure 6 shows a profiling current meter designed for a vertical fall rate of approximately 0.2 m s^{-1}. In use this instrument either falls freely, dropping a weight when reaching a specified depth, or it slides slowly down a hydrographic wire.

Figure 7 shows a general purpose three-axis current meter integrated into a general data acquisition system including battery, tape recorder, programmable clock, 12-bit A/D converter, etc. This current meter has a resolution of 1 mm s^{-1}, range 0 to ±2 m s^{-1}, and a storage capacity of approximately 30,000 observations of current, temperature, direction, and time. To control its performance in the field, the recorded information is simultaneously flashed through an optical window on the top and detected by an optical detector combined with a printer.

4.1 Experiences with Ultrasonic Current Meters

Good ultrasonic current meters are superior to mechanical current meters in bandwidth and sensitivity. Provided the hydrodynamic probe design is good, they will also be extremely linear over a wide dynamic range. A realistic comparison with existing instruments in the field is difficult. The best impression of instrument performance is achieved in towing tank experiments. Several such calibration experiments have been carried out.

Figure 8 shows the result of a linearity test with a two-axis probe of the reflector type as shown in Figure 7. With this probe the response is very linear up to ±1 m s^{-1}. At higher velocities the response falls off slightly, due partly to hydrodynamic effects and partly to electronic saturation. The latter effect has now been removed by changes in the design. Figure 9 shows the response when the same sensor is rotated while being towed at constant speed.

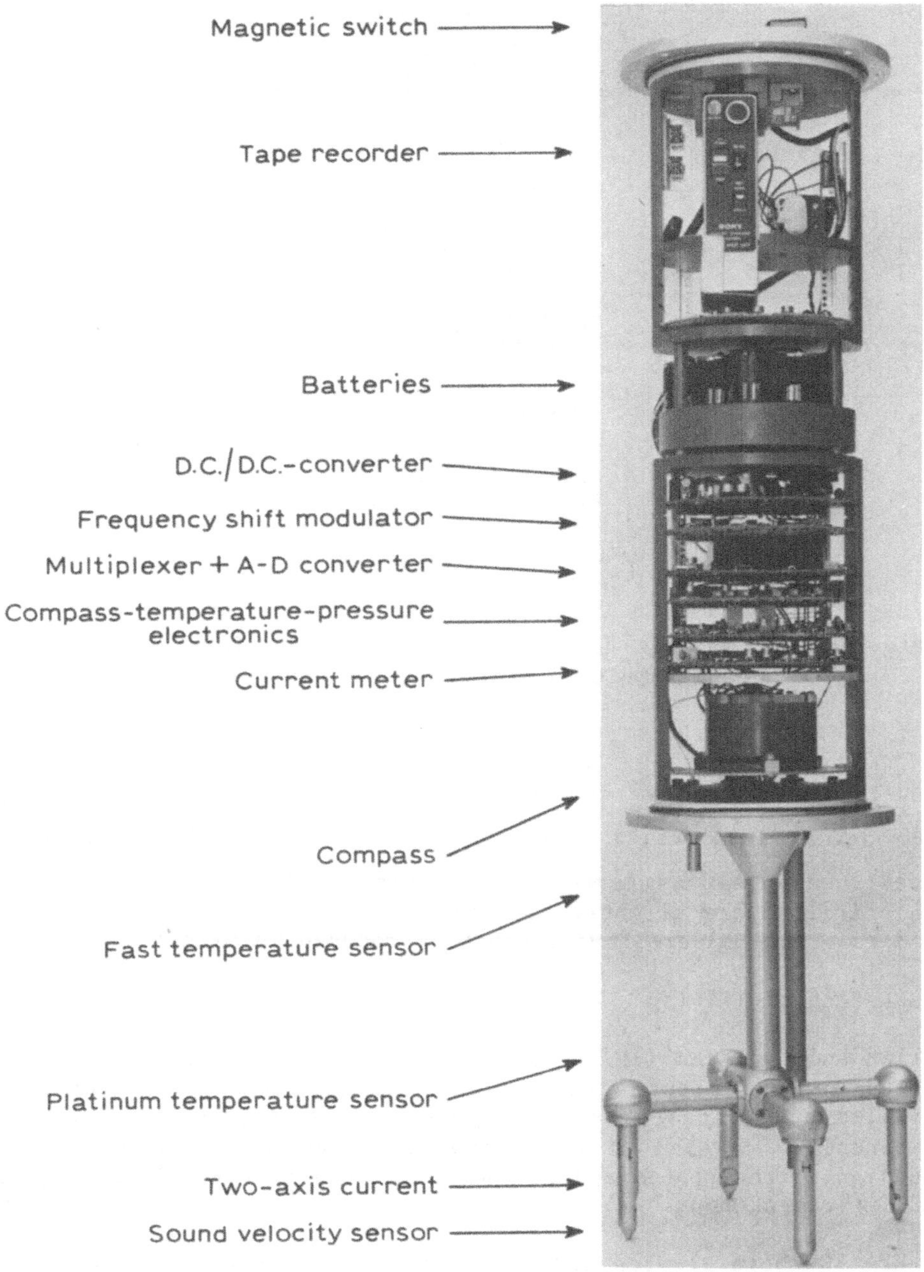

Fig. 6 Two-axis current meter integrated into a profiling instrument.

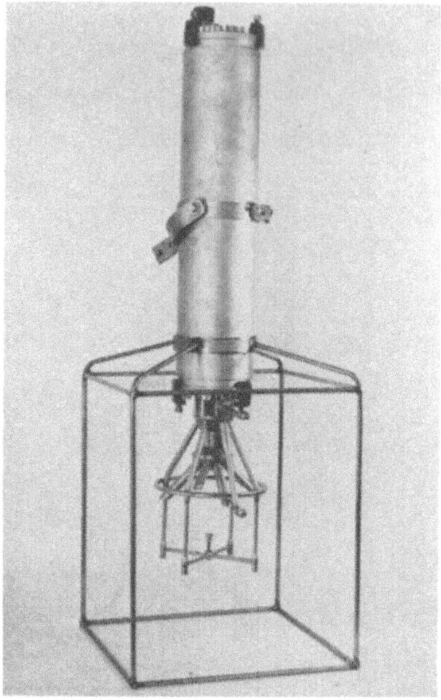

Fig. 7 Three-axis current meter integrated into a complete oceanographic data acquisition system.

Ideally the V_x and V_y current signals should follow perfect sine and cosine responses; a plot of $(V_x^2 + V_y^2)^{\frac{1}{2}}$ at different speeds would then appear as perfect concentric circles. As can be seen, the 'circles' are slightly squared due to the effect of the mirror holders. The holders have since been reduced in size.

4.2 Effect of Tilting

Collar and Gwilliam (1977) have carried out towing tank tests on a tilted current meter with a reflector-type sensor. Their tests revealed a non-ideal cosine response for strong vertical current components. Several of these problems have now been overcome by redesigning the probes and by increasing the distance between probe and instrument housing.

Figure 10 shows results from a profiling experiment in a Norwegian lake. The net current across the lake could be obtained independently by measuring the flow in the river leading out from the lake. The computed flow through the lake from Figure 10 is about 500 m^3 s^{-1}. On that day the river flow was estimated to be 450 m^3 s^{-1}.

4.3 Comparison of Different Ultrasonic Current Meters

Ultrasonic current measurements can be carried out using the travel time difference principle, the sing-around principle, the phase difference principle, or the Doppler principle (see Pinkel, Chapter 10).

Although all exploit the same physical phenomena, the different principles leave the designer with different degrees of freedom. In Norway the travel time difference principle has been selected due to its great versatility. It offers an unlimited dynamic range, no restrictions in choice of path length, and about the same bandwidth, resolution, zero stability, and linearity possible with the phase difference current meter. Due to its discontinuous sampling, the same electronic circuit can time-share several current sensor pairs, and any practical number of additional sensors that operate by expressing the variable as a time difference. As an example a complete three-axis current meter - or a two-axis current meter with an acoustic salinity sensor - will only occupy an 11 x 21 cm printed card. Presently this circuit consumes approximately 500 mW, but it can easily be brought down to, say, 100 mW.

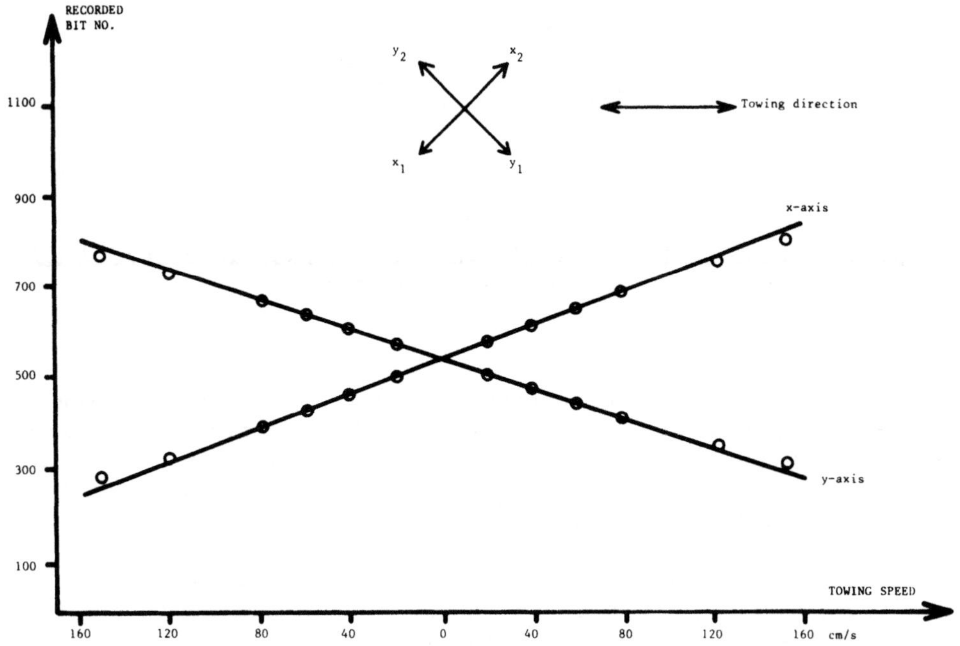

Fig. 8 Calibration curve for the horizontal component of the sensor in Figure 7.

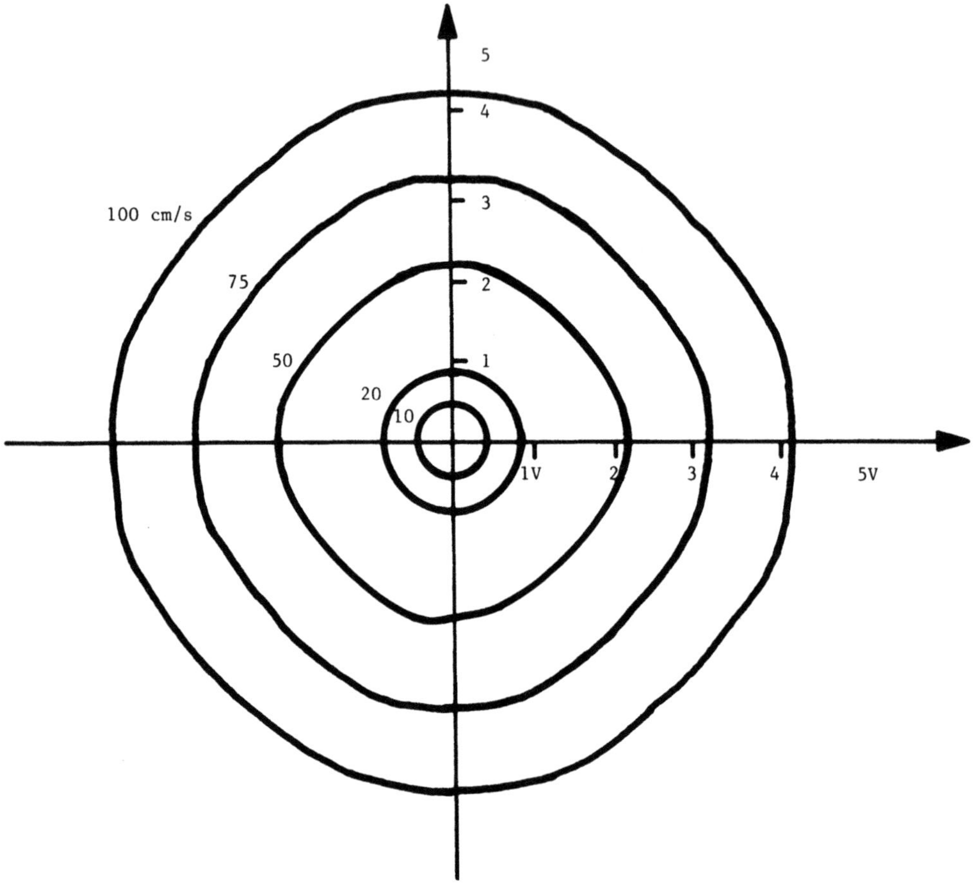

Fig. 9 Plot of $[V_x^2 + V_y^2]^{\frac{1}{2}}$ when the sensor in Figure 7 is rotated at different current speeds.

The sing-around principle offers an easy way of eliminating the need to determine the sound velocity, C. Combined with modern phase-locked loop techniques it may develop into a reliable solution. Presently the zero stability is not as good as with the travel time difference principle.

The phase shift principle offers low power consumption and stable zero level. The principle has, however, limitations in choice of dynamic range and of distance between probes. It cannot so easily be used in time-sharing of signals from several probe pairs and hence current meters with several sensors need more volume and cost more than corresponding travel time difference types.

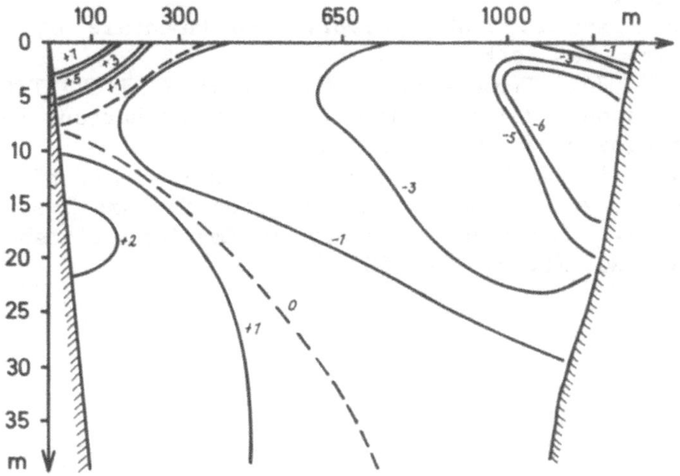

Fig. 10 Current profile of a Norwegian lake made with a two-axis current meter. + denotes current towards north (cm s^{-1}); - denotes current towards south.

Today the electronic problems associated with measuring currents acoustically have been satisfactorily solved. Before acoustic current meters can be more widely used, better hydrodynamic probe designs, and more 'field friendly' data acquisition systems are required, into which the ultrasonic current sensor can be harmoniously integrated. In particular the integration of acoustic current sensors into solid state recording data acquisition systems will make possible the design of a new generation of compact, low cost and highly versatile oceanographic instruments.

REFERENCES

COLLAR, P.G., and T.J.P. GWILLIAM. 1977. Some laboratory measurements on an acoustic current meter developed at Christian Michelsens Institute, Norway. Institute of Oceanographic Sciences, Wormley, Godalming, Surrey, England, Report 47.

FOSTER, F.S., and J.W. HUNT. 1978. The design and characterisation of short pulse ultrasound transducers. *Ultrasonics,* 16, (3): 116-128.

GYTRE, T. 1976. The use of a high sensitivity ultrasonic current meter in an oceanographic data acquisition system. *The Radio and Electronic Engineer,* 46: 617-623.

HARDIES, C.E. 1975. An advanced two axis acoustic current meter. 7th Annual Offshore Technology Conference, Houston, Texas.

KINSLER, L.E., and P. FREY. 1962. Fundamentals of acoustics. Wiley, New York.

LAUKHOLM, A. 1974. Some error sources common to the Transit Time Ultrasonic Velocity Meter. River and Harbour Laboratory, Trondheim, Norway. Internal Note 1.

LAWSON, K.D., N.L. BROWN, D.H. JOHNSON, and R.A. MATLEY. 1976. A three axis acoustic current meter for small scale turbulence. Instrument Society of America, Pittsburgh, Pa., ASI, No. 76269: 501-508.

10

Acoustic Doppler Techniques

R. Pinkel

1. INTRODUCTION

Doppler backscatter current meters have been used for the last 35 years to make relative velocity measurements in the ocean. While the initial widespread application of the principle was for ship speedometers (Harvard University Acoustic Research Laboratory, 1945), Doppler flowmeters have more recently been used by the oceanographic community. In general, these devices transmit high frequency sound in a narrow beam. The sound scatters off plankton and other drifting (and swimming) objects in the sea. The Doppler shift of the returning sound gives information about the component of scatterer velocity along the beam. Acoustic frequencies from several kilohertz to ~10 mHz have been used for Doppler sensors. The Doppler shift associated with a given relative velocity increases with increasing frequency. The higher the frequency, however, the more rapid is the acoustic attenuation with range.

Doppler sensors offer several advantages over alternative velocity measurement methods. They can measure current at some distance from the instrument. Thus the flow which is measured need not be affected by the presence of the sensor. They measure one component of the flow precisely. Little design effort is required to achieve the ideal 'cosine directional response', as compared to mechanical sensors. There is no inherent threshold velocity, below which the response of the instrument becomes uncertain. A zero hertz Doppler 'shift' can be detected with the same precision as any other frequency shift, within the limits of the data processing capability. The sensor is not degraded by biological fouling, since the frequencies often used in these devices are those used in commercial ultrasonic cleaners.

On the other hand, the precision of the measurement depends on the strength of the backscattered signal. Typically, at frequencies greater than 50 kHz, only 10^{-5} to 10^{-9} of the acoustic energy incident on a cubic meter of water is backscattered toward the transmitting source. There is an inherent inefficiency in the approach, in that the strength of the signal of interest is of the order of one-millionth of the energy required to produce that signal. Below 400 to 600 m depth the scattering strength is even lower. Also, scattering strength at a given site might vary with time of year (due to plankton blooms), surface conditions (stirring up silt), or other causes not related to the flow field being sensed. Thus, it is frequently not possible to assess the precision of a Doppler instrument prior to its actual use. This liability can cause difficulty in planning geophysical experiments.

A variety of Doppler sensors has been developed for oceanographic use. Rather than describe many particular systems, the major effort in this paper will be to review the general principles common to all. Single examples of the various types of systems are presented at the end. The discussion of general principles has been taken, in part, from standard works on radar, sonar, and detection theory, and applied to the case of oceanic Doppler sonar. It begins with definitions of incoherent backscattering, the scatterer autocorrelation function, and the Doppler spectrum. A discussion follows of sonar range and frequency resolution, in which the role of the signal to noise ratios is quantified. The possibility of improving resolution at a fixed signal-to-noise ratio by using coded transmissions is raised, and the liabilities of this approach are introduced. Finally, examples of several types of Doppler sensors are presented.

2. INCOHERENT BACKSCATTERING

Consider the situation when a pulsed signal of the form

$$S(t) = A\, e^{i\omega_0 t}, \quad 0 < t < T \tag{1}$$

scatters off a 'cloud' of many discrete 'targets' (analogous concepts apply for the case of continuous transmission). Following Serafin (1975), the received signal can be represented as a sum of the contributions from each scatterer.

$$S_r(t) = e^{i\omega_0 t} \sum_{\substack{\text{scattering} \\ \text{volume}}} a_j e^{-i\phi_j} = A_r e^{i\phi_r} \tag{2}$$

where $\phi_j = k \cdot 2r_j$ and k is the wavenumber of the radiation, r_j is the range to the jth scatterer, and a_j is the relative amplitude of the jth backscattered signal. 'Incoherent backscatter' results when the returns from the individual scatterers add randomly. If there is a sufficient number of scatterers in the scattering volume, the Central Limit Theorem applies. Then, A_r has a Rayleigh probability density function, and the statistics of ϕ_r are uniform on the interval (0, 2π). The functions $A_r \cos \phi_r$ and $A_r \sin \phi_r$ obey Gaussian statistics (Davenport and Root, 1968).

If the scatterers are moving, each perhaps with a different velocity, their contribution to the return phase will change with time, and

$$\partial\phi_j/\partial t = 2k\ \partial r_j/\partial t = 2kv_j \tag{3}$$

where v_j is the component of relative velocity parallel to the sonar beam. In this situation, the echoes from a given scattering volume evolve with time. If pulses are transmitted rapidly compared to the time in which scatterers change their relative positions, the phase of successive echoes at fixed time after transmission will vary deterministically. The rate of change of phase is proportional to the mean motion of the scatterers. However, if sufficient re-arrangement of the scatterers occurs between transmissions, the phase will vary randomly. Here, the information on scatterer motion is contained in the complex autocovariance function of the echo

$$R(\tau;x) = R(\tau;t) = \langle S_r(t)\ S_r^*(t + \tau)\rangle \tag{4}$$

The brackets imply pulse-to-pulse averaging, and the normal range is $x = ct/2$. The Fourier transform of the autocovariance function is the Doppler spectrum

$$S(\omega;x) = \int_{-\infty}^{\infty} R(\tau;x)\ e^{-i\omega\tau}\ d\tau \tag{5}$$

The frequency of the peak of the Doppler spectrum indicates an intensity or scattering amplitude weighted average velocity of the scatterers. Multiple peaks are possible. The Doppler spectrum corresponding to a single fish (strong scatterer) swimming upstream through a cloud of drifting plankton (weak scatterers) would have two peaks. The width of the Doppler spectrum is inversely proportioned to the time lag required for the envelope of $R(\tau;x)$ to become small, in the usual Fourier sense. Thus, the width is at least as wide as $1/T$. It can be wider, as discussed below.

3. RANGE AND VELOCITY RESOLUTION

It is convenient to examine the question of Doppler resolution in terms analogous to the discussion of frequency resolution in classical time series analysis. Recall that a linear estimate of the power spectrum of a stationary random process is related to the actual spectrum by a convolution integral

$$\hat{S}(\omega) = \int_{-\infty}^{\infty} S(\Omega)|\chi(\omega-\Omega)|^2 d\Omega \tag{6}$$

where $|\chi(\omega)|^2$ is termed the spectral window. If $|\chi(\omega)|^2 = \delta(\omega)$, then the estimate would be perfect. In any realizable situation the width of the central peak of the spectral window is finite, inversely proportional to the duration of the time series analyzed. This sets the resolution of the estimate. 'Side lobes' of the central peak result in 'leakage' between spectral bands. Aliased replications of the window occur at regular intervals in frequency as a result of the finite sampling rate of the process. The contribution to the estimate $\hat{S}(\omega)$ from the replicate 'aliased' peaks in the window cannot be separated from the contribution of the central peak; some prior knowledge of the form of $S(\Omega)$ is necessary in order that the data sampling scheme can be designed to minimize the effects of aliasing in the estimate $\hat{S}$.

All of these effects have direct analogy in the discussion of Doppler sonar resolution. The situation is slightly more complex, in that both frequency and range resolution must be considered simultaneously. The joint range-Doppler resolution for a linear processing method is again expressed in terms of an integral relationship between the Doppler spectral estimate $\hat{S}(\omega;x)$ and the 'actual' spectrum of the scatterer field

$$\hat{S}(\omega;x) = \int_{-\infty}^{\infty}\int_{-\infty}^{\infty} S(\Omega;X)\ |\chi(X-x,\ \Omega-\omega)|^2 d\Omega\, dX \tag{7}$$

$|\chi(X,\Omega)|^2$ is called the ambiguity function (Woodward, 1953). Note that it is traditionally defined such that $|\chi(-X,-\Omega)|^2$ is convolved with the actual spectrum, in contrast to the convention used in time series analysis.

The definition of the ambiguity function can be introduced as follows. Suppose a signal of the form $S(t) = s(t)\, e^{i\omega_0 t}$ is transmitted repeatedly at time intervals T_0, where we define

$$\frac{1}{T_0}\int_0^{T_0} s^2(t)\, dt = E \tag{8}$$

as the average transmitted power. At a later time echoes are received from two scatterers separated by a time lag $\tau = \Delta x/c$ and a Doppler frequency shift ω

$$S_R(t) = \sigma\left[s(t)\, e^{i\omega_0 t} + s(t+\tau)\, e^{i(\omega_0+\omega)(t+\tau)}\right] + n(t) \quad (9)$$

where σ is a measure of the strength of the return, accounting for the effects of attenuation, inverse square spreading loss, and the scattering cross section of the reflectors, n(t) is the noise, and

$$\frac{1}{T_0}\int_0^{T_0} n^2(t)dt = N, \text{ the average noise power.} \quad (10)$$

For a given transmitted power, noise power, time lag, and frequency shift, the ability to distinguish the separate arrivals can be quantified by

$$\Delta = \frac{\sigma}{T_0}\int_{-\infty}^{\infty} |s(t)\, e^{i\omega_0 t} - s(t+\tau)\, e^{i(\omega_0+\omega)(t+\tau)}|^2 dt \quad (11)$$

The ability to resolve the separate returns will depend on the size of Δ relative to the noise power N.

$$\frac{\Delta}{N} = \frac{2\,E\sigma^2}{N}\left[1-\mathrm{Re}\left(\frac{1}{ET_0}\, e^{-i(\omega_0+\omega)\tau}\int_{-\infty}^{\infty} s(t)\, s(t+\tau)\, e^{-i\omega t}\, dt\right)\right] \quad (12)$$

$$= \frac{2\,E\sigma^2}{N}\left[1-\mathrm{Re}\left(e^{-i(\omega_0+\omega)\tau}\, \chi(\tau,\omega)\right)\right]$$

where $$\chi(\tau,\omega) = \frac{1}{ET_0}\int_{-\infty}^{\infty} s(t)\, s(t+\tau)\, e^{-i\omega t} dt \quad (13)$$

is the 'ambiguity function' of the transmitted signal (Woodward, 1953). For a given transmitted signal s(t), and given signal-to-noise ratio, the ability to separate the arrivals is large if the time and frequency lags are such that $|\chi(\tau,\omega)|^2$ is small compared to unity.

It would be ideal if $|\chi(\tau,\omega)|^2 = \delta(\tau)\,\delta(\omega)$. However, as in the case of time series analysis, there are a number of constraints. In particular, the maximum value of the function is finite, equal to unity. The value is achieved at zero lag, $\chi(0,0) = 1$. For some signals, there are secondary peaks at other values of τ and ω, resulting in 'ambiguous' contributions to the estimate of Equation 7 which are indistinguishable from the 'true' contribution. Examples of this situation are presented below. The volume under the ambiguity surface is unity:

$$\int_{-\infty}^{\infty} \int_{-\infty}^{\infty} \chi(\tau,\omega)d\tau d\omega = 1 \tag{14}$$

For the simple case of a sinusoidal transmitted pulse of duration T, the 'width' of the peak of the ambiguity function in the time dimension is of order T; the width of the peak along the frequency axis is of order 1/T. Other examples are presented schematically in Figure 1 and discussed in greater detail in Rihaczek (1969) or other standard radar and sonar texts.

In discrete scattering situations, the combined range-Doppler resolution can be improved by increasing the average transmitted power, reducing the system noise, or narrowing the central peak of the ambiguity function. Efforts to increase the total energy by increasing peak power are ultimately limited by cavitation or nonlinear propagation of the acoustic wave; it is then necessary to lengthen the duration of the pulse to further increase the average transmitted power. To preserve range resolution while increasing average power, the transmitted signal, s(t), can be coded. Either phase or frequency discontinuities can be introduced into the carrier signal to divide the pulse into a series of 'bits'. If the return signal is suitably decoded, the frequency resolution will be inversely proportional to the total length of the transmission, while the range resolution will be proportional to the bit length, which in turn is inversely proportional to the code bandwidth, B. The technique of achieving the energy of a long pulse with the range resolution of a short one is frequently referred to as 'pulse compression'. The degree of compression, as quantified by the volume of the central peak of the ambiguity function $\chi(\tau,\omega)$, is proportional to the time-bandwidth product of the code T x B. Since the volume of the entire ambiguity function is unity, reducing the volume of the central peak produces corresponding increases elsewhere in the $\tau - \omega$ plane. These increases can take the form of a general rise in the ambiguity surface or the appearance of ambiguous secondary peaks in the function, which are associated with signals at spurious ranges and Doppler shifts in the decoded return information.

It should be noted that the bandwidth of the transmitted signal cannot be made arbitrarily large. For a given scatterer velocity, the higher frequency bands of a broad band transmission will be Doppler-shifted by a greater amount that the lower frequencies. As the ratio of bandwidth to carrier frequency approaches unity, this effect becomes significant. Equation 9 no longer represents the return from two scatterers; the ambiguity function does not describe the process.

While the theory relating range and velocity resolution with the form of the transmitted signal is straightforward, the actual data

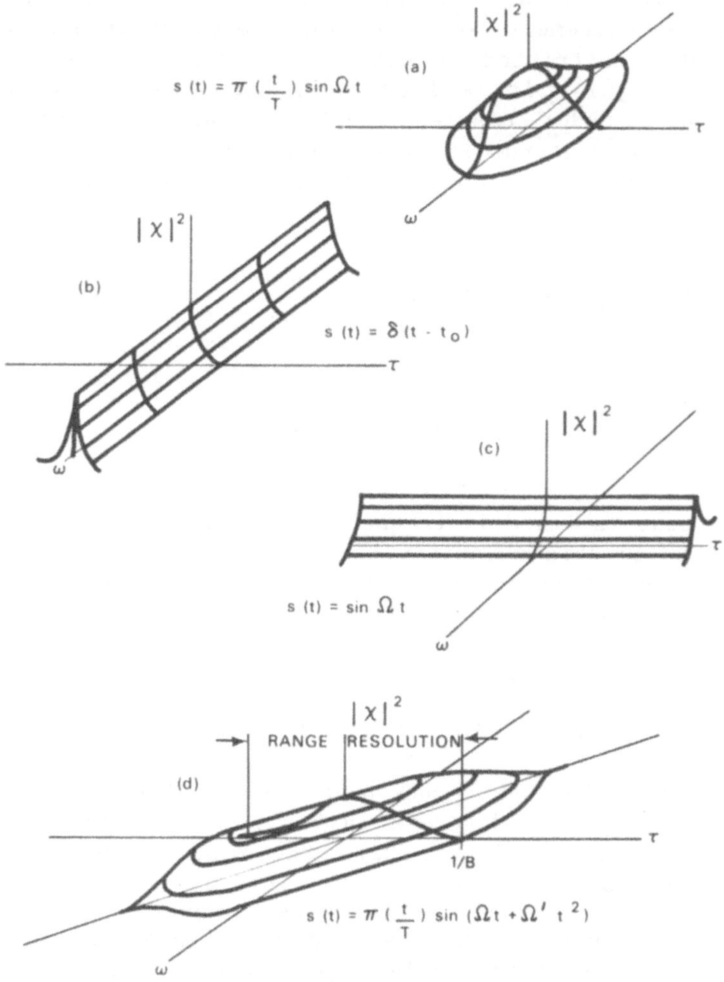

Fig. 1 Schematic drawing of the ambiguity functions associated with several different types of pulses. The 'gate function' is defined as $\pi(t) = e^{-t^2}$ for these examples:

(a) a short sinusoidal transmission of duration T;
(b) a delta function pulse; no frequency resolution;
(c) a continuous sinusoidal transmission; no range resolution;
(d) A 'chirp' transmission, in which the frequency changes linearly with time. The range resolution is inversely proportional to the bandwidth of the transmission, not proportional to its length.

processing required to achieve an optimal estimate of the Doppler spectrum can be extremely complex. The choice of transmitted code s(t) is greatly restricted by practical considerations. As with time series analysis, it is frequently more feasible to estimate parameters of a model Doppler spectrum than to attempt direct estimation of the spectrum itself. In particular, it is common to estimate the first and second moments of the Doppler spectrum directly from the return signal.

$$\hat{\Omega}(x) = \int_{-\infty}^{\infty} \omega S(\omega;x)d\omega / \int_{-\infty}^{\infty} S(\omega;x)d\omega \qquad (15)$$

$$\hat{B}(x) = \int_{-\infty}^{\infty} (\omega - \Omega)^2 \, S(\omega;x)d\omega / \int_{-\infty}^{\infty} S(\omega;x)d\omega \qquad (16)$$

The first moment estimate is frequently interpreted as 'the Doppler shift', and it is this quantity which is contoured in Figure 7 below. The second moment is frequently used as an index of the bandwidth of the Doppler spectrum. A brief comparison of spectral modelling techniques used in atmospheric sounding is presented in Sirmans and Bumgarner (1975).

4. RESOLUTION IN INCOHERENT SCATTERING SITUATIONS

The discussion in Section 3 has been presented from the point of view of scattering from discrete targets. Here, the form of the target echo is deterministic, with the random component of the return resulting solely from 'noise'. This is not necessarily the case when a 'cloud' of distributed targets occupies the scattering volume. Since many texts on radar and sonar theory do not address the incoherent backscatter problem, care must be taken in extending 'well known' results to this situation. For example, consider a pulse-to-pulse incoherent sonar (Section 6.5) operating at large signal-to-noise ratio. Doppler estimates formed from N pulses of a given intensity will be more precise than those formed from N/2 pulses of twice the intensity (Miller and Rochwarger, 1972). For coherent reflections, the Doppler precision is the same either way. Even for the case of zero noise, the return from a single pulse does not provide a precise estimate of the process Doppler spectrum in the same sense that a single realization (time series) of a physical process provides only a two degree-of-freedom estimate of the process spectrum.

It is not surprising that the concepts of the spectral window and ensemble averaging which are central to the spectral analysis of time series have direct analogies in the incoherent backscatter problem. However, the situation can exist where this analogy does not hold and the range-velocity relationship becomes dependent on the motion of the scatterers. Specifically, if the scatterers change their relative positions on the order of one acoustic wavelength while the pulse is passing through the scattering volume, the individual scattered return will be 'de-correlated'. The envelope of $R(\tau;x)$ will fall toward zero in a time, τ_c, shorter than the length of the transmitted pulse, as a result of the changing phase relationships among the echoes from the individual scatterers. τ_c can be called the scattering correlation time. It is a function of the geometry of the scattering volume, the acoustic wavelength, and the motion of the scatterers. Increasing the duration of a transmission beyond the scattering correlation time does not serve to further concentrate the return energy in the frequency domain; range resolution is lost and there is no corresponding increase in velocity precision. On the other hand, if it is desired to monitor the variations in Doppler spectral width it is necessary to use a long transmission. Here the associated ambiguity function is sufficiently narrow in the frequency dimension that the broadening effects of the velocity variability can be easily seen. It should be emphasized that many pulse compression codes, including the 'Barker codes' first used in atmospheric radar sounding, cannot be successfully decoded when reflected from scatterers whose correlation time is short compared to the code length (Grey and Farley, 1973). When this situation occurs, the interpretation of the measurement becomes uncertain.

Given the central role that the scattering correlation time plays, it is unfortunate that so little is known about its actual value as a function of acoustic frequency, system geometry, location in the sea, and time. Acousticians have generally concentrated on measurement of scattering strength alone, and not on correlation time, since it is the scattering strength which determines the level of 'reverberation noise' in sonar detection problems (a brief discussion of scattering strength variability is presented in the appendix). Additional measurements of correlation time as well as scattering strength, at many frequencies, times, and locations in the sea, are necessary before the performance of any given Doppler sonar can be predicted with precision.

5. RESOLUTION IN GEOPHYSICAL MEASUREMENTS

In Section 3, the concept of spatial resolution was quantified in terms of the ability to distinguish discrete targets. This is not necessarily the most relevant definition for many types of geophysical work. As an illustration, consider the situation where a pulsed incoherent sonar is used to sense one component of velocity at many ranges. This 'line array' of velocity measurements will be used to estimate the wavenumber spectrum of the current field (Fig. 2). In this context, achieving 'high spatial resolution' means having the ability to accurately estimate the wavenumber spectrum at large wavenumber. The factors involved are the spatial resolution of the sonar, the resolution (precision) of the velocity estimate, and the form of the wavenumber spectrum of the motions. Frequently in geophysics the spectrum of naturally occurring motions is 'red,' decreasing with increasing wavenumber. The noise level in the velocity estimate can be so large that the spectrum falls into the noise at a wavenumber lower than the spatial resolution limit of the sonar (Fig. 2a). In this situation it is necesssary to reduce the velocity estimate noise rather than increase the spatial resolution in order to estimate the spectrum at higher wavenumber. Increasing the length of the transmission will result in a more precise velocity estimate (if the pulse length is still less than the scattering correlation time), which will provide an accurate estimate of the wavenumber spectrum at higher wavenumber (Fig. 2b). An optimal pulse length can be defined such that the spectrum falls into the noise at exactly that wavenumber at which the return cuts off due to the spatial resolution limit of the sonar (Fig. 2c).

6. EXAMPLES OF DOPPLER CURRENT MEASUREMENT SYSTEMS

It is difficult to impose an overall structure on the many types of Doppler systems that have been developed. The most frequently discussed distinction is between pulse to pulse coherent and incoherent sensors. Systems in which scatterer motion is sensed from pulse to pulse variations in return echo phase are termed 'coherent'. Conceptually, the transmitted signal s(t) is considered to extend to ± infinity in time. It is intermittently zero between transmissions. The ambiguity function associated with this (pseudo) continuous transmission has zero width along the velocity axis. Thus, precise velocity measurements can be made with this type of system. Pulse length or beam geometry determines range resolution. Pulses can be coded to increase average power while maintaining range resolution.

Secondary ambiguous peaks (6.1, 6.2) or non-zero regions are characteristic of the ambiguity function of coherent systems. This is a consequence of the fact that the volume under the ambiguity surface is constrained to be unity, while the volume of the central peak is zero (as it has zero width in the frequency dimension). The key differences between the various types of coherent systems are seen in their associated ambiguity properties.

Efforts to develop coherent radar-sonar techniques have been concentrated in two areas. The first has been to create codes which maintain range resolution while increasing the transmission time, to increase the average power transmitted (Section 6.2). The second direction of development has been to improve overall system range. It was first thought that the maximum range of a coherent

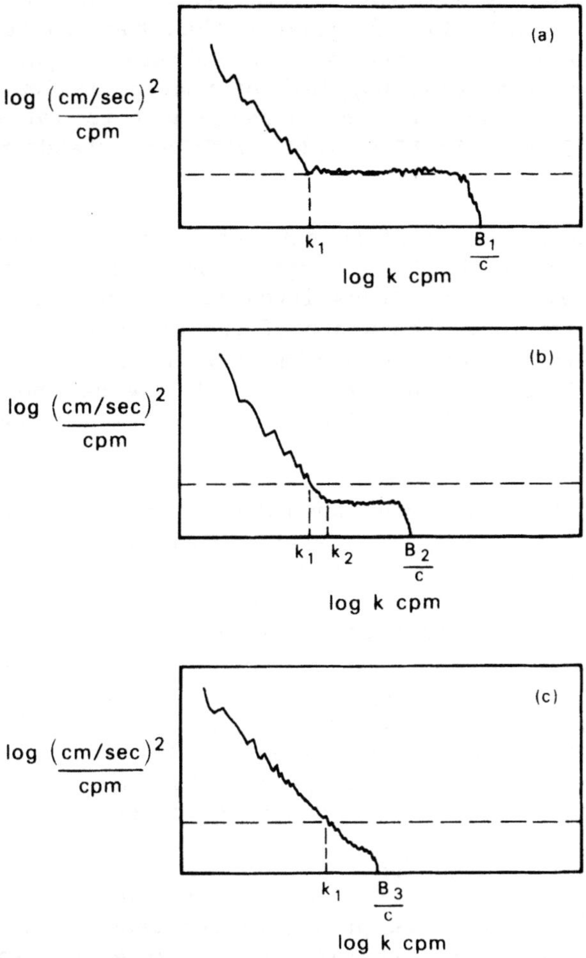

Fig. 2 Schematic wavenumber spectra of horizontal velocity. In this hypothetical example, it is assumed that the data, $U(x)$, are obtained using a narrow beam sonar which transmits a sinusoidal pulse of duration T, bandwidth $B = 1/T$. In case (a), the resolution of the sonar is given by $k = B_1/c$, but the wavenumber spectrum is only resolved out to k_1 before it falls into the spectral noise. In case (b) the pulse is lengthened (for fixed average power), the spectral noise level is reduced, and the spectrum is resolved to a higher wavenumber, k_2. An optimal situation exists when the spectrum falls into the noise exactly at the resolution limit of the sonar, $k_3 = B_3/c$, as shown in case (c). In these examples spatial resolution is increased, in the spectral sense, by increasing the transmitted pulse length (decreasing the bandwidth).

sonar was limited by the round trip travel time of the transmission. If the travel time is greater than the scattering correlation time, the return phase of the successive pulses (at fixed range) will vary randomly; Doppler information is lost. Within the last ten years methods for circumventing this range restriction have been developed (Section 6.3), with an attendant increase in system complexity.

Incoherent sonars have a different set of constraints. Here, motion is sensed from the time progression of phase within a single echo (Section 6.4). The transmitted signal s(t) is considered to be of finite duration. The width of the central peak of the ambiguity function in the frequency dimension is inversely proportional to the pulse length. The central peak now has non-zero volume. Secondary ambiguous peaks need not necessarily be present with this type of sonar.

In an incoherent system, the separate returns are considered to be independent realizations of a statistical process, as outlined in Section 2. The concept of a scatterer spatial distribution, which changes only slowly from pulse to pulse (for coherent sonar), is replaced by the notion of a statistical scattering process which is stationary from pulse to pulse. Here there is no restriction that the travel time be less than the scattering correlation time. On the other hand, because the ambiguity function has finite width in the velocity dimension, either power or statistical averaging have to be increased (at fixed noise level), to produce a velocity estimate as precise as that from a coherent sonar.

This distinction between coherent and incoherent systems is perhaps somewhat misleading. Even in 'incoherent' systems, the return from an individual echo must be processed coherently, to sense Doppler shift. Many systems are hybrids, in the sense that some aspects of the signal are processed coherently, some incoherently. For example, to increase average power while maintaining range resolution, the pulses transmitted by an incoherent sonar can be coded. The individual echoes are decoded coherently, then averaged incoherently with other returns from the same range. The coherent and incoherent approaches conceptually merge when the pulse length exceeds the scattering time. Beyond this point the return echo appears to consist of 'fragments' of the transmitted signal of duration τ_c which are available for coherent processing.

Given the central role that the scattering correlation time plays, it is convenient to use it as a basis for classifying the various types of Doppler systems that have been developed. For the purpose of classification, a system can be considered 'long range', if the round trip travel time to the range of interest is long compared to τ_c. Similarly, a 'long pulse' system transmits continuously for periods longer than τ_c. The converse is true for short range and short pulse systems.

Due to the lack of a priori knowledge of probable correlation times in most scattering situations, it is usually necessary to design Doppler systems by assuming that the transmission time and propagation time are either very long or very short compared to τ_c. Thus there is a tendency for existing instruments to fall clearly within one of these categories (see table). One example of each of the classes of instruments is presented below, starting with the 'short range' systems. A limited amount of data is presented to illustrate the capabilities and limitations of the instruments.

6.1 Short Range-Short Pulse Systems: The Pulsed Coherent Radar

Doppler sensors of this type transmit a succession of short pulses. The motion of the scatterers at a given range is determined by the pulse-to-pulse variation in return signal phase at fixed time lags

Table 1

Classification of Doppler Sonar Types

Propagation time / Correlation time	Transmission time / Correlation time $\geqslant 1$	Transmission time / Correlation time $\leqslant 1$
$\leqslant 1$	Atmospheric 'Pulse to Pulse Coherent Radars' Doppler frequency is determined by pulse to pulse variation in the return phase at fixed range. Range is determined by time after transmission.	Coded Continuous Transmission Multi Range Sonar Doppler frequency is determined by bit to bit return phase variation at fixed range. Range is selected by correlating the continuous return signal with time shifted replicas of the transmitted code.
$\geqslant 1$	Pulse to Pulse Incoherent Sonar Doppler frequency is determined by phase variations within a single return. Range is determined by time after transmission.	Continuous Transmission Current Meter Doppler frequency is determined by phase variation in continuous return. There is no range discrimination in the code; the sampling volume is determined by transducer orientation.

after transmission. The ambiguity function of these 'pulse-to-pulse coherent' systems has zero width in the Doppler dimension. The range resolution is proportional to the length of the transmitted pulse. The difficulty with the method is that there is an infinite number of peaks in the ambiguity function. At the same time the return from a given pulse is being received, the echoes from previous pulses, scattered back from greater ranges, are also being received. This results in range ambiguities. This problem is not significant if the attenuation of the pulses is sufficiently great that weak scattering from the most recent transmission produces a larger return signal than potentially stronger scattering at greater ranges from previous pulses. Ambiguous Doppler velocity estimates will result if the scatterers can change their range by an amount comparable to the wavelength of the transmitted signal during the interval between successive pulses. For example, if a scatterer moves exactly one wavelength during the interval between pulses, no pulse-to-pulse difference will be observed in the return, in spite of the fact that the scatterer is moving.

The interplay between pulse repetition rate, ambiguous range interval, and ambiguous velocity interval can be stated in a more fundamental form. If T_0 is the time between successive pulses, then the ambiguous ranges are separated by

$$\Delta R = cT_0/2 \qquad (17)$$

Ambiguous velocities are spaced by

$$\Delta V = \frac{\lambda}{2T_0} = \frac{c}{2T_0 f_0} \qquad (18)$$

where f_0 is the frequency of the radiation.

The product of the ambiguous range and velocity intervals is

$$\Delta V \Delta R = \frac{c^2}{4f_0} \qquad (19)$$

This result, independent of pulse repetition frequency, represents a basic performance limit for this type of system. It is possible to extend the limit somewhat by cleverly varying the pulse rate T_0, but this does not remove the problem.

Pulse-to-pulse coherent sonars have not yet been developed for oceanic use. Pulsed coherent radars have been used in atmospheric studies with great effectiveness. Because electromagnetic radiation travels at the speed of light, not sound, the ambiguity limits expressed in Equation 19 are not as restrictive. Microsecond pulses can be used, transmitted every millisecond. The first range ambiguity is 300 km removed from the range of interest (frequently outside the earth's atmosphere, depending on beam inclination). Similarly, ambiguous Doppler velocities correspond to unlikely physical situations.

6.2 Short Range-Long Pulse Systems: The Coded Continuous Transmission Sonar

A difficulty with the pulsed coherent system is that in order to achieve high range resolution, short pulses are required. This results in a very low duty cycle. High peak power is necessary to maintain an adequate signal-to-noise ratio. This problem can be avoided by using long coded pulses and determining the range by decoding the return appropriately. Although no research sonar of this type is in oceanographic use, a prototype has been built at the Scripps Marine Physical Laboratory to test the concept. This sonar uses a frequency stepped code. Five tones are transmitted sequentially, starting at 70 kHz and increasing by 104 Hz every 10 ms. The code is repeated continuously; i.e., the transmitter is always operating. The received signal is multiplied by time delayed in-phase and quadrature replicas of the transmitted code. This has the effect of frequency shifting the return from ~70 kHz to ±400 Hz. The component near zero hertz always comes from a fixed 7.5 m cell whose range depends on the time lag chosen. A ±50 Hz low-pass filter admits only the signal from the desired range cell.

In Figure 3, the results of a test of this sonar are presented. The test was conducted from a barge in Lake San Vicente, a city water reservoir east of San Diego. Spectra of backscatter intensity versus Doppler frequency and time are presented for a range cell centered 35 m from the barge. Note that the spectra are characteristically 1 to 2 Hz in width, which suggests that the scatterer correlation time is of order 0.5 to 1 s (Eq. 5). The signal-to-noise ratio is such that it is possible to estimate the Doppler peak to 1 Hz precision using only 5 s of data. At the operating frequency, a 1 Hz Doppler shift corresponds to approximately 1 cm s^{-1} of relative velocity.

The zig-zag nature of the velocity signal with time resulted from the barge being moved back and forth through the water. In this way, the accuracy of the sonar could be checked. The range discrimination of the system was verified by turning the transmitter on and off repeatedly. The signal was detected by the receiver only when the code was passing through the specified range cell.

Because the code repeats every 50 ms, it has associated frequency ambiguities every 20 Hz. Range ambiguities occur every code repetition length (which is about 37 m). The range-velocity ambiguity product is the same as that of the pulsed coherent sonar (Eq. 19).

In addition, there is a combined range-Doppler ambiguity. If a scatterer is changing range so as to cause a 104 Hz Doppler shift, the return signal will appear to come from the adjacent range cell, with 0 Hz Doppler shift.

The effect of the Doppler ambiguities can be seen in Figure 4. Here, the data in Figure 3 are duplicated, but the full frequency range passed by the receiver filter is included. In many upper ocean situations it would be difficult to guess which of the peaks, separated by 20 cm s^{-1}, corresponds to the true scatterer velocity. Nevertheless, the ability of this type of sonar to make remote measurements of water velocity very rapidly and accurately is impressive. In situations where precise measurement of small scale

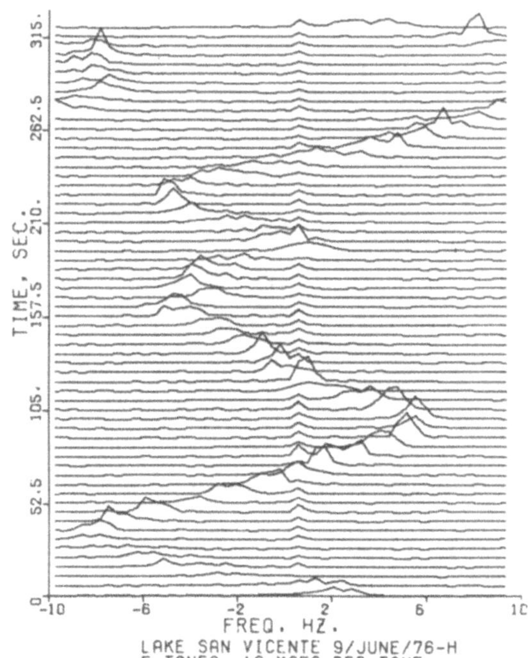

Fig. 3 Doppler backscatter spectral density is plotted in arbitrary units against Doppler frequency for a coded CW sonar. Each line is a spectral estimate based on 5 s of data. Successive estimates are vertically offset, to record velocity variations over a 5 min period. The range cell being monitored is 35 m from the transducers. Range resolution is 7.5 m. The small peak at 0.5 Hz is due to digital noise in the system.

velocity difference is important, such as in boundary layer or shear flow studies, this type of sonar could be used effectively.

6.3 Long Range-Long Pulse Systems: The Continuous Transmission Flowmeter

The fundamental ambiguity limit on coherent sonars, stated in Equation 19, results from the repetitive nature of the transmitted code. Each pulse or sequence of tones is identical to the previous pulse or sequence. If a scatterer moves one wavelength between repetitions, there is no way to distinguish that the motion has occurred. This problem can be avoided by either transmitting continuously or by using codes which never repeat, for example, by transmitting Gaussian noise. In the latter approach, random noise is

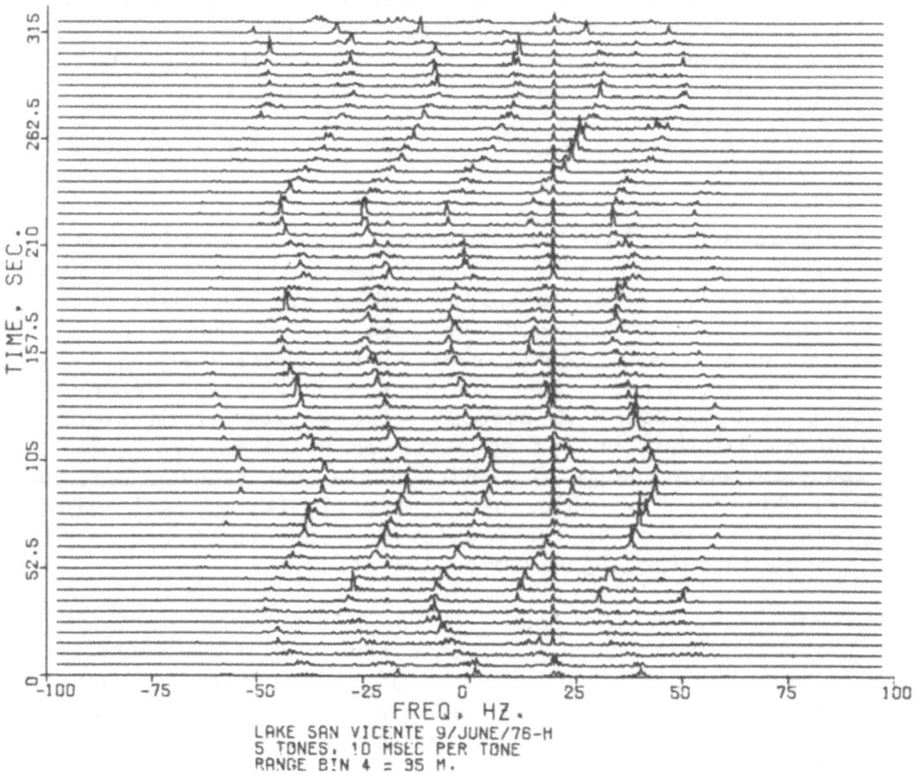

Fig. 4 Doppler spectra versus frequency and time from a coded CW sonar. These are the same data displayed in Figure 3, but the entire frequency band which is passed by the receiver filter has been plotted, revealing Doppler ambiguities.

generated in some broad frequency band. The noise sample is both transmitted (either pulsed or continuously) and stored in a delay line. The received echo is correlated with a replica of the transmitted waveform, delayed in time by an amount appropriate to the range of interest. The spectrum of the output signal is the Doppler spectrum at the range of interest. Range resolution is proportional to the bandwidth of the noise transmitted.

With a sonar of this type, many pulses can be in the water at the same time. The scattering volume is insonified (filled with sound) at a rate which is fast compared to the scattering correlation time, in spite of the fact that the travel time of an individual pulse is much greater than the correlation time. This can be done because each pulse is individually distinguishable from the others. Only the pulses passing through the range of interest will be detected in the correlation process. Also, there is little theoretical difference between a pulsed or continuously transmitting random signal sonar. The distinction has mainly to do with average power considerations, not ambiguity properties. This situation is analogous to the pulsed versus continuously transmitting coherent sonars, which have similar ambiguity constraints, even though transmitting quite differently. An example of a random signal sonar, developed for blood flow measurements, is described in Bendick and Newhouse (1974).

Random signal sonars have no ambiguous range or Doppler peaks; the ambiguity function of Gaussian noise has only a single peak (at zero lag in both range and Doppler). However, the value of the ambiguity function does not drop to zero away from the peak, since the volume under the ambiguity surface is constrained to be unity. The form of the ambiguity function is somewhat like a 'thumb tack', with small but significant skirts. Thus, there is a significant contribution to the Doppler spectral estimate at any given range and frequency from energy at other ranges and frequencies, as expressed in Equation 7. This contribution is termed 'self-clutter'. It represents an error in the Doppler estimate which occurs even in the limit of infinite signal-to-noise ratio.

The decorrelation of the echo due to scatterer motion introduces further error. The received signal can no longer be represented as a sum of many time delayed replicas of the transmitted code. The echo from the theoretical range of interest will not be perfectly detected by the correlator. In addition to Doppler broadening, range resolution will be degraded.

In some siutations self-clutter can be decreased by reducing the volume insonified by the sonar. This can be accomplished by using a spatially separated transmitter and receiver, so that the return signal comes from the volume common to the two transducer beams. The output of the correlator represents motion within the common volume, in a range cell of length inversely proportional to the transmitted beamwidth. However, the clutter contribution now comes only from the slightly larger volume common to the crossed beams.

At this point, it becomes attractive to consider abandoning codes altogether and transmitting a single frequency through narrow-beam transducers, with the crossed beams completely defining the scattering volume. While this is a 'long range' technique as defined in this chapter, most sonars of this type have been developed for local flow sensing. Most use continuous transmission (CW), with a separate sound source and receiver for each component of flow to be measured. The precision of the velocity estimate is very high; the central peak of the ambiguity function is infinitely narrow in the frequency dimension. It is also, however, infinitely wide in the time dimension, implying no range resolution. The volume of water sampled is determined by the transmitter and receiver beam pattern and the geometry of the instrument (Fig. 5).

Without a coded signal to assist in range discrimination, it is difficult to eliminate crosstalk, both electrical and acoustic, between the transmitter and receiver. Given the weak level of scattering which takes place in the scattering volume, very small side lobes of the transmitter beam can leak significant acoustic energy into side lobes of the receiver. The ratio of crosstalk noise to true signal could be acceptable in one environment, but unacceptable when the natural scattering strength drops. In

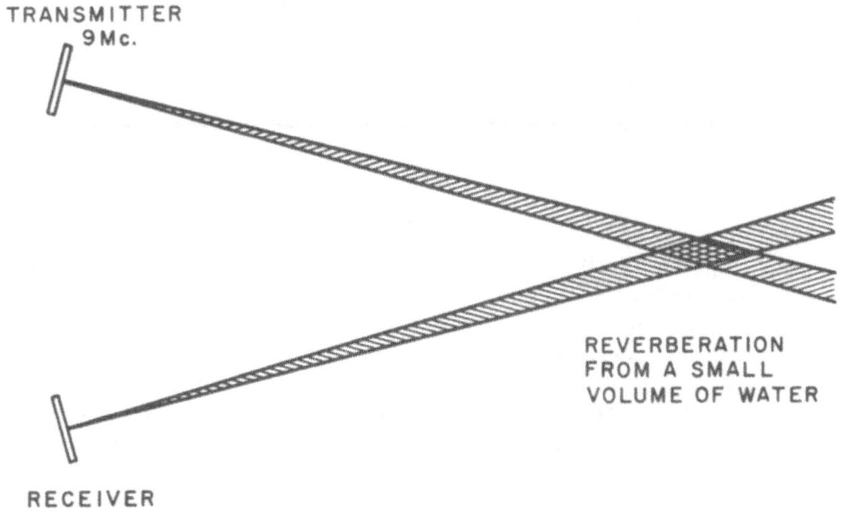

Fig. 5 Schematic configuration of CW Doppler current meter (from Squier, 1968).

marginal operating circumstances it is necessary to record data at a sufficient rate that an attempt can be made to separate the crosstalk (zero hertz Doppler) from the actual signal using spectral analysis techniques (Fig. 6). To the extent that crosstalk problems cannot be overcome, increasing the sonar power does not increase the signal-to-noise ratio. Also, at very high power levels, high frequency (megahertz) sonar transducers themselves can generate a local 'streaming' flow in the water (Squier, 1968). If this flow is large enough to affect the motion in the scattering volume, the measurement is suspect.

6.4 Long Range-Short Pulse Systems: The Pulsed Incoherent Sonar

As an alternative to crossed beam systems, Doppler measurements can be made at great range using pulse-to-pulse incoherent methods.

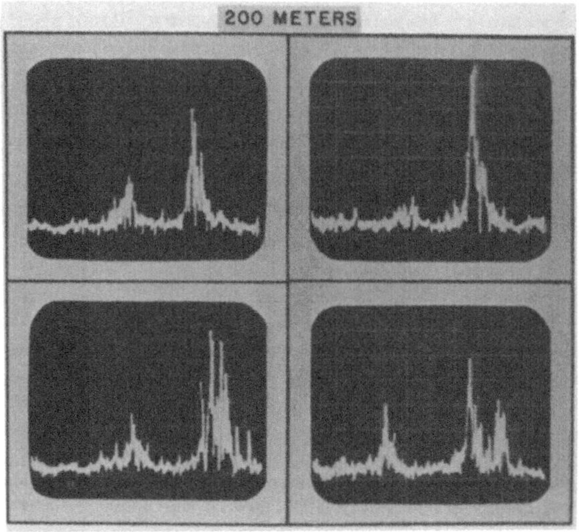

Fig. 6 Four representative reverberation spectra from the Squier CW Doppler current meter (from Squier, 1968). Backscatter spectral density is plotted in arbitrary units against Doppler frequency. Each division of the abcissa corresponds to 100 Hz, or 0.87 cm s^{-1}. Zero velocity is given approximately by the position of the crosstalk peak in the upper left spectrum. The fluctuation in size and frequency of the crosstalk peak suggests that it is acoustic rather than electrical in origin, and is caused by scatterers in the side lobes of the transducer beam patterns. Note the splitting of the Doppler peak in the lower spectra.

Simple or coded pulses can be used. Scatterer velocity is sensed by the phase variations within the return from each pulse. Maximum range is limited only by signal-to-noise limitations. However, the width of the ambiguity peak in the frequency dimension is now finite and inversely proportional to pulse length. More power is needed to achieve the Doppler precision of the previous systems for the same noise level.

Acoustic sounders of this type have proved effective in atmospheric research. They could prove useful in a large class of oceanographic problems. Velocity measurements can be made at many ranges, without recourse to separate moorings or arrays of discrete sensors. The spatial separation between measurements is precisely known, and the acoustic 'array' is not distorted by the currents it is trying to measure. Since the measurements are remote, there is no deformation of the flow by the sensor itself. There is no difficulty with spatial aliasing. Measurements are obtained continuously with range; spatial resolution is limited by the length of the acoustic pulse, not by the spacing between individual sensors.

On the other hand, a number of difficulties must be surmounted before long range measurements can be made. The sonar must be held very steady to prevent the outer reaches of the beam from gyrating excessively. Side lobes which point toward the sea surface must be suppressed, since volume reverberation caused by plankton is much weaker than scattering from the sea surface. Very small amounts of acoustic energy incident on the surface can scatter back and dominate the reverberation signal. As with conventional current meters, it is the relative velocity between the instrument and the water which is sensed. An absolute velocity measurement requires knowledge of the motion of the sensor. The sonar does, however, provide precise measurements of the strain rate $\partial u/\partial r$, where u is the component of velocity parallel to the beam and r is the range, even when the motion is not known. (Rapid time variations in sensor motion can produce spurious range variations in the velocity estimate. The magnitude of this effect is usually reduced when averaging the estimates from successive pulses.)

A possible concern with long range Doppler velocity measurements is that a time change of sound velocity along the acoustic path will introduce a frequency shift in the received signal which is not related to the relative speed of the scatterers. If, due to time changes of c, the 'leading edge' of the transmitted pulse travels with a different 'mean velocity' than the 'trailing edge', the pulse will be gradually distorted, and its frequency altered. This can be a significant effect in the thermocline, where depth variations in sound velocity are great. Vertical motions, perhaps associated with internal waves, can advect the sound velocity field through the propagation path, changing the mean sound velocity. The possible magnitude of the effect can be estimated from

$$\phi_j = k \cdot 2r_j = \frac{\omega}{c} \cdot 2r_j \tag{20}$$

$$\frac{\partial \phi_j}{\partial t} = 2\omega \frac{1}{c} \left(v_j - \frac{r_j}{c} \frac{\partial c}{\partial t} \right)$$

If we take $\partial c/\partial t = -w\, \partial c/\partial z$, where w is the vertical velocity, it becomes clear that the radial velocity at the scattering point has the same influence in the determination of the frequency shift as the mean vertical velocity along the path when

$$v_j \simeq \frac{r_j}{c} \frac{\partial c}{\partial z} w. \tag{21}$$

For typical thermocline values of $\partial c/\partial z = 0.1\ s^{-1}$ and $c = 1500$ m s^{-1}, a given change in mean vertical velocity will be just as effective in producing a 'Doppler shift' as the same change in scatterer radial velocity for ranges 15 km or greater. Since it is doubtful that vertical motion is coherent on a 15 km scale in the thermocline, this effect is probably not significant.

A final consideration is the refraction of the beam due to depth variations in the sound velocity profile. A beam transmitted horizontally in a sound velocity gradient $\partial c/\partial z = 0.1\ s^{-1}$ will be refracted by several degrees after ~750 m of travel. Also, the time variations in the sound velocity profile will cause distortions in the acoustic 'array', although the magnitude of this problem is small compared to that of a conventional moored array. Note that the magnitude of the Doppler velocity measurement is unaffected by refraction; the position of the beam and the direction of the velocity component sensed are altered.

An example of incoherent sonar data taken in the open ocean is presented in Figure 7. The measurements were made from the Research Platform FLIP in October 1978, 300 km west of San Diego. The sonar was mounted on FLIP's hull at a depth of 85 m, with the beam directed 45° down from horizontal. The width of the sonar beam was 2°. Thirty millisecond pulses were transmitted at 75 kHz. The peak transmitter power was 4 kW. Velocity was estimated from the sonar echo using an algorithm developed by Rummler (1968). The signal-to-noise ratio was such that a root mean square velocity precision of ~1 cm s^{-1} was achieved at ranges up to 400 m, after 30 s of pulse-to-pulse averaging.

The range-time variation in velocity is presented as a deformed surface, with positive vertical displacements of the surface corresponding to toward-FLIP velocity. The dominant features seen in the 11-hr record are internal gravity waves. From longer records of this sort much can be learned about the generation, propagation, and decay of these waves. Further oceanographic examples of incoherent sonar data are presented in Pinkel (1979).

7. FUTURE DEVELOPMENTS

The pulsed incoherent sonar is a particularly promising tool for upper ocean studies, because of its remote ranging capabilities. Ship-mounted sonars of this type are now available commercially. In several of these devices pairs of downward slanting sonars are used, in the so-called 'Janus' configuration (Fig. 8a). The difference in Doppler shift between the forward and aft slanting

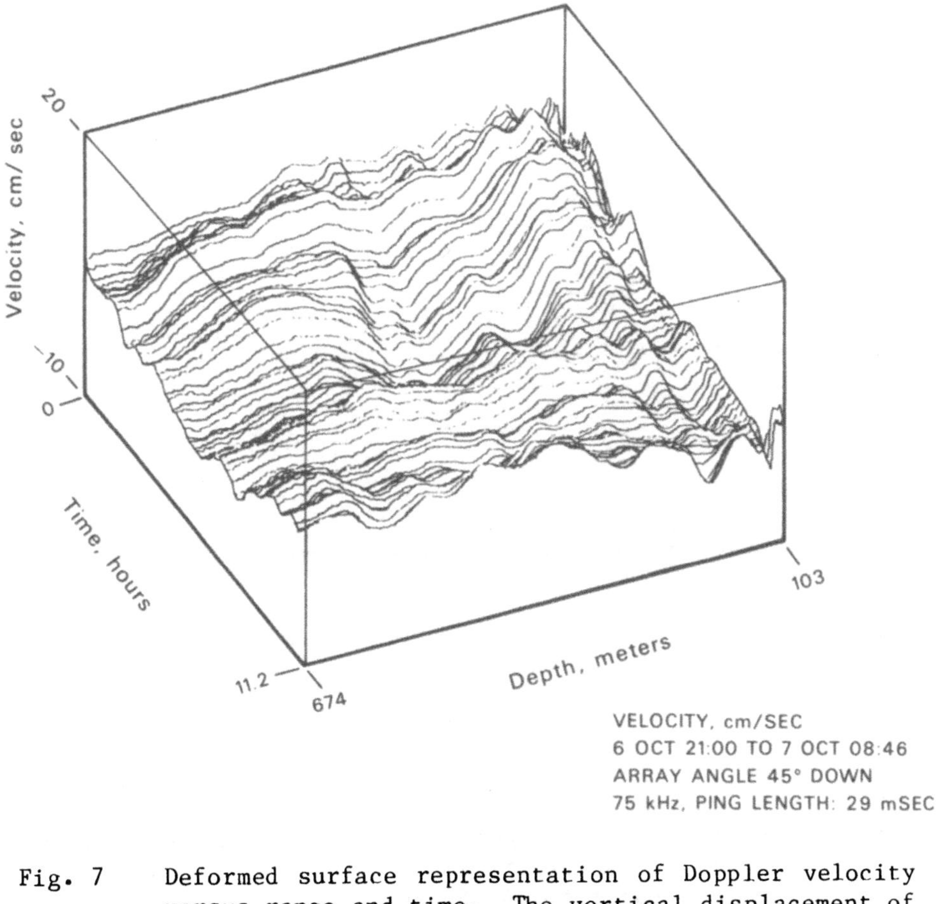

Fig. 7 Deformed surface representation of Doppler velocity versus range and time. The vertical displacement of the surface is proportional to the velocity perturbation parallel to the sonar beam. The dominant motions are due to internal waves.

beams gives an estimate of the average horizontal velocity of the water under the ship. The sum of the Doppler shifts is an estimate of the average vertical velocity. Errors are introduced if there is significant variation in the horizontal or vertical velocity over the distance between the two beams. The importance of these errors, and the overall performance of the available shipboard systems, have yet to be established. One alternative to the Janus

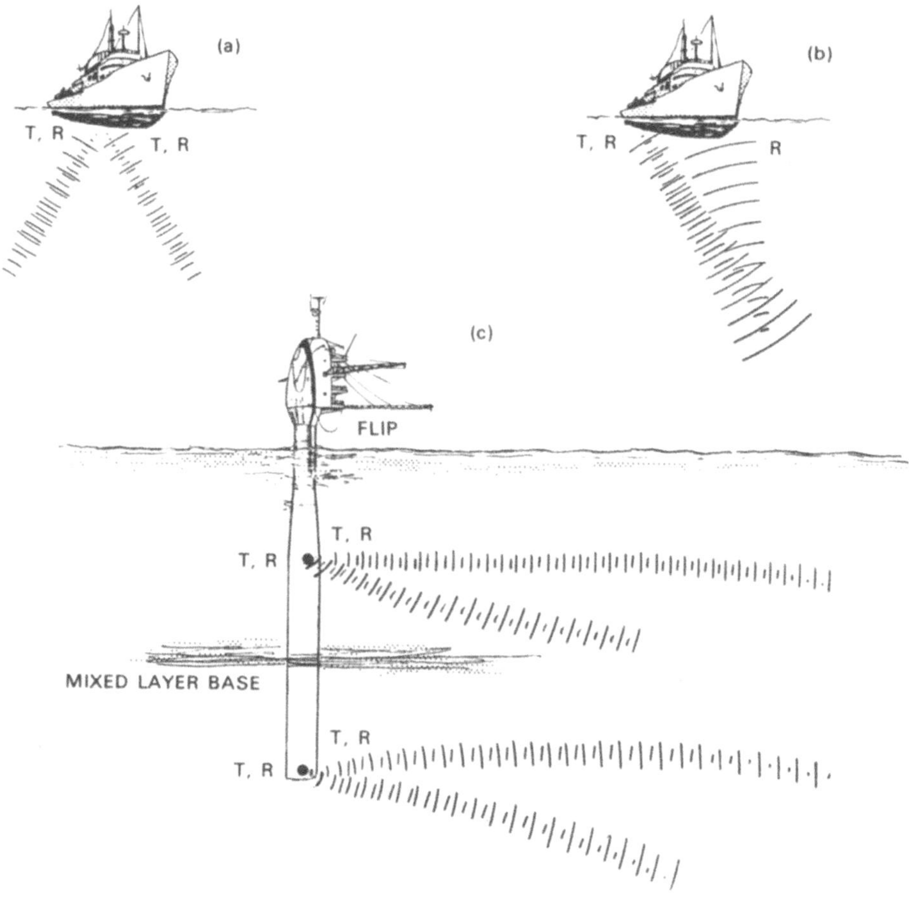

Fig. 8 Schematic configurations of possible pulsed incoherent sonar systems:

a. A dual beam JANUS-type sonar;
b. A bistatic sonar, with one transmitter and two receivers;
c. A multibeam FLIP-mounted system, for mixed layer and upper thermocline studies.

geometry is a 'bi-static' configuration, in which a downward slanted beam is transmitted from one end of the ship and received both at the transmitter location and at a second receiver at the other end of the ship (Fig. 8b). From the difference in Doppler shift between the two received signals, both horizontal and vertical relative velocity components can be estimated. The errors associated with this geometry resulting from ship motion have not been investigated.

The possibility exists of using moored Doppler instruments as an alternative to mechanical current meters for upper ocean studies. The relatively high power required by the sonar as compared to that needed by a mechanical instrument makes it difficult to design a device which will run unattended for long periods of time. However, the remote ranging ability of the sonar allows measurements in strongly oscillating flows near the sea surface where the wake from the body of conventional instruments interferes with the measurement.

This paper has concentrated solely on sonars which scatter from particles suspended in the water. Analogous principles apply to scattering from the sea surface. Generally, surface reverberation levels are much stronger than volume reverberation (Urick, 1975), and surface scattering sonar systems should not be particularly difficult to design. The difficulty is in interpreting the surface scattering data. Scattering can occur from the particles suspended in the water moving at nearly the water velocity, from bubbles (which do not necessarily track the water velocity), and from the underside of surface waves, resulting in a Doppler shift proportional to the phase velocity of the waves. Carefully designed experiments will be necessary in order that these effects can be separated.

Additional pulsed incoherent sonars are currently being constructed for use from FLIP. With several sonars operating simultaneously in different directions, a multidimensional array of velocity measurements will be obtained (Fig. 8c). The use of more portable spar buoys wold allow sonar deployment from conventional ships or helicopters, if azimuthal orientation could be controlled.

8. CONCLUSIONS

The use of Doppler sonar to map the upper ocean velocity field appears highly promising. Major questions regarding the ultimate capability of the sonar approach can only be answered with more operational experience. In particular, the worldwide variation in scattering strength, as well as the temporal variation at a given depth must be better understood. The species of organisms responsible for the scattering should be determined. Fish and other swimming scatterers surely contaminate the results some fraction of the time. Is their scattering signature distinct enough that this effect can be removed? The variability of the correlation time of incoherently backscattered acoustic codes must be better known. It is possible to create an acoustic code which is effective in one environment, yet degraded in another; this would introduce possible nonlinear behaviour in an otherwise linear approach to current measurement.

Fortunately, useful information can be obtained with present Doppler sonars, prior to fully answering all of these questions. The initial FLIP-mounted system will be used to monitor the shear across the mixed layer thermocline transition, and to measure horizontal strain rate near the sea surface. With the experience gained from these first applications, a great deal more can be determined about the future role of Doppler sonar measurements in air-sea interaction studies.

9. ACKNOWLEDGEMENTS

Dr. F.N. Spiess first suggested that a Doppler sonar might best provide the large 'array' of velocity information needed in air-sea interaction investigations, without the array. Discussions with Drs. V.C. Anderson, C.S. Cox, R.E. Davis, F.H. Fisher, W.H. Munk, and F.N. Spiess were most helpful. Dr. F.H. Fisher kindly loaned us his 87.5 kHz sonar system for initial incoherent sonar tests. Instrumentation was designed and operated by L.M. Occhiello, L.S. Tomooka, and E. Slater. This work was initially sponsored by the Advanced Research Projects Agency. Subsequent support for development and sea tests was provided by the Office of Naval Research, Code 480 and ONR Code 500.

APPENDIX

Biological Scattering and Sonar Design

The design of a sonar system involves balancing several factors. It is desirable to use higher frequencies, which scatter from small scale features, produce larger Doppler shifts for a given relative velocity, and require smaller arrays for good beam patterns. Acoustic attenuation, however, increases rapidly with frequency (Urick, 1975) and it is necessary to work at lower frequencies if great range capability is desired. The strength of a received sonar signal depends on the strength of the transmission, the scattering strength at the given angle of scatter, the geometry of the situation with respect to the beam patterns of the transmitter and receiver, and the inverse square spreading loss and attenuation.

The level of scattering strength is quantified by

$$S_s = 10 \log \frac{I_{scattered}}{I_{incident}}$$

where $I_{incident}$ is the intensity (power/area) of a plane wave incident on a cubic yard of water, and $I_{scattered}$ is the scattered intensity at a fixed reference distance from the scatterers (Urick, 1975). For typical acoustic frequencies, 25 to 100 kHz, S_s varies from about -60 dB to -100 dB for volume backscattering. Scattering strengths are even less below the top 500 m of the sea, where most marine life is concentrated. The low values of scattering strength pose great problems in the design of Doppler sonar systems. Very sensitive low-noise receivers must be used to record the returning sound. For a backscattering situation, the return signal intensity will start out at 10^{-6} - 10^{-9} of the transmitted level, at zero range, and decrease from there. Almost any hard target, such as a ship's hull, the cable lowering the instrument, or the sea surface, will have a greater 'target strength' than the volume reverberation. A related problem results from the distributed nature of the targets. If it is assumed that the scatterers are distributed uniformly, relatively few scatterers are found within a narrow beam as compared to the many that lie outside the small solid angle of the beam. Unless extreme care is taken to reduce side-lobe levels in the transmitter and receiver, the majority of the return at a given range will come from the myriad of scatterers outside the beam, rather than the few which are in the desired scattering volume. The requirement for small side-lobes in narrow beam scattering sonars is far more stringent than in search sonars, which have to detect whether a single hard target is in the beam or not, at a given range. The problem is particularly difficult if the scattering strength is greater outside of the desired scattering volume, due to instrument supports, sea surface, etc.

Progress is being made in determining the spatial variability in scattering strength. In general, the large-scale horizontal variability follows global patterns in biological productivity; scattering is greater in upwelling regions than in the centre of mid-ocean gyres, for example. The smaller scale horizontal variability, associated with the so-called 'patchiness' of planktonic communities has yet to be measured adequately. In the vertical, the tendency of scatterers to occur in distinct layers has been known since World War II. Some of the layering is thought to be associated with regions of high density gradient: the plankton are simply 'floating' on the density layers. The diurnal vertical migration of some zooplanktonic species, as well as many types of smaller swimmers, has been extensively studied. The migrators spend the night in the surface layers and the daylight hours at greater depth (200-400 m). Examples of diurnal variability in scattering strength profiles can be found in Urick (1975, p. 231).

The variation of scattering strength with frequency appears to be surprisingly small. Urick (1975) reports little variability above 20 kHz in his summary of the observations. Below 20 kHz, there does not appear to be any consistent relationship; the dependency varies from one measurement to the next. Castile (personal communication) has done simultaneous vertical scattering profiles at four frequencies, from 155 to 819 kHz in the coastal waters off California. He finds a slight tendency for scattering strength to increase with frequency, although there is great variability in the trend. The increase is generally less than 10 dB; 3 to 4 dB variation is more usual. Variations from one cruise to the next at the same location and frequency are generally greater than the cross frequency variation at the same time.

A more complete review of biological sound scattering can be found in Farquhar (1977). The volume 'Oceanic Sound Scattering Prediction', in which his paper occurs, represents an attempt to bring together most of the factors important in the understanding of scattering strength variation in the sea. Considerably more information is necessary, particularly with regard to the scattering correlation time, before Doppler sonar performance can be accurately predicted.

REFERENCES

DAVENPORT, W.B. and W.L. ROOT. 1958. *An Introduction to the Theory of Random Signals and Noise*. McGraw-Hill, New York, 393 pp.

FARQUHAR, G.B. 1975. Biological sound scattering in the oceans: A review. In: *Oceanic Sound Scattering Prediction*, edited by N.R. Andersen and B.J. Zahuranec, Plenum Press, New York, 859 pp.

GREY, R.W. and D.T. FARLEY. 1973. Theory of incoherent scatter measurements using compressed pulses. *Radio Science*, 8: 123-131.

HARVARD UNIVERSITY Acoustics Research Laboratory, Division of Engineering and Applied Physics. 1945. Acoustic Marine Speedometer Completion Report 6-1, ser. 287-2074. Harvard University, Cambridge, Mass.

PINKEL, R. 1979. Observations of strongly nonlinear internal motion in the open sea using a range gated Doppler sonar. *Journal of Physical Oceanography*, 9 (in press).

RUMMLER, W.D. 1968. Introduction of a new estimator for velocity spectral parameters. Technical Memo MM-68-4141-5, Bell Telephone Laboratories, Whippany, New Jersey.

RIHACZEK, A.W. 1969. *Principles of High Resolution Radar*. McGraw-Hill, New York, 498 pp.

SERAFIN, R.J. 1975. Information extraction from meteorological radars. *Atmospheric Technology*, 6: 46-65.

SIRMANS, D. and B. BUMGARNER. 1975. Numerical comparison of five mean frequency estimators. *Journal of Applied Meteorology*, 14 (6): 991-1003.

SQUIER, E.D. 1968. A Doppler shift flowmeter. Prepared for the ISA Marine Sciences Instrumentation Symposium, Florida, 16-19 January 1968. Report MPL-U-48/67, Marine Physical Laboratory, San Diego, California, 92152, U.S.A. 5 pp.

URICK, R.J. 1975. *Principles of Underwater Sound*. McGraw-Hill, New York, 391 pp.

WOODWARD, P.M. 1953. *Probability and Information Theory with Application to Radar*. Pergamon Press, New York. 136 pp.

11

Drifters

W.A. Vachon

1. INTRODUCTION

A Lagrangian drifter is a platform whose position is periodically determined as it drifts with the currents. Many platforms make other measurements such as temperature or barometric pressure. A Lagrangian drifter of any design is never perfectly locked to a particular water mass because its large size must react to forces from all water particles around it - particles which are constantly mixing and separating. Therefore, a Lagrangian drifter is only a quasi-Lagrangian sensor.

In its simplest form the position of a Lagrangian drifter is determined periodically such that trajectory (or streamline) data is derived. Drifters derive their name from the Lagrangian specification of a flow field which results from the platform specifying the position of a particular mass of water at a particular time. This method is opposed to the Eulerian specification of a flow field in which the flow quantities are defined as a function of position and time by the installation of sensors on fixed platforms.

There are three main reasons for the appeal of Lagrangian drifters:

(1) low cost

(2) expendability (reduces ship costs for retrieval if position determined remotely)

(3) horizontal patterns of flow readily visualized

Most recent applications have stressed the expendable nature of Lagrangian drifters for the acquisition of long-term trajectory data. These buoys have been deemed expendable because of their low life-cycle costs (or cost per data point) when remotely positioned by satellite, radar, or radio. The low cost per data point results from the minimal ship time (required only for deployments). In short-term applications nonexpendable drifters are often positioned by attending ships or aircraft. For specific applications, the latter approach has been found to be cost effective.

Lagrangian drifters are possibly the oldest method of measuring water currents - whether it be in bays, rivers, or in the deep ocean. A comprehensive history of their uses and application is summarized in Monahan et al. (1974). It appears that the first applications of drifters were to chart prevailing currents as an aid to maritime shipping as early as the sixteenth century. Even in the twentieth century there has been cost-effective use of drift cards and bottles for gross measurements of currents (Bumpus, 1965; Monahan et al., 1975). In the last 30 years, however, Lagrangian drifting buoys have seen their greatest use in the oceanographic studies of currents in coastal waters and deep oceans. The first practical uses of remotely-positioned Lagrangian buoys for the acquisition of oceanographic data was that of Stommel (1954) using a system developed at the Woods Hole Oceanographic Institution. Later studies by Cromwell et al. (1954), Volkmann et al. (1956), and Parker (1972) derived very useful oceanographic data from drifters. The study by Stommel was the first to employ a remotely positioned radio buoy.

Present technology divides Lagrangian drifters into two main categories. The surface-trackable Lagrangian drifting buoys are familiar to most people. There is, however, another class of Lagrangian sensors - neutrally buoyant floats - whose position is determined in most cases by acoustic means (see Swallow, 1955; Pochapsky, 1963; or Rossby and Webb, 1970). In the past, the latter type of float has not in general been employed for studies of near-surface phenomena (0 to 100 m). As a result, the technology will not be described in detail here, but a few key references to this work are given. One noteworthy case of the use of a neutrally buoyant float in near-surface studies is that of Bradley and Wells at M.I.T. (see Wells, 1973). A shallow water neutral density float was designed whose depth was servo-controlled and whose position was determined by variations in travel time of acoustic signals picked up by two receivers on an attending ship. This type of float has great potential for air-sea interaction studies and may see broader development and application in the near future. Many of the recent developments involving servo-controlled profiling instruments, such as these reported by Van Leer in this text (see Chapter 36), will have applicability to servo-controlled, near-surface neutral density Lagrangian drifters. The remainder of this chapter will be devoted to a description of the present technology of surface-trackable Lagrangian drifting buoys.

2. TYPES OF DRIFTING BUOYS

A large number of surface-trackable Lagrangian drifting buoy designs have emerged in recent years. Very few have seen widespread use over a prolonged period of time, however. The main feature which seems to classify such buoys is their cost and the manner by which they are positioned. The less expensive buoys are generally positioned by an attending ship which might use visual sighting or radar range and bearing of the buoy while determining ship position by other means such as Loran-C. Such buoys may cost as little as \$100 each if the drogue is a parachute and can be procured from government surplus. The more costly buoys, on the other hand, generally employ Buoy Transmit Terminals (BTTs) which are remotely positioned by a UHF signal to a satellite. Their position is then automatically computed at least twice per day. The satellites that recently became available for this purpose are the polar-orbiting NIMBUS-6 (1975 to present), the TIROS-N (late 1978 launch), and subsequent NOAA satellites. The plan is to keep two of the latter type of satellite in orbit until 1985. The single NIMBUS satellite gave a position, accurate to approximately ±2 km (rms), at least twice per day while the two TIROS-N or NOAA satellites give positions of comparable accuracy at twice this rate. The GOES satellite (Geostationary Operational Environmental Satellite) can also be employed for data transmission but not position. All such systems can accept data words from other onboard sensors [such as those for temperature (air, water), barometric pressure, and drogue tension to confirm the presence of the drogue]. With GOES satellites geostationary over the equator at 75°, 100°, and 135°W longitude, affording coverage over the Atlantic and most of the Pacific, it is possible to send data words whenever desired. It is also possible to send more words per transmission to the GOES satellite than to the NIMBUS, TIROS, or NOAA satellites. The present cost for a NIMBUS BTT is approximately \$1500, while that for a buoy-mounted GOES package is approximately \$3500. The average buoy power requirements of each are comparable although the peak for the GOES package is much higher. The total cost of a buoy, drogue, batteries, and BTT positioning system may be approximately \$4100 each (1978 dollars) with no additional onboard sensors. The price may reach approximately \$5000 with barometric pressure and temperature sensors aboard.

Other remote positioning methods have been employed including direct positioning by High Frequency radio direction finding and radio retransmission of LORAN-C or OMEGA signals received on the buoy. A study by Vachon and Yoerger (1977) indicates that more accurate position data, at higher frequencies than presently afforded by satellite systems, can be economically derived by retransmitting

LORAN-C signals. A program under way at the Woods Hole Oceanographic Institution will verify the feasibility through prototype construction and testing.

An HF direction-finding system described by Whelan et al. (1975) has claimed $\pm 0.5°$ accuracy using two shore stations and an onboard HF transponder. Inability to utilize the HF band at night and the personnel costs of attended shore stations were found to limit system utility, and so commercial production of the units was suspended in 1976. The concept, however, remains viable.

3. BUOY CONFIGURATIONS

The standard type of hull employed in deep ocean drifting buoy applications is an upright spar with the characteristics of a surface-following buoy. This feature is designed into the buoy by either a floatation collar at the water line or by ensuring that the spar buoy draft is small compared to the ocean wave length (i.e. $H/\lambda \ll (1/2\pi)$, H = draft, λ = wavelength). The equation describing the undamped heave response of a circular cylinder to sinusoidal wave inputs of frequency ω is given by (see Newman, 1963):

$$Z = A \sin \omega t \left(\frac{e^{-\frac{2\pi H}{\lambda}}}{1 - \frac{2\pi H}{\lambda}}\right) = A \sin \omega t \left(\frac{e^{-\frac{\omega^2 H}{\lambda}}}{1 - \frac{\omega^2 H}{g}}\right) \quad (1)$$

where A = wave amplitude, and ω = radian frequency. The deep-water dispersion relation ($\omega^2 = 2\pi g/\lambda$) is employed in order to derive the second form of Equation 1. For this undamped case, the heave will go to infinity if the wave frequency corresponds to the heave-resonant frequency of a uniform circular cylinder (i.e., $[g/H]^{\frac{1}{2}}$). Equation 1 does not adequately represent the more complex heave response of cylindrical spar buoys whose diameter is not uniform at the waterline (see Fig. 2).

In order for a drogue to exhibit a known droguing efficiency or Lagrangian performance (see Section 4) it is important that it not undergo large excursions forced by buoy motion and imparted through the tether line. Such motions may also give rise to loads which may limit the life of both buoy and drogue. Therefore, the use of a spar buoy, which is insensitive to passing waves, looks appealing. Equation 1 indicates, however, that in order for a spar to be wave insensitive, the buoy draft, H, must be greater than approximately one-sixth of the ocean wavelength. This criterion indicates, for example, that for modest seas with 100 m wavelengths a buoy draft of approximately 15 to 20 m is required. Such a buoy is costly to build, cumbersome to handle, and, because of its stability, may result in excessive antenna submergence in heavy seas.

Therefore, as a compromise, shallow-draft spar buoys, lumped floatation buoys, or buoys with non-uniform cross-section are often built.

The effective mass of a spar buoy can be increased by the addition of damping plates on the submerged portion of the buoy (e.g. Fig. 1). These plates will tend to entrain the buoy with surface wave motion, which may not be desirable for shallow draft buoys if vertical motions are to be minimized. For the case of deep draft spars (i.e., $H/\lambda > 1/2\pi$), damping plates on the bottom will tend to stabilize buoy motion.

The main desire with remotely positioned buoys is to keep the antenna (satellite or otherwise) from submerging too often while remaining close to a vertical orientation. The two main types of drifting drogued buoy hulls employed by investigators today are shown in Figures 1 and 2. Figure 1 is a drawing of the Scripps spar buoy (Kirwan et al., 1979) which has had considerable deployment experience in the rugged environment of the North Pacific from 1975 through 1978. Fabricated completely of fibreglass and sealed, this 3-m long buoy exhibits 107 kg of reserve buoyancy. It presently contains a BTT, an auxiliary sensor for water temperature, and a drogue indicator switch. The damping plates shown on the lower portion of the buoy will reduce the heave resonant frequency while minimizing heave excitation at the resonant frequency. This buoy, in particular, has demonstrated excellent ruggedness while

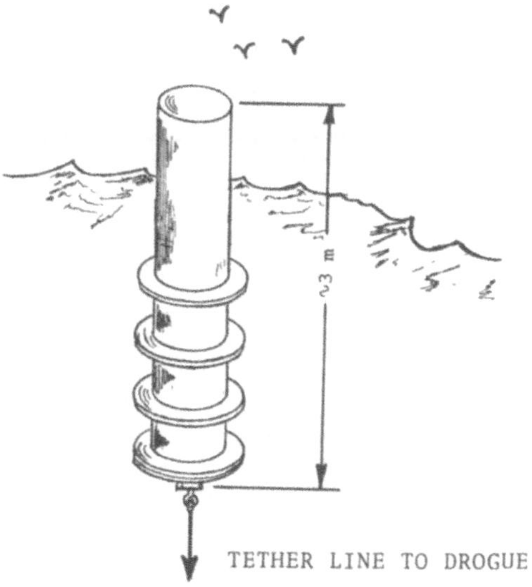

Fig. 1 Scripps drifting spar buoy.

presenting a relatively low profile to wind forces, which are a serious source of drogue slippage error.

Figure 2 depicts the type of hull that has been fabricated for the NOAA Data Buoy Office (NDBO) in recent years. It has been adapted from a design whose aim was to withstand moderately severe environments (Hall and Kerut, 1975). The present design exhibits approximately 90 kg of reserve buoyancy and employs a welded aluminum

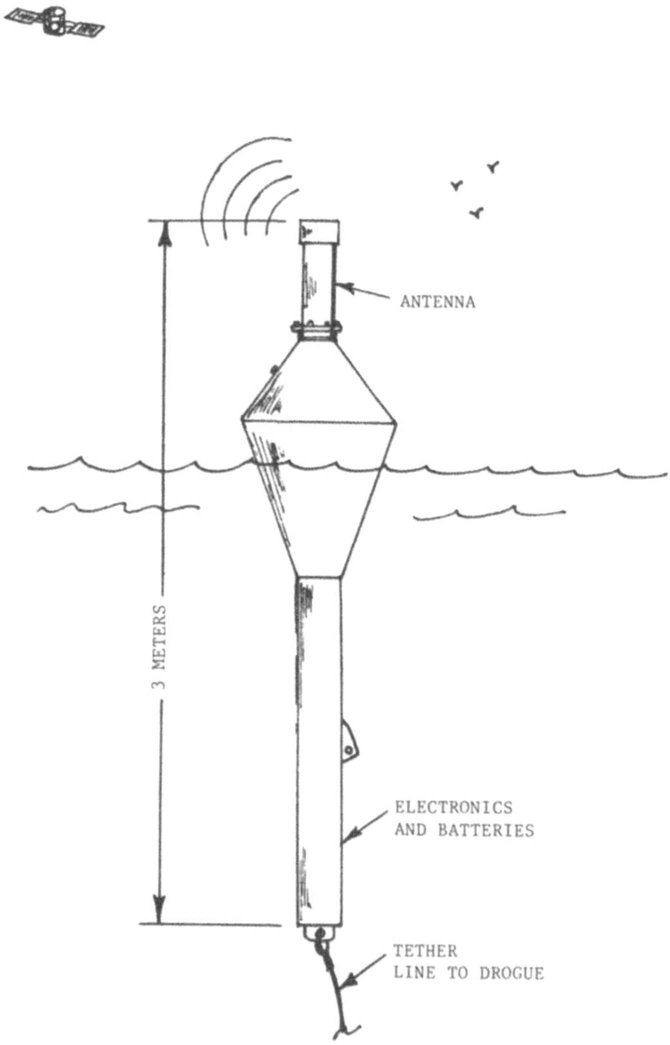

Fig. 2 NOAA Data Buoy Office (NDBO) drifting buoy hull.

hull, where formerly a PVC and fiberglass combination were used with limited success. The former design was of inadequate strength and ruggedness to survive for extended periods in the ocean with a drogue attached. The design shown in Figure 2 has been employed in numerous applications such as ring-tracking experiments (Richardson et al., 1977) and monitoring the flow of major currents. At the time of writing the buoy has had limited experience in the harsh environments of the southern oceans and more extensive experience in the North Pacific where Dr. Donald Hansen of the NOAA Atlantic Oceanographic and Meteorological Lab (AOML) has made numerous drogued buoy deployments. Because the Scripps buoy is heavier than the NDBO version it displaces more water and, as a result, has potentially a higher drag area to surface currents. On the other hand, the Scripps buoy has a drag area to wind forces comparable to that of the NDBO buoy. Both considerations are important sources of drogue slippage error.

A third and radically different type of drifting drogue buoy design is that employed by John Garrett of Environment Canada (Patricia Bay, Victoria, B.C.). The design, although not reported in the literature, has many desirable features. It includes a plywood, boat-shaped hull on the surface which houses batteries, sensors, and the BTT. A window shade drogue is tethered at some distance below the surface to one end of the hull via a line containing distributed buoyancy. The distributed buoyancy is derived by a series of clamp-on floats but it could also be a buoyant tether line. With such an arrangement it is hoped that the hull will 'weathervane' into the wind and present a low wind drag profile; the hull draft is so small that it should also present a low drag profile to surface currents. In addition, it is hoped that the distributed buoyancy between the buoy and drogue will minimize drogue dynamic loads and motion: the loads can limit survival time while drogue motion can possibly lead to serious drogue slippage errors. There are potential problems associated with this design, too. First, it is more difficult to package. Second, because hydrodynamic drag normal to the tether causes the water to act much like a pulley, horizontal buoy motion (i.e. surge) is converted to vertical motion and forces at the drogue. Last, there can be considerable dynamic and shock loading at the tether attachment point on the buoy.

4. DROGUES

A second major classification of Lagrangian surface-trackable drifters is the manner of droguing or locking to the current. Most systems employ a large drogue in order to maximize the ratio of drag area (i.e., $C_D A$) at measurement depth to that in the wind and surface current zone. There are various choices of drogues available, and with each there are serious questions, still unresolved, regarding drogue slippage and its impact on both drogue and overall system survivability. The choice of the proper drogue and its method of attachment is, in most cases, of key importance in addressing these questions. At the present time there is inadequate knowledge of the in situ performance of the various types of drogues, both from a current-locking or slippage and also from a dynamic-loading point of view. The former consideration has serious implications for the value of the derived data in oceanographic studies while the latter can seriously affect the long-term survival of both drogue and buoy. Both considerations are inherently linked, in that a drogue with little dynamic response to buoy or wave dynamics should generally last longer and behave in a more predictable manner than another which exhibits a large response to buoy motions.

At present the overall droguing efficiency of Lagrangian drifters is unknown and as a result the data derived from such systems cannot be used in oceanographic studies requiring a high degree of accuracy. It has long been hoped that the slippage-inducing forces, arising primarily from buoy forces and transmitted by tether line, can be adequately estimated so that the drogue slippage can be correspondingly estimated in the manner described by Kirwan et al. (1975). If it is estimated that the buoy and tether line give rise to a net drag force, F_D, a balancing of the forces in the system indicates that the drogue must supply an equal and opposite force. The simplest way in which to model this drogue force is through the hydrodynamic vector force relationship given by:

$$F_D = \tfrac{1}{2}\,\rho_w C_D A(\vec{V}_C - \vec{V}_B)\,|\vec{V}_C - \vec{V}_B| \qquad (2)$$

where $\vec{V}_B$ is the buoy (and drogue) vector velocity over the ground, $\vec{V}_C$ is the vector velocity of the true current at the drogue depth, ρ_w is water density, and $C_D A$ is the drogue drag coefficient times its lateral area. The velocity difference, $\vec{V}_C - \vec{V}_B$, is the velocity of the drogue with respect to the surrounding water; it is often called drogue slippage. If the drogue slippage can be measured or otherwise estimated in a vector sense, then the true current vector, $\vec{V}_C$, can be estimated by vectorially adding it to the measured buoy velocity: $\vec{V}_C = \vec{V}_B + (\vec{V}_C - \vec{V}_B)$. In this way it is (theoretically) possible to correct drifter trajectories in a piecewise linear manner between position data points. In reality complete trajectories are not correctable because slippage errors can move a drifter into a new body of water, whose flow may be wholly unlike that in which it was originally drogued. For this reason the corrected current vectors, $\vec{V}_C$, cannot honestly be attached end-to-end in the manner of a progressive vector diagram. Rather, the resultant measurement is a series of independent current vectors emanating from the position of the start of a data interval and laterally displaced from the previous one by an amount proportional to the drogue slippage.

The difficult parameters to obtain in the slippage equation are F_D and C_D. F_D is the vector sum of wind, wave, and current forces on the buoy and tether. Estimates of buoy wind drag can be made in the manner discussed in Kirwan et al. (1978) using steady flow drag coefficients. In many studies wind forces have been found to be the largest part of the error-inducing drag force, F_D (see Vachon, 1975). The wave-induced forces may include those due to Stokes drift velocities in the near-surface layer and nonlinear buoy-wave coupling forces and/or phasing which may lead to a rectification of the wave orbital motion by the buoy. The latter phenomenon can give rise to an additional net spar buoy wave force in the direction of wave travel for the case of $2\pi H/\lambda < 1$ (case 1) or opposite to wave travel for $2\pi H/\lambda > 1$ (case 2). The forces can result from the buoy being more immersed during the forward portion of wave orbital motion (case 1) [or less (case 2)] than the steady state draft would predict (see Equation 1). As a result of these considerations, the dynamic environment experienced by the buoy tends to make the steady flow drag coefficient data reported for standard shapes inapplicable for estimating buoy forces. Estimates of the average wave forces on buoys have been made by Vachon and Yoerger (1977) using the work of Sarpkaya (1976). Similar estimates can be derived in a manner described in Nath (1977b) where a computer program averages the instantaneous hydrodynamic and inertial forces on a buoy. This and other mathematical models of buoy forces and responses are presently unverified. Thus, at present, wave forces on buoys are not well understood.

Surface current forces can be estimated using steady drag coefficients and an estimate of the true vector current derived from independent data or in the manner of Wu (1975). Wu's study recommends a surface current of approximately 3.5% of, and colinear with wind velocity. Kirwan et al. (1978) have data to indicate that the mean surface current runs approximately 30° to the right of the wind direction. With these approaches, a first order estimate of slippage error forces can be obtained.

5. DROGUE PERFORMANCE ESTIMATES

In the same manner as buoys, the drag coefficient of drogues, C_D, is known only for steady flow conditions. It is a formidable design challenge to make the environment at the drogue sufficiently quiet to allow the use of steady-flow drag coefficients in drogue slippage estimates. A summary of steady-flow drag coefficients for most of the commonly used drogues is found in Vachon (1973). The drogue scale-model towing-test results described in this report were, in most cases, conducted under steady relative flow conditions with limited attempts to measure dynamic effects. The table is a summary of some of the drogue designs tested and their measured steady-flow drag coefficients derived for relative velocities from 0 to 10 cm s^{-1}.

It was found that parachutes, without spreader bars and added buoyancy at the canopy, hang downward unopened until acted on by a relative velocity which depends on the drogue wetted weight. With a parachute hanging down in such a fashion, there is an increased chance of both shroud line fouling and high dynamic loading in response to buoy heaving motion. It was felt, however, that with spreader bars and canopy buoyancy a parachute could potentially give satisfactory steady flow drag performance and minimize buoy-drogue dynamic interactions. This feature is theoretically achievable since the parachute is coupled to the buoy through a two-arm linkage consisting of the tether line and the parachute shroud lines which join at nearly a right angle at the ballast weight. Moderate seas should only result in the rise and fall of the ballast weight and rotations of the parachute canopy (the same

Table 1

Summary of Drogue Drag Coefficients Measured During Scale Model Tests
(Reynolds No. $\simeq 10^4$ to 10^5, Model Scale Factor $\simeq 1/16$)

Drogue Description	Average Drag Coefficient, C_D
Parachute	1.35
Conical muslin sea anchor*	1.53
Muslin sea bucket	1.54
Crossed vane, 2 solid walls	1.2**
Crossed vane, 3 solid walls	1.18**
Crossed vane, 3 plastic walls	1.4**
Buoyed fishing net, streaming parallel to flow	0.2^+
Cylinder	1.1
Window shade drogue	1.93

* Downstream spill hole area = 5% of inlet area.
** Based on full area of one wall.
\+ Based on full area of net.

comments apply to sea anchors and sea buckets). These features are yet to be verified.

Based on the numbers presented in the table, as well as on their simplicity and apparent behavior, window shade drogues have been used extensively over the past five years. In order to assess drogue slippage, a dimensionless plot (Fig. 3) of the estimated slip velocity of a window shade drogue as a function of horizontal force and ballast weight was presented in Vachon (1974). This curve, which incorporates the effects of area and ballast weight, is based on an analytical expression which seems to agree well with measured steady-flow drag data for forces up to approximately 75% of that of the ballast weight. If properly sized and ballasted a window shade drogue should not experience horizontal drag forces

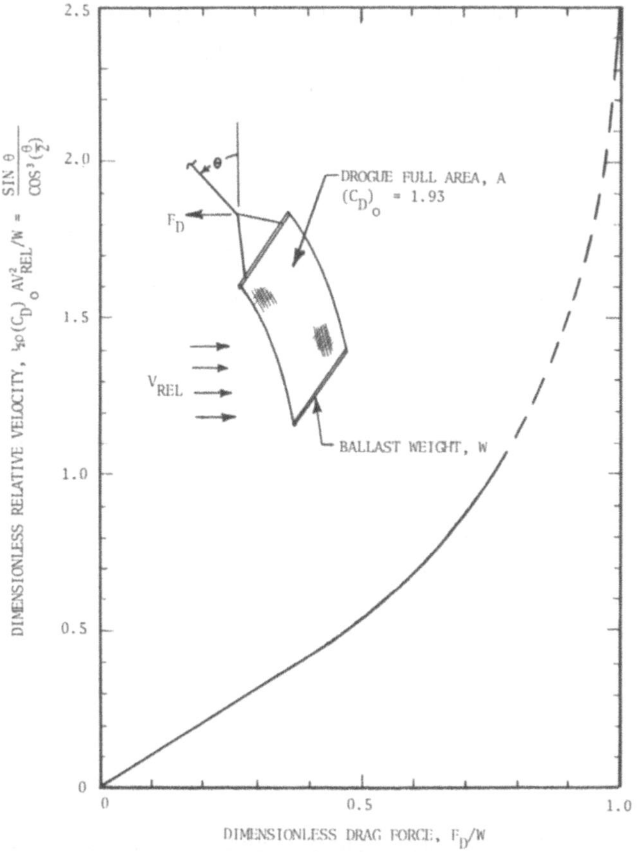

Fig. 3 Window shade drogue estimated performance curve (after Vachon, 1974).

which exceed one-half the ballast weight (i.e., linear region of curve) in order to keep the drogue from streaming upward in the flow.

More recent scale model wind tunnel tests of window shade drogues (Nath, 1977a) indicate that a steady drag coefficient of approximately 1.6 is applicable. If such were the case the slip indicated by the ordinate in Figure 3 would be approximately 9% higher. Subsequent laboratory tests and analytic model developments (Nath, 1977b, 1977c) have produced a readily available computer program that predicts drifting buoy motion and loads in waves, but does not adequately represent combined buoy and drogue motion and slippage. A key result of this work, however, is that there is considerable 'jerking' motion in the system and that extensible tethers are recommended for longer life.

Full-scale, nondynamic, drogue towing-test results, described in Vachon (1975), show that drag coefficient of a window shade drogue can be as high as 2.6. It is pointed out, however, that in the ocean environment this value may be too high due to buoy-drogue dynamics and non-weathervaning of the drogue. Further evidence of excessive drogue slippage is presented in Saunders (1976). He

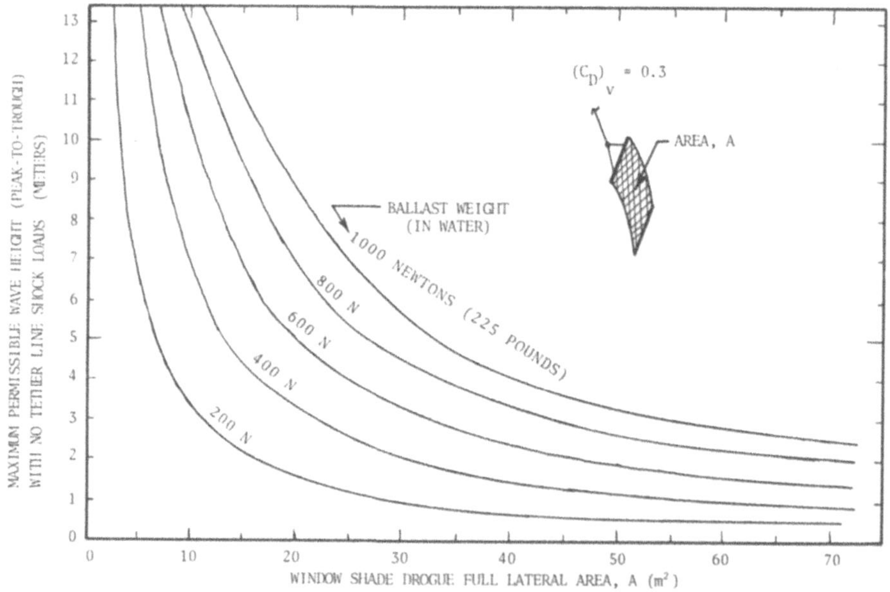

Fig. 4 Estimated window shade drogue dynamic shock load conditions based on full scale measurements of the drogue vertical drag coefficient, $(C_D)_V$ (after Vachon, 1975).

attributes the excess slippage to 'wave forces' of an undefined nature. In summary, it appears that if an estimate for the total error-inducing drag force, F_D, can be obtained, the curve shown in Figure 3 is still a good representation of window shade drogue slippage until more hard data are available.

6. BUOY-DROGUE DYNAMICS

As discussed earlier, a major limitation on the prolonged use of a Lagrangian buoy and window shade drogue in severe environments results from its sensitivity to tether line dynamics. The problem arises in the standard drogued buoy configuration, in which the drogue is suspended directly beneath a surface-following buoy. The vertical drag and inertia of such a drogue exhibits a relatively large downward force on the buoy as the buoy rises on the crest of a wave. On the subsequent buoy descent into a wave trough the tether line may go slack, because the drogue may not descend as rapidly as the buoy, leading to the possibility of severe shock loading when the buoy rises on the next wave crest. The combination of wave heights, drogue ballast weights, and drogue areas which may give rise to this condition (Fig. 4) are taken from Vachon (1975). The curves are based on full-scale measurements of

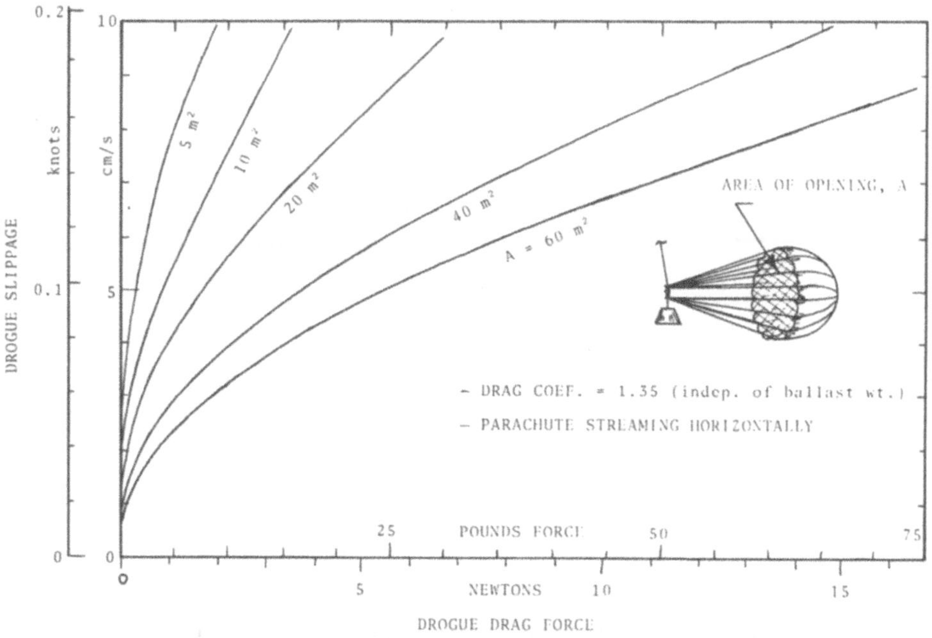

Fig. 5 Parachute drogue performance curves (after Vachon, 1973).

the vertical drag coefficient of a window shade drogue, $(C_D)_V$, and mathematical estimates of the mechanism taking place using sinusoidal wave inputs with amplitudes corresponding to the Pierson-Moskowitz spectrum for fully developed wind-driven seas. The curves are of practical value in determining the size of window shade drogue ballast weights as a function of lateral area for increased system survival. For a more detailed anaysis of this problem, a computer simulation, such as that used by Nath (1977b), should be useful.

At present there seem to be the four major drogue candidates for Lagrangian drifter applications: window shade, parachute, crossed vane, and a 'holey sock' or porous vertical axis fabric cylinder (Nath, 1977a). The estimated slippage performance of the window shade drogue is shown in Figure 3. Equivalent steady-flow slippage curves for both the parachute and bi-planar crossed vane are shown in Figures 5 and 6 respectively, which use the drag coefficients shown in the table. The measured drag coefficient of a scale model 'holey sock' was found to be approximately 1.6; this value compares favourably with values given in the table. The benefit of this design is that is presents the same vertical projected area to relative flow, regardless of flow direction. The merits of a cylinder

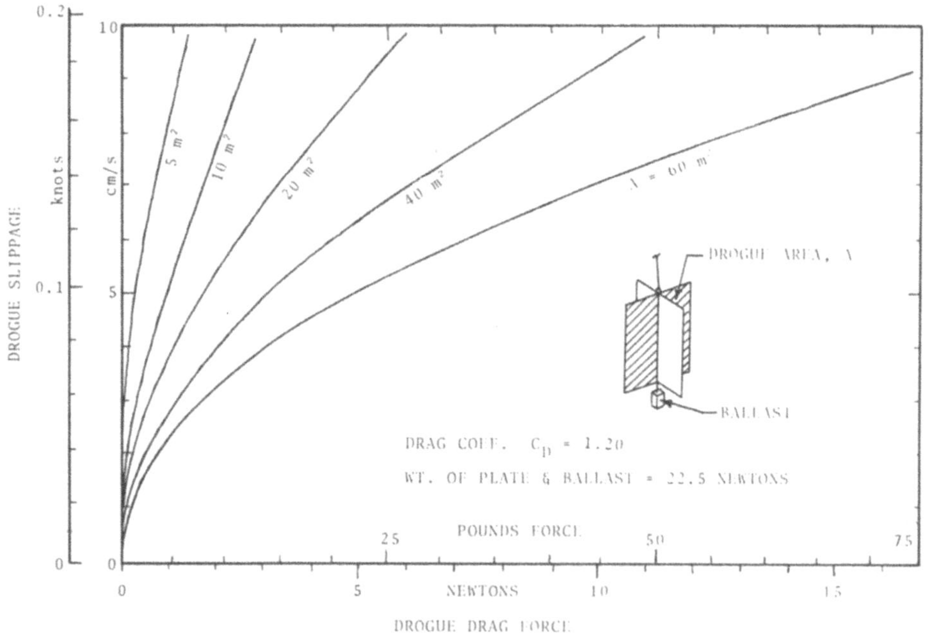

Fig. 6 Bi-planar crossed vane performance curves (after Vachon, 1973).

versus a bi- or tri-planar crossed vane, which exhibit similar directional properties, are yet to be fully established. It does appear that such drogues may be better adapted to near-surface droguing than a window shade or parachute because their drag area (i.e., $C_D A$) is less contaminated by the complexities of wave action and the resulting variations of drogue orientation.

7. DROGUE DEPTH STABILITY

The depth stability of drogues can most easily be estimated by neglecting tether line drag (good for near-surface drogues) and balancing drogue drag and ballast weight moments about the buoy. The straight-line tether assumption greatly simplifies the geometry. The results of a more detailed analysis, involving the lift of a window shade drogue, are given in Vachon (1974). The analysis can be easily modified for a crossed vane, which also exhibits lift.

8. SUMMARY AND CONCLUSIONS

As a result of recent advances in electronic systems for remote positioning and data collection, drifting buoys can be a cost-effective tool for the acquisition of oceanographic data in all regions of the earth. Estimates of drogue slippage errors are, however, unverified at present. Therefore, the measurement of near-surface currents by drogued buoy trajectories is still relatively crude. Recent analytical studies are providing a basis for quantifying these errors. The real deficiency at present is the lack of instrumented, full-scale engineering test data for verifying the estimated errors for different designs and conditions.

While lacking a complete description of the errors and their sources, it is still possible to recommend design approaches. As a result of early scale model and full scale tests by the author and others, the following general conclusions have been reached:

(1) Parachutes and other similar shapes, which depend on a relative velocity for deployment, can be poor drogue choices because they perform most poorly at low relative velocities. Lagrangian buoys, on the other hand, should aim for a zero relative velocity - a poor match between goals and observed performance. If such designs are used as drogues they should be held open by rings or spreaders in order to enhance their deployment at low relative velocities and should be made as neutrally buoyant as possible or even exhibit a slight positive buoyancy in order to minimize buoy-drogue dynamic interactions.

(2) Simple shapes, such as the window shade drogue or crossed vane, are not nearly as susceptible to fouling as are shapes like the parachute.

(3) Under steady-flow conditions a weighted window shade or sail drogue, when tethered in line with the centre of pressure, will align itself perpendicular to the prevailing flow due to its hydrodynamic characteristics. Under dynamic conditions induced by buoy and wave interactions, a window shade may not align itself perpendicular to the net relative velocity. Therefore, where a conservative design or more reliable error estimates are needed, a crossed vane or 'holey sock' drogue are recommended - both of which provide nearly uniform C_DA product regardless of direction.

(4) Elasticity or compliance should be installed in the system, primarily to extend the life of the drogue in severe environments.

(5) Where possible, component failures should be retrieved and analyzed so that design improvements can be made in subsequent deployments.

REFERENCES

BUMPUS, D.F. 1965. Bottled oceanography. *Oceanus*, 11: 20-23.

CROMWELL, T.R., R.B. MONTGOMERY and E.D. STROUP. 1954. Equatorial undercurrent in the Pacific revealed by new methods. *Science*, 119: 648-649.

HALL, J.M. and E.G. KERUT. 1975. Development of a meteorological and oceanographic drifting buoy system. Oceans 75; Record of the Combined Meeting of 1975 IEEE Conference on Engineering in the Ocean Environment and 11th Annual Meeting of the Marine Technology Society, San Diego, Calif., 1975: 56-69.

KIRWAN, A.D., JR., G. McNALLY, M.S. CHANG and R. MOLINARI. 1975. The effect of wind and surface currents on drifters. *Journal of Physical Oceanography*, 5: 361-368.

KIRWAN, A.D., JR., G. McNALLY and S. PAZAN. 1978. Wind drag and relative separations of undrogued drifters. *Journal of Physical Oceanography*, 8: 1146-1150.

KIRWAN, A.D., JR., G. McNALLY, S. PAZAN and R. WERT. 1979. Analysis of surface current response to wind. *Journal of Physical Oceanography*, 9: 401-412.

MONAHAN, E.C., P.C. HAWKINS and E.A. MONAHAN. 1974. Surface current drifters: evolution and application. Michigan Sea Grant Program Report MICHU-SG-74-603, Department of Atmospheric and Oceanic Science, 4072 E. Engineering Bldg., University of Michigan, Ann Arbor, Michigan 48104.

MONAHAN, E.C., B.J. HIGGINS and G.T. KAYE. 1975. A comparison of vertical drift - envelopes to conventional drift bottles. *Limnology and Oceanography*, 20: 141-147.

NATH, J.H. 1977a. Wind tunnel tests on drogues. Final Report to NOAA Data Buoy Office on Contract No. 03-6-038-128, NOAA Data Buoy Office, NSTL Station, Mississippi 39529. 71 pp.

NATH, J.H. 1977b. Laboratory validation of numerical model drifting buoy - tether - drogue system. Final Report to NOAA Data Buoy Office on Contract No. 03-6-038-128, National Space Technology Laboratories, NSTL Station, Mississippi 39529. 180 pp.

NATH, J.H. 1977c. Laboratory model tests of drifting buoy and drogue. Environmental Fluid Dynamics Laboratory, Wave Research Facility, Oregon State University, Corvallis, Oregon 97331. 86 pp.

NEWMAN, J.N. 1963. The motions of a spar buoy in regular waves. Report No. 1499, David W. Taylor Naval Ship Research and Development Center, Bethesda, Maryland 20084. 27 pp.

PARKER, C.E. 1972. Some direct observations of currents in the Gulf Stream. *Deep-Sea Research,* 19: 879-893.

POCHAPSKY, T.E. 1963. Measurement of small-scale oceanic motions with neutrally-buoyant floats. *Tellus,* 15: 352-362.

RICHARDSON, P.L., R.E. CHENEY and L.A. MANTINI. 1977. Tracking a Gulf Stream ring with a free drifting surface buoy. *Journal of Physical Oceanography,* 7: 580-590.

ROSSBY, T.E. and D. WEBB. 1970. Observing abyssal motions by tracking Swallow floats in the SOFAR Channel. *Deep-Sea Research,* 17: 359-365.

SARPKAYA, T. 1976. In-line and transverse forces on cylinders in oscillating flow at high Reynolds numbers. *Proceedings of the Eighth Annual Offshore Technology Conference, 1975.* Paper No. OTC 2533: 95-108.

SAUNDERS, P.M. 1976. Drifting buoy Lagrangian test. Final report of Woods Hole Contract NAS13-5 with NOAA Data Buoy Office, NSTL Station, Mississippi 39529.

STOMMEL, H. 1954. Serial observations of drift currents in the Central North Atlantic Ocean. *Tellus,* 6: 203-214.

SWALLOW, J.C. 1955. A neutral-buoyancy float for measuring deep current. *Deep-Sea Research,* 3: 74-81.

VACHON, W.A. 1973. Scale model testing of drogues for free drifting buoys. Report No. R769, C.S. Draper Laboratory, 555 Technology Square, Cambridge, Mass. 02139. 131 pp.

VACHON, W.A. 1974. Improving drifting buoy performance by scale model drogue testing. *Marine Technology Society Journal,* 8: 58-62.

VACHON, W.A. 1975. Instrumented full scale tests of a drifting buoy and drogue. Report No. R-947, C.S. Draper Laboratory, 555 Technology Square, Cambridge, Mass. 02139. 163 pp.

VACHON, W.A. 1977. Current measurement by Lagrangian drifting buoys - problems and potential. Oceans 77 Conference Record; Third Annual Combined Conference, Los Angeles, 1977: 46B1-46B7.

VACHON, W.A. and D.R. YOERGER. 1977. A drifting buoy positioning system for the study of western boundary currents. Report No. R-1094, C.S. Draper Laboratory, 555 Technology Square, Cambridge, Mass. 02139. 66 pp.

VOLKMANN, G., J. KNAUSS and A. VINE. 1956. The use of parachute drogues in the measurement of subsurface ocean currents. *EOS, Transactions of the American Geophysical Union,* 37: 573-577.

WELLS, R.K. 1973. The design and construction of a constant depth current tracking buoy. M.S. thesis, M.I.T. Department of Ocean Engineering, Cambridge, Mass. 02139.

WHELAN, W.T., H.G. TORNATORE, S.P. MURRAY, H.H. ROBERTS and W.J. WISEMAN. 1975. An over-the-horizon radio direction-finding system for tracking coastal and shelf currents. *Geophysical Research Letters,* 2: 211-213.

WU, J. 1975. Wind-induced drift currents. *Journal of Fluid Mechanics,* 68: 49-70.

12

Electromagnetic Current Meters

I.S.F. Jones

1. INTRODUCTION

Electromagnetic current meters produce a magnetic field and then measure the electric potential generated by flow through this magnetic field. With two sets of orthogonal electrodes, the magnitude and direction of the water velocity in a plane perpendicular to the magnetic field can be measured and various configurations, such as those shown in Figure 1, have been developed to house the magnetic coils and electrodes.

One would like to be able to calculate the expected sensitivity, i.e. voltage output as a function of flow velocity that would have occurred in the absence of the sensor, but we are limited to estimating the relative importance of different regions around the sensor in determining the output voltage. An adequate description of the flow around a sensor has yet to be combined with the weighting functions so that our understanding to date of the response of electromagnetic current meters to different fluid speeds and directions has come from flow calibration tests.

2. THEORY

For sea water and a slowly fluctuating magnetic field Shercliff (1962) shows that the voltage around an electromagnetic current meter is given by

$$\nabla^2 E = \text{div}\ (\underset{\sim}{V} \times \underset{\sim}{B}) \qquad (1)$$

where $\underset{\sim}{V}$ is the fluid velocity and $\underset{\sim}{B}$ is the magnetic field vector.

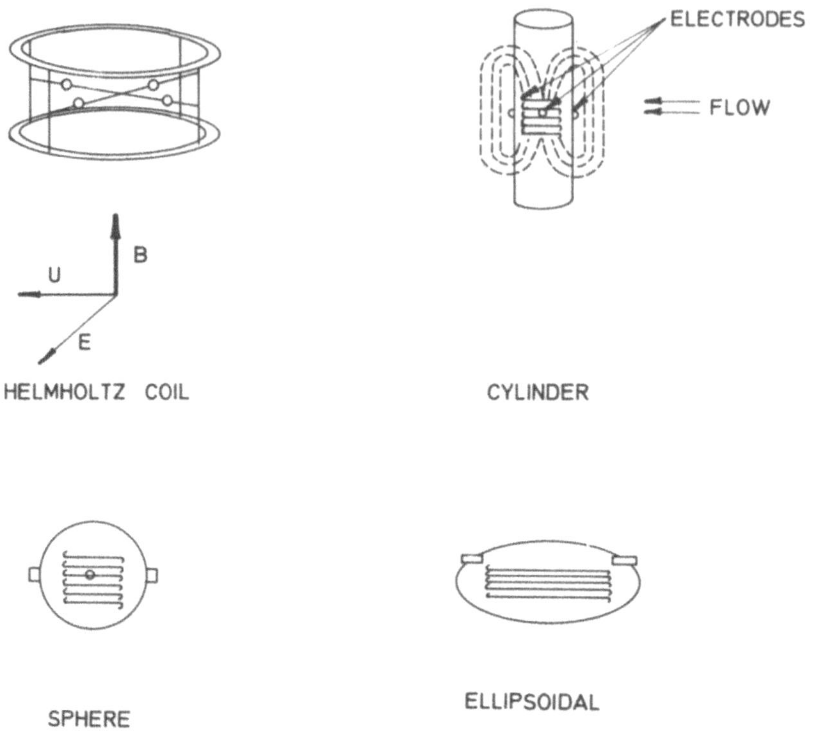

Fig. 1 A number of electromagnetic current meter shapes that have been developed.

While the flow meter is usually arranged to draw no current, the sea water is a conductor and as a result an electrical current, j, described by McCullough (see Chapter 6) flows in the fluid around the sensor.

Bevir (1970) shows that a solution to Poisson's equation can be found in terms of the current $\mathbf{i}$ that flows as a result of unit current through the electrodes of the sensor when no fluid motion is present, since this current vector $\mathbf{i}$ takes into account the boundary conditions imposed on Equation 1. The solution to Equation 1 for the voltage difference between the two electrodes of the current meter is

$$E_1 - E_2 = \int \mathbf{V} \cdot (\mathbf{B} \times \mathbf{i})\ d\ (\mathrm{vol})$$

where the volume integral is over all space. The vector $\mathbf{G} \times \mathbf{i}$ can be considered as a weighting function for the volume integral of the velocity and can be determined from a knowledge of the configuration of the current meter. As an example, an almost spherically shaped current meter with an external magnetic field that can be approximated by a dipole at the centre of the sensor has weighting functions W_1 and W_2 that are shown in Figure 2. These weighting functions are for the plane that contains the electrodes and is perpendicular to the magnetic dipole axis. The induced potential difference due to the flow within a slab dx_3 wide about the plane that contains the electrodes is given by

$$\frac{d}{dx_3}(E_1 - E_2) = \int\int (V_1W_1 + V_2W_2)\, dx_1\, dx_2$$

The weighting functions decrease rapidly with x_3 as one moves away from the plane shown, so that one can see that the region of flow

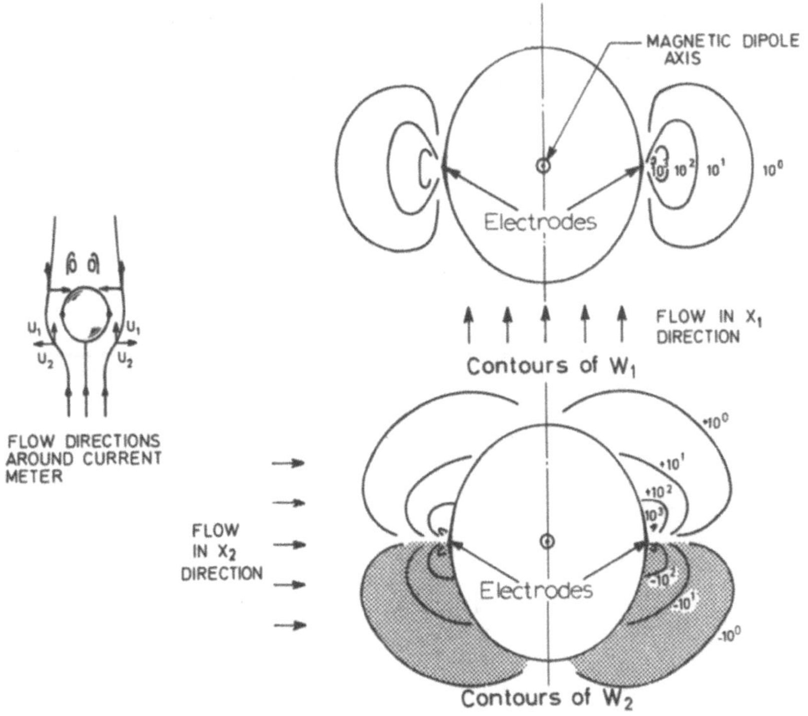

Fig. 2 Weighting functions for an electromagnetic current meter with a dipole magnetic field. W_1 is component of $\mathbf{B} \times \mathbf{i}$ in x_1 direction.

that affects the sensor is confined to about two characteristic diameters of the flow meter. For the configuration in Figure 2 the 'sensitive' axis is the free stream direction x_1 and one can see that the response of the sensor is zero for axisymmetrical flow perpendicular to the sensitive axis since the positive and negative values of W_2 cancel. The question of flow in directions other than x_1 and x_2 is discussed later.

3. FLUID FLOW

If the pattern of fluid flow around the sensor remains similar for all velocities (that would have been present in the absence of the sensor) the output of the device will be a linear function of velocity and the electrical voltage changes sign with the velocity. However, the flow around bodies is not similar, in general, for different Reynolds numbers, VD/ν, defined in terms of a characteristic length of the sensor, D, and the kinematic viscosity, ν. As the Reynolds number increases, the flow in the boundary layer on the sensor becomes turbulent and this point of transition depends upon the free stream turbulence and the roughness of the body. The influence of transition on the calibration of a spherical sensor is shown in Figure 3 where the velocity was changed to vary the Reynolds number. Transition of the boundary layer occurs between Reynolds numbers of 0.8×10^5 to 3×10^5 according to Hoerner (1958) and this is the range where the sensitivity of the current meter took a

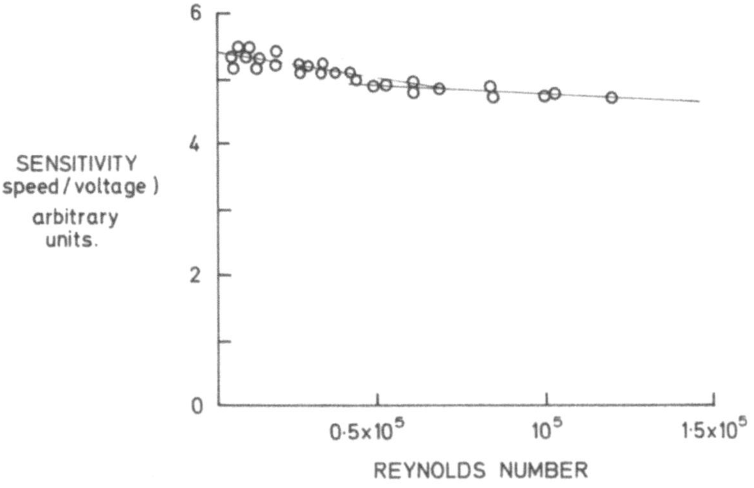

Fig. 3 The change in sensitivity of a spherical electromagnetic current meter of diameter 4 cm as the velocity increases. Data from tests in Sydney University towing tank.

sharp step. Vibration, surface roughness and free steam turbulence all affect transition which may be the cause of the change in sensitivity reported by some experimenters. It seems desirable, at present, to choose characteristic dimensions for the required velocity range that ensure Reynolds numbers well below transition. This restricts sensors to less than about 10 cm for mixed layer work.

A velocity component detector in a uniform field has an output proportional to $\cos\theta \cos\phi$ where θ and ϕ are the angles of the flow to the 'sensitive' axis. In the previous example θ is the angle to the x_1 axis while $90 + \phi$, the angle to the magnetic axis, is usually termed the tilt angle. (Most current meters are operated with the magnetic axis vertical so the sensor detects horizontal currents.) If the fluid flow around the sensor is a function of θ and ϕ, there is little chance of good cosine response. In this regard, a spherical sensor appears promising.

One of the most successful devices at achieving a good cosine response appears to be the Helmholtz coil meter developed by Olsen (1972). This device detects the induced EMF within the 'uniform' magnetic field produced by two coils arranged as shown in Figure 4. The excellent cosine response due to variations in both θ and tilt is shown in Figure 4 which is the result of a calibration in an oscillating flow (the downfall of the Savonius rotor). Unfortunately, even this result must be treated with some caution as the size of the tank used for the unsteady calibrations may not have been large enough. Various techniques have been tried to improve the cosine response of sensors that rely on external flow. One such technique is to raise the electrodes above the sensor boundary layer so that the flow that contributes most strongly to the voltage is less dependent on the boundary layers. Presumably on cylindrical and spherical sensors raising the electrodes keeps them clear of the separated wake of the body for larger flow angles. The time for the flow to reach equilibrium around the sensor is important in turbulent flows where the sensor calibration relies on steady state tests. One suspects a suitable time scale is 5D/V where V is the free stream velocity. Vortex shedding is a related feature that can potentially lead to 'flow noise' from the sensor, especially for an electrode that is towards the rear of the body. The characteristic shedding frequency is of the order of 0.2V/D and so is above the frequency often required for mixed layer work.

Changing temperature in sea water affects viscosity and conductivity. Both these variables should have a negligible effect on the sensitivity of the electromagnetic current meter.

4. ELECTRONIC DESIGN

The induced voltage across the electrodes depends upon the magnetic

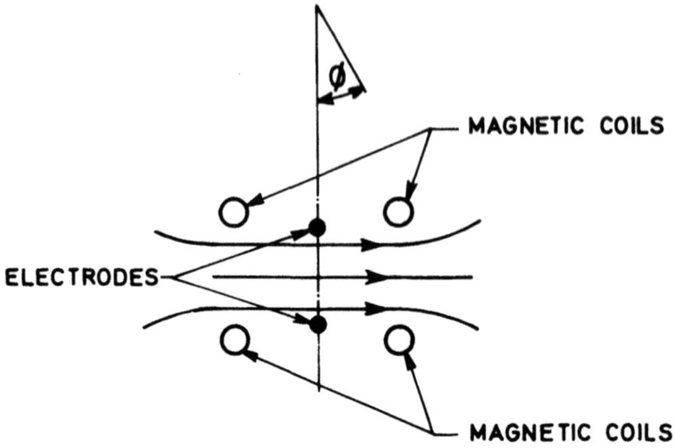

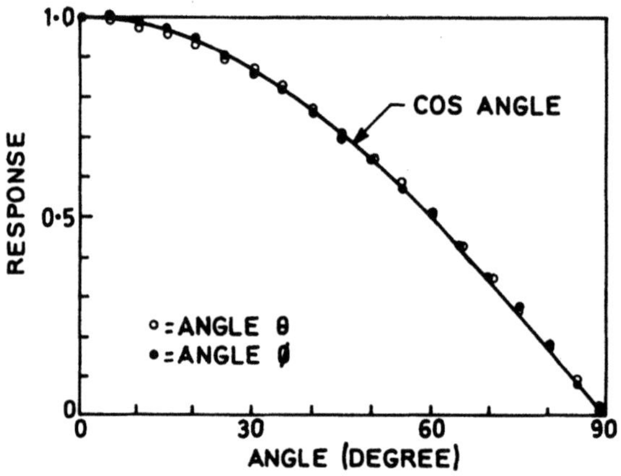

Fig. 4 The angular response of a Helmholtz coil current meter after Olsen (1972).

flux generated by the sensor. At the magnet power consumption of the order of 200 mW, commercial electromagnetic current meters have sensitivities of the order of 10^{-2} to 10^{-1} μV s cm^{-1}. These low signal levels mean that electrochemical reaction at the electrodes can induce voltage drifts of significant magnitude. To overcome this difficulty the sign of the magnetic field is varied and a phase sensitive detector is used to eliminate the effects of drift due to the electrodes and the first stage amplifiers. Most readily available operational amplifiers have input noise levels of about 10^{-3} (μV^2) Hz^{-1}, a value which sets the output noise. The electromagnetic current meter has no threshold in the usual sense but the signal does have a noise component that limits the accuracy at which one can determine the mean current. Since the sensitivity increases with magnetic flux density, and so magnet power consumption, the noise floor can be lowered by expending more power. Figure 5 shows the noise from a commercially available cylindrical current meter (magnet power 200 mW) together with an estimate from Jones and Kenney (1977) of the 'turbulence' in the ocean mixed layer under a number of values of surface stress. Also shown in this figure are a number of estimates of the orbital motion that correspond to surface wave spectral values, $\phi_\eta(f)$, at a frequency of 0.16 Hz (i.e. near the peak in the wind wave spectra). A 10 m s^{-1} wind of long duration produces a power spectral density of surface elevation of about 10^4 cm^2 Hz^{-1} at a frequency of 0.16 Hz.

Since the sensitivity of electromagnetic current meters is so low, fluctuating voltage gradients in the water due to the power mains can cause noise in the form of low frequency beats. As the phase sensitive detector reverses the sign of the output from the electrodes at twice the magnet frequency, the low frequency beating can be eliminated if the current meter is operated synchronously with and at a subharmonic of the noise field.

Ground loops due to the current needed to operate the magnet can be a more serious problem as this form of noise is at the same frequency as the magnet and so can appear as a shift in the zero

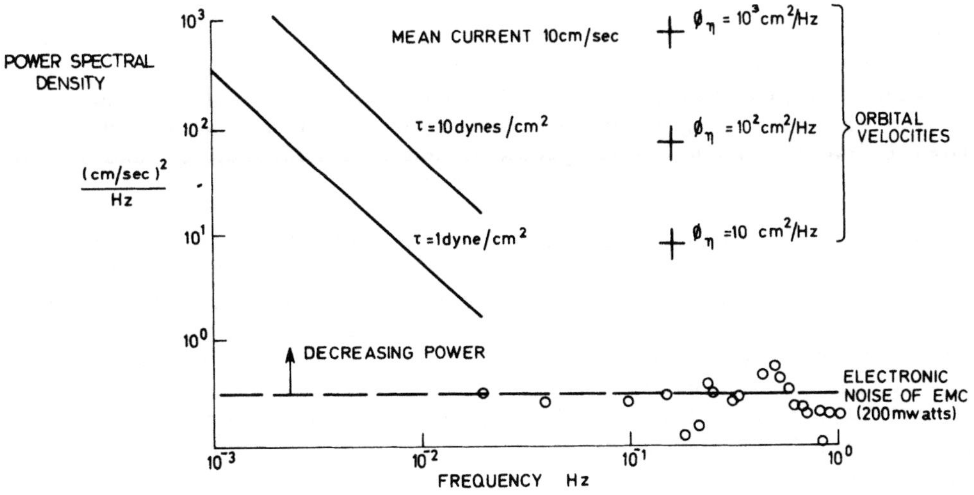

Fig. 5 The noise level from a Marsh McBirney Inc. electromagnetic current meter compared with the expected power spectral density of velocity in the ocean mixed layer at a depth of 10 m.

output of the instrument.

Another potential problem is the long term drift of the zero flow point. Changes in zero of the instrument can be due to two causes:

(i) offset drift of the electronics,
(ii) variations in impedance between the sensor magnets and the electrodes.

Using modern design the electronic drift can be kept low. With the electrodes shorted together, a commercial instrument, using low-drift, low-noise operational amplifiers, had a zero drift over 14 days that was less than the rms variation between individual 30 second averages of the output. At a magnet power of 200 mW, a 30 second average of the output had a value within ±1 cm s^{-1} of zero flow (with a 90% confidence level). Griffiths, Collar and Braithwaite (1978) have found that water absorbed into the sensor can lower the leakage between the magnet and the electrodes in some sensor head designs and since the magnet voltage is in phase with the detector, the leakage current appears as an erroneous output voltage. Attention to the sensor head design to eliminate this problem is required.

Finally there are time variations in the gain of the current meter but to date there appears to have been little consideration given to this potential difficulty.

5. ADVANTAGES

The principal advantage of the electromagnetic current meter over most other sensors is that it contains no moving parts and, in most designs, no protruding components that are liable to damage. The sensor can be arranged to be a strength member and so avoid an external 'cage' which always produces some degradation of the sensor performance due to the wake of the cage.

Some designs of electromagnetic current meters perform adequately when the flow comes from any direction including that of the support. In the upper ocean where the surface wave orbital motions can induce both horizontal and vertical flows for portions of the wave period, this is a very desirable feature. Since there are no moving parts, the sensor can be coated with anti-fouling material on all surfaces except the electrodes. This is important since the marine growth in near surface waters can be very rapid. The electromagnetic current meter, since it has an output proportional to a volume integral, should not be very sensitive to small changes in surface roughness (away from flow transition). Electromagnetic current meters have been used for a number of years in oceanography with some success c.f. Thorpe et al. (1973).

6. DISADVANTAGES

The power supply of the electromagnetic current meter can be substantial in relation to modern remote data collection systems since a large magnetic field is required to produce an adequate signal to noise ratio for detecting oceanic turbulence. Another minor disadvantage is that electromagnetic current meters require nontrivial electronics, necessitating calibration prior to use.

Commercially available instruments have marginal cosine response for direct use in the wave orbital fields of the oceanic mixed layer although there is promise of better response (c.f. Figure 4). In this regard, the integration of the sensor with the support and electronics case must be carried out carefully to ensure that these additional structures do not cause large flow distortions to degrade the carefully obtained near ideal 'tilt' response.

7. MIXED LAYER MEASUREMENTS

Electromagnetic current meters have been used for a number of surface mixed layer studies. Thornton and Krapohl (1974) have used a cylindrical electromagnetic current meter to show that in the frequency band 0.1 to 0.5 Hz the velocity fluctuations are dominated by surface wave orbital motions. Medwin used a similar instrument to show that surface wave induced fluctuations were important in determining the nature of bubble resonant frequencies (see Antonia et al., 1974). Bruzzone and Jones (1976) measured the three components of mean and fluctuating velocity from a large spar buoy in order to calculate the shear stress in the mixed layer. Jones and Kenney (1977) presented horizontal velocity spectra in the wind driven layer of a lake and show how at frequencies below 0.1 Hz the velocity fluctuations scale on inner law boundary layer variables. Scott and Csanady (1976) used cylindrical current meters from semi-rigid spars attached to the sea floor on the coastal shelf to measure mean velocities in the presence of surface wave motions and to relate the velocities to the wind stress.

8. FUTURE DEVELOPMENTS

Three problems that have received attention in the past are the noise floor, the cosine response, and the power requirements. Noise floors adequate for many applications can already be achieved and improved amplifier design promises gains in the future. Cosine response presents a more difficult problem to overcome when one considers the very high accuracies required to extract mean velocities and, in particular, mean velocity shear in the presence of surface orbital motions. The power required for the current meter does not appear to be capable of being changed dramatically, given that a certain level of signal to noise is necessary.

To date sensors have contained a single magnet and detected two flow components. A three-axis sensor shown in Figure 6 is presently being tested to find the cosine response to both steady and unsteady flows. Here the sensor consists of two magnetic coils with perpendicular axes. By switching the magnets alternately and measuring the voltage between two pairs of electrodes, the three components of current can be continuously measured. A future development could include a microprocessor to allow mean velocities to be determined in the presence of very large fluctuating velocities. The microprocessor could contain the experimentally determined cosine response curve to make first order corrections to the instantaneous voltages.

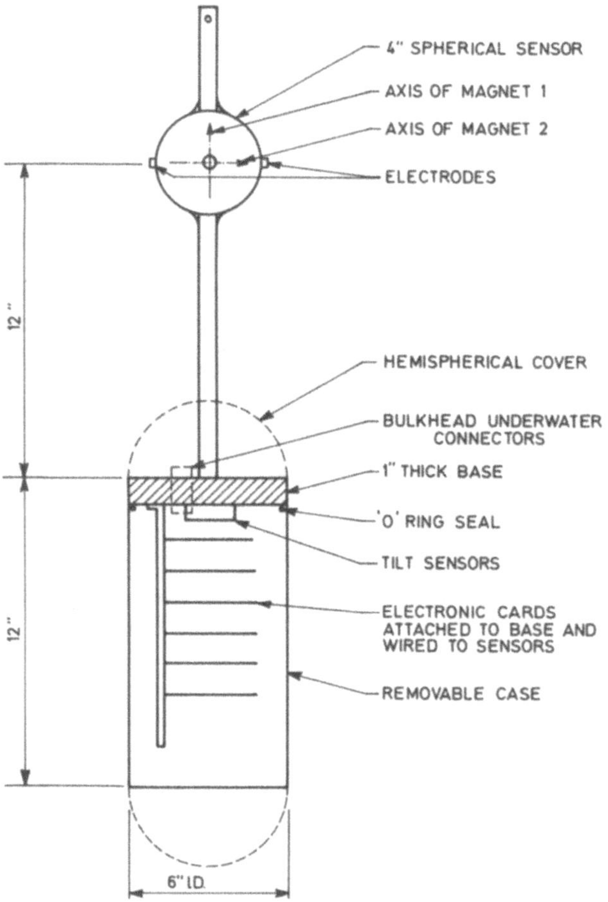

Fig. 6 A three-component electromagnetic current meter under test. The meter is designed to measure Reynolds stress and so tilt sensors are included to provide accurate reference for the current meter axes.

REFERENCES

ANTONIA, R.A., D.R. BLACKMAN, I.S.F. JONES, R.E. LUXTON, and H. MEDWIN. 1974. Project BASS, environmental measurements. RAN Research Laboratory T.M. 18/74.

BEVIR, M.K. 1970. The theory of induced voltage electromagnetic flowmeters. *Journal of Fluid Mechanics,* 43: 577.

BRUZZONE, F. and I.S.F. JONES. 1976. A compilation of mixed layer current meter and wind observations. Woods Hole Oceanographic Institution Tech. Rep. 76-101.

GRIFFITHS, G., P.G. COLLAR, and A.C. BRAITHWAITE. 1978. The characteristics of electromagnetic current sensors in laminar flow conditions. Institute of Oceanographic Sciences Report 56, Wormley, Godalming, Surrey, England.

HOERNER, S.F. 1958. *Fluid-Dynamic Drag.* S.F. Hoerner, Brick Town N.J., U.S.A.

JONES, I.S.F. and B.C. KENNEY. 1977. The scaling of velocity fluctuations in the surface mixed layer. *Journal of Geophysical Research,* 82: 1392.

OLSEN, J.R. 1972. Two-component electromagnetic flowmeter. *Marine Technology Society Journal,* 6: 19.

SCOTT, J.T. and G.T. CSANADY. 1976. Nearshore currents off Long Island. *Journal of Geophysical Research,* 81: 5401.

SHERCLIFF, J.A. 1962. *The theory of electromagnetic flowmeasurements.* Cambridge University Press, Cambridge.

THORNTON, E.B. and R.F. KRAPOHL. 1974. Water particle velocities measured under ocean waves. *Journal of Geophysical Research,* 79: 847.

THORPE, S.A., E.P. COLLINS, and D.I. GAUNT. 1973. An electromagnetic current meter to measure turbulent fluctuations near the ocean floor. *Deep-Sea Research,* 20: 933.

13

Air Pressure Measurement Techniques

F.W. Dobson

1. INTRODUCTION

Air pressure is defined as the force acting normal to a unit area. Its SI unit is the Pascal, or Newton m^{-2} *. In the fluid momentum equations fluid velocities are linked with pressure gradients. The effect of pressure is to transport energy and to redistribute the velocity components towards isotropy, since its effects are independent of direction.

2. THE PRESSURE SPECTRUM

A schematic spectrum of air pressure (Fig. 1) after Gossard (1960) can be rather artificially divided into three regions of interest in this discussion: the 'synoptic' region, with frequencies of 10^{-6} to 10^{-4} Hz (periods of 4 days to 4 hr), the 'mesoscale' region, from 10^{-4} to 10^{-2} Hz (periods of 4 hr to a few minutes), and the 'microscale' region, from 10^{-2} to 10 Hz. The spectrum is seen to be 'red' - that is largest at the lowest frequencies.

Because of the large range of frequencies and dynamic range of the pressure field, a large number of different instruments have been developed for the synoptic and microscale regions, with instruments of both types being utilized in the mesoscale region. The synoptic instruments are called 'barometers', and usually measure the total pressure exerted by the atmosphere. The mesoscale and microscale instruments are simply referred to as 'pressure gauges' or

---

* 1 Pa = 10 dyne cm^{-2} = 10^{-2} mbar = 9.87×10^{-6} atm = 7.50×10^{-3} mm Hg (Torr) = 1.450×10^{-4} psi.

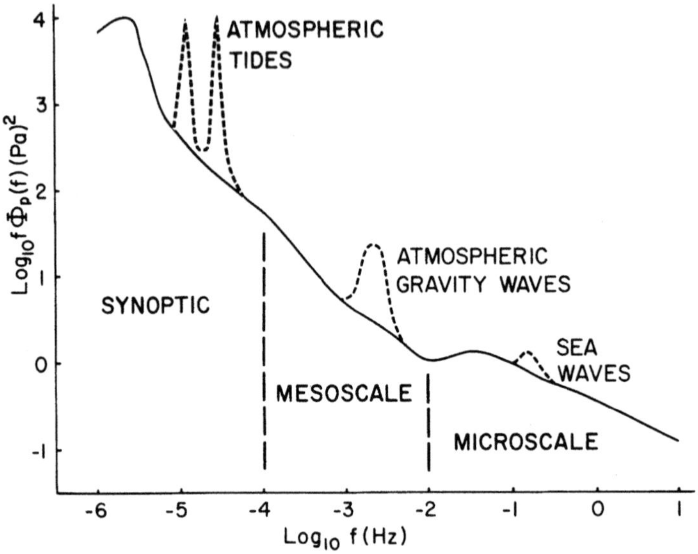

Fig. 1 Spectrum of pressure fluctuations in the atmospheric boundary layer (after Gossard, 1960).

sometimes 'microbarographs'. They are usually differential instruments; that is they measure differences between the instantaneous and the mean atmospheric pressure.

3. AREAS OF INTEREST

3.1 Geostrophic Flow, and Barometry

The approximate momentum equations (the terms left out are typically at least one order of magnitude smaller) for horizontal flow in the atmosphere are

$$\frac{dU}{dt} - fV = -\frac{1}{\rho}\frac{\partial P}{\partial x}$$

$$\frac{dV}{dt} + fU = -\frac{1}{\rho}\frac{\partial P}{\partial y}$$

where $f = 2\Omega\sin\phi$ is the 'Coriolis parameter', Ω is th earth's rotation rate, ϕ is latitude, and ρ is air density. The acceleration terms, important in weather prediction, are normally a factor of 10 smaller than the other two terms in each equation, and so to first order the balance is

$$-fV_g = -\frac{1}{\rho}\,\partial P/\partial x$$

$$fU_g = -\frac{1}{\rho}\,\partial P/\partial y$$

where U_g and V_g are the two components of the 'geostrophic' wind velocity vector $\vec{U}_g$. In practice, $\vec{U}_g$ is estimated from pressure differences measured with barometers and their separations.

If, for example, the barometers used to measure the pressure have an accuracy of ±10 Pa (0.1 mbar) and are $\Delta y = 100$ km apart, then

$$\delta(\partial P/\partial y) \simeq \pm 10^{-4}\ \text{Pa m}^{-1}$$

$$\delta(U_g) = (1/\rho f)\,\delta(\partial P/\partial y) = \pm 0.7\ \text{m s}^{-1}$$

where $\rho f \simeq 1.5\times10^{-4}$ kg s^{-1} m^{-3} at 45° latitude. If the measurement error is caused by dynamic pressures (see Section 4 below), it varies with the wind speed squared, and the fractional error in the geostrophic wind speed is given by

$$\frac{(U_g)_{measured} - U_g}{U_g} = \frac{U_g\ \delta C\ G^2}{2f\ \delta x}$$

where δC is the difference in pressure coefficients (C = fraction of $\frac{1}{2}\rho U^2$ measured by the sensing head) between the two barometers' sensing heads, and G is a function relating the geostrophic wind to that measured at the sensing head (see, for instance, Hasse, 1973). For $C \sim 0.2$, $\delta C \sim 0.1$, $G^2 \sim 0.5$, $U_g = 20$ m s^{-1}, $\Delta x = 100$ km, for instance, the error is ~10%.

Barometry is also used in the study of atmospheric tides (e.g., Siebert, 1961) and of the effect of air pressure on oceanic tides (Proudman, 1953).

3.2 Mesoscale Studies

A large body of literature exists on the use of pressure sensors to study 'infrasound' (1-10 Hz), caused by surf, earth disturbances, and man-made explosions (see, for instance, Gossard and Hooke, 1975). Low-frequency pressure fluctuations have also been associated (Cunning, 1974) with internal waves in the atmosphere. Cunning shows typical amplitudes of 20 to 40 Pa, in the frequency range 10^{-2} to 10^{-1} Hz.

3.3 Microscale Studies

An understanding of turbulent pressure fluctuations is fundamental to the understanding of the atmospheric turbulent flow equations (e.g. Lumley and Panofsky, 1964, p. 161 ff.). Many studies have been published (see Willmarth, 1975) of turbulent pressure fluctuations in wind tunnel boundary layers, but few have been done in the atmospheric boundary layer (Priestley, 1966; Elliott, 1972b; McBean and Elliott, 1975). The turbulent energy balance in the atmospheric boundary layer (the 'friction' layer) can be approximated by (see Lumley and Panofsky, 1964, p. 67 ff.)

$$\frac{1}{\rho}\frac{\partial q}{\partial t} = 0 = -\overline{u'w'}\frac{\partial U}{\partial z} + \frac{g}{T}\overline{w'\theta'} - \frac{\partial}{\partial z}[(q + \frac{p}{\rho})w'] + \varepsilon$$

where $q = u'^2 + v'^2 + w'^2$, U is the mean downwind velocity component, T and θ' are the mean and fluctuating temperatures, and ε is dissipation. Energy enters the turbulence as large-scale motions via the first term on the right-hand side, and about the same amount of energy is dissipated to heat via molecular viscosity as small-scale motions by the last term on the right. The second term on the right can add or subtract turbulent energy, depending on the sign of the $\overline{w'\theta'}$correlation; in stable conditions (air warmer than sea) the turbulence loses energy to buoyancy forces, while in unstable conditions (sea warmer than air), the turbulence gains energy. The second-last term has two parts, and both represent a divergence of kinetic energy transport. Neither, summed over all three components, is a net source or sink of turbulent energy. The first, on the average, redistributes energy from large to small scales and from place to place, via nonlinear interactions; the pressure transport term, because pressures act in all directions, redistributes the energy of the downwind component to the other components, tending to make the flow isotropic. The total energy balance in near-neutral conditions over land and over the sea has been investigated by Elliott (1972b,c), who finds that the two middle terms in the equation are about equal with opposite signs, and are about 1/10 of the first and last terms. The non-neutral case has been studied by McBean and Elliott (1975).

3.4 Wave Generation Studies

If a field of wave-coherent pressure fluctuations exists in the air flow over gravity waves at sea, then work is done on the waves according to

$$\overline{\partial E_w/\partial t} = \overline{p\partial\zeta/\partial t}$$

where E_w is the wave energy per unit area and p(t) and $\zeta(t)$ are the pressure and water level fluctuations. Associated with the energy flux $\partial E_w/\partial t$ is a momentum flux

$$\overline{\underset{\rightarrow}{\tau}} = -\overline{p\nabla\underset{\rightarrow}{\zeta}}$$

From a measurement of the pressure-wave height directional cross spectrum $X_{p\zeta}(\omega,\theta) = Co_{p\zeta}(\omega,\theta) + iQu_{p\zeta}(\omega,\theta)$ the downwards wave-induced energy and momentum fluxes can be computed (see, for instance, Snyder, 1974):

$$\overline{\frac{\partial E_w}{\partial t}} = -\int_0^\infty d\omega \int_\theta d\theta \; \omega \; Qu_{p\zeta}(\omega,\theta)$$

and

$$\overline{\underset{\rightarrow}{\tau}_w} = -\int_0^\infty d\omega \int_\theta d\theta \; \underset{\rightarrow}{k} \; Qu_{p\zeta}(\omega,\theta)$$

where ω is radian frquency, θ is wave direction, and k is (vector) wavenumber. The inner integrands are then the spectra of the wave-supported energy and momentum fluxes.

There have been two approaches to the measurement problem. The first (Longuet-Higgins et al., 1963; Shemdin and Hsu, 1967; Dobson, 1971) has been to attempt to measure the pressure as near the water surface as possible; since the water surface moves this has necessitated mounting pressure sensors on wave-following devices (see Shemdin, Chapter 33). The second approach (Elliott, 1972c; Snyder et al., 1978) has been to measure the pressure fluctuations from fixed sensors at a variety of heights out of range of the wave crests, and attempt to extrapolate the resulting pressure-wave quadrature spectra to the surface, assuming the wave-induced pressures vary exponentially with height ($e^{-\alpha kz}$, where α is empirically determined), and their phase with respect to the waves is constant with height.

When growing waves are present Elliott (1972c) finds a subtle but dramatic difference in the behavior of the pressure-velocity correlation from that observed over land, which is related to energy transfer within the turbulence. Over land, for scale sizes larger than the height of measurement, the pressure and velocity are in phase; for smaller scales they are in quadrature and energy transfer occurs among the turbulent components, mostly at scales just smaller than the phase transition to quadrature. Over water, the phase transition occurs at much larger scale sizes - in fact, the transition scale size is determined by the length of the dominant water waves

$$\lambda_{\text{transition}} = \frac{U}{c}\,\lambda_{\text{dominant waves}}$$

The reasons are not clear for this disparity. It appears certain that the wave generation process is involved, but it remains unclear whether the effect is important in the energy budgets of either the turbulence or the wave-coherent motions.

4. MEASUREMENT PROBLEMS

Measurements of air pressure are of interest over an enormous dynamic range - from hundreds of kilopascals in dynamical meteorology to tenths of micropascals in acoustics. Because the spectrum of atmospheric pressure is so red (Fig. 1), considerable care needs to be exercised when designing pressure sensors. Meteorologists have long recognized the inherent difficulties involved in pressure measurements. Users of mercury barometers on board ship, for example, are forced, in addition to using gymbals to keep their instrument upright, to install constrictions in the mercury column which act as low-pass filters to prevent the 'pumping' action of the ship's vertical motion on the column. Large temperature gradients in the room containing the barometer can also cause significant errors, both at sea and in other harsh environments (such as the Arctic; see Martin, 1973, for typical problems and solutions). If frequencies of more than 10^{-2} Hz are to be considered, the applicable Gas Law for an enclosed volume changes from isothermal at low frequencies, $PV = nRT$, to adiabatic at high frequencies $(PV)^{\gamma}$ = constant, where $\gamma = c_p/c_v$ is the specific heat ratio. In what frequency interval the isothermal-adiabatic transition occurs depends on the thermodynamics of the particular experimental setup; it normally occurs between 10^{-3} and 10^{-2} Hz (see Section 8 on calibration of pistonphones).

Another problem, just as basic as the thermodynamic one, is that any sensing head placed in an air flow disturbs the very flow it is trying to measure. The Bernoulli relation states that in a steady, incompressible, vorticity-free flow at constant height

$$P + \tfrac{1}{2}\rho U^2 = \text{constant}.$$

For undisturbed uniform flow U_0 upstream of an obstacle, if $P = P_0$ at some distance from the object, then the constant is $P_0 + \tfrac{1}{2}\rho U_0^2$; if at some point the flow speed U drops to zero (as it normally does on the upwind side of an obstacle), the pressure at that point equals $P_0 + \tfrac{1}{2}\rho U_0^2$ - that is, the ambient pressure is augmented by an amount $\tfrac{1}{2}\rho U_0^2$, which is called the 'stagnation pressure'. In fact, everywhere in the vicinity of the object there exists an excess pressure, caused by the presence of the object in the flow, which is normally called the 'dynamic pressure' (such pressures vary with the wind speed squared, and result in the exertion of wind forces on obstacles) which is some fraction C of the stagnation pressure P_s; the so-called 'pressure coefficient' C varies with position relative to the object and with the orientation of the object with respect to the flow direction. Figure 2 gives the computed potential-flow excess, or blockage, pressure, upstream of a cylinder and a sphere, expressed in terms of fractions of the stagnation pressure versus the number of radii upstream of the object. Such curves may be used for estimating the expected size of blockage pressures caused by structures which support pressure sensors.

Dynamic pressures can be measured and converted to flow speed, and this is the basis of one of the earliest of the modern wind tunnel instruments. The pitot-static, or Prandtl, tube (see, for instance, Prandtl and Tietjens, 1957) produces two output pressures, $P_0 + \frac{1}{2}\rho U_0^2$ and P_0, and is used as an anemometer (see Smith, Chapter 3). The device is particularly interesting because to obtain P_0 at its static ports it uses the technique of balancing the positive blockage pressure from its stem with the negative dynamic pressure associated with the sensing head. This technique, discussed later, is widely used in sensing head design.

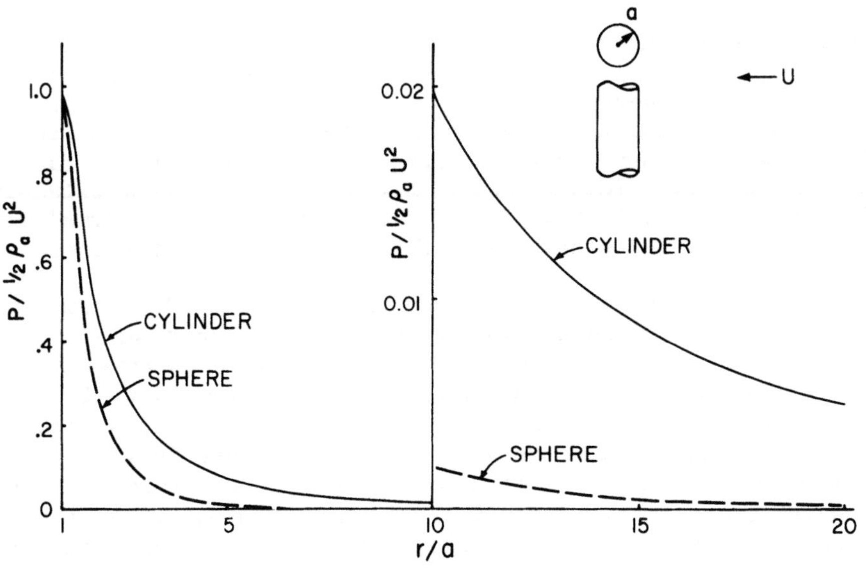

Fig. 2 Dimensionless blockage pressure caused by a cylinder and a sphere

5. INSTRUMENTS

5.1 The Aneroid Barometer

The design of aneroid barometers and sources of error are described in Middleton and Spilhaus (1953), and are not repeated here. The major difficulty with using aneroid sensors is maintaining their absolute calibration over time periods of more than a few days (Martin, 1973, p. 89). A typical field accuracy for a well maintained precision mechanical aneroid barometer is ± 50 Pa; in the laboratory, devices which do not drive a pointer or pen can be kept to ± 10 Pa.

Aneroid capsules are also used as the pressure-sensing elements in various electronically-controlled barometers. The device types are aneroid capacitance gauges (Pike and Bargen, 1976), aneroid capsules which are servo-driven to oscillate at their resonant frequency, and aneroid capsules with servo force balances. The last two designs are commercially available. None of the devices includes a separate spring to balance the pressure forces on the evacuated capsule; the ability to sense small deflections of a metal plate has increased so enormously that the capsules have become their own springs. The 'electronic aneroids' have typical overall accuracies of ± 50 Pa; they can, if temperature-corrected, be kept to within ± 10 Pa.

5.2 Capacitance Gauges

Whereas aneroid gauges have been historically connected with barometry, capacitance microphones have been connected with acoustic, that is, high-frequency, low-amplitude pressure sensing. They have been adapted for the measurement of turbulent pressures by the provision of an air leak of known time constant around the diaphragm (see Section 7). The control of the leak has been extensively studied (Rasmussen, 1960). FM carrier techniques are normally used with microphones for sensing turbulent pressure fluctuations.

If small size and weight are unimportant, the diaphragm of the transducer may be made larger, resulting in higher sensitivities. Units are commercially available with full-scale ranges of 10^3 to 10^5 Pa absolute and differential, with resolutions down to 10^{-3} Pa and accuracies over periods of a few days of $\pm 10^{-2}$ Pa. They are particularly well suited for use with Pitot tubes in wind tunnels, and as field sensors of turbulent pressure fluctuations. Commercial units are also available with two diaphragms, one a dummy that is used to minimize the acceleration sensitivity of the transducer.

5.3 Crystal Resonance Gauges

Two types of pressure gauges have been developed, both for measuring pressure variations in the deep ocean, which have a sufficient resolution to record barometric and even turbulent pressure fluctuations. Both are commercially available. The first type consists of a piezoelectrically-driven quartz disc which forms an integral part of a quartz cylinder filled with an inert gas. The device has a pressure range of $7x10^4$ kPa, stability over a week of ±70 Pa, and over 100 s of 7 Pa and hysteresis of ±70 Pa. The second resonant crystal device consists of a specially-shaped and loaded crystal bar, enclosed in a vacuum and driven piezoelectrically at resonance in a shear mode. Such devices, when sampled four times per second, have typical resolutions of 4 Pa and measured long term stability (after three months aging) of about ±50 Pa. It is worth noting that in carefully controlled laboratory conditions the device is capable of attaining a long term stability of ±10 Pa.

5.4 Strain Gauge Transducers

Strain gauges measure the deformation of spring material under the action of pressure forces. They have excellent frequency response (typically DC - 200 Hz or more), low volume change for a given pressure change, require little power, and are rugged. They suffer from the problems associated with all springs: hysteresis and sensitivity to temperature, particularly with regard to drift and sudden changes in offset with temperature cycling. In general, they have less dynamic range than the crystal devices, and less sensitivity than the aneroids and capacitance gauges. The best quoted long term accuracies are ±100 Pa.

5.5 Variable Reluctance Sensors

If the deflection of a spring by pressure forces is sensed by a change in magnetic field, the sensor falls into the class of variable reluctance pressure transducers. Variable reluctance transducers are in general less expensive for a given resolution than the strain gauge types, but suffer from the same problems, since they rely for their output on the deflection of a spring. They have relatively high hysteresis, drift, sudden changes in offset due to shock and temperature change. They consume more power, and have in general lower sensitivity, than aneroid or capacitance types.

6. SENSING HEADS

6.1 Sensing Head Design

The objective of sensing head design is to reject as large a fraction as possible of the stagnation pressure $\frac{1}{2}\rho U^2$ which is induced on the surface of the sensing head.

The wind velocity $\underset{\rightarrow}{U}$ can be expressed as the sum of a mean wind speed $\overline{U}$, with which the pressure sensing head is aligned, and a fluctuating vector of length $q = u'^2 + v'^2 + w'^2$. The total stagnation pressure P_d in such a flow is

$$P_d = \tfrac{1}{2}\rho|U|^2 = \tfrac{1}{2}\rho[\overline{U}^2 + 2\overline{U}q^{\frac{1}{2}} + q],$$

and a probe will have a pressure error given by

$$P_e = \tfrac{1}{2}\rho C_p(\psi, \phi)[\overline{U}^2 + 2\overline{U}q^{\frac{1}{2}} + q]$$

where ψ and ϕ are respectively the pitch and yaw angles the sensing head makes with $\vec{U}$. The first term in the brackets exceeds the second by about one order of magnitude, and the second exceeds the third by about the same factor (Lumley and Panofsky, 1964). The 'pressure coefficent' C_p is the fraction of the stagnation pressure $\frac{1}{2}\rho\overline{U}^2$ measured by the sensing head. The purpose of sensing head design is to minimize $C_p(\psi, \phi)$. Yaw dependence is eliminated by making the head horizontally omnidirectional, and there are two techniques for minimizing pitch dependence: (i) the use of the blockage pressure from an obstacle placed downstream from the sensing port, or (ii) by changing the direction of the disturbed flow by careful head shaping. An example of (i) is the pitot-static tube (Prandtl and Tietjens, 1957), and of (ii) is any one of the microscale sensing heads described below.

Static heads for barometry are presently being tested which can be mounted on small buoys for use in the open sea. The design specifications call for mean pressure errors less than ±0.5 mbar (50 Pa) up to sea state 9 (30 m s^{-1}); this calls for $C_p < 0.1$ for buoy pitch angles of up to ±30°. The British Meteorological Office (1962) has designed a vane-mounted sensing head which has a C_p of less than 0.1 at pitch angles less than ±20°. It is unfortunately not possible at time of writing to refer to published literature on other heads, since patents are being applied for. The most important features of the heads are that they have circular symmetry, eliminating yaw dependence, and their sensing ports look downwards, minimizing blockage by water. Typical variations of $C_p(\psi)$ are ±0.1 over pitch angle variations of ±30°, and typical mean C_p values are -0.1. The dependence of C_p on buoy freeboard has been investigated but no published work is known to this writer; it is in general equal in importance to the pitch dependence. It can be approximately computed from potential flow arguments for simple geometries.

Two further points are worth making. Firstly, the full stagnation pressure only reaches 100 Pa (1 mbar) at a mean wind speed of about 41 m s^{-1}. Therefore, before considering the need for a well designed sensing head it is first advisable to consider the accuracies required in the designed pressure measurement. Often an open port inside a building or in an enclosed ships's bridge will do (in fact, an inside location protects the barometer from the elements and averages out dynamic pressure variations due to wind direction changes). If the port must be outside it can be placed (Robertson, 1972) in a region of parallel flow, shielded from precipitation between two parallel discs with rounded edges (to reduce flow separation in downdrafts or updrafts).

The second point is that if accuracies of better than ±1 mbar are desired, then it is not enough to design a good sensing head; its mounting must be carefully considered. All supporting structures produce their own flow distortion pattern (see Wucknitz, Chapter 32). The pressure errors reproduced by an otherwise ideal sensing head depend greatly on its mounting location. If placed on a boom at some distance from its supporting structure, the head will 'see' the pressure disturbance produced by the entire obstacle (ship, large mast, or building); the errors can be estimated by wind tunnel testing or potential flow solutions in the vicinity of bodies with similar geometry. If the sensing head is mounted directly on its supporting structure, the resulting pressure errors will be the sum of the errors induced by the large structure and those induced by nearby surface irregularities (the latter errors are typically as large as the overall obstacle errors and may be of the same or opposite sign). For an example of the type of errors involved in shipboard locations see Thompson (1975).

The magnitude $|p'|$ of the turbulent pressure fluctuations at the bottom of the atmospheric boundary layer (Elliott, 1975) is typically ten times smaller than the dynamic pressure contamination $\rho\bar{U}u'$ measured by any sensing head placed in the turbulent flow. Therefore to obtain a signal-to-noise ratio of 10 over most of the microscale frequency range the sensing head must have a pressure coefficient C_p of 0.01 or less. Variations of C_p with pitch angle are compared in Figure 3 for three recently developed microscale sensing heads, due to Elliott (1972a), Snyder et al. (1974), and Miksad (1976).

The Elliott and Snyder probes were designed for work in turbulent boundary layers over the sea, where pitch angles rarely exceed 10° and the typical turbulent pressure fluctuations are ~1 Pa rms (see Fig. 1). The Miksad head was designed for measuring 0.001 to 0.1 Hz 'mesoscale' pressures accurately over a wide range of pitch angles at an exposed location. Whereas the Miksad probe is meant to be omnidirectional, the Snyder probe must be mounted on a vane

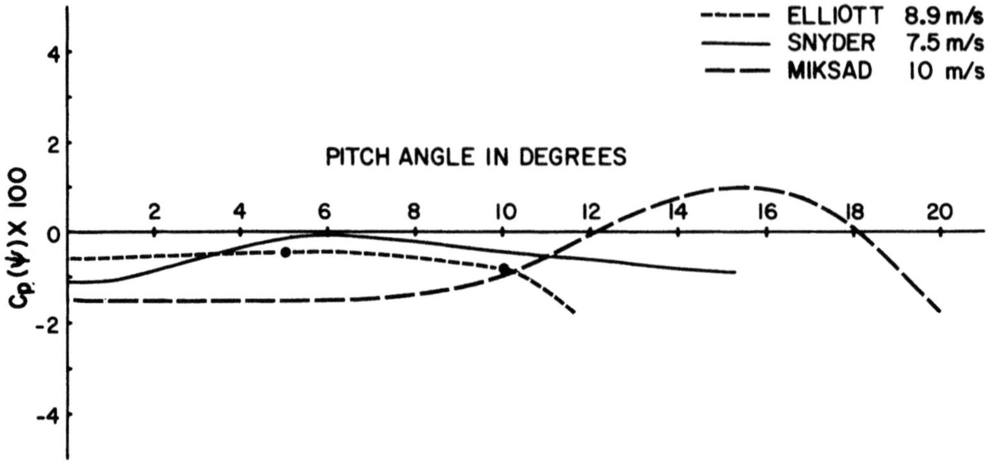

Fig. 3 Pressure coefficient versus pitch angle in degrees for three sensing heads: (a) Elliott (1972a); (b) Snyder (1974); (c) Miksad (1976).

to keep it within 15° of the instantaneous wind direction, and the Elliott probe has a fixed-head version and a vertically mounted vane-directed version (Elliott, 1975) for use in turbulent boundary layers on land, in convective conditions.

6.2 Port Blockage, Drainage, and Size

One of the most serious problems for work at sea is how to detect the presence of port blockage. Often the signal from a partially-blocked head is indistinguishable from the real pressure signal. If the sensor is attended, the head must be carefully kept clean and dry prior to each measurement, and checked for blockage afterwards. If the probe is not attended, some way must be found either of accepting dynamic contamination or of monitoring the head's pressure coefficient. If the sensing head is on a buoy the buoy pitch may be monitored and correlated with the measured pressure fluctuations in the same frequency band. If the head is fixed, then the wind speed may be measured and correlated with the measured pressure. To the writer's knowledge no in situ monitoring of sensing head pressure coefficient has been attempted, either for buoys or for microscale sensors.

Two rather crude approaches may be used to minimize the effects of port blockage: either (i) shelter the port (the shelter, of course, will influence the head's dynamic pressure sensitivity) and allow for drainage, or (ii) cover it completely. If approach (i) is taken, the drainage path must not itself be a source of pressure for the sensor, and the only solution is on land, to drain it into the ground, and at sea, to the water. The pressure-sensing orifice should be made as large as possible, the upper limit being set by the permissible dynamic pressure sensitivity of the orifice itself (Franklyn and Wallace, 1970). If a flexible flush diaphragm is used to close the pressure port of a barometer [approach (ii) above] it considerably reduces the barometer's pressure sensitivity, and encloses a volume of air between it and the transducer. This volume of air manufactures spurious pressure fluctuations p_s of its own, according to

$$\Delta p_s = \frac{\Delta T_a P_a}{T_0}$$

where ΔT_a is any externally-driven temperature fluctuation and P_a is the average pressure of the enclosed volume. Materials with low temperature coefficients of expansion must be used, and the enclosed volume must be sufficiently large that volume changes caused by diaphragm movement may be ignored. If the diaphragm is mounted on a moving base such as a buoy, it will respond to accelerations along its axis. This effect can be filtered out for barometric measurements, but can cause large errors in systems designed to study pressures coherent with the source of motion of the sensor (such as occur in wave generation problems).

7. PNEUMATIC FILTERING

Because most differential sensors lack the dynamic range to resolve turbulent pressure fluctuations with amplitudes of 10^{-2} Pa in the presence of barometric variations with amplitudes of up to 10^3 Pa, it is necessary to suppress the comparatively low-frequency barometric variations with pneumatic high-pass filters. Such filters also serve to suppress pressure fluctuations caused by slowly-varying temperature changes of the air in the reference volume and transducer.

Pressure signals may be pneumatically filtered by building into the plumbing of the sensor narrow tubing (resistance) and air volumes (capacitance). The analogue with an electrical circuit is exact for isothermal conditions, with pressure corresponding to voltage and volume flow rate to current; pneumatic resistance has units of pressure/volume flow rate or Pa s m^{-3}, and pneumatic capacitance has units of volume/pressure, or m^3 Pa^{-1}. A typical pneumatic bandpass filter, after Priestley (1966), is shown in Figure 4. Filter characteristics (the time constants $\tau_i = R_iC_i$) are determined by making accurate measurements of the volume flow rate-pressure relation of the orifice 'resistors' and of the volumes of the 'capacitors', including the interior of the transducer.

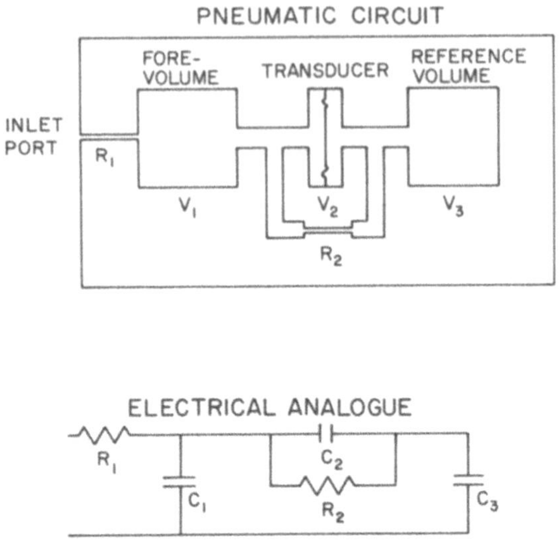

Fig. 4 Pneumatic low-pass filter and its electrical analogue for turbulence pressure transducers (after Priestley, 1966).

'Resistors' are typically capillary tubes of a material, such as stainless steel, with a low temperature coefficient of expansion. They are trimmed by changing their length. Volumes must be well insulated thermally; for low-frequency measurements they can be kept isothermal by filling with fine steel wool. Dewar flasks are often used as volumes, but they are fragile. To prevent the volume they contain from being subjected to large temperature fluctuations in the field they must be shielded from the sun (or other sources of radiant heat).

8. CALIBRATION

Pressure sensors are normally calibrated by comparison with a standard. In the case of barometers, that standard is the mercury barometer, in which the height is measured of the column of pure mercury which is supported by the pressure to be measured. Middleton and Spilhaus (1953) quote the long-term absolute accuracy of a carefully corrected and regularly calibrated mercury barometer of the 'Fortin' type to be about ±10 Pa, and the error in individual readings to be two or three times that. Martin (1973) found it impossible to obtain even this accuracy with his Fortin transfer standards in the Arctic, in spite of elaborate precautions. The mercury barometer is nevertheless still regarded as the standard against which all other barometers are calibrated.

Differential sensors may be calibrated statically using a U-tube manometer. Many such devices are commercially available with resolutions down to 10^{-2} Pa and full scale ranges of up to 10^3 Pa. They are used extensively in wind tunnel work with pitot-static tubes, and in measurements of small mean pressures.

All other methods of measuring pressure or of producing an accurately known pressure depend on some other instrument for knowledge of the calibration pressure. The methods can be divided into static and dynamic, and are discussed briefly below in that order.

The deadweight tester consists of a piston loaded with accurately-known weights, and supported by fluid pressure on its lower side. Cross (1964) gives a good account of the various corrections which must be applied. The static calibration of high-frequency pressure transducers may be accomplished using any of the techniques already described; manometers are most commonly used for this purpose.

Some excellent techniques (Brüel, 1964) for dynamic (frequency response) calibration were originally designed for use with microphones. The best of these is the so-called 'piston-phone'. A motor-driven cam drives a piston which oscillates through a seal, sinusoidally varying the volume of a calibration chamber to which the test transducer is coupled. With care the pressure amplitude in the calibration volume can be calculated with an accuracy of ±0.2 Pa. The accuracy is almost wholly dependent on the accuracy of the piston amplitude measurement.

A simpler variation of the piston-phone is often used in the 0.01 to 10 Hz frequency band, for relative dynamic calibration of sensing systems used for the measurement of turbulent pressures. The calibration volume is varied sinusoidally with either a mechanical or an electronic driver (motor-driven cam or permanent-magnet speaker coil), and the interior pressure is measured by a 'standard' transducer and the system to be calibrated. By varying the driver frequency a complete response curve can be drawn over the frequency range of interest. The only constraint on amplitude is that to obtain a useful signal/noise ratio it be much greater than ambient air pressure variations, which can amount to 10 Pa rms or more in the 0.01 to 10 Hz frequency band in a building on a windy day. A typical system of this type is described in Elliott (1972a). Relative accuracies of ±1% in amplitude and 5° in phase are obtainable with care. The technique is limited at high frequencies by Helmholtz resonances in the volume, and at low frequencies by leakage from the volume.

The frequency response of mesoscale sensing systems, in the 10^{-4} to 10^{-2} Hz frequency band, can be calculated by forming the electrical analogue of the system (Section 7) or measured by recording time series of ambient air pressure measured by the sensing system and by an absolute pressure gauge (the technique can in fact be used for any relative dynamic calibration). The frequency response is then, in the absence of appreciable instrumental noise, given as the ratio of the cross spectrum between the time series of standard and system and the power spectrum of the standard sensor (see, for instance, Jenkins and Watts, 1968).

9. Error Signals and Noise

Error signals may be divided into two parts, incoherent noise and coherent but spurious error signals; the term 'coherent' means that the error signal will form part of the cross spectrum between the pressure and any other variable with which it is being correlated.

One of the most serious sources of incoherent and coherent error signals, particularly for turbulence sensors, is dynamic pressure contamination, and ways and means of minimizing it are discussed in detail in Section 6. A typical example applies to pressure sensors mounted on buoys. If the sensing head or pressure port becomes encrusted with ice or salt crystals or corrosion products, its pressure coefficient will be changed in an unknowable way. Other sources of error excluding electronics noise (offset and gain fluctuations in amplifiers, etc.), are hysteresis, thermal zero shift, and drift of the transducer springs, ambient temperature and radiation, temperature and strain sensitivity of the reference volume of differential sensors and strain and acceleration sensitivity of sensor tubing. Noise can be considerably reduced, in springy systems, by rapidly cycling the instrument through wide ranges of temperatures and pressure before it is used in the field (sometimes called 'burn-in' or 'aging').

Sources of coherent error signals include electronic cross-talk from capacitive or inductive pickup or from mutual sensitivity to power supply variations, and for pressure-wave correlations, dynamic pressures caused by wave-induced variations of the pitch and roll angles of the sensing head, and acceleration sensitivity of the transducer and of fluids in internal tubing (Section 10). Low-pass filtering will not completely remove the dynamic pressure offset $\frac{1}{2}\rho C_p\overline{U}^2$ due to the action of the mean wind U on a sensing head with mean pressure coefficient C_p; nor will it completely remove the errors due to fluctuations $\overline{\frac{1}{2}\rho C_p(\psi)U^2}$ in dynamic pressure caused by a buoy's tilting and the variation of C_p with tilt angle ψ, since the tilt sensitivity of the sensing head is normally nonlinear (see, for instance, Miksad, 1976).

After having expended all reasonable efforts to minimize the noise of the sensing system from all sources, the residual noise within the frequency range of interest must be measured. The proper way to make a noise measurement is to arrange the system exactly as it is to be used in the field and measure its output in the presence of no input pressure fluctuations, say by blocking the instrument's sensing ports. This is generally very difficult to do with high-resolution pressure sensors, since blocking the ports affects the instrument's frequency response and it is next to impossible to make an enclosure in which there are no fluctuations of pressure, temperature, or volume. If the experimenter has access to a standard transducer with known low noise, then it is simple to measure the same pressure field with standard and unknown sensors and correlate their signals; the incoherent residual is then the noise of the unknown sensor.

10. THE USE OF PRESSURE SENSORS ON WAVE-FOLLOWING DEVICES

10.1 Techniques

Both buoys (Longuet-Higgins et al., 1963) and servo-driven devices (e.g., Peep, 1972) have been used to support pressure sensors. Servo-driven wave-following devices are discussed in Shemdin (Chapter 33). The most important practicalities for pressure measurements from such devices are discussed below.

10.2 Sources of Error

The single most important source of error in the measurement of wave-coherent pressures is generated by the waves themselves. The potential flow solution for the wave-induced pressures at a fixed point in a uniform air flow U_0 above waves ζ of speed c is (Lamb, 1945)

$$P_w(t) = -\rho g\zeta(1 - U_0\cos\theta/c)^2.$$

where θ is the angle between wind and wave directions. In a wind-driven sea the wavelength increases until c approaches $U_0\cos\theta$, and since U_0 and c have the same sign, the term in brackets is small. If other wave fields are present, for which c is negative (waves, such as swell, travelling against the wind), the term in brackets can become > 1, completely masking the pressures from growing waves. It is remarkably difficult to avoid this problem; see Snyder et al. (1974), Latif (1974), and Dobson and Elliott (1978). If successful measurements of wave-coherent pressures are to be made no effort must be spared to find a site where the presence of back-scattered wave components and swell is a minimum. In addition it is considered essential to obtain careful measurements of the directional spectrum of the wave field and, if possible, of the pressure field in the frequency band of the wind-driven sea as well as at swell frequencies.

Potential flow theory shows (Longuet-Higgins et al., 1963) that in a uniform air flow of speed U above waves with elevation $\zeta(t)$ and phase speed c the wave-coherent pressure signal $\underset{\sim}{p}(t)$ which would be observed from a sensor moving with the water surface, such as a surface-following buoy, is

$$\underset{\sim}{p}(t) = -\rho g\zeta(t)\{1 + (1 - U_0\cos\theta/c)^2\}$$

since to first order the buoy moves through a vertical air pressure gradient

$$\frac{\partial \underset{\sim}{p}}{\partial z} = -\rho g$$

Therefore the magnitude of the signal from the moving sensor is larger than that of a fixed sensor by $-\rho g\zeta(t)$. This turns out to be a major limitation on the accurate measurement of the amplitude of the wave-coherent pressure (see Dobson and Elliott, 1978). Over most of the spectral region in which the pressure and wave height are coherent, the residual cospectrum after $-\rho g\zeta(t)$ is corrected for, is typically one order of magnitude less than the original (see 'Error Correction'). Calibration errors in the wave and pressure sensors become critically important, and in practice only poor estimates can be made of the in-phase component of the pressure, and therefore of the phase of the pressure with respect to the waves. To first order the wave height-pressure quadrature spectrum is fortunately unaffected, and accurate estimates may be made of the wave-supported energy and momentum fluxes.

Although it produces only small corrections to the pressure signal, wave-follower tracking error can make a substantial contribution to the wave height pressure quadrature spectrum. If the tracking error is measured by a servo system, a correction term $\rho g\varepsilon$, where ε is the measured tracking height error, may be added to the pressure signal. If tracking errors are large and it is felt the pressure gradient near the wave surface is really not $-\rho g$, it is possible to keep the pressure-sensing head at nearly constant height by mounting it a short distance Δx down-wave from the wave sensor to introduce a phase lead of

$$\Delta\phi(f) = \frac{\omega^2 \Delta x}{2\pi\ g} \text{ radians.}$$

This phase lead approximates the generally nonlinear phase response of the servo system and reduces the resultant tracking phase error of the pressure-sensing head to a few degrees, over a frequency range determined by Δx and the frequency (phase) response of the wave follower.

Pressure sensors also measure acceleration perpendicular to the plane of the sensing diaphragm. It is common practice to mount the pressure sensor on a wave-following device with its most sensitive axis horizontal and pointed in the cross-wave direction. Because buoy tilts or piston flexure can cause significant wave-coherent accelerations in the cross-wave direction it is good practice to measure the accelerations along the sensitive axis and correct the measured pressure signal accordingly. Any enclosed vertical air columns within the sensor system, when accelerated, will produce a pressure signal

$$\rho \Delta h \partial^2 \zeta(t) / \partial t^2$$

where Δh is the effective length of the air columns leading from sensing head and reference volume to the transducer.

10.3 Error Correction

All of the effects listed in the preceding section must be corrected for in the analysis of data from a wave following device. If the pressure sensor is mounted in a buoy, then tracking errors are small but flow blockage becomes more significant since the attack angle of the buoy varies in a wave-coherent way (see Dobson, 1971).

When correcting the pressure and wave signals one must be careful to take frequency response into consideration. It is simple, using either analogue or digital techniques, to add signals $-\rho g \pi(t)$ and $+\rho g \varepsilon(t)$ to the recorded pressure time series where $\pi(t)$ and $\varepsilon(t)$ are respectively wave follower vertical position and tracking error. If, however, the pressure sensor has a frequency response which must be corrected for, all the corrections are best done after spectra of the time series are produced.

11. PRESENT DEVELOPMENTS

11.1 Barometry

Design work is in progress which should result in omnidirectional sensing heads for use at sea, which do not easily become clogged by spray or affected by icing, and which have pressure coefficients less than 0.1 over all expected angles of pitch and yaw. The next step must be to produce designs which also allow in some way for changes in buoy freeboard (the effect is large: the pressure coefficient can be a strong function of the amount of the structure beneath it which is above the waterline).

11.2 Micrometeorology

The pressure-velocity correlation divergence term in the turbulent energy balance equation has been little studied, particularly in various conditions of atmospheric stability (see Section 4, Wyngaard and Coté (1971), and McBean and Elliott, 1975). Although the term appears to be small in conditions of neutral stability, its importance in the turbulent energy budget appears to increase with increasing instability. Since the pressure-velocity correlations are the only mechanism by which turbulent energy can be redistributed among the velocity components, it is fair to say that our understanding of the turbulent flow in the lower atmospheric boundary layer is not satisfactory at the moment. The turbulent energy budget and the budgets for the variances of the individual velocity components in the turbulent momentum equations need to be studied in the large variety of boundary-layer flows encountered in nature.

The problem in the past has always been to design a pressure sensing head which maintains its low pressure coefficient in the face of large fluctuations in the incident wind direction (for instance, in highly convective situations in the atmospheric boundary layer). To get around the problem Elliott (1975) and Dobson and Elliott (1978) have used a pressure-sensing disc mounted vertically on a vane, leading the pressure signal via an 'O' ring seal to a fixed transducer; they have found it very difficult to maintain a good pressure seal while allowing the vane sufficient freedom to respond to the most energetic of the wind direction fluctuations, in order to keep the angle of the sensing head within its stall angle (typically ±15°) with respect to the wind vector. Reliable, well understood sensing heads are required, in order that the measurement of pressure in the atmospheric boundary layer will become accessible to anyone but the experienced specialist.

11.3 Wave Generation

Recent developments (Snyder et al., 1978) indicate that before the energy and momentum balance of the wave field is understood, it will be necessary to study wave-pressure correlations over very 'young' (short fetch or duration) and over 'fully-developed' fields of waves in the open ocean. The interactions of turbulence with the wave-induced flows near the sea surface have been tantalizingly little studied (see Elliott, 1972c). An investigation of the reasons for the different scaling relationships in the turbulence over land and over growing waves would undoubtedly help to clarify the effect of wave-turbulence interactions on wave growth rates (see Miles, 1967).

The instruments to make the above-mentioned studies are being developed at present or already exist. A wave-following device, operated at low servo gain so that (a) high-frequency vibrations do not swamp the signal being sought but (b) the sensor package is maintained at a safe distance from the surface of the larger, low-frequency waves, would be ideal for investigating, in fetch-limited conditions, wave-turbulence interactions in the air, and the high-frequency energy and momentum fluxes to the wave field. For fully-developed waves, a buoy similar to that used by Longuet-Higgins et al. (1963) is being developed by this writer, and has already been used in the JASIN (Joint Air-Sea INteraction) 1978 experiment to make some preliminary wave growth measurements.

It is hoped that the almost mystical fear expressed by fluid dynamicists of making measurements of pressure fluctuations in the air will soon be overcome so that the field becomes available to a wider variety of investigators; there are many doors left to be opened.

REFERENCES

BRITISH METEOROLOGICAL OFFICE. 1962. A static pressure head for use with precision aneroid barometers. BMO 16, Design Study 15, File M22997/62. British Meteorological Office, London Road, Bracknell, Berkshire, RG12 2S2, England, 3 pp.

BRUEL, P.V. 1964. Accuracy of condenser microphone calibration methods. Part I. Brüel and Kjaer, Copenhagen, Denmark. Technical Review 4: 1-29.

CROSS, J.L. 1964. Reduction of data for piston gauge pressure measurements. U.S. National Bureau of Standards, Monograph 65, 8 pp.

CUNNING, J.B. 1974. The analysis of surface pressure perturbations within the mesoscale range. *Journal of Applied Meteorology,* 13: 325-330.

DOBSON, F.W. 1971. Measurements of atmospheric pressure on wind-generated sea waves. *Journal of Fluid Mechanics,* 48: 91-127.

DOBSON, F.W. and J.A. ELLIOTT. 1978. Wave-pressure correlation measurements over growing sea waves with a wave follower and fixed-height pressure sensors. In: *Turbulent Fluxes Through the Sea Surface, Wave Dynamics, and Prediction,* edited by A. Favre and K. Hasselmann, Plenum, New York: 421-432.

ELLIOTT, J.A. 1972a. Instrumentation for measuring static pressure fluctuations. *Boundary-Layer Meteorology,* 2: 476-495.

ELLIOTT, J.A. 1972b. Microscale pressure fluctuations measured within the lower atmospheric boundary layer. *Journal of Fluid Mechanics,* 53: 351-383.

ELLIOTT, J.A. 1972c. Microscale pressure fluctuations near waves being generated by the wind. *Journal of Fluid Mechanics,* 54: 427-448.

ELLIOTT, J.A. 1975. The measurement of pressure fluctuations in the atmospheric boundary layer. National Centre for Atmospheric Research (NCAR), *Atmospheric Technology,* 7: 30-32.

FRANKLIN, R.E. and J.M. WALLACE. 1970. Absolute measurements of static-hole error using flush transducers. *Journal of Fluid Mechanics,* 42: 33-48.

GOSSARD, E.E. 1960. Spectra of atmospheric scalars. *Journal of Geophysical Research,* 65: 3393-3351.

GOSSARD, E.E. and W.H. HOOKE. 1975. *Waves in the Atmosphere.* Elsevier, Amsterdam, 456 pp.

HASSE, L. 1973. Note on the surface-to-geostrophic wind relationship from observations in the German Bight. *Boundary-Layer Meteorology,* 6: 197-201.

JENKINS, G.M. and D.G. WATTS. 1968. *Spectral Analysis and Its Applications.* Holden-Day, San Francisco, 525 pp.

LAMB, H. 1945. *Hydrodynamics.* Dover Publications, Inc., New York, 1945 edition, 738 pp.

LATIF, M.A. 1974. Acoustic effects on pressure measurements over water waves in the laboratory. Department of Coastal and Oceanographic Engineering, University of Florida, Gainesville, Technical Report 25, 123 pp.

LONGUET-HIGGINS, M.S., D.E. CARTWRIGHT and N.D. SMITH. 1963. Observations of the directional spectrum of sea waves using the motions of a floating buoy. In: *Ocean Wave Spectra,* Prentice-Hall: 111-132.

LUMLEY, J.A. and H.A. PANOFSKY. 1964. *The Structure of Atmospheric Turbulence.* John Wiley and Sons, New York, 239 pp.

MARTIN, P. 1973. Barometric measurements from buoys during AIDJEX 1972. Arctic Ice Dynamics Joint Experiment Bulletin 22. AIDJEX, Division of Marine Resources, University of Washington, Seattle, Wash., 98105, U.S.A.: 89-111.

McBEAN, G.A. and J.A. ELLIOTT. 1975. The vertical transports of kinetic energy by turbulence and pressure in the boundary layer. *Journal of Atmospheric Sciences,* 32: 753-766.

MIDDLETON, W.E. and A.F. SPILHAUS. 1953. *Meteorological Instruments.* University of Toronto Press, 286 pp.

MIKSAD, R.W. 1976. An omni-directional static pressure probe. *Journal of Applied Meteorology,* 15: 1215-1225.

MILES, J.W. 1957. On the generation of surface waves by shear flows. *Journal of Fluid Mechanics,* 3: 185-204.

MILES, J.W. 1967. On the generation of surface waves by shear flows. Part 5. *Journal of Fluid Mechanics,* 30: 163-175.

PEEP, M. 1972. A wave follower for field study of air-sea interactions. *Institute of Electronic and Electrical Engineers (IEEE) Transactions on Geoscience Electronics GE-10:* 24-32.

PIKE, J.M. and D.W. BARGEN. 1976. The NCAR digital barometer. *Bulletin of the American Meteorological Society,* 57: 1106-1111.

PRANDTL, L. and O.G. TIETJENS. 1957. *Applied Hydro- and Aero-Mechanics.* Dover Publications, Inc., New York, 311 pp.

PRIESTLEY, J.T. 1966. Correlation studies of pressure fluctuations on the ground beneath a turbulent boundary layer. U.S. Department of Commerce, National Bureau of Standards Report 8942, 92 pp.

PROUDMAN, J. 1953. *Dynamical Oceanography.* Methuen and Co., London, 409 pp.

RASMUSSEN, G. 1960. Pressure equalization of condenser microphones and performance at varying altitudes. Brüel and Kjaer, Copenhagen, Denmark, Technical Review 1 (1960): 1-23.

ROBERTSON, P. 1972. A direction-insensitive static head sensor. *Journal of Physics E, Scientific Instruments,* 5: 1080-1083.

SIEBERT, M. 1961. *Atmospheric Tides. Advances in Geophysics 7.* Academic Press, New York: 105-182.

SHEMDIN, O.H. and E.Y. HSU. 1967. Direct measurement of aerodynamic pressure above a simple progressive gravity wave. *Journal of Fluid Mechanics,* 30: 403-416.

SNYDER, R.L. 1974. A field study of wave-induced pressure fluctuations above surface gravity waves. *Journal of Marine Research,* 32: 497-531.

SNYDER, R.L., R.B. LONG, J. IRISH, D.G. HUNLEY and N.C. PFLAUM. 1974. An instrument to measure atmospheric pressure fluctuations above surface gravity waves. *Journal of Marine Research,* 32: 485-496.

SNYDER, R.L., R.B. LONG, F.W. DOBSON and J.A. ELLIOTT. 1978. The Bight of Abaco pressure experiment. In: *Turbulent Fluxes Through the Sea Surface, Wave Dynamics, and Prediction,* edited by A. Favre and K. Hasselmann, Plenum, New York: 433-444.

THOMPSON, N. 1975. Shipboard pressure measurements during JASIN 1972. *Meteorological Magazine,* 104: 157-179.

WILLMARTH, W.W. 1975. Pressure fluctuations beneath turbulent boundary layers. *Annual Reviews of Fluid Mechanics,* 7, Annual Reviews, Inc., Palo Alto, CA: 13-38.

WYNGAARD, J.C. and O.R. COTE. 1971. The budgets of turbulent kinetic energy and temperature variance in the atmospheric surface layer. *Journal of the Atmospheric Sciences,* 28: 190-201.

14

Slow-Response Temperature Sensors

E.L. Deacon

1. INTRODUCTION

In the study of heat, moisture and momentum transfers between sea and atmosphere relevant observations involving slow-response temperature sensors are those of: air temperatures, dry bulb and wet bulb, sea surface temperature, especially their air-sea differences, and temperature profiles (to ~10 m), dry bulb and wet bulb. The air-sea differences are particularly important as the flux of sensible heat is directly related to the difference in temperature between the sea surface and the air, while the total flux of heat (sensible plus latent) is directly related to the difference between the sea surface temperature and the wet bulb temperature of the air (Montgomery, 1948). For the absolute values of air and sea temperatures an accuracy within ±0.2°C is sufficient. However, for the air-sea differences ±0.1°C should be aimed at and profile observations are more exacting still. This follows from the fact that observations in the lowest few metres over the sea show that

$$\frac{\Delta\theta}{\Delta(\ln z)} \simeq 0.1\ (\theta_a - \theta_s) \qquad (1)$$

where θ is potential temperature, $(\theta_a - \theta_s)$ is the air-sea difference and z is height above the surface. Profile studies therefore require the temperature differences between the various observation levels (1, 2, 4, 8 m etc.) to be measured to ±0.01°C. Such an accuracy is likely to be unattainable in the case of unattended installations or buoys and is not easily achieved under more favourable conditions. In practical applications the relationships between the transfers and wind speed and the air-sea differences are of prime importance and the latter are fortunately much more easily observed with sufficient accuracy than the vertical gradients in the air.

Air-sea differences in temperature over the ocean range mainly between 3 and -8°C. However, within a few kilometres of the land greater differences can be experienced with offshore winds.

2. TYPES OF SENSORS

2.1 Thermocouples

Thermocouples are frequently used for the measurement of temperature and may therefore be conveniently applied to the measurement of air-sea temperature differences and to temperature gradient measurement. For recording the absolute value of a temperature they are less convenient owing to the need for a reference junction held at a known temperature. The output is relatively small, e.g. 40 μV/°C for copper-constantan, but this is no great disadvantage now that stable dc amplifiers are readily available. In measuring the temperature difference over a fairly small spatial interval the output can be increased by employing a number of junction pairs in series, thermopile fashion. The output μV/°C varies somewhat with temperature, as may be seen from the data given for various thermocouple combinations in Smithsonian Physical Tables and also by Benedict and Ashby (1962). Between 0 and 30°C the increase amounts to 7% for copper-constantan and 4% for Chromel-Alumel.

Copper leads can be used when recording at a distance by following the arrangement shown in Figure 1. The junctions of the leads to each thermoelectric pair A and B are potted and clad in such fashion that the temperature differences within the pots are negligibly small. Alternation of layers of conducting and insulating material

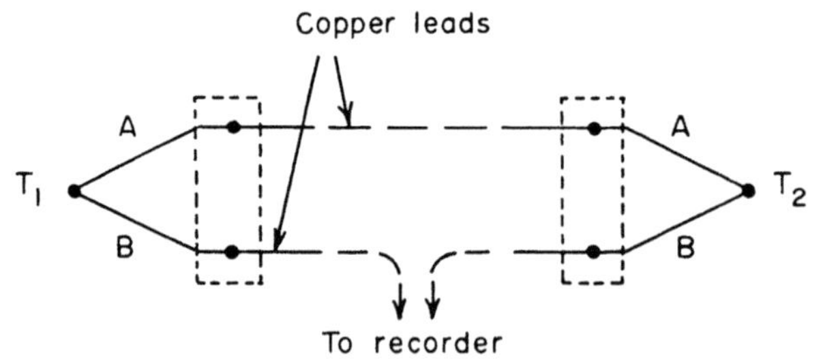

Fig. 1 Thermocouple arrangement using thermoelectric pair A and B with copper leads to measure temperature difference $T_1 - T_2$.

gives added assurance of isothermal conditions in the interior. Switches should be selected not to introduce unwanted emfs.

2.2 Resistance Thermometers

Resistance thermometers are also frequently used for both temperature and temperature difference measurement. For the latter, resistance elements, usually of platinum, are used in a bridge circuit as in Figure 2a. The arms AB and BC are equal fixed resistors

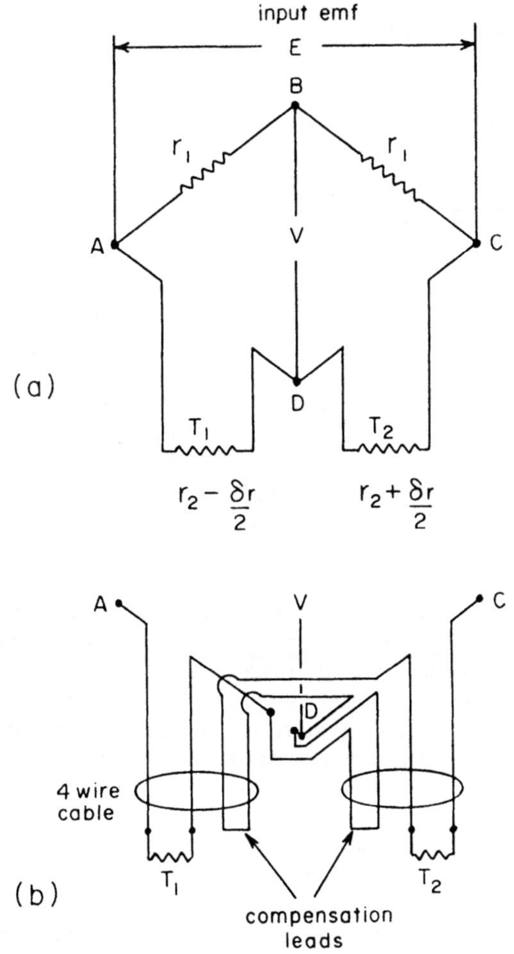

Fig. 2 a. Basic resistance-thermometer bridge for temperature difference measurement.
b. Arrangement of lower half of bridge circuit with the inclusion of compensation leads.

while arms AD and DC are the thermometer elements of resistance r_2 at temperature $(T_1 + T_2)/2$. In the usual application BD is a virtually infinite impedance (e.g. with negative feedback amplification of the output emf) and so the bridge may be viewed as two voltage dividers whose difference is the bridge output voltage V. This gives

$$V = E\{r_1/2r_1 - (r_2 - \tfrac{1}{2}\delta r)/2r_2\} = E\delta r/4r_2 \qquad (2)$$

Then with the usual approximate expression for the variation of resistance with temperature

$$r_2 = r_0\ (1 + \alpha T) \qquad (3)$$

in which α is the temperature coefficient of resistance, T is temperature, °C, and r_0 is the resistance at 0°C, one derives

$$V = \frac{E\ \alpha\ (T_2 - T_1)}{4(1 + \alpha\ T_a)} \qquad (4)$$

where T_a is the ambient temperature $(T_1 + T_2)/2$. So for a given resistance material the sensitivity of the bridge varies with ambient temperature as $(1 + \alpha\ T_a)^{-1}$. Over the range 0 to 30°C this amounts to a decrease of 12% in the case of platinum elements (α = 0.0039). With nickel elements α increases with temperature sufficiently for it to be possible to secure a constant sensitivity (Grant and Hicks, 1962). However, it is more usual in meteorological work to use commercially available platinum elements and correct for the variation in sensitivity of the bridge with ambient temperature.

Where long leads are involved compensation leads as shown in Figure 2b can counterbalance temperature and length differences. However, it is often possible to avoid long leads in the critical parts of the bridge circuit by keeping the fixed arms together with the elements on the temperature gradient mast. Then only long output leads are required and these present no problem (e.g. Collins, 1965).

For the measurement of temperature differences it is necessary to use closely matched pairs of thermometer elements. These may either be selected out of a suitably large batch or matching may be effected by shunt and parallel resistances applied to the member of a pair having the larger temperature coefficient. As an example of the degree of variation found with commercially available elements it was found that a batch of 18 platinum elements (nominal resistance 100 Ω at 0°C) were of equal resistance at 15°C to within ±0.1%. The α values varied within the limits ±1.2% (Collins, 1965).

The voltage applied to the bridge is limited by the maximum permissible rise in temperature of the elements caused by joule heating. For cylindrical elements in a transverse current of air of velocity v, the temperature rise can be calculated from Equation 6 below. With axial ventilation the temperature rise will be nearly twice as great. As the dissipation in the element is proportional to E^2/r_2, it follows from Equation 4 that, with a given size of element limited to a given temperature rise, the bridge sensitivity may be increased by a factor, f, by increasing the element resistance by the factor f^2. With high resistance elements lead resistance variations and unwanted thermal emfs become less important. The latter can be minimized by using fixed resistors wound with wire having both a low temperature coefficient and a low thermal emf against copper.

2.3 Thermistors

Thermistors are resistance elements with a very high negative temperature coefficient of resistance. Their characteristic variation corresponds approximately to

$$R = a \exp(b/T) \tag{5}$$

where T is in K. At 15°C the percentage change in resistance per degree is about 10 times that of platinum. This combined with a relatively high resistance, typically in the range 10^3 to 10^5 Ω makes them attractive for temperature measurement in many geophysical applications.

Thermistors are available in a variety of different forms: beads, rods, or discs. The dissipation of heat is better from rods than from beads, so rods can give a greater sensitivity. However, a coating impervious to gas is necessary otherwise traces of such gases as H_2S and NH_3 have a deleterious effect causing a progressive change in calibration.

A thermistor bridge arrangement, capable over some 30°C range of giving a closely linear relationship between the balance point of a linear potentiometer and temperature, is that described by Droms (1962).

A group of thermistors can be adjusted to a common calibration curve over a limited range of temperature by means of shunt and series resistors. Details of the procedure are given by Anderson (1949). Only at two temperatures is exact coincidence obtained, but within the limited temperature range involved in work at sea the departures elsewhere are acceptably small.

2.4 Quartz Crystal Thermometers

The resonant frequency of a suitably cut quartz crystal is temperature dependent and can therefore be used for temperature sensing (Wade and Slutsky, 1962). The output beat frequency is adapted to counting circuits, an advantage in obtaining the mean temperature over a period.

Some sources of the various types of thermometer element and associated equipment for recording are given in Monteith (1972).

3. CALIBRATION

3.1 Stability of Calibration

Long years of service in science and industry have proved the dependability of platinum resistance thermometers and thermocouples for temperature measurement. For a really high degree of stability of calibration platinum resistance thermometry is the method of choice. It has the further advantage where small temperature differences are to be measured, as in the observation of vertical temperature gradients over the sea, of readily providing an output emf an order of magnitude greater than with thermocouples. Thermistors are less stable than metal resistance elements. It is advisable to 'age' new thermistors by passing a suitable current through them for several days; an aging program can be established by calibrating some specimens at intervals during this treatment.

The calibrations of quartz temperature sensors have been found to be very sensitive to mechanical shock, which may disqualify them from many applications at sea.

3.2 Calibration System

For temperature difference work using platinum resistance elements some such procedure as the following is adopted. With other types of sensor the procedure is little different. A batch of resistance thermometers is carefully calibrated over a range of 0 to 30°C against a good mercury-in-glass thermometer certificated by an accredited standards laboratory. Well matched pairs may then be selected and, if necessary, final more exact matching achieved by providing the more sensitive element of each pair with suitable series and parallel resistances.

The selected pairs of thermometers are then connected to the bridge circuits and put in a well stirred water bath at about 20°C. Small series resistors are then inserted as trimmers to bring the recorder reading to zero for each pair. Two Dewar flasks are then employed as calibrating baths to calibrate each pair against a standard thermometer scaled to 0.01°C for temperature difference. The elements are interchanged between baths to give positive and negative differences. Before and after each calibration run, recordings are made with test resistors switched into the circuit in place of the thermometer elements. A small immersion heater is used to change the bath temperatures as required.

4. RADIATION SHIELDING

A thermometer element unshielded from radiation indicates a temperature considerably above the true air temperature. The magnitude of the effect for a cylindrical element (diameter d) can be estimated from the empirical relation for the forced convection of heat from cylinders.

For Reynolds number of the order of 1000 this is approximately

$$H/\Delta T = 1.5\ K(vd/\nu)^{\frac{1}{2}} \qquad (6)$$

where H is the rate of heat transfer per unit length of cylinder, ΔT is the temperature excess of the cylinder above air, K and ν are the thermal conductivity and the kinematic viscosity of air, and v is the air velocity across the cylinder. With $K = 0.26$ mW $cm^{-1}\,°C^{-1}$ and $\nu = 0.15\ cm^2\ s^{-1}$ this gives

$$\Delta T \simeq 1.0\ H(vd)^{-\frac{1}{2}} \qquad (7)$$

with H in mW cm^{-1}, v in cm s^{-1}, and d in cm. So for a cylinder with d = 0.5 cm and reflection coefficient of 0.8 in full sunshine (100 mW cm^{-2}) H is 10 mW cm^{-1}. In a 2 m s^{-1} wind, this gives ΔT = 1.0°C and even at 20 m s^{-1} it amounts to 0.3°C. It follows that efficient shielding is needed, not only from downcoming radiation, but also from radiation emitted by and reflected from the surface.

Radiation shields are often similar in design to those employed in the Assmann psychrometer; that is, the sensing elements are mounted axially inside two concentric metal sleeves which are brightly plated so as to have a high reflectivity and are aspirated at about 5 m s^{-1}. At sea aerosols from spray quickly dull a polished surface, so it would be advantageous for the outer sleeve to consist of a material of low thermal conductivity sheathed internally and externally with plated metal. With an arrangement of this general type, with the axis vertical and the intake facing downward, Slob (1978) found by wind tunnel tests that the intake air is mingled with some air which has been in contact with the exterior of the outer sleeve whenever the wind speed exceeds the aspiration rate. With normal apsiration rates of 3 m s^{-1} or more, the maximum error with his arrangement amounts to 3 or 4% of the temperature difference between the exterior of the outer sleeve and the true air temperature. Without the thermal insulation in the outer sleeve the effect is a round twice as great. When experimental conditions permit, this type of error may be avoided in all but very light winds by having the intake facing into the wind.

Heat conduction between the inner and outer sleeves and between mounting and the sensitive part of the element needs to be minimized by suitable design.

Radiation shields based on the Assmann type have been used by the Hamburg team on a special meteorological buoy (Dunckel, 1966). Each pair of elements had an aspiration fan-motor unit and the wet and dry bulb elements were in separate housings. However, with a tubular mast it is often convenient to use the mast as an aspiration duct to each level and so need only one centrifugal fan unit. Collins (1965) describes equipment of this type which economizes aspiration by housing both elements, wet and dry, in the one tubular housing, the wet bulb element downstream of the dry bulb. Equality of aspiration rate as between the various housings is secured by having an orifice plate and pressure tapping points built into each housing along with a simple throttle valve.

If observations under light wind conditions are not required then aspiration may be dispensed with by using some such arrangement as that described by Sheppard et al. (1972). Equation 7 shows that radiation effects will be reduced by using elements of small diameter.

When fine-wire thermometer elements (see Larsen et al., Chapter 15) are used to give mean air temperatures, as well as statistics of the turbulent fluctuations, then calculations of the temperature rises caused by radiation absorption and by the current through the wire need more accurate relationships than Equations 6 and 7. Those found experimentally by Collis and Williams (1959) are available for Reynolds numbers ranging from 0.02 to 140. The small difference in temperature between wire and air introduces a trace of hot-wire anemometer effect which causes some distortion of temperature fluctuation statistics (Wyngaard, 1971).

5. WET BULB TEMPERATURES

To obtain wet bulb temperatures of the high accuracy needed in profile work certain precautions need to be taken to reduce to a minimum extraneous sources of heat from:

(a) radiation,

(b) conduction of heat along the thermometer mountings and leads, and

(c) water being fed to the wet bulb sleeve at higher than wet bulb temperature.

In addition, an adequate rate of aspiration is needed and an adequate water supply.

Heat flow along the mountings can be minimized by making them of material of low thermal conductivity and of small cross section. The wet bulb sleeve should also be extended to cover more than the element proper.

To be sure of an adequate supply of water at wet bulb temperature Collins (1965) used a positive water feed via stainless steel hypodermic tubing so arranged that, for a major part of its length, it is under an extension of the wet bulb sleeve. With the higher relative humidities at sea a capillary water feed is adequate but suitable precaution is needed to keep the feed at wet bulb temperature in some such manner as that used by Sheppard et al. (1972).

Cellulose tissue is suitable for the wet bulb sheath as it holds more water than muslin and gives a rapid and even distribution of water. When wet it clings in position without tying and so is easier to renew.

The variation of wet bulb depression with air speed and element diameter has been investigated by Wylie (1949). Figure 3 summarizes his results. The air flow in these experiments was

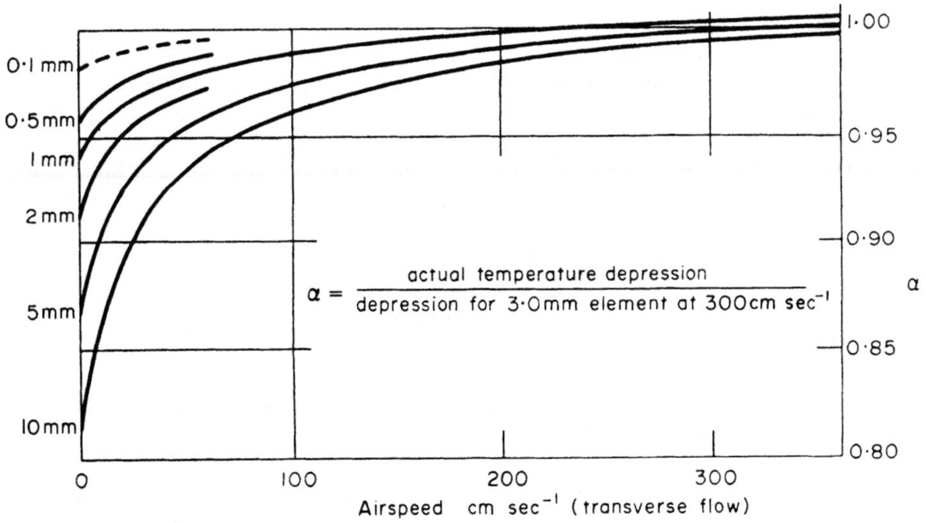

Fig. 3 Curves showing the effect of air speed and diameter on the temperature depression of a wet element (after Wylie, 1949).

transverse to the element axis. The corresponding air speeds for flow along the element would be somewhat greater.

The response time of a thermometer element is changed when it is used as a wet bulb. When it is desirable to match the time constants of dry and wet bulb element pairs, this may be effected by increasing the thermal inertia of the more responsive element, either by fitting a sleeve or by dipping it in a suitable coating medium. However, as slow-response temperature sensors are normally used to obtain mean values over periods of 10 minutes or more, a small degree of mismatch causes negligible error except when conditions are changing rapidly.

For work under conditions of below freezing wet bulb temperature the psychrometric method of measuring humidity is unsuitable for use at sea and an alternative method needs to be employed.

6. SEA SURFACE TEMPERATURE

The sea surface temperature T_s usually measured is that of the water some 0.5 m or so below the surface. The true surface temperature (see Katsaros, Chapter 16) usually differs slightly from this temperature owing to heat transfer between sea and air giving rise to an appreciable temperature gradient across the viscous sublayer on the sea surface. This difference has been investigated by Hasse (1971) whose results provide a means of making an approximate correction whenever the total heat transfer can be measured or estimated.

With stationary installations or buoys T_s may readily be measured by methods already mentioned. With ships under way several methods are available:

(a) taking bucket samples of the surface water and measuring their temperatures,

(b) taking the condenser intake temperature,

(c) using a suitable thermometer element housed in a recess in the ship's side plating, or

(d) towing a suitable thermometer element in the sea surface from a boom rigged out from the ship's side.

The relative merits of the first three methods are discussed by Roll (1965). The last mentioned has been used by the author on several cruises when a continuous record of sea surface temperature was desired. A thermistor element enclosed in a cylindrical metal housing some 100 to 200 g in weight was found to ride satisfactorily in the sea surface with the ship under way. A comparison of such sea surface temperatures bracketing temperatures taken on station by the bucket method gave agreement within 0.1°C in the mean and an rms difference of 0.2°C. The sea temperature bucket used was a vacuum flask surrounded by sponge rubber in a stout steel cylindrical bucket.

7. ENVIRONMENTAL HAZARDS

The marine atmosphere is a harsh environment for scientific equipment and particular attention needs to be given to designing sensors, mountings, etc., to be as robust and as resistant to corrosion as possible. Salt nuclei are abundant in the marine atmosphere with the result that exposed surfaces soon become coated with salt even in the absence of spray. Electrical insulation needs to be given particularly careful attention. Where routine attention is not available, the radiation shielding of thermometer elements needs to be designed to allow for deterioration in the reflectivity of polished metal surfaces.

During periods of rain and fog thermometer elements are likely to become wet and register too low. Furthermore, even after the wetting has ceased, drying may take an hour or more if the relative humidity is high. Slob (1978) has reported experiments in which a shielded and aspirated wet and dry bulb pair of elements is supplemented by a third similar dry bulb element with a heater winding enabling it to be subjected to a cycle of heating and cooling. After heating to some 30°C above air temperature, it is allowed to cool to within 0.1°C of its equilibrium temperature before its temperature is recorded along with those of the other two. From comparison of the recorder traces periods of precipitation are clearly detected, as well as periods during which the 'dry bulb' is drying off.

At relative humidities above 65% the temperature indicated by a salt-coated sensor will depart somewhat from the true air temperature whenever the relative humidity is changing appreciably. This is a result of the hygroscopic nature of salt and the heat exchanges accompanying change in hydration. With well aspirated elements this will generally have negligible effect on mean air temperatures but somewhat inaccurate temperature gradients may be recorded at such times.

Accurate wet bulb thermometry is difficult in the marine environment unless wet bulb wicks can be washed or changed daily or even more frequently under adverse conditions. With unattended installations some automatic washing of the wet bulb should be arranged for. It would also be helpful to provide a cover to the wet bulb with means for automatic removal and replacement before and after the observation period.

Unattended installations in coastal regions are welcomed by cormorants as providing convenient perches for their lengthy periods of meditation and digestion. They occur in all parts of the world and can be quite troublesome. Their habits should be borne in mind at the design stage. It may be advisable to provide perches where they can do little harm. Cup anemometers and wind vanes are tempting targets for the bored marksman and other sensors may suffer as a consequence.

REFERENCES

ANDERSON, L.J. 1949. Compensation methods for thermistor beads. *Bulletin of the American Meteorological Society*, 30: 192-193.

BENEDICT, R.P. and H.F. ASHBY. 1962. Improved reference tables for thermocouples. In: *Temperature: Its measurement and control in science and industry*, 3, part 2, Applied methods and instruments, edited by A.I. Dahl, Reinhold Publishing Co., New York: 51-64.

COLLINS, B.G. 1965. An integrating temperature and humidity gradient recorder. In: *Humidity and Moisture: measurement and control in science and industry*, 1, edited by A. Wexler, Reinhold Publishing Co., New York: 83-94.

COLLIS, D.C. and M.J. WILLIAMS. 1959. Two-dimensional convection from heated wires at low Reynolds number. *Journal of Fluid Mechanics*, 6: 357-384.

DROMS, C.R. 1962. Thermistors for temperature measurement. In: *Temperature: Its measurement and control in science and industry*, 3, part 2, Applied methods and instruments, edited by A.I. Dahl, Reinhold Publishing Co., New York: 339-346.

DUNCKEL, M. 1966. Eine Apparatur zur Messung des vertikalen Wind-Temperatur- und Feuchteprofils über dem Ozean. *"Meteor" Forschungsergebnisse, Reihe B*, 1: 45-53.

GRANT, D.A. and W.F. HICKS. 1962. Industrial temperature measurement with nickel resistance thermometers. In: *Temperature: Its measurement and control in science and industry*, 3, part 2, Applied methods and instruments, edited by A.I. Dahl, Reinhold Publishing Co., New York: 305-315.

HASSE, L. 1971. The sea surface temperature deviation and the heat flow at the sea-air interface. *Boundary-Layer Meteorology*, 1: 368-379.

MONTEITH, J.L. 1972. Survey of instruments for micrometeorology. *International Biological Programme Handbook No. 22*. Blackwell Scientific Publications, Oxford, 263 pp.

MONTGOMERY, R.B. 1948. Vertical eddy flux of heat in the atmosphere. *Journal of Meteorology*, 5: 265-274.

ROLL, H.U. 1965. *Physics of the Marine Atmosphere*. Academic Press, New York, 426 pp.

SHEPPARD, P.A., D.T. TRIBBLE and J.R. GARRATT. 1972. Studies of turbulence in the surface layers over water (Lough Neagh). Part 1. Instrumentation, programme, profiles. *Quarterly Journal of the Royal Meteorological Society,* 98: 627-741.

SLOB, W.H. 1978. The accuracy of aspiration thermometers. Scientific Report W.R.78-1, Koninklijk Nederlands Meteorologisch Inst., De Bilt.

WADE, W.H. and L.J. SLUTSKY. 1962. Quartz crystal thermometer. *Review of Scientific Instruments*, 33: 212-213.

WYLIE, R.G. 1949. *Psychrometry.* Division of Physics, CSIRO, Sydney, Report No. PA-4, 55 pp.

WYNGAARD, J.C. 1971. The effect of velocity sensitivity on temperature derivative statistics in isotropic turbulence. *Journal of Fluid Mechanics*, 48: 763-769.

15

Fast-Response Temperature Sensors

S.E. Larsen, J. Højstrup, and C.H. Gibson

1. INTRODUCTION

Measurements of temperature fluctuations in the atmospheric boundary layer are important for estimating many parameters including: heat flux, thermal stability, variability of refractive index, and temperature dissipation. In this paper we discuss the most commonly applied methods and some recently discovered difficulties that may be encountered in performing temperature fluctuation measurements above the ocean surface.

Since the responses of individual measuring systems are controlled by a multitude of environmental parameters, we have chosen a set of 'typical' conditions for which relevant properties are estimated. measuring height 5 m, mean speed 5 m s^{-1}, mean ambient temperature 20°C on a clear sunny day.

2. CHARACTERISTICS OF THE FLUCTUATING AIR TEMPERATURE

The frequency n(Hz) of turbulent fluctuations is related to wavenumber k(rad m^{-1}) and wavelength λ(m) by:

$$n = \frac{u}{2\pi} k = \frac{u}{\lambda} , \tag{1}$$

where u is the mean air velocity (m s^{-1}) and Taylor's hypothesis is used.

The dimensionless frequency f, used in connection with Monin-Obukhov scaling of spectra, is given by:

$$f = \frac{nz}{u} = \frac{z}{2\pi} k = \frac{z}{\lambda} , \qquad (2)$$

where z is the measuring height.

Of special importance for describing the behaviour of the smallest scales of turbulence are the Kolmogorov length scales for velocity η and the diffusion length scale for temperature η_θ given by

$$\eta_\theta = \eta Pr^{-\frac{3}{4}}, \; \eta = (\nu^3/\varepsilon)^{\frac{1}{4}}, \; Pr = (\nu/\kappa) \qquad (3)$$

where ν and κ are the molecular diffusivities of momentum and heat respectively ($m^2 \; s^{-1}$) and ε is the dissipation of velocity fluctuations ($m^2 \; s^{-3}$). The Prandtl number, Pr, can usually be considered a constant of about 0.7 in the atmospheric boundary layer. η_θ will typically be of the order of 1 mm.

Associated with η, the Kolmogorov wavenumber k_s and wavelength λ_s are given by:

$$\lambda_s = \frac{2\pi}{k_s} = 2\pi\eta \; . \qquad (4)$$

In Figure 1 representative normalized spectra and co-spectra involving boundary layer temperature fluctuations are plotted for neutral conditions as functions of the dimensionless frequency f. Temperature spectra over the ocean may range from curve 1, describing over-land data and some over-ocean data, to curve 2, describing most over-ocean data. This difference is, of course, also reflected in the co-spectra, but usually all co-spectra are found within the hatched area 3. Curve 4 finally displays typical behaviour of the so-called dissipation spectrum, i.e. the spectrum of $\partial T/\partial x$.

It should be noted that the spectra in the figure pertain to neutral conditions only and that their behaviour will vary with stability. The most important change in this context is that the spectra for increasing stability will slide along the frequency axis to higher frequencies, e.g. Busch (1973).

From spectra such as those in Figure 1 one can estimate (Table 1) the contribution to the total (cross) variances from different Δf-intervals.

The fluctuation level of temperature over the open ocean is usually less than over land. For design purposes, a standard deviation of about 0.1°C can be used. Furthermore, it should be noted that both over land and over water the distribution function of temperature is found to be highly skewed.

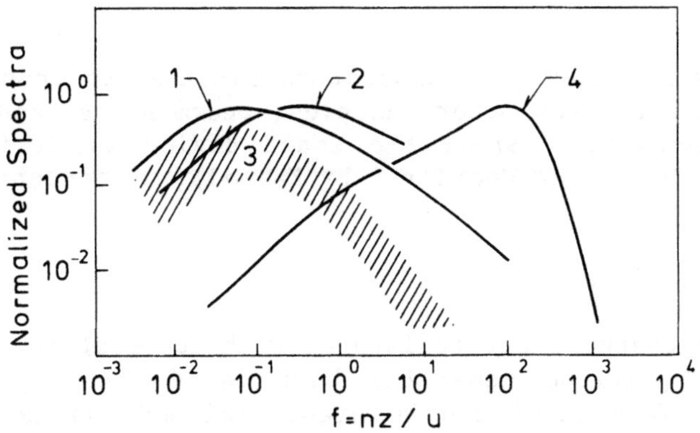

Fig. 1 Spectra under neutral conditions, normalized according to the Monin-Obukhov hypotheses (1,2) and the Kolmogorov hypotheses (3,4).

1: $nS_T(n)/T_*^2$ for over-land measurements and some over-ocean measurements (Kaimal et al., 1972; Phelps and Pond, 1971).

2: Alternative over-water $nS_T(n)/T_*^2$, compiled from Leavitt (1975), Phelps and Pond (1971), and Kruspe (1974).

3: Absolute values of logarithmic co-spectra involving temperature and vertical velocity, horizontal velocity and humidity (Kaimal et al., 1975; Friehe et al., 1975).

4: 'Dissipation spectrum', i.e. spectrum of $\partial T/\partial x$ (Williams and Paulson, 1977). The curve is related to curve 1 through multiplication by f^2 and some rescaling.

Table 1.

Relative contribution of temperature variance and $\overline{wT}$-covariance from different Δf intervals, calculated from curve 1 of Figure 1 and Kaimal et al. (1972), valid for neutral stratification; the low frequency contributions are very sensitive to stability changes.

Δf-range	0-0.01	0.1-1	1-10	10-30	30-100	100-200	300-1000
$\overline{T^2}$ (%)	52	34	11	1.7	0.9	0.37	0.03
$\overline{wT}$ (%)	50	41	8.0	-	-	-	-

3. SENSORS

The most commonly used air temperature fluctuation sensors are metallic resistance wires, thermocouples, thermistors, and sonic anemometer/thermometers. Since the sonic device is discussed elsewhere in this book (see Chapter 4 by Kaimal), we concentrate on the other three types.

3.1 Resistance Wires

These sensors change their resistance with temperature. Of all usable metals, platinum generally best meets the requirements of thermometry. This metal can be highly refined and is very resistant to corrosion. It is mechanically and electrically stable and its temperature-resistance relationship is well established. Occasionally, other metals such as tungsten and nickel are used in thermometry because of their higher strength and sensitivity, respectively.

A uniformly cylindrical wire of length, L, and diameter, D, has a resistance given by

$$R = \frac{4L\sigma^{-1}}{\pi D^2} \tag{5}$$

where σ^{-1} is the material resistivity.

The temperature variation of R for a given resistance element is usually given in terms of second- or third-order polynomials, of which the best known is the Callendar-Van Dusen equation for platinum, valid for $-185°C < T \leq 1000°C$

$$R(T) = R(0)\{1 + \alpha[T - \delta\left(\frac{T}{100} - 1\right)\frac{T}{100} - \gamma\left(\frac{T}{100} - 1\right)\left(\frac{T}{100}\right)^3]\} \tag{6}$$

where T is in °C and α, β, and δ are constants for each individual element. Typical values are $\alpha = 3.92\times10^{-3}\,°C^{-1}$, $\gamma = 0.1$ for $T < 0°C$, and $\gamma = 0$ for $T \geq 0°C$, $\delta \simeq 1.49$.

For measurements of small temperature variations in the vicinity of the temperature T_0, it is customary to use a simplified expression that can be obtained from equations such as (6) through a Taylor expansion

$$R(T) = R(T_0)\left(1 + \alpha_T(T - T_0) + \beta_T(T - T_0)^2\right) . \tag{7}$$

However, of more direct interest for measurements and for comparisons between different sensors, are the sensitivities S_1, S_2 defined by the following equation

Table 2

Typical properties of different temperature sensors at 20°C. Thermistors can have resistance values in the range indicated, independent of the geometry of the sensors. In the thermocouple group, the upper material parameters relate to copper, the middle to constantan, and the lower to chromel. The sensitivity values for the resistance sensors are based on a current-resistance product, RI = 0.07 ΩA. This value is chosen arbitrarily but it is characteristic of the values used in practice. β_T and S_2 are not given for W- and Ni-wire, since they vary greatly between individual wires. R(T) is generally less linear for those materials than for Pt.

		Temperature Coefficient α_T	β_T	Sensitivities S_1	S_2	Density ρ	Resistivity σ^{-1}	Heat Capacity C	Heat Conductivity k
Sensor Type	Sensor	(K^{-1})	(K^{-2})	($\mu V\ K^{-1}$)	($\mu V\ K^{-2}$)	(kg m^{-3})	(Ω m)	(Joule $kg^{-1}\ K^{-1}$)	(W $m^{-1}\ K^{-1}$)
Metal Wires	Pt-wire	3.7×10^{-3}	-1.1×10^{-6}	260	-0.08	2.1×10^{4}	1.2×10^{-7}	140	70
	W-wire	4.5×10^{-3}	--	315	-	1.9×10^{4}	5.5×10^{-7}	140	150
	Ni-wire	6.0×10^{-3}	--	420	-	8.8×10^{3}	6.8×10^{-7}	440	59
Thermistors	β = 2000 K	-2.3×10^{-2}	3.5×10^{-4}	1600	25	4.9×10^{3}	$10^{2}-10^{7}$ Ω	590	2
	β = 6000 K	-7.0×10^{-2}	2.7×10^{-3}	-4900	190				
Thermocouples	Copper	3.9×10^{-3}	--			8.9×10^{3}	1.9×10^{-8}	390	380
				41	0.04				
	Constantan (55% Cu + 45% Ni)	5×10^{-6}				8.9×10^{3}	6×10^{-7}	410	23
				61	0.05				
	Chromel-P (90% Ni + 10% Cr)	3.2×10^{-3}	--			8.8×10^{3}	8.5×10^{-7}	440	190

$$V(T) = IR(T) = V(T_0) + S_1(T - T_0) + S_2(T - T_0)^2 \tag{8}$$

where I is the current and V(T) the corresponding voltage drop across the sensor, and $S_1 = IR(T_0)\alpha_T(T_0)$, $S_2 = IR(T_0)\beta_T(T_0)$. Table 2 presents properties of resistance wires.

3.2 Thermistors

Thermistors (thermal resistors) are semiconductors, usually coated by glass or a metal, made of ceramic material produced by sintering mixtures of metallic oxides. They are characterized by (a) a high negative resistance sensitivity to temperature changes, and (b) a wide range of resistances (10 Ω to 10 MΩ) with all other physical properties being constant.

The resistance-temperature relation for thermistors is usually given as

$$R(T) = R(T_0)\ \exp[\beta(\frac{1}{T} - \frac{1}{T_0})]\ , \tag{9}$$

where T_0 is a reference temperature and β is a constant for each thermistor element, usually ranging in the interval 2000 to 6000 K.

Corresponding to Equation 7 for metallic resistance wires, for thermistors we can derive a small perturbation expression with temperature coefficients given by

$$\alpha_T(T_0) = \frac{1}{R(T_0)} \left.\frac{\partial R}{\partial T}\right|_{T_0} = -\frac{\beta}{T_0^{\,2}} \tag{10}$$

$$\beta_T(T_0) = \frac{1}{2R(T_0)} \left.\frac{\partial^2 R}{\partial T^2}\right|_{T_0} = \frac{\beta}{T_0^{\,3}}\left(1 + \frac{\beta}{2T_0}\right)$$

with corresponding sensitivities (Eq. 8). Properties relevant to thermistors are also listed in Table 2. The material parameters are taken from Lueck et al. (1977).

3.3 Thermocouples

Thermocouples are based on the principle of thermoelectricity illustrated in Figure 2. Let two different materials, A and B, be in contact at two points kept at the temperatures T_1 and T_2. Between the two points there will then be a difference in the electric potential, which is a function of the materials involved and the temperature difference.

It has furthermore been shown that the introduction of a third material, C, into the circuit will not change the thermoelectrical

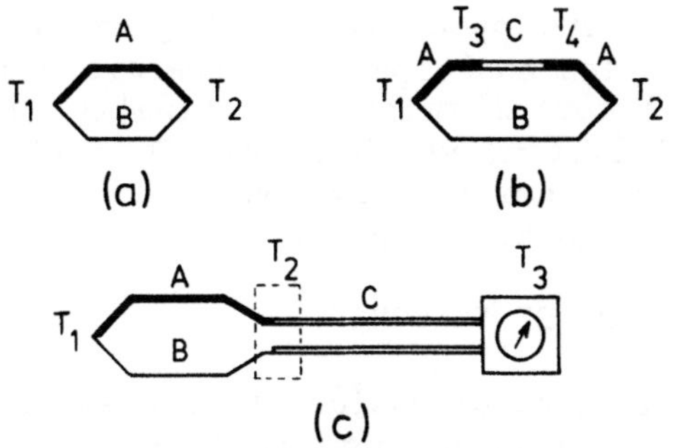

Fig. 2 Thermocouple measurement. (a) Thermoelectrical voltage between the two touching points is a function of $T_1 - T_2$. (b) Introduction of a third metal in the circuit does not change the voltage between points 1 and 2 provided that $T_3 = T_4$. (c) Principle for practical measurements.

voltage, provided that the two endpoints of materical C are kept at the same temperature.

By means of manufacturer's tables we have estimated the sensitivity coefficients S_1 and S_2, defined in Equation 11 and corresponding to Equation 8, for two commonly used thermocouples

$$V(T - T_0) = S_1(T - T_0) + S_2(T - T_0)^2 , \qquad (11)$$

where T_0 is the temperature at the reference junction.

4. PROBE CONFIGURATIONS

Figure 3 illustrates the appearance of different types of temperature sensors. Figure 3a shows the usual construction of simply spanning the wire between two prongs. The sensing element may be a metal wire used as a resistance sensor, or two different wires, welded together at the middle and constituting the one junction of a thermocouple. In the latter case, the prongs are of the same material as the corresponding wire halves. Part of the wire is plated in some resistance wire sensors (Fig. 3b), thus providing a mechanical and thermal buffer between the sensitive middle part of the wire and the prongs (Maye, 1970). This configuration is especially common for the so-called Wollaston wires, where the thin Pt wire is supplied by the manufacturer embedded in a silver coating. After the silver-coated wire has been soldered to the prongs, the silver is etched away from the middle part (Sandborn, 1972). In Figure 3c and d the resistance of a wire sensor, and thereby its sensitivity, has been increased by increasing the length of the wire. For strength and spatial resolution, the wire has been wound around a rod or a more open frame. Figure 3e shows the common shapes of thermocouples and especially thermistors. The bead shape is the most common for thermistors. Finally, Figure 3f shows a shielded aspirated system, which is rarely used for fast measurements, but which is commercially available and claimed to have better than 10 Hz frequency response.

5. SPATIAL RESOLUTION OF SENSORS

For a sensor with largest dimension, L, the smallest wavelength, λ, that can be resolved without noticeable distortion is about 10L. For decreasing λ, the signal becomes increasingly distorted, until it is finally suppressed for $\lambda \lesssim L$.

The signal is distorted partly because the sensor distorts the flow and thereby the temperature field, and partly because it averages the incoming temperature fluctuations spatially over its volume. For thin wire sensors, used in a field with velocity larger than about 1 m s^{-1}, it is a good approximation to characterize the signal distortion as a line averaging along the wire. Wyngaard (1971a) used this assumption to derive approximate expressions for the spectral attenuation of the signal for $\lambda \gtrsim L$. Most thermocouples and thermistors used for fast temperature measurements can be considered very close approximations to point sensors, since

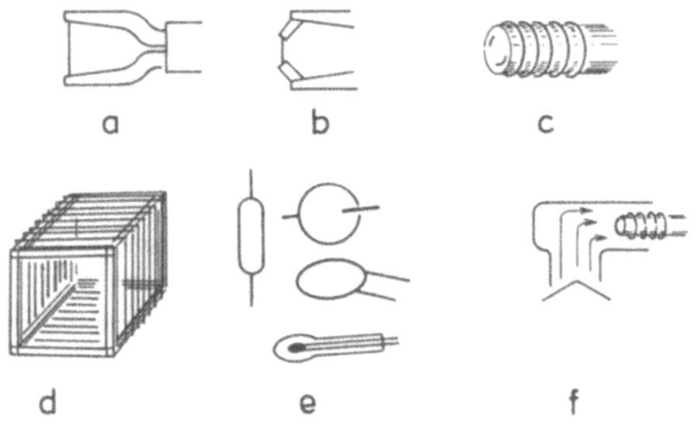

Fig. 3 Different sensor configurations. a and b show a sensor spanned between two prongs. Dimensions are typically 0.5 mm to 2 cm. In b the outer part of the sensor is coated. c and d are wires wound around frames of dimension 1 to 30 cm. e shows the shapes that thermocouples and thermistors can have: rod, bead, disc, and glass-coated sensor, dimensions 0.5 to 5 μm. Finally, f illustrates a shielded, aspirated system.

their relevant physical dimensions are less than or comparable to the Kolmogorov scale.

6. FREQUENCY (OR TEMPORAL) RESOLUTION

While spatial resolution is dependent on the sensor dimensions, temporal resolution depends on the sensor heat balance, which in turn depends strongly on the detailed characteristics of the flow field and the probe geometry and material characteristics.

We shall here summarize the theory for a simple probe configuration such as that in Figure 3a. It can be thought of as a resistance wire, a thermocouple or a rod-shaped metal coated thermistor. The heat balance for a unit length wire can be written

$$q_s = q_j - q_k - q_c - q_r \,. \qquad (12)$$

For a velocity above approximately 1 m s^{-1}, we can assume the convective heat loss, q_c, to be forced and thus avoid the other difficulties associated with very low wind speeds (Højstrup et al., 1976; 1977). The wire is placed along the x-axis perpendicular to the mean wind speed. This means that the terms in Equation 12 are:

stored heat, $q_s = \rho_w C_w \dfrac{\pi D^2}{4} \dfrac{\partial T_w}{\partial t}$.

electrical heating, $q_j = \dfrac{4}{\pi D^2} \sigma_0{}^{-1}\ I\ [1 + \alpha_T(T_w - T_0)]$,

conduction along wire, $q_k = -k_w \dfrac{\pi D^2}{4} \dfrac{\partial^2 T_w}{\partial x^2}$

convective heat loss, $q_c = \pi\ k_a\ Nu(T_w - T_a)$

radiative heat loss, $q_r = q_r{}'\ \pi\ D$,

where w, a, and 0 refer to wire material, air and reference conditions, respectively. D is wire diameter, ρ is density, C is heat capacity, k is heat conductivity, and $q_r{}'$ is the radiative heat loss per unit area. For forced convection, the following expression can be used for the Nusselt number, Nu (Collis and Williams, 1959),

$$Nu = [(0.24 + 0.56\ Re^{0.45})^{-1} + 2\ \frac{\lambda}{D}]^{-1} \qquad (13)$$

where $Re = uD/\nu$, u is the velocity and ν the kinematic viscosity of the air, and λ is the mean free path of the air molecules (at the sea surface $\lambda \simeq 0.07$ µm). Note that $\alpha_T > 0$ for resistance wires and $\alpha_T < 0$ for the thermistors, and that $I \simeq 0$ for thermocouples.

Equation 12 can be written as

$$\tau \frac{\partial T_w}{\partial t} = \frac{1}{\eta^2} \frac{\partial^2 T_w}{\partial x^2} - T_w + \gamma T_a - K - \frac{q_r' D}{Nu\ k_a} \tag{14}$$

where

$$\gamma = \left(1 - \frac{4\ I^2\ \sigma_0^{-1}\ \alpha_T}{\pi^2\ D^2\ Nu\ k_a}\right)^{-1}, \quad \eta^2 = \frac{4\ Nu\ k_a}{D^2 k_w \gamma}$$

$$\tau = \frac{\rho_w\ C_w\ D^2}{4\ Nu\ k_a} \gamma, \quad K = \frac{4\ I^2\ \sigma_0^{-1}}{\pi^2\ D^2\ Nu\ k_a} (\alpha_T\ T_0 - 1).$$

Neglecting radiation, Equation 14 can be solved by separating the temperatures into a static and a fluctuating part (Højstrup et al., 1976) as $T = T_s + \theta\ e^{j\omega t}$. The boundary conditions at the wire ends, $x = \pm L/2$, are $T_{ws}(\pm L/2) = T_{as}$ and $\theta_w(\pm L/2) = \theta_a/(1 + j\omega\tau_p)$, where the response of the prongs is approximated by a first-order filter with time constant τ_p, which like τ is a function of air velocity, prong material and dimensions (Sandborn, 1972). For the mean temperature averaged along the wire, we find

$$\langle T_{ws} \rangle_L = T_{as} + [T_{as}(1 - \gamma) + K] \left[\frac{\tanh(\eta L/2)}{\eta L/2}\right]. \tag{15}$$

For $\tau_p >> \tau$, the frequency response for a resistance wire $H_{Rw}(\omega)$ and for the thermocouple sensor, $H_{Tc}(\omega)$, can be approximated by

$$H_{Rw}(\omega) = \frac{\langle\theta_w\rangle_L}{\theta_a} = \frac{1}{1 + j\omega\tau} \left(1 - \frac{j\omega\tau_p}{1 + j\omega\tau_p} \frac{\tanh\left(\frac{\eta L}{2}\sqrt{1+j\omega\tau}\right)}{\eta L/2}\right)$$

$$H_{Tc}(\omega) = \frac{\theta_w(x=0)}{\theta_a} = \frac{1}{1 + j\omega\tau} \left(1 - \frac{j\omega\tau_p}{1 + j\omega\tau_p} \frac{1}{\cosh\left(\frac{\eta L}{2}\sqrt{1 + j\omega\tau}\right)}\right). \tag{16}$$

In deriving $H_{Tc}(\omega)$, we have neglected the different characteristics of the two different materials constituting the thermocouple. A more thorough discussion of the response characteristics of bare-wire thermocouples is given by Scadron and Warshawsky (1952), where it is shown that Equations 16 are good approximations, provided that average material parameters are used and the two sensor halves are not too dissimilar.

The difference between the two frequency response functions in Equations 16 stems from the fact that the resistance sensor measures an average temperature along the wire, while the thermocouple measures the temperature at the contact point between the two sensor halves. This difference is illustrated by considering

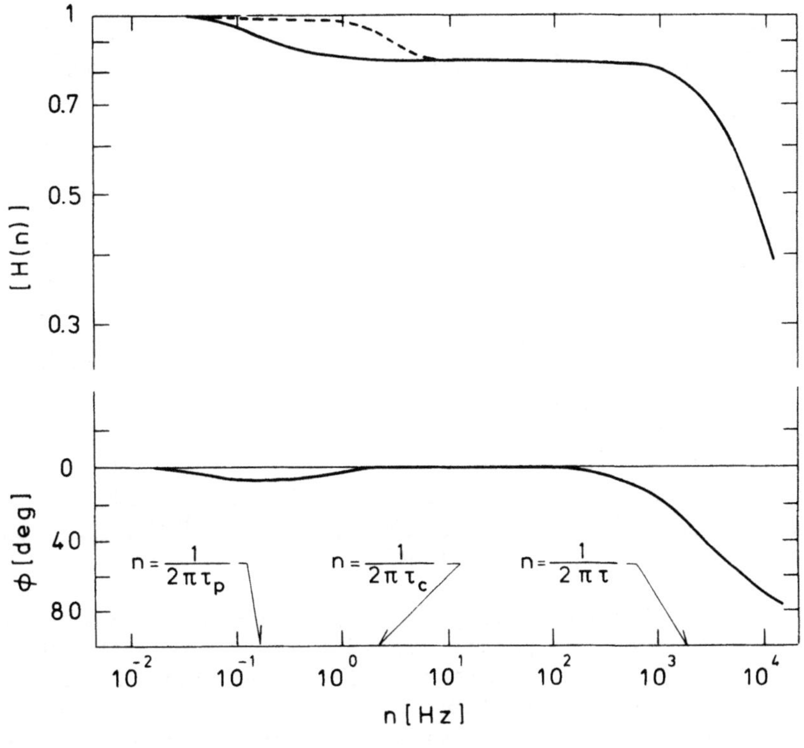

Fig. 4 Amplitude and phase response for a Pt wire sensor, D = 1 μm, L/D = 400, wind velocity 5 m s^{-1}. The solid curve pertains to a sensor such as that depicted in Figure 3a with a prong time constant, τ_p, corresponding to a prong tip diameter equal to approximately 100 μm. The broken curve corresponds to the sensor in Figure 3b with L/D = 400 for the uncoated wire, τ_c is the time constant for a 30 μm coating diameter.

the attenuation, δ, of the energy transfer function $|H(\omega)|^2$ for $\tau_p^{-1} << \omega << \tau^{-1}$.

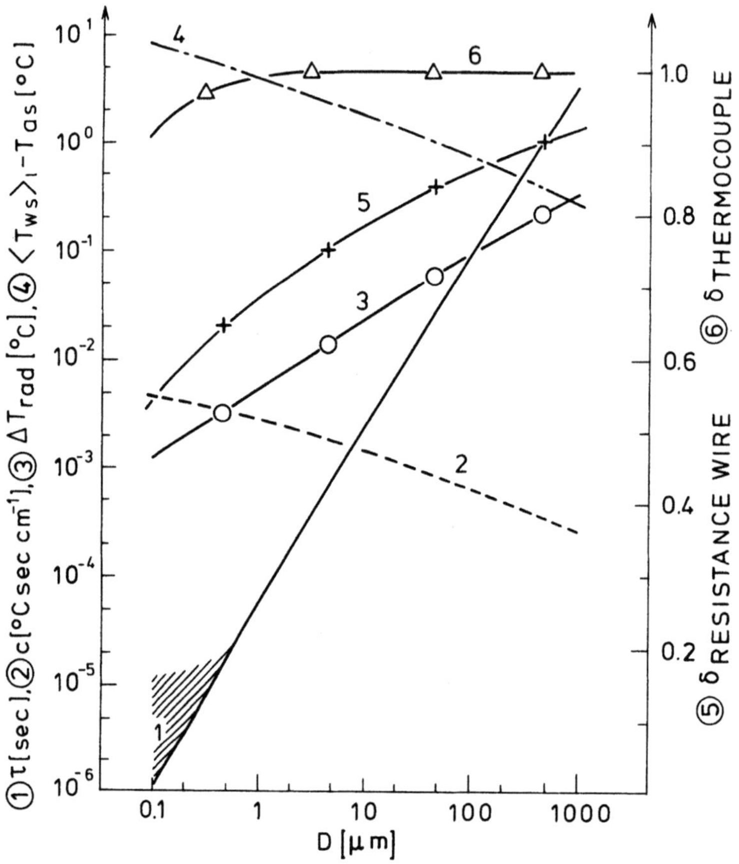

Fig. 5 The figure displays characteristics of sensors with Pt material parameters (for simplicity, the thermocouple curve is also calculated with Pt parameter values) as functions of the wire diameter, D, with L/D = 400, wind velocity, u = 5 m s^{-1}, and, when relevant, I • R = 0.07 AΩ as in Table 2. The curves describe (with reference to equations in the text): (1) time constant, with uncertainty indicated for very thin wires (14); (2) velocity sensitivity (20); (3) radiation influence (18,19); (4) overheat (15); (5) and (6) high frequency attenuation (17). The displayed characteristics can be estimated for other parameter values using the curves and the formulae referred to in the parentheses.

$$\delta = \begin{cases} \left(1 - \tanh(\eta L/2)/(\eta L/2)\right)^2 & \text{for resistance sensors} \\ \left(1 - [\cosh(\eta L/2)]^{-1}\right)^2 & \text{for thermocouples} \end{cases} \quad (17)$$

Characteristic values for δ as well as τ and $\langle T_{ws}\rangle_L - T_{as}$ as functions of the diameter D are shown in Figure 5. The phase and amplitude behaviour of $H_{Rw}(\omega)$ is illustrated in Figure 4.

7. RADIATION

Approximate methods for including the radiative term in Equation 14 are given in Sandborn (1972) and in Scadron and Warshawsky (1952). However, this term is usually neglected; it is often quite unimportant and its interpretation is complicated. Its influence can be estimated from Equation 14

$$\Delta T_{rad} \simeq -\frac{q_r' D}{Nu\, k_a} , \quad (18)$$

where we write q_r' as

$$\begin{aligned} q_r' = {} & q_r' \text{ (emitted from wire)} - q_r' \text{ (received from air)} \\ & - q_r' \text{ (received from the sea and other surfaces)} \quad (19) \\ & - q_r' \text{ (short wave)} \\ \simeq {} & 21 - 15 \quad 12 - 69\ \mathrm{Wm^{-2}} = -75\ \mathrm{Wm^{-2}} , \end{aligned}$$

where the numbers are estimated for the typical condition described in the introduction for a vertical Pt wire with a solar inclination of 45° (e.g., Kondrat'yev, 1965).

With this value for q_r', a characteristic variation of ΔT_{rad} with D is shown in Figure 5. The relative magnitude of the different terms in Equation 19 shows that the largest rapid variation of ΔT_{rad} may result from clouds passing the sun with associated time scales of the order of minutes. As seen from Figure 5, the radiative term is quite small for thin sensors; however, radiation will also influence the larger probe supports and (through conduction) the thinner sensors. This may be especially important for sensor configurations such as those in Figures 3c and d.

8. RESPONSE TO VARIATIONS IN THE NUSSELT NUMBER

In the above description of the sensor response it is implicitly assumed that temperature is the only flow quantity that varies. This is obviously an approximation only and the Nusselt number can be shown to vary with many flow variables (Larsen and Busch, 1974).

However, the variation that affects the temperature measurements most is the velocity variation, which affects all parameters in Equation 14.

Using the expression for Nu in Equation 13 (with $D \gg \lambda$), Wyngaard (1971b) derived the following relation between the measured temperature T_m and the 'true' temperature T, when measurements are performed in a velocity field with a vertical wire sensor

$$T_m = T - cu'$$

$$c \simeq \frac{Re^{0.45}(IR)^2}{16\sigma^{-1} \, k_a Nu^2 \, \bar{u}} \frac{D^2}{L^2} \tag{20}$$

Figure 5 shows characteristic values of c as a function of D for a constant L/D value.

9. CONTAMINATION AND CORROSION

Sensors are unavoidably contaminated when used close to the sea surface. This problem can be most serious for bare wire thermocouples, where corrosion may aggravate existing inhomogeneities in the materials to the point that they themselves start functioning as small thermocouples. For noncorroding sensors, such as platinum wires and properly coated thermocouples and thermistors, it is generally found that even quite a thick salt coating does not change their static characteristics, such as α_T and R(T) (see, e.g., Schacher and Fairall, 1976).

A layer of contaminants around a sensor will influence its thermal properties in several ways, affecting both the radiative and the convective heat transfer. The convective heat transfer from the probe will be suppressed by the insulating layer and enhanced by the greater surface area of the sensor. On balance, it is generally found that the efficiency of convective heat transfer is decreased by contamination over land, and increased by contamination over the ocean.

The time constant, τ in Equation 14, of a probe increases with contamination. The problem becomes progressively worse the smaller the sensor. This is both because small sensors act as more efficient collectors of aerosols and because the relative change in their characteristics will be greater for each particle collected. Indeed, while the formula for τ, given in connection with Equation 14, compares well with experiments for $D > 0.5$ μm, the experimentally determined τ for smaller diameters is 3 to 5 times greater than predicted by theory, as illustrated in Figure 5 (LaRue et al., 1975; Højstrup et al., 1976). This may be due to an almost instantaneous contamination of the very small sensors. Finally, it is generally found that a used but cleaned sensor has a larger time constant than an unused sensor (LaRue et al., 1975).

Salt particles act as condensation nuclei for water vapour under subsaturated conditions (e.g., for NaCl for a relative humidity around 70%). Under normal humidity conditions, salt particles on a sensor therefore can result in contamination of the temperature signals with humidity variations, since the heat associated with condensation/evaporation is transferred to and from the sensor as the relative humidity varies around the condensation point for a given salt particle.

The influence of salt cover on a temperature sensor has recently been discussed by Holmes and O'Brian (1975) and by Schmitt et al. (1978). The first used a typical slow response sensor, which due to its larger surface, is more liable to radiation errors. This could have led to the observed difference in readings from salt covered and clean sensors. They suggest that the condensation/evaporation at a salt covered temperature sensor is responsible for the difference between over-land and over-ocean temperature data that is described in Section 2. Figure 6 (Schmitt et al., 1978) shows the typical spiky behaviour of over-ocean temperature signals associated with a step change in humidity.

Much research is still needed to obtain a satisfactorily quantitative theory for this phenomenon. However, the seriousness of the problem can hardly be overestimated, and at the present moment we can only advise the experimenter to keep the probes as clean as possible and to beware of signal patterns such as that shown in Figure 6. During the experiments described in Schmitt et al. (1978), it was generally found that a probe could stay clean, in the sense that it showed no temperature spikes, for 5 to 60 minutes after exposure to a marine environment.

Finally, one should mention that under arctic conditions, rimed ice on the sensor can be a problem. However, this can be reduced by periodically passing a heating current through the sensor.

10. PROBES WITH COMPLICATED CHARACTERISTICS

The theory in Section 6 is easily extended to a configuration such as that in Figure 3b (see Fig. 4). For the configuration as in Figures 3c and d, the convection and conduction terms become more complicated. However, the time constant τ will again be given by essentially the same expression as in Equation 14, and the transfer function will show an attenuation at frequencies greater than τ_p^{-1}, τ_p being the response time of the supporting frame. However, for such configurations, the wire support can be made of materials with very small heat conductivities, thus considerably reducing the influence of heat conductivity.

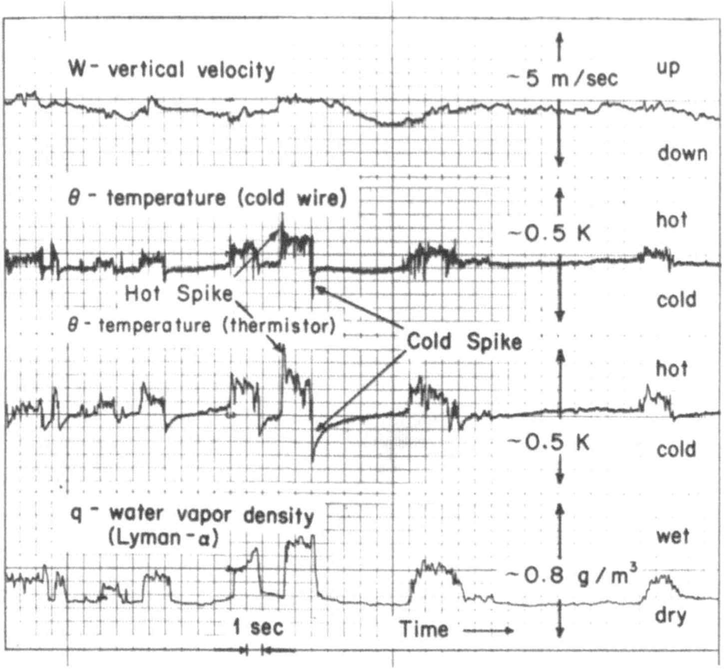

Fig. 6 The figures show the relation between abrupt humidity changes and temperature spikes (Schmitt et al., 1978). In the reference it is argued that the spikes are associated with the condensation/evaporation around salt particles with the fluctuating humidity.

For coated sensors and film sensors, the heat conductivity and response times of the different layers constituting the sensor interplay in complicated ways, e.g., Lowell and Patton (1955), Sandborn (1972), and Lueck et al. (1977). For such sensors and the shapes shown in Figure 3e, it is often better to treat each sensor as a whole, rather than break the response down into the response of the various elements, and either determine the response characteristics experimentally, or rely on information provided by the manufacturer. For thermocouples, τ is usually specified for a given velocity. For thermistors, the overheat corresponding to a given current and a given velocity is specified. By use of an approximate heat balance

$$m_w c_w \frac{\partial T_w}{\partial t} = R(T_0)I^2[1 + \alpha_T(T_w - T_0)] - Ah(T_w - T_a) \qquad (21)$$

the probe overheat and τ can be estimated at other velocities. m_w, c_w, A and h are the mass, heat capacity, surface area, and heat transfer coefficient, respectively, for the probe. Information on h in terms of Nusselt number can be found in the literature for many probe shapes (see McAdams, 1954).

However, in view of inaccuracies both in using Equation 21 and the manufacturer's specifications, the best approach to determination of response characteristics for complicated probes is probably to compare the response to that of a simple resistance wire system (see, e.g., Lueck et al., 1977).

Table 3 summarizes the minimal dimensions and response times for temperature sensors presently in use in atmospheric boundary layer work.

11. BRIDGES

Almost all thermometers for measurement of fast temperature fluctuations use a bridge at the input stage. We first consider a dc bridge for thermistors or resistance wires, such as shown in Figure 7 (without the thermocouple or capacitor). For a given probe current I_p, the voltage across the sensor is given by

$$V(T) = I_p R(T). \tag{22}$$

Table 3

Minimal dimensions and response time for temperature sensors currently used in boundary layer work

Sensor Description	Dimensions	Time Constant for $u = 5\ m\ s^{-1}$
Pt wire (Fig. 3b)	$L \simeq 0.2$ mm	$\simeq 10^{-5}$ s
Bare wire thermocouple constantan-chromel (Fig. 3a)	$D \simeq 12.5\ \mu$	$\simeq 2 \times 10^{-2}$ s
Thermistor, bead shape (Fig. 3e)	$D \simeq 25\ \mu$	$\simeq 0.3$ s

In the bridge configuration I_p is usually generated by either a constant current or a constant voltage generator G across the bridge. The corresponding expressions for I_p and the bridge output are easily deduced, and are only given here for a constant voltage supply, V_0.

$$I_p = \frac{V_0}{R_1 + R(T)} \tag{23}$$

$$V_B = \frac{V_0 R_2 R_3}{[R_1 + R(T)](R_2 + R_3)} \left(\frac{R(T)}{R_3} - \frac{R_1}{R_2}\right)$$

With a proper choice of resistance values, the bridge offers a convenient method for subtracting the mean value of V(T) and permits increased amplification of the relatively low level fluctuating signal component. However, a bridge also embodies a number of drawbacks, (a) for the same I_p, it is seen that $\Delta V_B < \Delta V(T)$, (b) due to the R_T in the denominator, I_p varies and V_B is more non linear in $(T-T_0)$ than V(T) (this nonlinearity is slightly less for a constant current supply), and (c) the noise level of the bridge is higher than for R(T) alone, as will be discussed below.

A problem with thermocouples is the need for a reference junction. One commercially available solution is shown on Figure 7. The two reference junctions and R(T), a Pt element, are placed in good thermal contact (note that only the leads from the thermocouple probe to the reference junctions need to be thermocouple material, see Fig. 2). With proper scaling of the circuit it follows from Figure 7 that any change in the thermocouple voltage resulting from a change in temperature at the reference junctions will be compensated by an opposite change in the bridge voltage.

A simpler method is to embed the reference junctions in a body with a large thermal response time (~1 cm^3 of glue has a time constant of around 10 min). The output of such a system is proportional to the temperature fluctuations high-pass filtered with the larger time constant and, it should be noted, contaminated by the radiation error on the larger body.

12. NOISE, DRIFT AND AMPLIFICATION

The electronics for temperature-measuring systems appear deceptively simple, their only role being to amplify the output of the bridge and, for resistance sensors, to provide the sensing current through the sensor. Owing to the low signal level from many sensors (see Table 2), however, the noise and drift requirements can be very severe. We shall here only touch upon the concepts involved, and refer readers to books on electronics for further information.

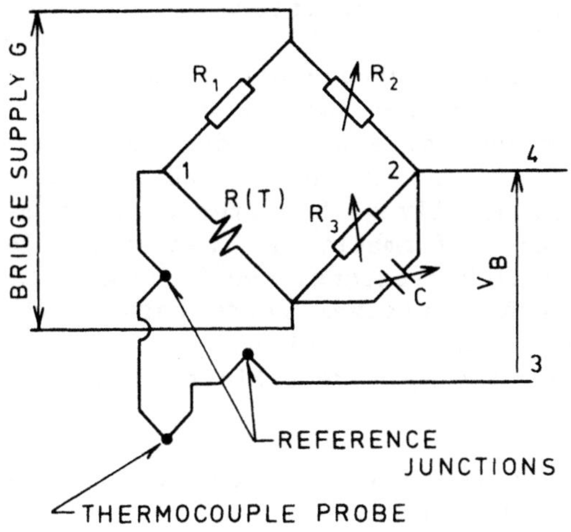

Fig. 7 A temperature bridge. R(T) is a resistance sensor. When the figure describes a resistance sensor bridge, the lead between points 1 and 3 contains no thermocouple junctions. C is a variable capacitor used for ac-bridges only to balance capacitance in the sensor cable.

All resistors in the system generate thermal noise, which is a white noise process with a constant spectral density, $e_N^2(n)$, given by.

$$e_N^2(n)/R = 4\ kT \simeq \frac{1}{64}\ (nV)^2\ (Hz\ ohm)^{-1}, \tag{24}$$

where k is the Boltzmann constant and the approximate expression pertains to T ≃ 283 K. With R equal to the resistance between points 3 and 4 on Figure 7, Equation 24 describes the lower bound for the noise content of a signal recovered from such a bridge.

Another noise source is related to all material interfaces in the system, but the main contribution comes from the active components, i.e. mainly from amplifiers and current sources. It is referred to as the 'f^{-1}' noise (f = frequency). For example, the noise spectrum from a good commercial amplifier might presently appear as:

$$\overline{N_A^2(n)} \simeq (1 + b/n)\ \overline{a^2(n)} \qquad \text{for } n > 10^{-5}\ \text{Hz} \tag{25}$$

where $a^2(n) \simeq 1\ (nV)^2\ Hz^{-1}$ is the white noise component and the parenthesis with $b \simeq 200$ Hz describes the 'f^{-1}' noise contribution. The drift due to temperature changes originates mainly in current sources and amplifiers. Presently, the drift is 0.2 to 1 μV °C^{-1}, referred to input, for good amplifiers.

For many purposes, the magnitude of the noise and drift described here is not serious for the quality of the data, and the bridge output can be increased by a simple dc amplification. If greater resolution is desired, one can use an ac system, where the sensor current oscillates, usually with a frequency of between 25 and 100 kHz. The temperature information is then available as an amplitude modulation on this high frequency oscillation, and therefore it is present in a frequency interval where both drift and 'f^{-1}' noise are absent. After amplification, the temperature signal is retrieved either by a simple rectification and low-pass filtering or by synchronous detection.

In practice, such ac systems have made it possible to reduce the total system noise to only two to three times the thermal noise given by Equation 24 (LaRue et al., 1975). The price for this is that the general degree of complication is increased and that the frequency characteristics of the system are more critical.

A compromise between an ac system and a dc system can be achieved by using a dc probe current combined with chopper amplifiers. Through a gating technique these break the input into a high frequency signal before amplification, and the signal is thereafter recovered as described above. Such amplifiers can also be used in connection with thermocouples.

Due to the low signal level, radio frequency interference has proven to be a problem both on ship and aircraft, for temperature sensors (and hence also for radiation instruments). Since radio transmitters have become more powerful, this problem has become even more troublesome and should be considered in an early design stage. Digitizing at a short distance from the sensor is recommended.

13. CALIBRATION PROCEDURES

Calibration procedures where the system output is measured for known air temperatures and known velocities are the basic method of static calibration. Recently, however, procedures have been extended to dynamic calibration with good results by Højstrup et al. (1976, 1977), who generated a sinusoidal temperature variation at a frequency which was variable between 2 and 10000 Hz by means of a strong sound field.

Also within this category of calibration methods is on-line

comparison of mean values and the low-frequency variances from the fast-responding system and a slowly responding system located nearby at the same height. This method has the advantage of being simple to carry out and of taking into account changes in the low-frequency response of the fast system. This method is recommended in any case as a valuable tool to check the instrumentation. There are difficulties since the natural variation of mean temperature may be small and slow response temperature sensors are more liable to radiation errors.

Another method of calibration used for resistance sensors relies on changes of the sensor current. First, it can provide a check on the dc gains of the system: a relative change in the current of magnitude α_T, neglecting the rather small change in overheat, will give a change in system output corresponding to 1°C. A larger change in current will also change the overheat. If the larger change is a step change, it has been shown (Sandborn, 1972) that the response of the heat balance Equation 14 to the first order will be of the type

$$T_w = C_1 + C_2 e^{-t/\tau}, \qquad (26)$$

where C_1 and C_2 are constants for given ambient conditions and τ is the time constant in Equation 14. A modification of this method employs a sinusoidal variation of I^2, instead of a step change in current, to study the relative amplitude changes of the output for increasing frequency.

The advantage of using the 'current change' calibration method is that it is easy to do and can be built into the system as a special mode of operation, thereby allowing for continuous checking of the time constant of the sensor, which may also be important since the time constant is a sensitive indicator of sensor contamination. An extensive description of this calibration method is given in LaRue et al. (1975). Note that this way of checking the time constant of the sensor seems to be valuable in air, while it is of doubtful utility in water [see Chapter 18 by Gibson and Deaton, and Lueck et al. (1977)].

14. DISCUSSION

Most measurements of fast temperature fluctuations in the atmospheric boundary layer have been performed by means of resistance wires (usually made of platinum), and if response times of less than ~0.01 s are important, these wires are still the only choice. Thermocouples are relatively new to the field, but they are now used routinely by several groups (see Leavitt and Paulson, 1975). They are easy to operate and have a number of attractive characteristics, as shown in Section 6. One of these - point measurement -

makes them well suited for use as wet bulb thermometers where time constants of about 0.3 s have been realized (Tillman, 1973). For theoretical response of such thermocouple psychrometers, see Sano and Mitsuta (1968).

The response time of thermistors is still rather large. Owing to their simple electronics and high sensitivity they are useful back-up instruments, especially for measurements at greater heights, where the signal is small and of rather low frequency.

Further improvement is likely in the response times of thermocouples and thermistors and in the performance of electronic circuitry, all of which will simplify matters for the experimenter. Less pleasant is the fact that salt contamination may prove to present such difficulties for reliable measurements of temperature fluctuations over the ocean that the development of different techniques may be required. At the present stage of knowledge we can only urge that experimenters should be aware of this problem, and use sensors that are easy to replace and not smaller than necessary for their purposes, as they seem to be more susceptible to contamination the smaller they are.

ACKNOWLEDGEMENTS

The authors wish to acknowledge discussions with Fred Weller, Jim Tillman, Buck Williams, Carl Friehe, Kurt Schmitt, Tom Deaton, John LaRue and Dennis Thomson. This research was supported by National Science Foundation Grants No. ATM76-00855, ATM76-23856 and DES75-07223 and U.S. Office of Naval Research Grant No. N00014-14-75-C0152. This work is contribution No. 435 from the Department of Atmospheric Sciences, University of Washington, Seattle, Washington, and was carried out while the senior author was a visitor there.

REFERENCES

BUSCH, N.E. 1973. On the mechanics of atmospheric turbulence. In: *Workshop on Micrometeorology*. Edited by D.A. Haugen (American Meteorological Society, Boston): 1-65.

COLLIS, M.D. and M.J. WILLIAMS. 1959. Two-dimensional convection from heated wires at low Reynolds numbers. *Journal of Fluid Mechanics*, 6: 357-384.

FRIEHE, C.A., J.C. LaRUE, F.H. CHAMPAGNE, C.H. GIBSON and G.F. DREYER. 1975. Effects of temperature and humidity fluctuations on the optical refractive index in the marine boundary layer. *Journal of the Optical Society of America*, 65: 1502-1511.

HOLMES, J.F. and J.F. O'BRIAN. 1975. Marine sensors and their interface with the environment. In: *Proceedings of the Instrumentation Society of America, Marine Sciences Divsion,* held at Milwaukee, Wisconsin, Oct. 6-9, 1975. Advances in Instrumentation, 30, part 3, 758/1-758/4.

HØJSTRUP, J., K. RASMUSSEN and S.E. LARSEN. 1976. Dynamic calibration of temperature wires in still air. DISA-Information No. 20: 22-30.

HØJSTRUP, J., K. RASMUSSEN and S.E. LARSEN. 1977. Dynamic calibration of temperature wires in moving air. DISA-Information No. 21: 33.

KAIMAL, J.F., J.C. WYNGAARD, Y. IZUMI and O.R. COTE. 1972. Spectral characteristics of surface layer turbulence. *Quarterly Journal of the Royal Meteorological Society,* 98: 563-589.

KONDRAT'YEV, K.Ya. 1965. *Radiative Heat Exchange in the Atmosphere.* Pergamon, Oxford, 411 pp.

KRUSPE, G. 1974. Measurements of fluctuations of vertical wind velocity, temperature and radio-refractive index above the sea. *Boundary-Layer Meteorology,* 6: 257-267.

LaRUE, J.C., E. DEATON and C.H. GIBSON. 1975. Measurement of high-frequency turbulent temperature. *Review of Scientific Instruments,* 46: 757-764.

LARSEN, S.E. and N.E. BUSCH. 1974. Hot-wire measurements in the atmosphere. Part I. Calibration and response characteristics. DISA-Information No. 16: 15-34.

LEAVITT, E. 1975. Spectral characteristics of surface layer turbulence over the tropical ocean. *Journal of Physical Oceanography,* 5: 157-163.

LEAVITT, E. and C.A. PAULSON. 1975. Statistics of surface layer turbulence over the tropical ocean. *Journal of Physical Oceanography,* 5: 143-156.

LOWELL, H.H. and N. PATTON. 1955. Response of homogeneous and two-material laminated cylinders to sinusoidal environmental temperature change, with application to hot-wire anemometry and thermocouple pyrometry. NACA Technical Note 3514, 143 pp.

LUECK, R.G., O. HERTZMAN and T.R. OSBORN. 1977. The spectral response of thermistors. *Deep-Sea Research,* 24: 951-970.

McADAMS, W.H. 1954. *Heat Transmission.* 3rd edition, McGraw-Hill, New York, 532 pp.

MAYE, J.P. 1970. Error due to thermal conduction between the sensing wire and the supports, when measuring temperature with a wire anemometer used as resistance thermometer. DISA-Information No. 9: 22-26.

PHELPS, G.T. and S. POND. 1971. Spectra of the temperature and humidity fluctuations and of the fluxes of moisture and sensible heat in the marine boundary layer. *Journal of Atmospheric Sciences,* 28: 918-928.

POND, S., G.T. PHELPS, J.E. PAQUIN, G. McBEAN and R.W. STEWART. 1971. Measurements of the turbulent fluxes of momentum, moisture and sensible heat over the ocean. *Journal of Atmospheric Sciences,* 28: 901-917.

SANDBORN, V.A. 1972. *Resistance Temperature Transducers.* Metrology Press, Fort Collins, Colorado, 545 pp.

SANO, Y. and Y. MITSUTA. 1968. Dynamic response of the hygrometer using fine thermocouple psychrometer. Special Contributions of the Geophysical Institute, Kyoto University, Japan, No. 8: 61-70.

SCADRON, M.D. and I. WARSHAWSKY. 1952. Experimental determination of time constants and Nusselt numbers for bare-wire thermocouples in high velocity air streams and analytic approximation of conduction and radiation errors. NACA-Technical Note 2599, 81 pp.

SCHACHER, G. and C.W. FAIRALL. 1976. Use of resistance wires for atmospheric turbulence measurements in the marine environment. *Review of Scientific Instruments,* 47: 703-707.

SCHMITT, K.F., C.A. FRIEHE and C.H. GIBSON. 1978. Humidity sensitivity of atmospheric temperature sensors by salt contamination. *Journal of Physical Oceanography,* 8: 151-161.

TILLMAN, J.E. 1973. Wet- and dry-bulb thermocouple psychrometry. Atmospheric Technology, NCAR, Boulder, Colorado, No. 2, p. 77.

WILLIAMS, R.M. and C.A. PAULSON. 1977. Microscale temperature and velocity spectra in the atmospheric boundary layer. *Journal of Fluid Mechanics,* 83: 547-567.

WYNGAARD, J.C. 1971a. Spatial resolution of a resistance temperature sensor. *Physics of Fluids,* 14: 2052-2054.

WYNGAARD, J.C. 1971b. The effect of velocity sensitivity on temperature derivative statistics in isotropic turbulence. *Journal of Fluid Mechanics,* 48: 763-769.

16

Radiative Sensing of Sea Surface Temperature

K. Katsaros

1. INTRODUCTION

1.1 Usefulness and Advantages of Remote Sensing

The technique of inferring sea surface temperatures from radiation observed in the infrared region of the spectrum has been used extensively for more than a decade. Remote sensing has successfully been employed from airplanes, notably in upwelling studies of the Pacific Ocean off the Oregon coast (Holladay and O'Brien, 1975) where the sea surface temperatures revealed the response of the ocean to varying wind stress conditions. Numerous studies have employed such data to study eddies of the Gulf Stream. The time scale of the feature of interest is in many of these cases days or weeks, which is too short for an adequate survey by a ship. The ability of airplanes or satellites to survey a large area in a short period of time and to repeat the measurements within a short time interval has revealed mesoscale circulation features in the ocean not imagined previously. The new technique has provided much informaton, and also posed new questions related to it. Operational uses of satellite data for weather forecasting have been particularly valuable over the oceans. Sea surface temperatures may be used increasingly as indicators of long range weather changes (e.g., Bjerknes, 1969; Namias, 1969).

The regions of the spectrum employed for remote sensing are the so-called 'window-regions', where the atmospheric gases are relatively transparent (see Fig. 1). There is, however, some absorption and emission which makes observed radiances ('radiance' is radiant flux per unit solid angle per unit projected surface area) dependent on atmospheric temperature and humidity structure if the path lengths are greater than a few tens of metres.

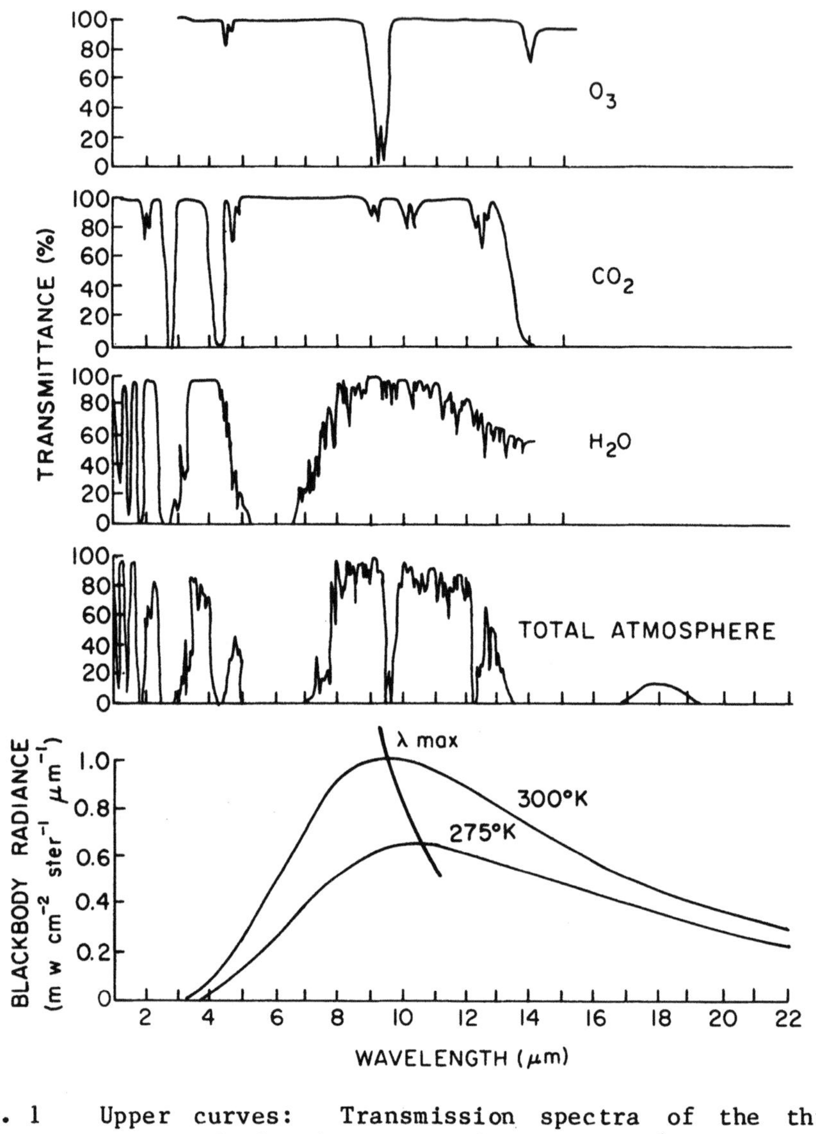

Fig. 1 Upper curves: Transmission spectra of the three principal absorbing atmospheric gases (ozone, carbon dioxide, and water vapour) and the transmission of the whole atmosphere at vertical incidence. The atmosphere is seen to present 'windows' between 3 and 4.5 μm and between 8 and 12 μm except for the ozone band between 9 and 10 μm.
Lower curve: Thermal radiation as a function of temperature and wavelength. 275 K and 300 K approximates the temperature range between polar and equatorial waters.

Deviation of water emittance and reflectance from unity is a function of the angle of emission and of the polarization. This must be considered when the reflected portion of the received radiation is eliminated. Since infrared radiation emanates from the top 50 μm or less of a water surface, while conventional bucket and ship-intake temperatures give an average for the top 0.2 m or more, it is important to understand the factors which influence the difference between these two measures.

1.2 Platforms

Remote measurements of sea surface temperatures have been obtained very close to the target of the observation and very far from it, from distances of about 1 m to satellite heights of 40,000 km. Remote in the first instance simply signifies that there is no direct contact between target and sensor. There is, however, exchange of radiant energy between them and the influence of the instrument on the observation needs to be considered when the distance is small.

The platform chosen naturally depends on the intended purpose of the measurement. Radiometers have been simply hand-held or have been mounted on booms attached to docks, masts, and ships. They have been carried in airplanes and in satellites. For such diverse uses the technology differs, and varies over a wide range in accuracy and sensitivity, and in spatial and temporal resolution.

From close range surface temperatures can be inferred with a spatial resolution of the order of centimetres (e.g., Ewing and McAlister, 1960; McAlister and McLeish, 1969; Clauss et al., 1970; Katsaros, 1973, 1977; Grassl, 1976).

The use of airplanes to observe sea surface temperatures has been extensive (e.g., Saunders, 1967a, 1971, 1973; McLeish, 1968; Holladay and O'Brien, 1975). Temperature patterns with a spatial resolution of centimetres to dekametres can be obtained from airplanes. Thermal resolution is typically of the order of 0.5°C but accuracy of 0.1°C can be obtained with care. Airborne measurements constitute a valuable complement to ship-based measurements by providing information about horizontal variability.

Satellite measurements, both in the 3 to 4 μm and 8 to 12 μm regions, have provided large quantities of sea surface temperature data. However, without sophisticated techniques to correct for the atmospheric interference, the information content has often not exceeded that of climatological data. Work in the future may benefit from the recent use of multi-wavelength or coincident atmospheric profiling which has greatly improved the accuracy.

2. INSTRUMENTAL DESIGN PRINCIPLES

Several types of instruments are used for measuring infrared radiance. They can be broken down, on the one hand, into point sensors and imagers, and, on the other, into heat-sensing and photon-sensing devices. Imagers are either made up of numerous point sensors in an array, or consist of a single sensor which scans the area of observation by optical means or in the case of some satellites by the axial spin and forward motion of the satellite itself.

Radiometers contain as essential parts a lens or mirror that focuses the image, a detector, filters, a chopper, and electronic circuitry. Lenses and detectors as well as filters have limited wavelength response, and can provide the desired cutoff on the long or short wavelength side. Detectors are the transducers, which convert radiant energy into an electrical signal.

2.1 General Description of a Radiometer

An infrared radiometer works on the principle of comparing the radiance in a specified wavelength band of a target area with that of an internal reference cavity (see Fig. 2). The entering radiation comes alternately from the target and from the cavity via a rotating toothed chopper wheel in front of the lens. The radiation from the target enters the openings between the chopper teeth and when the chopper covers the lens its gold coating acts as a mirror, reflecting the radiation from the cavity back onto the detector. This chopping of the radiation produces an alternating signal which can be readily amplified.

2.2 Detectors

There are two basic types of detectors: heat sensors and photon detectors. Heating of a material or the absorption of photons by a semiconductor can both produce changes which are observable electrically. The various principles of detection in common use are listed below with a short description. The reader is referred to textbooks on infrared radiation for complete details (e.g., Kruse et al., 1962).

Thermal detectors respond to the total energy of the radiation impinging on them irrespective of its spectral distribution. The photon-detector responds to the rate at which quanta of radiation are absorbed. Since these detectors only respond to quanta with energy greater than a minimum level, they have a long wavelength cutoff, which is temperature dependent. Since a quantum of short wavelength radiation contains more energy than a quantum of longer wavelength, the energy response for photon detectors increases linearly as a function of wavelength.

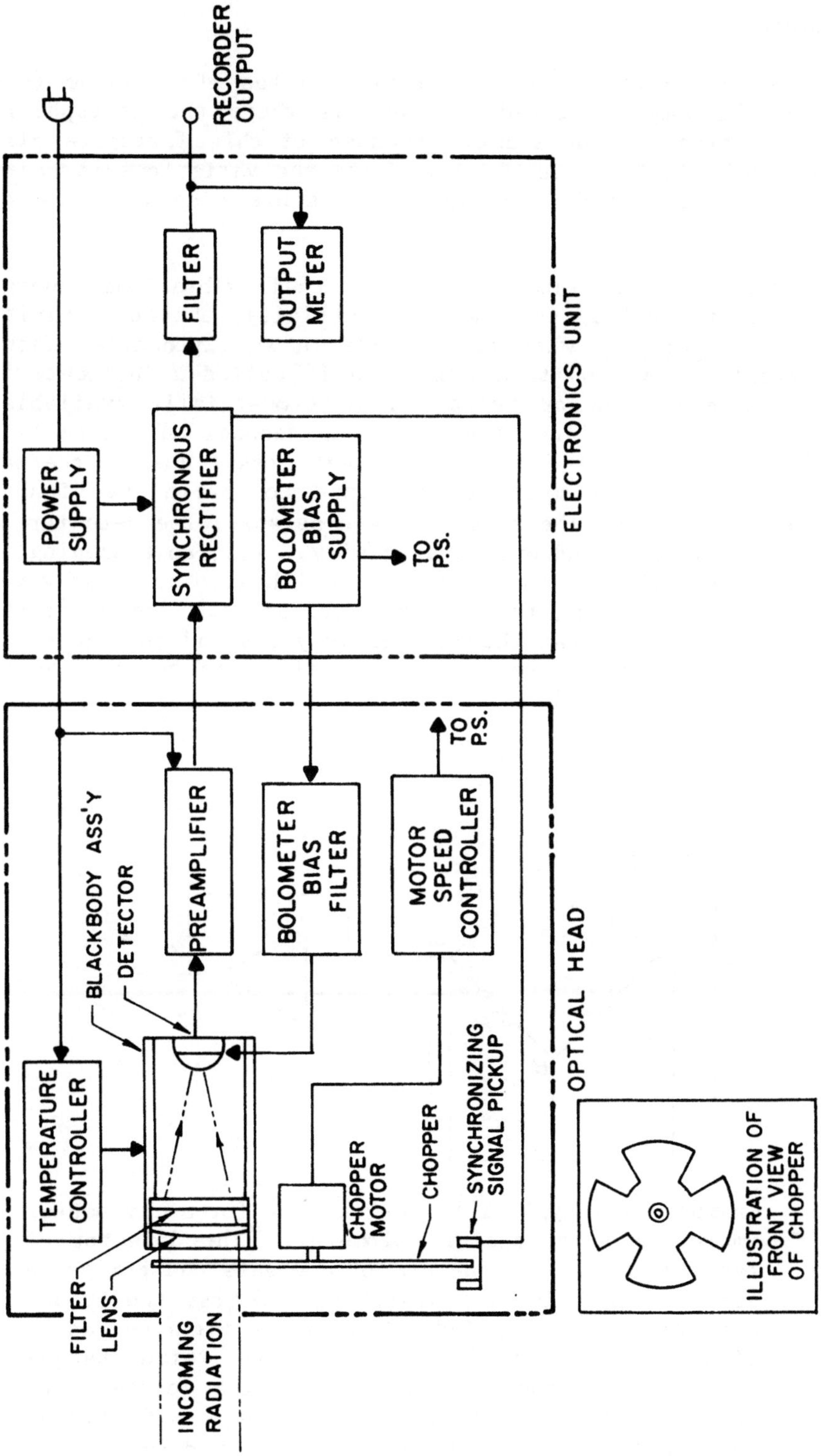

Fig. 2 An example of the physical arrangement of the various components in an infrared radiometer (courtesy of Barnes Engineering Company).

(a) Bolometers

The energy spectrum of terrestrial radiation has its maximum (see Fig. 1) at the same infrared wavelengths where the atmospheric gases are relatively transparent. Because of this fortunate circumstance simple instruments which measure the variations in heating produced by this radiation typically achieve an accuracy of 0.1°C.

Heating of semiconductor materials changes their vibrational energy and thereby their ability to conduct electricity; the conductivity of thermistors increases with temperature in an exponential fashion. A detector based on this principle is called a 'bolometer', and is the common detector mechanism in commercially available, hand-held radiometers. Since the detection depends on the heating effect of absorbed infrared radiation, heat losses due to thermal conduction or convection must be optimized to obtain the highest possible sensitivity and the needed time constant. The detector is mounted in an insulated and evacuated cavity. The front opening is covered by a 'window' or a focusing lens, which also provides wavelength limits by filtering the incoming radiation. The detectors are made very small, so that they can quickly respond to changes in the incident radiation; thermistor flakes that are 50 μm x 50 μm and 10 μm thick are typical.

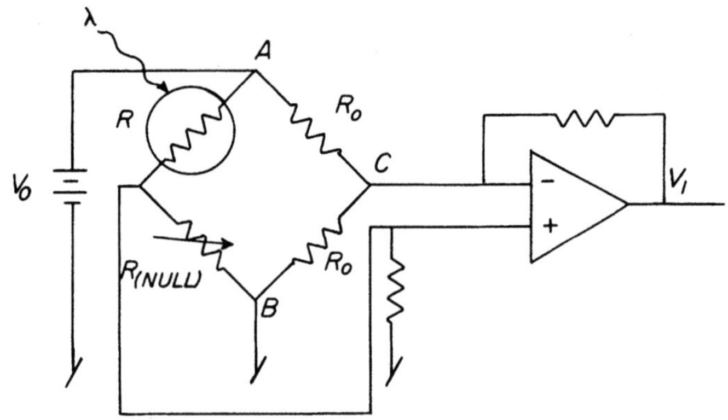

Fig. 3 Example of a typical bridge circuit for a radiometer employing a thermistor bolometer. The detector of resistance R forms one arm of a wheatstone bridge in series with balance resistor R_0. Across the bridge a voltage V_0 is applied. Heating by infrared radiation causes the resistance to change which changes the current through the bridge and the voltage drop between points A and B and between C and ground. The amplified output voltage V_1 varies with the intensity of the radiation.

In order to minimize effects on the signal due to spurious thermal drift in the detector housing, two identical detectors are usually used in a bridge circuit, one being shielded from the radiation, and the other one alternately exposed to the external radiation and the internal radiation of the cavity by a 'chopper wheel'. A typical bridge circuit is seen in Figure 3.

(b) Photon-detectors

In the near infrared window at 3.5 to 4 μm individual photons have more energy than in the 8 to 12 μm region. Thus transducers which employ various effects produced as quanta of electromagnetic radiation interact with semiconductor materials are often used in this wavelength region.

The photoconductive effect is commonly used in photon detectors. Increase in conductivity results when the energy of a photon produces a freeing of electrons or 'holes' or electron-hole pairs within the crystal lattice. The detection of excitation by infrared radiation is enhanced if the background of thermal excitation is minimized. For this reason semiconductor photon detectors are usually operated at low temperatures. Both detectivity, which will be defined below, and wavelength response are affected (see Fig. 4). The normal cooling technique is to mount the sensor flake in contact with a dewar containing liquid nitrogen (-196°C). Electrical (Peltier) cooling can also be used. In satellites the back of the sensor is mounted so that it loses heat to space by radiation.

(c) Quantitative Measures of Detector Properties

A measure of the performance of a detector is called its 'detectivity', D*. It is defined here as (beware, since definitions are in use which differ slightly):

$$D^* = \frac{1}{(\text{NEI})\, A_D^{\frac{1}{2}}} \quad (1)$$

where A_D is the detector area, and NEI (noise equivalent input) is defined by:

$$\text{NEI} = J \left(\frac{V_N}{V_S}\right) \frac{1}{\Delta f^{\frac{1}{2}}} \, . \quad (2)$$

V_S is a measure of the voltage produced by an rms value of irradiance J on the detector of area A_D and V_N is the noise voltage; both voltages are measured in an electrical bandwidth Δf.

When detectivities are to be compared, the conditions for obtaining the D* value should be carefully specified. Thus the black body radiation source for J is typically 500 K; the reference bandwidth is usually 1 Hz or 5 Hz; the centre frequency is 90, 400, 800, or 900 Hz. The reference temperature of the detector is either the ambient (i.e., 22°C) or the operating temperature (if the detector is to be used cooled). The intensity of the irradiance and the angular opening of the detectivity determination should also be specified. When a detectivity is given, the values of source temperature, centre frequency, and bandwidth are given in parenthesis; e.g., D* (500 K, 500, 5). Sometimes the subscript λ is added to D* when performance at one wavelength is specified, D^*_λ. This is usually the wavelength of maximum performance; operating the same detector at different temperatures will result in different detectivity. Both the magnitude of the response and the wavelength maximum are affected (see Fig. 4). The time constant of a detector must also be known if the radiation is to be chopped. The variation of the response with frequency is therefore often given with D*.

Another quantity used to describe the merits of a detector is the responsivity, R, which gives the signal voltage per unit of radiant power:

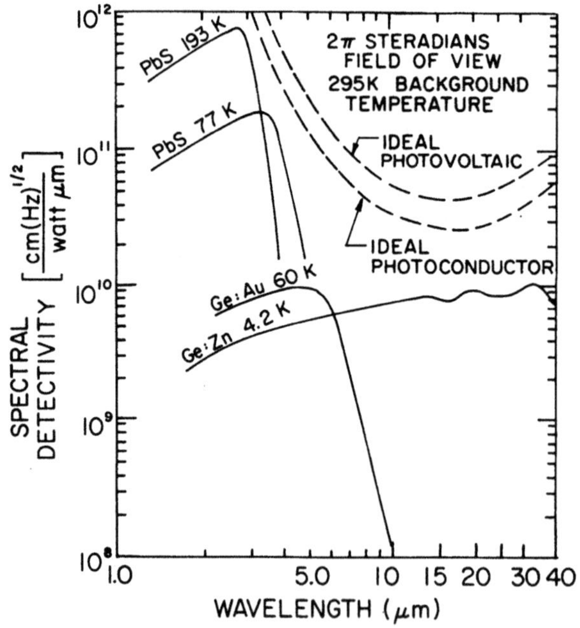

Fig. 4 Spectral detectivity and wavelength response of some semiconductor detectors as a function of temperature (courtesy of Santa Barbara Research Center).

$$R = \frac{V_S}{JA_D} . \tag{3}$$

Units are volts/watt per area and the temperature of the black body is usually 500 K. The spectral responsivity, R_λ, is often specified.

2.3 Optics, Filters, and Choppers

Several optical components are used in detection of infrared radiation. Among them are a focusing lens or mirror, windows or domes that separate the detector from the environment, and filters that limit the wavelength response to the desired interval.

To observe surface temperatures a narrow field of view is desirable in order to resolve horizontal variations. The linear spot size for a 2° angular opening is 35 cm at a distance of 10 m and 350 m at a distance of 10 km. This wide angle has been used from ships. Satellites have resolution elements, so-called 'footprints', of the order of 2 km, which require much narrower angular openings. The angular opening specified by manufacturers typically represents the angle where the intensity is reduced by 50%. If the observed target is seen against a contrasting background the extended region of response must be kept in mind.

The sensitivity of the radiative measurement depends on the angular opening. With a larger opening more radiant energy is focused on the detector, and the relative error due to noise is smaller. Wide angle bolometers may give 0.01°C resolution when used from ships while ones with a 2° opening typically provide 0.1°C. Spatial resolution can thus be sacrificed for temperature resolution.

(a) Windows, Lenses, Mirrors, Choppers

Since detectors, especially cooled ones, must be enclosed in a vacuum, infrared windows that transmit well throughout the desired spectral interval must be found. The window material must also bond well to the rest of the detector housing. The requirements of transmittance are the same for domes as for windows. Sometimes the function of the window is combined with those of a focusing lens and a filter.

Metal mirrors and choppers are also used extensively in the optical system. They usually have gold coatings because gold is a good reflector in the infrared. Chopping the incoming radiation to produce an alternating electrical signal and a reference radiance was mentioned above. Choppers are often operated at 50 or 60 Hz and may then depend on the power line frequency. This must be considered when the power source is a small generator on board a ship or airplane or at a field site. Any deviation in the chopper frequency will affect the calibration.

All of the materials exposed to the atmospheric environment must have chemically stable compositions that are not sensitive to moisture or abrasion; otherwise the output will drift.

(b) Filter Types

Interference filters are in common use. They consist of dielectric films that interact constructively or destructively with the electromagnetic radiation, such that only certain wavelengths are transmitted. Transmission can be quite high, nearly 100% at the peak wavelength, and the edges of the transmission band can be extremely sharp.

Semiconductor filters exhibit strong absorption at wavelengths shorter than the cutoff limit and are therefore used to eliminate undesired short wavelength radiation.

The optical lens used to focus the radiation on the detector element can be made of a material which also filters. A germanium lens provides a useful optical bandpass between 8 and 12 μm. The detector element is often bonded directly to the lens (hyperimmersed). This reduces reflection losses at the surfaces between lens and detector.

(c) Imagers

There are many advantages in obtaining a visible image since the human brain is particularly apt at pattern recognition. Two-dimensional images can be produced by an array of detectors or by a scanning radiometer. The former converts the radiances received by each element in an array of detectors in the focal plane of the lens into a visible image by light modulation. The image can then be photographed.

In satellites the scanning technique has been employed. Only one detector is used and the target area is scanned either by a moving mirror or by the spin and advance motion of the satellite. In the spin-stabilized TIROS-series satellites, detectors view the earth from horizon to horizon in lines perpendicular to the path of advance. In the Nimbus series the same side of the satellite always faces the earth, and the scan is provided by a rotating mirror. An image is later produced by having the electrical signal modulate a light source which scans the screen or photographic plate.

(d) Calibration

Calibration is performed in the laboratory by positioning a black body source of known temperature T_{B1} and known emittance ε_λ (near

1), so that it fills the field of view of the instrument. The radiance $B(T_{Bl})$ can then be calculated from Planck's radiation law:

$$\bar{J}_e = \int_0^\infty \phi_\lambda \, [\varepsilon_\lambda B(T_{Bl},\lambda) + (1 - \varepsilon_\lambda) \, J_R(\lambda)] \, d\lambda \qquad (4)$$

where J_R is the radiance of the environment being reflected, and ϕ_λ is the transmittance of the filters and lenses on the instrument. J_e is measured as a voltage, and one obtains a calibration curve of J_e in W m^{-2} versus voltage. It is important that the emittance of the black body is well known, and that the measured temperature T_{Bl} corresponds to the surface temperature of the radiating surfaces in the black body cavity (for a discussion of 'black bodies' see Hudson, 1969).

Interpretation of the radiance of a target is given in terms of the brightness temperature, T_B (also called equivalent black body temperature; the term brightness is simpler and therefore preferable). The brightness temperature is obtained from:

$$\bar{J}_e \, (T_B) = \int_0^\infty \phi_\lambda \, B(T_B, \lambda) \, d\lambda \qquad (5)$$

Equation 5 is identical to Equation 4 for $\varepsilon_\lambda = 1$. Since the target is not necessarily a black body, one must not casually interchange T_B with its true surface temperature, T_S.

Many radiometer systems employ in-flight calibration to compensate for electronic drifts. A scan mirror is utilized to bring the cavities of one or more black bodies of known temperature into the field of view.

(e) Accuracy, Sensitivity, and Stability

There are several sources of error which limit the accuracy of an infrared radiometer. A 'leak' in the filters will, for instance, admit spurious radiation (often visible light) from outside the desired infrared spectral region. The accuracy of the temperature control of the cavity will determine the ultimate accuracy obtainable with the instrument. Care must be taken when the radiometer is exposed to cold or wind, since normally the heating element of the temperature-controlled cavity has limited power. If the chill factor is very large, differences between lens and chopper temperature and the internal cavity temperature may also become too large and the emission of these two components may enter differently than during instrument calibration. Extra insulation around the sensing head and a long baffled tube in front of the optics will limit these problems. The extended tube also keeps sea spray or precipitation particles from entering the instrument.

Since the infrared surface temperature depends on inverting Equation 4, the accuracy will depend on knowledge of the emittance, reflected radiance, and ε_λ of the calibration black body source.

Sensitivity is mainly determined by the various sources of noise and the field of view and can therefore range widely; it is a function of the quality and temperature of the detector material. Two sources of noise are the following:

(1) Bolometer flakes are limited by thermal noise caused by the random motion of the molecules due to the elevated temperature.

(2) Photon detectors have their special noise sources due to thermal excitation creating random fluctuations in concentration of charge carriers, rates of generation and recombination of electron-hole pairs ('gr'-noise) etc.

It is common practice to average several data points together. This reduces spatial or temporal resolution, but takes advantage of the randomness of several of the noise sources.

Electronic drift is no longer a major problem but one should be aware that amplifiers and other components have limited lifetimes. Optical materials deteriorate in time. Mirrors and choppers may become scratched, dirty, corroded, or wet in an adverse environment, all of which will decrease sensitivity. Frequent calibration checks of a radiometer are therefore recommended. For air-sea interaction studies a well stirred water bath whose temperature can be accurately read with a mercury thermometer has often proved to be a handy and adequate substitute for clumsy black bodies to make quick checks (care must, of course, be taken that reflected signals are minimized and that the thermal boundary layer in the bath is completely destroyed). The outputs from satellite radiometers are tested against data obtained at sea level.

3. SIGNAL INTERPRETATION

In addition to instrumental considerations for interpreting the voltage output from an infrared radiometer, the effects of atmospheric absorption, emission and scattering, and water surface characteristics must be taken into account. The presence of a thermal boundary layer in the water must also be considered when comparing remotely sensed sea surface temperatures with 'bucket' measurements.

3.1 Effects of Imperfect Transparency of the Atmosphere

The absorption and emission of infrared radiative energy by atmospheric gases is mainly due to vibrational and rotational energy bands of the triatomic molecules: water vapour (H_2O), carbon dioxide (CO_2), and ozone (O_3) (see Fig. 1). Treatises have been written on the absorption by weak absorption bands, exemplified by the water vapour bands in the 8 to 12 μm region, and on the absorption by strong bands, exemplified by the carbon dioxide absorption at 15 μm; both are of concern here (see Goody, 1964, for details). The absorption and emission both depend on the concentration and temperature of the gas.

Redirection of the electromagnetic radiation (scattering) occurs when the radiation interacts with particulates in the atmosphere. The effect depends on the ratio of particle diameter to wavelength of the radiation, and on the refractive index of the particle.

As long as the path length is short enough that the atmosphere is nearly transparent, the attenuation follows the Beer-Lambert Law

$$\frac{J_\lambda}{J_{\lambda_0}} = e^{-\gamma_\lambda u} = \tau_\lambda \tag{6}$$

where J_λ and J_{λ_0} are the transmitted and incident spectral radiances respectively. γ_λ is the extinction coefficient, and u is optical thickness defined through $u = \int_0^x \rho \, dx$, where ρ is the density of the absorbing gas and x is path length. γ_λ is the linear sum of the extinction due to absorption, α_λ, and due to scattering, β_λ.

In the main infrared window, 8 to 12 μm, absorption is usually negligible for path lengths < 300 m; for long paths, the ozone absorption band at 9.6 μm degrades this window somewhat. In the long wavelength region carbon dioxide absorption is mainly responsible for reducing the atmospheric transmittance, while at short wavelengths water vapour absorption is the responsible agent (Fig. 1). In tropical regions the high water vapour content in the atmosphere may reduce the transmittance sufficiently to render the determination of sea surface temperature impossible. Recent satellite sensors and some commercial radiation thermometers have employed a narrower window, between 10.5 and 11.5 μm. Although the incident radiance is reduced accordingly, advances in detector technology are making the use of this window more attractive.

Calculations of atmospheric extinction for varying atmospheric path lengths above a target are given by Saunders and Wilkins (1966). Saunders (1970) has also measured the change in observed radiance with altitude.

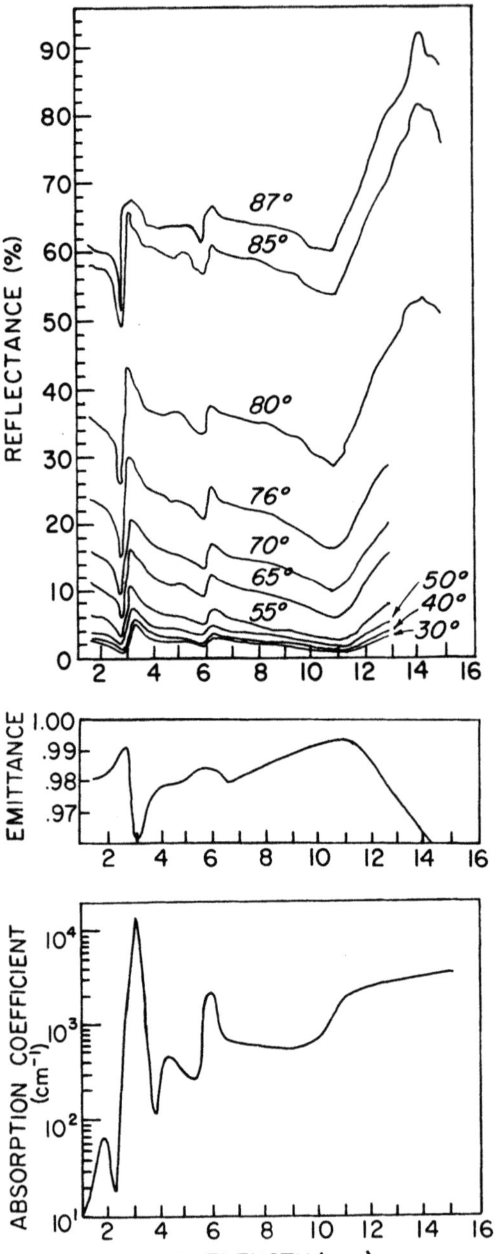

Fig. 5 Top: Reflectance of water as a function of wavelength and of the angle from the normal (after Mc-Swain and Bernstein, 1961).
Middle: Emittance of liquid water (after Mikhaylov and Zolotarev, 1970).
Bottom: Absorption coefficient of liquid water (after Irvine and Pollack, 1968).

3.2 Effects of 'Non-blackness' of a Water Surface in the Infrared

Beyond 30° of normal, water reflectance is as strong a function of angle in the infrared as it is in the visible. Near the vertical, however, water emittance is almost constant and >0.95 for infrared wavelengths >4 μm out to 13 μm [see Fig. 5 middle (after Mikhaylov and Zolotarev, 1970); see also Querry et al. (1977) for recent measurements of the effects of salinity]. Figure 5 top (after McSwain and Bernstein, 1961) shows variations of water reflectance with angle and wavelength. Transmittance of water is small at all middle infrared wavelengths, so that $r_\lambda = 1 - \varepsilon_\lambda$. This is not an adequate approximation at shorter wavelengths, however, as can be seen in Figure 5 bottom, where values for the absorption coefficient are taken from Irvine and Pollack (1968). The reflectance of an oil film is at vertical incidence a few percent greater than the reflectance of clean water.

When converting measured infrared radiances to surface temperature one needs to consider deviations from unity of the emittance of water. Since emittance and reflectance are functions of angle of emission (i.e., the angle between the surface normal and the line of sight to the instrument), one must consider (1) the nadir angle of the radiometer, and (2) the wave slopes (i.e. the roughness, of the sea surface). Including considerations of the reflected signal one obtains the following equation for the observed radiance from an infinitesimal surface area

$$\overline{J}_e(T_B) = \int_0^\infty \phi_\lambda \varepsilon_\lambda B(\overline{T}_s,\lambda)\tau(\lambda)d\lambda + \int_0^\infty \phi_\lambda(1 - \varepsilon_\lambda)J_{sky}(\lambda)\tau_\lambda d\lambda$$

$$+ \text{ atmospheric emission from below} \qquad (7)$$

where $\overline{J}_e(T_B)$ is the observed radiance, T_s is average water surface temperature, and $\overline{\varepsilon}_\lambda$ is spectral emittance for the angle of emission from the area. T_{sky} is the temperature of the sky (integrated atmospheric emission), in the area of the sky reflected by the water facet into the radiometer (see sketch in Fig. 6). If the radiometer is more than 300 m above the water, atmospheric emission in the path between the water surface and the radiometer must also be calculated (see Fleagle and Businger, 1963). Equation 7 must further be integrated over the solid angle of the instrument. The size of the sky cone being reflected at a certain viewing angle also depends on the slope distribution. Cox and Munk (1955) studied the down- and cross-wind slope distribution by analyzing photographs of the sun's glitter pattern observed from an airplane. Their results showed rms slope, σ, to be directly proportional to wind speed. Since for winds <15 m s^{-1} the rms wave slopes are <16°, the roughness of the sea does not affect emittance strongly (see their paper for a more complete discussion of this problem). The most important consideration with respect to emittance is the nadir angle of the instrument.

Saunders and Wilkins (1966) discuss the effect of varying sea surface roughness and sky conditions (clear-summer, clear-winter, and overcast) on the radiance sensed by a radiometer at different nadir angles. Variable cloudiness presents a very difficult problem since the whole sky dome is reflected if the sea is rough. An upward facing radiometer has been used to provide a measure of the reflected radiance (see also discussion of Brewster angle measurements below). Saunders and Wilkins also calculated the reflected solar radiation. Even in a calm sea with the radiometer pointed to the sun's image at a large nadir angle, the change in brightness temperature is only 0.2°C for the 8 to 12 μm window. However, at 3 to 4 μm, the sun's image can produce a 20°C increase.

A radiometer mounted on a boom of a ship or from a dock may see the ship's hull or its own image by reflection. A similar consideration must be made for the reflection of the hull of an airplane, when it flies 100 m or less above the sea. Saunders and Wilkins (1966) found for a radiometer of small angular field of view that the presence of the aircraft is not sensed as long as its angular diameter as observed from the surface is less than the rms slope of the waves. This implies for aircraft of 30 m wingspan and 15 m s^{-1} surface winds a minimum height of 40 m. For weak winds the minimum height will increase to about 150 m.

3.3 Corrections for Variations in Water Surface Emittance and in Atmospheric Transmittance and Emittance

Correcting the signal for variations from unity in water emittance

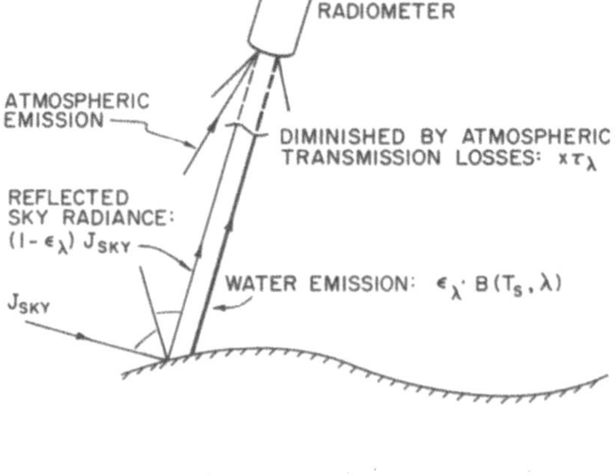

Fig. 6 Illustration of the three sources of the radiance received by a downward looking infrared radiometer over the sea.

can be accomplished in various ways. The first is to measure the sky radiance (J_{sky}) (usually at zenith, with the surface temperature measurement made vertically downward) and invert Equation 7 using known values for $\phi(\lambda)$, $\varepsilon(\lambda)$. and $\tau(\lambda)$. A second technique is to obtain the measurement at the Brewster angle. At this angle (for water it is about 57° from zenith) the reflection is at a minimum for both horizontally and vertically polarized radiation (Grassl and Hinzpeter, 1975, and Grassl, 1976). A technique which employs measurements at the Brewster angle with a polarizer, whose axis is alternatingly perpendicular and parallel to the plane of incidence, is described by Lecomte et al. (1973). This technique has been tried from a ship but is not in general use. It may reduce the error due to sky reflection by a factor of four. With a third technique, the effect of sky reflectance can be eliminated by looking at a stirred water bath of known temperature at the same angle under which the measurement of sea surface temperature is being obtained and in the same direction. Both measurements are then affected by the same sky reflection. This technique has been successfully employed on a ship (Clauss et al., 1970).

Correcting the signal for varying atmospheric absorption and emission has proved to be a difficult problem. In the past, information about the atmospheric temperature and humidity structure was obtained from the radiosonde network; however, over most of the oceans such data are not adequate and climatological data were then often substituted. An improved method uses atmospheric temperature and humidity profiles obtained simultaneously on the satellite (see Allison et al., 1975, for a review). Radiance measurements are obtained in several narrow wavelength intervals within the absorption bands of carbon dioxide and water vapour. As the satellite radiometer sees through a sequence of narrow spectral regions with decreasing absorption (emission), layers of the atmosphere at greater and greater depths are being sampled. An 'inversion' calculation produces the profiles of temperature and humidity. The carbon dioxide band at 15 μm is used for temperature on the assumption that the CO_2 is well mixed and variations in emitted radiance only reflect the temperature of the gas. For water vapour the radiance is both a function of temperature and the water vapour mixing ratio. A very simple but effective technique to correct for the interference of the atmosphere is to use the radiance at neighbouring wavelengths, where the absorption by atmospheric water vapour varies rapidly with wavelength (Prabhakara et al., 1974). These authors use spectral data from the Nimbus 3 Iris radiometer and compare radiances in bands of 1 μm width. The accuracy of their sea surface temperature is about 1 K.

A similar technique is proposed by Saunders and Wilkins (1966), who suggest that a radiometer in an airplane could view the surface at 0° and 60° nadir angles and thereby double the path length. Even though the temperature and humidity structure is unknown, the net effect of the atmosphere can be deduced (the reflected signal must, however, also be included). In all case studies the ultimate test has been comparison with surface-based measurements: so-called 'surface truth'. For airplanes this check allows one to correct the absolute scale when the atmospheric conditions are fairly uniform, since the area of resolution is not too diverse for the two measures. For satellites the check is still of importance, but a point measurement from one ship is not adequate for the large areas covered by satellites. However, after a correction for atmospheric degradation has been applied, satellite data have been found to agree with ship-based T_w temperature measurements to ±1 K.

3.4 Effects of the Thermal Boundary Layer in the Water on Interpretations of Radiometric Surface Temperatures

The absorption coefficient of water is very large in the entire infrared region (see Fig. 5 bottom), which implies that the radiation in the relevant 'window' regions of the spectrum emanates mainly from water above 50 μm depth. The remotely sensed surface temperature is therefore truly representative of the interface, and is the relevant temperature for evaluating air-sea temperature differences and saturation vapour pressure near the interface.

Near the surface there exist thin boundary layers both in the air and in the water where molecular processes dominate. For this reason a large fraction of the total temperature difference between the surface and the interior of the fluid occurs over a very small distance, of the order of 1 mm. The remotely sensed surface temperature can therefore differ by a significant amount from the 'surface temperature' obtained by conventional means, at depths of 20 cm to 1 m. The difference becomes of interest when using both kinds of data simultaneously. Studying how it varies with heat flux, wind stress, and waves has provided additional insights (see Katsaros, 1979, for a more comprehensive review).

On the open sea the surface is typically colder than lower layers by a few tenths of a degree Kelvin, because the solar heat source is distributed over about 100 m depth, while all the losses occur from the interface. The losses consist of a net upward heat flux due to sensible and latent heat fluxes by turbulent exchange as well as a net upward radiative flux due to exchange with the atmosphere and clouds. With weak winds and strong insolation the surface layer will occasionally be warmer than lower strata. Because the aqueous thermal boundary layer is typically colder than lower strata it has been called 'the cool skin'. The use of 'film' for 'skin' should be discouraged, since a 'film' more properly refers to thin layers of surface contamination.

Figure 7 illustrates the establishment of a cold thermal boundary layer by showing the depths where the sources and sinks of heat are located, i.e. short and long wave radiation and the turbulent fluxes of sensible and latent heat. A scaling length for the depth of the thermal boundary layer, δ, and the temperature drop across the layer, ΔT, are defined by the conduction equation:

$$Q_N = -k \left(\frac{\partial T}{\partial z}\right)_0 = -k \frac{\Delta T}{\delta} \tag{8}$$

where $(\partial T/\partial z)_0$ is the slope of the temperature profile at the interface, and k is molecular conductivity of heat. Q_N is the net result of the fluxes of heat across the interface excluding solar radiation.

$$Q_N = +Q_{IR} + Q_G + Q_H + Q_E \tag{9}$$

The individual terms are defined on Figure 7. During daylight a correction must be made for the short wave radiance absorbed within the thermal boundary layer (see Saunders, 1967b).

In turbulent water, eddies will occasionally penetrate the thermal boundary layer and remove the cooler water to greater depths, upon which the re-establishment of a cool layer by conduction will start again. This intermittent process, referred to as 'surface renewal', limits the growth of the boundary layer and both δ and ΔT (e.g., Liu and Businger, 1975). That δ should depend on the intensity of the turbulent shear stress τ_w is predicted by Saunders (1967b) using the dimensional argument

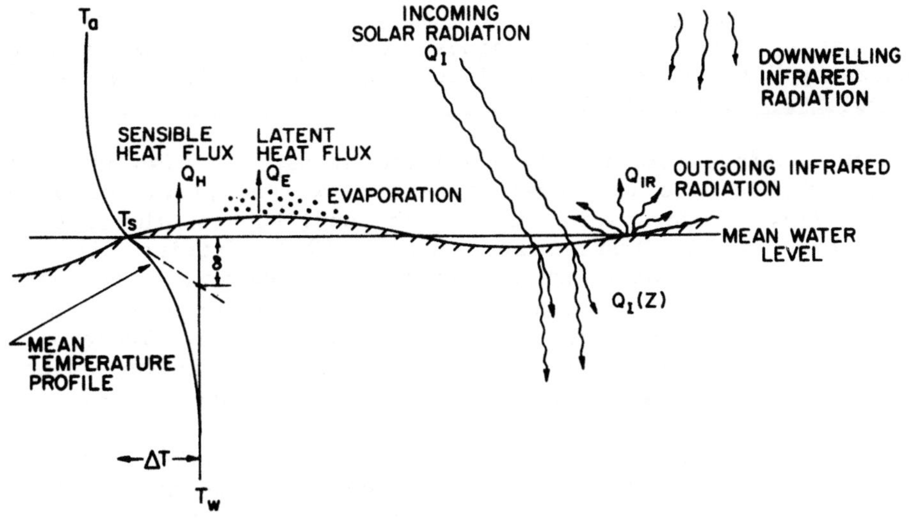

Fig. 7 Illustration of the factors influencing the temperature drop across the aqueous thermal boundary layer.

$$\delta \propto \frac{\nu}{\sqrt{\tau_w/\rho_w}} \tag{10}$$

where ν is the kinematic viscosity. Current shear near the surface is mainly induced by the atmospheric shear stress. Thus $\tau_w \propto \tau_a$, but the constant of proportionality is not well established. It also appears to be dependent on the stage of wave growth.

Substituting Equation 10 into Equation 8 Saunders (1967b) obtained a prediction for ΔT:

$$\Delta T = \lambda \frac{Q}{\sqrt{\tau_w/\rho_w}} \tag{11}$$

where λ is a proportionality factor containing physical constants, the ratio between the thicknesses of the viscous and thermal boundary layers, and the constant of proportionality between τ_w and τ_a. Since Q_E and Q_H (contained in Q_N) are both linear functions of the mean wind speed $\overline{U}$ (or almost so) and $\tau_w \propto \overline{U}^2$, while Q_{IR} is independent of $\overline{U}$, ΔT is expected to be almost independent of wind speed, except when $Q_{IR} + Q_G$ is an important term in the heat budget.

During conditions of free convection in the water ΔT can be predicted to increase as $Q_N^{\frac{3}{4}}$ (e.g., Katsaros, 1977). Further studies of ΔT during forced mixing are reported by Hasse (1971), who derived a relationship similar to Saunders', and found excellent agreement with field data. Paulson and Parker (1972) calculated values of λ ranging from 4 to 17 using available field and laboratory data. The smallest values were derived from laboratory studies by Hill (1972) and McAlister and McLeish (1969) where wind-driven waves were present.

That the surface temperature deviation should diminish when waves perturb the surface (at constant heat flux) was predicted by Witting (1971). Papers which predict the amplitude of the associated temperature waves (O'Brien and Omholt, 1969; Witting, 1972) are contradictory regarding their magnitude. However, a temperature wave in general agreement with Witting's prediction was found by Chang and Wagner (1975) and by Miller et al. (1975). The amplitude is very small $[O(10^{-2}\,°C)]$. Effects due to the varying emittance (and reflectance) are usually an order of magnitude larger.

An interesting application of the existence of a strong temperature gradient near the surface was made by McAlister and McLeish (1970). They developed a specialized two-wavelength radiometer to measure the temperature gradient within the thermal boundary layer. The radiometer measures the emitted radiances in two narrow wavelength intervals at 4.8 and 3.8 μm. Between these wavelengths the absorption coefficient of water varies strongly, while atmospheric transmission is nearly constant (Fig. 5 bottom, and Fig. 1). The effective depths of emission are 25 and 75 μm respectively. The measured temperature gradient can then be related to the net upward heat flux by Equation 8.

Effects of contaminant films on the total temperature drop between the interior and the surface have been observed. The films often collect in 'slicks' of alternating smooth and rough water, which align themselves with the wind. That organic material collects in the smooth regions has been documented (e.g. Sutcliffe et al., 1963); the smooth appearance is due to destruction of capillary waves at the surface tension discontinuity represented by the slick. Such films, of natural and anthropogenic origin, have received detailed study in recent years (see MacIntyre, 1974, for an excellent review).

Existence of a surfactant film may influence the remotely sensed radiation in several ways: (a) the emittance of oils is typically less than that of water, so the correction for sky reflection will be greater and the surface will appear cooler; (b) some but not all organic materials decrease evaporation; (c) since capillary waves will not form in the slick and since waves reduce ΔT, other factors being constant, one would expect a greater ΔT in the area of a slick; (d) if the oil film is fairly thick, but not so thick that convection occurs within it, its effect is to provide an additional layer through which heat is conducted primarly at the molecular rate. There is therefore an additional temperature drop augmenting the ΔT in the water. With so many conflicting possibilities for the effect of a slick the net result on the total ΔT cannot readily be predicted. However, McLeish (1970) noted from airplane observations large values in horizontal spectra of temperature at 10^{-5} cycles cm^{-1}, corresponding to a horizontal scale of 1 km. He interprets this 'cooler' surface temperature as only a superficial manifestation of the convergence associated with subsurface turbulence. He suggests that surface films collect in the convergence zone, and that their inhibition of the transport of heat results in colder brightness temperatures.

Since values of $\Delta T = T_s - T_w$ are typically of the order of -0.2 to -0.5°C and $T_a - T_w$ is often of the order of -1°C to -2°C, ΔT represents an important consideration (a) when values of T_w are sought from the remote measurements of T_s or (b) when heat and water vapour fluxes are calculated on the basis of T_a and T_w measurements.

Evaluation of ΔT can be accomplished by stirring the water. Vigorous stirring increases τ_w in Equation 10 and thereby decreases δ; T_w is exposed to the radiometer rather than T_s. With this procedure the reflected sky signal is the same for the two measurements.

Since this is not always possible (as from a ship at sea) the two values T_s and T_w must be obtained separately and corrections for the sky reflection can be obtained as described in Section 2.3.

4. INHERENT LIMITATIONS - FUTURE EXPECTATIONS

Inherent limitations are caused by the natural variability of the sea and atmosphere and of the sensing system. The variations in concentration of atmospheric absorbing gases and their horizontal and vertical temperature distribution can only be corrected to a degree which depends on the resolution of the spectral sensors used. Unless sophisticated cooled detectors are used, random thermal fluctuations limit the accuracy of the common bolometer. The simple commercial instruments have not been improved beyond 0.1 K for the narrow beam versions over the last decade.

The high moisture content of tropical air masses severely attenuates the signal and degrades the contrasts in surface temperature. This limitation affects satellite observations most strongly. Many observed variations in ΔT are caused by variations in the upper ocean dynamics, and therefore slick concentrations may introduce an apparent surface temperature fluctuation, which is not related to subsurface temperature changes.

One of the greatest limitations to infrared sensing of sea surface temperature from satellites is the obstruction by clouds in the field of view. The narrower the beam the greater the likelihood that the radiometer will be able to peek between clouds. A technique involving 18 channels in the infrared to correct for cloud distribution at three levels in addition to the atmospheric transmittance is reported by Chahine et al. (1977). With more sophisticated data collection and analysis techniques one may hope eventually to obtain sea surface infrared temperatures accurate to $\pm 0.5°C$ from satellites in the absence of clouds.

In the Gulf of Alaska and many other parts of the ocean clouds are too prevalent; the hope is that microwave radiometers will provide a solution. Cloud droplets do not interfere substantially with radiation of millimetre to centimetre wavelength. Raindrops will, however, strongly emit and absorb at these wavelengths.

Multispectral microwave radiometers will be carried on the new suite of experimental satellites to be launched in 1978 and beyond. The surface resolution obtainable with such instruments is about 100 km. The need for large resolution elements is due to the very weak emission at these long wavelengths. To interpret microwave radiances in terms of temperature a correction for the low emittance of water (as low as 50%) must be made. The emittance at microwave wavelengths depends on temperature and salinity and also on the extent of white caps on the ocean surface.

ACKNOWLEDGEMENT

The author is grateful for support from the Office of Naval Research under Contract N00014-75-C-0502. Contribution number 450, Department of Atmospheric Sciences, University of Washington, Seattle, WA 98195.

REFERENCES

ALLISON, L.J., A. ARKING, W.R. BANDEEN, W.E. SHENK and R. WEXLER. 1975. Meteorological satellite accomplishments. *Reviews of Geophysics and Space Physics,* 13: 737-745.

BJERKNES, J. 1969. Atmospheric teleconnections from the equatorial Pacific. *Monthly Weather Review,* 97: 163-173.

CHANG, J.H. and N.N. WAGNER. 1975. Laboratory measurement of surface temperature fluctuations induced by small amplitude surface waves. *Journal of Geophysical Research,* 80: 2677-2687.

CHAHINE, M.T., H.H. AUMANN and F.W. TAYLOR. 1977. Remote sounding of cloudy atmospheres. III. Experimental verifications. *Journal of Atmospheric Science,* 34: 758-765.

CLAUSS, E., H. HINZPETER and J. MUELLER-GLEWE. 1970. Messungen zur Temperaturstruktur im Wasser an der Graenzflaeche Ozean-Atmosphaere. *'Meteor' Forschungsergebnisse, Reihe B.*: 90-94.

COX, C. and W. MUNK. 1955. Some problems in optical oceanography. *Journal of Marine Research,* 14: 63-78.

EWING, G.C. and E.D. McALISTER. 1960. On the thermal boundary layer of the ocean. *Science,* 131: 1374-1376.

FLEAGLE, R.G. and J.A. BUSINGER. 1963. *An Introduction to Atmospheric Physics.* Academic Press, New York, 346 pp.

GOODY, R.M. 1964. *Atmospheric Radiation.* Clarendon Press, Oxford, 436 pp.

GRASSL, H. 1976. The dependence of the measured cool skin of the ocean on wind stress and total heat flux. *Boundary-Layer Meteorology,* 10: 465-474.

GRASSL, H. and H. HINZPETER. 1975. The cool skin of the ocean. GATE Report 14, Vol. II, World Meteorological Organization, International Council of Scientific Unions, Geneva: 229-236.

HASSE, L. 1971. The sea surface temperature deviation and the heat flux at the sea-air interface. *Boundary-Layer Meteorology,* 1: 368-379.

HILL, R.H. 1972. Laboratory measurement of heat transfer and thermal structure near an air-water interface. *Journal of Physical Oceanography,* 2: 190-198.

HOLLADAY, C.G. and J.J. O'BRIEN. 1975. Mesoscale variability of sea surface temperatures. *Journal of Physical Oceanography,* 5: 761-772.

HUDSON, R.D., JR. 1969. *Infrared System Engineering.* Wiley Interscience, New York, 642 pp.

IRVINE, W.M. and J.B. POLLACK. 1968. Infrared optical properties of water and ice spheres. *Icarus,* 8: 324-360.

KATSAROS, K.B. 1973. Supercooling at the surface of an arctic lead. *Journal of Physical Oceanography,* 3: 482-486.

KATSAROS, K.B. 1977. The sea surface temperature deviation at very low wind speeds; is there a limit? *Tellus,* 29: 229-239.

KATSAROS, K.B. 1979. The aqueous thermal boundary layer. *Boundary-Layer Meteorology* (in press).

KRUSE, P.W., L.D. McGLACHLIN and R.B. McQUISTAN. 1962. *Elements of Infrared Technology.* John Wiley & Sons, Inc., New York, 448 pp.

LECOMTE, P., P.Y. DESCHAMPS and J.C. VANHOUTTE. 1973. Ameliorations apportées à la mesure de la température de surface de l'ocean par l'utilisation d'un radiomètre infrarouge polarisant. *Applied Optics,* 12: 2115-2121.

LIU, W.T. and J.A. BUSINGER. 1975. Temperature profile in the molecular sublayer near the interface of a fluid in turbulent motion. *Geophysical Research Letters,* 2: 403-404.

MacINTYRE, F. 1974. The top millimeter of the ocean. *Scientific American,* 230: 62-77.

McALISTER, E.D. and W. McLEISH. 1969. Heat transfer in the top millimeter of the ocean. *Journal of Geophysical Research,* 74: 3408-3414.

McALISTER, E.D. and W. McLEISH. 1970. A radiometer system for airborne measurement of the total heat flux from the sea. *Applied Optics,* 9: 2697-2705.

McLEISH, W. 1968. On the mechanism of wind slick generation. *Deep-Sea Research,* 15: 461-469.

McLEISH, W. 1970. Spatial spectra of ocean surface temperature. *Journal of Geophysical Research,* 75: 6872-6877.

McSWAIN, B. and J. BERNSTEIN. 1961. Specular reflectance of water in the 1.5 to 15 micron region as a function of wavelength and incidence angle. Navweps Rept. 7162, Quarterly Report Fundamental Research Project, Naval Ordnance Laboratory, Department of the Navy, Corona, California.

MIKHAYLOV, B.A. and V.M. ZOLOTAREV. 1970. Emissivity of liquid water. *Atmospheric and Oceanic Physics,* 6: 52.

MILLER, A.W., R.L. STREET and E.Y. HSU. 1975. The structure of the aqueous thermal sublayer at an air-water interface. Department of Civil Engineering, Technical Report No. 195, Stanford University, 193 pp.

NAMIAS, J. 1969. Seasonal interactions between the North Pacific ocean and the atmosphere during the 1960's. *Monthly Weather Review,* 97: 192-193.

O'BRIEN, E.E. and T. OMHOLT. 1969. Heat flux and temperature variation at a wavy water-air interface. *Journal of Geophysical Research,* 74: 3384-3385.

PAULSON, C.A. and T.W. PARKER. 1972. Cooling of a water surface by evaporation, radiation, and heat transfer. *Journal of Geophysical Research,* 77: 491-495.

PRABHAKARA, C., G. DALU and V.G. KUNDE. 1974. Estimation of sea surface temperature from remote sensing in the 11- to 13-μm window region. *Journal of Geophseyical Research,* 79: 5039-5044.

QUERRY, M.R., W.E. HOLLAND, R.C. WARING, L.M. EARLS and M.D. QUERRY. 1977. Relative reflectance and complex refractive index in the infrared for saline environmental waters. *Journal of Geophysical Research,* 82: 1425-1433.

SAUNDERS, P.M. 1967a. Aerial measurement of sea surface temperature in the infrared. *Journal of Geophseyical Research,* 72: 4109-4117.

SAUNDERS, P.M. 1967b. The temperature of the ocean-air interface. *Journal of Atmospheric Sciences,* 24: 269-273.

SAUNDERS, P.M. 1970. Corrections for airborne radiation thermometry. *Journal of Geophysical Research,* 75: 7596-7601.

SAUNDERS, P.M. 1971. Anticyclonic eddies formed from shoreward meanders of the Gulf Stream. *Deep-Sea Research,* 18: 1207-1219.

SAUNDERS, P.M. 1973. Tracing surface flow with surface isotherms. *Memoires Societe des Sciences de Liege*, 6 ser., tom VI: 99-108.

SAUNDERS, P.M. and C.H. WILKINS. 1966. Precise airborne radiation thermometry. *Proceedings of the 4th Symposium on Remote Sensing of Environment,* Institute of Science and Technology, University of Michigan, Ann Arbor, Michigan: 815-826.

SUTCLIFFE, W.H., E.R. BAYLOR and D. MENZEL. 1963. Sea surface chemistry and Langmuir circulation. *Deep-Sea Research,* 10: 233-243.

WITTING, J. 1971. Effects of plane progressive irrotational waves on thermal boundary layers. *Journal of Fluid Mechanics,* 50: 321-334.

WITTING, J. 1972. Temperature fluctuations at an air-water interface caused by surface waves. *Journal of Geophysical Research,* 77: 3265-3269.

17

High Resolution Salinity Measurement Techniques

M.C. Gregg and A.M. Pederson

1. SCIENTIFIC OBJECTIVES REQUIRING SALINITY MEASUREMENTS

Salinity measurements are of fundamental importance to physical studies of the ocean because variations in salt content have strong effects upon the density field and in addition are conservative tracers of water movements below the sea surface. In the following discussion, some of the scientific objectives of near surface work are outlined and then used to estimate observational requirements. Since salinity is not measured directly, these requirements are compared to the sensitivities of measurable parameters to changes in salinity, temperature, and pressure. Finally, the observational techniques currently in use or under discussion are evaluated against the scientific requirements with emphasis on measurements resolving structures smaller than one metre. Many of these measurements are made from vehicles that move freely through the ocean rather than from platforms tethered to a ship.

Three-dimensional time histories of the major structures in the upper ocean are essential to studies of air-sea interactions. The most basic information is the static stability of the surface water; is it well mixed in the mean or are there weak gradients that indicate either a slight stratification or convective instability? Over what lateral scales do well-mixed surface layers extend, and how are heterogeneities in the surface water related to the formation of subsurface intrusions? How far do these subsurface well-mixed fluid parcels intrude before they are dissipated as identifiable features? Within surface mixed layers, do either individual large scale turbulent eddies or Langmuir cells extend across the depth of the layer or are they limited to the upper parts of the layers? Salinity and temperature data systems capable of detecting characteristic signatures of these structures, having vertical scales of 10 to 200 m and lateral dimensions up to 1000 times greater, are necessary to advance beyond the meager information now available.

Measurements over much smaller scales are required to obtain direct evidence for mixing of heat and salt. In stratified profiles, mixing events occur principally as occasional overturns, with vertical scales ranging from 1 to 2 metres to as little as a few centimetres (Gregg, 1977). The downward growth of a surface mixed layer by entrainment is shown by density instabilities at the base of the layer. To assess fully the response of the ocean to surface forcing, it is necessary to obtain statistics of the occurrence of these events for a wide variety of conditions. To obtain quantitative information about the rate of entropy generation associated with the mixing, it is necessary to fully resolve the variance of the salinity gradients, $\overline{(\nabla S)^2}$, as well as the coarser scales associated with the overturn. Within the mixed layer, the shape of the salinity spectrum can be compared with the forms expected for fully developed homogeneous turbulence.

2. SPECIFICATIONS FOR SALINITY DATA

2.1 Definitions

In translating these general scientific objectives into specifications for salinity measurements, it is necessary to consider the spatial resolution, precision and sensitivity required for the different objectives. The spatial resolution is determined by the scales in the salinity field of the processes considered. The minimum frequency response of a sensor, in relation to its speed through the water, and the maximum size follow from the required spatial resolution. The latter involves the dimensions of the volume being sampled, the separation of the probe from other sensors necessary to determine salinity, and the distortion of the structures in the water by the probes and their mountings. The sensitivity is the minimum signal level necessary to see an unambiguous signal of the structures being studied in a given record. Precision, or confidence limits, are determined by the need to use the salinity data in determining density gradients or doing lateral mapping.

For some of the structures considered in Section 1, little is known about what signal levels are to be expected and undoubtedly these levels will vary by several decades with different locations and forcing conditions. In the following, the observations reported by Gregg (1976) will be used as a basis for estimation. The requirements for salinity are expressed in parts per thousand (ppt) and parts per million (ppm).

Table 1

Estimated Specifications for Salinity Measurements in Near Surface Observations

Scientific Objective	Trace Lateral Variability and Intrusions below the Mixed Layer	Detect Entrainment and Overturns in Thermocline	Determine Mean Stability of Mixed Layer and Signature of Large Scale Processes	Resolve $\overline{(\nabla S)^2}$
Sensitivity (ppm)	10 to 1	6 to 0.1	Gradients of 0.03 ppm m^{-1} or net ΔS of 6 to 0.3	3×10^{-2} to 3×10^{-4}
Precision (ppm)	10 to 1 over periods of hours to days	2 to 0.2 over periods of minutes	6 to 0.3 over periods of minutes. Not important for spectral approach or for gradients	not important
Spatial Resolution (m)	50 to 2	1 to 0.1	1 m for gradients 200 to 10 for net ΔS	10^{-3} to 10^{-4}

2.2 Tracing Lateral Variability Below the Mixed Layer

The delineation of gross lateral heterogeneities and the tracing of intrusions are the least demanding of salinity measurements. The major intrusions have temperature and salinity signatures of several tenths in terms of °C and ppt, over vertical scales of 50 m or more. To trace unambiguously the TS characteristics of these features, a sensitivity of about 10 ppm, with corresponding precision, is sufficient. However, this level of sensitivity frequently reveals intrusions with vertical scales down to several metres. To map them would require an increase of sensitivity and precision by a factor of 10.

2.3 Detecting Entrainment and Overturns in Vertical Profiles

In the upper ocean, the temperature gradient averaged over scales of a few metres is most frequently in the range of 20 to 80 m°C m^{-1}. From the mean TS relation for the Central North Pacific, $\Delta S/\Delta T = 0.08$ ppt $°C^{-1}$, giving salinity gradients of 1.6 to 6.4 ppm m^{-1}. A linear equation of state, estimated from Tables V through IX of Neumann and Pierson (1966), can be used to express the gradient of potential density as:

$$\frac{d\rho_\theta}{dz} = -0.22\ (dT/dz + \Gamma) + 0.77\ dS/dz\ (\text{kg m}^{-4}) \qquad (1)$$

where z is positive upward, Γ is the adiabatic temperature gradient ($\simeq -10^{-4}$ °C m^{-1}), and T and S are in °C and ppt, respectively. Thus, the potential density gradient in the North Pacific varies between -3 to 13×10^{-3} kg m^{-4} with salinity differences accounting for 25%. Precisions of ±2 m°C and ±2 ppm are necessary to determine this gradient to $\pm2\times10^{-3}$ kg m^{-4} over a scale of 1 m. This would show only the largest and least frequent of the instabilities. To detect most of the instabilities, resolution to 0.1 m with an increase in both sensitivity and precision of a factor of 10 would be necessary. The signals associated with entrainment at the base of the mixed layer are of roughly the same scales and amplitudes.

2.4 Detecting Low Wavenumber Structures within Mixed Layers

The detection of structures within nearly well mixed surface layers is a much more difficult task than those discussed above. For a well mixed layer, the temperature and salinity profiles would be adiabatic and isohaline, respectively. Since there is little information about how large a departure from the neutral state has significant dynamical effects, it would be desirable to measure temperature gradients of 10^{-4} °C m^{-1}, i.e. the adiabatic gradient, over scales greater than 1 m. A salinity gradient of 0.03 ppm m^{-1} has an equivalent effect upon the stratification. Lacking this, the minimum capability is the measurement of net differences across the layers corresponding to these gradients. For layer depths of 10 to 200 m, this corresponds to both sensitivities and precisions of 1 to 20 m °C and 0.3 to 6 ppm.

Mixed layers that are nearly adiabatic in the mean are found to have rms temperature fluctuations of between 10 and 1 m°C; for the latter case, the rms salinity fluctuations have been investigated and found to be about 0.1 ppm. Most of the variance was contributed from scales of 10 m and greater. The levels of these larger scale salinity fluctuations will vary greatly, both in an absolute sense and with respect to the temperature fluctuations, depending upon whether there are active sources of fluctations due to heat transfer and evaporation at the surface or to entrainment at the base of the layer (the low value mentioned above occurred in a layer that appeared to have only weak sources). To extract the signature of large turbulent eddies driven from the surface, or of Langmuir cells, may take sensitivities to at least these rms levels with spatial resolution ranging from 10 to 200 m. If TS relations are used in investigating these structures, precisions equivalent to the sensitivity will be required; but, on the other hand, if a spectral signature is sufficient, then precision is not important.

2.5 Detecting Turbulent Structures within Mixed Layers

During a mild storm, temperature spectra over scales of several metres and less have been found to closely resemble the forms predicted for homogeneous, isotropic turbulence. For a passive scalar variable, such as temperature or salinity, the scale at which the

Table 2

Scales of diffusive cutoff for temperature and salinity for values of the rate of energy dissipation, ε, expected in turbulent mixed layers. 99% of the variance of the gradients will be obtained by measurements that resolve the spectra to k_B.

ε (W kg^{-1})	k_B^T (cpm)	k_B^S (cpm)
10^{-5}	756	7560
10^{-6}	425	4250
10^{-7}	239	2390
10^{-8}	134	1340

spectra change from a slope of −5/3 in the inertial subrange to −1 in the viscous-convective subrange depends upon ε, the rate at which kinetic energy is dissipated by viscosity. The length scale at which diffusion cuts off the fluctuations of a scalar, such as temperature or salinity, is given by the Batchelor scale, $L_B \sim (\nu K^2/\varepsilon)^{1/4}$, where ν is the kinematic viscosity and K is the scalar diffusivity. The units of the scale are generally considered to be radians per metre. The corresponding wavenumber scale, $k_B \sim (2\pi L_B)^{-1}$ in cpm (cycles m^{-1}), is shown in Table 2 for the dissipation rates expected in turbulently stirred mixed layers. Measurements that resolve the spectrum to k_B obtain 99% of the gradient variance, while half of the variance is resolved to scales of $k_B/3$. The levels of the high wavenumber spectra will vary depending upon the source strength of the low wavenumber fluctuations previously

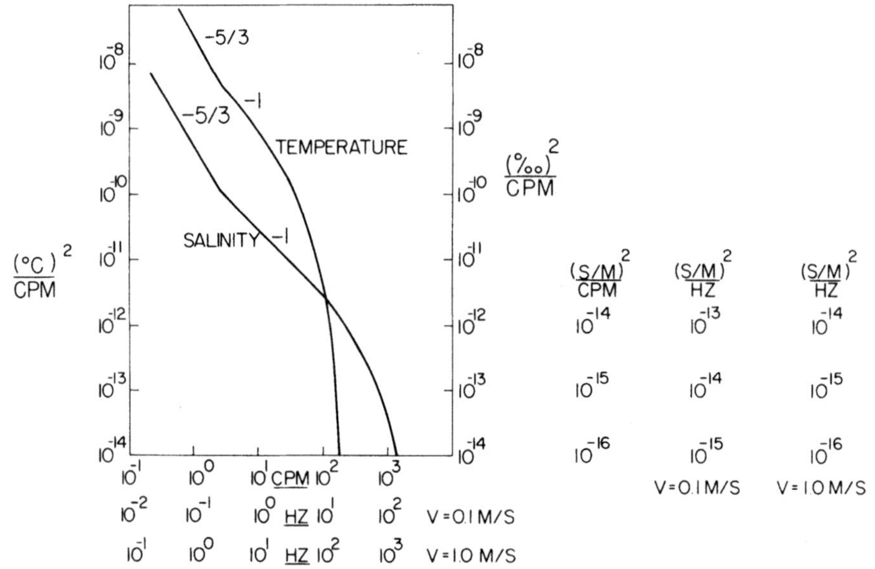

Fig. 1 Temperature and salinity spectra expected for a surface mixed layer in a mild storm. The salinity spectrum is based upon the mean TS relation for the Central North Pacific and upon the temperature spectrum, which was measured to 80 cpm. The two scales on the bottom express the spatial spectra in terms of the frequencies sensed by probes moving at 0.1 and 1.0 m s^{-1}. The three scales on the right express the conductivity spectra for the high wavenumber portion of the spectra where the temperature fluctuations will have negligible effect upon the conductivity.

discussed. As an example of the spectral levels expected, a temperature spectrum measured in a turbulent mixed layer has been plotted in Figure 1, together with a salinity spectrum, which was inferred by using the break-in-slope from the -5/3 to the -1 range to obtain ε, and the mean TS relation of the surface waters (the measured salinity spectrum had a somewhat higher level and followed the same slopes but was resolved to only 10 cpm). The value of ε was 10^{-8} W kg^{-1}. Since this was a mild storm and ε was low, the requirement for spatial resolution was about the least that can be expected for mixed layer work. Figure 1 is scaled to give both the spatial spectra and the frequency spectra observed by probes moving at 0.1 and 1.0 m s^{-1}. These spectra are related by $\Phi(f) = \Phi(k)/U$, where U is the speed in m s^{-1} and f = kU. Thus, assuming that the measurements are limited by inherent system noise, the sensitivity criteria, as well as the frequency resolution, become more severe for fast moving sensors. As shown in Figure 2, where equal areas under the curves contribute equally to the respective variances, the significant contributions to the salinity variance come at wavenumbers greater than the diffusive cut-off of the temperature fluctuations. In terms of the summary of salinity specifications in Table 1, resolution of $\overline{(\nabla S)^2}$ in turbulent mixed layers will require spatial resolutions between 1 and 0.1 mm and sensitivities of 10^{-2} to 10^{-4} ppm; a formidable prospect!

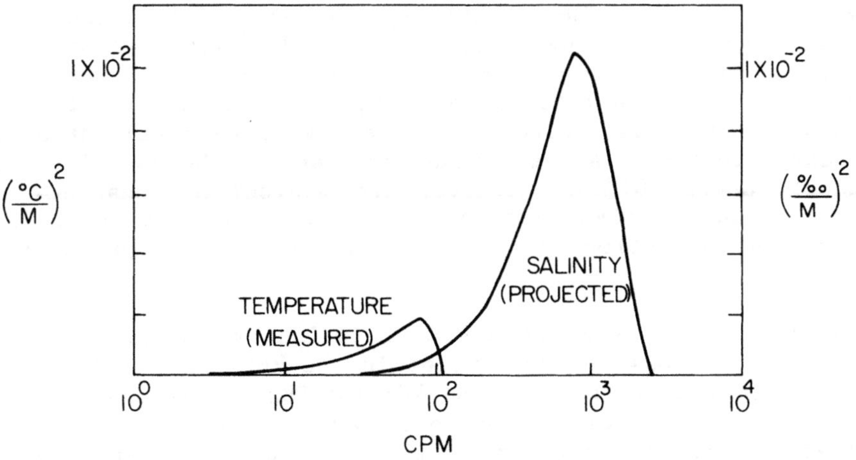

Fig. 2 Variance preserving plots for the spectra in Figure 1. Significant contributions to the salinity variance occur at wavenumbers past the range of appreciable temperature variance. Thus, it is not necessary to make simultaneous temperature and conductivity measurements at these scales.

3. SENSITIVITY OF MEASURABLE PARAMETERS TO SALINITY VARIATIONS

3.1 Parameters That Are Affected by Salinity

A general discussion of the physical properties of sea water is given by Horne (1969). Of these properties, dielectric constant, $\hat{\varepsilon}$, index of refraction, n, electrical conductivity, c, and velocity of sound, U_s, are among the parameters that have been used for measurements in various fields of technology. Each of these exhibits an appreciable variation with changes in salinity and has been considered as the basis for an in situ salinometer.

3.2 Dielectric Constant and Index of Refraction

The dielectric constant, which is discussed in detail by Owen et al. (1961), is a measure of the polarizability of a substance by an applied electric field, and this in turn affects the propagation characteristics of electromagnetic radiation in the material. The polarizability varies with the frequency of the field depending upon the natural periods of the electric dipoles in the substance. For a given frequency, the dielectric constant has both real and imaginary, or lossy, parts whose values are determined by the characteristics of those dipoles having nearly equal or greater natural frequencies; those dipoles with significantly lower natural frequencies cannot respond to the external field. For pure water below the GHz range there is little variation of $\hat{\varepsilon}$ with frequency and the value of 80 for the static dielectric constant, $\hat{\varepsilon}_s$, can be used for water at 20 °C. The relatively massive water molecules have the lowest natural frequencies, about 1.7×10^{10} Hz, and produce an appreciable imaginary component of $\hat{\varepsilon}$ at frequencies greater than 10^8 Hz. The detailed behavior at frequencies beyond the GHz range is not well known but there is apparently another relaxation process near 3×10^{12} Hz. At yet higher frequencies the oscillations of electrons, about 10^{16} Hz, control the dielectric constant. For frequencies in the visible range, e.g. 6×10^{14} Hz for green light, the index of refraction is related to the dielectric properties by $n^2 = \hat{\varepsilon}$.

For an electrolyte, such as a NaCl solution, there is an appreciable dielectric loss at frequencies in the kHz range and lower due to ionic conduction. Between the kHz and GHz ranges the dielectric constant is lower than that of pure water ($\hat{\varepsilon} \simeq 74$ for concentrations equivalent to sea water at 20 °C), and varies with ion concentration due to the volume effect of the added nonpolar molecules and the interaction of the electric fields of the ions, which reduces the ability of the water molecules to orient to an external field. Polarization of the ions in an electrolyte increases n above the value for pure water (about 1% for sea water), and produces a salinity dependence in the visible range.

3.3 Electrical Conductivity

The electrical conductivity is a measure of ion migration due to an applied field and has little frequency dependence below the GHz range; at those frequencies dielectric losses due to the water molecules produce an apparent conductivity. The salinity dependence is thus directly related to ion strength, rather than indirectly related as in the case of dielectric constant and index of refraction.

3.4 Sound Speed

The phase velocity of compressional waves in water is given by the adiabatic compressibility, $U_s = (\partial\rho/\partial p)_{AD}^{-\frac{1}{2}}$, and increases with the decrease in compressibility caused by increases in T, S, and p. The sound speed has a significant frequency dependence only for frequencies close to the structural relaxation frequency, which is 160 GHz.

3.5 Comparison of the Salinity Sensitivities of the Parameters

The ability to determine salinity by the in situ values of these parameters is complicated by variations of the parameters with

Table 3

Sensitivity of electrical conductivity (c), sound speed (U_s), index of refraction (n), and static dielectric constant (ε_s) to per unit changes in temperature (°C), salinity (ppt) and pressure (P_a) (10^4 Pa = 1 dB). The values have been obtained from: Hasted (1973) and Owen et al. (1961) for ε_s, Stanley (1971) for n, Utterback et al. (1934) for c, and Albers (1960) for U_s.

	C	ppt	P_a
ε_s	-3.7×10^{-1}	-1.6×10^{-1}	3.7×10^{-8}
n	-8.4×10^{-5}	1.8×10^{-4}	1.4×10^{-10}
c(S m^{-1})	9.6×10^{-2}	9.9×10^{-2}	4×10^{-9}
U_s(m s^{-1})	3.7	1.1	1.8×10^{-6}

temperature and pressure as well as with salinity, as shown in Table 3. Although the pressure effects are weak, those due to temperature are not; only the index of refraction has a significantly greater sensitivity to salinity (in ppt) than to temperature (in °C). The importance of the temperature sensitivity varies with the TS relation, but typically the change of temperature in the ocean in °C is 10 times the salinity variation in ppt. Thus, extraction of the salinity variation from the measured change in one of the parameters requires subtraction of the temperature change; salinity is effectively the small difference obtained when one large number is subtracted from another. This requires great precision in both of the measured values. Since temperature is basically a point measurement and the other parameters are volume averages, the problem is even more severe at scales approaching the size of the probe dimensions when the differing phase and amplitude responses of the two sensors must be considered. Only at scales less than the diffusive temperature cut-off is the problem one of salinity alone.

The measurement problem posed by the scientific requirements for salinity can be seen in Table 4, which lists the changes in temperature, salinity, and pressure required to produce a 1% change in the measurable parameters. Conductivity, the parameter that depends directly upon salinity, is a factor of 10 more sensitive to salinity than the other parameters. To meet the various

Table 4

Changes in temperature, salinity, and pressure required to produce a 1% change in the parameters on the left. One MPa corresponds to nearly 100 m in depth. Electrical conductivity is the most sensitive to changes in T and S and index of refraction is the least sensitive. Typical values for the parameters in sea water are $\hat{\varepsilon}_s = 74$, $c = 4$ S m^{-1}, $n = 1.34$, and $U_s = 1500$ m s^{-1}.

	°C	ppt	MPa
$\hat{\varepsilon}_s$	-2.0	-4.7	-
n	-160	75	96
c(S m^{-1})	0.45	0.43	11
U_s(m s-1)	4.1	14.0	8.4

sensitivity requirements of Table 1 requires conductivity measurements that can detect fractional changes of $2x10^{-4}$ to $7x10^{-9}$. When problems of removing temperature effects are considered, the precision requirements are at least as formidable.

In the next section the conductivity sensors that are presently in use or under development are described. Further consideration of the other parameters is given in the last section.

4. CONDUCTIVITY SENSORS

4.1 Basic Design Problems

Inductive conductivity sensors were the first to be used successfully for continuous profiling as part of commercially developed STD systems. However, as pointed out by Brown (1974) transformer errors increase if the sensor dimensions are reduced; consequently subsequent developments, which have sought to improve spatial resolution, have used direct resistance measurements between electrodes exposed to sea water. Due to electrode polarization these have been alternating current devices operating in the kHz range, where c is not frequency dependent. However, even with the use of ac exciting voltages, the exposure of electrodes to sea water results in broadband noise, including low frequency shifts in the cell constant, resulting from electrochemical effects at the electrode surfaces.

One line of development, pioneered by Dauphinee (1968), sought to minimize these electrode effects by the use of four electrodes and geometric configurations that make the cell constant independent of electrode characteristics. One pair of electrodes is used to impress a controlled ac current through the cell, while a second pair of electrodes is used to sense the induced voltage. The ratio of impressed current to sensed voltage is directly proportional to conductivity. If the sense amplifier has a sufficiently high input impedance, negligible current will flow through the sense electrodes, and hence the sensed voltage will be independent of changes in electrode impedance due to fouling or electrochemical effects. High resistance paths and well defined sensing volumes occur when the electric streamlines between electrodes are constricted by a tube or orifice. Such constrictions can result in a large hydraulic resistance, and hence long flushing lengths, when the probe velocity through the water is small. The small cells are also sensitive to heating of the water sample as it passes through the cell and to dimensional changes in the constriction due to even slight fouling.

Another approach has been to obtain very small sensing volumes by the convergence of electric streamlines near a point or needle-like electrode that is used with a distant ground plane. These have the

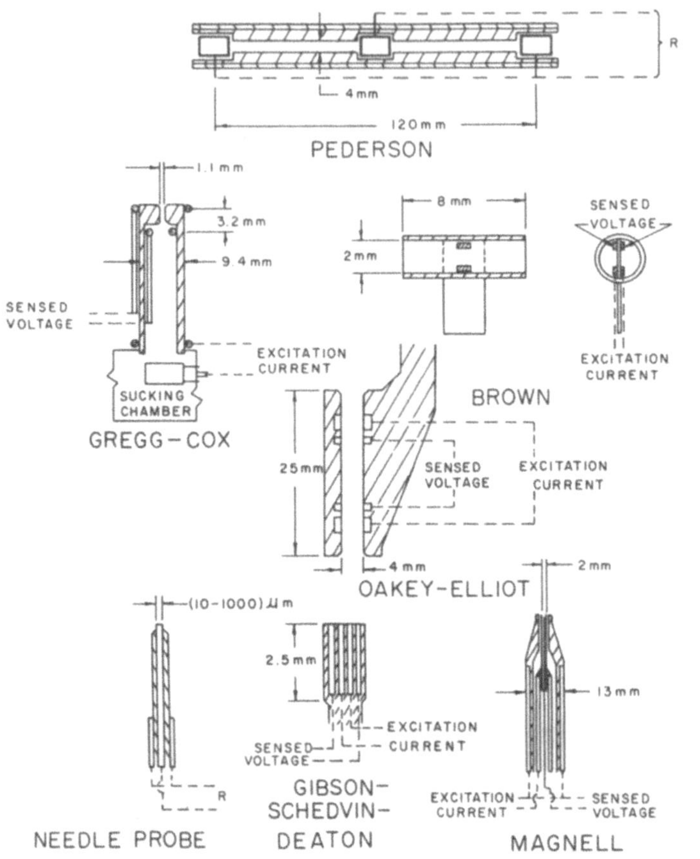

Fig. 3 Schematics of the principal high resolution conductivity cells. Pederson's cell and the needle probe are two electrode designs while the others are variations of a four-electrode configuration. The Gibson-Schedvin-Deaton probe is completely free-flushing. Suction is used on Gregg-Cox and Magnell sensors and can be applied to the Pederson and Brown units.

Table 5

Characteristics of some of the conductivity sensors. $K_C = cR$ is the cell constant. Gregg-Cox probe data are for the configuration used in November 1975. The Brown data represent one of several probe versions in use.

Characteristic	Pederson	Gregg-Cox	Brown	Gibson Schedvin Deaton	Oakey Elliott
K_C (m^{-1})	2000	2690	720	196	300
E (V rms)	0.035 (across bridge)	1.35	0.1	0.022	0.07
Volume (m^3)	2.2×10^{-6}	3×10^{-9}	2.5×10^{-8}	not applicable	3×10^{-7}
Flushing Characteristic	190 mm at 1 m s^{-1}	1.2 mm (for typical sucking rate and 80 mm s^{-1} fall rate)	20 mm at 1 m s^{-1}	boundary layer limited	60 mm at 0.5 m s^{-1}
Self-heating (°C)	8×10^{-9} at 1 m s^{-1}	55×10^{-3} at typical sucking rate	1.3×10^{-5} at 1 m s^{-1}	not known	6×10^{-6}
rms noise (S m^{-1}) bandwidth	1.1×10^{-5} 0–2.5 Hz (C dependent)	2.2×10^{-6} 0–25 Hz (C dependent)	1×10^{-4} 0–30 Hz (least count)	1.8×10^{-4} 0.1–500 Hz	4×10^{-4} 1–100 Hz
Estimated Repeatability over a few days time (S m^{-1})	$\frac{10^{-3}}{\text{month}}$	$(6\text{–}15) \times 10^{-3}$ (absolute precision, not repeatability)	$\leq 1 \times 10^{-3}$	not known	not known
Output Signal	frequency	rectified voltage; high and low pass	16-bit digital	frequency	voltage

advantage of being free-flushing but have all the electrode problems discussed above. A more recent approach is the use of four electrodes in an open configuration, without a constriction.

In evaluating the different conductivity probes it is necessary to consider the dynamic range of the data system with which they have been used, since the performance of some is limited by least count rather than by sensor or electronic noise. Presently the analogue-to-digital (A/D) converters with the widest dynamic range can obtain 16 bits for an input range of ±10 V, and a sample rate of 50 Hz or greater. If this were used to span the conductivity range for the open ocean 2.5 to 6.2 S m^{-1} (the units are Siemens m^{-1}; 1 S m^{-1} = 10 mmho cm^{-1}), the least count would be 5.7×10^{-5} S m^{-1}. The corresponding spectral level, for a 50 Hz sample rate, is 1.1×10^{-11} (S $m^{-1})^2$ Hz^{-1}. This value for the spectral level of the digitization noise assumes that both the signal and the noise have white spectra. To obtain maximum precision, in some systems the conductivity range of 0 to 6 S m^{-1} has spanned a digital range of either 16 or 15 bits. Since these least count levels are much greater than the levels of the conductivity signal at high wavenumbers, a separate data channel has been used in some cases to record high-gain high-pass data, thus increasing the dynamic range of the system. Another approach has been to use a sensing circuit that yields frequency, so that an A/D converter is not needed. It should be noted here that the variance of the error of a frequency modulated signal that is either period or frequency counted is twice the variance of the error of an ideal A/D converter (Irish and Levine, 1977).

4.2 Specific Systems

Although none of the conductivity systems in use has been fully documented under all of the important test configurations, relatively complete information exists for some. Schematic drawings of the principal small-scale probes are shown in Figure 3 and those characteristics that could be obtained are given in Table 5. Other probes have been reported but we have not been able to obtain sufficient information about them.

Pederson (1973) used a standard three-electrode Beckman conductivity cell as a two-electrode device by holding the two outer electrodes at the same potential. The path between the outer and inner electrodes forms a resistance element in a Wein bridge oscillator circuit, giving a frequency output: $f = M_C(R)^{-\frac{1}{2}}$, where M_C is a constant depending upon the bridge parameters, and the cell constant K_c relates the cell resistance, R, to the conductivity. The dynamic range of the digital data depends upon the method of period counting. As used on the free-fall microstructure recorder (MSR), 1408 periods of the data signal are used to count a 5 MHz reference

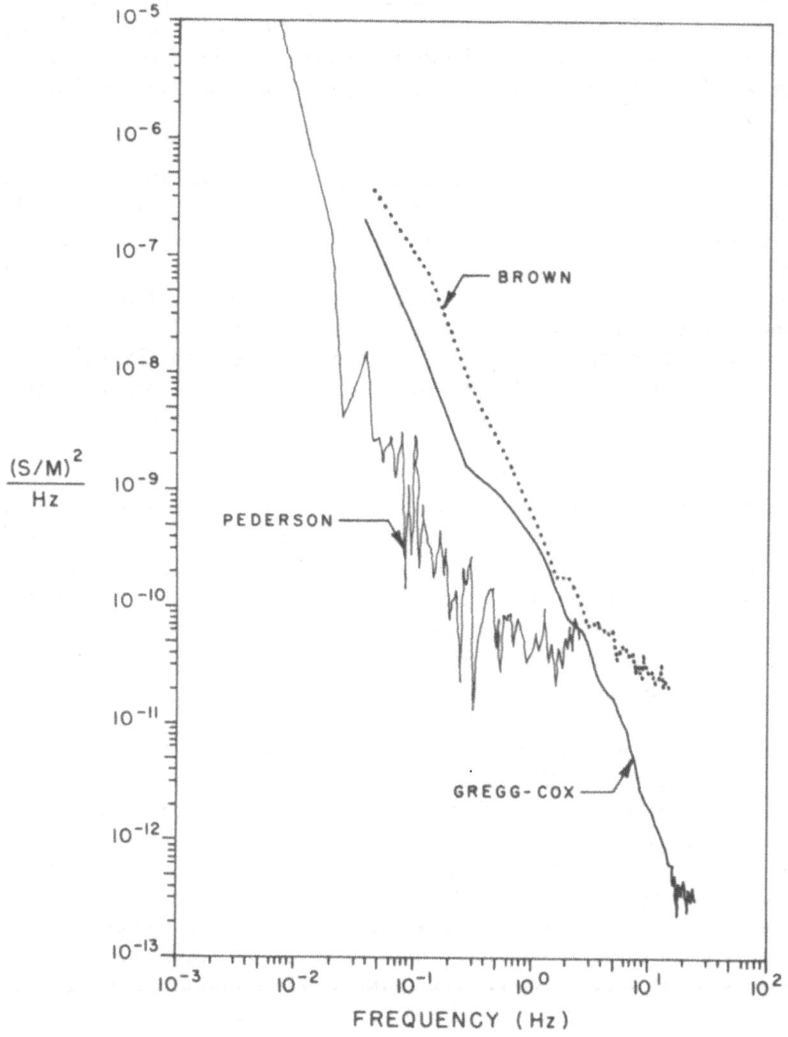

Fig. 4 Spectra representing the lowest activity records that were available from data runs with these sensors. The record with the Pederson cell was taken at a depth of 360 m in a weakly stratified basin in Jervis Inlet, B.C.; the data with the Gregg-Cox probe were obtained in a surface mixed layer in the North Atlantic. Both had fall speeds of 0.08 m s^{-1}. The data taken with the Brown CTD were obtained from 3124 to 3412 dB in the North Atlantic with a descent rate of 1 m s^{-1}. The rapidly falling portions of the Pederson and Brown spectra at low frequency are signal, while the flat sections at high frequency are noise. The full spectrum of the Gregg-Cox probe is signal for $f < 15$ Hz.

clock. At a 5 Hz sample rate the least count varies from 5.7×10^{-6} S m^{-1} for c = 2.5 S m^{-1} to 2.2×10^{-5} S m^{-1} at 6.2 S m^{-1}. A pump has been used to give a 0.13 s flushing time on the slowly moving MSR while the estimated free-flow flushing length is 0.19 m for a speed of 1 m s^{-1}.

The noise level of this two-electrode probe is shown in Figure 4 by the spectrum of a quiet section of MSR data taken in a fjord. The spectral level of the noise is 5×10^{-11} (S m^{-1})^2 Hz^{-1}, corresponding to an rms value of 1.1×10^{-5} S m^{-1} over a 2.5 Hz bandwidth. If the conductivity signal were due entirely to temperature this would correspond to 0.1 m °C. Since the least count is expected to be 8.6×10^{-6} S m^{-1} for this data, the noise level in Figure 4 represents the inherent noise of the probe. Successive data from a calibration facility have yielded differences of 0.001 S m^{-1} over one month.

A two-electrode configuration with a large distant ground plane and a wire as small as 0.5 μm in diameter have been used by physiologists for many years. The convergence of the electric streamlines near the sensing electrode produces signals that are attenuated by 3 dB at distances of five times the wire diameter. Gibson and Schwarz (1962) used the same design, but with 10 to 50 μm wires, in a laboratory turbulence experiment that verified the existence of the viscous-convective subrange in a dilute salt solution; but, due to a relatively poor signal-to-noise ratio they were not able to resolve the variance. The excellent spatial resolution makes this an attractive configuration for studies in the high wavenumber portion of the spectrum. In addition to turbulent studies in dilute solutions, many investigators (e.g. Mied and Merceret at the Johns Hopkins University) have used it in tanks that are strongly stratified by salt. Unfortunately, the small electrode size and complicated electrochemistry of concentrated solutions result in high noise levels and strong drift of the cell constant, even in relation to the large signal levels in the tanks. The weaker signals and more complicated electrochemistry of the ocean have frustrated attempts to make a seagoing instrument. Elliott and Oakey (J. Elliott, Bedford Institute of Oceanography, personal communication) used the needle probe in an attempt to obtain high wavenumber information from a tethered free-fall vehicle. Due to the high noise level, the conductivity signal could not be obtained at wavenumbers beyond the temperature cutoff. Even when a commercial CTD was used for the low frequency information, changes in the cell constant during individual casts resulted in unacceptable errors in salinity. They have abandoned the two-electrode configuration for a four-electrode device.

Gregg and Cox (1971) developed a small four-electrode probe by using a mechanical device to suck water through a 1.1 mm diameter, 3.2 mm long hole. Flow distortion due to the size of the probe tip limits resolution to scales of 10 mm. Since the current driving electrodes are outside the sensing volume great care must be taken to contain the water that has been sucked through the hole within an electrically insulated chamber to prevent the existence of a conductivity path in parallel with the relatively high resistance hole. Presently the back resistance is about 15 MΩ, compared to hole resistances of 600 to 1200 Ω.

Since this probe is presently being redesigned, the configuration used on a cruise during October 1975 will be described. To achieve a stable gross conductivity signal, the rectified ac level across the sensing electrodes was digitized at 5 Hz with a least count of $3x10^{-4}$ S m^{-1}, which limited the sensitivity of this data channel. Another channel of high-pass data, with an RC time constant = 0.1 s, was amplified with a gain of 1000 and recorded with a resolution of 16 bits, at 50 Hz. The least count, in the passband, varies with conductivity and is given by $dc/dN = (1.1x10^{-10}\ c^2)\ I^{-1}$, where I is the driving current in amperes. With I = 2.84 mA this corresponds to ($2.4x10^{-7}$ to $1.5x10^{-6}$) S m^{-1} over the oceanic range of conductivity, corresponding to spectral levels of ($1.9x10^{-16}$ to $7.5x10^{-15}$) $(S\ m^{-1})^2\ Hz^{-1}$. The lowest signal level obtained on this cruise was in a surface mixed layer and is shown by the spectrum in Figure 4; at 25 Hz the spectral level was ($3.5x10^{-13}$) $(S\ m^{-1})^2$ Hz^{-1}. A laboratory test, Figure 5, subsequent to the cruise revealed a noise level less than the signal level in the mixed layer. The sharp spikes in the spectrum at 10, 15, and 20 Hz are due to a grounding problem and are correctable. The broad noise spike from 6 to 9 Hz is believed due to jitter in the sucking device.

To prevent blockage of the small hole the probe tip has been covered until the beginning of the MSR data cycle. Except in a local fjord we have not yet had problems with blockage, even in surface waters. The other price of the high signal-to-noise levels obtained is significant ohmic heating of the water passing through the hole (about 5.5 m°C). This level of heating corresponds to the two least significant bits of the gross conductivity, but can produce spurious signals in the high-gain high-pass data if the sucking mechanism produces an irregular flow rate. This problem has not been fully solved and, together with the necessity to trap the sampled water, limits the overall usefulness of the configuration.

By using a vertical plate and symmetric placement of the electrodes, Brown (1974) was able to confine the electric field of a four-electrode configuration within a 2 mm diameter and 8 mm long ceramic tube. The probe has been used on wire-lowered, free-fall and self-propelled vehicles and is estimated to have a flushing length about 2½ times the tube length. The cell is excited by a 10 kHz sine wave with an average amplitude of 0.1 V rms at the sense electrodes. In the system the ratio of the current across the exciting electrodes to the sensed voltage is linear with conductivity and is digitized to 16 bits. Since full-scale is set to 6.5 S m^{-1} the least count is $1x10^{-4}$ S m^{-1}. The sensitivity is limited by the least count, as can be seen by the high frequency flattening of the spectrum of a deep cast, shown in Figure 4. In this case the high frequency spectral level is a factor of 2 lower than the theoretical digital noise, $5x10^{-11}$ (S $m^{-1})^2$ Hz^{-1}, due to the -2 slope of the signal spectrum (M. Briscoe, Woods Hole Oceanographic Institution, personal communication). The cell itself is believed capable of a factor of 4 improvement in sensitivity for a reduced dynamic range or slower digitization rate (N. Brown, N. Brown Instrument Systems, Catoumet, Mass., 02534, personal communication). The low-noise high-accuracy digitization has been achieved by use of an ac A/D converter of the successive-approximation type. However, the use of this type of A/D converter without sample-and-hold circuitry

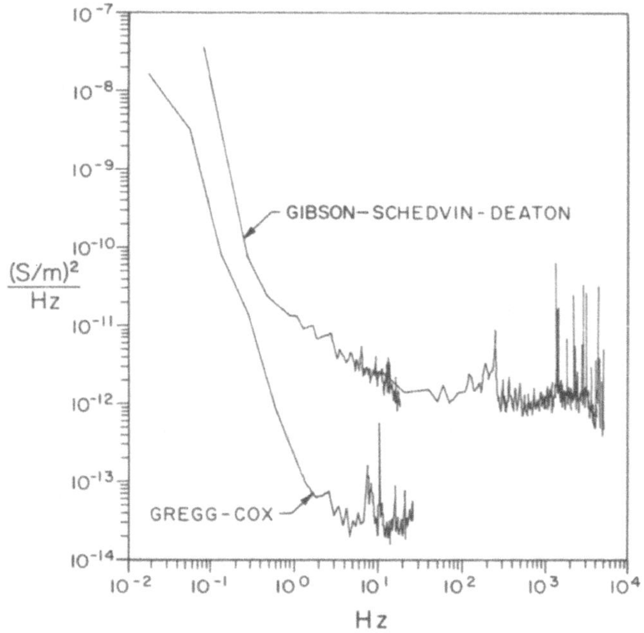

Fig. 5 Laboratory noise spectra for the Gibson-Schedvin-Deaton and Gregg and Cox probes. In both cases the low frequency levels are thought to be real signals in the laboratory baths. The sharp noise spikes are due to grounding problems or are in the tape recording system and are not inherent in the probes. These spectra may be translated into wavenumber spectra by $\Phi(k) = U\ \Phi(f)$, where U is the probe velocity.

results in an uncertainty in the exact time within the aperture (≈10 ms) when the signal is actually digitized. The digitization takes place in the instrument and, when used on a wire-lowered platform, the data are transmitted in frequency shift key (FSK) format. The precision of the measurement is sensitive to minute changes of the inside diameter of the tube due to fouling resulting from chemical deposition, but can be restored to the original calibration by an acid flush. The repeatability obtained corresponds to 1×10^{-3} S m^{-1} over periods of a few days (R. Millard, Woods Hole Oceanographic Institution, personal communication). Because of the small cell size, self-heating of the water within the cell by the driving current has proved to be a problem in some applications. For probe velocities greater than ≈0.3 m s^{-1} the self-heating is negligible compared with the least count of the 16-bit data system. However, when the flow of water through wire-lowered probes stops due to ship roll, ohmic heating of 10 m°C has been experienced. This produces a spurious conductivity reading of 1×10^{-3} S m^{-1}. When used with a slowly descending free-fall vehicle a detectable heating effect has been observed which is attributed to the thermal lag of the cell wall (A.J. Williams III, Woods Hole Oceanographic Institution, personal communication).

It should be pointed out that there are a number of different configurations of the Brown instrument in use, representing changes due to ongoing development. In fact, some instruments have been built at both Woods Hole Oceanographic Institution and Scripps Institution of Oceanography. Early models used about three times as much cell-driving current as the present model, and the geometry of the cell has also changed. Thus the published data do not necessarily represent the present capabilities of the instrument.

Other variations of the four-electrode configuration are shown in Figure 3. Magnell's design was used on a towed body and sucked water between the inner and outer electrodes. A common ground plane for the driving and sensing circuits necessitated the use of only three electrodes.

Oakey (1977) uses four electrodes on the inside surface of a 4 mm diameter, 25 mm long, free-flushing tube. The output voltage from an ac bridge is directly proportional to c and has been found to be linear to 1%. The noise level has been found to correspond to an rms value of 4×10^{-4} S m^{-1} over a bandwidth of 1 to 100 Hz. When used on a free-fall device with a speed of 0.5 m s^{-1} the flushing characteristics of the cell have been found, by comparison with simultaneous temperature measurements, to result in a signal attenuation of 3 dB at 15 cpm.

Gibson, Schedvin, and Deaton of Scripps Institution of Oceanography (personal communication) are using four electrodes on a thin plate that is 2.5 mm long. The device is intended to be mounted on a towed body and oriented end-on into the flow. The spatial resolution has not been documented but will be determined by the boundary layer formed by the plate and the three-dimensional structure of the potential field. The excitation current is placed across the inner two electrodes in an effort to improve the spatial resolution. The laboratory noise spectrum, Figure 5, is relatively flat in the range of 5 to $5x10^3$ Hz, with levels of the order $(1-4)x10^{-12}$ $(S\ m^{-1})^2\ Hz^{-1}$.

5. SENSOR CALIBRATION AND THE COMPUTATION OF SALINITY

To obtain salinity, and to evaluate the quality of the results, it is necessary to consider the complete system that also measures temperature and pressure as well as conductivity. In the following a few remarks are made of the effects of temperature sensors on salinity data. A more complete description of the characteristics of temperature sensors is given by Gibson and Deaton (see Chapter 18).

5.1 Static Calibration of Conductivity and Temperature Sensors

The static calibration of a sensor is the relationship between its output when it has come to equilibrium with the environment and the environmental parameters of interest, e.g. T, c, and p. It is also necessary to know the calibration of the electronic system, which is given by a relationship between digital counts, or frequency or volts, and the sensor output. Since the electronic system may vary with environmental conditions, particularly temperature, it is best to calibrate the complete sytem in a bath. However, frequently this is not possible and the calibration is done in two steps: the sensors are calibrated in the bath to give the cell constant for the conductivity probe and R = R(T) for the temperature probe, and the electronics are calibrated using known inputs, such as a resistance box, to simulate the probes. By repeating the calibrations under different conditions the temperature and pressure dependence of the cell constant and the temperature drift of the electronics can be obtained.

Since the procedures we have used for calibration are typical, a brief description may be of value to investigators unfamiliar with salinity measurements. The Pederson conductivity probe, and the corresponding temperature sensor have the electronics and sensors mounted as an integral unit. The calibration bath contains artificial sea water that is made by mixing sea salt (a commercially available product) in distilled water. For a given bath salinity readings of sensor frequency are taken for a series of temperature values, typically seven points. The bath is well mixed and regulated to about 1 m°C; the temperature is read to a precision of 1 m°C with a platinum resistance thermometer and Mueller Bridge. The accuracy of the reading is U.S. National Bureau of Standards (NBS) traceable to ±5 m°C, while the long term repeatability is about

2 m°C. For each calibration point a water sample is taken and, using a commercially available standard bench salinometer with a reference arm of Copenhagen water, the ratio

$$R_T = \frac{C(S,T,0)}{C(35,T,0)}$$

is determined. The International Oceanographic Tables (UNESCO 1966), extended for temperatures below 10°C using the data of Brown and Allentoft (1966) and corrected for the 1968 adjustment to the temperature scale, are then used to compute the salinity and conductivity of the bath to an accuracy of better than ±0.01 ppt and ±0.001 S m^{-1}, respectively (the bench salinometer is capable of ±0.003 ppt accuracy, and Copenhagen water has a variability of about 0.003 ppt). Calibrations are performed at two different salinities so that any temperature dependence of the conductivity electronics and cell constant can be determined. A least squares fit of f_C(Hz) = g(S,T) is then made to the calibration data. The Gregg-Cox probe has been calibrated in the same bath and has similar stability but less accuracy.

The Northwest Regional Calibration Center (NRCC), where our calibrations are performed, routinely dilutes sea salt to produce standard salinities in increments of 5 ppt from 5 to 40 ppt. Twelve samples are taken of each salinity and sent to the National Oceanographic Instrumentation Center (NOIC) and oceanographic laboratories in the United States and abroad for standardization tests.

Due to the added difficulty of obtaining laboratory measurements of conductivity and temperature at elevated pressures, sensors are routinely calibrated only at atmospheric pressure. Some investigators have performed special tests or have done calculations to investigate the pressure dependence of conductivity cell constants and thermistors and have found that the effects are relatively small. Although this subject needs more work for observations in the deep ocean the effects at shallow depths are believed to be negligible in relation to the signal levels.

5.2 Dynamic Calibrations

The response of temperature sensors to a change in the ambient temperature is limited by the time required for diffusion through the thermal boundary layer around the probe, the insulating coating, and through the temperature sensitive material. The response of many of the conductivity sensors is limited by the flushing time of the sensing hole.

For probes moving in linear gradients the effect of the lag of the probes is to produce a constant offset in the measurements. This can be seen by assuming that the transient responses are adequately modelled by single pole low pass filters. Then the impulse response functions have the form

$$h(t) = \frac{1}{\tau} e^{-t/\tau} \quad t > 0$$
$$= 0 \qquad t < 0$$

where τ is the time constant. The temperature sensed by the probe, $\hat{T}(t)$, is the convolution of the temperature passing the probe, $T(t)$, and the impulse response. For a sensor that acts as a simple low-pass filter falling through a linear temperature profile

$$\hat{T}(t) = T(t) - \overline{(\partial T/\partial Z)}\, W\tau(1 - e^{-t/\tau})$$

where $\overline{\partial T/\partial Z}$ is the mean gradient and W is the fall speed. Thus the steady state response is a constant offset.

For $\overline{(\partial T/\partial Z)} = 0.1°C\ m^{-1}$, which is frequently found in the seasonal thermocline at the base of the mixed layer, $W = -1\ m\ s^{-1}$ and $\tau = 1$ s, the offset is 0.1°C. A similar effect occurs in conductivity and hence in salinity.

For up and down casts of a CTD, or a cycling towed body, the offsets are in the opposite direction and produce a separation of the two TS diagrams. Since good laboratory determinations of the transient responses of most probes have not been made, Dantzler (1974) empirically determined the difference in the time constants for an STD system to be the value that brought the up and down TS relations into best agreement when a simple lag correction was applied to the observed profiles.

Due to the existence of finestructure (changes in the gradients over scales from about 0.5 m to tens of metres) in the profiles, simple lag corrections are not adequate to reconstruct the true profile from the data. This is illustrated schematically in Figure 6, where two sections with linear profiles are separated by a step change. The difference in the phase and amplitude responses of the temperature and conductivity sensors to this transient produces a salinity spike, which is not removed by a lag correction. To correct for such transients, digital filters based on accurate determinations of the frequency response functions of both temperature and conductivity sensors must be used to deconvolve the data. To date this has not been done, although the problem is being worked on. The principal difficulty has been in obtaining the transfer functions of the sensors. The values published by many manufacturers are not adequate since they were obtained by plunging the probes from air into water and thus do not simulate the boundary layers present on probes as used in the ocean.

In situ comparisons of CTD data with higher resolution instruments as well as laboratory studies of sensor responses to impulse and step structures are being conducted in several laboratories. Preliminary results of work by this author and T.B. Sanford of Woods Hole, by R. Millard and coworkers at WHOI, and by Paige (1979) show that the response of the Brown CTD to temperature structures of about one metre is significantly attenuated. Since even a slight attenuation can seriously affect salinity calculations, caution should be exercised in using salinity and density data over scales of a few metres and less. The caution should, of course, be extended to larger scales if the CTD is being towed or dropped through the water at speeds of more than 1 or 2 m s^{-1}.

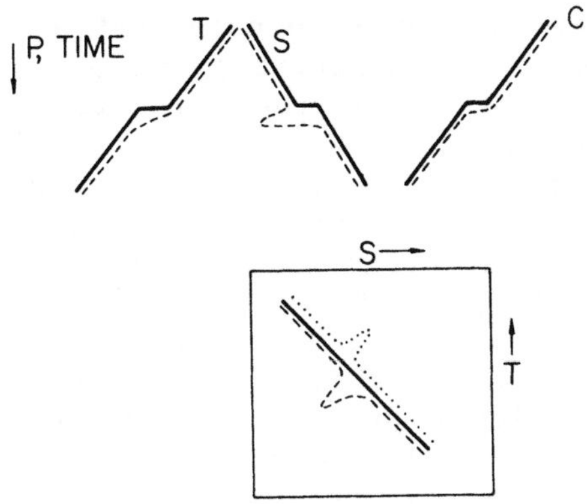

Fig. 6 Schematic representation of the effects of sensor time constants on measured T,c data and upon the computed salinity, S. In the upper portion of the figure the solid lines represent the true profiles and the dashed lines the measured data from a descending instrument. It has been assumed that the conductivity probe has a faster response than does the temperature sensor. In a linear gradient the measured profiles are offset by a constant amount from the true profile. When a rapid change in the gradient occurs the differing transient responses of the temperature and conductivity probes produce a spike in salinity. In the TS diagram in the lower portion of the figure an up profile, indicated by the dotted trace, through the same structure produces an offset in the opposite direction in the linear region. Using a simple lag correction can correct for the offsets in the linear regions but not for the spikes.

5.3 Computation of Salinity

Several algorithms have been developed for computing salinity from measured values of temperature, conductivity, and pressure. Although none has been accepted for standard use there is discussion of redefining salinity in a manner that would facilitate a standard means of computation. A thorough review of the present and proposed methods of computing salinity is given by Lewis and Perkin (1978). Their article also gives references which contain the algorithms in current use.

6. ASSESSMENT OF THE ABILITY TO MEASURE SALINITY

6.1 Comparison of the Capabilities of Current Sensors with the Scientific Objectives

To compare the characteristics of these probes with the requirements in Table 1 it is useful to note that a salinity fluctuation of 1 ppm will produce a conductivity change of 10^{-4} S m^{-1}. Thus the four electrode probes and Pederson's have the raw sensitivity to trace lateral features outside of the mixed layer and to partially meet the needs for sensing structures and the mean stability within mixed layers, but only the Brown system has a precision that meets even the least stringent requirements of Table 1. Since the needle probe does not appear to be useful in the ocean, 10 mm-scale spatial resolution is the most that can be expected from the present generation of sensors. The Gregg-Cox approach has been found to offer the sensitivity required for detecting overturns and to be partially in the range required for the salinity variance but it does not have the necessary spatial resolution. Hence it does not appear that conductivity probes will be able to resolve fully the salinity gradients since even the needle probes are large enough that they would distort the salinity structures near the scale of the diffusive cutoff.

To translate the raw conductivity sensitivity into an equivalent salinity depends upon the characteristics of the full data system and not just upon the conductivity. This involves both the temperature and pressure data, as well as the dynamics of the vehicles as they move through the water. The difficult task of matching the phase and amplitude responses of the temperature and conductivity sensors involves details specific to each system such as the relative locations of the probes, filtering, and the algorithms applied in processing. The pressure dependence of conductivity is sufficiently strong that errors in the pressure data used to convert T,C,p to T,S,p can produce significant salinity errors. For example, the Neil Brown CTD uses a strain gauge that has an uncertainty of ±65 kPa. The corresponding conductivity and salinity errors are $\pm 2.6 \times 10^{-4}$ S m^{-1} and ±2.6 ppm, respectively. These gauges usually have additional temperature dependent offsets of ±6.5 kPa °C^{-1} which, if not corrected for, can produce errors of ±0.26 ppm $°C^{-1}$ in salinity. These errors are slowly varying during a vertical profile and result in errors in precision but not in sensitivity. For measurements at very high wavenumbers the pressure effect can also produce a degradation in sensitivity. The rms conductivity level between 90 and 110 cpm in Figure 1 is 7×10^{-7} S m^{-1}. An uncorrected rms pressure fluctuation of ±175 Pa would produce such a level. This could be produced in the conductivity data by undetected depth oscillations of ±17.5 mm at frequencies of 0.1 to 10 Hz, for velocities of 0.1 to 1.0 m s^{-1}. An equivalent effect could also be produced in the salinity data using noisy pressure values. These effects are in addition to the production of spurious signals by probe oscillations in a salinity gradient. To measure such low level conductivity signals it is also necessary to use accelerometers or very high resolution pressure gauges to know the vertical motions of the sensors at the same frequencies.

Since the basic scientific questions need to be answered many times under a wide variety of external forcing conditions, the performance of both the data collection and analysis systems under non-optimum situations is more important than the results under the best of circumstances. In our opinion, the best prospect for upper ocean studies over the next several years lies in the refinement and careful documentation of the various four-electrode sensors that are now being used or developed. Different configurations should be considered for specific tasks, e.g., the larger cell, 4 mm x 4 mm x 30 mm, being constructed by Neil Brown offers better precision for routine use than the current model. Much more effort will be required to optimize the salinity data, even given the same basic sensitivities as shown in Table 5.

6.2 Method for Estimating the Effect of Calibration Errors and System Noise on Quantities Computed from Salinity, Temperature, and Pressure Data

The measurement objectives that were obtained in Section 1 are adequate to give an order-of-magnitude reference for comparison with probe characteristics. However, once a sensor is selected it is necessary to evaluate the quality of the data, generally on a point-by-point basis, since some of the effects of noise are dependent upon the signal level. Gregg (1979) has developed expressions for estimating the uncertainty and noise in quantities derived from T,C,p data due to calibration errors and the inherent system noise. Since this analysis does not include the effects of salinity spiking and other problems specific to individual data systems it represents the most optimistic assessment of the data.

The uncertainty in salinity data due to calibration uncertainties in the measured values of T,c,p is $\pm\gamma_s$ where

$$\gamma_s = \left|\frac{\partial S}{\partial T}\,\varepsilon_T\right| + \left|\frac{\partial S}{\partial c}\,\varepsilon_c\right| + \left|\frac{\partial S}{\partial p}\,\varepsilon_p\right| \tag{2}$$

For the typical calibration uncertainties of ±0.01 °C, $\pm 10^{-3}$ S m^{-1}, and $\pm 10^5$ Pa, γ_s = ±(14-30) ppm. The range of variation is due to variations of the salinity derivatives over the range of normal oceanic conditions.

The uncertainty in salinity, as well as in temperature, is not a problem if only one data system is used and it maintains its calibration. However, if individual sensors must be replaced during a cruise or if a survey is done with several ships then the estimates of $\pm\gamma_s$ are the minimum uncertainty in salinity between the different data systems; as Lewis and Perkin (1978) point out, the differences between calibration baths are frequently significant and will increase the uncertainty. In addition to limiting the ability to trace salinity features these calibration errors also restrict the accuracy of maps of tracers on potential density surfaces as well as maps of dynamic height anomalies.

The system noise in the individual data channels limits the sensitivity of the salinity data, and of derived quantities such as N^2, the stability frequency. The variance of the salinity noise is given by

$$\sigma_s^2 = \left(\frac{\partial S}{\partial T}\right)^2 \sigma_T^2 + \left(\frac{\partial S}{\partial c}\right)^2 \sigma_c^2 + \left(\frac{\partial S}{\partial p}\right)^2 \sigma_p^2 \tag{3}$$

For a Brown CTD that is limited only by the quantization noise, σ_s = 0.2 to 0.4 ppm. Gregg (1978) also obtained similar expressions for the noise of N^2 and Ri, the gradient Richardson number, in terms of basic system noise values. It is important that the system noise values be monitored periodically and that expressions such as Equation 3 be used to determine the statistical significance of the data. In most cases the calibration errors and noise in conductivity are the major source of the resulting salinity errors and noise.

Until the salinity spiking problem is solved, the actual quality of the data will continue to be much worse than the rms noise values given by the above expressions. How much worse is difficult to quantify and varies with the system and the sharpness of the finestructure. During some recent work in the Pacific we found salinity spiking of several tenths of a ppt, with both the Pederson cell and the Brown CTD, when there was vigorous overturning at the base of the mixed layer.

7. MEASUREMENT OF OTHER PARAMETERS

Since the other parameters are less sensitive to salinity variations than is conductivity their potential usefulness lies in using measurement techniques that compensate for the weaker signal. A commercially manufactured sound velocimeter quotes sensitivities of 15 to 1.5 mm s^{-1} over a path 0.15 mm long by 26 mm in diameter, with sample rates of 10 to 1 Hz, respectively. The quoted precision is ±40 mm s^{-1} over six months. If these specifications are verified, the sound speed sensitivities achieved correspond to salinity fluctuations of 1 to 10 ppm depending upon the data rate, and would be suitable for use in tracing features and the stability of deep surface layers. However, neither the stability nor the precision offer an advantage over conductivity.

The techniques for measuring dielectric constant depend upon the frequency of interest; impedance bridges are most commonly used in the kHz band, resonant circuits in the MHz band, and transmission-line techniques for frequencies greater than 1 GHz. For a 0.01 μF capacitor in a 1 MHz circuit we estimate a sensitivity of about 1 kHz/ppt, a factor of 10 greater than obtained with Pederson's Wein bridge oscillator. The capacitor would have a ratio of area to plate separation of 14.5 m and could be built in several configurations of interleaved plates. The use of a pump for flushing would be required for the small sizes. The direct frequency output, which can be period-counted with great sensitivity, is an advantage as is the absence of electrochemical effects encountered by direct conductivity cells. Among the problems to be overcome is the dimensional stability of the capacitor.

In the optical range of frequencies, the most common method of measuring changes in the index of refraction is by interferometry. Changes in the optical path length of one of two superimposed beams of coherent light produce shifts of the interference fringes, which can be measured to at least 0.01 λ, i.e., the sensitivity is given by $\Delta(nL) = 0.01\ \lambda$, where L is the path length. For green light, $\lambda = 0.5$ μm, and a 0.1 m long path, $\Delta n = 5\times10^{-8}$, corresponding to $\Delta S = 0.3$ ppm. More sophisticated heterodyning techniques with two light beams or a combination of an acoustic and a light source can produce very sensitive measurements.

The most obvious liability is that interferometry will measure relative changes of n once the sensor is in the water but cannot easily be made into an absolute device. Also, to obtain the sensitivities above it is necessary to compensate for changes of L; even with the use of Vycor glass, which has a linear coefficient of expansion of $8\times10^{-7}\,°C^{-1}$, the temperature changes expected in the ocean, i.e. ±15°C, would change L by ±24 λ.

The development of operational instruments to make useful measurements of the index of refraction or the dielectric constant may thus be possible but will be very difficult. In addition, a large effort would be required to determine ε(T,S,p) and n(T,S,p). Therefore, it seems that such efforts should be postponed until the data from forthcoming upper ocean programs (such as MILE and JASIN*) are assessed and the estimates in Table 1 revised. If a need for a sensitive instrument exists and cannot be satisfied by conductivity probes then the use of these parameters should be seriously considered.

ACKNOWLEDGEMENTS

We are indebted to R. Millard and A.J. Williams III for spectra and information on the performance of the Brown conductivity cell and to N. Brown, N. Oakey, J. Elliott, C. Gibson, J. Schedvin, and R. Huggett for very useful discussions. The preparation of this paper was supported by the U.S. Office of Naval Research Contribution No. 987 from the Department of Oceanography, University of Washington, Seattle, Washington, 98105.

---

* MILE = MIxed Layer Experiment;
JASIN = Joint Air Sea INteraction experiment.

REFERENCES

ALBERS, V.M. 1960. *Underwater Acoustics Handbook*. Pennsylvania State University Press, 290 pp.

BROWN, N. 1974. A precision CTD microprofiler. *Proceedings of Ocean 74 Institute of Electrical and Electronic Engineers Conference on Engineering in the Ocean Environment*, Halifax, Nova Scotia, 1974. IEEE, Inc., 345 East 47th St., New York, N.Y. 10017.

BROWN, N.L. and B. ALLENTOFT. 1966. Salinity, conductivity, and temperature relation of sea water over the range 0 to 50°/₀₀. Final report on contract NOnr4290(00) M.J.O., Number 2003 Bissett-Berman, United States Office of Naval Research, Washington, D.C.

DANTZLER, JR., H.L. 1974. Dynamic salinity calibrations of continuous salinity/temperature/depth data. *Deep-Sea Research*, 21: 675-682.

DAUPHINEE, T.M. 1968. In-situ conductivity measurements using low frequency square wave AC. *Instrument Society of America Marine Sciences Instrumentation*, 4: 555-562.

GIBSON, C.H. and W.H. SCHWARZ. 1963. Detection of conductivity fluctuations in a turbulent flow field. *Journal of Fluid Mechanics*, 16 (3): 357-364.

GREGG, M. 1976. Fine and microstructure observations during the passage of a mild storm. *Journal of Physical Oceanography,* 6: 528-555.

GREGG, M.C. 1977. Variations in the intensity of small-scale mixing in the main thermocline. *Journal of Physical Oceanography,* 7: 436-454.

GREGG, M.C. 1979. The effect of bias errors and system noise on parameters computed from T,C,PV profiles. *Journal of Physical Oceanography,* 9: 199-217.

GREGG, M.C. and C.S. COX. 1971. Measurements of the oceanic microstructure of temperature and electrical conductivity. *Deep-Sea Research,* 18: 925-934.

HASTED, J.B. 1973. *Aqueous Dielectrics.* Chapman and Hall, London, 302 pp.

HORNE, R.A. 1969. *Marine Chemistry.* Wiley-Interscience, New York, 568 pp.

IRISH, J.D. and M.D. LEVINE. 1978. Digitizing error from period and frequency counting techniques. *Deep-Sea Research,* 25: 211-219.

LEWIS, E.L. and R.G. PERKIN. 1978. Salinity: Its definition and calculation. *Journal of Geophysical Research,* 83: 466-478.

MAGNELL, G. 1976. Salt fingers observed in the Mediterranean outflow region (34°N, 11°W) using a towed sensor. *Journal of Physical Oceanography,* 6: 511-523.

MIED, R.P. and F.J. MERCERET, JR. (Unpublished Manuscript). The construction of a simple conductivity probe, Department of Mechanical Engineering, Johns Hopkins University, Baltimore, Md., U.S.A. 21218.

NEUMANN, G. and W.J. PIERSON, JR. 1966. *Principles of Physical Oceanography.* Prentice-Hall, Englewood Cliffs, N.J., 545 pp.

OAKEY, N.S. 1977. An instrument to measure oceanic turbulence and microstructure. Bedford Institute of Oceanography, Report Series BI-R-77-3, 52 pp.

OWEN, B.B., R.C. MILLER, C.E. MILLER and H.L. COGAN. 1961. The dielectric constant of water as a function of temperature and pressure. *Journal of Physical Chemistry,* 65: 2065-2070.

PAIGE, M.A. 1979. Response characteristics of the NBIS sensors to step changes in temperature and conductivity. *Exposure* (a newsletter published by the School of Oceanography, OSU), 6(6): 7-12.

PEDERSON, A. 1973. A small in-situ conductivity instrument. *Proceedings of Ocean 73 Institute of Electrical and Electronic Engineers Conference on Engineering,* Seattle, Washington, September 25-28, 1973. IEEE, Inc., 345 East 47th St., New York, N.Y. 10017.

STANLEY, E.M. 1971. The refractive index of seawater as a function of temperature, pressure and two wavelengths. *Deep-Sea Research,* 18: 833-840.

UNESCO. 1966. Table 1 in International Oceanographic Tables, National Institute of Oceanography of Great Britain and UNESCO, Paris.

UTTERBACK, C.L., T.G. THOMPSON and B.A. THOMAS. 1934. Refractivity-chlorinity temperature relationship of ocean waters. *Conseil Permanent International pour l'Exploration de la Mer, Journal du Conseil,* 9: 35-38.

18

Hot / Cold Sensors of Oceanic Microstructure

C.H. Gibson and T.K. Deaton

1. INTRODUCTION

During the 50-year history of laboratory and atmospheric experimental turbulence studies the preferred method of detecting small scale velocity and temperature fluctuations has beeen to observe the fluctuations in resistance of small elements which are temperature sensitive, such as platinum wires and films or metal oxide thermistors. Large electrical heating of such elements causes them to be sensitive to fluctuations in fluid velocity since the temperature, and therefore resistance, depends on the convective heat transfer. If the electric current through the sensing element is small, its temperature will adjust to the fluctuating ambient value, although with the thermal lag depending on the size of the element, the thickness of the fluid boundary layer, and their thermal diffusivities.

The frequency response to velocity fluctuations can be extended by orders of magnitude using a feedback circuit which maintains the element at constant temperature by adjusting the necessary heating current. The time period needed to sense and correct for a small resistance error signal is much less than the thermal lag period of the element, which accounts for the higher frequency response. An example of such a circuit is given in Section 3. Most laboratory turbulence studies are done with constant temperature anemometer circuits.

Frequency response to temperature fluctuations may be extended by circuit or computer compensation for the thermal lag of the passive element if it is known and constant. Variable contamination may be compensated by operating two sensors at different constant overheat temperatures. Since the sensitivities of both circuits to velocity and temperature fluctuations are known and different, two independent equations with two unknowns are produced which can be solved for the fluctuating velocity and temperature, with a high frequency response for both quantities. For a review of such hot and cold element 'anemometry' and 'thermometry' techniques, see Bradshaw (1971) or Corrsin (1963). Most laboratory/atmospheric temperature measurements have been done with cold wires so small that frequency response corrections can be neglected.

Unfortunately, a simple extrapolation of hot/cold element technology from the laboratory to the ocean has not been possible. Two factors are major obstacles. First of all, oceanic signal levels are astoundingly small by laboratory standards. Secondly, they are generally absent from a large volume fraction of any particular ocean layer; rather than the homogeneous, isotropic turbulence of high Reynolds number theory and laboratory/atmospheric practice, oceanic turbulence is strongly damped by stable stratification and occurs in intermittent bursts in thin layers or isolated patches where the local shear instability overcomes the local density stratification.

For comparison, a typical viscous dissipation rate ε in the upper ocean is 10^{-3} cm^2 s^{-3} (10^{-7} W kg^{-1} or $\rho\varepsilon = 10^{-4}$ W m^{-3}). This is three orders of magnitude less than atmospheric values and five orders of magnitude less than most wind tunnel experiments. Temperature variance dissipation rates χ might be 10^{-6} °C^2 s^{-1} in the upper ocean, compared to 10^{-4} in the atmosphere and 10^{-1} in a wind tunnel. Kolmogorov velocity scales $(\nu\varepsilon)^{\frac{1}{4}}$, where ν is the kinematic viscosity, are of order 10^{-2} cm s^{-1} compared to 1 cm s^{-1} in wind tunnels. Small scale turbulent temperature fluctuations are measured by the Batchelor temperature scale $\Sigma_B = (\chi/\gamma)^{\frac{1}{2}}$, $\gamma = (\varepsilon/\nu)^{\frac{1}{2}}$, defined by Gibson (1968), which typically has values of 10^{-5}°C or less; this is smaller than a good laboratory 'isothermal' bath, let alone a wind tunnel experiment where Σ_B might be 10^{-1}°C. However, despite the fact that turbulence in the ocean is weak, it is generally assumed that turbulent mixing is crucial to vertical diffusion of heat and chemical species in most of the ocean.

Little is known at present of the statistical description of oceanic intermittency and patchiness, but these factors must be taken into account in selecting microstructure detection sensors, platforms and sampling patterns to ensure adequate sensor dynamic range and spatial resolution, and adequate platform stability, range, and speed of coverage. Bodyakov et al. (1977) present evidence that the intensity of turbulence and mixing may be positively correlated with the size of the patch, which may mean that very large volume samples must be obtained for the average to be representative, especially if the volume fraction of the active patches is small and the level of activity large compared to the rest of the fluid.

The requirement for rapid sampling of a large fluid volume is in conflict with the fact that signal levels are low and the frequency response of the sensors is limited. In this volume, Nasmyth (see Chapter 38) presents a discussion of high speed sensor platforms and Gregg and Pederson (see Chapter 17) describe the scientific objectives and difficulties of small scale salinity detection. In the following, the fluid processes which determine sensor requirements for small scale oceanic temperature and velocity fluctuations are discussed in Section 2. Various velocity sensors and circuits are described in Section 3, and temperature sensors and circuits are given in Section 4; these have been designed to meet the requirements for use in the ocean. The sensors discussed are intended for use either on high speed towed bodies or submersibles, or on lower speed (and therefore much quieter) dropsondes as described in articles in this volume by Nasmyth, and by Lange (see Chapter 37) and Osborn and Crawford (see Chapter 19).

2. SENSOR REQUIREMENTS

The scientific objectives of small scale velocity and temperature measurements are similar to those for small scale salinity described by Gregg and Pederson in the preceding chapter; namely, to provide a qualitative and quantitative statistical description of the mixing and diffusive transport of conserved quantities such as heat, chemical species, mechanical energy, and momentum in the ocean. Small scale measurements are necessary because three-dimensional turbulence and turbulent mixing are confined by stable stratification to scales smaller than a buoyancy length $L_R = (\varepsilon/N^3)^{\frac{1}{2}}$, where N is the Brunt-Väisälä frequency $[g(\partial\rho/\partial z)/\rho]^{\frac{1}{2}}$, g is the acceleration of gravity, ρ is the density averaged over some vertical length larger than L_R, and z is the depth; the requirement for turbulent mixing to occur is that L_R must be larger than about 15 times the Kolmogorov scale $L_K = (\nu^3/\varepsilon)^{\frac{1}{4}}$. For typical N values of 10^{-2} rad s^{-1}, L_R varies from 1 to 0.1 m for the ε range of 10^{-1} to 10^{-3} cm^2 s^{-3}. In the absence of turbulence, the small scale velocity field probably consists primarily of shear and internal wave motions, with a minimum 'buoyant-viscous' length scale due to viscous damping proportional to $L_{KV} = (\nu/N)^{\frac{1}{2}}$ for such stratified flows.

For active turbulent mixing, where 'active turbulence' means a random fluid motion where inertial forces dominate both viscous and buoyancy forces for a finite range of length scales, temperature and salinity fluctuations should be homogeneous and isotropic over length scales ranging from L_R to L_B, where L_B is the Batchelor diffusive length $(D/\gamma)^{\frac{1}{2}}$, D is the molecular diffusivity, and γ is the turbulent rate of strain $(\varepsilon/\nu)^{\frac{1}{2}}$. However, homogeneous, isotropic fluctuations of temperature and salinity do not necessarily indicate the presence of active turbulence. Stable stratification may damp out the turbulent kinetic energy of an active patch, leaving small scale internal waves and a remnant patch of temperature and salinity microstructure activity. Microstructure activity in temperature and salinity caused by turbulence in nonturbulent fluid is called fossil temperature turbulence and fossil salinity turbulence. Such fossils may persist for long periods and give valuable indications of previous mixing intensities. They may also give misleading values for vertical diffusivity if interpreted as if they were actively turbulent (for example, using the model of Osborn and Cox, 1972) and may cause overestimates of the volume fraction of active turbulence if all microstructure activity is identified with active turbulence regions (for example, as assumed by Gargett, 1976). Consequently, it is necessary for the sensors to distinguish between active and fossil turbulence regions. Temperature measurements by Nasmyth are interpreted by Stewart (1969) as possible examples of turbulent temperature fossils based on simultaneous velocity fluctuation measurements which show a very low level of activity. More sensitive velocity sensors (Crawford, 1976) show that active microtemperature fluctuations are apparently always accompanied by increased microvelocity fluctuations. It then becomes necessary to determine whether such microvelocity fluctuations are active turbulence or the small scale internal waves (fossil vorticity turbulence) formed when the active turbulence has been damped out by the stable stratification. The question of how to recognize the 'signature' of fossil turbulence in spectra of temperature and velocity signals measured in the ocean is a controversial one; its exposition lies beyond the scope of this article.

The length scales in active and fossil turbulence regions in the ocean are substantially larger than those found in laboratory studies, which is a distinct advantage for frequency response and spatial resolution requirements for velocity and temperature sensors. The peak of the universal active turbulent temperature dissipation spectrum occurs when $k(D/\gamma)^{\frac{1}{2}} = 0.24$, which for a Prandtl number ν/D of about 9 corresponding to ocean water and $\varepsilon = 10^{-3}$ cm^2 s^{-3} is at a wavelength of $2\pi(D/\gamma)^{\frac{1}{2}}/0.24 = 1.8$ cm. For salinity with Schmidt number 1000, the peak wavelength is only 0.17 cm. Clearly, measurements with such fine resolution will be difficult even for temperature, and it may be necessary to rely on universal similarity hypotheses to infer the small scale structure for both temperature and salinity from large scale measurements in many cases. The peak of fossil temperature spectra appears to occur (Schedvin, 1979) at a wavelength of $2\pi(D/N)^{\frac{1}{2}}/0.3$ or about 8 cm for $N = 10^{-2}$ rad s^{-1}.

The peak of the turbulent velocity dissipation spectrum occurs at about $k(\varepsilon^{-1}\nu^3)^{\frac{1}{4}} = 0.09$, corresponding to wavelength $\lambda_A = 14.9$ cm at $\varepsilon = 10^{-3}$ cm^2 s^{-3} for active turbulence. In the absence of turbulence, buoyancy-dominated 'fossil vorticity turbulence' (small-scale internal waves; the buoyancy range of turbulence) should be limited by a small scale viscous cutoff wavelength proportional to $L_{KV} = (\nu/N)^{\frac{1}{2}}$. Very little is known about this regime of oceanic fluid motion at this time, primarily because it is extremely weak and few if any relevant measurements have been possible; see Gibson (1979) for a review. For example, $\varepsilon_F = (15/2)\ \nu(\partial u/\partial z)^2 \simeq 30\nu N^2 = 3.6 \times 10^{-5}$ cm^2 s^{-3} for $N = 10^{-2}$ rad s^{-1}. Up to this time heated element sensors have generally been too noisy to measure ε values less than about 10^{-4} cm^2 s^{-3}. Fortunately, the length scales are sufficiently large that very sensitive instruments such as the thrust velocimeter described by Osborn and Crawford (see Chapter 19) or the acoustic velocimeters described by Gytre (see Chapter 9) may be effective for this application even though their spatial resolution might be inadequate to measure the smaller scale motions produced by active turbulence.

3. HEATED ELEMENT VELOCITY SENSORS

As previously mentioned, the principle of heated element velocity sensors is that as the heat transfer rate increases with increasing convection velocity past the element, so the heating current required to maintain the element at constant temperature must also increase. The empirical relations describing forced convective heat transfer from heated element velocity sensors are generally expressed as $Nu = f(Re,Pr)$, where $Nu = (Q/A)/(k\Delta T/\Delta x)$ is the Nusselt number, $Re = U\Delta/\nu$ is the Reynolds number, $Pr = \nu/D$ is the Prandtl number, Q is the heat transfer rate through area A, k is the thermal conductivity of the film, ΔT is a characteristic temperature difference through thickness Δx, U is the velocity past the film, ν is the kinematic viscosity of the fluid film and D is the thermal diffusivity. For hot wires (Corrsin, 1963), Kramers (1946) and van der Hegge Zijnen find

$$Nu = 0.42\ Pr^{0.20} + 0.57\ Pr^{0.33}\ Re^{0.50}$$

for the range $0.71 < Pr < 1000$ and $0.01 < Re < 1000$, which is consistent with the earlier semi-empirical form proposed by King (see Corrsin, 1963), $Nu = A + B\ Re^{0.50}$, known as 'King's Law'. More recent measurements suggest a similar form, but with the exponent reduced to 0.45 for hot wires in air or 0.33 for hot films in water and with 'temperature loading factors' $(T_{max}/T_{film})^{-0.17}$ applied to the Nusselt number to correct for the fact that for very high overheat temperatures the fluid properties will vary (Bradshaw, 1971).

The precise form of the heat transfer law actually depends on a great variety of factors not represented by the most refined semi-empirical 'laws' (for example, geometrical effects, edge losses and surface contamination) so that even for laboratory studies it is necessary to rely on direct calibration against a standard instrument such as a pitot tube or sonic anemometer in the range of velocities and temperatures to be encountered in the experiment. If the calibration is likely to change due to 'aging' of the probe or to surface contamination, it is best to record the output of the standard instrument simultaneously during the experiment and compare joint probability density functions or power spectral levels at low frequencies to be sure the calibration has not drifted, at least for frequencies within the bandwidth of the instrument used as a standard.

For atmospheric turbulence work it has become standard practice to have at least double redundancy of all signals recorded in order to monitor intercalibration drifts. For oceanic turbulence work this is perhaps even more important, since the possibility of sensor contamination for heated element sensors is high, and, even if the laboratory calibration is reproduced after the experiment, it does not provide the certainty that goes with an independent simultaneous measurement. For towed measurements we have used a ducted current meter with approximately ±1% absolute accuracy and frequency response to about 2 Hz as our standard, and found satisfactory agreement with laboratory calibrations. The propeller blades produce pulses in a magnetic sensor as they rotate, with frequency proportional to the mean velocity.

Note that since the heated element anemometer output voltage V is usually proportional to the heating current I, and $I^2 = A + BU^n$, the sensitivity $\partial I/\partial U \simeq (Bn/2IU^{1-n}) \simeq (n/U^{0.75})$ decreases with increasing velocity and decreasing n. To make $V \sim U$ for the range of U values, either a 'linearizer' circuit must be used or (possibly expensive) computer operations carried out if large percentage changes in U are to be measured. A tradeoff between the advantages of increased linearity at high speed and the disadvantages of increased heating current requirements and reduced sensitivity is necessary to select the optimum platform speed for a given turbulence activity.

Particular care must be taken to account for the effects of temperature sensitivity in oceanic turbulence measurements. It follows from the heat transfer relation that $I^2 = \Delta T(A + BU^n)$, where ΔT is the overheat temperature difference ($T_{element} - T_{ambient}$). Clearly, fluctuations in the heating current I can result from fluctuations in either U or the ambient temperature. From the heat transfer relation we also find that

$$(2I\delta I/n\Delta TBU^{n-1}) = \delta U\{1 - (1 + A/BU^n)(U/n\Delta T)(\delta T/\delta U)\}$$

$$\simeq \delta U\{1 - (U/n\Delta T)(\delta T/\delta U)\}$$

assuming $A << BU^n$. For temperature fluctuations to be negligible it is necessary for the second term in the right hand brackets to be small. If the field is actively turbulent, $\delta T/\delta U$ in the inertial subrange is of order $(\beta_K \chi \varepsilon^{-1/3} k^{-5/3}/\alpha \varepsilon^{2/3} k^{-5/3})^{\frac{1}{2}} \simeq (\chi/\varepsilon)^{\frac{1}{2}}$ and the output fluctuations will be approximately

$$\delta U\{1 - (U/n\Delta T)(\chi/\varepsilon)^{\frac{1}{2}}\} \text{ (inertial subrange).}$$

The second term is the 'temperature noise' and must be small if the velocity signal is to dominate. If $U = 100$ cm s^{-1}, $n = \frac{1}{2}$, and $\Delta T = 40°C$, then $(\chi/\varepsilon)^{\frac{1}{2}}$ must be much less than 0.2°C cm^{-1} s. Typical values of χ and ε are 10^{-6} °C^2 s^{-1} and 10^{-3} cm^2 s^{-3} for active oceanic turbulence, which gives 0.03°C cm^{-1} s and only a 15% 'temperature noise'. However, if the region is well mixed or if the turbulence has been damped by stratification, as in a turbulent temperature fossil, the ratio may be only 0.1°C cm^{-1} s (as in Schedvin, 1979), so the sensitivity of the heated element sensor to velocity and temperature fluctuations in this example would be equal in the inertial range, and dominated by temperature at smaller scales. Examination of available oceanic 'velocity' spectra (e.g. Vega, 1975) typically shows high frequency content above that expected from universal similarity spectral forms. These spectra may actually be contaminated by temperature sensitivity at high frequencies.

It would seem that heated element velocity signals in the ocean will generally be somewhat contaminated and sometimes dominated by temperature signals, especially for high velocities or small overheat temperatures. However, temperature and velocity fluctuations are also likely to be correlated, especially at large scales, so spectra and correlation functions of such signals will be a mix of spectra and cospectra, correlations and cross correlations.

Sensitivity of the heat transfer to salinity fluctuations will occur through the effects on viscosity, density and thermal conductivities; from an inspection of the heat transfer equation, it can be seen that the resulting salinity signal should be small compared to velocity and temperature fluctuation signals.

The first successful measurements of oceanic turbulent velocity fluctuations were made using millimetre size heated platinum films deposited on the tip of rounded conical quartz probes, mounted on a torpedo-like body towed from a ship (Grant et al., 1962; see also Nasmyth, Chapter 38). Initial tests were in a tidal passage with dissipation rates ε of order 10^{-1} cm^2 s^{-3} (10^{-5} W kg^{-1}), which is 2 to 4 orders of magnitude larger than open ocean values found in later measurements (see Gargett, 1976, for a review). The Kolmogorov inertial subrange constant and spectral form observed in the tidal passage were almost identical to those observed in wind and water tunnel grid turbulence, which supports the Kolmogorov (1941) first and second universal similarity hypotheses. However, open ocean spectra are frequently low level, often contaminated by noise and in relatively poor agreement with universal forms (Grant et al., 1968), either for velocity or temperature. Subsequent attempts to measure oceanic turbulence with heated element sensors on towed bodies have also met with limited success in the open ocean (Gargett, 1976; Williams and Gibson, 1974; Gibson et al., 1974; Vega, 1975) except in strong current systems (Belyaev et al., 1975a, 1975b). Biological fouling is a severe problem. The Canadian group solves this by a remotely controlled jet washer. Sensitivity to mechanical vibration and ambient temperature fluctuations may dominate the output velocity signals.

Lange (personal communication) has recently made measurements of velocity fluctuations from a dropsonde using a microbead thermistor operated with constant heating current. Heated thermistors have the advantage that their resistance decreases with increasing temperature and their burnout temperature is higher than platinum film probes, hence they are more rugged and can be operated at higher temperatures. Constant current circuits are very simple compared to constant temperature bridges; their primary disadvantage is that the frequency response is determined by the thermal lag of the heated element.

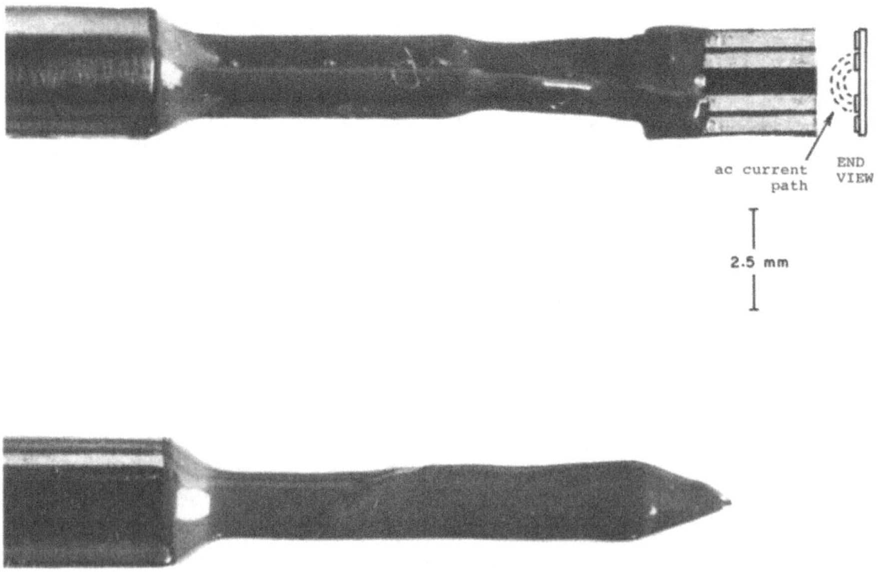

Fig. 1 (top) Open flow, 2 mm, 4 electrode microconductivity probe. (bottom) 0.0127 cm microbead thermistor probe.

Figure 1 shows a photograph of a microbead thermistor probe, along with a four-element microconductivity probe (see Gregg and Pederson, Chapter 17) both of which have been developed for use on our towed body (described by Nasmyth, see Chapter 38). A 0.13 mm diameter thermistor is mounted on the conical tip of a 4 mm diameter quartz cylinder using a commercial platinum film probe, and may be heated to become velocity sensitive or used as a temperature sensor in circuits described in following sections.

Heated thermistor elements have the advantage that their temperature coefficient of resistance $\partial \ln R/\partial T$ is about -4% $°C^{-1}$ compared to only 0.2% $°C^{-1}$ for typical platinum film probes; their resistance values cover a range from hundreds to thousands of ohms, compared to only 5 to 10 ohms for platinum films. The low impedance

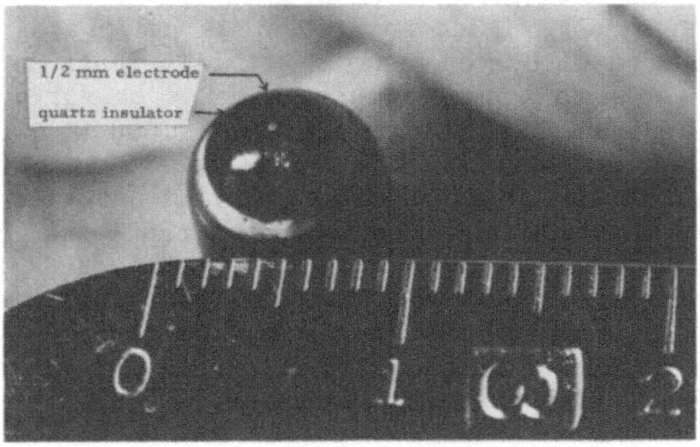

Fig 2a USSR hydroresistance anemometer - ½ mm Pt electrode.

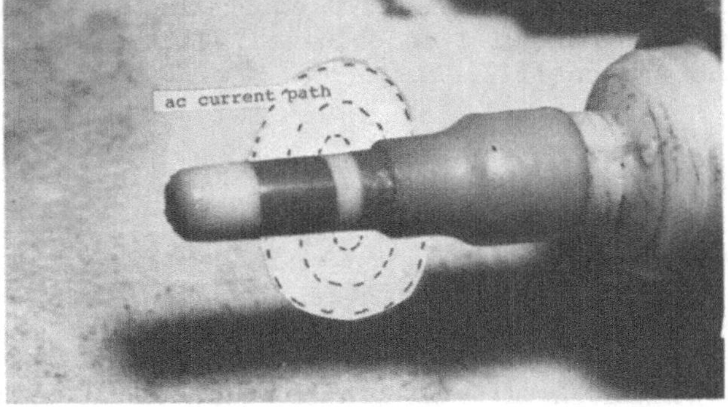

Fig. 2b USSR ac coupled conductivity probe - 3 cm resolution.

of platinum film sensors requires special care to avoid any fluctuations in the lead and contact resistance between the bridge and probe: the slightest mechanical flexing can introduce significant noise.

A unique velocity sensor developed by the Moscow Institute of Oceanology (Belyaev et al., 1975a, 1975b) is called a 'hydroresistance anemometer', and is shown in the photograph of Figure 2a. A 0.5 mm platinum electrode in the centre of a hemispheric quartz insulator forms a circuit through the sea water to the stainless steel sheath. High frequency current density (2 MHz) sufficient to heat the fluid inside and outside the stagnation point boundary layer is imposed by a constant current circuit mounted in the probe sting; the resistance fluctuations caused by the variable convective heat transfer and large ($-2\%\ ^{\circ}C^{-1}$) temperature coefficient of sea water conductivity is used to detect velocity fluctuations based on much the same principles as described previously for other heated element sensors. The instrument is calibrated on board the ship by comparison with a commercial hot film instrument in a small flow channel and with a ducted current meter on various tow bodies at speeds of 1.5 to 3 $m\ s^{-1}$ to depths of 1000 m. The sensor is rugged compared to platinum film or microbead thermistor probes; the particular sensor shown in Figure 2a had survived two years of service in frequent sea tests, including a nondestructive encounter with a mud bottom, prior to its use in the U.S.-USSR intercomparison experiment (Schedvin, 1979) when the photograph was taken. Its frequency response should be comparable to that of platinum film probes with similar dimensions used as resistance thermometers (Fabula, 1968a, 1968b) as described below.

Any of the previously described heated resistance element velocity sensors, or any of the resistance element temperature sensors to be described in the next section, can be used in a constant temperature (constant resistance) bridge to obtain increased frequency response to fluctuations in both velocity and temperature. Using two probes at two temperatures, two equations of the form $dE = A\delta U + B\delta T$ with two unknowns will be obtained which can be solved for δU and δT. Since the coefficients A and B will be different for each circuit,

$$\delta T = [A_2/(B_1A_2-B_2A_1)]dE_1 - [A_1/(B_1A_2-B_2A_1)]dE_2$$

and $$\delta U = [B_2/(A_1B_2-A_2B_1)]dE_1 - [B_1/(A_1B_2-A_2B_1)]dE_2,$$

with $A_1 \neq A_2$, $B_1 \neq B_2$.

Maximum sensitivity results when the overheat temperature ΔT is maximum for one circuit and minimum for the other. The minimum ΔT must always be greater than δT, and the maximum may be limited by the boiling or degassing temperatures, or by nonlinearities resulting from variable fluid properties.

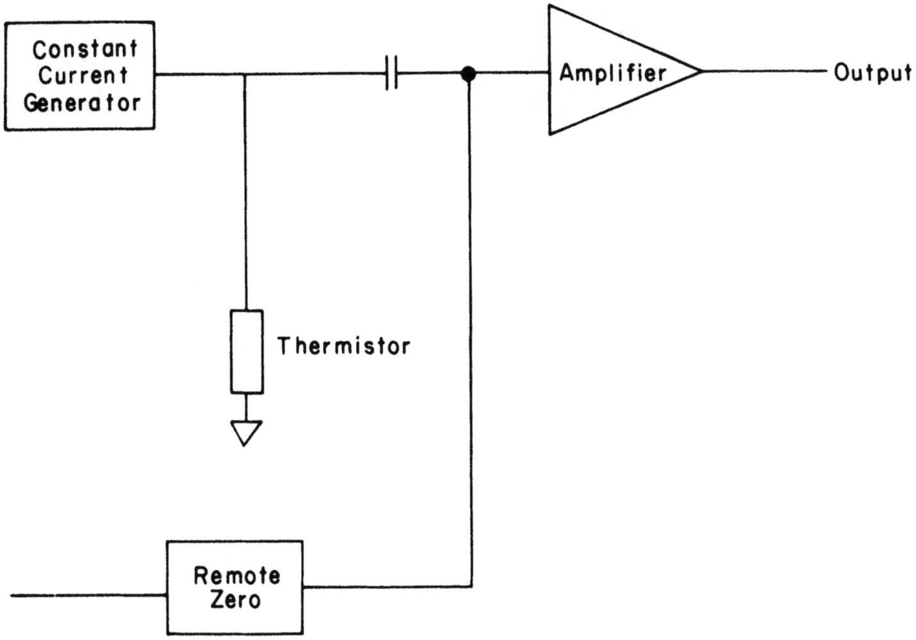

Fig. 3 Constant current anemometer circuit.

Figure 3 is an example of a constant current anemometer bridge, designed by one of us (Deaton), which we use on our towed body. The circuit is ac coupled with a 200 s time constant to avoid the need to rebalance the bridge while still responding to the lowest frequencies of interest. Should the mean velocity change, a remotely controlled circuit charges the coupling capacitor on command to avoid a prolonged off-scale condition. Constant current circuits have the advantage of simplicity of design and are not subject to instabilities common to constant temperature bridges. However, frequency response to temperature or velocity fluctuations corresponds to the thermal lag of the sensor, which may be inadequate to resolve the viscous and thermal diffusive scales except for very slowly moving sensor platforms.

Figure 4 shows a schematic diagram of a constant temperature anemometer circuit with remote control of balance, operation, standby, and on-off. Remote control is necessary, since the bridge must be close to the sensor probe to minimize noise. Water tunnel comparison of velocity spectra measured with this circuit, given in Gibson et al. (1974), shows that its frequency response and noise levels are as good as or better than a commercial bridge. Our bridge is first balanced by remote control signal 2. If the output voltage

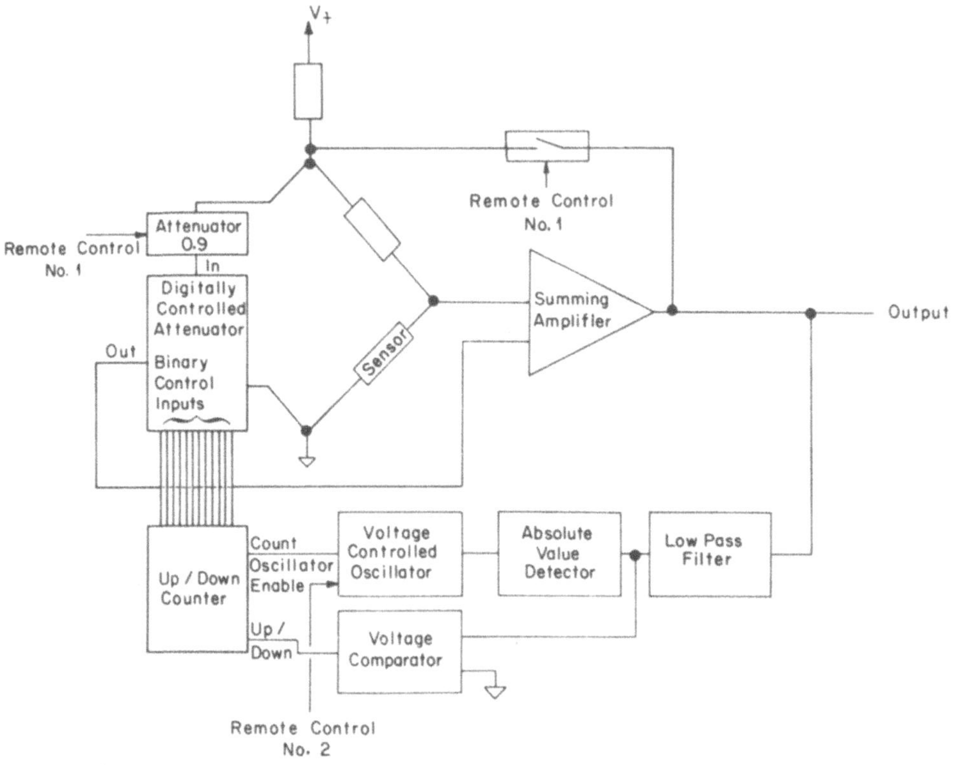

Fig. 4 Constant temperature anemometer circuit.

from the summing amplifier is not zero, indicating bridge balance, this activates the voltage-controlled oscillator whose frequency activates step changes through the up/down counter of a digitally-controlled attenuator. Operation of the bridge is caused by remote control signal number 1, which disengages the 0.9X attenuator and engages the summing amplifier to the top of the bridge. The bridge balances by increasing the bridge current and hence heating the sensor. For a platinum resistance probe a 10% resistance change corresponds to a ΔT value of about 40°C. For negative temperature coefficient sensors, such as thermistors or conductivity probes, a different circuit arrangement would be required.

Heated platinum film and conductivity sensors used in the ocean seem to be able to measure turbulence with minimum reported ε values of about 10^{-3} cm^2 s^{-3}. We may express this as a velocity sensitivity based on the peak of the universal turbulent dissipation spectrum at $k_{peak} = 0.09\ (\nu^3/\varepsilon)^{\frac{1}{4}}$, to give

$$u' = (k\phi)^{\frac{1}{2}}_{peak} = 1.6(\varepsilon\nu)^{\frac{1}{4}},$$

$$= 0.09 \text{ cm s}^{-1} \text{ for } \varepsilon = 10^{-3} \text{ cm}^2 \text{ s}^{-3}.$$

If sensitivity to velocity fluctuations increases by a factor of 20 for heated thermistors (which is proportional to the increase in their temperature coefficient of resistance) then $u' = 4.5 \times 10^{-3}$ cm s^{-1}, and $\varepsilon_{min} = 6 \times 10^{-9}$ cm^2 s^{-3}: both substantially less than the minimum small scale turbulence or internal wave signals expected in the ocean. Whether or not this greatly improved sensitivity to small scale velocity fluctuations using heated thermistors can actually be achieved remains to be demonstrated, but it is certainly needed to improve the present inadequate experimental description of small scale oceanic velocity fluctuations. Conductivity sensors also have a high temperature coefficient of resistance, so it is possible that lower-noise, heated-conductivity-cell velocity

Table 1

Heated Element Velocity Sensors

Sensor	Advantages	Disadvantages	Application
Platinum film (conical)	- shape resists fouling - high frequency response constant temperature circuits documented	- low sensitivity - low impedance sensor increases sensitivity to contact and lead resistance noise	- high turbulence regions
Thermistor	- high sensitivity - negative temperature coefficient resists burnout in constant current operation	- low frequency response in constant current mode - subject to fouling	- dropsonde or slow moving towed bodies - medium turbulence regions
Conductivity	- rugged - high sensitivity - burnout impossible	- behavior poorly documented - salinity sensitive - possibly noisy	- medium speed towed bodies - medium to high turbulence regions

sensors may be developed in the future which may be more rugged and have a higher frequency response than thermistors.

Table 1 summarizes various heated element velocity sensors, compares advantages and disadvantages, and suggests possible preferred applications. The list is by no means exhaustive, but represents the sensors most frequently used in ocean turbulence experiments.

4. TEMPERATURE SENSORS

Compared to velocity, small scale temperature fluctuations in the ocean are relatively easy to detect with a satisfactory signal-to-noise ratio. Consequently, most oceanic 'microstructure' and 'turbulence' data reported are actually microtemperature fluctuations, from which inferences about the density, velocity, and other fields may or may not be possible. Gregg et al. (1978) report an rms noise level of 3 μ°C over a bandwidth of 20 to 100 cpm, which is less than 1% of the signal levels observed at 3500 m in the South Pacific.

However, Lueck et al. (1977) have studied the frequency response of thermistor probes commonly used on dropsonde devices. Flat plates, ellipsoidal beads, and rods were compared to a conical platinum film probe for response to temperature fluctuations produced by injecting bursts of heated fluid in a water tunnel flow at 125 cm s^{-1}. The conical film probe spectral response had been carefully studied by Fabula (1968a) using a diffusion layer stagnation flow model and observations of response when the probe is rapidly plunged through a thin heated sheet of water rising from a heated wire in quiescent fluid. It was found that the thermistors had 3 dB attenuation frequencies of 6 to 15 Hz. In Section 2 we found that resolution of a temperature gradient spectrum for fossil temperature turbulence requires spatial and frequency response resolution to about 1 cm or less, which would constrain platform velocities for the probes studied by Lueck et al. (1977) to less than 6 to 15 cm s^{-1}. Such low velocities have been possible using carefully designed dropsondes (Gregg et al., 1972, 1978) but may be impractical for towed sensor platforms.

More effective frequency response can be obtained by compensation for the thermal lag of the sensor if the response function is known. Schedvin (1979) used Fabula's (1968a, 1968b) results to correct the temperature spectrum from a 0.15 mm conical platinum film on the tip of a probe, such as that shown on the bottom of Fig. 1, by factors up to 4; compensated response was obtained to 175 Hz with resolution to wavelengths of 1 cm. It is necessary to assume that the probe is clean and the velocity is constant. For sensor platform speeds above about 2 m s^{-1} it would seem that thermistor or platinum film probe temperature sensors require more sophisticated methods to extend frequency response than simple thermal lag compensation: for example, the 'constant temperature thermometer' technique described previously.

Both platinum films and thermistor beads are subject to damage because of their small size, and their already limited frequency response is sensitive to fouling. Conductivity probes with the same spatial resolution and sensitivity, but with much better frequency response and strength, can be used as temperature sensors where the conductivity is dominated by temperature and not salinity (much of the ocean). For scales smaller than the thermal diffusive cutoff, conductivity fluctuations will be dominated by salinity.

Figure 2b shows an ac coupled conductivity probe used by the Russian oceanic turbulence group on towed bodies. Two 0.6 cm diameter silver films, cylindrical about the probe axis and closely spaced, are coupled as capacitor plates through dielectric material and sea water in a 2 MHz bridge mounted in the probe body. Fluctuations in sea water resistivity within the 3 cm sample volume, shown as the current path between the plates, are sensed by the bridge. Since the boundary layer on the probe occupies a relatively small portion of the sample volume, the frequency response and spatial resolution of such a probe should not be greatly affected by it.

The open flow conductivity probe shown at the top of Figure 1 is also designed to detect small scale conductivity fluctuations on a high speed platform. It is described by Gregg and Pederson (see Chapter 17). The sample volume is essentially the 2 mm diameter cylindrical shell with axis along the space between the inner (constant current) electrodes shown by the current path in the end view sketch. Frequency response is limited by the flushing of the fluid boundary layers on the inner electrodes; these boundary layers occupy a relatively small fraction of the total sample volume ($\delta \simeq (\nu x/U)^{\frac{1}{2}} \simeq 30\ \mu m$).

Both the microconductivity probe and the microbead thermistor probe shown in Figure 1 are used in similar bridge circuits, shown schematically in Figure 5. The 22 kHz ac bridge is balanced remotely by a control command, and a 'calibrate' square wave signal of about 0.1°C (or equivalent for conductivity) is imposed on the output by a remotely controlled unbalance impedance activated by a 'calibrate' command. Balance occurs in 0.1 s. For operation with a microbead thermistor, the high impedance voltage amplifier is attached to the constant current generator leads to the thermistor rather than to the outer electrodes of the microconductivity probe.

Due to biological fouling at sea, laboratory measurements of temperature sensor frequency response corrections must be regarded as a lower bound. If such corrections are necessary it is preferable to use direct measurements of the frequency response of the fouled sensor. A technique for in situ frequency response measurements has been used for contaminated platinum wire temperature sensors in the marine atmosphere by LaRue et al. (1975). A high frequency ac heating signal is amplitude modulated at various frequencies and the resulting output resistance fluctuations measure the probe response at the modulation frequency. Application of in situ oceanic frequency response measurements should, in principle, be no more difficult than in the atmosphere, although comparison with universal spectra in documented flow fields (LaRue et al., 1975) or higher frequency response sensors in laboratory flows (Lueck et al., 1977) are needed to confirm this assumption.

Table 2 summarizes advantages, disadvantages, and possible applications of various temperature sensors. It should be repeated that proper circuitry can offset apparent disadvantages such as low frequency response or low sensitivity.

Intercalibration drifts and loss of high frequency sensitivity can be tested for temperature sensors if a high degree of redundancy exists between sensors. This is desirable for temperature, just as it is for velocity. One possibility is to use a suitably accurate, low-drift, mean temperature sensor such as a commercial platinum wire resistance thermometer or a pressure-protected thermistor bead, which has frequency response to about ½ Hz and absolute accuracy to 0.01°C, and then compare spectral levels for other

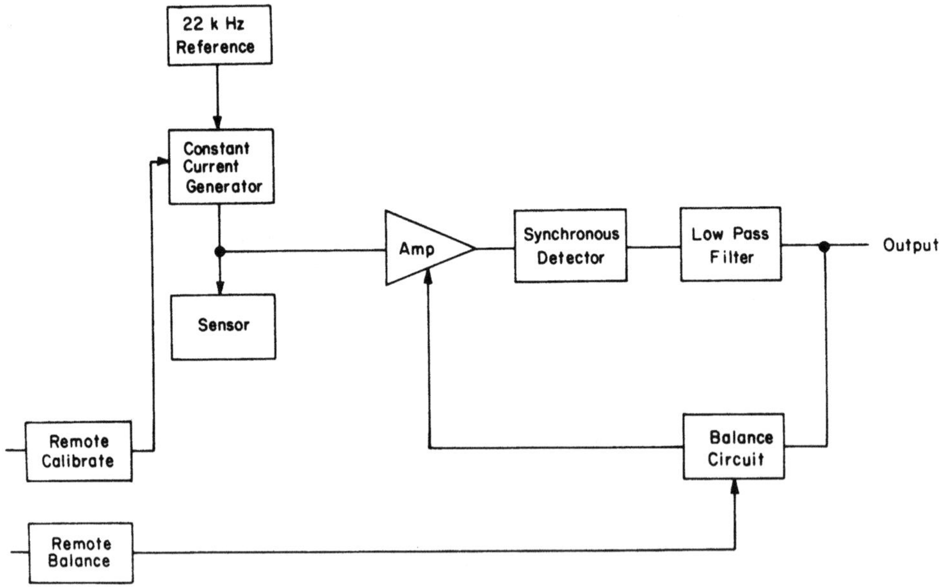

Fig. 5 Small scale conductivity/microbead thermistor circuit.

Table 2

Temperature Sensors

Sensor	Advantages	Disadvantages	Application
Platinum film (conical) (cylindrical coated quartz fibre)	- resists fouling - documented frequency response - high frequency response	- low sensitivity - low impedance - fouls rapidly	- towed bodies - upper ocean - active turbulence
Thermistor	- high sensitivity	- low frequency response	- dropsondes - deep ocean - towed bodies (with circuit compensation for frequency response)
Conductivity	- high sensitivity - high frequency response - resists fouling	- salinity sensitive	- towed bodies - dropsondes - all depths

temperature sensors at frequencies less than ½ Hz with the mean temperature sensor as a standard. Deterioration of the frequency response of a thermistor or platinum film probe would be indicated by comparison with the high frequency spectrum of a microconductivity probe in regions of the ocean where the conductivity is dominated by temperature.

Since the degree of activity in the ocean varies widely, it is usually possible to find a region which is so quiet that the output signal is dominated by noise. Assuming the instrument noise is constant, one can then measure the spectrum from the quietest portion of the record and possibly identify and correct the data from the more active regions for the indicated noise.

5. SUMMARY

Small heated and cold velocity and temperature sensors are needed to detect active and fossil turbulent microstructure or double diffusive processes in the ocean with spatial resolution to scales of molecular diffusion. Because velocity signals are so low, heated velocity sensors are marginally capable of detecting the turbulence levels or the residual small scale internal wave motions present in the ocean, and have substantial sensitivity to temperature fluctuations even at high overheat temperatures. Thermistor beads have been shown to have adequate sensitivity to detect most oceanic temperature and possible velocity signal levels, and may be operated at constant temperature to extend their frequency response. Preferably, two beads should be operated close together at two very different, constant overheat temperatures so that simultaneous temperature and velocity measurements can be made.

Platinum film and conductivity probes have also been used in the ocean to measure velocity fluctuations, but seem more likely to be limited by noise than thermistors. Frequency response corrections are possible for slow moving thermistors if the probe remains clean, but for speeds over about 10 cm s^{-1} the uncertainty due to fouling and variable velocity make frequency response corrections impractical for presently available thermistors. This may be a severe limitation, since the patchy, intermittent nature of most oceanic microprocesses seems to require higher sensor platform velocities to provide an adequate statistical description in a reasonable time.

Conductivity sensors offer an alternative means of detecting temperature at high speeds when salinity fluctuations are small and may be heated to detect velocity. They are relatively rugged, resist fouling, are nearly as sensitive to temperature fluctuations as thermistors, and have high frequency response. A combination of small scale conductivity, temperature, and heated velocity sensors operating in close proximity to each other is more useful than the same sensors operated separately because the mixed sensitivities permit redundant measurements of the same quantities and improved confidence in the results. The combination is also needed to infer small scale fluctuations of salinity and density, which will be needed for any complete description of the thermodynamic and dynamic state of the fluid.

AKNOWLEDGEMENTS

Support for this work was provided by ONR N00014-75-C-0152, NSF ATM 76-23856 and NSF ENG 27398 (C.I.T. P.O.#28-464865). Several useful suggestions by reviewers are acknowledged.

REFERENCES

BELYAEV, V.S., A.N. GEZENTSVEY, A.S. MONIN, R.V. OZMIDOV and V.T. PAKA. 1975a. Spectral characteristics of small-scale fluctuations of hydrophysical fields in the upper layer of the ocean. *Journal of Physical Oceanography,* 5: 492-498.

BELYAEV, V.S., M.M. LUBIMTZEV and R.V. OZMIDOV. 1975b. The rate of dissipation of turbulent energy in the upper layer of the ocean. *Journal of Physical Oceanography,* 5: 499-505.

BODYAKOV, G.I., Y.A. GERMAN and R.V. OZMIDOV. 1977. Statistical analysis of intermittent ocean turbulence. *Atmospheric and Oceanic Physics,* 13: 74-76.

BRADSHAW, P. 1971. *An Introduction to Turbulence and Its Measurement.* Pergamon, Oxford, 218 pp.

CORRSIN, S. 1963. Turbulence: experimental methods. In: *Handbuch der Physik,* 8/2, Springer, Berlin: 524-590.

CRAWFORD, W.R. 1976. Turbulent energy dissipation in the Atlantic equatorial undercurrent. Ph.D. Dissertation, University of British Columbia, 156 pp.

FABULA, A.G. 1968a. Theoretical frequency response of stagnation-region thin-film thermometers. *Journal of Scientific Instruments,* 1: 1200-1206.

FABULA, A.G. 1968b. The dynamic response of towed thermistors. *Journal of Fluid Mechanics,* 34: 449-464.

GARGETT, A.E. 1976. An investigation of the occurrence of oceanic turbulence with respect to fine structure. *Journal of Physical Oceanography,* 6: 139-146.

GIBSON, C.H. 1968. Fine structure of scalar fields mixed by turbulence: I. Zero-gradient points and minimal gradient surfaces. *Physics of Fluids,* 11: 2305-2315.

GIBSON, C.H. 1979. Fossil temperature, salinity and vorticity turbulence in the ocean. In: *Marine Turbulence,* edited by J.C.J. Nihoul, Elsevier Oceanography Series, to be published.

GIBSON, C.H. and W.H. SCHWARZ. 1963. Detection of conductivity fluctuations in a turbulent flow field. *Journal of Fluid Mechanics,* 16: 357-364.

GIBSON, C.H., L.A. VEGA and R.B. WILLIAMS. 1974. Turbulent diffusion of heat and momentum in the ocean. *Advances in Geophysics,* 18A: 353-370.

GRANT, H.L., R.W. STEWART and A. MOILLIET. 1962. Turbulent spectra from a tidal channel. *Journal of Fluid Mechanics,* 12: 241-268.

GRANT, H.L., B.A. HUGHES, W.M. VOGEL and A. MOILLIET. 1968. The spectrum of temperature fluctuations in turbulent flow. *Journal of Fluid Mechanics,* 34: 423-442.

GREGG, M.C., C.S. COX and P.W. HACKER. 1973. Vertical microstructure measurements in the Central North Pacific. *Journal of Physical Oceanography,* 3: 458-469.

GREGG, M.C., T. MEAGER, A. PEDERSON and E. AAGAARD. 1978. Low noise temperature microstructure measurements with thermistors. *Deep-Sea Research,* 25: 843-856.

KOLMOGOROV, A.N. 1941. The local structure of turbulence in incompressible viscous fluid for very large Reynolds numbers. In: *Turbulence - classical papers on statistical theory,* edited by S.K. Friedlander and L. Topper, Interscience, New York (1961): 151-156.

KRAMERS, H. 1946. Heat transfer from spheres to flowing media. *Physica,* 12: 61-80.

LaRUE, J.C., T.K. DEATON and C.H. GIBSON. 1975. Measurement of high-frequency turbulent temperature. *Review of Scientific Instruments,* 46: 757-764.

LUECK, R.G., O. HERTZMAN and T.R. OSBORN. 1977. The spectral response of thermistors. *Deep-Sea Research,* 24: 951-970.

OSBORN, T.R. and C.S. COX. 1972. Ocean fine structure. *Geophysical Fluid Dynamics,* 3: 321-345.

SCHEDVIN, J.C., 1979. Microscale measurements of temperatures in the upper ocean from a towed body. Ph.D. Dissertation, University of California at San Diego, 400 pp.

STEWART, R.W. 1969. Turbulence and waves in a stratified atmosphere. *Radio Science,* 4: 1269-1278.

VEGA, L.A. 1975. Small-scale measurements of velocity and temperature in the upper layers of the ocean. Ph.D. Dissertation, University of California, San Diego, 171 pp.

WILLIAMS, R.B. and C.H. GIBSON. 1974. Direct measurements of turbulence in the Pacific equatorial undercurrent. *Journal of Physical Oceanography,* 4: 104-108.

19

An Airfoil Probe for Measuring Turbulent Velocity Fluctuations in Water

T.R. Osborn and W.R. Crawford

1. INTRODUCTION

The study of oceanic turbulence on the scale of tenths and hundredths of metres is limited by the difficulty of measurement. Small-scale velocity fluctuations are of interest for they demonstrate the presence of turbulence and mixing. Temperature, salinity, and density fluctuations can be measured, albeit with difficulty, to the scale of tens of centimetres and these data used to infer mixing and turbulence. Unfortunately, measurements of a scalar field like temperature or salinity do not give much insight into the velocity field.

Atmospheric velocity measurements are usually made with heated anemometers such as hot wires and films, which are highly nonlinear. In water there are added problems due to the higher heat capacity of the fluid and the possibility of particulate matter contaminating the probes. This paper discusses another sensor - the airfoil probe - which has proven useful in the ocean. The probe is a pointed body of revolution in which the cross force on the axisymmetric nose is sensed with a piezoceramic transducer. The probe measures the cross-stream component of velocity whereas heated anemometers usually measure the downstream component.

The original idea for the airfoil probe was due to H.S. Ribner at the University of Toronto and was developed there in conjunction with T.E. Siddon. Siddon (1969) describes a combined pressure/velocity measuring probe designed to operate in conjunction with an analogue analysis system and to correct the pressure measurement for the effect of the interaction of the velocity field with the stationary probe. The paper includes measurements in several specially designed flows to test the combined system.

Siddon (1971) describes a probe made for operation in air. That paper includes a comparison of the probe with a pair of crossed hot wires in a round jet. The energy-containing eddies are on a scale of 20 times the probe diameter. The two techniques give close agreement over most of the range with the airfoil probe falling below the crossed wires at wavelengths smaller than four times the probe diameter. This effect is attributed to the spatial averaging of the probe. Other papers by Siddon and Ribner are listed in the bibliography for the reader who is interested in the complete selection.

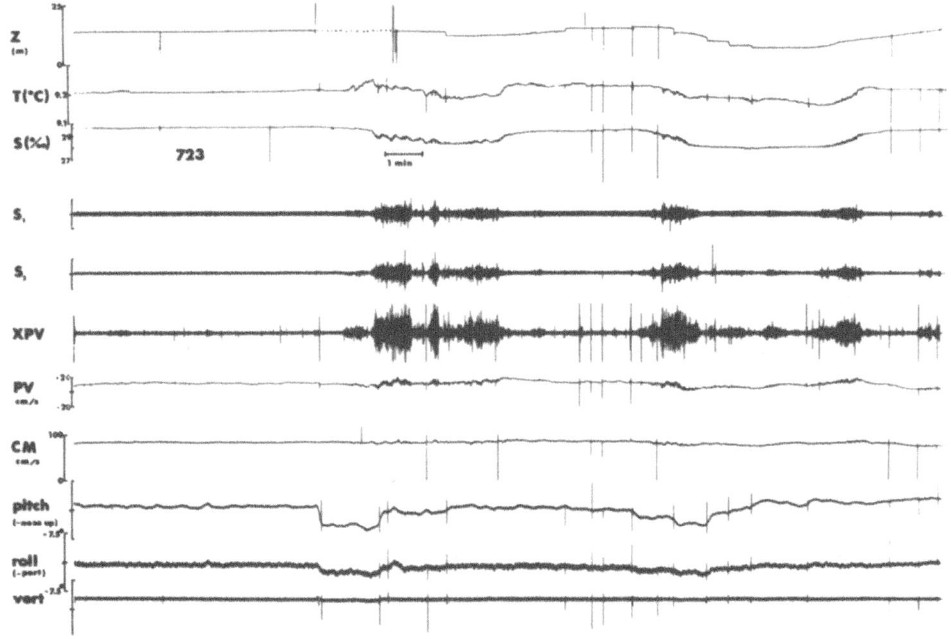

Fig. 1 Data collected by the Institute of Ocean Sciences, Patricia Bay, showing the turbulent interface between the Fraser River Plume and the underlying water in th Strait of Georgia. Time progresses from left to right. S_1 and S_2 are the velocity shears from two perpendicular channels of the airfoil probe. PV is the velocity signal from a platinum hot film anemometer. XPV is the PV data which have been bandpassed in the frequency range 1 to 100 Hz and then amplified.

Vehicles for oceanic microstructure measurements are discussed in two other chapters of this book (17 and 18) so this chapter will only discuss the characteristics of the airfoil probe. Where the probe limitations place restrictions on the vehicle behavior we note the problem, but we do not discuss the performance of a complete instrument system.

Airfoil probes are currently being used for oceanic measurements at three institutions. The Institute of Ocean Sciences, Patricia Bay, B.C. (IOS), uses them in a measurement program where the vehicle is a 'Pisces' submersible; the Bedford Institute of Oceanography (BIO) and the University of British Columbia (UBC) use the probes on free-fall vehicles. Manuscript reports are available on the latter two systems (see bibliography for details) and can be procured by writing the institution concerned.

Before looking at the details of the probe, we present some data taken with airfoil probes in two different applications. Figure 1 shows data from a system mounted on the 'Pisces IV' submersible of the Institute of Ocean Sciences at Patricia Bay, B.C. Dr. A.E. Gargett kindly provided the figure and details of the operation; the data were taken in November 1976 off the mouth of the Fraser River in Georgia Strait. The base of the river plume was 10 to 12 m, where a region of high-frequency velocity fluctuations marked the interface between colder, fresher river water and the underlying warmer, saltier water. It should be noted that only the interface was turbulent, not the entire river plume. The salinity trace showed that the submersible first penetrated into, but not through, the interface from below, sank down, rose up through the interface into the river plume, and then sank through the interface again.

The similarity is very strong between the relative intensity of the downstream turbulent velocity fluctuations as monitored by the platinum film probe and the cross-stream velocity shear fluctuations derived from the airfoil probe. Problems in these data include frequent isolated spikes due to dropouts in the data recording system and step-like offsets in the depth (pressure) record because the pressure transducer shared a common hydraulic reservoir with the submersible operation systems. The difference in noise level between the two channels of the airfoil probe is attributed to the difference between the horizontal and vertical vibrations of the mounting structure.

The data in Figure 2 are from Crawford and Osborn (1979a) and were taken using a free-fall vehicle developed at UBC. The figure shows temperature, its gradient, and two perpendicular velocity shear channels in the depth interval from 10 to 160 m near 32°59'S, 0°2'N in July 1974. At this time and location the core of the Atlantic Equatorial Undercurrent was at approximately 90 m depth. The temperature and velocity microstructure were associated with the

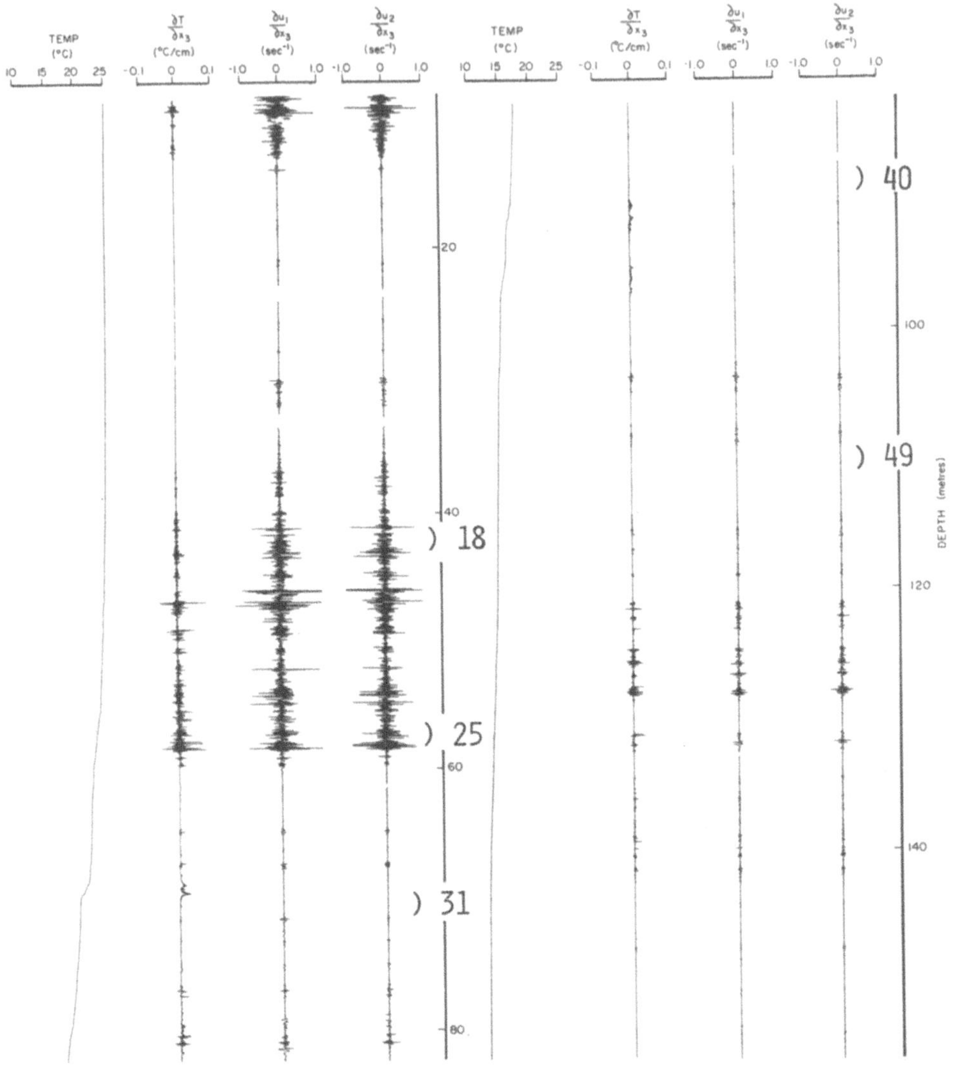

Fig. 2 Data collected with a free-fall instrument in July 1974 at 32°59'W, 0°2'N. The data represent temperature, its gradient, and two perpendicular shear channels from an airfoil probe. The five numbers between 18 and 49 indicate the location of spectra of 2 m intervals of the data from the $\partial u_1/\partial x_3$ channel that are shown in Figure 7.

regions of relatively high mean shear above and below the current. The temperature and velocity microstructure generally occurred simultaneously except in regions where the temperature gradient signal was uniformly of one sign; e.g., 69 to 70 m and 91 to 96 m. In these cases no velocity microstructure was observed. We will return to this point in the section on thermal response of the probe.

The airfoil probes presently used in the ocean are all similar (Fig. 3). The nose is a rubber compound, with silicon rubbers the most popular at the moment. The sensing element is a piezoceramic bimorph beam, the sensitive element from a ceramic phonograph cartridge. (The piezoceramic beam is a capacitively-coupled voltage source; thus the dc portion of the response is not measured; oceanic uses of the probe include a differentiating circuit before recording, hence the data are presented as velocity shear.) Much development work is taking place in the manufacturing so specific details should come directly from the active researchers. The BIO probes are single-channel. The UBC probes (which are also used at IOS) have been two-channel probes in the past (e.g., Fig. 1 and 2) but we are now making only single-channel probes and will use these for future measurements. The present design uses some features of the BIO probes plus other improvements suggested by the work of Nowell at IOS (Patricia Bay, B.C., personal communication).

Let us now look at the characteristics of the probe starting with the theoretical response. The next section will describe calibration, linearity, and sensitivity. Then the temperature, pressure,

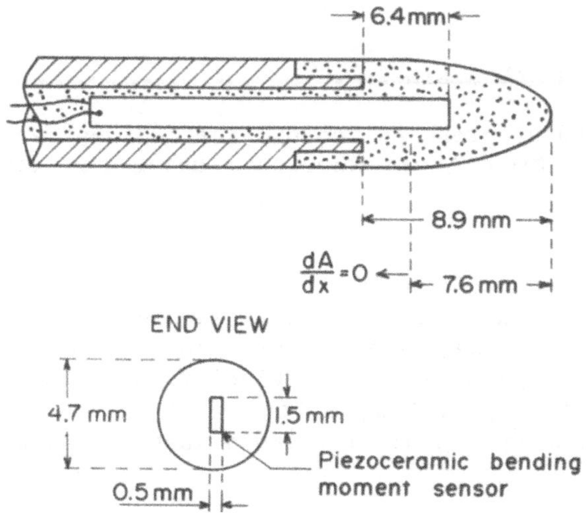

Fig. 3 Airfoil probe as manufactured in early 1978 at UBC.

and acceleration sensitivities are outlined. Finally, the noise level will be examined by looking at a series of spectra from a single profile.

2. THEORETICAL RESPONSE

The probe is an axisymmetric airfoil of revolution (Fig. 3) mounted so that the mean velocity (V) is aligned with the axis of revolution. In oceanographic applications the mean axial velocity is due to the vehicle motion through the water, and u is the fluctuating cross flow in the water column. It is u which the probe is designed to measure.

For a slender body of revolution in an inviscid flow of speed U and angle of attack α (assumed small), Allen and Perkins (1952) derive f_p, the cross force per unit length due to the potential flow as

$$f_p = (\tfrac{1}{2}\rho U^2)\,\frac{dA}{dx}\,\sin 2\alpha \qquad (1)$$

where ρ is the density of fluid, and dA/dx is the rate of change in body cross-sectional area with longitudinal distance along the body (the lift force is perpendicular to the mean velocity vector; note that in this discussion it is the cross force, i.e., the force perpendicular to the airfoil, and not the lift force that is being discussed).

This formula is correct to the second power in α for inviscid flow. The total cross force exerted on the probe by the potential flow is derived by integrating Equation 1 along the length of the probe from the tip where A = 0, to the base of the probe where dA/dx = 0. Equation 1 becomes

$$F_p = \int f_p dx = \tfrac{1}{2}\rho U^2 A \sin 2\alpha \ . \qquad (2a)$$

The ability of the shear probe to measure the cross-stream velocity u is evident if Equation 2a is rewritten using the double-angle relation for sin 2α:

$$F_p = \rho A V u \qquad (2b)$$

The importance of two aspects of vehicle motion becomes apparent immediately. First, a large mean axial velocity is necessary to reduce the relative contribution of velocity fluctuations in the axial direction. Second, tilting of the axis of the probe mixes the cross-stream velocity signal with the downstream velocity fluctuations.

If the cross force due to the pressure distribution were the only force, then the output would be linear in the cross-stream velocity. However, there is a second cross force, F_v, which is due to viscous effects. This force can be modelled as proportional to the square of the cross-stream velocity.

$$F_V = B \cdot \tfrac{1}{2}\rho U^2 \sin|\alpha| \sin\alpha \ , \qquad (3)$$

where B is a constant of proportionality.

The effect of this force is to increase the slope of the cross force curve with increasing angle of attack. Therefore the slope of the force curve should be monotonically increasing with increasing magnitude of the angle of attack. In this and many other respects an airfoil of revolution is like a delta wing. Both can operate at large angles of attack without stalling. We will now look at the calibration procedure and results to get some information about sensitivity and linearity.

3. CALIBRATION TECHNIQUE, VELOCITY SENSITIVITY AND LINEARITY

Our present calibrator is shown in Figure 4. A jet is discharged vertically into a tank of water, with the tip of the probe mounted just above the centre of the outlet at an angle α. Water for the jet passes through a mesh-and-honeycomb arrangement to reduce the scale size and intensity of turbulence. Details can be found in Crawford (1976) and Osborn and Crawford (1977).

The probe is tiled successively at nominal $2\frac{1}{2}°$ intervals from 20° left to 20° right, and rotated about its axis; thus a sinusoidal voltage is generated. Slip rings bring supply voltages to the preamplifier, and transmit the signal from the preamplifier to a bandpass filter and an rms meter; the output of the rms meter is denoted by E_{rms}.

As the probe rotates in the jet, the total cross force on the probe remains constant but the force on the beam varies sinusoidally. The values of $E_{rms}/\rho U^2$ for one of the new UBC single-channel probes are plotted versus $\sin 2\alpha$ in Figure 5. The flow rate U is measured with a pitot tube and an inclined manometer.

The data have been examined in two ways: first, by fitting in a least squares manner against a function

$$X \sin 2\alpha + Y \sin|\alpha| \sin\alpha \qquad (4)$$

the slope of which with respect to $\sin 2\alpha$ is

$$X + Y/2 \tan|2\alpha| \ , \qquad (5)$$

and second, by using a cubic spline routine to fit the data and then plotting the fit and the slope. The second method has the advantages of not specifying the functional form and not requiring

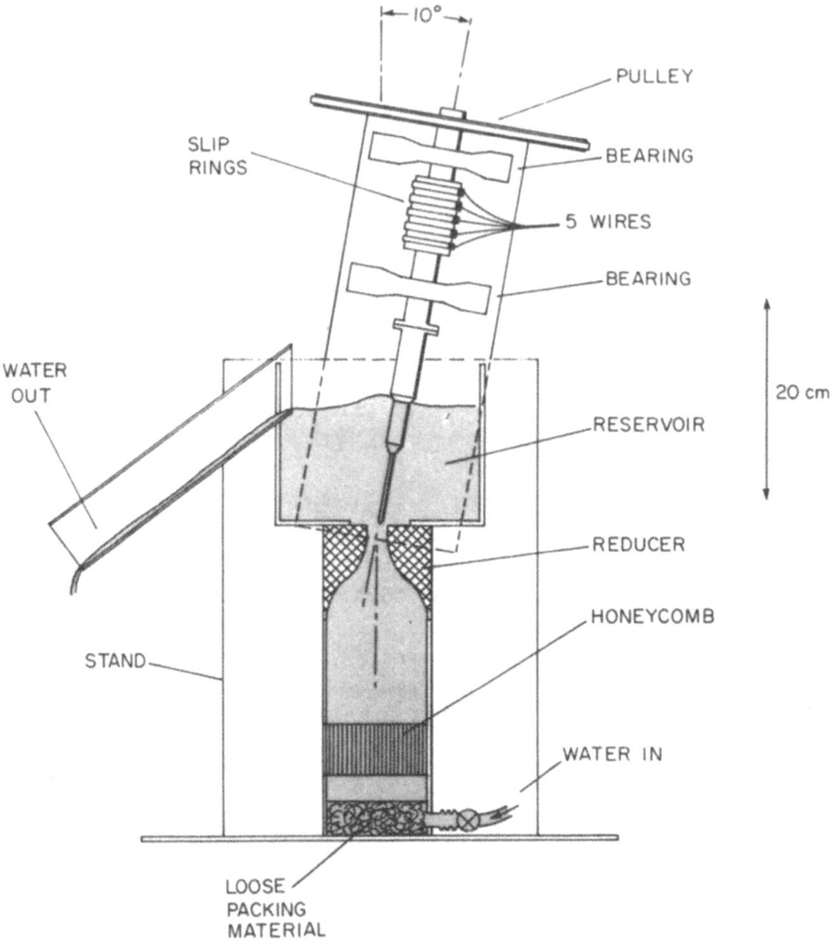

Fig. 4 The present UBC calibrator.

symmetry, but has the disadvantage of forcing the curvature toward zero at the end of the interval. The tension in the spline is adjusted so that the slope is monotonically increasing from the minimum. Both of these techniques can allow for a variation in the weighting assigned to each data point. Figure 5 shows both fits (and their slopes) with the spline fit as a solid line and the least squares fit as a dashed line.

The variation in the slope with angle of attack is apparent in Figure 5. In most oceanic operations the maximum angle of attack is determined by the vehicle and the mean flow, and can be kept to less than 10°, or at most 15°, by suitable vehicle design. Since it would be difficult to monitor the mean angle of attack, the sensitivity is taken as a constant and the nonlinearity included in the error bounds for the small-scale shear measurement. For example, calculating the geometric mean of the slope over the ±10° range by using the maximum and minimum values gives a 2% higher value for the spline fit than for the least squares fit.

A constant multiplicative error in the calibration arises from errors in the measurement of ρU^2. Crawford (1976) gives a value of 4% or less for the uncertainty in the dynamic pressure of the jet. Considerable time has been required to make the measurements as accurate as possible by minimizing noise, adjusting alignments, and

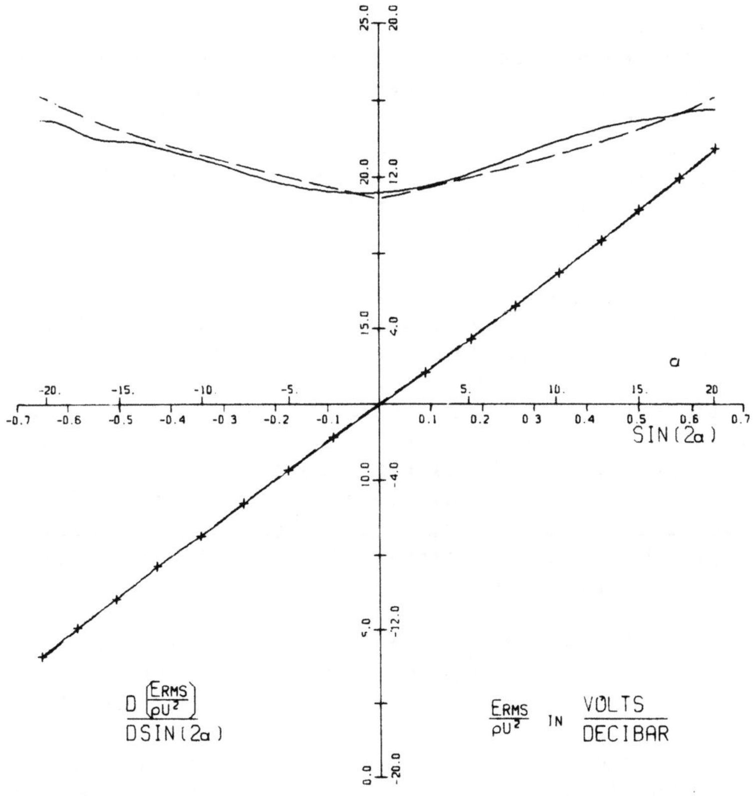

Fig. 5 Calibration curve for a single-channel UBC probe. The data points are indicated with an x. The spline fit to the data is shown as a solid line with the least squares fit of Equation 4 indicated in a dashed line. The slopes of the two fits are also indicated in a similar fashion.

improving the angle measurements. The 0° setting of the calibrator is determined by adjusting the axis of rotation of the probe in two perpendicular planes until the minimum value is achieved. Since symmetry requires the cross force to be zero at $\alpha = 0$, the value measured is an estimate of the error, and the data values for 0° angle of attack are not used in the fitting scheme. Rather, the data are used to indicate the quality of the calibrator setup. The output at this setting is probably due to a combination of residual misalignment, vibration induced from the bearing, etc., and turbulence in the flow. It is only 3% of the output at 2.5° and the percentage decreases with increasing angle.

If the output at 0° is considered to be totally due to the angular misalignment then that error in alignment would be less than 0.1°. The various angles at which the probes are calibrated are set relative to the minimum. Error in the measurement of the angular adjustment is less than ±0.1°. The asymmetry of the slope of the spline fit suggests that there is a systematic error which is believed to be associated with the calibrator in that the tip of the probe is not along the line perpendicular to the plane of rotation through the pivot point. In the calibration shown, the difference in output between ±20° corresponds to a difference in the angles of 0.2°; using ±0.2° as the error in α, the error in $\sin 2\alpha$ is ±4% at 2.5° decreasing to ±0.4% at 20°.

The errors in the calibration procedure are associated with the measurement of flow in the calibrator, adjustment of the calibrator, and noise. An improved but similar calibrator has been constructed to reduce the problem of alignment and to simplify the adjustment procedure. A more accurate system for measuring the dynamic pressure is being tested. However, the small inherent nonlinearity in the probe limits the accuracy of measuremet when using a constant value for the calibration coefficient. The experience to date indicates that the minimum value for the overall error in sensitivity (including nonlinearities) is between ±5% and ±10%, depending on the range of α.

The increase in sensitivity with increasing angle of attack affects the vehicle design via four different mechanisms. First the mean velocity must be sufficiently larger than the turbulent velocity fluctuations in order to limit the maximum angle of attack. [In oceanic applications, the fall speed necessary to produce a reasonable signal level ($V > 40$ cm s^{-1}) is large enough that the angle of attack fluctuations due to the turbulent velocity fluctuations is less than 1°]. Second, pitching in a horizontally moving vehicle or tilting in a free fall must be restrained to a limited angular excursion. Third, the frequency of any oscillation of the support vehicle is also important since there is an apparent cross-stream velocity at the probe due to the rotation of the vehicle about its centre of mass, and this can lead to large effective angles of attack. Fourth, a large mean shear in the water can induce a significant angle of attack variation over the instrument and may cause a large value to occur at the probe tip, depending on the mean axial speed and the configuration of the vehicle.

4. THERMAL SENSITIVITY

There are three aspects of the effect of temperature on the probe that we wish to discuss. The first is the change in the calibration coefficient with a change in the mean temperature. The second is the response of the probe to relatively low frequency changes in the mean temperature. The third is the response of the probe to relatively high frequency changes in the temperature.

The change in sensitivity with mean temperature has been examined by calibrating at different temperatures and by fixing the calibration at one angle (10°) while slowly varying the temperature of the water with the rms output voltage recorded as a function of temperature. The results are consistent with a change in sensitivity of +1% $°C^{-1}$. More work is needed to see if this value varies over the oceanic temperature range, and to determine the variation between probes.

Piezoelectric devices also have an output due to thermal variations; this phenomenon is called the 'pyroelectric effect'. Thus, the probes have a low-frequency response to temperature changes. In Figure 6 the data show a temperature change of 1°C between 69 and 70 m depth (these data are a subset of those shown in Fig. 2). The response in the velocity shear is of low frequency with an amplitude on the order of 2.5×10^{-2} s^{-1}. This level of response is typical. Some probes are ten times better and some probes have been almost five times worse. The variations are believed to be due to differences between individual beams. Also, some construction techniques accentuate the problem. This thermal sensitivity limits the useful low-frequency response of the probe to somewhere between 1 and 0.1 Hz.

We now come to the question of the high-frequency response of the probes to temperature fluctuations. Are the signals that we interpret as velocity fluctuations contaminated by the high-frequency temperature fluctuations? Figure 6 shows a large temperature change just above 70 m depth with many small-scale temperature gradient fluctuations of up to 1°C m^{-1} associated with it, but no corresponding small-scale velocity shear fluctuations, indicating the probe's insensitivity to small-scale temperature fluctuations.

After use, some probes start having large noise spikes associated with temperature changes. Probes exhibiting this phenomenon are retired immediately after the drop in which it begins. The phenomenon is considered to be a mechanical failure of the probe, but has

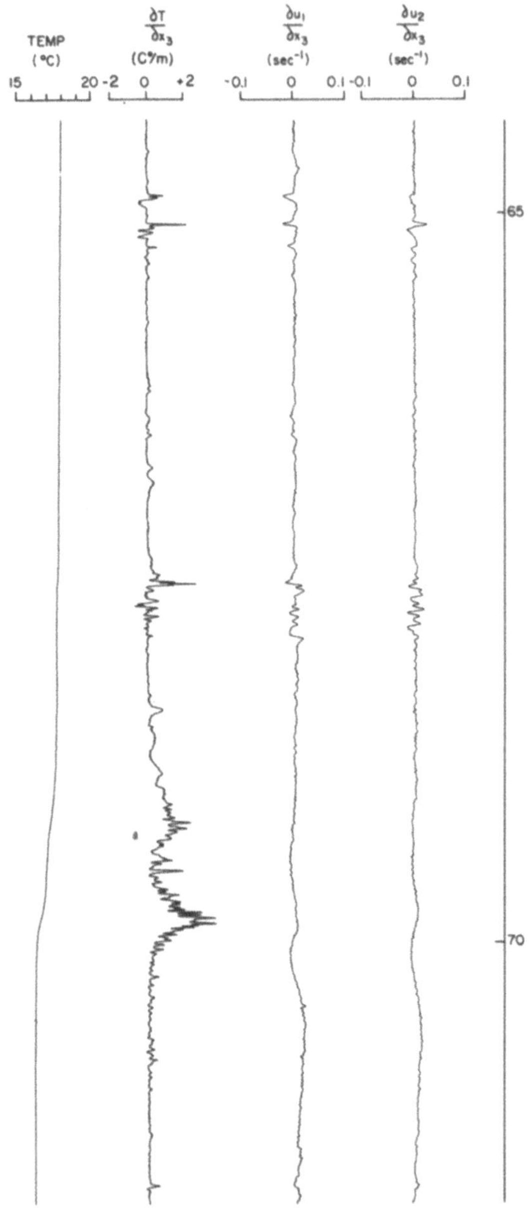

Fig. 6 Expanded trace from Figure 2 showing the response to the relatively broad temeprature features above 70 m depth.

not been serious lately due, we believe, to a change in construction technique involving careful selection and testing of the beams prior to encapsulation and a change in the nose material (which has resulted in a better bond to the stainless steel base).

In summary, the water temperature affects the probe by changing the sensitivity. Large mean temperature changes induce a low-frequency signal which limits the usefulness of the low-frequency portion of the data. High-frequency signals due to temperature fluctuations do not appear in the data. Thermally-induced spiking during large temperature changes can occur and denotes the end of useful life for a probe.

5. PRESSURE EFFECTS

There are two possible pressure effects: pressure-induced noise and a change in sensitivity with pressure. Noise spikes occur intermittently on our velocity shear data records. To determine the cause we tested a probe in a pressure tank by itself and in conjunction with our Richardson-type stretched pin releases. The noise appeared only when the probe was pressurized with the release. Keeping the releases and all other moving parts of the free-fall instrument well greased has reduced the occurrence of these noise spikes.

We also tested the probe for a change in sensitivity with pressure by mounting a 12.7 mm diameter bar 0.3 m in front of the probe tip on our free-fall instrument. Drops were then made in excess of 200 m depth through isothermal water in Howe Sound (a local B.C. inlet). A spectral analysis revealed less than a ±5% change in sensitivity over that depth interval and this uncertainty is due to the uncertainty in the fall speed. Oakey (1977) also reports the effect of hydrostatic pressure to be negligible at the same 5% level.

6. SPATIAL AND TEMPORAL RESOLUTION

Measurements of the velocity shear can be used to estimate the local rate of viscous energy dissipation. If one has all the necessary derivatives, the dissipation can be calculated exactly (Lamb, 1945). Using the relation for isotropic turbulence $\varepsilon = 7.5 \nu[\partial u/\partial z]^2$, where u is a horizontal velocity component and z is the vertical coordinate, one can estimate the dissipation from considerably less information. A crucial question is whether the variance of the velocity shear is completely resolved by the probe. Just as in the situation of estimating the turbulent heat flux from the small-scale temperature gradient, spatial averaging of the signals at the scale size of the peak of the spectrum can lead to seriously underestimating the variance.

There are actually two problems being studied at once. The first is a resolution problem in which a velocity fluctuation with a varying value of cross-stream velocity impinging on the probe modulates the angle of attack at different positions on the probe. The second is a frequency response problem, similar to that of the pitching airfoil (von Kármán and Sears, 1938) in which the angle of attack is uniform but the flow is not in equilibrium with the instantaneous position of the airfoil. The response of the probe to any real feature will be a combination of the two effects.

Siddon (1969) intercompared the airfoil probe with a set of crossed hot wires in a jet. The results showed agreement between the two systems to wavelengths as small as four times the diameter of the probe. One can calculate the transfer function associated with averaging a sine wave of wavelength λ over an interval L. The result is the familiar $\sin(x)/x$ function with $x = \pi L/\lambda$. This relation implies a half-power point (-3 dB) at $\lambda = 2.25$ L and a -1 dB point at $\lambda = 4L$. Siddon's result shows the crossed wire and airfoil probe just starting to diverge at $\lambda = 4d$. The desired linear lift force due to the flow arises from the portion of the probe where the diameter changes and is in fact weighted by the rate of change of cross-sectional area. Thus the length over which the probe changes diameter rather than the diameter itself might be a more appropriate length scale. The change in area occurs in about 1.5 diameters for both of our probes and that described by Siddon.

Crawford (1976) has analyzed data collected with the probe (mounted on our 'Camel' dropsonde) along the Atlantic Equatorial Undercurrent in order to study the energetics of the current. In the course of that work it was necessary to estimate the dissipation rate from the measured variance of the small-scale shears measured by the airfoil probe. As a check on the measured variances the velocity spectra were also computed by dividing out digitally, in the frequency domain of the spectra, the electronic time differentiation. The velocity spectra were fitted against a universal spectrum for cross-stream fluctuations derived from the downstream spectra of Nasmyth (1970). The fitting was done in the fashion described by Stewart and Grant (1962).

The shape of the measured spectra relative to that of the universal spectrum suggests that the probes have a poorer spatial resolution than the extrapolation of Siddon's results would lead one to expect. The -1 dB point occurs at about 4 cm and the -3 dB point at around 3 cm. These values suggest averaging lengths, L, of 1 cm and 1.3 cm, respectively, when used in the relations for the linear running average. These are longer than the 0.75 cm over which the probe changes cross-sectional area or the 0.475 cm diameter. But this result is consistent with Nowell's analysis (IOS, Patricia Bay, B.C., personal communication) of wind tunnel results from a probe model where the slope of the lift force curve increases and becomes more nonlinear as the length of the cylindrical after section is increased. The implication is clear that a shorter probe would be better and the present UBC probes (Fig. 3) have been modified accordingly.

A major aspect of the probe's future development will have to be the documentation of the spatial and temporal resolution. The difficulty is in finding the appropriate method. Comparisons with a universal curve are not definitive, and cross-stream velocity measurements in water with heated anemometry are not sufficiently documented to be satisfactory. Using a series of successively smaller probes might work; however, the range of probe sizes that could be compared in most water tunnels is quite limited and the test is only a relative one. Laser Doppler Anemometry may be suitable and this possibility is presently under examination.

7. VIBRATION SENSITIVITY AND NOISE LEVEL

The airfoil probe is sensitive to accelerations. Being a force sensor the probe interprets a lateral acceleration as a force due to the inertial loading of the probe tip. Also, a vibrating probe sees a varying angle of attack of the velocity vector. Thus even a set of crossed wires which have no acceleration sensitivity, when vibrated perpendicular to a steady flow, would respond to the time-varying cross flow. Similarly, a hot film vibrated in the downstream direction senses the apparent fluctuation in the flow. Shaker table tests of the airfoil probes manufactured at UBC indicate that the apparent cross-stream flow dominates over the probe's acceleration sensitivity. Hence the noise level from vibrations is determined by the vehicle and probe mount, and not the probe itself.

Figure 7, taken from Crawford (1976), shows a series of spectra from 2-m vertical sections of the data shown in Figure 2. The dissipations have been estimated from the variance of the shear signal (Crawford and Osborn, 1979a). All spectra are from the $\partial u_1/\partial x_3$ signal of the data displayed in Figure 2. The high frequency region contains the vibrations and the electronic noise; the turbulence signal is at the lower frequencies, generally below 20 Hz. Especially noticeable is the vibration peak at 23 Hz. Complete elimination of vibrations is impossible. The instrument has subsequently been redesigned so that vibrations would be outside the frequency band of interest; the modification has significantly reduced the 23 Hz noise.

The spectral noise level of the probe is shown with an associated value of $\varepsilon = 4\times10^{-6}$ cm^2 s^{-3} for a bandwidth of 0.5 to 10 Hz. For higher values of ε the bandwidth of the integration is increased to include the region of the spectrum where the signal exceeds the noise level. The bottom panel shows a low dissipation region in

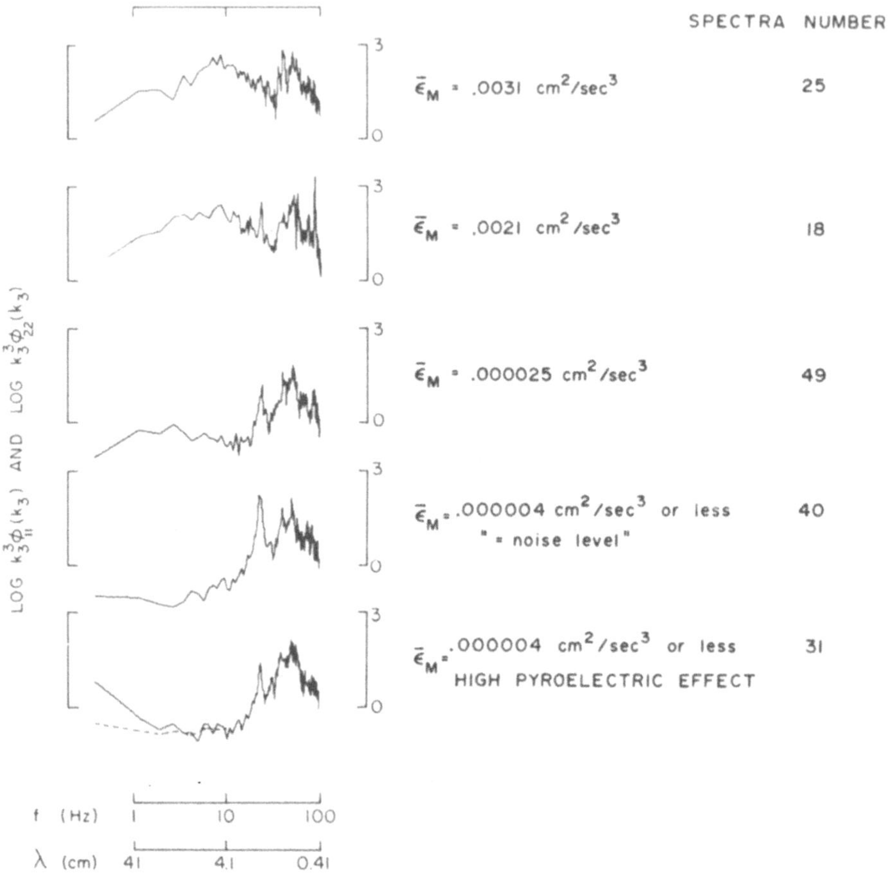

Fig. 7 Spectra of velocity shear (Crawford, 1976) showing the change in the spectra with changing signal level. The dotted line in the bottom spectrum shows the effect on the spectra of removing the low-frequency temperature-induced signal with a digital filter. Locations of spectra are marked on Figure 2.

which there was significant thermally-induced variation. High-pass digital filtering was used to reduce the low-frequency content and produce the spectral shape shown by the dotted line.

8. SUMMARY AND CONCLUSIONS

With any probe one is interested in:

(1) resolution - spatial and temporal
(2) linearity
(3) sensitivity
(4) calibration
(5) signal-to-noise ratio - noise level
(6) innate complexities - manufacture, lifetime, specialized electronics

This chapter has endeavored, for one instrument, to outline the presently available information about the first five items. In general, the characteristics are quite suitable for certain oceanic applications. The probe is undergoing a period of rapid development with improvements in its characteristics, as well as other aspects such as lifetime and ease of manufacture. The operational procedures and the electronics are considerably simpler than those associated with heated anemometry. With continued effort the probe can be an increasingly useful tool in the study of oceanic turbulence.

ACKNOWLEDGEMENTS

This work has been supported by the National Research Council of Canada, the Department of Fisheries and the Environment, and the U.S. Office of Naval Research on Contract N00014-76-6-0446-NR083-207. This work has benefited from many people who have given aid through the years. Dr. P.W. Nasmyth first suggested the probe to one of the authors; Dr. T.E. Siddon gave help, encouragement, and information; Dr. R.W. Stewart gave encouragement to the early work, and Drs. A.E. Gargett, A. Nowell, and S. Pond have been helpful in discussions.

REFERENCES

ALLEN, H.J. and E.W. PERKINS. 1952. A study of effects of flow over slender, inclined bodies of revolution. U.S. National Advisory Committee for Aeronautics, Report No. 1048, 15 pp.

CRAWFORD, W.R. 1976. Turbulent energy dissipation in the Atlantic Equatorial Undercurrent. Ph.D. Thesis, University of British Columbia, 149 pp.

CRAWFORD, W.R. and T.R. OSBORN. 1979a. Microstructure measurements in the Atlantic Equatorial Undercurrent during GATE. In press, *Deep-Sea Research*, GATE Supplement, edited by W. Duing.

CRAWFORD, W.R. and T.R. OSBORN. 1979b. Energetics of the Atlantic Equatorial Currents. In press, *Deep-Sea Research*, GATE Supplement, edited by W. Duing.

LAMB, H. 1945. *Hydrodynamics*. Sixth edition. Dover Publications, 738 pp.

OAKEY, N.S. 1977. An instrument to measure oceanic turbulence and microstructure. Bedford Institute of Oceanography Report Series/BI-R-77-3, Dartmouth, N.S., Canada, 52 pp.

OSBORN, T.R. 1974. Vertical profiling of velocity microstructure. *Journal of Physical Oceanography*, 4: 109-115.

OSBORN, T.R. 1977. The design and performance of free fall microstructure instruments at the Institute of Oceanography, University of British Columbia. Manuscript Report No. 30, University of British Columbia, 37 pp.

OSBORN, T.R. 1978. Measurements of energy-dissipation adjacent to an island. *Journal of Geophysical Research*, 83: 2939-2957.

OSBORN, T.R. and W.R. CRAWFORD. 1977. Turbulent velocity measurement with an airfoil probe. Manuscript Report No. 31, University of British Columbia, 39 pp.

OSBORN, T.R. and T.E. SIDDON. 1975. Oceanic shear measurements using the airfoil probe. Turbulence in Liquids: *Proceedings of the Third Symposium on Turbulence in Liquids*, September 1973, edited by G.K. Patterson and J.L. Zakin, University of Missouri-Rolla.

NASMYTH, P.W. 1970. Oceanic turbulence. Ph.D. Thesis, University of British Columbia, 69 pp.

SIDDON, T.E. 1965. A turbulence probe utilizing aerodynamic lift. University of Toronto, Institute for Aerospace Studies, Technical Note 88, 15 pp. (Available in the libraries of University of Toronto and University of British Columbia.)

SIDDON, T.E. 1969. On the response of pressure-measuring instrumentation in unsteady flow. University of Toronto, Institute for Aerospace Studies, Report No. 136, 32 pp. (Available in the libraries of the University of Toronto and the University of British Columbia.)

SIDDON, T.E. 1971. A miniature turbulence gauge utilizing aerodynamic lift. *Review of Scientific Instruments*, 42: 653-656.

SIDDON, T.E. 1974. A new type turbulence gauge for use in liquids. Symposium on Flow - Its Measurement and Control in Science and Industry. *Proceedings of the 1971 Symposium*, Pittsburgh: 435-439.

SIDDON, T.E. and H.S. RIBNER. 1965. An aerofoil probe for measuring the transverse component of turbulence. American Institute of Aeronautics and Astronautics. *AIAA Journal*, 3: 747-749.

STEWART, R.W. and H.L. GRANT. 1962. Determination of the rate of dissipation of turbulent energy near the sea surface in the presence of waves. *Journal of Geophysical Research*, 67: 3177-3180.

VON KARMAN, T. and W.R. SEARS. 1938. Airfoil theory for non-uniform motion. *Journal of Aeronautical Science*, 5: 379-390.

20

Expendable Measuring Devices

T.P. Barnett and R.L. Bernstein

1. INTRODUCTION

In recent years oceanographers have begun to study the large scale variability that occurs in the world's oceans. This represents a significant departure from the classical methods of oceanographic observation wherein a single ship conducts relatively long cruises to define the 'mean' properties of a large area. The emphasis on variation in the ocean now requires that large areas be sampled in a time short compared with that of the fluctuations to be studied and/or the background noise levels that might alias a larger scale signal. These requirements clearly dictate measurement capabilities that the older methods cannot realistically satisfy.

Variations of the thermal structure have been among the first large scale oceanic fluctuations to be studied. One of the main reasons for this, aside from the scientific justification, is that the basic instrumentation necessary to observe the temperature field is relatively simple. Thus, a set of expendable temperature-measuring devices has been developed that now enable oceanographers to sample large regions of the oceans quickly and inexpensively. The major portion of this article will describe the general properties of these instruments: the expendable bathythermograph (XBT) and the airborne expendable bathythermograph (AXBT). Both of these systems have been in use for approximately a decade and their general characteristics are relatively well known. However, it is only recently that a detailed analysis has been made of the operational capabilities of some of the available probes. It is also true that only in the last few years have the instruments been used to study, in a concentrated scientific way, large scale oceanic variations.

In this article we will discuss the operational characteristics of both devices as well as their strong and weak points. We do not intend this to be an exhaustive treatise on all available XBT/AXBTs since neither space nor present information make this possible. Particular emphasis will be placed on the AXBT system since it is less well known. Examples will be given on some of the scientific uses to which the expendable devices have been put. The final sections will suggest future improvements in the instrumentation as well as a summary statement indicating the future potential for expendable instrumentation.

2. XBT/AXBT OPERATIONAL CHARACTERISTICS

2.1 XBT Operational Characteristics

The operation of the original commercially-available XBT system is generally well known and is only briefly summarized here. The system consists of three parts: probe, launcher, and recorder. As the probe sinks through the water its thermistor sensor changes resistance in proportion to water temperature. This change is transmitted to a strip chart recorder through a fine wire connecting the probe and recorder. Twin de-reeling spools (one in the probe, another retained at the launcher) ensure that little stress is exerted to break the wire or slow the fall rate of the instrument. Probe depth is derived from timing, using a drop rate relation of the form

$$d = 6.472t - 0.00216t^2 \quad (1)$$

where d is depth in metres and t time in seconds. The second order time term accounts for the linear decrease of sinking velocity with time, which is caused by the steady wire weight loss in the probe while de-reeling (see Section 3.1 for additional comments on drop rate). The thermistor in the probe has a 0.1 s time constant, which when combined with Equation 1 yields a 65 cm depth resolution. The two most standard probes provide temperature data down to 450 m and 700 m, respectively.

2.2 AXBT Operational Characteristics

The airborne expendable bathythermograph (AN/SQQ-36) used by the U.S. Navy is an instrument dropped from an aircraft to obtain a temperature profile of the upper 305 m of the ocean (Fig. 1). After being launched from an aircraft, its rate of descent through the air is slowed by a spin-stabilizing rotochute (or, alternatively, a parachute). Upon striking the sea surface, a bottom plate and rotochute come off, an antenna deploys, and salt-water batteries are activated. About 5 s later a continuous unmodulated radio frequency carrier begins to be transmitted. Approximately 10 to 100 s later a sensor probe with a thermistor located on its lower edge is released from the floating AXBT housing. Simultaneously the carrier is modulated with an audio frequency proportional to the temperature sensed by the thermistor according to the following manufacturer-specified equation:

$$f = a + bT \tag{2}$$

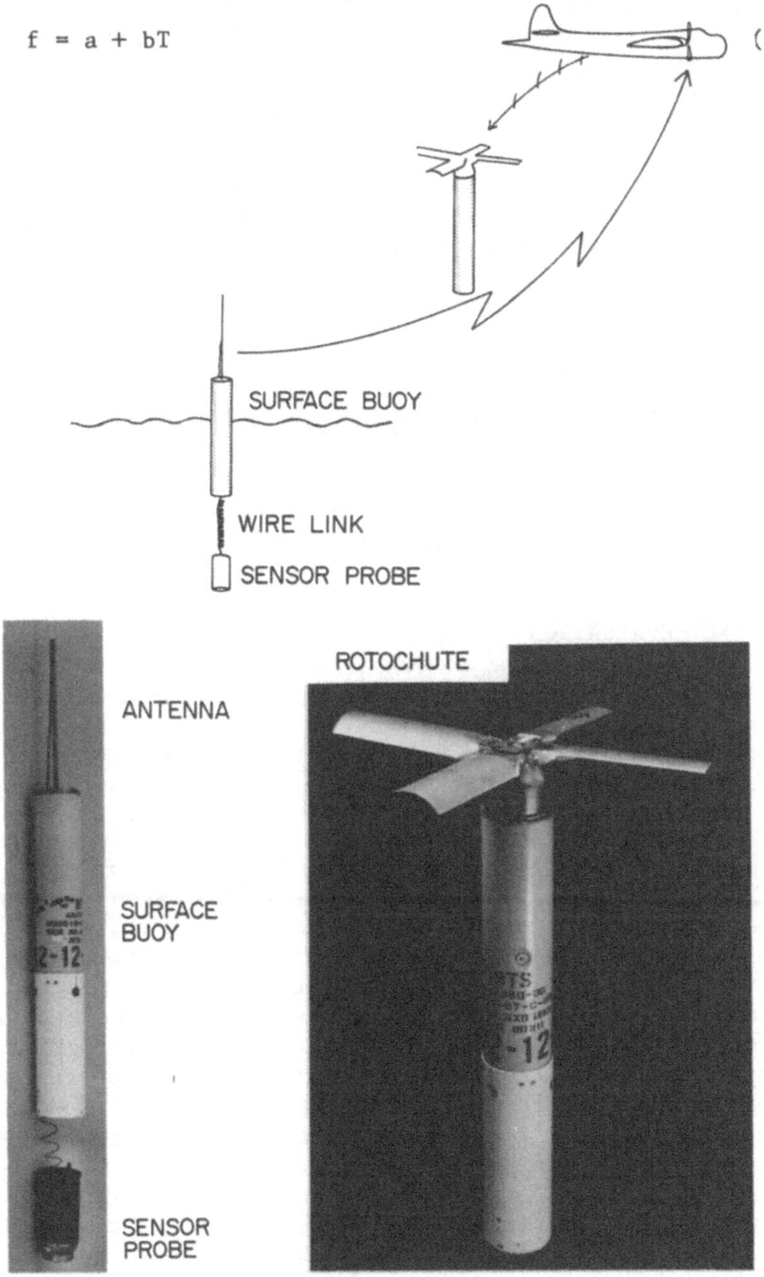

Fig. 1 The U.S. Navy airborne expendable bathythermograph (AXBT).

where f is the audio frequency in hertz (Hz), T is temperature in degrees Celsius, and 'a' and 'b' are constants. This audio frequency is generated in the probe and transmitted to the AXBT housing via a wire link. With a specified fall rate of 1.52 m s^{-1} for the probe, it takes about 200 s for the 300 m temperature profile to be completed. Approximately 7 min after striking the water the AXBT housing is flooded and the entire unit sinks.

Aboard the aircraft the modulated carrier is received and the demodulated audio frequency is converted into a dc voltage which is plotted on a strip-chart recorder versus time. The recording device used is not suitable for most scientific studies due to the poor temperature resolution (~0.5°C). More modern naval aircraft record the desired data in high resolution format (adequate for most scientific purposes) on analogue tape recorders.

Each AXBT is 92 cm in length, 12.5 cm in diameter, and 8.2 kg in weight. Performance specifications for the AXBT include a 95% reliability (95% of the units will provide a full 305 m temperature profile), a fall rate of 1.52 m s^{-1} ±5%, and a temperature accuracy of ±0.56°C over the range -2°C to +35°C (which also is the resolution of the strip-chart recorder). The newer AXBTs have potentially better temperature accuracy than the required ±0.56°C.

3. XBT/AXBT EVALUATION

3.1 XBT Evaluation: Advantages/Disadvantages/Characteristics

The primary advantages of the XBT system are low cost and its ability to operate from virtually any ship travelling at any speed, in any sea state.

The primary disadvantage of the system lies in its reliance on a dc signal path through a long, fine wire link. When problems arise they are usually from one of three sources. First, since the wire link depends on a sea-water ground return path, improper grounding of the recorder can lead to apparent malfunctions. Second, if the wire should rub against some object, such as the side of the ship while de-reeling, its enamel insulation may be worn off, inducing what appear to be warm spikes or bumps in the temperature record. This problem also occurs if the insulation is faulty for other reasons, thereby leading to an 'apparently' good temperature record ... but one that is too warm at depth. Third, if the wire should temporarily 'hang up' on some object and thereby stretch a little before freeing itself, this also induces an apparent warming in the record. The effects of insulation wear and wire stretch can be difficult to detect in oceanic regions containing natural temperature inversions. A fourth, more minor, disadvantage is the system susceptibility to the ship's radio interference.

The XBT system overall accuracies are quoted by one manufacturer at ±2% of depth (or 15 m, whichever is greater). One investigation by R. Scarlett (personal communication) found that the ±0.2°C temperature error could be reduced to ±0.1°C by improving the recorder system. The ±2% drop rate error is quoted by the manufacturer as a conservative figure. Recent experimental work indicates that this drop rate error could be halved by taking into account the variable viscosity of sea water, which is mainly a function of temperature (Sippican, 1976).

The above results may be contrasted with those obtained through 'at-sea' experience. Work by McDowell (1977) has identified a previously unrecognized systematic error in the drop rate (Eq. 1) below 400 m. However, when this is accounted for, XBT agreement with an STD comparison standard was within 0.1°C and 1% of depth. McDowell utilized XBT probes of recent manufacture.

Much more serious is the recent evidence that systematic errors occur in batches of the older probes (e.g., Dugan and Schuetz, 1977; Barnett et al., 1977). The errors may be traced to distributed electrical leakage through the wire insulation. The result is a temperature trace which is visually acceptable yet turns out to have a 0.5 to 1.0°C, and sometimes much greater, warm bias relative to a standard of comparison. It is important to point out that 300,000 of these older probes are available, principally through a stockpile of the U.S. Navy. A study in 1977 by the Navy has identified one particular production run as the principal offender with this insulation problem. It has been eliminated from the Navy distribution system. The problem occurs in less than 5% of all probes from other production runs. Furthermore, the manufacturer has made changes in the method of insulating the wire since these older probes were built, so this problem does not occur in probes of recent manufacture.

The scientific user is well advised to be cautious if he uses the older probes, as the quoted accuracies (0.2°C, 2% depth) may not be achieved with them. On the other hand, when working with recent probes and by exercising care it may be possible to achieve accuracies several times better than those quoted.

3.2 AXBT Evaluation: Advantages/Disadvantages/Characteristics

Perhaps the greatest virtue of the AXBT, besides its relatively low cost, is that it may be deployed from an aircraft thus offering the oceanographer a truly synoptic picture of large regions of the ocean. The relative cost per mile of each observation making up the synoptic picture is lower than if the data were obtained from, say, a research vessel. Finally, the deploying aircraft can gain useful information from AXBTs in almost any sea state.

A major disadvantage of the AXBT system is that it must be deployed in a somewhat restricted altitude/air speed 'window' (<10,000 ft and/or 250 knots indicated air speed). These constraints may not allow the deploying aircraft to operate in a most efficient manner, thus limiting its effective range and unnecessarily increasing the length of the data collection operation. Another disadvantage is that the devices return temperature information on only the upper 300 m of the ocean. For many studies, data to considerably greater depth may be required. Finally, as we shall see below, AXBTs made by different manufacturers may vary widely in quality. In some cases, it may be necessary to individually calibrate each instrument in order to obtain temperature data reliable to ±0.2°C.

The general characteristics of the AXBT and its component parts have been intensively studied (Sessions et al., 1974; Sessions et al., 1976). Here we briefly summarize the principal results of these two publications and add a new result on the fall rate of the sensor probe.

TABLE 1

Results of AXBT Temperature Calibrations

Manufacturer	T_r	Δ	Standard Deviation
Motorola	8°C	0.55°C	0.24°C
"	10	-0.37	0.45
"	14	-0.29	0.45
"	18	0.60	0.70
Magnavox	0	-0.10	0.14
"	10	-0.37	0.11
"	25	0.23	0.4

Δ is the mean AXBT deviation obtained by subtracting the reference bath temperature (T_r) from the temperature obtained from the manufacturer's standard curve (Eq. 1). Data from 291 Motorola and 59 Magnavox units were used to develop this table.

- Calibration - The AXBTs are constructed to a standard manufacturers' specification so that a single calibration curve is supposed to relate output frequency (f) to temperature (T). The consistency of this curve has been checked for older units made by Motorola and newer units provided by Magnavox. The results are shown in Table 1. The Motorola units varied so widely about the 'standard' calibration equation that it was necessary to individually calibrate each unit prior to deployment to ensure adequate accuracy. The Magnavox devices, on the other hand, were homogeneous enough to be represented by a 'universal' calibration curve (see below).

- Linearity - The linearity of each of the above systems between 5° and 20°C was investigated. The Motorola system, after adjustment of the calibration constants for each instrument, was found to follow Equation 2 to within ±0.1°C. The Magnavox instruments followed a slightly modified calibration curve over the same interval and beyond (to 0°C and 25°C) to ±0.15°C. The modified curve fit to all Magnavox systems tested is given by

$$T = a' + b'f + c'f^2$$

where

$$a' = -45.11°C, \quad b' = 0.03381°C\ Hz^{-1},$$

$$c' = -1.676 \times 10^{-6}\ °C\ Hz^{-2} \qquad (3)$$

- Temporal Response - Dynamic calibrations were conducted by placing a (calibrated) Motorola AXBT in a reference temperature bath and rapidly decreasing the bath temperature. After a 168 s simulated fall time the two temperatures differed by -0.6°C. This temperature difference can be converted into estimates of isotherm depth error provided one knows something about the vertical temperature gradient. In the central Pacific the error in isotherm depth at 200 m would be approximately 40 m.

  More extensive tests (M.H. Sessions, personal communication, 1978, Scripps) of the Magnavox AXBTs indicate they are a 'two-time constant' system with characteristic time of 2 and 15 s. The relatively slow instrument response can induce large errors (~2°C) in regions of high vertical temperature gradient, i.e., the bottom of the mixed layer.

- Fall Rate - AXBTs have specified fall rates of 1.52 m s^{-1}, a number supposedly accurate to ±5%. The uncertainty of this number is of crucial importance since it is used in conjunction with time to determine the depth that a reported temperature represents. In a recent experiment,

Magnavox AXBTs have been equipped with pressure transducers and the fall rate actually measured. Care was taken in the fitting of the pressure transducers so that the weight, moments of inertia, etc., of the AXBT were not significantly altered. The average fall rate observed for 10 instruments was 1.58 m s^{-1}, a value that was independent of depth. This latter fact suggests that the designed neutrally buoyant character of the wire does negate the need for a second order correction term (cf. Eq. 1). Also, viscous effects were within the experimental scatter of the data. Thus observed values are consistent but slightly biased with respect to design specifications.

- Voltage Sensitivity - The power supplied to the system by the salt-water battery was found to fluctuate by up to 0.7 V during the course of a simulated drop. In the Motorola system this affected the primary oscillation circuit in the probe, inducing typical errors of order 0.2°C. The corresponding error in the Magnavox system was much less than 0.1°C.

4. EXAMPLES OF USE

During the last few years XBT and AXBT systems have played an increasingly important role in oceanographic studies. We mention below four examples from the North Pacific, which happened to occur during the month of November 1975. Figure 2 displays the locations of BT lines that were occupied, which include one swath of XBT data in the western Pacific, six east-west XBT transects of the entire central North Pacific, two north-south AXBT transects between Hawaii and Alaska, and two Hawaii-west coast U.S. XBT sections. The swath was conducted by Wilson and Dugan (1978) utilizing seven U.S. Navy vessels sailing in parallel with 40 km track separation. The east-west transects were carried out by special observers aboard American President Lines freighters (Bernstein and White, 1977). The north-south transects were done by a U.S. Navy P-3C aircraft (Barnett, 1976). The Hawaii-west coast sections were performed by ship's personnel aboard Matson line freighters (Saur, 1975).

Recalling our remarks in the introduction on the shifting emphasis by oceanographers toward the study of larger and larger regional ocean variability, it is clear from Figure 2 that XBT/AXBT systems can provide very cost-effective means for collecting the necessary data.

To illustrate this point we mention the distribution of data collected during July, August, and September 1955 by the NORPAC program (NORPAC Atlas, 1960). Here, 20 ships occupied 1002 hydrographic stations across the North Pacific, an effort that required roughly 20 ship months and would consume approximately 2.5 million

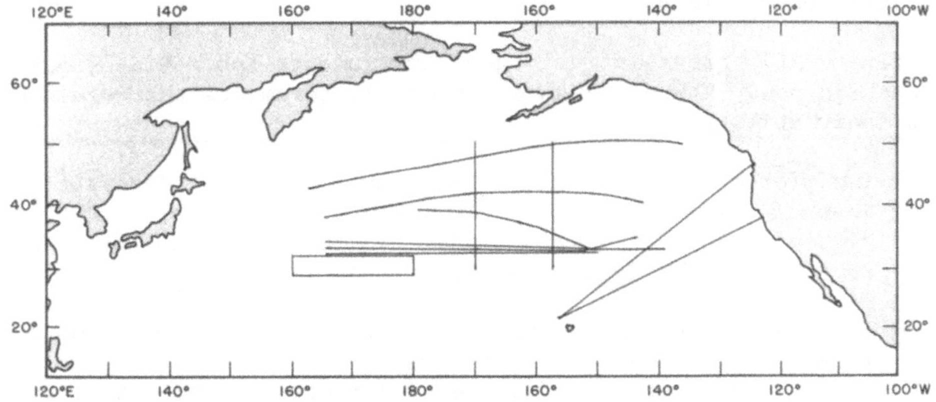

Fig. 2 Expendable BT sections in the central North Pacific Ocean, taken during November 1975.

research dollars at present rates. In comparison, by using XBT and AXBT systems on platforms of opportunity, elements of the NORPAX program during three months of 1976 collected about 1200 BT stations at an estimated cost of $50,000. While the comparison of the two programs is somewhat unfair (differing program objectives, hydrographic versus BT data, etc.) the striking cost ratios do make the point that expendable systems can make important contributions for future large area studies.

5. IMPROVEMENTS IN XBTs/AXBTs

5.1 Improvements in XBTs

There are a few areas where the present XBT system could be significantly upgraded. Most of the problems with the system arise in the dc signal path through the wire link, as discussed in Section 3.1. Yet it is this link which allows low cost and wide sea-state operating ranges. To retain the wire link while upgrading the signal quality might require some microelectronic circuitry in the probe to permit a digital data transmission mode up the wire to the recorder. Developments in microelectronic chip circuitry make such a development economically feasible today. Digital transmission schemes tend to be inherently more noise-free than any others, and might avoid effects of wire stretch and insulation abrasion.

Other areas for improvement include a new generation of recorders developed to match a digital probe, incorporating solid-state circuits, and digital cassette recording. It would also be desirable to conduct a detailed drop rate calibration through engineering tests.

5.2 Improvements in AXBTs

The newer AXBTs are relatively good instruments for a wide range of scientific use. There are some obvious improvements that would extend their utility.

- The probes presently are capable of measuring temperatures down to only 300 m. An improved probe, perhaps using a system similar to that of the normal XBT, could be added to increase the depth capability substantially.

- The present operating window for the AXBTs is not compatible with the optimum operating window of long-range aircraft. Thus surveys of large and/or distant regions of the ocean cannot be performed in an optimum manner. The AXBTs should be modified so that they may be dropped from higher altitudes and at higher air speeds.

- Scientific use of the AXBT requires substantiation of its accuracy. In the past this has meant that the individual scientist must conduct a calibration program to ensure that the probes are returning data as advertised. With a small amount of effort at the manufacturer's assembly line, suitable calibration procedures could be instituted such that the AXBT reliability could be clearly substantiated.

- The present failure rate of AXBTs dropped by standard naval procedures and personnel under Scripps supervision is 11 to 15% (the advertised failure rate is 4%). This number is based on a sample of nearly 2000 drops. This reliability factor is of crucial importance when a series of stations is to be occupied. Failure of the system at one station then requires that the aircraft return to that station to redrop another BT. This is time-consuming and can represent a severe limit on the plane operating range. In terms of efficiency and actual cost, the reliability of the AXBT should be improved.

- One could envision a number of 'local' studies that would use AXBTs. Such studies might be conducted from relatively small aircraft that have a limited payload. The present size and weight of the standard AXBT system would preclude their extensive use in such studies. It seems advisable to consider the development of a 'mini' AXBT that would be suited to deployment from small aircraft that are less expensive to operate.

6. CONCLUSION

The increasing scientific interest in the study of large scale ocean variability has been partially responsible for the development of a first generation of expendable devices for measuring ocean temperature. These devices have proved themselves to be reliable, cost-effective, and, with proper use, capable of returning a picture of ocean structure that has heretofore been impossible to obtain.

In the future, further advancements in the development of expendable measuring devices could revolutionize the field of oceanography. For instance, it is not farfetched to envision a standardized expendable unit capable of returning one or more signals to either the deploying ship or aircraft. The standardized base unit could have accompanying it a suite of sensor transducer elements of plug-in, modular type construction. These transducers could provide a variety of measurements by sensing not just temperature but other oceanic parameters, such as conductivity, pH or oxygen. Most of the transducers exist to make these additional measurements. With regard to velocity, Sanford and Drever at Woods Hole (personal communication) are currently investigating the feasibility of linking an electromagnetic probe capable of measuring velocity shear with the standard XBT sytem.

The development of the above expendable systems seems an attractive alternative to the high (and rising) costs of traditional oceanographic measurement methods. Ships-of-opportunity, i.e., those that cost nothing to the scientific user, offer tremendous platform savings that make even the cost of several hundred dollars per expendable probe seem economically sensible. By the same token, a single, dedicated research aircraft could perform a large portion of the standard oceanographic surveys presently carried out by an entire fleet of research vessels.

REFERENCES

BARNETT, T.P. 1976. Large scale variations of the temperature field in the North Pacific Ocean. Naval Research Reviews, XXIX(3), U.S. Department of the Navy, Office of Naval Research, Arlington, VA. 22217: 36-51.

BARNETT, T.P., R.A. KNOX and R.A. WELLER. 1977. Space/time structure of the near-surface temperature field during the NORPAX POLE experiment. *Journal of Physical Oceanography*, 7: 572-579.

BERNSTEIN, R.L. and W.B. WHITE. 1977. Zonal variability in the distribution of eddy energy in the mid-latitude North Pacific Ocean. *Journal of Physical Oceanography*, 7: 123-126.

DUGAN, J.T. and A.F. SCHUETZ. 1977. Subtle T4 XBT malfunction. NRL Memorandum Report 3612, U.S. Naval Research Laboratory, Washington, D.C. 20375, 30 pp.

McDOWELL, S. 1977. A note on XBT accuracy. POLYMODE News, 29, unpublished MS, Woods Hole Oceanographic Institution, Woods Hole, Mass. 02543, pp. 1, 4.

NORPAC Atlas. 1960. *Oceanic observations of the Pacific, 1955.* University of California Press, Berkeley, California, 123 plates.

SAUR, J.F.T. 1975. Changes in position and character of the transition zone between Hawaii and California, inferred from surface salinity gradients, 1972-75 (abstract). EOS, 56: 1007.

SESSIONS, M.H., T.P. BARNETT and W.S. WILSON. 1976. The airborne expendable bathythermograph. *Deep-Sea Research,* 23: 779-782.

SESSIONS, M.H., W.R. BRYAN and T.P. BARNETT. 1974. AXBT calibration and operation for NORPAX POLE experiment. SIO Reference Series 74-31, Scripps Institution of Oceanography, La Jolla, California, 27 pp.

SIPPICAN CORPORATION. 1976. The effect of manufacturing tolerances, fluid density and viscosity variations on XBT depth accuracy. Sippican Internal Report R-752. Sippican Corporation, Barnabas Road, Marion, Mass. 02738, U.S.A., 38 pp.

WILSON, W.S. and J.P. DUGAN. 1978. Mesoscale thermal variability in the vicinity of the Kuroshio Extension. *Journal of Physical Oceanography,* 8: 537-540.

21

Slow-Response Humidity Sensors

M. Coantic and C.A. Friehe

1. INTRODUCTION

In addition to specifying the general state of the air mass, the measurement of the mean humidity distribution in the boundary layer over the sea is used to determine the rate of evaporation of water from the sea surface, a variable of basic importance in air-sea interaction studies. The large heat of vaporization of water plays important roles in the earth's oceans and atmosphere by cooling the upper ocean layers during evaporation and heating the atmosphere when water vapor condenses. The presence of water vapor in the atmosphere also causes refractive index effects on the propagation of acoustic, microwave, and optical radiation. Humidity measurements are also needed for computing atmospheric stability, radiation flux divergences, forecasting fog, etc.

Many different units are used to specify the water vapor content of air (see e.g. List, 1949); here we shall primarily use absolute humidity, ρ_w (mass/unit volume), or specific humidity, q (mass of water/mass of moist air), as they arise naturally when considering the turbulent flux of water vapor from the sea, Q (mass/area x time). By definition:

$$Q = \overline{\rho_w' w'} \simeq \rho \, \overline{q' w'} \qquad (1)$$

where w' is the instantaneous vertical velocity and ρ is the average air density. Fast response q' instrumentation for direct covariance measurements is discussed by Hay (see Chapter 22). We consider here techniques whereby Q can be obtained from simpler measurements of the mean distributions of humidity and other variables. Two such methods are currently used: the profile method (e.g. Paulson et al., 1972; Dunckel et al., 1974), and the bulk transfer coefficient method. For the evaporation flux, the latter is described by:

$$Q = \rho \, C_r \, U_{10} \, (q_s - q_{10}) \tag{2}$$

where U_{10} and q_{10} are the average 10 m wind speed and specific humidity, respectively, and q_s is the surface specific humidity (obtained by assuming saturation at the sea surface temperature). The empirical coefficient C_r is about 1.3×10^{-3} (e.g. Kondo, 1975; Friehe and Schmitt, 1976).

The average flux of water vapor over the oceans is about 2.5×10^{-5} kg m^{-2} s^{-1} (~ 0.8 m yr^{-1}; Monin, 1972). An average humidity difference of about 1.7×10^{-3} kg kg^{-1} is obtained by applying the bulk formula, Equation 2, and the assumption of a 10 m wind speed of 10 m s^{-1}. The upper limit of the air humidity is between 4×10^{-3} and 27×10^{-3} kg kg^{-1}, corresponding to saturation at temperatures from 0 to 30°C. In the Arctic or near arid land regions the air can be as dry as approximately 5×10^{-4} kg kg^{-1}.

In general, humidity measurements are required over a wide range of environmental conditions: air temperatures from below freezing to 30°C, wind speeds to 50 m s^{-1}, intense solar radiation, and the usual motion, vibration and electrical noise fields if the instrument is used on a ship, airplane or buoy. Also, a humidity instrument should be protected from rain, ice, fog, and salt spray, and should recover from exposure to saturated conditions without loss of calibration.

An estimate of the accuracy required in the measurement of q_{10} of $\pm 4 \times 10^{-4}$ kg kg^{-1} is obtained by accepting a 20% error in the estimation of Q, neglecting errors in U_{10}, and assuming the error in q_{10} is uncorrelated with that in q_s. This is a very high accuracy requirement, and it is not met by most humidity sensors, especially when used over the sea. For the profile method, relative accuracy requirements are even more stringent, but they may possibly be overcome by restricting measurements to large evaporation rates, and by cycling a reference humidity calibration instrument over a range of different heights. To obtain the mean value of the fluctuating humidity field, averaging times of about 10 to 15 minutes should be used. This may be achieved by using very slow response sensors, or filtering or averaging the outputs of fast response sensors.

2. EXPERIMENTAL TECHNIQUES

A large number of different techniques exist to measure humidity in the atmosphere. For experiments over the ocean, we will consider methods based upon psychrometry, dew-point measurement, equilibrium with hygroscopic materials, and the use of electromagnetic radiation. For detailed reviews of humidity instrumentation, see Wexler (1965 and 1970).

2.1 Psychrometers

The psychrometer is extensively used in meteorology because of its simplicity and low cost. It consists of two identical ventilated temperature sensors, one covered with a porous coating, or 'wick', which is saturated with pure water.

The saturation specific humidity q_w at the surface of the wick, which is at temperature T_w, is larger than the specific humidity q of the incoming air. Therefore, evaporation and cooling of the wetted thermometer occurs, so that the 'wet bulb' temperature is finally less than the 'dry bulb' temperature, T. The idealized psychrometer theory assumes (Wexler, 1965) an isenthalpic process so that one can write:

$$L_v(q_w - q) + c_p(T_w - T) = 0 \qquad (3)$$

where L_v is the latent heat of vaporization and c_p the specific heat capacity of air.

A physically more realistic derivation of a psychrometer equation is to equate the amount of heat required for evaporation from the wet bulb, H, with that provided by convection from the airstream:

$$H = L_v\, C_r\, \rho\, U(q_w - q) = c_p\, C_\theta\, \rho\, U(T - T_w) \qquad (4)$$

where U = free-stream velocity. This gives

$$q = q_w - \frac{C_\theta}{C_r}\frac{c_p}{L_v}(T - T_w) \qquad (5)$$

where C_θ and C_r are heat and mass exchange coefficients between the wet wick and the airflow; for a given geometry their ratio is known to depend on the air velocity and 'Lewis' number (ratio of thermal to mass diffusivity).

Semi-empirical turbulent flow heat and mass transfer formulations give the dependence of the ratio of coefficients on the Lewis number as the Lewis number to the 2/3 power. For the water-air system the Lewis number is 0.81, so that the coefficients are predicted to be nearly equal and Equation 3 is a reasonable approximation for an air-water psychrometer.

For a truly realistic analysis of psychrometer accuracy, a number of additional temperature differences and parasitic heat sources must be considered. Therefore, the psychrometer Equation 5 is generally written

$$q = q_w - A'(T - T_w) \tag{6}$$

where $A = A' M_a/M_v$ is the 'psychrometric constant', determined by calibration (e.g., List, 1949), and where M_a and M_v are the molecular weights of air and water vapor respectively.

The value of A from Equation 3 is 6.5×10^{-4} °C^{-1}; experiments cited by Harrison in Wexler (1965) gave $A = 10 \times 10^{-4}$ to 6.5×10^{-4} °C^{-1} as the ventilation wind speed was increased from 0.2 to 6 m s^{-1}. It is generally recommended that the ventilation wind speed be at least 3 m s^{-1} for thermometer-element psychrometers. For small thermocouple elements, Harrison cites measurements which gave $A = 6.15 \times 10^{-4}$ °C^{-1} to 6.24×10^{-4} °C^{-1} for ventilation speeds of 0.1 to 3 m s^{-1}.

Psychrometric instruments using thermometers range from the well known inexpensive sling psychrometer to fan-aspirated, radiation-shielded units such as the Assmann type. Practical considerations include the need for adequate ventilation velocity, proper shielding from solar radiation (see Deacon, Chapter 14), precise and accurate thermometry since small temperature differences are encountered, and care in exposing the thermometers for a sufficient time for equilibrium to be reached. Research applications have used small thermocouples or thermistors for temperature sensors so that radiation effects are lessened and continuous electrical output signals for mean and fluctuating humidity are obtained (Paulson et al., 1972; Polavarapu and Munn, 1967).

According to Wexler (1970), the best absolute accuracy for psychrometers is ±1% in relative humidity, i.e., ±4 to 25×10^{-5} kg kg^{-1}. Some authors claim relative accuracies of 1×10^{-5} kg kg^{-1}, so that mean vertical gradient measurements are in principle possible (Collins, 1965; Seck and Perrier, 1970).

A major source of error at sea, in addition to failure to observe the considerations given above, is the contamination of the wetted wick by salt. Salt will lower the vapor pressure or the specific humidity q_w at the surface of the wick (Raoult's law) and consequently the wet bulb temperature will be higher for equilibrium to be achieved. Woolcock (1949) investigated this effect for several salts (except NaCl) and found significant errors. If the salt concentration is high, the wet bulb temperature can actually exceed the dry bulb temperature. There are also indications that salt deposition on dry bulb thermometers can cause erroneous readings due to changed radiational and thermal properties (Holmes and O'Brien, 1975; Larsen et al., see Chapter 15).

2.2 Dew-Point Instruments

Dew-point hygrometry is, in principle, an absolute method since, at a given pressure, liquid (or solid) and gaseous water can be in equilibrium only at the 'saturation' temperature given by the Clausius-Clapeyron equation. Partial water vapor pressure being thus known from the temperature of a dew (or frost) layer on a surface in contact with air, absolute or specific humidity can be obtained from a simultaneous measurement of air temperature or pressure. For surface temperatures below 0°C, as encountered in the Arctic, the possible occurrence of either frost or subcooled liquid water can, however, lead to some ambiguity.

In micrometeorological instruments, the condensation surface is a small polished metal membrane or 'mirror,' cooled by a Peltier-effect element. The presence of dew or frost is detected by a variety of techniques (Wexler, 1965): the optical method is widely used, wherein the ratio of scattered to reflected light from the mirror surface is measured. An appropriate feedback circuit controls the surface temperature, through the Peltier current or by means of additional heating, to maintain a constant small dew layer thickness (see Fig. 1). The surface temperature is measured with a thermocouple, thermistor, or platinum thermometer. The device is generally housed in an aspirated radiation shield, and an air temperature sensor is often included.

Sources of, and procedures for, reducing errors are discussed in the literature (Wexler, 1965):

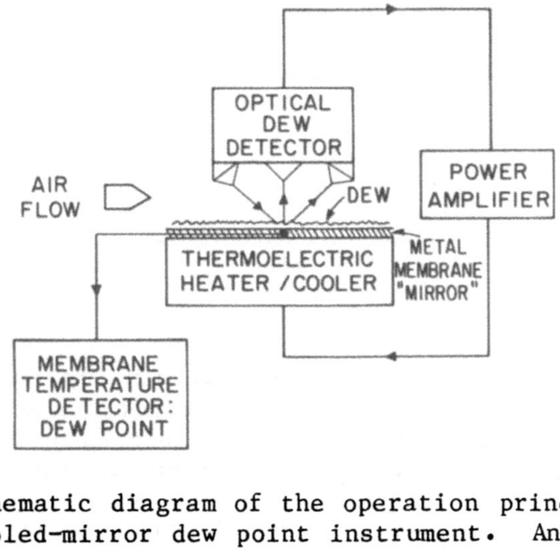

Fig. 1 Schematic diagram of the operation principle of the cooled-mirror dew point instrument. An ambient air temperature sensor is usually included.

- Differences between the temperature of the dew deposit and the true dew-point of air are due to the Kelvin effect (droplet curvature reduces the measured dew-point) and to the Raoult effect (soluble contaminants within the dew increase it). Relatively thick dew layers, and adequate airflow paths and screens, have to be used.

- Uncertainties arise from the detection procedure (mainly because of deposition of dirt or salt over the mirror) and from the finite sensitivity and stability of the feedback system. Efficient air cleaning and adjustment of the feedback loop and the airflow rate are necessary.

- Errors in the temperature measurement result from temperature gradients and calibration uncertainties in the sensor and read-

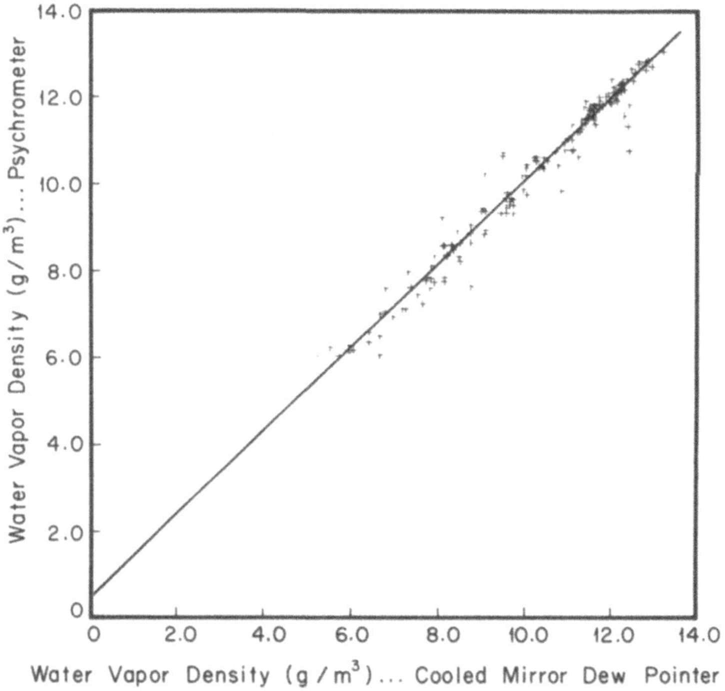

Fig. 2 Comparison of water vapour density obtained with a precision cooled-mirror dew point instrument and an air aspirated psychrometer obtained at sea (Friehe and Schmitt). The equation of the least-squares line is Y (psychrometer) = 2.6×10^{-4} kg m^{-3} + 0.972x (cooled-mirror dew point) with a standard deviation of 3.6×10^{-4} kg m^{-3}.

out system: adequate design and calibration procedures are needed.

Absolute accuracies to ±0.2°C are quoted for research-grade instruments. A large source of error arises from temperature sensor calibration, and relative accuracies better than ±0.05°C are attained in the laboratory. In terms of specific humidity for the 0 to 30°C saturation range, this leads to figures of $(5\text{-}30)\times10^{-5}$ and $(1.2\text{-}7.5)\times10^{-5}$ kg kg^{-1}, respectively. Over the ocean, salt spray contamination is naturally the major source of error.

In Figure 2 we present about 150 simultaneous observations of absolute humidity, made with care under open ocean conditions with a shielded, aspirated thermometer psychrometer and a research-grade dew-point unit. (The measurements were made with the assistance of K.F. Schmitt, UCSD, from the R/P FLIP in the North Pacific.) A least-squares fit of the data gave a correlation coefficient of 0.97 and a standard deviation of 3.6×10^{-4} kg m^{-3} or about 3×10^{-4} kg kg^{-1}, approximately equal to the intercept term. The agreement indicates that accuracies of the order specified for flux estimations from the bulk transfer formula can be obtained under field conditions with careful psychrometer measurements.

2.3 Hygroscopic Devices

In addition to the classical hair hygrometer, there exists a wide variety of devices which make use of various materials that undergo physical-chemical reactions when the humidity of the surrounding air changes, in such a way that one of their physical properties can be calibrated in terms of moisture and, in general, temperature. We shall cite some of the most widely used.

The lithium chloride hygrometer, invented by Dunmore in 1938 (Wexler, 1965), is still one of the most widely used instruments. An aqueous solution of lithium chloride, dispersed in a plastic binder, is applied to a dielectric substrate. Changes in relative humidity produce corresponding changes in the concentration of the solution which, in turn, result in substantial changes in resistance. Resistance also varies with temperature, so that a simultaneous temperature measurement is needed. The Dunmore hygrometer, the standard radiosonde sensor for many years, has been the object of much developmental work leading to a well proven method.

The carbon element, now the standard U.S. radiosonde sensor, makes use of the changes in size with humidity of hydroxyethylcellulose (mixed with powdered carbon), which results in changes in the resistance of a thin sensitive layer deposited on an inert substrate. The carbon 'Hygristor' is now a well proven instrument, but is also temperature sensitive (Morrissey and Brousaides, 1970). Numerous chemical film coatings undergo resistance changes as a function of moisture, and have been used for hygrometry: barium fluoride, potassium chloride, potassium metaphosphate, styrene copolymer.

Salt solution phase-transition hygrometers are based on the fact that the saturation relative humidity over a saturated salt solution depends only on temperature. If the temperature of such a solution is above that corresponding to the ambient air humidity, evaporation, and therefore drying, will occur, and conversely wetting will be observed for lower temperatures. Because the transition from the liquid to the solid phase considerably increases electrical resistivity, a self-regulating instrument can be realized; by simply applying a given voltage between two electrodes (Viton, 1970). Lithium chloride is currently used for relative humidities in the range 11 to 100%, with an estimated accuracy around ±1.5°C dew-point. The instrument is often called a 'Dewcel'.

The aluminium oxide hygrometer operates by adsorption of water molecules within a highly porous oxide layer, which results in a change in the dielectric properties of the layer. Manufacturing is delicate: the porous oxide layer is obtained by anodization of a pure aluminium substrate, and has then to be covered by an outer electrode of sufficient porosity to allow water molecules to travel in and out of the oxide layer. This outer conducting layer is generally obtained by vacuum deposition of an aluminum or gold layer, and its electrical connection is difficult (Guizouarn, 1970; Hasegawa et al., 1974).

In the piezoelectric sorption hygrometer, the weight of water sorbed by an hygroscopic coating is sensed by the change in the oscillating frequency of a crystal (Gjessing et al., 1968).

The 'bulk effect' sensor, known as the 'Brady array', is a proprietary, humidity-sensitive, solid-state semiconductor device. Its advantages are small size and low power consumption. Calibration tests have revealed hysteresis effects, non-linearity, and temperature sensitivity (Little et al., 1974).

Most of these techniques are relatively cheap and easy to use. The time response is generally adequate for the measurement of mean values. Relative accuracy can be as good as ±2% relative humidity, $\pm 8\times10^{-5}$ to 50×10^{-5} kg kg^{-1}. There are problems, however, in making measurements with some of the hygroscopic devices when the relative humidity is high. Also, they suffer from a common failing: the adsorption-desorption mechanism is responsible for substantial hysteresis effects, and is severely affected by surface contamination. To reduce contamination effects, sensors can be put inside protective permeable membranes (see e.g. Viton, 1970), but only at the cost of increased time response and hysteresis. These methods are consequently of limited interest for fixed-point air-sea interaction measurements. But they are widely used for measuring profiles with expendable instrument packages such as the NCAR Dropsonde (Cole et al., 1973), when small weight and cost are essential factors. Carbon films and some other fast-response sorption sensors are also being used for humidity fluctuation measurements.

2.4 Optical and Microwave Methods

For local point measurements of average humidity, use of the averaged output of one of the fast-response, short-path remote instruments (microwave, infrared or Lŷman-alpha) as described by Hay (see Chapter 22) would have the advantage that salt spray in the path would not seriously degrade the measurement. Spray accumulation on the window or cavity surfaces would, however, have to be avoided or circumvented.

Remote sensing methods are generally in the development stage at present, but offer promise for two reasons: they are free from salt contamination and are capable of profiling humidity over large distances.

Little (1973) cites integrated water content measurements from passive microwave radiometry and profiling by means of two-wavelength lidars. Another promising although difficult method is laser-induced Raman scattering (Penner and Jerskey, 1973). Remote methods will develop raidly in the future in conjunction with satellite-based air-sea interaction studies.

3. DISCUSSION AND CONCLUSIONS

The above sections have indicated the need for humidity measurements near the ocean surface, estimated typical values and accuracy requirements, and described most of the available instruments. Humidity is one of the more difficult properties of the atmosphere to measure, and many of the instruments are not suitable for the marine environment. Salt spray contamination is the chief source of trouble in most of the devices.

For the non-marine environment, Wexler (1970) has compared the 'relative uncertainty' (estimated percentage maximum error plus three times the standard deviation) of many humidity instruments, and the results are presented in Figure 3. Only the dew point and microwave refractometer methods possess accuracies greater than 10^{-4} kg kg^{-1}. For dew points of -10 to +20°C, the relatively inexpensive methods, such as the lithium cell, treated hair, carbon, and saturated lithium chloride cells have uncertainties of 5 to 20%. McKay (1978) has reported results of long-term tests of several types of humidity sensors, also for the non-marine environment. Considerable deviations in dew point (up to 5°C) and relative humidity (15%) were found between various instruments and reference cooled-mirror and lithium chloride instruments. Thus the data of Figure 3 only apply to the performance of sensors under conditions where the instruments can be periodically maintained. Figure 3 should be used with caution, since only part of the necessary information is included. For example, problems of drift make the microwave refractometer useless as an absolute instrument although it is a valuable tool for fast response measurements (see Hay, Chapter 22). On the other hand, aspirated psychrometers, which have a stable calibration, have proved to be a useful instrument for profile measurements (e.g. Dunckel et al., 1974).

In view of the accuracy requirements, high-quality initial and subsequent calibrations are desirable for air-sea interaction studies. High accuracy absolute humidity chambers are available in some of the major laboratories for initial calibrations (Wexler, 1968), but are not portable for at-sea use. Small chambers using saturated salt solutions are suited for field use, but the accuracies are

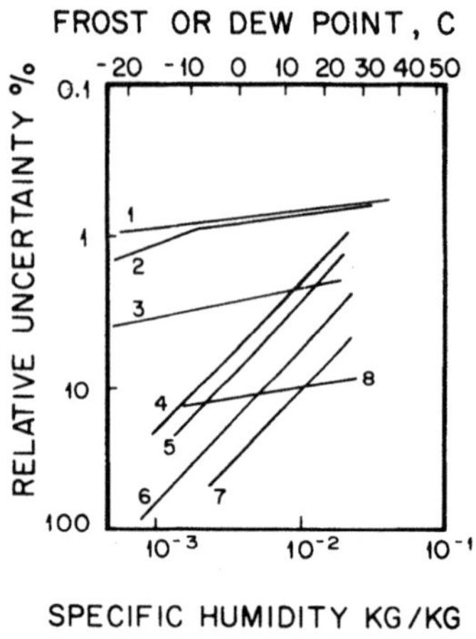

Fig. 3 Range and relative uncertainty of some of the hygrometric techniques discussed in this chapter (after Wexler, 1970) for a non-marine environment.

1. cooled mirror dew point
2. microwave refractometer
3. infrared absorption meter
4. aspirated psychrometer (25°C)
5. lithium chloride 'Dunmore' cell (25°C)
6. treated hair (25°C)
7. carbon or bariumfluoride cell (25°C)
8. saturated LiCl 'Dewcel'

low. A recommended procedure is to use a well calibrated and maintained accurate instrument, such as a cooled-mirror dew point system, as a transfer standard in the field. Aspirated psychrometers are another alternative; they need only a clean wick and a temperature calibration.

In conclusion, the choice of a humidity instrument will depend upon the accuracy desired, installation conditions, and funds available. For continuous measurements, a properly attended cooled-mirror dew point system is superior to the other methods. Lower-cost saturated lithium chloride cells can also be used if accuracy requirements are not high. For spot observations, a radiation shielded, aspirated, properly used psychrometer can provide reasonably accurate results, as shown in Figure 2. Finally, in all methods considerable effort will be required to maintain calibration accuracy, especially with regard to salt spray contamination.

ACKNOWLEDGEMENTS

We would like to thank Dr. A. Wexler, U.S. National Bureau of Standards, for his comments on this work. In France, investigations on humidity sensors at IMST have been supported by DRME, CNEXO, and CNRS. At UCSD, support was received from the North Pacific Experiment (NORPAX), through the Office of Naval Research and National Science Foundation.

REFERENCES

COLE, H.L., S. ROSSBY, and P.K. GOVIND. 1973. The N.C.A.R. wind-finding Dropsonde. *Atmospheric Technology*, No. 2.

COLLINS, B.G. 1965. An integrating temperature and humidity gradient recorder. In *Humidity and Moisture*, Edited by A. Wexler, 1: 83-94, Reinhold, New York.

DUNCKEL, M., L. HASSE, L. KRUGERMEYER, D. SCHRIEVER, and J. WUCKNITZ. 1974. Turbulent fluxes of momentum, heat and water vapor in the atmospheric surface layer at sea during ATEX. *Boundary-Layer Meteorology*, 6: 81-106.

FRIEHE, C.A., and K.F. SCHMITT. 1976. Parameterization of air-sea interface fluxes of sensible heat and moisture by the bulk aerodynamic formulas. *Journal of Physical Oceanography*, 6: 801-809.

GJESSING, D.T., C. HOLM, T. LANES, and A. TANGERUD. 1968. A simple instrument for the measurement of small-scale structure of temperature and humidity, and hence also the refractive index, in the troposphere. *Journal of Scientific Instruments, Ser. 2*, 1: 107-112.

GUIZOUARN, L. 1970. Hygromètre à oxyde d'aluminium, In: *Techniques d'étude des facteurs physiques de la biosphère.* Institut National de la Recherche Agronomique, Paris: 235-241.

HASEGAWA, S., L. GREENSPAN, J.W. LITTLE, and A. WEXLER. 1973. A laboratory study of some performance characteristics of an aluminium oxide humidity sensor, Technical Note No. 824. U.S. National Bureau of Standards, Washington, D.C.

HOLMES, J.F., and J.F. O'BRIEN. 1975. Marine sensors and their interface with the environment. *Instrument Society of America,* 15 (4): 76-81.

KONDO, J. 1975. Air-sea bulk transfer coefficients in diabatic conditions. *Boundary-Layer Meteorology,* 9: 91-112.

LIST, R.J. 1949. *Smithsonian Meteorological Tables,* Smithsonian Institution Press, Washington, D.C., 527 pp.

LITTLE, C.G. 1973. Remote sensing of the atmosphere. *Atmospheric Technology,* 2: 51-56.

LITTLE, J.W., S. HASEGAWA, and L. GREENSPAN. 1973. Performance characteristics of a 'bulk effect' humidity sensor. Report NBSIR 74-477; U.S. National Bureau of Standards, Washington, D.C.

McKAY, D.J. 1978. A sad look at commercial humidity sensors for meteorological applications. Preprint Volume, Fourth Symposium on Meteorological Observations and Instrumentation, American Meteorological Society, Denver, Colorado, 10-14 April 1978: 7-14.

MONIN, A.S. 1972. *Weather forecasting as a problem in physics.* M.I.T. Press, Cambridge, Mass.

MORRISSEY, J.F., and F.J. BROUSAIDES. 1970. Temperature-induced errors in the ML-476 humidity data. *Journal of Applied Meteorology,* 9 (5): 805-808.

PAULSON, C.A., E. LEAVITT, and R.G. FLEAGLE. 1972. Air-sea transfer of momentum heat and water determined from profile measurements during BOMEX. *Journal of Physical Oceanography,* 2: 487-497.

PENNER, S.S., and T. JERSKEY. 1973. Use of lasers for local measurements of velocity components, species densities, and temperatures. *Annual Review of Fluid Mechanics:* 9-30.

POLOVARAPU, R.J., and R.E. MUNN. 1967. Direct measurement of vapour pressure fluctuations and gradients. *Journal of Applied Meteorology,* 6: 699-706.

SECK, M., and A. PERRIER. 1970. Description d'un psychromètre à thermocouples. Son application à la mesure des gradients d'humidité. In: *Techniques d'étude des facteurs physiques de la biosphère.* Institut National de la Recherche Agronomique, Paris: 223-234.

VITON, P. 1970. Utilisation des capteurs hygrométriques à sorption thermocontrôlée. In: *Techniques d'étude des facteurs physiques de la biosphère.* Institut National de la Recherche Agronomique, Paris: 219-222.

WEXLER, A., Editor-in-Chief. 1965. *Humidity and moisture, measurement and control in science and industry.* Volumes I, II, III, IV, Reinhold, New York.

WEXLER, A. 1968. Calibration of humidity measuring instruments at the National Bureau of Standards. *Instrument Society of America Transactions,* 7: 356-362.

WEXLER, A. 1970. *Measurement of humidity in the free atmosphere near the surface of the earth.* Meteorological Monographs, Vol. 11, No. 30, American Meteorological Society, Boston, Mass: 262-282.

WOOLCOCK, M.E. 1949. Some possible sources of error in operating the Assmann psychrometer. *Instrument Practice,* 3: 13-17.

22

Fast-Response Humidity Sensors

D.R. Hay

1. INTRODUCTION

Changes in the environment above the earth's surface encompass a wide range of time- and length-scales. When averaged over a period of several years and over the surface of the earth, the mean air temperature and density decrease from 15°C and 1.23 kg m^{-3} at mean sea level to -56°C and 0.36 kg m^{-3} at altitude 11 km. The mean molecular weight remains constant at 29 g-mol^{-1}. Humidity is a highly variable component. A mean profile at mid-latitudes shows specific humidity decreasing from 6.2 to 0.027 g kg^{-1} over the same height interval. The changes that are superimposed upon this mean background extend in space from circulation on the global scale to molecular dissipation of eddy scales on the order of millimetres. The corresponding time-scales range from annual and seasonal variations to minute fluctuations occurring within a fraction of one second.

The present chapter will refer to the measurement of atmospheric changes near the small scale (or high wavenumber) end of this range of eddies. Specifically, it will deal with rapid response devices for measuring fluctuations in atmospheric humidity in eddies whose scale sizes are as small as a few hundred metres down to a fraction of one centimetre, and with response times as short as 10^{-3} s. The measurements here will apply to humidity in the gaseous state. Ideally, the technique should introduce no significant change in the observed medium, apply at temperatures above and below the freezing point, and be insensitive to water in the non-gaseous state.

Three techniques for humidity measurement will be reviewed. These involve the absorption of infrared radiation by water vapour (the infrared hygrometer), the absorption of ultraviolet radiation by water vapour (the Lyman-α hygrometer), and the simultaneous measurement of refractivity, pressure, and temperature fluctuations to yield information on humidity fluctuations (microwave refractometers - in this review, the term 'microwave' will be interpreted broadly to include radio wavelengths in the range 0.01 to 30 m).

Measurements of this type are important to three aspects of air-sea interaction. One is the study of vapour transport in both the vertical and horizontal. In practice, the 'eddy flux' technique is used, in which simultaneous measurements are made of humidity and velocity fluctuations with sensors responding to the same range of wavenumbers. The flux is derived through the covariance of these parameters; further details may be found, for instance, in Busch (1973).

A second application is in the ducted transmission of centimetre and millimetre radio waves over water. Although this phenomenon has been known for about 40 years, and the theory relating vertical gradients of humidity (and refractivity) to the ducting radio modes has been developed extensively, the prediction of ducting conditions over water and land-water boundaries remains a topic of current concern (see, for example, Group II papers in Zancla, 1973).

A third application is found in the study of the energetics of vapour turbulence in the air. This deals with turbulence spectra, the scale sizes of eddies and turbulence structure that transfer vapour most efficiently, and their dependence upon air stability (see, for example, Busch, 1973).

The remainder of this chapter includes the nature of humidity fluctuations in the environment, the requirements imposed by the environment upon the technique of humidity measurement, and progress in the design and application of the infrared hygrometer, the Lyman-α hygrometer, and microwave refractometers during the past decade.

2. THE NATURE OF HUMIDITY FLUCTUATIONS IN THE ENVIRONMENT

One of the earliest reports on humidity fluctuations at the wavenumbers of present interest was provided by Elagina in 1963. Since that time the observing techniques have been developed substantially, as noted below. While these investigations are far from complete, numerous measurements on stationary, shipborne, and airborne platforms have indicated the trends in humidity spectra. Figure 1 represents the main features of these spectra. At heights within a few hundred metres of the surface, the spectral estimates conform to the '-5/3 law' of Kolmogorov-Obukhov within an inertial subrange. The wavenumber dependence changes at the low wavenumber end of this subrange, where maximum contribution to humidity variance occurs. A low-frequency limit for the spectrum E(f) may be defined as the frequency where fE(f) falls to one-tenth of its maximum value. In the lower boundary layer over the sea it depends upon air stability and other meteorological conditions, with typical values around $3x10^{-3}$ in dimensionless frequency (f = nz/U). Here, n is frequency in hertz, z is height above the surface, and U is mean wind speed. There is some suggestion of a departure from the '-5/3 law' near the high wavenumber end of this subrange, beginning at scale sizes some ten times the Kolmogorov microscale (Coantic and Leducq, 1969; Miyake and McBean, 1970; Elagina and Koprov, 1971; Martin, 1971; Phelps and Pond, 1971; Bean et al., 1972; Krechmer et al., 1972; Martin, 1972; Bolgiano and Warhaft, 1973; Bean and Emmanuel, 1973; Smedman-Högström, 1973; Warhaft, 1973; Warner, 1973; Kruspe, 1974).

The spectrum of Figure 1 is placed in perspective by noting the magnitudes of the humidity fluctuations involved. It will be seen that the '-5/3 law' trend encompasses a range of 10^7 in spectral density, or a range of 3300 in variation of specific humidity. For an air temperature of 20°C, the maximum is saturation humidity at 15 g kg^{-1}; hence the minimum is around 0.005 g kg^{-1}. These limits decrease rapidly with decreasing air temperature.

There is evidence also of horizontal stratification of humid air within the general turbulence at altitudes exceeding a few tens of metres. This stratification may extend a few kilometres horizontally, but only a few metres or less in the vertical, with local enhancement in specific humidity of about 3 g kg^{-1}. These layers

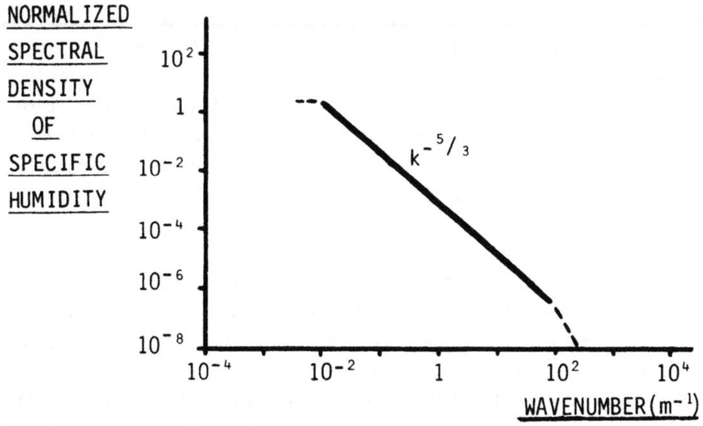

Fig. 1 Spectral characteristics of humidity fluctuations near the surface.

have been associated with anti-cyclonic subsidence, Kelvin-Helmholtz instabilities, and other phenomena (Lane, 1965, 1969; Gjessing et al., 1969; Bean and Emmanuel, 1973; Warhaft, 1973).

Further insight into the physical nature of vapour fluctuations may be obtained from reports on vertical vapour flux. Through measurements involving the eddy correlation technique, Bean et al. (1972) and Bean and Emmanuel (1973) have found overwater fluxes in the BOMEX project to be approximately constant within the lower subcloud layer; they average about 6×10^{-5} kg m^{-2} s^{-1} with some day-to-day variations and some variations within the day. The differences in the spectra of these fluxes when observed along wind and across wind suggest that the flux depends upon organized convective structures such as thermal plumes. Vertical gradients in specific humidity within 1 m of an open water surface tend to be very strong, with representative values around 1.2 g kg^{-1} m^{-1}.

3. DESIGN REQUIREMENTS IN THE HUMIDITY MEASUREMENT

Numerous factors will govern the design of an instrument for observing atmospheric humidity fluctuations. These include the required spatial resolution, recording bandwidth, functional range, sensitivity, robustness, instrument platform, observing interval, and environmental contamination. The present section will refer to each of these.

The physical dimensions of the humidity sensor and the type of air flow through it govern its spatial resolution. The latter is the minimum distance along the direction of mean flow for which a complete change of sampled air occurs within the sensor. Analysis of the flow through a microwave refractometer cavity shows that this minimum distance is of the order of the cavity length for laminar flow; but when the flow is turbulent, the flushing efficiency deteriorates and the resolution length increases (Hartman, 1960; Gilmer et al., 1965). For example, at wind speeds exceeding 5 m s^{-1}, the resolution length increases to some ten times the cavity length. The same principle may be applied to other types of sensors in which air flow is impeded. Thus, a conservative estimate of the minimum sensor length for the range of wavenumbers in Figure 1 is about 10 m at the low wavenumber end and 1 mm at the high wavenumber end.

The bandwidth of a hygrometer recording circuit is dictated by the rate at which eddies pass the sensor. For an eddy of wavenumber k moving with the mean wind speed U, Taylor's hypothesis gives the associated spectral period $T = (k\ U)^{-1}$. For each spectral period, the desired bandwidth of the circuit is $(2T)^{-1}$. Thus, in a wind of 5 m s^{-1} and at a height of 10 m, the spectral periods extend from 500 s to 2 ms if the production range of humidity variance and the -5/3 range are to be covered. The required passband of the hygrometer circuit would extend from 0.001 Hz to 250 Hz.

Recording and analysis of the sensor signals is aided by modern electronic techniques. Encoded transcription on magnetic tape is desirable, not only because of the wide dynamic range of signal amplitudes involved, and the usual requirement for simultaneous recording of two or more fluctuating quantities (e.g., q' and w'), but also because the computing of auto-and cross-correlations generally is required. An example of the sensitivity required in such recordings has been given in the previous section.

Robust construction is essential in the humidity sensor. Current applications involve mounting of the hygrometer on aircraft travelling some 100 m s^{-1}, on buoys that are buffeted by waves and wind, and on ship masts that are subject to large accelerations.

Exposure of the sensor to the environment presents problems of design that have not yet been fully resolved. These include vapour adsorption and condensation on the sensor surfaces (Turner and Hay, 1970), wetting of the sensor by precipitation and spray, and contamination by salts and dust. These are capable of modifying the sensor calibration, its thermal lag, and reversibility.

The mechanical mounting and configuration of the sensor must take into account several requirements. Free flow of air through the sensor is important to its response time (Miyake and McBean, 1970; Phelps et al., 1970). In covariance studies, the paired sensors must be separated by a distance that is less than the smallest significant eddy size, and the response times of the two sensors must be similar over the desired wavenumber range. The atmospheric volume that is monitored by the sensors must be sufficiently large to include important spatial variations, e.g. natural thermals (Kruspe, 1974). The aspect sensitivity of the sensor to wind-borne humidity fluctuations also must be considered (Bean, 1971; Bean et al., 1972). For measurements near an open water surface, the sensor platform should be stationary for observing the effects of wave-coherent motions (Vershinskiy, 1970; Laevastu, 1973). If the hygrometer is to be mounted on a balloon-borne platform, weight and positioning of the sensor outside of the balloon wake are important design factors (Lane, 1965; Readings and Butler, 1972). The mounting arrangements should provide for continuous observation over a representative interval of time, normally about 30 minutes in a stationary regime of turbulence.

4. THE INFRARED HYGROMETER

Humidity measurement by the techniques described in this and the following section is based upon the Bouguer-Beer-Lambert law (Johns, 1965; Zuev, 1974). This relates the transmittance of the medium (τ) to its extinction coefficient (γ) by the equation:

$$\tau = \exp \left[\int_0^{D_a} - \gamma_\lambda(r)\, dr \right] \tag{1}$$

$$= \exp (-\bar{\gamma}_\lambda D_a)$$

where the spatial coordinate r extends through the depth D_a of the humid medium, and γ is wavelength dependent. In general,

$$\gamma_\lambda = \alpha + \beta \tag{2}$$

where β represents the extinction component that results from scatter by particulates and molecules, and α refers to the absorption loss.

Equations 1 and 2 are applied in two ways. In one, τ is recorded at two adjacent wavelengths. Since β is a relatively slow function of wavelength in Rayleigh and Mie scattering, it is approximately constant over this narrow wavelength interval. By appropriate choice of the spectral region, α is negligible at one selected wavelength but subject to strong absorption by water vapour molecules at the other wavelength. Hence, the differential measurement of τ at two adjacent wavelengths indicates the humidity concentration along the test path, through the instrument calibration.

The second method requires a measurement of τ at only one wavelength. The latter is chosen for α much greater than β. Hence, fluctuations in τ correspond to fluctuations in α (and in humidity). The calibration is completed by referring the mean of these fluctuations to the corresponding mean of an independent humidity measurement with a slow-response hygrometer, over the recording range of the instrument.

Applications of this technique in the infrared region have been reported for both long and short test paths. Tomasi and Guzzi (1974) describe measurements using a test path that extends through the full depth of the atmosphere, which give the integrated humidity concentration along the path. Evaluation of γ in this application is very difficult since it depends upon gas temperature and the concentrations of other gaseous components. The calibration procedure is greatly simplified if D_a is less than 1 m, where the temperature and pressure may be assumed constant along the path. The present discussion will be limited to short-path hygrometers of the latter type.

Appropriate regions of the infrared spectrum for this hygrometer technique are noted in Figure 2. Absorption by various gases is included in the figure. The strong water vapour absorption lines

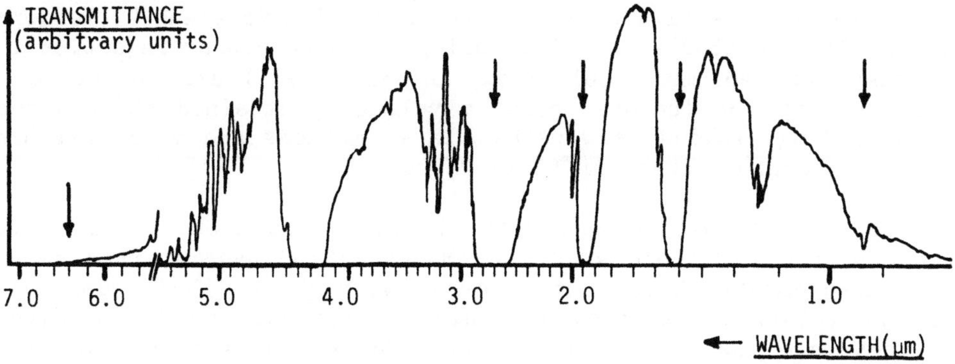

Fig. 2 Atmospheric infrared absorption spectrum, using the sun as a source (adapted from GATES, 1960, and Gates and Harrop, 1963).

associated with vibrational and rotational transitions are centered at 0.94, 1.4, 1.9, 2.7, and 6.4 μm, as indicated by the arrows. These have been used in hygrometers described by Staats et al. (1965), Wood (1965), Bardeau (1971), Elagina and Koprov (1971), and Hyson and Hicks (1975).

Figure 3 illustrates the elements of a short-path hygrometer. The beam from the infrared energy source is modulated by a chopper to facilitate stable high-gain amplification of the detected signal. The filter transmits alternately at the absorbing and nonabsorbing wavelengths, and comparison of the two transmitted signals yields α through Equations 1 and 2. Since the lengths of the test paths lie within the range 0.20 to 1.0 m, the calibration relating α to atmospheric humidity is obtained conveniently with the aid of standard, saturated salt solutions at controlled temperature and pressure.

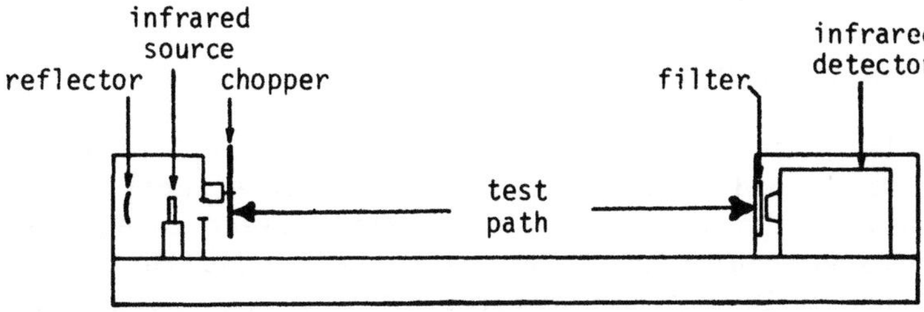

Fig. 3 Basic components of the short-path infrared hygrometer.

For a more precise calibration at the absorbing wavelength, Henderson (1970) describes a technique using measured vapour injection in the test path volume. Golubitsky et al. (1968) also describe a correction at the 'nonabsorbing' wavelength, obtained through the use of a long, folded test path. These methods are more suitable for the laboratory than for field work.

Two infrared hygrometers described by Hyson and Hicks (1975) have been designed for field studies. Moore (1976) has reported eddy-flux measurements above a pine forest with them. Their construction is relatively simple, mechanically rugged, and inexpensive, and they have low power consumption. One operates at 6.3 μm; because of the restricted passband of its recorder, it covers only the range of wavenumbers up to 0.4 m^{-1}. The other operates at 2.7 μm; assuming a mean flow of 2.5 m s^{-1}, its upper wavenumber range of 5 m^{-1} is governed by laminar flow through its sensor and by the recorder passband. The recorder noise level of approximately 0.1 g kg^{-1} of vapour corresponds to the expected humidity fluctuation at this upper wavenumber limit for air temperatures around 20°C. A temperature correction is provided for the calibration, and calibration instabilities associated with aging of the signal source and contamination of its windows are eliminated through signal ratioing. The mean output is referred to a slow-response hygrometer.

Two limitations of these hygrometers should be noted, in addition to the restricted range of wavenumbers described above. The calibration of the hygrometer becomes invalid at humidities approaching saturation, where β changes significantly through vapour adsorption and absorption on atmospheric particles (Elagina et al., 1970; Filippov and Mirumiants, 1971). With regard to the two-wavelength hygrometers, the latter authors also have pointed out that transmittance at the 'nonabsorbing' wavelength is affected by the wings of the adjacent absorption lines; thus, precise humidity measurements by this method require knowledge of each spectral region in more detail than currently is available. See also related papers by Blum et al. (1972), Breckinridge and Hall (1973), Viktorova and Zhevakin (1973), and Cox (1973).

5. THE LYMAN-α HYGROMETER

The principle of this hygrometer is similar to that of the infrared hygrometer described in the previous section, except for the spectral region in which it operates. In the present case, transitions between the electronic levels of water molecules in the air yield absorption bands in the ultraviolet region (Garton et al., 1957; Randall et al., 1965; Tillman, 1965; Martini et al., 1973). The only significant absorbers of ultraviolet radiation in the atmosphere are oxygen, water vapour and ozone, and the last is important only in the stratosphere [photochemical reactions in some types of smog may occasionally create ozone layers in the lower atmosphere (see, for example, Lea, 1968)]. Figure 4 illustrates the relative absorption by oxygen and water vapour. At the Lyman-α emission line of atomic hydrogen (0.12156 μm), the absorption

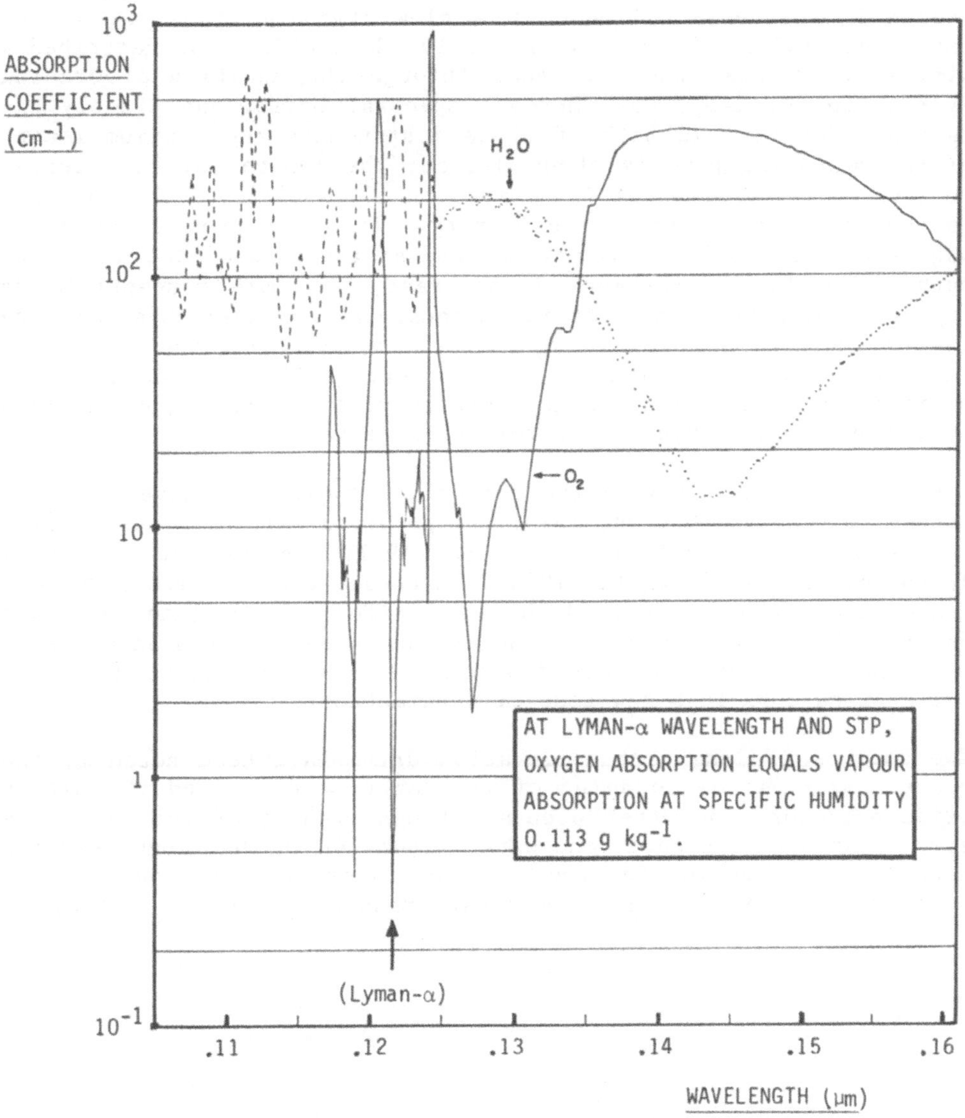

Fig. 4 Spectral absorption by atmospheric water vapour and oxygen in the ultraviolet region [adapted from K. Watanabe, M. Zelikoff and E.C. Inn (Tillman, 1965)].

coefficients are 0.34 and 387 cm^{-1}, respectively. Hence, an instrument with only modest spectral resolving power is required to measure the transmittance of the Lyman-α line through humid air, where the absorption is due almost entirely to water vapour.

The essential components of the hygrometer are represented in Figure 5. Various types of Lyman-α source have been investigated for the desired strength and spectral purity. A commercial hygrometer uses a hydrogen-neon mixture in a glow discharge tube, with life expectancy 100 to 200 hours. Recently, Buck (1976) has described a similar hygrometer whose source, through the addition of uranium hydride to the lamp, has improved spectral purity and greatly extended lifetime. In both of these instruments the minimum length of the measuring path is about 0.2 cm; the length may be extended to 1 cm in the commercial instrument or 10 cm in Buck's hygrometer, as required for increased sensitivity. Either lithium fluoride or magnesium fluoride windows may be used at the ends of the measuring path. The former are more transparent, but more susceptible to vapour etching than the latter. Because of the small physical size of its sensing head, this hygrometer may be calibrated in a bell jar. More commonly, it is used to observe relative changes in humidity; its mean indication is compared with the corresponding mean of a slow-response hygrometer for absolute calibration.

Several field trials with the commercial hygrometer have been reported (Miyake and McBean, 1970; Phelps et al., 1970; Martin, 1971; Phelps and Pond, 1971; Pond et al., 1971; Krechmer et al, 1972; Thorpe et al., 1973; Smith, 1974). In these, the hygrometer was operated on stationary and shipborne platforms over land and over water. The measured humidity spectra and vapour fluxes agreed well with those measured by other rapid-response techniques for eddy scale sizes down to a few tens of centimetres.

Two sources of long-term calibration drift have been noted by the users. One is due to aging of the Lyman-α source and the nitric oxide detector; the other occurs through etching of the cell windows by atmospheric gases and water. Buck (1976) has suggested two methods of correcting for the calibration drift: continual reference of this hygrometer to a slower-response dewpoint hygrometer

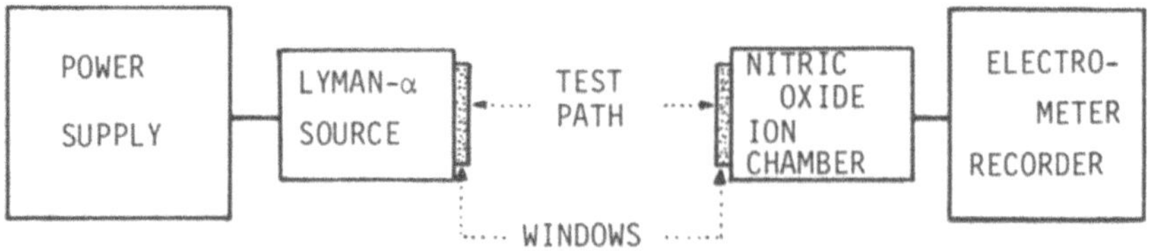

Fig. 5 Basic components of the Lyman-α hygrometer.

for calibration revision, and the use of the variable path technique. In the latter, the Lyman-α hygrometer record is supplemented by measurements of air pressure and temperature at the test path, as the length of the test path is altered. Using this information to solve the instrument equations yields the revised calibration. This procedure normally is followed a few times per day. The absolute accuracy in measured specific humidity is about 4 g kg^{-1}. However, relative accuracy is much better, 0.1 g kg^{-1} or less. Turbulence in the flow at the sensor may limit the maximum observed wavenumber to about 50 m^{-1}. Noting that the bandwidth of the recording circuit is 80 Hz, the criteria of Section 3 (above) indicate that these limits of accuracy and wavenumber are roughly compatible at wind speeds less than a few metres per second and temperatures around 20°C.

6. THE MICROWAVE REFRACTOMETER AS A HYGROMETER

The radio wave refractometer has been used widely in the measurement of atmospheric humidity fluctuations. This device has assumed various forms during its 28 years of development and application: the microwave refractometer, the UHF (or coaxial cavity) refractometer, and the HF (or capacitor) refractometer. Their basic similarities and differences will be noted below.

In principle, this method of humidity measurement depends upon the relationship between refractivity (N_a), specific humidity (q), temperature (T), and pressure (p) of the sampled air (McGavin and Vetter, 1965). Assuming that condensation is absent and that the air behaves as a perfect gas,

$$q(\text{g kg}^{-1}) = 1.67\text{x}10^{-3}\,\frac{N_a\,[T\,(K)]^2}{p(\text{mb})} - 0.129\,T(K) \qquad (3)$$

(for air pressure, 1 mbar = 10^2 N m^{-2} = 10^{-1} kPa). Thus q may be derived from simultaneous measurements of N_a, T and p. For the measurement of humidity fluctuations, the differential form of Equation 3 is applied:

$$q' = \left(\frac{1.67\text{x}10^{-3}\,T^2}{p}\right)N_a' + \left(\frac{3.34\text{x}10^{-3}N_a T}{p} - 0.129\right)T' - \left(\frac{1.67\text{x}10^{-3}N_a T^2}{p^2}\right)p' \qquad (4)$$

where the units of Equation 3 are implied, and the parameters in the brackets are the local mean values about which the fluctuations (primed quantities) occur. Slow-response sensors may be used to obtain the coefficients of N_a', T', and p'. The special requirements for these fluctuation measurements may be illustrated by an application under average sea-level conditions with 75% relative humidity. Here, Equation 4 becomes:

$$q'(\mathrm{g\ kg^{-1}}) = 0.137\ N_a' + 0.184T'(\mathrm{K}) - 0.0445p'(\mathrm{mbar}). \quad (5)$$

If q' is 0.005 g $\mathrm{kg^{-1}}$ (corresponding to the high-wavenumber limit in Section 2), then appropriate values of Na', T', and p' are 0.02 N-unit, 0.015K, and 0.06 mbar respectively. It is clear that refractivity must be observed with a device like the microwave refractometer, that temperature must be measured with very fine resistance wire or micro-bead thermistors, and that pressure must be recorded with acoustic sensors. Techniques for observing these rapid fluctuations of temperature and pressure are described elsewhere in this volume (see Larsen et al., Chapter 15, and Dobson, Chapter 13). The remainder of the present section will deal with refractivity instruments.

All radio wave refractometers operate upon the same basic principle. The air sampling sensor is part of a frequency sensitive network. When the sampled air alters the electrical characteristics of the sensor, the resulting shift in frequency response (Δn) of the network indicates the change in refractive index (Δn_a) of the air, according to the relation:

$$\Delta n = K\ \Delta n_a = 10^{-6}\ K\ \Delta N_a. \quad (6)$$

In most refractometer applications, only relative refractivity is recorded; the absolute refractivity may be derived from auxiliary measurements with slow-response sensors through Equation 3, if required. Hence, it is desirable to maintain K constant over the dynamic range of the refractometer. In the microwave refractometer, the sensing element is a resonating microwave cavity with ventilating ports (see, for example, Vetter and Thompson, 1971). The HF refractometer uses an air capacitor as the sensing element (Hay, 1971), and the UHF refractometer employs a ventilated coaxial cavity for the air sensor (Deam, 1962).

Departures from ideal response in the refractometer arise from several causes. As noted earlier, the spatial resolution is governed by flushing efficiency in the sensor; this may be some ten times the cavity length in turbulent flow at 5 m $\mathrm{s^{-1}}$ or greater (Gilmer et al., 1965), which corresponds to wavenumbers around 1.5 $\mathrm{m^{-1}}$. The same authors found also that the cavity response to turbulent air was sensitive to the aspect angle between the cavity axis and the wind direction. Substantial reduction in the aspect sensitivity occurred when side ports were introduced into the cavity. Vapour adsorption on the untreated surfaces of the sensor at humidities above 60% will introduce errors in the apparent refractivity that increase towards saturation pressure (Turner and Hay, 1970; Hay, 1971). The application of a smooth hydrophobic coating (e.g. teflon or beeswax) to the sensor surfaces effectively eliminates this source of error, except at saturation where condensation appears.

Different responses to environmental temperature in the sensing and reference elements of a refractometer may result in large errors in apparent refractivity (Hay, 1971). This is due to the different patterns of heat exchange and dissimilar thermal compensation in these elements. Figure 6 illustrates these characteristics. The upper graph represents the step-change in air refractivity to which the dual-cavity refractometer was subjected. The refractometer response is indicated in the lower graph, where the apparent refractivity undergoes an instantaneous change that is followed by a complex pattern of slow transients of different lag times.

Recent design improvements in microwave refractometers have provided reliable operational instruments (M.C. Thompson, Jr., private

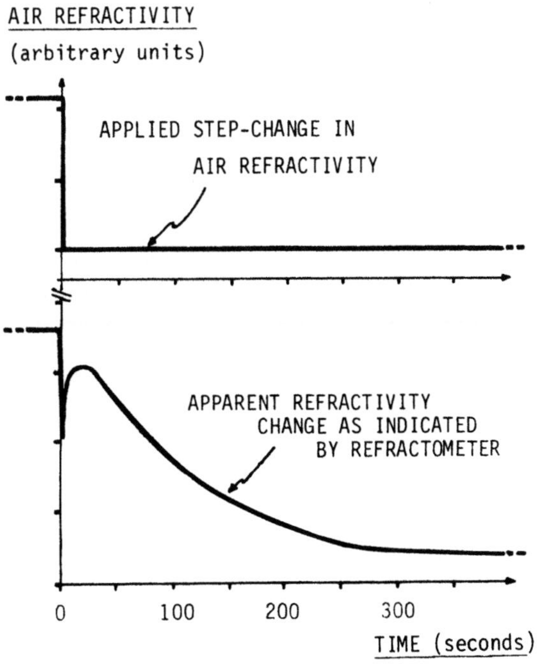

Fig. 6 Response of a dual-cavity refractometer to a step-change in air refractivity, illustrating the effects of different thermal lags and coefficients of thermal expansion in the sensing and reference cavities.

communication, Institute for Telecommunication Sciences, U.S. Department of Commerce, Boulder, Colorado; R. Gilmer and B.R. Bean, private communication, Environmental Research Laboratory, NOAA, U.S. Department of Commerce, Boulder, Colorado). Such instruments have been used extensively in field observations. Figure 7 illustrates the form of the sensing cavity that is used on aircraft. Openings are provided only in the end plates for this application, where air flow is essentially in one direction. For a stationary platform, slots are cut in the side plates to reduce aspect sensitivity. Readings et al. (1973) have used a microwave refractometer with solid state circuitry for tethered balloon ascent; aspect sensitivity of the cavity flushing was reduced by maintaining the desired heading with a drag chute. Special techniques of temperature compensation in this cavity reduce thermal instabilities to the equivalent of 0.03 N-unit. Differential response between the sensing and reference cavities is effectively eliminated by housing the latter in a temperature-controlled chamber, or by replacing it with a stable crystal oscillator. A hydrophobic coating is applied to the inner surfaces of the cavity to eliminate adsorption errors. In practice, the refractivity error at wind speeds below 13 m s^{-1} is less than 0.1 N-unit; this increases to about 1 N-unit at wind speeds around 50 m s^{-1} (Gilmer et al., 1965).

Field experiments with the capacitor refractometer have been

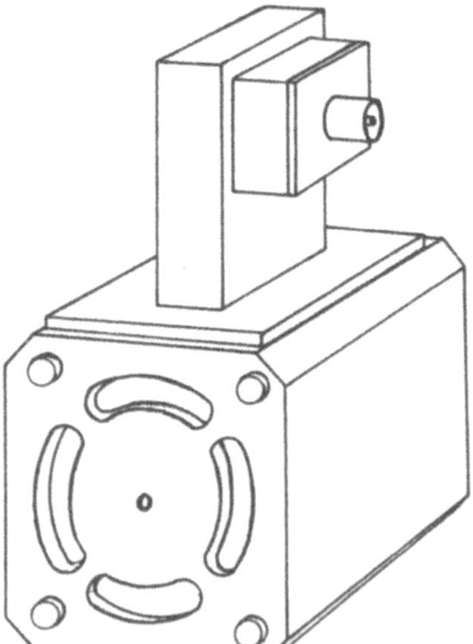

Fig. 7 The air-sensing cavity of a microwave refractometer, surmounted by a detector termination.

reported by Martin (1971, 1972, 1973) and Hicks and Martin (1972). The open construction of the capacitor sensor, and its short length along the direction of air flow (approximately 1 cm), provide for resolution to wavenumbers up to 100 m^{-1}. Little information is available on the electrical stability of the instrument.

7. COMMENTS

Three types of hygrometer have been considered as devices for monitoring fluctuations in air humidity at the air-sea interface. They have been selected because of their extensive use in field experiments and because they operate readily at subfreezing and higher temperatures. These are the infrared hygrometer, the Lyman-α hygrometer, and the microwave refractometer (supplemented by temperature and pressure sensors). As rapid-response hygrometers, they must be capable of observing humidity fluctuations at wavenumbers exceeding 10^{-2} m^{-1}, and as high as 10^{2} m^{-1} or more.

The design requirements have been given in Section 3. In field operations, these hygrometers best meet the design requirements when they are used as dependent instruments; they provide high relative accuracy in observing humidity fluctuations, but absolute calibration requires reference to a stable, slow-response hygrometer. However, each is capable of precise and absolute measurements in the laboratory, and design improvements are continuing. The major design limitation is the failure of instrument calibration at humidities approaching saturation vapour pressure, when vapour adsorption merges into condensation on the sensor surfaces and on airborne particulates in the test path. Field experience has provided working solutions to most of the other design problems, involving robust construction, operating in corrosive sea spray, and dealing with precipitation. At the present time, these instruments require attended operation. The sensing heads of the Lyman-α hygrometer and the microwave refractometer are only a few centimetres in length, and the associated electronics with solid state circuits weighs less than a few kilograms. Each hygrometer is a sophisticated instrument, and the writer cannot overemphasize the importance to a prospective user of contacting those with extensive experience in their design and field applications.

Currently, microwave hygrometers operating in aircraft observe turbulence at wavenumbers up to 1 m^{-1}. This limit is imposed by turbulent flow in the sensor and by system noise. Stationary instruments may respond to higher wavenumbers. The smaller physical size and open sensing path of the Lyman-α hygrometer limits its wavenumber range to 50 m^{-1} or less. Probably, redesign of the source and detector (including their configurations) of this last instrument offers the best prospect for extending the wavenumber limit towards the Kolmogorov microscale.

ACKNOWLEDGEMENT

The preparation of this contribution has been supported in part through Grant A5378 from the National Research Council of Canada.

REFERENCES

BARDEAU, H. 1971. The instrumentation of an aircraft intended for meteorological research. In *Statistical Methods and Instrumentation in Geophysics,* Teknologisk Forlag, Oslo: 277-291.

BEAN, B.R. 1971. Comparison of remote and in-situ measurements of meteorological parameters and processes. In *Statistical Methods and Instrumentation in Geophysics,* Teknologisk Forlag, Oslo: 181-196.

BEAN, B.R. and C.B. EMMANUEL. 1973. The dynamics of water vapor flux in the marine boundary layer. In *Modern Topics in Microwave Propagation and Air-Sea Interaction,* D. Reidel Publishing Co., Dordrecht-Holland: 51-64.

BEAN, B.R., R. GILMER, R.L. GROSSMAN, R.E. McGAVIN and C. TRAVIS. 1972. An analysis of airborne measurements of the vertical water vapor flux during BOMEX. *Journal of Atmospheric Sciences,* 29: 860-869.

BLUM, F.A., K.W. NILL, P.L. KELLEY, A.R. CALAWA and T.C. HARMAN. 1972. Tunable infrared laser spectroscopy of atmospheric water vapor. *Science,* 177: 694-695.

BOLGIANO, R., JR. and Z. WARHAFT. 1973. Preliminary observation of the evaporation duct in the I.M.S.T. Air-Sea Wind Tunnel. In *Modern Topics in Microwave Propagation and Air-Sea Interaction,* D. Reidel Publishing Co., Dordrecht-Holland: 37-50.

BRECKINRIDGE, J.B. and D.N.B. HALL. 1973. Absorption spectrum of atmospheric water vapour in the vicinity of the He 10830 Å triplet. *Solar Physics,* 28: 15-21.

BUCK, A.L. 1976. The variable-path Lyman-alpha hygrometer and its operating characteristics. *Bulletin of the American Meteorological Society,* 57: 1113-1118.

BUSCH, N.E. 1973. On the mechanics of atmospheric turbulence. In: *Workshop on Micrometeorology,* American Meteorological Society, Boston: 1-65.

COANTIC, M. and D. LEDUCQ. 1969. Turbulent fluctuations of humidity and their measurement. *Radio Science,* 4: 1169-1174.

COX, S.K. 1973. Infra-red heating calculations with a water vapour pressure broadened continuum. *Quarterly Journal of the Royal Meteorological Society,* 99: 669-679.

DEAM, A.P. 1962. Radiosonde for atmospheric refractive index measurements. *Review of Scientific Instruments*, 33: 438-441,

ELAGINA, L.G. 1963. Measurement of the frequency spectra of pulsations in absolute humidity in the near-ground layer of the atmosphere. *Izvestia Academy of Sciences*, U.S.S.R. Geophysical Series, 12: 1133-1137 (English translation).

ELAGINA, L.G., V.I. GORSHKOV and E.T. MIRONENKO. 1970. The measurement of turbulent vapor flux by infrared hygrometer. *Izvestia Atmospheric and Oceanic Physics*, 6: 49-51.

ELAGINA, L.G. and B.M. KOPROV. 1971. Measurement of eddy fluxes of humidity and their frequency spectra. *Izvestia Atmospheric and Oceanic Physics*, 7: 83-85.

FILIPPOV, V.L. and S.O. MIRUMIANTS. 1971. Attenuation of I.R. radiation by atmospheric haze in spectral regions coinciding with the absorption band of liquid water. *Izvestia Atmospheric and Oceanic Physics*, 7: 60-63.

GARTON, W.R.S., M.S.W. WEBB and P.C. WILDY. 1957. The application of vacuum ultraviolet techniques to the continuous monitoring of trace concentrations of water in several gases. *Journal of Scientific Instruments*, 34: 496-500.

GATES, D.M. 1960. Near infrared atmospheric transmission of solar radiation. *Journal of the Optical Society of America*, 50: 1299-1304.

GATES, D.M. and W.J. HARROP. 1963. Infrared transmission of the atmosphere to solar radiation. *Applied Optics*, 2: 887-898.

GILMER, R.O., R.E. McGAVIN and B.R. BEAN. 1965. Response of the NBS Microwave refractometer cavities to atmospheric variations. *Journal of Research, National Bureau of Standards*, 69D: 1213-1217.

GJESSING, D.T., H. JESKE and N.K. HANSEN. 1969. An investigation of the tropospheric fine-scale properties using radio, radar and direct methods. *Journal of Atmospheric and Terrestrial Physics*, 31: 1157-1182.

GOLUBITSKY, B.M., S.O. MIRUMIANTS, M.V. TANTASHEV and V.I. FILIPPOV. 1968. System for studying the spectral transparency of the atmosphere in the infrared spectral region. *Izvestia Atmospheric and Oceanic Physics*, 4: 676-678.

HARTMAN, W.J. 1960. Limit of spatial resolution of refractometer cavities. *Journal of Research, National Bureau of Standards (Radio Science)*, 64D: 65-72.

HAY, D.R. 1971. Some problems of calibration instability in radio refractometers. In *Statistical Methods and Instrumentation in Geophysics*, Teknologisk Forlag, Oslo: 247-268.

HENDERSON, W.F. 1970. Hygrometer calibration system. *Journal of Physics E., Scientific Instruments*, 3: 984-986.

HICKS, B.B. and H.C. MARTIN. 1972. Atmospheric turbulent fluxes over snow. *Boundary-Layer Meteorology*, 2: 496-502.

HYSON, P. and B.B. HICKS. 1975. Single beam infrared hygrometer for evaporation measurement. *Journal of Applied Meteorology*, 14: 301-307.

JOHNS, J.W.C. 1965. The absorption of radiation by water vapour. In *Humidity and Moisture,* Reinhold Publishing Corp., New York, Vol. 1: 417-427.

KRECHMER, S.I., G.N. PANIN and V.V. IPATOV. 1972. Measurement of humidity pulsations above the sea. *Izvestia Atmospheric and Oceanic Physics,* 8: 448-450.

KRUSPE, G. 1974. Measurements of fluctuations of vertical wind velocity, temperature and radio-refractive index above the sea. *Boundary-Layer Meteorology,* 6: 257-267.

LAEVASTU, T. 1973. Present state of knowledge of evaporation from the sea. In *Modern Topics in Microwave Propagation and Air-Sea Interaction,* Reidel Publishing Company, Dordrecht-Holland: 22-36.

LANE, J.A. 1965. Some investigations of the structure of elevated layers in the troposphere. *Journal of Atmospheric and Terrestrial Physics,* 27: 969-978.

LANE, J.A. 1969. Some aspects of the fine structure of elevated layers in the troposphere. *Radio Science,* 4: 1111-1114.

LEA, D.A. 1968. Vertical ozone distribution in the lower troposphere near an urban pollution complex. *Journal of Applied Meteorology,* 7: 252-267.

MARTIN, H.C. 1971. The humidity microstructure: a comparison between a refractometer and a Lyman-alpha humidiometer. *Boundary-Layer Meteorology,* 2: 169-172.

MARTIN, H.C. 1972. Humidity and temperature microstructure near the ground. *Quarterly Journal of the Royal Meteorological Society,* 98: 440-446.

MARTIN, H.C. 1973. Some observations on the microstructure of dry cold fronts. *Journal of Applied Meteorology,* 12: 658-663.

MARTINI, L., B. STARK and G. HUNSALTZ. 1973. Electronic Lyman-α humidity measuring device. *Zeitschrift für Meteorologie,* 23: 313-322.

McGAVIN, R.E. and M.J. VETTER. 1965. Radio refractometry and its potential for humidity studies. In *Humidity and Moisture,* Reinhold Publishing Corp., New York, Vol. 2: 553-560.

MIYAKE, M. and G. McBEAN. 1970. On the measurement of humidity transport over land. *Boundary-Layer Meteorology,* 1: 88-101.

MOORE, C.J. 1976. Eddy flux measurements above a pine forest. *Quarterly Journal of the Royal Meteorological Society,* 102: 913-918.

PHELPS, G.T., S. POND and V. GORNER. 1970. Simultaneous measurements of humidity and temperature fluctuations. *Journal of Atmospheric Sciences,* 27: 343-345.

PHELPS, G.T., and S. POND. 1971. Spectra of temperature and humidity fluctuations and of the fluxes of moisture and sensible heat in the marine boundary layer. *Journal of Atmospheric Sciences,* 28: 918-928.

POND, S., G.T. PHELPS, J.W. PAQUIN, G. McBEAN and R.W. STEWART. 1971. Measurement of the turbulent fluxes of momentum, moisture and sensible heat over the ocean. *Journal of Atmospheric Sciences,* 28: 901-917.

RANDALL, D.R., T.E. HANLEY and O.K. LARISON. 1965. The NRL Lyman-Alpha humidiometer. In *Humidity and Moisture,* Reinhold Publishing Corp., New York, Vol. 1: 444-454.

READINGS, C.J. and H.E. BUTLER. 1972. The measurement of atmospheric turbulence from a captive balloon. *Meteorological Magazine,* 101: 286-298.

READINGS, C.J., E. GOLTON and K.A. BROWNING. 1973. Fine-scale structure and mixing within an inversion. *Boundary-Layer Meteorology,* 4: 275-287.

SMEDMAN-HÖGSTRÖM, A.S. 1973. Temperature and humidity spectra in the atmospheric surface layer. *Boundary-Layer Meteorology,* 3: 329-347.

SMITH, S.D. 1974. Eddy flux measurements over Lake Ontario. *Boundary-Layer Meteorology,* 6: 235-255.

STAATS, W.F., L.W. FOSKETT and H.P. JENSEN. 1965. Infrared absorption hygrometer. In: *Humidity and Moisture,* Reinhold Publishing Corp., New York, Vol. 1: 465-480.

THORPE, M.R., E.G. BANKE and S.D. SMITH. 1973. Eddy correlation measurements of evporation and sensible heat flux over Arctic sea ice. *Journal of Geophysical Research,* 78: 3573-3584.

TILLMAN, J.E. 1965. Water vapor density measurements utilizing the absorption of vacuum ultraviolet and infrared radiation. In *Humidity and Moisture,* Reinhold Publishing Corp., New York, vol. 1: 428-433.

TOMASI, C. and R. GUZZI. 1974. High precision atmospheric hygrometry using the solar infrared spectrum. *Journal of Physics E., Scientific Instruments,* 7: 647-649.

TURNER, H.E. and D.R. HAY. 1970. Atmospheric refractometry at high relative humidities. *Canadian Journal of Physics,* 48: 2517-2536.

VERSHINSKIY, N.V. 1970. Humidity pulsation spectra above the Mediterranean Sea. *Doklady of Academy of Sciences,* USSR, Earth Sciences Section, 193: 20-22.

VETTER, M.J. and M.C. THOMPSON, Jr. 1971. Direct-reading microwave refractometer with quartz-crystal reference. *IEEE Transactions on Instrumentation and Measurement,* 20: 58.

VIKTOROVA, A.A. and S.A. ZHEVAKIN. 1973. Rotational spectral lines of water vapour dimers in the upper troposphere. *Izvestia Atmospheric and Oceanic Physics,* 9: 145-154.

WARHAFT, Z. 1973. The relation between temperature and humidity in the free atmosphere under conditions of stable stratification and strong thermal intermittency . . . . a case study. *Quarterly Journal of the Royal Meteorological Society,* 99: 89-104.

WARNER, J. 1973. Spectra of the temperature and humidity fluctuations in the marine boundary layer. *Quarterly Journal of the Royal Meteorological Society,* 99: 82-88.

WOOD, R.C. 1965. The infrared hygrometer - Its application to difficult humidity measurement problems. In *Humidity and Moisture,* Reinhold Publishing Corp., New York, Vol. 1: 492-504.

ZANCLA, A. (editor). 1973. *Modern Topics in Microwave Propagation and Air-Sea Interaction.* D. Reidel Publishing Company, Dordrecht-Holland, 364 pp.

ZUEV, V.E. 1974. *Propagation of Visible and Infrared Radiation in the Atmosphere.* J. Wylie and Sons, New York, 405 pp.

23

Gas Exchange

E.P. Jones

1. INTRODUCTION

For most gases, the surface region of the ocean, especially the top 100 μm or so, is undoubtedly the relevant region for gas exchange between the ocean and atmosphere. Since the surface region is so thin and delicate, it has been very difficult to study, and therefore the details of chemical and physical processes involved in gas exchange are quite elusive. Much of what is known about the chemical nature of the surface has been learned using various skimmers which can sample the top few hundred micrometres of the ocean (Garrett and Duce, see Chapter 25). Few, if any, studies have tackled measurements at the surface relevant to gas exchange.

This section will focus on the problem of gas exchange measurements. Marine aerosols can sometimes be involved in gas exchange; for example, organic particles may be ejected into the atmosphere where they produce SO_2 gas (Cuong et al., 1974), and, conversely, gases in the atmosphere can form particles which eventually reach the oceans. The following discussion will not deal with these more involved processes.

A major objective of gas flux measurements is to find out what environmental quantities control the gas exchange process so that gas exchanges can be predicted using a reasonable number of fairly easily measured quantities. The reason this objective is important is that although we can measure concentrations of many naturally occurring and man-produced gases, we cannot make very reliable estimates of their exchange between the ocean and atmosphere. The barrier to achieving this objective is that until recently there have been no direct measurements of gas fluxes over a short enough time period and a local enough region to preclude major changes in the quantities which may affect gas exchange. These include air and water temperatures, gas concentration differences between the air and water, wind velocity, wave state, humidity, and other physical and chemical parameters which can change over a local region in a short time.

Until recently, there have been only two types of measurements of gas fluxes between the ocean and atmosphere, both involving radioactive nuclei and both suffering important limitations in relating measured fluxes to environmental quantities. A new application of the eddy correlation technique to measure gas fluxes directly has been recently demonstrated and holds considerable promise for tackling this problem. Since the first two techniques have been recently reviewed (Broecker and Peng, 1974), they will be only briefly discussed here. The eddy correlation technique will be discussed more fully.

2. CARBON-14 AND RADON-222 METHODS

Carbon-14 is a radioactive isotope generated in the atmosphere by cosmic rays and recently by nuclear explosions. It enters the ocean as part of the CO_2 molecule and decays largely within the ocean with a half-life of 5770 years. By measuring the relative concentration of C^{14} and the stable C^{12} isotopes one can determine the exchange rate of CO_2 (Broecker and Peng, 1974). It is assumed in this method of determining gas exchange that there is a steady state in the C^{14} and C^{12} cycles and a uniform C^{14}/C^{12} ratio in air and sea surface. Because of a lack of complete data in different parts of the ocean corrections due to non-uniform distributions cannot be made, but have been judged to have at most a 40% effect. The CO_2 flux determined in this way is, of course, a long time, global average value and cannot be related to environmental parameters mentioned earlier. The measurement may be quite useful, however, in determining fluxes of other gases which are more or less uniformly distributed throughout the atmosphere and sea surface and whose solubility in the sea is not markedly different from CO_2.

Radon-222 is a daughter of radium-226. Its half-life is 3.8 days, short compared with C^{14}, and thus is suitable for measurements on an entirely different time scale. The usual, though not essential, assumption for flux measurements is that the parent, Ra^{226}, is uniformly distributed in the upper regions of the ocean. With this assumption, if there were no radon flux out of the ocean, the Rn^{222} isotope should be uniformly distributed also since it would be produced and would decay at fixed rates. The deficiency between the concentrations of radon in the upper 100 m or so and the concentrations at greater depths is thus assumed to be due to what has escaped to the atmosphere. Broecker and Peng (1974) reviewed the basic method and measurements.

Radon-222 flux determinations, in contrast to carbon-14 flux determinations, are local rather than global and reflect relatively short duration events. There are at least two problems associated with these measurements, however. First, no account has been taken of the possible effects of advection on the radon profile which could explain some rather scattered results of radon flux dependence on wind speed (Peng and Broecker, 1976). Second, weather conditions can change in very short times (hours) and the mixed layer can too. Thus, a particular radon profile may be far from a steady state condition, since a period of stable conditions two or three times longer than the Rn^{222} half-life is required to approximate one. It may prove possible to work with non-steady profiles, but a careful history of atmospheric and mixed layer conditions would be required to do so.

3. THE EDDY CORRELATION METHOD

The eddy correlation method can make direct measurements of local fluxes over short time periods, less than one hour, and thus is well suited to relating fluxes to the environmental quantities listed earlier. The technique has been applied to determine moisture, heat, and momentum fluxes (e.g. Pond, 1971; Businger, 1975) over a wide variety of conditions, and recently it has been applied for the first time to determine a gas flux, carbon dioxide, between the ocean and atmosphere (Jones and Smith, 1977).

The eddy correlation technique applied to gas fluxes correlates fluctuations in gas concentration in the atmosphere with the vertical wind velocity. The eddy correlation method for gas flux measurements is so far the only direct measurement and it is hard to foresee another possible direct method to compete with it. There are no special assumptions involved beyond those already well established in boundary-layer meteorology theory. The average flux F is given by

$$F = \overline{w'q'} \tag{1}$$

where w' is the vertical wind velocity, q' is the fluctuation of the concentration (mass/vol) of the gas about its average value in the atmosphere, and the overbar denotes a time average. Thus, to determine the gas flux, one has to measure fluctuations in vertical wind velocities and gas concentrations. For very small fluxes of gases, where the average gas concentration is perhaps four orders of magnitude larger than the rms average concentration fluctuation, it is necessary either to take account of a very small average vertical wind velocity or instead to measure the fluctuations of the gas of interest relative to the instantaneous air concentration (mass of gas/mass of air). One of these procedures is necessary to overcome the effect of air density fluctuations caused primarily by a heat flux (Bakan, 1978; Jones and Smith, 1978a; Smith and Jones, 1979).

Vertical wind velocities can be measured quite reliably and accurately (see Smith, Chapter 3; and Kaimal, Chapter 4) from a stable platform such as a special buoy, FLIP, or a flat beach. An ordinary ship is unfortunately not as suitable both because of ship's motion and air flow distortion. The key to a successful measurement of a gas flux between the ocean and atmosphere using the eddy correlation method is a sensor with a high sensitivity and short response time. Carbon dioxide was chosen as the gas to try this technique because it plays a major role in ocean chemistry and in life processes and, especially in this context, it is a strong infrared light absorber. It may be possible to use the eddy correlation technique for other gases if suitable detectors are developed. However, for most gases, estimated fluxes are extremely small (Liss and Slater, 1974).

Because there is almost no information on the processes associated with gas exchange, it is difficult to make a precise estimate of the sensitivity required to detect gas concentration fluctuations. For the frequency response, one can take as a guide frequencies observed for temperature and moisture fluctuations at 10 m above the sea surface. These lie between about 0.05 and 10 Hz. A reasonable estimate of the CO_2 flux assuming a concentration difference between the air and sea of 10% and an exchange rate (see Section 5) of 3×10^{-5} m s^{-1} is perhaps 10^{-7} mol m^{-2} s^{-1}, although this could be either too small or too large for particular conditions of wind, sea state, concentration differences in the air and water, etc. To convert this flux to concentration fluctuations, following a common practice, we rewrite Equation 1 in terms of the friction velocity u_*

$$F = u_* q_* \qquad (2)$$

where q_* is a concentration fluctuation scale and $u_* = (-\overline{u'w'})^{\frac{1}{2}}$ with u' the horizontal wind velocity fluctuation. For typical values of u_* (0.2 m s^{-1}) the detector should be able to detect concentration fluctuations of about 5×10^{-7} mol m^{-3} or about 0.01 ppm. This imposes an extremely stringent noise requirement for a sensor to meet when the signal can be averaged for only 0.1 s corresponding to the 10 Hz bandwidth. However, the eddy correlation method does not require nearly such stringent conditions be met for at least two reasons. First, the flux is determined by the correlation between the vertical wind velocity and the gas concentration fluctuations. Thus, since the vertical wind velocity can be measured with a good signal-to-noise ratio, the time over which the signal is averaged will be the duration of the measurement, typically more than half an hour, rather than the 0.1 s response time

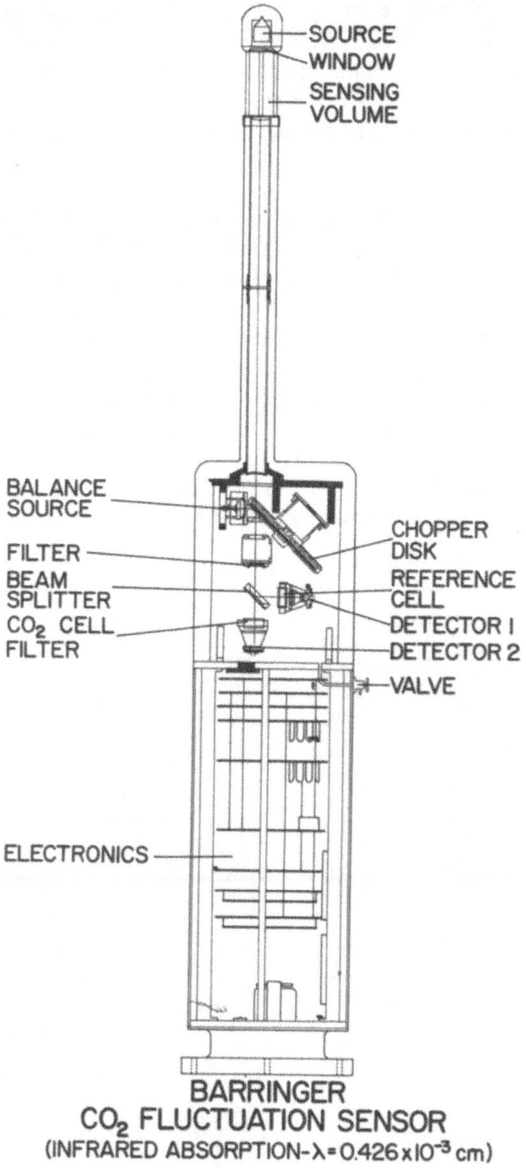

Fig. 1 The CO_2 sensor. The sensing path near the top is 10 cm long (after Jones and Smith, 1978b).

of the detector alone. Second, using the fluctuation magnitude obtained from Equation 2 ignores the distribution of frequencies of fluctuations. Although some fluctuations will occur up to about 10 Hz, the spectrum is not uniform but has a lower frequency peak so that the magnitude of maximum fluctuations is greater than indicated by Equation 2. Also, since there are almost no fluctuations below about 0.01 Hz, the sensor need not be exceptionally stable. Baseline drifting over time intervals of more than a few minutes does not present a problem, since spurious high correlations can be filtered out.

Because of the rather unusual requirements outlined above, a new CO_2 sensor was developed (Jones et al., 1978) since previously available devices were either not fast enough or not sensitive enough. The sensor is a variant of a nondispersive gas cell analyzer developed commercially to detect infrared absorption from trace quantities of gases in the atmosphere. A radiant source black body is housed at the top of the sensor (Fig. 1 and 2). The source radiant energy passes via a germanium lens and window through a sampling region 10 cm long and 1 cm in diameter where some energy is absorbed by the CO_2 absorption band centered on 4.3 μm. This source signal is then focused down a tube to the main

Fig. 2 Instrumented 10 m tower on the beach at Sable Island (after Jones et al., 1978)

sensor body where it passes through an optical filter, is chopped at about 1 kHz, and finally is directed by a beam splitter to two thermoelectrically cooled PbSe detectors. The whole measuring path except for one gas cell is evacuated to remove CO_2. This one gas cell is filled with enough CO_2 to make it opaque in the CO_2 absorption band so that the detector behind the cell is insensitive to CO_2 fluctuations. The output signal from one detector is electronically subtracted from the other so that absorption signals not due to CO_2 will cancel. The difference signal is thus proportional to the CO_2 concentration in the sampling region. A second black body inside the main sensor body provides a constant balance signal which is about the same magnitude as the source signal and which reaches the detectors when the sample signal is blocked by the chopper. Its purpose is to provide a background signal well above any ambient or spurious source within the detector. Over a 10 Hz bandwidth, determined by electrical filtering, the rms noise

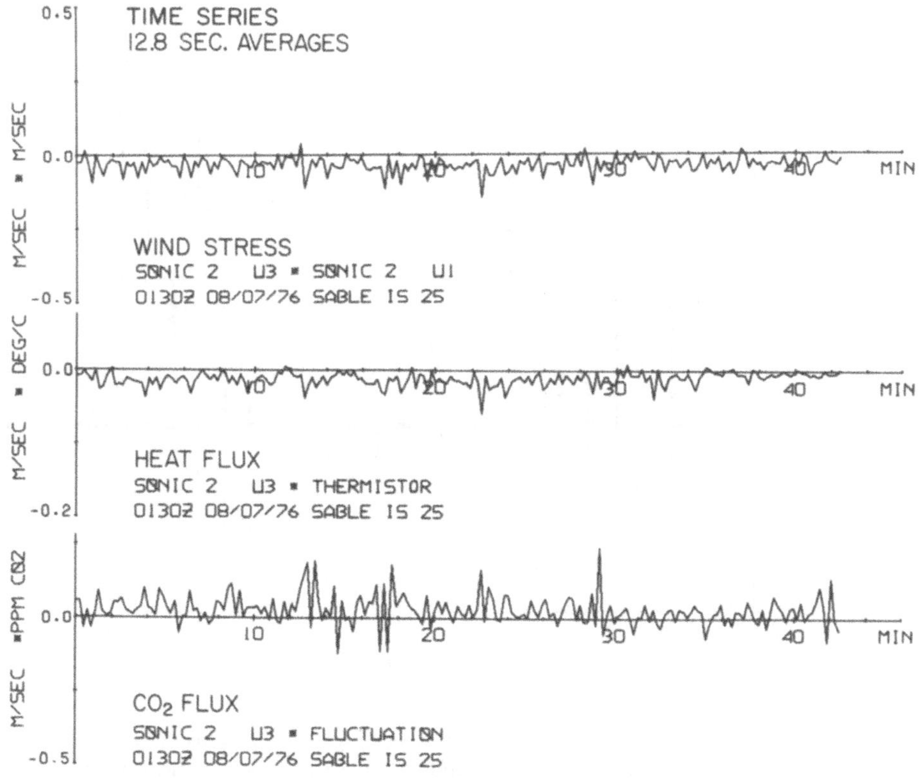

Fig. 3 Time series of CO_2 flux, heat flux, and wind stress averaged over intervals of 12.8 s (after Jones and Smith, 1978b)

corresponds to about 0.3 ppm or about 0.1% of the ambient CO_2 concentration.

Eddy correlation gas flux measurements between the ocean and atmosphere have been limited to one set performed for CO_2 on a beach with onshore winds on Sable Island (300 km east-southeast from Halifax, Nova Scotia). Results have been reported elsewhere (Jones and Smith, 1977). In summary, CO_2 fluxes measured were about 10^{-6} mol m^{-2} s^{-1}, somewhat higher than the estimate made above, but reasonable especially in view of relatively rapidly rising sea temperatures. For a typical run, time series of products of CO_2 concentrations, temperatures, and downwind velocities with vertical wind velocities (Fig. 3) give a qualitative indication of the averaging time required for representative estimates of the fluxes.

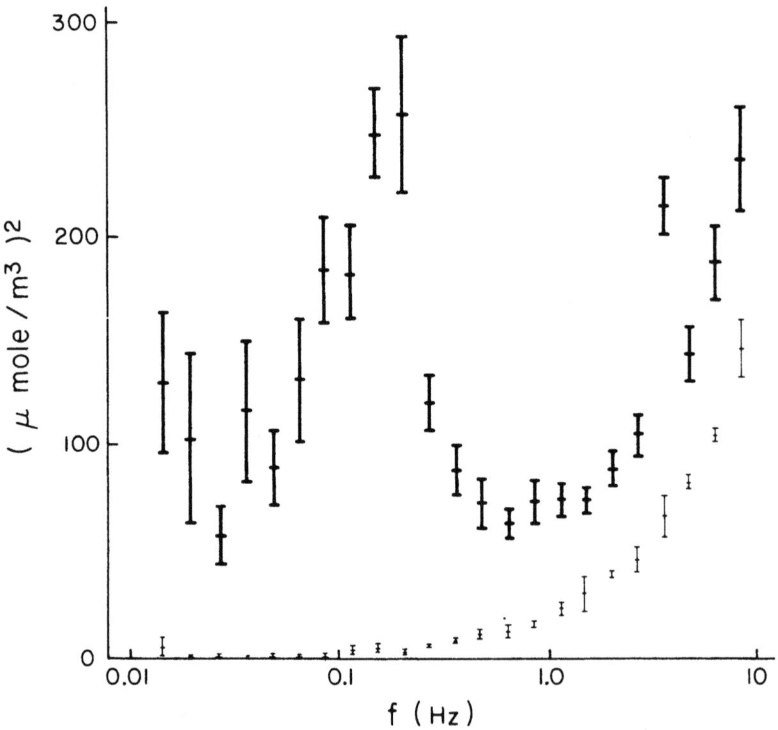

Fig. 4 Upper plot: CO_2 concentration spectrum times frequency versus logarithmic frequency. Lower plot: Noise of the instrument. Height of symbols gives standard error calculated from variability during data run (after Jones et al., 1978).

Individual events typically last 1 to 4 min, so we must average for much longer times, in this case 43 min. The fluctuation frequency spectrum (Fig. 4) shows a fairly sharp maximum at a lower frequency than the spectra for other quantities (heat, moisture, momentum) which, if typical, allows for somewhat less stringent sensor requirements than expected because the bandwidth required is less than anticipated. There is no obvious reason why the spectrum is more sharply peaked than is usual for the fluctuation spectra associated with heat and moisture fluxes.

4. LABORATORY STUDIES OF GAS EXCHANGE

Many laboratory studies of gas exchange done by people interested in industrial processes are reviewed by Danckwerts (1970). Some laboratory studies have been designed in particular to elucidate gas exchanges between the atmosphere and ocean. Two earlier ones (Kanwisher, 1963; Liss, 1973) indicated that the exchange rate (see Section 5) varied as the square of the wind speed. Another focused on the effect particularly of capillary waves on the exchange rate, and qualitative results were consistent with a strong enhancement of the exchange rate as the capillary wave field developed (MacIntyre, 1974). Recently, Broecker et al. (1978) found a linear relationship between the exchange rate and the wind velocity. Flothmann et al. (1979) found a linear relationship between the exchange rate and the wind velocity followed by an abrupt increase at a wind velocity of about 7.5 m s^{-1} to another steeper relationship between the two. These studies are most appropriate for gases which are only slightly soluble in sea water; i.e., gases for which the major resistance to exchange is in the surface boundary region of the water. Merlivat and Coantic (1975) have done laboratory studies on moisture fluxes which are very analogous to fluxes of highly soluble gases, since, in both these cases, it is the boundary region of the atmosphere rather than of the ocean which controls the exchange rate.

Because field conditions are hard to simulate in the laboratory and because we do not know which are the most important environmental quantities affecting the exchange rate, one must be cautious in trying to apply laboratory results to field situations. Nevertheless, these studies help point the way to understanding which mechanisms are involved in gas exchange and permit examination of mechanisms which are very difficult to study at sea.

5. MODELS OF GAS EXCHANGE

Models of gas exchange between the ocean and atmosphere are required to predict exchanges of gases which cannot be measured directly. A model derived from basic principles which can be applied to field conditions without recourse to adjustable parameters would be most desirable, but since such a model is unlikely to be found soon, empirical models into which one can incorporate environmental parameters such as wind speed, etc., must suffice. Such models may be derived from laboratory work, but need verification from field measurements to have real credibility.

Several models for gas exchange between liquid and gas phases are reviewed by Danckwerts (1970). Recently, Deacon (1977) developed a model which seems appropriate for moderate wind conditions. In general, oceanographers have relied primarily on a diffusive sub-layer model whose major appeal is its simplicity. The model assumes that at the interface between the atmosphere and ocean, molecular diffusion provides the transport mechanism both in the liquid and gas. Above and below this region turbulence is assumed to be strong; eddy diffusion is the dominating transport mechanism and the model assumes dissolved gases are uniformly distributed. In ocean-atmosphere exchanges, molecular diffusion through the air layer is thought to be much faster than through the sea layer, so the sea layer is the rate limiting layer and the only one which has to be considered for all but highly soluble gases. Thus, for most gases, including radon and CO_2, the flux, F, is given by

$$F = \frac{D}{\delta} \Delta C$$

where D is the molecular diffusion constant of the gas in the sea layer, δ is the thickness of the diffusive sublayer in the sea, and ΔC is the gas concentration difference on either side of the layer.

A second model is a surface renewal model which involves the replacement at intervals of elements of liquid at the surface by liquid from the interior which has the local mean bulk composition. While at the surface, the elements exchange gas as though they were quiescent and infinitely deep; the flux is a function of the average time, τ, the elements spend at the surface. This leads to the expression

$$F = \left(\frac{D}{\tau}\right)^{\frac{1}{2}} \Delta C$$

Both of these models are single parameter models so that without being able to measure δ or τ directly there is little to choose between them. Their value lies in the possibility of being able to make relative predictions of gas fluxes from concentration difference measurements and from estimates of the 'exchange rate' $k = F/\Delta C$. The fact that k depends in one model on D and in the other on $D^{\frac{1}{2}}$ is not particularly important for these predictions because diffusion constants of different gases in sea water are similar enough that the predictions are not much affected and estimates of global fluxes between the ocean and atmosphere can be attempted (Liss and Slater, 1974).

From carbon-14 and radon-222 isotope studies, δ has been estimated for the world ocean to be 50 ±30 μm (Broecker and Peng, 1974). This corresponds roughly to a value of τ of one second in terms of the surface renewal model. Since studies of surface layer temperatures indicate lifetimes of about one second (Saunders, 1974), there is certainly room for skepticism regarding the laminar layer model. But more important is the difficulty of a direct measurement of the single parameter, δ or τ, or indeed whether one parameter is enough to characterize exchange processes.

A model for gas exchange of highly soluble gases was developed for the situation in which the boundary layer in the atmosphere rather than in the ocean determines the gas exchange rate (Hicks and Liss, 1976). Models and laboratory experiments for moisture fluxes, which are very analogous to highly soluble gases, are also relevant to this case; in particular, an isotopic method (Merlivat and Coantic, 1975) could be a fruitful one to pursue further.

6. CONCLUDING REMARKS

The problem of gas transport between the ocean and atmosphere is obviously not a simple one. Both reliable field measurements and suitable theories are in very short supply. We have some understanding of what occurs in the atmosphere near the ocean surface, but only a vague comprehension of the relevant processes in the ocean. To understand gas exchange, we probably need to know much more about mixed layer processes as well as those at the surface. The eddy correlation method for measuring gas fluxes is so far the only direct method, but it is a difficult technique and cannot be used for all conditions and locations for which field measurements are needed.

Considerable insight into gas transport within the surface region of the ocean can come from considering heat transport. Right at the ocean-atmosphere interface, the total heat flux has contributions from radiation, conduction, and evaporation, but below the top 20 μm or so, at least at night, heat is transported primarily by molecular and eddy diffusion, just as dissolved gases are. Thus, any successful heat flux model should also be applicable to gas fluxes, and, since temperature profiles are generally easier to measure than gas concentration profiles, heat transport measurements both in the laboratory and in the field may show the way to understanding gas exchange processes. One model (Hasse, 1971) developed to explain heat flux measurements at sea has no adjustable parameters and is in remarkable agreement with the experimental results. Two recent papers (Street and Miller, 1977; Katsaros et al., 1977) on heat transport show studies in this area have progressed farther than similar ones for gases.

REFERENCES

BAKAN, S. 1978. Note on the eddy correlation method for CO_2 flux measurements. *Boundary-Layer Meteorology*, 14: 597-600.

BROECKER, H.Ch., J. PETERMANN and W.SIEMS. 1978. The influence of wind on CO_2 exchange in a wind-wave tunnel, including the effects of monolayers. *Journal of Marine Research*, 36: 595-610.

BROECKER, W.S. and T.H. PENG. 1974. Gas exchange rates between air and sea. *Tellus*, 26: 21-35.

BUSINGER, J.A. 1975. Interactions of sea and atmosphere. *Reviews of Geophysics and Space Physics*, 13: 720-726 and 817-822.

CUONG, N.B., B. BONSANG and G. LAMBERT. 1974. The atmospheric concentration of sulfur dioxide and sulfate aerosols over Antarctic, subantarctic areas and ocean. *Tellus*, 26: 241-249.

DANCKWERTS, P.V. 1970. *Gas-liquid reactions*. McGraw-Hill Book Company, New York, 276 pp.

DEACON, E.L. 1977. Gas transfer to and across an air-water interface. *Tellus*, 29: 363-374.

FLOTHMANN, D., E. LOHSE and K.O. MÜNNICH. 1979. Gas exchange in a circular wind/water tunnel. *Naturwissenschaften*, 66: 49-50.

HASSE, L. 1971. The surface temperature deviation and the heat flow at the sea-air interface. *Boundary-Layer Meteorology*, 1: 368-379.

HICKS, B.B. and P.S. LISS. 1976. Transfer of SO_2 and other reactive gases across the air-sea interface. *Tellus*, 28: 348-354.

JONES, E.P. and S.D. SMITH. 1977. A first measurement of sea-air CO_2 fluxes by eddy correlation. *Journal of Geophysical Research*, 82: 5990-5992.

JONES, E.P. and S.D. SMITH. 1978a. The air density correction to eddy flux measurements. *Boundary-Layer Meteorology*, 15: 357-360.

JONES, E.P. and S.D. SMITH. 1978b. Eddy correlation measurement of sea-air CO_2 flux. In *Turbulent fluxes through the sea surface, wave dynamics and prediction*, edited by A. Favre and K. Hasselmann, Plenum Press, 677 pp.

JONES, E.P., T.V. WARD and H.H. ZWICK. 1978. A fast response atmospheric CO_2 sensor for eddy correlation flux measurement. *Atmospheric Environment*, 12: 845-851.

KANWISHER, J. 1963. On the exchange of gases between the atmosphere and sea. *Deep-Sea Research*, 10: 195-207.

KATSAROS, K.B., W.T. LIU, A. BUSINGER and J.E. TILLMAN. 1977. Heat transport and thermal structure in the interfacial boundary layer measured in an open tank of water in turbulent free convection. *Journal of Fluid Mechanics*, 83: 311-335.

LISS, P.S. 1973. Process of gas exchange across an air-water interface. *Deep-Sea Research,* 20: 221-238.

LISS, P.S. and P.G. SLATER. 1974. Flux of gases across the air-sea interface. *Nature,* 247: 181-184.

MacINTYRE, F. 1974. Chemical fractionation and sea-surface micro-layer processes. In *The Sea,* edited by E.D. Goldberg, John Wiley and Sons: 245-299.

MERLIVAT, L. and M. COANTIC. 1975. Study of mass transfer at the air-water interface by an isotope method. *Journal of Geophysical Research,* 80: 3455-3464.

PENG, T.H. and W.S. BROECKER. 1976. The rate of gas exchange across the sea-air interface in the Atlantic Ocean. *Bulletin of the American Meteorological Society,* 57: 144.

POND, S. 1971. Air-sea interaction. *Transactions of the American Geophysical Union,* 53: 389-394.

SAUNDERS, P.M. 1974. The skin temperature of the ocean. A review. *Memoires Sociêtê Royal des Sciences de Liège,* 6e série, tome vi: 295-303.

SMITH, S.D. and E.P. JONES. 1979. Dry air boundary condition for correction of eddy flux measurements. In press, *Boundary-Layer Meteorology.*

STREET, R.L. and A.W. MILLER, JR. 1977. Determination of the aqueous sublayer thicknesses at an air-water interface. *Journal of Physical Oceanography,* 7: 110-117.

24

Ocean Wave Measurement Techniques

R.H. Stewart

1. INTRODUCTION

Ocean waves have been measured for many reasons in many places, and there exists a large body of literature on techniques for obtaining these measurements. I have no intention of surveying this work. Rather I wish to outline the important techniques, their accuracy and applicability, with some emphasis on the new methods of studying the fine structure of the sea surface and the directional distribution of all waves from centimetre to hectometre wave lengths.

I will begin by discussing measurements at a point, progress to measurements at several points, and end with a discussion of measurements over areas. But before beginning it is useful to define what is to be measured and the reference frame in which it is to be measured.

The ocean surface is taken to be $z = \zeta(x,y,t)$ in a coordinate system in which z is vertically upward, t is time, and x,y are horizontal, such that $\bar{\zeta} = 0$. In general, the statistical properties of this surface vary slowly in time and space. As such, it can be described locally by a three-dimensional Fourier transform, $X(k,\alpha,n)$. That is, the sea surface can be considered as a superposition of waves of all wavelengths, $L = 2\pi/k$, and periods, $T = 2\pi/n$, travelling in all possible directions α.

Usually it is assumed that the longer waves ($L > 1$ m) obey the dispersion relation for small amplitude gravity waves, $n^2 = gk$. This reduces the dimension of the spectrum by one and the resulting function is the directional spectrum $\psi(k,\alpha) = \psi_1(n,\alpha)$. The sea surface is now described as a superposition of plane waves having varying wavelengths and directions. Integration of ψ over all angles yields the one-dimensional spectrum $\Phi(n)$. This is the same as the spectrum of sea-surface elevation measured at a point. Integration of Φ over n gives the variance of wave elevation $\overline{\zeta^2}$. If ζ is a Gaussian random process, and if ψ is confined to a narrow band of frequencies, then the square root of the wave variance is one-quarter of the 'significant wave height' (Longuet-Higgins, 1952). Originally, this term was used to denote the height of the highest one-third of the larger ocean waves, $H_{1/3}$, because this corresponds well with visual estimates of wave height, but its relationship to wave variance is now used more often.

For shorter waves ($L < 1$ m) the dispersion relation is usually not known. Such short waves are strongly influenced by surface currents, and their Doppler shift must be accounted for. In general, their frequency on a fixed coordinate system is

$$n = (gk + \gamma k^3)^{\frac{1}{2}} + k\,\overline{U}$$

where γ is the ratio of surface tension to water density, and $\overline{U}$ is the mean surface current. If larger, longer waves are present, these, too, influence the dispersion relation (Phillips, 1966, Chapter 3). The uncertainty in knowing $\overline{U}$ and the influence of the longer waves limits the accuracy with which measurements of $\psi(k,\alpha)$ produced by some instruments can be related to measurements of $\psi_1(n_1,\alpha)$ produced by other instruments. The notation used here and in the rest of this chapter is identical to that in Phillips (1966), and the relationships between the various spectra and their interactions can be found there.

Measurements of ζ must be related to some fixed coordinate system x,y,z,t. Near shore this is easy, the coordinate system is fixed by securing the instrument to the bottom. In mid-ocean it is not. The difficulty of finding a frame of reference often exceeds the difficulty in detecting the water surface. There two references are used: the average sea level, and still water at depth where wave motion is small. Typically this latter depth is taken to be 0.5 to 1.0 wavelengths of the longest wave to be measured since wave motion attenuates rapidly with depth, as exp[2πdepth/wavelength]: wave motion at 3/4 wavelength depth is less than 1% of the surface motion.

2. MEASUREMENTS AT A POINT

The commonest wave measurement is of water surface elevation at a point, and many ingenious devices have been used to obtain this. These generally fall into three classes: surface-piercing instruments which directly measure wave height relative to mean water level, pressure measuring devices, and inertial measurements of wave acceleration which are then integrated to obtain wave height.

2.1 Surface-piercing Instruments

These instruments are usually mounted on a stable platform and sense water height in several ways: by changes in the resistance or capacitance of a vertical wire suspended in the water, by photographs of a pole, by the deflection of a laser beam shining through the surface, or by the reflection of laser light from the surface. The measurement of resistance is the preferred oceanic method, while the surface-penetrating laser technique measures the slope of short capillary-gravity waves difficult to measure by other methods.

Changes in the resistance of a wire dangling into the sea result from sea water shunting that part below water. The total resistance of the wire above the water plus the sea-water path is a function, usually linear, of the water level; it is measured by exciting the wire with an audio-frequency oscillator and measuring the induced current. Sensitivities of a few millimetres and precision of a few centimetres of water level can be obtained using nickel-chromium wire with 0.5 $\Omega\ cm^{-1}$ resistance, or with tungsten or conducting plastic wires having diameters of 0.2 to 0.5 mm. An apparatus as simple as a weighted length of wire hanging into the water works well for many purposes; but vibration of the wire by wind, and the sideways drag by waves, can contribute errors. These can be avoided by wrapping the wire around a rigid support, or by anchoring the wire to a rigid structure. Farmer (1963) provides a more detailed analysis of the accuracy of the technique and shows that the instrument should work well in theory and in practice.

The resistance need not be primarily in the wire. One widely used gauge uses sturdy cables of high conductance to measure the resistance of the sea water path between them; the resistance decreases or increases as the water level rises and falls. The gauge is rugged, but the interpretation of its signal requires an understanding of the changes of conductance of sea water with temperature and salinity, and the geometry of the conducting path.

If the wire is insulated, its capacitance is a function of water level, and this, too, can be used to measure wave height. Unfortunately, the measurement is often inaccurate. Water films left as the water drains downward contribute to the capacitance and distort the measurement of wave height. And the slow absorption of water into the insulation changes its dielectric properties, and produces a slow drift in the instrument response. For these reasons the technique is now seldom used. However, a detailed description of a particular instrument of this type, and a fine account of the problems of calibrating any wave instrument, are given by Kinsman (1960).

Ultimately, the surface-piercing wire can be eliminated completely. The height of the sea surface can be sensed sonically or with a radio-frequency measurement of distance. These methods are perhaps a little easier to use from large platforms, but still require the presence of the platform to support the instruments in a fixed position relative to mean water level.

2.2 Surface-piercing Gauges for Short Waves

Steep sea-surface slopes are produced by the short gravity and capillary waves forming the high frequency tail of $\Psi(n)$. They are responsible for the glitter of reflected sunlight and for the scatter of microwave radar signals; but, being short, they are difficult to measure by the usual techniques discussed above. In a typical wind sea, wave height decreases as wavelength squared, while wave slope decreases only in proportion to wavelength, and one seeks to measure slope directly. For example, if a narrow beam of laser light pierces the surface from below it is deflected at an angle proportional to the sea-surface slope. If it then strikes a horizontal screen a known distance above the sea surface, its position can be observed and recorded by a television camera. Such an instrument described by Hughes et al. (1978) was used to measure the modulation of short waves by internal wave velocities. Instruments of this type are primarily research tools, and suffer from practical difficulties. They must work close to the sea surface in order that the laser beam can be caught and imaged by a lens just above the surface; accordingly the waves must be of low amplitude, or the instrument must follow the surface of the longer waves. Stray sunlight and varying laser intensity, modulated by the curvature of the waves being studied, often fool the television. Weak intensities are not recorded, strong intensities saturate the tube producing persistent spots, and sun glitter adds extra spots. Altogether, perhaps 10 to 20% of the data points can be lost, with a comparable reduction in the usefulness of the data.

Despite these problems, the technique provides a useful means of measuring short, high-frequency waves. Alternatively, derivatives of the signal from a resistance wire (Mitsuyasu, 1977) or radio scatter, discussed below, can also be used to measure these waves.

2.3 Pressure Measurements

Ocean waves produce measurable pressure fluctuations beneath them. Under certain circumstances these can be uniquely and accurately related to wave height; but past confusion of the nature of the hydrodynamics producing the pressure signal has led to some unnecessary controversy about the accuracy of the technique.

The amplitude of the pressure signal varies as

$$\Delta p \simeq \rho g a \cosh k(z + h)/\cosh kh \qquad (1)$$

where ρ is water density, g is acceleration of gravity, a is height of the wave with wavenumber k, and h is water depth. In shallow water, $\Delta p \simeq \rho g a$ and is independent of depth; in deep water, $\Delta p \simeq \rho g a \exp(kz)$ and rapidly decreases with depth. The theory assumes $ak \ll 1$ or that $a/h \ll 1$. The former is satisfied for most studies, but the second can be false near the surf zone where $a/h = O(1)$. But to the extent that these assumptions hold, each Fourier component in the pressure spectrum can be linearly related to the same frequency component in the wave height spectrum $\Phi(n)$ (see Fig. 1; Esteva and Harris, 1970).

Controversy over the usefulness of the technique arises indirectly from attempts to simplify the measurement. To avoid calculating the spectrum, many early pressure records were converted to wave height by simply using a single value of k appropriate for the dominant wave. This is not very accurate, but is not now a problem because of the development of the Fast Fourier algorithm. However, placement of gauges in shallow water to avoid the cost of long cables is a continuing source of error. There, horizontal velocities contribute to the pressure field through the Bernoulli equation, and they must either be accounted for or shown to be small. The latter is the easier course, and requires placing the gauges in sufficiently deep water. 'How deep' is estimated from linear theory by comparing the mean square velocity at the depth of the gauge with wave pressure: for a shallow water wave this ratio is $a/(4d)$, and is found by using Equation 2.45 in Wiegel (1964), evaluated at the bottom, and compared with Equation 1 above. Substitution of one-half the significant wave height for wave amplitude gives an estimate of the total error in computed wave height.

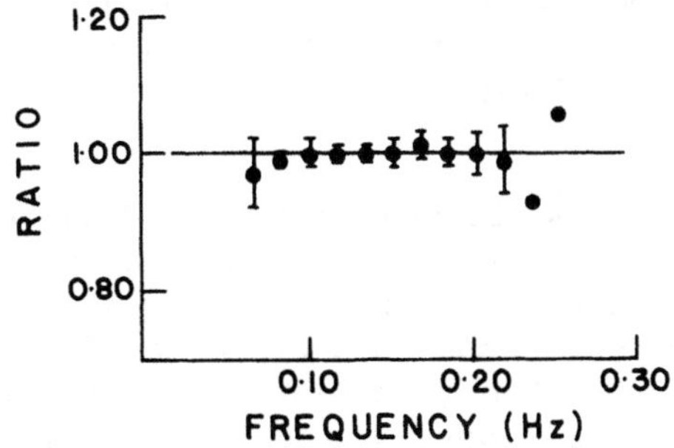

Fig. 1 Average and standard deviation of ratio (n) of wave spectra observed by resistance wire to those measured by a pressure gauge - from data recorded at Atlantic City whenever average wave height exceeded 30 cm (after Esteva and Harris, 1970).

A different pressure technique makes simple wave measurements in deep water under relatively calm conditions. A pressure gauge is dangled from a surface float to a depth of perhaps 100 m where there are no appreciable internal pressure fluctuations due to waves on the surface. As the surface float moves up and down, so does the pressure gauge, and it measures a change in hydrostatic pressure directly proportional to wave height. But work at the Institute of Oceanographic Sciences (Barber and Tucker, 1962) shows it does not perform well in storms because the wire fails to hang vertically and false readings are made.

2.4 Inertial Measurements

These measurements of the vertical acceleration of the water surface avoid the difficulty of finding a suitable reference frame in which to measure wave height in the open ocean. On average the ocean surface is fixed and has no mean acceleration; so vertical acceleration, when integrated twice in time, yields sea-surface elevation relative to mean sea level. The measurement is not without difficulty. Low frequency waves, especially swell, tend to have weak accelerations. To measure these accurately in the presence of large, high-frequency accelerations requires an instrument that responds linearly to vertical accelerations and that ignores horizontal motion. Furthermore, low-frequency drifts in the accelerometer appear, after integration, as large but fictitious excursions in sea height. Nevertheless, it is easier to build a good vertical accelerometer than to build a stable platform, and inertial instruments are the preferred wave measurement device in deep water.

The vertical reference for the acceleration measurement is obtained from a vertical gyro which aligns itself with downward acceleration averaged over a few minutes. During this time wave accelerations average to zero, and the gyro aligns itself to true vertical. Alternatively, a damped pendulum with a small unbalanced mass and long period, when supported at a point close to its centre of mass, will respond to the average acceleration, and it too will remain vertical.

The accelerometer in the instruments should be aligned to vertical by being attached to the gyro or the pendulum. Otherwise the signal from a strapped-down accelerometer requires correction for two components of both tilt and horizontal acceleration. For highest accuracy this requires two additional accelerometers, or, correct to third order in wave slope, the horizontal accelerations can be calculated from the wave tilt.

Neither the gyro nor the pendulum referenced accelerometer is cheap and simple to make, although the pendulum is closer to the ideal.

A much simpler but less accurate wave gauge consists of an accelerometer rigidly attached to a flat plate floating on the sea surface. Because the slopes of gravity waves longer than, say, a metre are small, on the order of 10°, the accelerometer is always close to vertical. Nevertheless, even these small deviations introduce errors. Tucker (1959) investigated this instrument in detail, and finds that all wave frequencies contribute to the error at a particular frequency, and that the acceleration error is nearly independent of frequency. Weak, low-frequency acelerations such as those produced by swell may be obscured, but the main wave components are not seriously affected.

The accuracy of vertical gyros and accelerometers is only a function of cost; the accuracy ranges from 1% of g and 1° from vertical, for cheap accelerometers and gyros, to very expensive inertial guidance systems capable of measuring the earth's rotation. Typical oceanic systems measure acceleration to an accuracy of 0.1% of g, and can be referenced to vertical within 0.25°. Such vertical accelerometers have low-frequency noise in the power spectrum of acceleration that is 0.1% of the acceleration in the most energetic part of the spectrum (Stewart, 1977).

The integration of the accelerometer signal is easily done if the spectrum of acceleration, $\Phi_A(n)$, is available because

$$\Phi(n) = n^{-4}\Phi_A(n)$$

This integration in frequency avoids the practical difficulties of double integration in time and simplifies removal of acceleration drifts.

A useful extension of the accelerometer technique results from mounting them on ships in combination with pressure meters. The accelerometers, placed next to pressure recorders on either side of the ship's hull, measure the vertical motion of the ship and thus the longer waves. The pressure gauges measure the shorter waves relative to the accelerometer position. Combining the two measurements gives $\Phi(n)$. The response of the system depends on the particular form of the ship on which it is mounted, and the system may be calibrated by comparison with wave spectra measured by a small accelerometer buoy deployed next to the ship (Cartwright, 1963). This has been done for the Ocean Weather Ship Weather Reporter and its response is now accurately known. All-in-all, this particular 'shipboard wave recorder' has been very successful. Its development contributed to the solution of many problems in wave measurement, and the instruments themselves, when placed on weather ships, have provided the only routine, accurate, and long-term measurements in mid-ocean (Neumann and Pierson, 1966, p. 330).

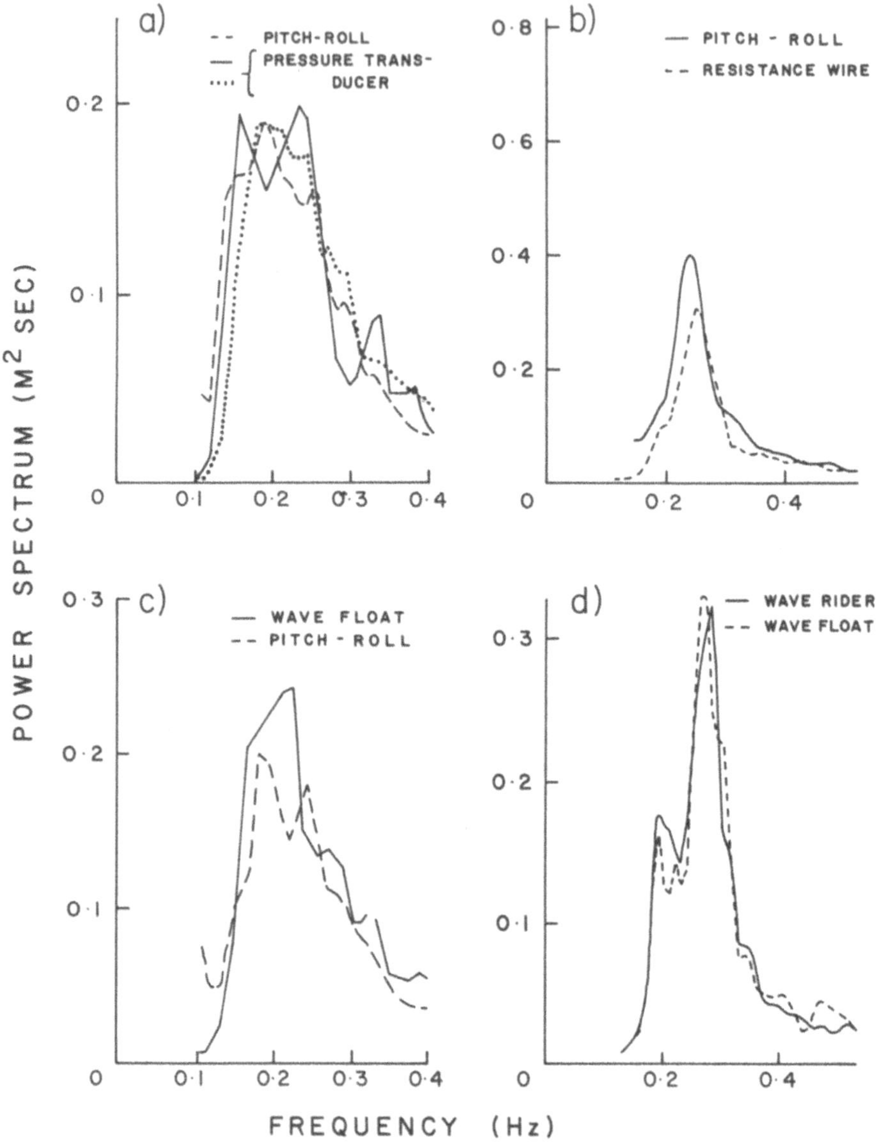

Fig. 2 Intercomparisons of wave measurements: (a)-(d) comparisons of accelerometer buoys - Wave Rider, wave float, pitch-roll with resistance wire and pressure gauges (after Hasselmann et al., 1973).

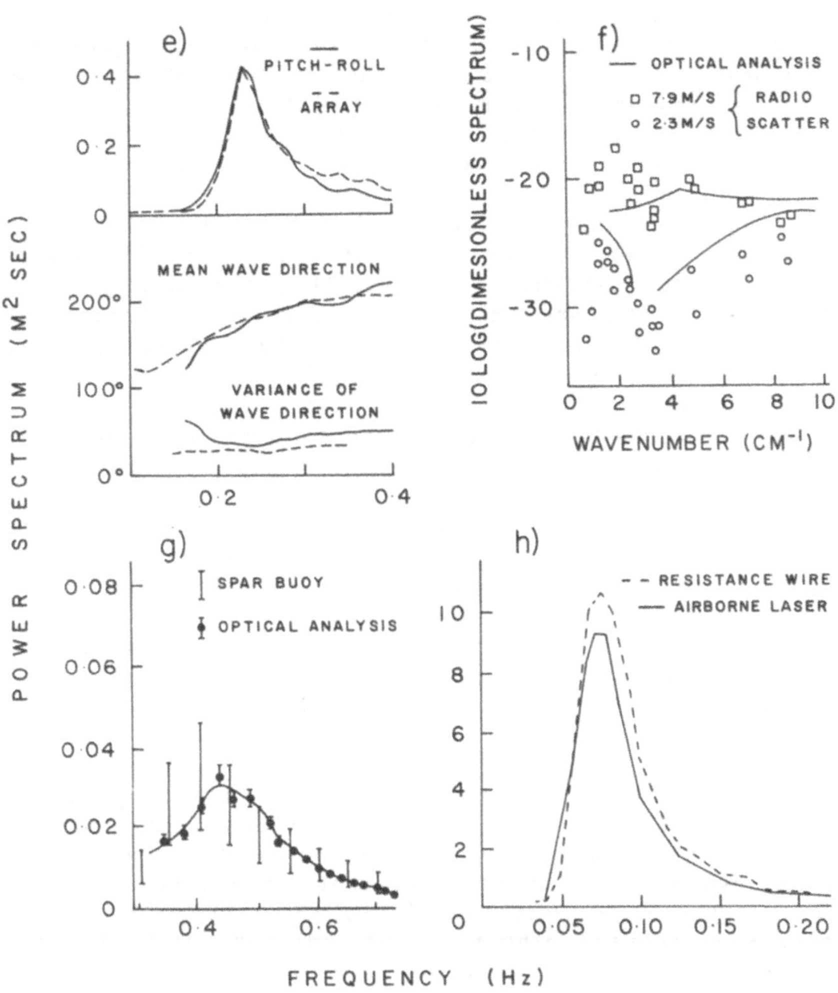

Fig. 2 Intercomparisons of wave measurements: (e) comparison of wave spectra and first two moments of directional distribution of waves measured by a six-element array and by a pitch-and-roll buoy (after Hasselmann et al., 1973); (f) wavenumber spectra measured by Bragg-scattered microwaves and by photometric techniques (after Wright and Keller, 1971); (g) comparison of photographic measurements of waves with resistance wire measurements (after Stilwell, 1969); (h) comparison of laser profilometer with resistance wire (after Ross et al., 1970).

2.5 Intercalibrations

A number of comparisons, reported in the literature, of the relative accuracy of the various wave measuring techniques validate their accuracy. In general, the comparisons agree on the 10% level, and all careful measurements appear to be equally accurate. Figure 2 shows typical comparisons of wave measurements at a point by wave staffs, pressure recorders, and accelerometers.

The favourable comparisons imply that all techniques measure waves to the same relative accuracy. This statement must be applied with caution. Not all instruments work in the same environment. Surface-piercing instruments must be supported on a stable platform, and measurements of large waves require large expensive platforms. Accelerometer buoys are generally useful except in severe storms when breaking waves can force small buoys under water or upside down. The shipborne wave recorder provides the only surface measurements under these conditions in deep water.

The intercalibrations apply only to well calibrated instruments. But many instruments are rarely calibrated; an initial or manufacturer's calibration is assumed to hold for months or years. Few reports of wave measurements state how and when the instrument was calibrated. And even fewer reports state the instrument was calibrated against known waves, say in a wave channel. Frequently the calibration is no more than a raising or lowering of the instrument in still water.

I wish to conclude this section with a plea that applies here as well as to the following sections. If you make wave measurements, carefully document the calibration and accuracy of your instrument, preferably dynamically, under conditions similar to those in which the instrument will be used. Don't rely upon its theoretical performance; many subtle, practical difficulties which can seriously degrade wave measurements have not been discussed here for lack of space.

Let me just cite several examples to accentuate the warning. A buoy floating on the sea surface does not measure wave height or acceleration at a point; rather, it measures wave height along a trajectory determined by the orbital velocities of the largest waves. Thus even a carefully calibrated accelerometer can still give a false measure of the wave spectrum. Nor is a fixed sensor any better. Short waves are carried by the long wave's orbital velocity and their frequency is Doppler shifted; and both long and short waves are Doppler shifted by mean currents. In addition, animal and plant life must be considered. Slime coats resistance wires, small animals live in the orifice of pressure transducers, and large animals bite lines. Lipids, oils, and other surface-active materials may be released by the structure supporting a wave gauge, or may exist in calibrating tanks, and seriously influence measurements of short or capillary waves.

These examples are applicable to wide classes of wave measurements. Their influence is usually small, but not always; so think carefully about a particular wave gauge. Determine exactly what is being measured, be it wave height along some trajectory or pressure at some depth, and then think carefully about how this can be related to what you really wanted to measure before beginning to use or to interpret wave information.

3. MEASUREMENTS AT SEVERAL POINTS

For many purposes knowledge of the distribution of wave directions as well as frequencies is required. This requires measurements of wave height over many wavelengths to obtain $\psi(\underset{\sim}{k})$ in the same way that measurements over many wave periods are required to obtain $\Phi(n)$; this greatly complicates the measurement. Usually the data over space are much sparser than data over time. The oceanographer then faces a problem shared among radio astronomy, communications, seismology, and other disciplines: how to efficiently determine the direction of signals using a small number of perhaps noisy receivers? The solution is not easy, but the problem is attacked most directly by measuring wave height at a number of points, or along a line.

3.1 Pitch-and-Roll Buoy

The first timid but very effective approach to the problem is to measure wave height and two components of wave slope at a point as a function of time. This Taylor Series expansion of the wave surface yields directly the first five coefficients a_i, b_i of the Fourier Series expansion of the distribution of wave directions at each wave frequency or length. That is, $\psi(k,\alpha)$ is written as $\psi(k,\alpha) = g(k,\alpha)\Phi(n)$, where

$$g(k,\alpha) = a_0(k) + a_1(k)\cos\alpha + b_1(k)\sin\alpha + a_2(k)\cos^2\alpha$$

$$+ b_2(k)\sin^2\alpha$$

such that

$$\int_0^{2\pi} g(k,\alpha)d\alpha = 1$$

Alternatively it is often assumed that $g(k,\alpha) = \cos^s[(\theta - \theta_0)/2]$ and the data used to calculate $\theta_0(k)$ and $s(k)$ (Tyler et al., 1974), or the data can be used to calculate the first five moments of the directional distribution.

The height and slope measurements are easily made by recording the tilts of a symmetric, disc-shaped buoy supporting a vertically-stabilized accelerometer such as that used to measure $\Phi(n)$. The technique, described fully by Longuet-Higgins et al. (1963), is especially attractive. The buoy is compact and easily deployed. It accurately follows the wave surface for wavelengths greater than twice its own diameter (Stewart, 1977). In contrast to wave arrays, it measures, with an angular resolution of $\pm 50°$, the directional distribution of all waves regardless of wavelength; analysis of data is relatively simple and straightforward; and the height and slope data are partially redundant, thus providing a check on the accuracy of the sensors used to measure these quantities. For example, the sum of the two slope spectra times wavenumber gives the height spectrum, and wave height and slope, being in quadrature, have a small cospectrum. For these reasons the buoys are popular and frequently used so long as measurements of large, breaking, storm waves are not required, although suitable inertial instruments on large, discus-hulled meteorological buoys may provide even these measurements.

Another technique, analogous to the pitch-roll buoy, has been used near shore. There, measurements of pressure and two components of horizontal velocity made from a fixed platform are used to calculate the first five moments of the directional distribution of wave energy (Nagata, 1964; Bowden and White, 1966).

3.2 Wave Arrays

Measurement of wave height as a function of time at a small number N of points is also used to calculate wave directions, but now all the difficulties of discrete time measurements carry over into the space domain. Probes must be close enough to avoid aliasing short waves into longer waves, but they must extend over sufficiently large distances to provide good wavenumber or directional resolution. Fortunately, the problem is common to many disciplines and a large literature on array theory exists.

Two questions must be answered: Given N probes, where should they be placed? And how should the data be analyzed? The general class of solutions is generally constrained. Ocean waves travel on a plane; they have a well known dispersion relation for wavelengths greater than a few metres; the waves are incoherent; and wave probes are noisy.

Within these constraints, Haubrich (1968) describes how probes may be placed in linear or planar arrays with variable excitation to form a good array. In general the co-array, the set of points which includes all the difference spacings of the array, must be maximized. Only linear arrays with N = 2,3,4 (Moffet, 1968) and planar arrays with N = 3,4 or 6 are 'perfect' (see Fig. 3); and imperfect but 'optimum' arrays can be found by direct search for low N if the probes lie on a prescribed grid. The more general problem of designing arrays with variable N, space distribution, amplitude, and phase excitation, for linear, rectangular, circular, and elliptical grids is the subject of a book by Ma (1974), but Haubrich's paper answers many of the simpler problems posed by oceanographers.

Schemes to analyze data depend critically on exactly what questions are to be answered by the data. The fundamental ones are: What is 'good' resolution? What trade-off will be made between noise rejection or low sidelobe levels and resolution? and how important is computational simplicity? Answering these questions requires an explicit mathematical statement of what is desired, and the

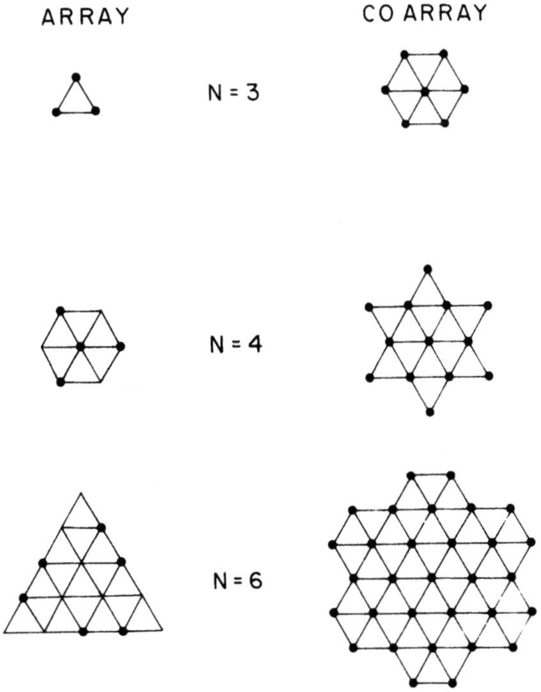

Fig. 3 The perfect arrays on a plane and their co-arrays (after Haubrich, 1968). The perfect uniformly-spaced arrays on a line have probes at coordinates: N = 2 (0,D), N = 3 (0,D,3D), and N = 4 (0,D,4D,6D). No other perfect linear arrays are known.

penalties to be paid for wrong answers. Once stated the problem is solved, at least in principle, by standard techniques of inverse theory (Parker, 1977).

Information on ocean-wave directions is used for many quite different purposes: the tracking of distant storms, the determination of the beamwidth of wind-generated seas, or the finding of dominant directions of large waves along coasts; so it is not possible to outline possible schemes for analyzing data. Davis and Regier (1977) formulate the problem of obtaining optimum algorithms to process various types of wave data, and evaluate the resolution and statistical reliability of both 'a priori' and data-adaptive schemes. The former rely only on prior criteria for optimum analysis; the latter are specifically designed for each particular set of observed data. These data-adaptive schemes frequently provide

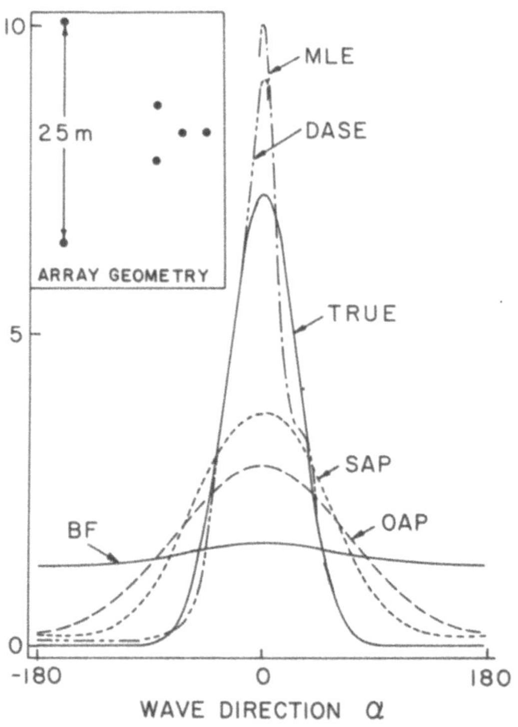

Fig. 4 Tests of the ability of a 25 m array of wave probes (inset) to resolve 150 m waves with a directional distribution of $\cos^{16}(\alpha/2)$ when data are analyzed using various 'a priori' - SAP, OAP, BF, and data-adaptive MLF, DASE algorithms (after Davis and Regier, 1977).

higher resolution than do 'a priori' techniques, especially for waves much longer than the length of the array (Fig. 4), but at a significant increase in the amount of computation.

3.3 Measurements Along a Line

Continuous observation of wave height over large distances avoids the difficulties inherent in the sparse spatial coverage from wave arrays. The simplest extensive measurement is of wave height along a line. The Fourier Transform of such a profile in the x direction yields

$$F(k_x) = \int_{-\infty}^{\infty} \psi(k_x, k_y)\, dk_y$$

From this, $\Phi(n)$ is calculated provided that the directional distribution of the wave is known beforehand.

The measurement of height is typically made from a low-flying aircraft; and it is not instantaneous. The transformation of this measured spectrum from the reference frame of the aircraft to a fixed coordinate system and the calculation of $\Phi(n)$ from these data are described in Barnett and Wilkerson (1967). They show that the motion of the platform complicates the calculation, but introduces no additional ambiguity; and that the calculation is not sensitive to the assumed directional distribution, provided the distribution is unimodal. In practice it is sufficient to assume that the waves travel with the wind and that they have a broad beamwidth, usually of the form $\cos^2(\theta)$, where θ is the angle relative to the wind. Of course, the presence of swell at an angle to the sea complicates the analysis.

Both lasers and radars have been used. The former is now preferred because it is more compact, has a narrower beamwidth, and so observes a smaller spot on the sea surface. Presently a commercial Geodelite using a red-light laser operating at a wavelength of 632.8 nm is routinely used from aircraft flying a few hundred metres above the sea. The light beam is modulated at a maximum radio frequency of 49 MHz, and it measures wave height with a resolution of around 5 cm by observing specular reflections from waves in a spot 15 cm in diameter.

Practical experience rather than theory is the basis for using the laser profilometer. Certainly, light of this wavelength penetrates many metres into clear sea water. But in practice the instrument track is large enough to contain some specular reflection, and this dominates the subsurface scatter. Comparisons of the signals from profilometers mounted on a tower and on an aircraft with that from a wave staff (Fig. 2) verify the operation and accuracy of the laser profilometer (Ross et al., 1970). The aircraft test also demonstrates that the vertical motion of a Lockheed Super Constellation was essentially below an equivalent wave frequency of 0.07 Hz and did not distort the measurement of ocean waves.

4. MEASUREMENTS OVER AN AREA

The sparsity of data from measurements at several points or along a line introduces ambiguities in their interpretation and complicates their analyses. Continuous measurements over an area, by providing complete spatial coverage, avoid these difficulties. Rather than wave height or slope, they often measure some quantity such as density variation on a photographic film that is only indirectly related, and this compromises the usefulness of the techniques. Only Bragg-scattered HF radio waves unambiguously measure $X(\vec{k},n)$ over large areas, but at the cost of large antennas on land.

4.1 Photographic Techniques

It is clear that waves can be seen in daylight by an observer with a clear view of the sea. It is equally clear that the process is selective and some wave components are enhanced, others suppressed, and that converting photographs of the ocean into wave spectra is not simple. One need only observe the relative prominence of different wave components in the glitter pattern of the sea to be convinced of the difficulties. Furthermore, because time variations are not observed, approaching waves cannot be distinguished from receding waves, and the observation is inherently ambiguous.

The technique uses geometrical optics to associate the brightness of every point on the sea surface with the slope that reflects the source of illumination toward the observer; the illumination is either the sun, or the diffuse skylight of a clear blue sky, or a uniformly overcast sky.

Two methods are employed. In the first, the average glitter of reflected sunlight is used to measure the probability distribution function of surface slopes and the second harmonic of the directional distribution of waves (Cox and Munk, 1954).

In the second method, the spatial variation of the brightness of the sea is related to the slope spectrum of longer waves. If every wave were resolved, if the source of illumination were sufficiently simple, and if the photographic process well controlled, then photographs of the sea would be directly and uniquely related to wave slope. But these conditions are difficult to satisfy. Usually short waves are unresolved, and it is assumed, perhaps not explicitly, that they have a uniform, measurable, slope distribution. This is not true, since the waves are modulated by longer waves, and the measurement is slightly nonlinear.

The illumination is usually satisfactory. Cox (1955) and Sugimori (1975) use the sun and observe the modulation of the glitter pattern. This measures the spectrum of that component of slope which increases away from the centre of the glitter, and two photographs are required to obtain the elevation spectrum $\psi(\underset{\sim}{k})$. In practice, because of the broad beamwidth of the wind-generated sea, one photograph may suffice (Sugimori, 1975).

Stilwell (1969) uses diffuse sky illumination. To avoid ambiguity the photograph is taken at an oblique angle well away from vertical and is used with photographs of the sky illumination to obtain slope spectra. The distortion of the oblique photograph can usually be minimized, but because of atmospheric haze, the technique is only used from low elevations and observes the spectra of shorter waves.

Neither method of measuring wave slope spectra has been well tested against independent measurements, and their accuracy and area of applicability are not established. The two published tests, one by Stilwell (1969, see Fig. 2) and one by Sugimori (1975) demonstrate their usefulness in two instances, but more tests are needed before the technique can be routinely used.

In addition to these indirect photographic methods, stereophotography has been used to contour the sea surface in two dimensions, from this the wavenumber spectrum is calculated. The computations are laborious and difficult, but have been used on a few occasions over the past four decades (Dietrich et al., 1975, p. 350). Again, because the measurement is made at a single time, wave directions have a 180° ambiguity.

4.2 Radio Measurements

Radio waves incident on the sea at angles away from the vertical are coherently scattered by ocean waves that match the Bragg scattering equation, and the scattered signal is Doppler-shifted by exactly the frequency of the ocean waves. This allows $X(\underset{\sim}{k},n)$ to be measured directly; the $\underset{\sim}{k}$ dependence is a function of geometry and radio wavelength through the Bragg equations, n is the Doppler shift, and the total backscattered radio power is directly proportional to $k^4X(\underset{\sim}{k},n)$. Because $X \sim k^{-4}$ the total received power is nearly independent of radio frequency or wavelength, and so this resonant scatter provides an accurate and sensitive technique to measure sea-surface roughness (Barrick, 1972).

Two frequency bands are used: that at HF with dekametre wavelengths which scatter from the dominant ocean waves; and microwaves with centimetre wavelengths which scatter from short waves. Both measure waves directly, but the latter is used most often to infer wind speed at the sea surface.

The long wavelength of HF signals poses a problem. Determining their propagation directions, and thus the direction of the scattering ocean waves, requires large antennas. For example, 2 MHz radio waves have a length of 150 m and are backscattered at grazing incidence from 75 m, 7 s waves. Determination of their direction within 6° requires an aperture of 10 wavelengths or an antenna 1.5 km long. Even a simple quarter-wave transmitting antenna is 40 m high, and the problem is to find simple directional antennas.

Fortunately, these long radio waves can be used in simple experiments to measure $X(k,\alpha,n)$ at fixed k. Typically, a pulsed transmitter is placed on the sea surface, an island, or a coast, and the backscattered signals are received by a directional antenna steered in azimuth. Such a radar can measure k, α, and n with an accuracy of 1%, and one was used on Wake Island to measure the directional distribution of 7 s waves with a resolution of 6° (Tyler et al., 1974). There, a directional antenna was synthesized by carrying a

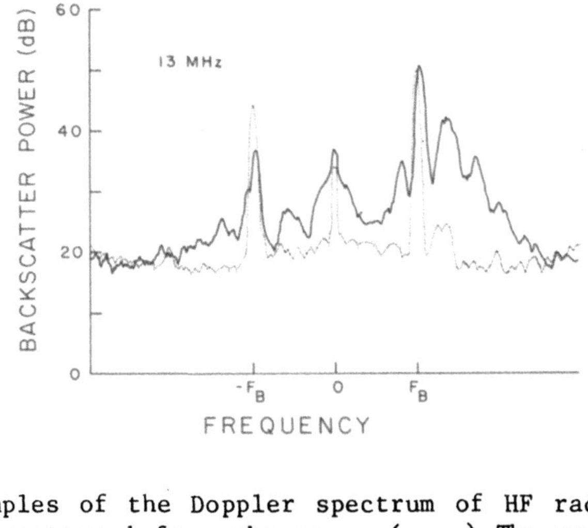

Fig. 5 Examples of the Doppler spectrum of HF radio waves backscattered from the sea. (——) The radio wavelength is much greater than the significant ocean-wave height, and the radar measures only two components of the wave spectrum. (....) Radio wavelength is less than ten times the wave height, and the spectrum contains information about all components of the ocean-wave spectrum. Echo at zero frequency is from land (after Teague et al., 1977); $\pm F_B$ are the ocean wave frequencies which match the resonant scatter.

simple antenna at constant velocity along an aircraft runway. In general these direct measurements are best at measuring X at constant k; multifrequency measurements are difficult to calibrate accurately and are usually avoided.

As the ratio of wave height to radio wavelength increases the scatter becomes more complex but contains more information (Fig. 5). The scatter of shorter HF wavelengths is modulated by longer ocean waves, and the Doppler spectrum of this signal is a function of the entire ocean-wave spectrum. Lipa (1977) has investigated the inverse calculation of using the Doppler spectrum to infer $\psi(k,\alpha)$ and finds that the calculation is possible under some limiting but useful conditions (Fig. 6). The technique is particularly attractive because a single radio frequency can be used to measure the entire wave spectrum.

At microwave frequencies local tilting of the surface by long waves as well as Doppler shifts produced by the long-wave orbital velocity greatly complicate the scatter. As at HF, this information can be used, at least in theory, to observe the long-wave spectrum, particularly with small air- or satellite-borne instruments. In practice this is yet to be proven because the relative importance of the various effects is not yet understood, particularly the amplitude modulation of short waves by the divergence of the long-wave velocity field, and the influence of Doppler shifts on the

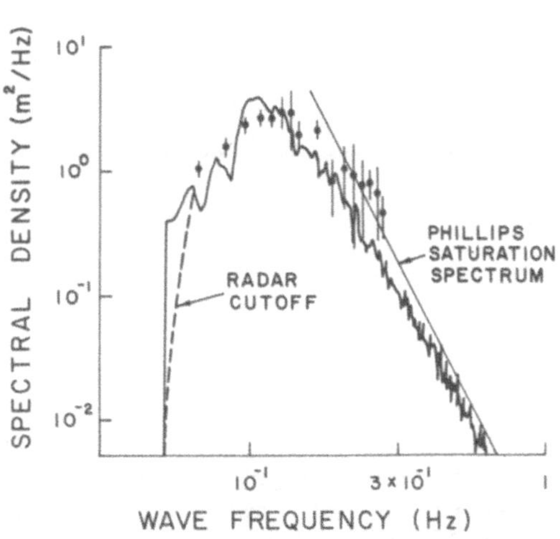

Fig. 6 Comparison of wave spectrum measured by a pitch-roll buoy (solid line) with that calculated from HF radio scatter using sidebands to the resonantly-scattered 22 MHz signal (from Lipa, 1977).

synthetic-aperture radar processing. Maps, with a resolution of a few metres, of the microwave power scattered from the sea, obtained by real and synthetic-aperture radars flown on aircraft, show modulations clearly related to the long waves, but interpretation of the data is controversial and these indirect techniques continue to be developed.

On the other hand, the direct measurement of short waves by microwave radars is well established (Wright, 1966; Wright and Keller, 1971) and is the primary source of information about short waves on the ocean. Examples include the observations by the Naval Research Laboratory (Daley, 1973) that these waves are not saturated but depend on the local wind velocity, Keller and Wright's (1975) measurements of the modulation of $X(k,n,\alpha)$ at fixed k by long waves, and Larsen and Wright's (1975) measurements of the rate of growth of short waves with time.

While Bragg scatter is a powerful tool for measuring waves, so, too, are specularly reflected microwaves. A short radio pulse

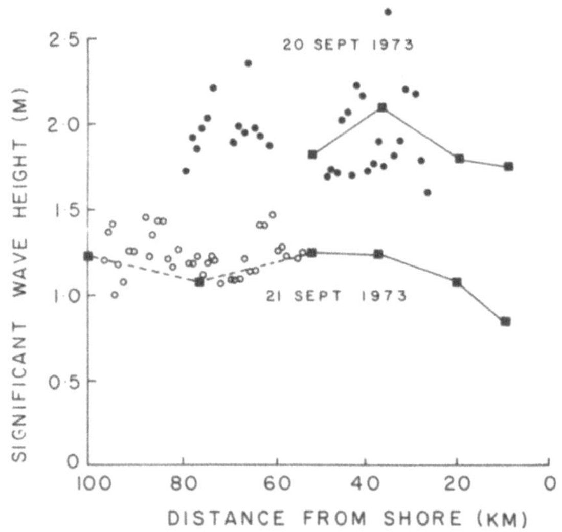

Fig. 7 Significant wave height measured by a short-pulse radar altimeter (o) compared with wave buoy (solid line) and laser profilometer (dashed line) measurements (after Walsh et al., 1977). The scatter in the radio data result from the statistical variability of the scatter and not from instrumental error.

striking a kilometre-sized area at vertical incidence scatters first from wave crests, then from their troughs, and the initially sharp pulse is broadened on reflection by an amount directly proportional to wave height. The technique allows radar altimeters on spacecraft such as GEOS-III to measure wave height, and Walsh et al. (1977), comparing airborne measurements with wave buoy measurements (Fig. 7), shows that the technique can be accurate.

5. AN OVERVIEW OF WAVE MEASUREMENTS

Dominant trends and clear perspectives are easily lost in a listing of techniques, so it is helpful to retreat and ponder these aspects. Clearly, the new techniques are becoming increasingly important and may soon dominate. The radar altimeter on GEOS-III has made hundreds of thousands of measurements of wave height over the oceans. A similar instrument on the now defunct SEASAT-A was capable of making millions of observations over all the world's oceans in winter as well as summer, even in the strongest storms. Such data, augmented by estimates of wave direction, will perhaps in the future provide the routine climatology of waves needed by most users of wave information.

Radio scatter techniques, because they can simultaneously measure wave frequency, direction, and wavelength to high accuracy, are providing many measurements previously difficult or impossible to obtain. Already they have provided most of the information about the high frequency waves on the sea surface. Because of their high precision and because they can be used from aircraft, they are increasingly used to study stormy seas, thus contributing to the most interesting and important area in air-sea interaction studies.

6. CONCLUSION

Simple accurate techniques exist to measure the dominant components of the ocean-wave frequency spectrum. Equally accurate, but much less simple, techniques exist to measure the wavenumber-frequency spectrum. But reports of wave measurements must be viewed with scepticism until it is clear that the measurements come from calibrated instruments because this essential aspect of the measurement is often ignored or quickly passed over.

ACKNOWLEDGEMENTS

This work was supported by the U.S. Office of Naval Research under Contract N00014-75-C-0152.

REFERENCES

BARBER, N.F. and M.J. TUCKER. 1962. Wind waves, In *The Sea,* Vol. 1, edited by M.N. Hill, Wiley-Intersciences, New York: 664-699.

BARNETT, T.P. and J.C. WILKERSON. 1967. On the generation of ocean wind waves as inferred from airborne radar measurements of fetch-limited spectra. *Journal of Marine Research,* 25: 292-328.

BARRICK, D.E. 1972. First-order theory and analysis of MF/HF/VHF scatter from the sea. *IEEE Transactions on Antennas and Propagation,* 20: 2-10.

BOWDEN, K.F. and R.A. WHITE. 1966. Measurements of the orbital velocities of sea waves and their use in determining the directional spectrum. *Geophysical Journal,* 12: 33-54.

CARTWRIGHT, D.E. 1963. The use of directional spectra in studying the output of a wave recorder on a moving ship. In *Ocean-wave Spectra,* Prentice-Hall: 203-218.

COX, C. 1955. Optical measurements of sea surface roughness. *Proceedings of the First Conference on Coastal Engineering Instruments,* Berkeley, 1955, Council on Wave Research: 1-15.*

COX, C. and W. MUNK. 1954. Statistics of the sea surface derived from sun gitter. *Journal of Marine Research,* 13: 198-227.

DALEY, J.C. 1973. Wind dependence of radar sea return. *Journal of Geophysical Research,* 78: 7823-7833.

DAVIS, R.E. and L.A. REGIER. 1977. Methods for estimating directional wave spectra from multi-element arrays. *Journal of Marine Research,* 35: 453-477.

DIETRICH, G., K. KALLE, W. KRAUS and G. SIEDLER. 1975. Allgemeine Meereskunde, 3. Auflage Gebrüder Borntraeger, Berlin, 593 pp.

ESTEVA, D. AND D.L. HARRIS. 1970. Comparison of pressure and staff wave gauge records. *Proceedings of 12th Coastal Engineering Conference,* Washington, D.C., 1970. New York, American Society of Civil Engineers: 101-116.

FARMER, H.G. 1963. A data acquisition and reduction system for wave measurements. In *Ocean Wave Spectra,* Prentice-Hall: 227-233.

---

* Copies available on interlibrary loan through the library of the Scripps Institution of Oceanography, or by writing to the author of this chapter.

HASSELMANN, K., T.P. BARNETT, E. BOUWS, H. CARLSON, D.E. CARTWRIGHT, K. ENKE, J.A. EWING, H. GIENAPP, D.E. HASSELMANN, P. KRUSEMAN, A. MEERBURG, P. MULLER, D.J. OLBERS, K. RICHTER, W. SELL and H. WALDEN. 1973. Measurements of wind-wave growth and swell decay during the joint North Sea wave project (JONSWAP). *Deutschen Hydrographischen Zeitschrift*, *Reihe A8,* 12, 95 pp.

HAUBRICH, R. 1968. Array design. *Seismological Society of America Bulletin,* 58: 977-991.

HUGHES, B.A., H.L. GRANT and R.W. CHAPPELL. 1977. A fast response surface-wave slope meter and measured wind-wave moments. *Deep-Sea Research,* 24: 1211-1223.

KELLER, J.C. and J.W. WRIGHT. 1975. Microwave scattering and the straining of wind-generated waves. *Radio Sciences,* 10: 139-147.

KINSMAN, B. 1960. *Surface waves at short fetches and low wind speeds - a field study.* Vol. 1, Johns Hopkins University, Chesapeake Bay Institute Technical Report 19, 581 pp.

LARSEN, T.R. and J.W. WRIGHT. 1975. Wind-generated gravity-capillary waves: laboratory measurements of temporal growth rates using microwave backscatter. *Journal of Fluid Mechanics,* 70: 417-436.

LIPA, B. 1977. Derivation of directional ocean-wave spectra by integral inversion of second-order radar echoes. *Radio Sciences,* 12: 425-434.

LONGUET-HIGGINS, M.S. 1952. On the statistical distribution of the heights of sea waves. *Journal of Marine Research,* 11: 245-266.

LONGUET-HIGGINS, M.S., D.E. CARTWRIGHT and N.D. SMITH. 1963. Observations of the directional spectrum of sea waves using the motions of a floating buoy. In *Ocean Wave Spectra,* Prentice-Hall: 111-132.

MA, M.T. 1974. *Theory and application of antenna arrays.* Wiley-Interscience, New York.

MITSUYASU, H. 1977. Measurement of the high frequency spectrum of ocean surface waves. *Journal of Physical Oceanography,* 7: 882-891.

MOFFET, A.T. 1968. Minimum-redundancy linear arrays. *IEEE Transactions on Antennas and Propagation,* 16: 172-175.

NAGATA, Y. 1964. The statistical properties of orbital wave motions and their application for the measurement of directional wave spectra. *Journal of the Oceanographical Society of Japan,* 19: 169-181.

NEUMANN, G. and W.J. PIERSON. 1966. *Principles of physical oceanography.* Prentice-Hall, Inc., Englewood Cliffs, N.J., 545 pp.

PARKER, R.L. 1977. Understanding inverse theory. *Annual Review of Earth and Planetary Sciences,* 5: 35-64.

PHILLIPS, O.M. 1966. *The dynamics of the upper ocean.* Cambridge University Press, 261 pp.

ROSS, D.B., V.J. CARDONE and J.W. CONWAY. 1970. Laser and microwave observations of sea-surface conditions for fetch-limited 17- to 25-m/s winds. *IEEE Transactions of Geoscience Electronics,* 8: 326-336.

STEWART, R.H. 1977. A discus-hulled wave measuring buoy. *Ocean Engineering,* 4: 101-107.

STILWELL, D. 1969. Directional energy spectra of the sea from photographs. *Journal of Geophysical Research,* 74: 1974-1986.

SUGIMORI, Y. 1975. A study of the application of the holographic method to the determination of the directional spectrum of ocean waves. *Deep-Sea Research,* 22: 339-350.

TEAGUE, C., G. TYLER and R.H. STEWART. 1977. Studies of the sea using HF radio scatter. *IEEE Transactions on Antennas and Propagation,* 25: 12-19.

TUCKER, M.J. 1959. The accuracy of wave measurements made with vertical accelerometers. *Deep-Sea Research,* 5: 185-192.

TYLER, G.L., C.C. TEAGUE, R.H. STEWART, A.M. PETERSON, W.H. MUNK and J.W. JOY. 1974. Wave directional spectra from synthetic aperture observations of radio scatter. *Deep-Sea Research,* 21: 989-1016.

WALSH, E.J., E.A. ULIANA and B.S. YAPLEE. 1978. Ocean wave heights measured by a high-resolution pulse-limited radar altimeter. *Boundary-Layer Meteorology,* 13: 263-276.

WIEGEL, R.L. 1964. *Oceanographical Engineering.* Prentice-Hall, Inc., Englewood Cliffs, N.J., 532 pp.

WRIGHT, J.W. 1966. Backscattering from capillary waves with applications to sea clutter. *IEEE Transactions on Antennas and Propagation,* 14: 749-754.

WRIGHT, J.W. and W.C. KELLER. 1971. Doppler spectra in microwave scattering from wind waves. *Physics of Fluids,* 14: 466-474.

25

Surface Microlayer Samplers

W.D. Garrett and R.A. Duce

1. INTRODUCTION

1.1 Background

Increasing interest in the chemistry, biology, and physics of the air-water interface has developed in the past decade. Research on natural and man-made surface films and other chemical accumulations near this phase boundary has been engendered by a realization that interfacial chemistry plays an important role in the following timely oceanographic problems. (1) Organic surface films modify air-water interfacial properties and hence influence waves, air-water transport processes, and other physical boundary layer properties. Consequently, ocean operations such as shipping or off-shore industrial production influence and are in turn affected by the nature of the air-sea interface. (2) Natural organic sea surface films or inadvertent petroleum spills act as accumulation sites for lipophilic pollutants such as chlorinated hydrocarbons or organic forms of heavy metals. (3) Organic surface films have a strong influence on the exchange of liquids, solids and gases between water and air. However, the magnitude and exact nature of this role have not been well characterized by 'in situ' experimentation. (4) Both natural and pollutant films modify water surface properties in ways which are detectable by various remote sensing systems; thus it is possible to detect and track oil spills and municipal wastes from airborne platforms. Scientific interest in the chemistry of the sea surface and air-sea interfacial exchange processes is demonstrated by two recent conferences devoted exclusively to this research area (JGR, 1972; JRA, 1974).

To provide data for these areas of interest, specialized surface sampling techniques are required to obtain samples for the determination of the chemical and physical constitution of the surface films and microlayers involved. For example, in the field of remote sensing, knowledge of the chemical nature and physical properties of the sea surface microlayer is essential to provide 'ground truth' for the proper interpretation of the sensing-system data. Sensible signals produced by organic surface films are usually caused by a modification of the physical properties of the air-sea interface and associated boundary layers. The magnitude of the physical effect is a function of the chemical nature of the film, its thickness, and the surface concentration of the chemical species involved. Many of the effects produced by surface films arise from a layer only one molecule in thickness (approximately 30 Å), such as a natural sea slick. These organic films attenuate and resist the formation of capillary waves, modify breaking waves and the resulting distribution of entrained bubbles, influence bubble bursting phenomena, and produce temperature anomalies by various mechanisms. In other instances involving petroleum spills and municipal effluents, thicker films are implicated in the production of sensible signals. In general, interfacial effects that are observed by remote sensors occur within a few millimetres of the air-water boundary. The special problems associated with sampling thin interfacial layers will be addressed in this chapter.

1.2 The Monomolecular Sea Slick versus the Surface Microlayer

The term 'surface film' has taken on a number of interpretations depending upon the scientific discipline and area of research. It may connote a truly monomolecular layer of organic molecules adsorbed at the air-water interface, a thin layer of interfacial water in which neuston dwell, or perhaps a hydrodynamically defined boundary layer which may or may not be under the influence of coherent organic films. In recent literature the term 'surface microlayer' seems to refer to the interfacial layer removed by a particular surface sampler and to imply no specific thickness, chemical composition, or concentration.

If the air-water interface is considered in terms of molecular dimensions, we can define a 'clean' aquatic surface as one where the adsorbed organic molecules, biological organisms, or detrital particles are too widely spaced to influence the air-water interfacial properties to a sensible extent. The concentration of polar, surface-active molecules adsorbed at the interface, however, may be increased by surface convergences, upward transport mechanisms (Fig. 1), or by atmospheric deposition. When the surface concentration reaches a level where the surface molecules are in close proximity, the interface is partially immobilized and the compressibility of the surface film markedly decreases. When this condition is attained, dynamic interfacial processes, such as capillary wave attenuation, bursting bubble phenomena, foaming properties, etc., are significantly modified (Garrett, 1972).

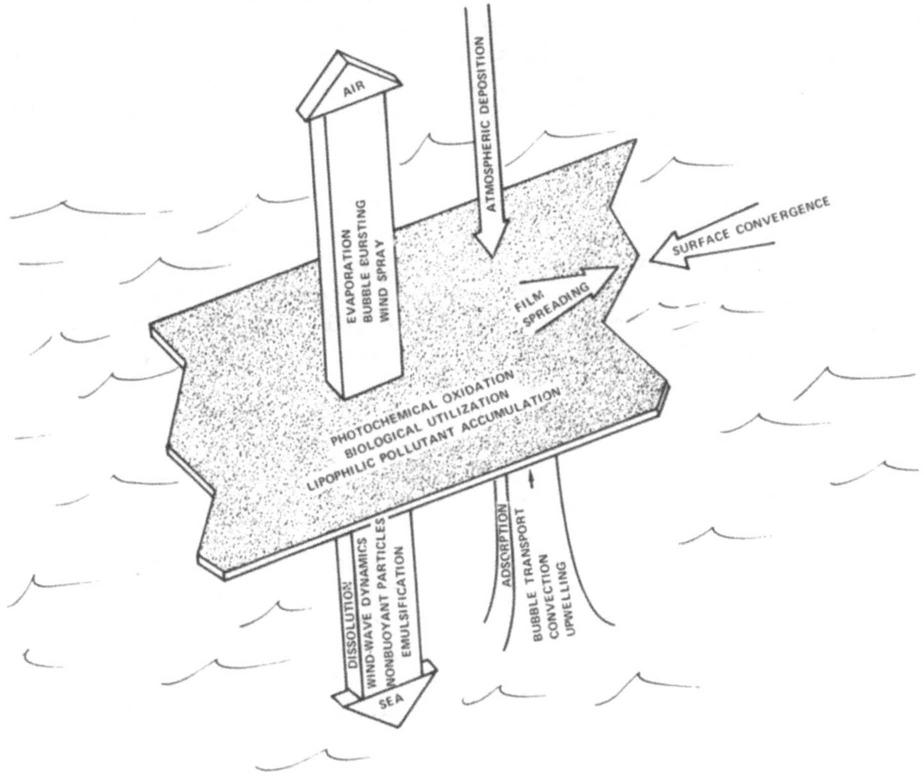

Fig. 1 Physical and chemical processes involved in the life history of an aquatic organic surface film (from Garrett and Smagin, 1976, reproduced with the permission of the World Meteorological Organization).

Natural sea slicks are monomolecular layers of polar organic material physically adsorbed at the air-water boundary. Films such as this, only one molecule in thickness, modify a number of interfacial properties and processes. Specialized samplers have been developed to sample the air-water microlayer, but none of these meets the ideal goal of removing only the adsorbed layer for subsequent chemical analyses and study.

In addition to the adsorbed surface film itself, consideration must be given to the chemical nature of the adjacent subsurface water. The underlying water is, of course, the principal source of the chemical constituents and biological organisms for the sea surface. The dynamic interplay of physical processes in a wind/wave situation constantly disrupts, disperses, and renews the constituents of the interface, such that its chemical composition is controlled by the sticking tendency of the chemical species which come into contact with the water surface. The more strongly adsorbed and less water-soluble species will have longer surface residence times, greater surface persistence, and consequently greater influence on surface properties, than the more oxygenated and water-soluble oceanic constituents.

Organic substances are involved in a number of air bubble processes at the air-water interface. For example, organics probably influence the chemical fractionation of inorganic constituents in bubble fragments ejected into the atmosphere when the bubbles burst. These polar organic materials originate below the surface, adsorb at the interfaces of rising bubbles and are transported to the surface. Thus, devices which duplicate these complex processes and can sample the ejected jet and film drops from bursting bubbles are of vital importance (see below, Section 2.3).

2. SURFACE MICROLAYER SAMPLERS

The water-surface microlayer samplers discussed in the following section were designed primarily to remove thin slices of water from the surface for the subsequent analysis of their chemical constituents. The advantages and limitations listed for the various sampling methods are based on present knowledge and depend on the type of surface sample desired; future research may modify these judgments.

Since ships are a source of both organic and inorganic contamination, surface microlayer sampling is usually performed at a suitable distance from a mother ship in the open ocean or from small boats near shore. Consequently, the methods are sea-state dependent and sampling is seldom possible in rough weather, unless the sampler is attached to booms off the bow of a ship. Sampling with some of the techniques is possible in high amplitude swells in the absence of wind-generated chop and spray. The inability to sample in high sea states is not a serious limitation, because strong wind-wave interactions disperse most substances from the sea surface, and essentially eliminate the physical-chemical effects of the surface microlayer.

2.1 Screen

The first approach to surface microlayer sampling (Garrett, 1962; 1965) involved the use of a metal screen built into a framework fitted with detachable handles. Research indicated that window screen (16 wires per 2.54 cm) had an optimum mesh size. In practice, the screen is brought into contact with the water parallel to its surface, withdrawn, and drained into an appropriate collection vessel. As the screen is raised through the water surface discrete segments of the surface layer between the wires are removed. This removal mechanism is nonselective since adsorptive processes are not involved except for contact with the screen wires. However, initial adsorption of polar chemical species onto the wires makes them hydrophobic, and after several contacts with natural waters, little additional adsorption takes place. The thickness of the surface layer sampled is a function of the wire diameter. However, additional water is collected by draining from the framework which holds the screen. Because there has been no standard screen sampler configuration adopted to date, the thickness of the layer sampled must be determined for each device.

After contact with the aquatic surface, the screen is placed over the collection bottle (usually fitted with a large-diameter funnel) in a tilted position so that the collected water leaves the screen assembly from one corner. If the sampler has a framework to support the screen the first water to drain off is that associated with this frame. Consequently, the water draining off the sampler during the first five seconds after removal from the water is not collected. Draining is allowed until most of the water films between the wires have broken.

The screen method has been used widely to obtain samples for the chemical analyses of both natural and pollutant microlayer constituents in the surface layer. Cleaned metal screens are used when organic constituents are to be studied, and plastic screens have been used for the collection of microlayer samples for inorganic analyses (Duce et al., 1972; Balashov et al., 1974). In addition, sterilized screens have been used extensively for the collection of surface-dwelling organisms (neuston) for identification and determination of their population densities (Sieburth, 1965, 1971; Tysban, 1971). The screen method provides a sample of known layer thickness, and the surface area sampled can be accurately measured and controlled. A screen sampler is simple to construct, operate, and clean. Thus, it may be standardized for use in monitoring programs as well as for research. Disadvantages include the care which must be taken to avoid sample contamination by floating debris, petroleum residues, jellyfish, etc., since these will be collected by the screen.

2.2 Solid Adsorbers and Collectors

A number of microlayer samplers utilize a solid surface which is touched to or drawn through the air-water interface. Collection of surface constituents is by the mechanism of adsorption, but usually some water is associated with the collected material. Such methods emphasize the adsorbable surface-active species and may not recover completely the nonpolar or highly water-soluble components in the surface microlayer, which are less competitive for adsorption sites. Solids dipped vertically have the additional disadvantage that the first surface-active material to be adsorbed is derived from subsurface water well below the interface.

(a) Rotating Drum

A rotating hydrophilic cylinder was developed by Harvey (1966) as a surface microlayer sampler. The motor-driven cylinder is rotated at a surface speed which is slightly less than that of the small boat on which it is mounted. The direction of rotation of the cylinder is such that a thin film of water is lifted upward from the surface. Near the top of the cylinder a 'doctor' blade pressed tightly to the rotating surface directs the collected water into a cup. The sample subsequently flows into a collection bottle via a connecting hose. In the models built to date, use is restricted to relatively calm waters, usually in coastal areas. The rotating drum method is capable of sampling large areas of a water surface (hundreds of square metres per hour) and eliminates the need for manual dippping. However, its greater complexity, sea state limitations, and restricted portability, in comparison with other methods, may limit it to research activities.

The screen and rotating drum samplers have been intercompared on the basis of chemical and biological analyses of field collected samples (Daumas et al., 1976; Roy et al., 1970). In general, the results indicated that a thinner layer (~60 μm) is sampled by the drum than by the screen. Surface enrichments of particulate matter were greater in the drum-collected samples, whereas dissolved species were distributed throughout a thicker surface layer and were more efficiently collected by the screen. To explain these facts, it was concluded by Daumas et al. that water very near and at the air-water interface contained particles and bacterial populations, while the water at slightly greater depth is characterized by dissolved substances and photosynthetic cells. In fact, Roy et al. reported considerably greater surface enrichments of phytoplankton in water collected by the screen method compared with drum samples. These intercomparison studies clearly show that the use of two or more collectors which sample different thicknesses of the surface microlayer can provide a valuable picture of chemical and biological distributions in the vicinity of the air-water boundary.

It should be noted that both the screen and drum methods will collect thin films and monomolecular layers of organic material adsorbed at the air-water interface. However, both methods collect a layer thickness which is 4 to 5 orders of magnitude greater than the dimensions of a monomolecular layer (30 Å). This represents an unwieldy quantity of subsurface water if interest is only in the interfacially adsorbed species. A sampler employing hydrophilic Teflon (Garrett et al., 1974) has been developed recently to collect only the adsorbed constituents of the air-water interface without recovering large quantities of subsurface water. This sampler is discussed in an ensuing section.

(b) Glass Plate

The glass-plate sampler described by Harvey and Burzell (1972) may suffer from the previously discussed limitations of solid adsorbers, e.g. selective adsorption and extensive subsurface contact. However, there has not been sufficient controlled research to determine accurately the collecting characteristics of this sampler. The glass plate is the ultimate in simplicity, consisting of a clean, wettable glass plate and a neoprene (or other noncontaminating material) wiper blade. The plate is inserted vertically through the air-water interface, withdrawn slowly, and the water wetting both sides of the plate is directed into a collection vessel with the wiper blade. A microlayer of thickness from 60 to 100 μm is collected by this technique, and 45 min is required for the collection of a litre of water with a square plate 20 cm on a side by 0.4 cm thick.

(c) Teflon

Teflon has been used as a surface microlayer sampler for lipids and hydrocarbons. Miget et al. (1974) utilized a disk of teflon touched lightly to the water, parallel to the surface. Following contact with the water the disk is held vertically while the adsorbed materials are washed free with a stream of solvent. Controlled laboratory studies indicated good collection efficiencies for various fatty acids, while lower efficiencies were reported for alkanes and aromatic hydrocarbons (Ledet and Laseter, 1974). Larsson et al. (1974) collected surface samples with a teflon plate perforated with conical holes which opened downward toward the water surface. Samples were analyzed for lipids and associated organochlorine residues. When this collector was dipped horizontally through the water surface, monomolecular layers of pure fatty acids and esters were transferred to the plate at from 70 to 100% area efficiency.

The effectiveness of teflon as an adsorber of surface films from water is difficult to explain. According to Garrett (1962), smooth, clean sheets of teflon and polyvinylfluoride did not remove any surface film or water when placed horizontally on a water surface covered with a monomolecular film of a fatty acid. In addition, hydrophilic surfaces such as glass ceased to retrieve surface films on horizontal contact, once they had adsorbed a closely packed monomolecular layer which covered all hydrophilic adsorption sites. It is possible that the success of the teflon samplers may be due to a less-than-clean surface condition giving rise to hydrophilic adsorption zones, or that the lipids and hydrocarbons were in particulate form and were physically removed along with an adherent water film.

(d) Hydrophilic Teflon

In some sampling situations it is desirable to reduce contact with subsurface water in order to minimize the quantity of nonsurface constituents collected. This goal has been achieved in part through the development of a surface sampler which takes advantage of the inertness of teflon and the high collection efficiency of a hydrophilic surface (Garrett and Barger, 1974). A sheet of teflon whose surface has been made hydrophilic by etching with a solution of sodium in liquid ammonia is used to adsorb surface film constituents from the air-water interface. This material proved to be the most effective of many solid adsorbers examined. It not only recovered monolayers of fatty acids, fatty esters, and nonpolar alkanes efficiently from water, but also continued to collect surface film upon repeated contact with a monolayer-covered surface. Laboratory and field research demonstrated that monolayers can be recovered by the technique, and proved that the collected organic material could be recovered by solvent extraction in sufficient yield for chemical analysis. Collection of subsurface organic matter is minimized by draining a thin film of collected water which wets the sampler. This is done between each contact with the water surface. Only the slightly wet adsorber sheet containing the adsorbed constituents of the surface film is returned to the laboratory for processing. The collector sheet is attached to a framework fitted with detachable handles and makes contact with the water parallel to the surface. This sampler was designed to collect organic material adsorbed at or floating on the water surface, to the exclusion of subsurface materials. Additional experience in using this new sampling technique is required as well as intercomparisons with other microlayer collectors.

(e) Internal Reflection Prisms

Surface films as thin as a monomolecular layer can be analyzed by infrared spectrophotometry using multiple internal reflection prisms (Baier, 1970). The surface films may be adsorbed from water solution or an interface onto optically polished prisms made of a suitable material, such as germanium. The film coatings are analyzed in an internal reflection mirror assembly which directs an infrared beam into the beveled end of the coated prism. The prism acts like a light pipe and causes multiple internal reflections which amplify the infrared absorption effects of the adsorbed film.

Aquatic surface films are transferred to a clean, wettable prism by dipping it through the air-water interface (Fig. 2). The mechanisms of film transfer to a completely hydrophilic solid have been reviewed by Baier (1970). There is no film transfer to a completely hydrophilic solid introduced vertically downward through a monolayer-covered surface. Upon removal a single molecular layer is transferred to each side of the plate. A single dip technique has been used to characterize the chemical nature of the surface microlayer (Baier, 1972; Baier et al., 1974). Multilayers of organic materials such as petroleum films are also transferred in a suitable condition for infrared analysis. In addition to internal refraction IR spectrophotometry, the adsorbed films may be subjected to other nondestructive analyses to further characterize their chemical composition. Contact angles of various liquids on the film-coated plate can provide information on the constitution of the outermost layers of the film. Ellipsometric measurements will give the film thickness, and the electrical properties of the film may be determined with a vibrating-reed electrometer.

Caution should be applied to the interpretation of the results obtained by this approach to surface film analysis. First, it is obvious from the film-transfer mechanisms that the hydrophilic prism

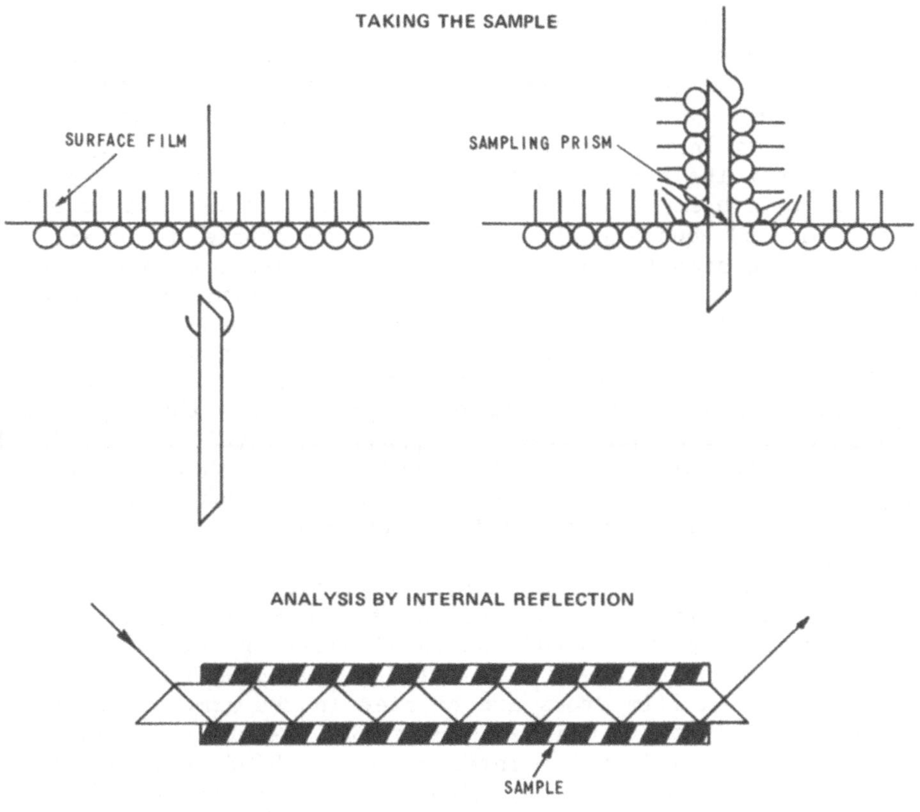

Fig. 2 Surface film recovery, and analysis by internal reflection (from Baier, 1970, reproduced with the permission of the Journal of Great Lakes Research).

does not pick up surface film on its entry into the water. Thus, its first contact is with water-soluble subsurface substances, many of which are surface active and will readily absorb. Although some contact will be made with lipids of low water solubility, the soluble proteinaceous material, carbohydrates, and humic substances predominate in natural water. When the prism is withdrawn, it will remove any surface film present, but the IR spectrophotometer will respond to both surface and subsurface material adsorbed onto the collector's surfaces. Secondly, the spectra obtained from the single-dip procedure show only the major adsorption bands. There may be insufficient detail in the form of second-order bands to provide an unambiguous identification of the chemical classes present. On the other hand, when used for the collection of multilayers such as thick natural foams and scums or films of pollutant petroleum products, spectra are obtained from which valuable information can be derived (Baier, 1970; 1972).

2.3 Bubble Interfacial Microlayer Sampler

A completely different type of surface microlayer collector is the Bubble Interfacial Sampler (BIMS). This instrument utilizes the bursting bubble as a microtome to skim off a very thin layer of surface water and inject it into the atmosphere as jet and film droplets. MacIntyre (1968) has calculated that the top droplet produced from the central jet in the bubble cavity is composed of material originally spread over the interior of the bubble surface at a thickness equal to about 0.05% of the bubble diameter. Thus for a 200 μm diameter bubble, the top jet droplet is composed of material from a surface layer of the bubble interior only 0.1 μm thick. The second jet drop is produced from the next 0.1 μm, etc. With a bubble size distribution in the sea ranging from approximately 50 μm to perhaps 1500 μm diameter, the top jet drops produced from these bubbles strip off approximately the top 0.025 to 0.75 μm of the air/water interface. MacIntyre (1972) suggests that the shattering bubble cap, which produces atmospheric film drops, is probably no less than 2 μm thick when it breaks, so film drops are probably produced from a thicker layer than jet drops.

The BIMS is shown in Figure 3 (Fasching et al., 1974). The device is suspended between the twin hulls of a 4 m long catamaran; bubbles approximatey 500 to 1000 μm in diameter are produced at variable depths down to 50 cm. The bubbles rise and burst at the sea surface; jet and film drops are created in the atmosphere enclosed within the truncated pyramid of the BIMS. The ambient marine atmosphere is excluded from the interior of the BIMS by clean air curtains in the front and back, wind screens on either side, and a slight positive pressure inside. The artificially produced sea salt particles are collected from the enclosed atmosphere at the top of the truncated pyramid on 20x25 cm Whatman 41 filters, using a commercial high volume air sampling pump.

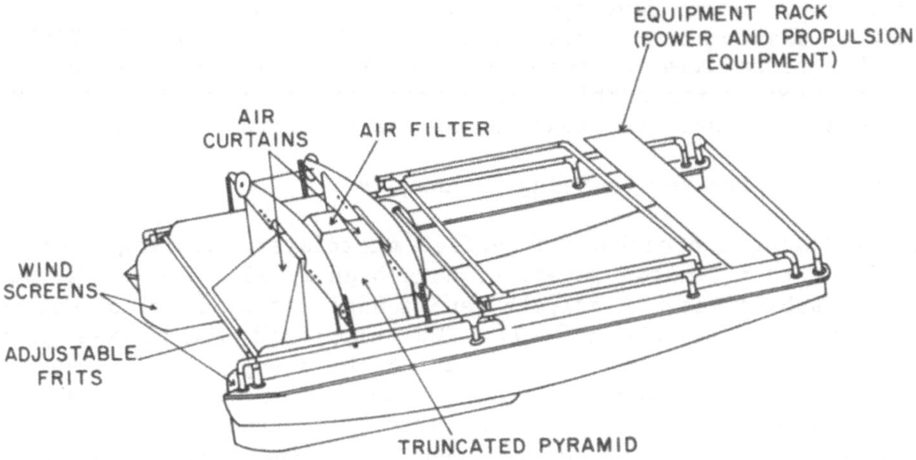

Fig. 3 The bubble interfacial microlayer sampler suspended between the twin hulls of a 4-m catamaran (from Fasching et al., 1974, reproduced with the permission of the Journal de Recherches Atmosphériques).

It must be emphasized that particles produced by the BIMS are not necessarily representative of the chemical composition of only the surface microlayer of the water being sampled, because the bubbles will scavenge a considerable amount of surface active material, and associated substances, and transport them to the surface. This bubble scavenged material and the surface microlayer material present before the bubble arrives at the surface are thus combined in the bubble skin when it bursts. This scavenging effect can be minimized by generating the bubbles very close to the sea surface. While the thickness of the microlayer thus sampled by the BIMS is rather uncertain, varying from a few hundredths of a micrometre up to a few micrometres, it does sample a layer at least one or two orders of magnitude thinner than the screen, glass plate, or rotating drum.

The foregoing description indicates that the BIMS is not strictly a surface microlayer sampler; rather it attempts to duplicate a sequence of natural processes which occur when air bubbles are entrained into the sea by breaking waves, rise through the water and break through the air-water interface. In this sense it is an important device for the study of chemical enrichments in the aerosols ejected from the bursting bubbles.

The system used to date is limited to coastal waters in near calm conditions, although an open ocean version is being built. In general the system is rather cumbersome to use and difficult to transport. It is being developed as part of a research program for the study of sea-to-air material exchange. The realism of the BIMS approach will be examined through a comparison of the number and size distribution of the bubbles and aerosol produced by the system with bubble and aerosol distributions at sea under various conditions.

2.4 Other Microlayer Sampling Techniques

The following approaches to surface microlayer sampling have been developed to a limited extent or have been merely proposed as workable possibilities. A polyethylene funnel was used to collect surface films by Morris (1974). The stoppered funnel was pushed through the water surface, after which the stopper was removed, and the funnel was allowed to fill with subsurface water. It was then moved sideways to an undisturbed area and brought slowly upwards; an area of film, equal to the open end of the funnel, was entrapped. The surface film adhered to the inside of the funnel as it was slowly pulled from the water. Solvent extraction removed the adhering film constituents from the collector. This method appears to be tedious, and the area sampled is only as large as the mouth of the funnel. As with any solid adsorber, the possibility exists of selective adsorption which could bias results in favour of the more strongly adsorbed species.

Several investigators have used the previously discussed screen and rotating drum methods to collect surface water for biological examination. A more recent technique uses sterile 'Nuclepore' membranes as adsorbers for microbes in the surface layer (Crow et al., 1975). The membranes are floated on the water surface and retrieved with sterile plastic dishes raised upwards through the water surface. Under calm conditions forceps were used for recovery of the floating membranes. Collected populations of bacteria and fungi were several orders of magnitude greater than were previously reported for surface slicks.

A method proposed for the collection of lipophilic organic substances deposited onto the sea surface from the atmosphere employs an artificial sea slick as a collector film (Garrett and Smagin, 1976). A large sea surface area can be covered with small quantities of oleyl alcohol, a liquid fatty alcohol which spreads spontaneously into a monomolecular layer. After a period of time the monolayer and collected lipophilic fallout from the atmosphere can be retrieved with an appropiate surface microlayer sampler. Sample calculations were made using existing data on nonvolatile hydrocarbon concentrations in the marine atmosphere and an assumed deposition velocity to determine the feasibility of the approach. Although the calculations proved encouraging, there are numerous logistical and analytical problems in such a scheme. For example, assuming that recovery of the collector film is possible, there would be 100 times as much oleyl alcohol as atmospheric hydrocarbons in the sample. It was also pointed out that since the incorporation of subsurface species into the organic collector film is possible, the experiment should be conducted in areas which are as free as possible of hydrocarbons from sources other than the atmosphere.

3. 'IN SITU' AIR-WATER INTERFACIAL MEASUREMENTS

3.1 Surface Tension

The surface tension of a natural water body may be determined in situ by a refinement of the spreading oil method of Adam (1937), a technique for the measurement of the tension of a dynamic water surface. In a typical application of this method by the Naval Research Laboratory, Washington, D.C., a large number of spreading oils were prepared and calibrated to ±0.6 dyn cm^{-1}, so that surface tension could be determined in approximately 1 dyn cm^{-1} increments between 74 and 62 dyn cm^{-1} and every 2 dyn cm^{-1} for lower values of surface tension. Calibration was performed on a hydrophil balance by the piston monolayer technique described in detail by Adam (1937) and Zisman (1941). A natural sea water substrate was used in the calibration procedure. Its surface was cleaned of interfering film-forming organic material prior to use.

The in situ surface tension measurement is performed as follows. A buoyant spreading oil applicator (toothpick) is dipped into a spreading oil and tossed onto the water surface. Oils with decreasing surface tension values are dispensed until spreading of the oil is observed as indicated by an obvious patch of iridescent colours and/or a silvery sheen. In this manner the surface tension can be bracketed between the values of the last nonspreading oil applied and the first oil which spreads. Alternating dilations and contractions of the water surface due to passing waves sometimes cause the oil to oscillate between a spread film and unspread lens due to actual surface tension changes in the sea surface. When this condition occurs, the in situ surface tension is very nearly that of the spreading oil used.

The table lists measured surface tension values at which the spreading oils will just spread onto clean natural sea water and the corresponding concentrations of 1-dodecanol in clean, light paraffin oil. Computed values are based on a hyperbolic expression derived from the data in the table by the method of least squares. The difference between these values and the actual surface tension as measured by the piston monolayer method are included. The relationship between the surface tension of the spreading oils and their composition is depicted in Figure 4.

Table 1

Spreading Oil Composition and Calibration

Concentration of* 1-dodecanol in Paraffin Oil ($ml\ l^{-1}$)	Measured** surface tension of Oil ($dyn\ cm^{-1}$)	Computed *** surface tension of Oil ($dyn\ cm^{-1}$)	Difference Measured-Computed ($dyn\ cm^{-1}$)
0.660	----	74.5	----
0.860	----	73.6	----
1.080	----	72.6	----
1.320	72.8	71.6	1.2
1.580	71.3	70.6	0.7
1.584	69.4	70.6	-1.2
1.776	69.3	69.9	-0.6
1.860	69.6	69.6	0.0
1.992	68.6	69.1	-0.5
2.256	68.2	68.3	-0.1
2.520	67.5	67.5	0.0
2.832	66.9	66.5	0.4
3.192	65.8	65.6	0.2
3.600	64.5	64.6	-0.1
4.032	63.4	63.6	-0.2
4.536	62.0	62.6	-0.6
5.760	60.3	60.5	-0.2
7.272	57.9	58.3	-0.4
9.192	56.5	56.3	0.2
11.559	54.7	54.3	0.4
14.712	52.8	52.4	0.4
18.600	51.7	50.8	0.9
23.520	48.5	49.3	-0.8

Filtered sea water substrate 73.0

* 1-dodecanol = lauryl alcohol, Eastman Kokak No. 873 dissolved in paraffin oil, N.F., Saybolt viscosity 125/135, Fisher Scientific No. 0-119

** By piston monolayer technique

*** Based on best fit of measured values to $r = [k/(x+a)] + b$ by the method of least squares where r = surface tension value, x = concentration in $ml\ l^{-1}$, and a = 6.08, b = 41.84, k = 220.34

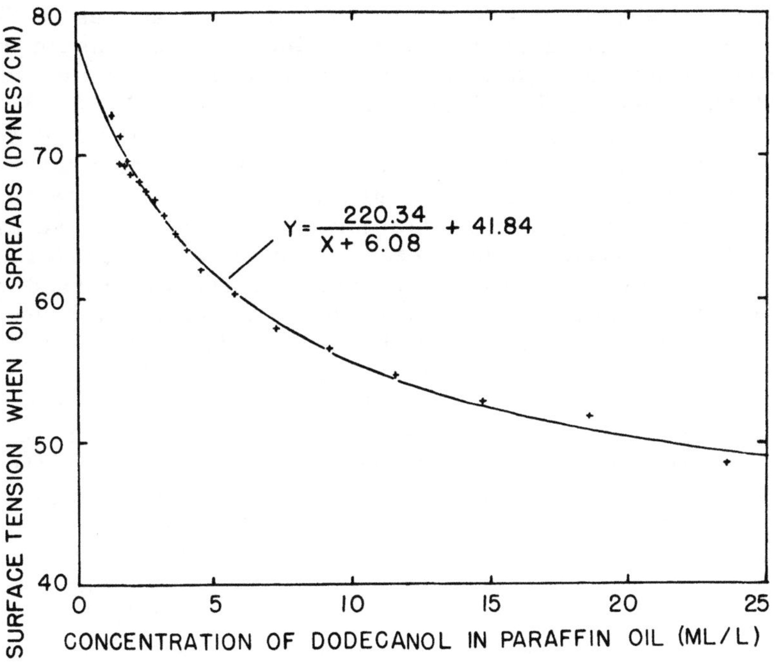

Fig. 4 Relationship between the surface tension at which a spreading oil just spreads and its composition. Solid line is best fit hyperbola; data points represent actual measurements.

3.2 Concentration of Film-forming Material at the Air-water Interface

Because of the dynamic nature of an air-water interface, the analysis of collected film-forming material does not provide sufficient data for the calculation of the concentration of surface-active molecules adsorbed at a particular instant. Information of this kind is required to predict the extent of surface convergence necessary to increase the concentration of film-forming molecules to a level where a coherent surface film forms. A device called a 'floating film balance' (Barger et al., 1974) has been successfully employed to determine in situ the relative surface concentration of slick-forming constituents of the sea surface on both slick-covered and clean rippled waters.

An improved approach involves the use of a floating plastic hoop which encompasses about 0.5 m^2 of water surface (Garrett and Barger, 1974). A plastic skirt projects 5 cm below and 10 cm above the ring to eliminate disruption of the captive surface by waves or spray. Prior to contact of the hoop with the water, powdered clean white paraffin wax is dusted onto the water to act as a floating indicator of surface motion. The inside of the plastic ring is coated with oleyl alcohol, a pure surface-active compound whose surface-film pressure (surface tension depression) is accurately known. When the coated hoop is placed onto the powdered water surface, the oleyl alcohol film spreads, pushing the natural film and the indicator powder toward the centre. When equilibrium is reached (in a few seconds), the powder marks the area occupied by natural film at the maximum film pressure (equilibrium spreading pressure) of the oleyl alcohol. The area of the compressed natural film is adjusted upwards by calculation to a value which would correspond to a film pressure of 1 dyn cm^{-1}. It is at this pressure that surface effects become appreciable and sensible sea slicks are first produced. Thus, it is possible to determine the extent of surface compression or surface area reduction required to form a coherent, visible slick at a particular sea site. In addition, if one assumes that the average composition of the surface film is that of a particular compound, it is possible to calculate a surface concentration in mass per unit area of water surface. Additional assumptions necessary for this calculation are: (1) that the surface films at a particular site are similar in properties to those used for the area adjustment calculations (Barger et al., 1974), and (2) that the natural surface films are essentially water insoluble, so that losses by dissolution under the pressure of the driving film are minimal. While this calculation may be slightly in error because of the assumptions, the described technique nevertheless provides an estimate of previously unobtainable concentration values of surface-active organic material at an air-water interface.

4. GUIDELINES FOR THE SELECTION OF SAMPLING TECHNIQUES

4.1 Normal Water Surfaces

At present no surface microlayer sampler can meet all of the requirements of the various research efforts in the field of air-sea interactions. It is unlikely that any of the described samplers will be universally adopted, although a few of them have been judged to be suitable for monitoring projects (UNESCO, 1976). The selection of a device for a particular project, whether it be for research, baseline studies, or monitoring, depends upon the microlayer constituents under study and the portion of the microlayer which is being emphasized. For example, studies of the monomolecular film adsorbed directly at the sea surface will require a different sampling approach than studies of photosynthetic organisms which live near but not at the interface. A broad research project which examines both chemical and biological species in the microlayer may require two or more sampling methods to derive realistic information. In addition, concentration and surface enrichment gradients may be obtained by utilizing several samplers which remove different thicknesses of the surface microlayer.

A judgment on sampler selection may be obtained from the descriptions of the devices outlined in this chapter; the following criteria are listed as additional guidelines for selection.

(1) The sampler should be inert and not modify or contaminate the collected chemical or biological constituents of the microlayer samples.

(2) The water surface area sampled or the layer thickness must be known if it is desired to calculate (a) the surface concentration of a particular component, or (b) the enrichment of a constituent in the surface layer.

(3) The sampler should have a known collection efficiency for the components sought if concentration data are desired.

(4) For monitoring programs the device should be simple to use, easy to clean, and portable enough for small-boat operation away from sources of contamination.

(5) The sampler should be used in a manner which avoids contact with water beneath the microlayer to the greatest extent possible.

4.2 Surface Covered with Petroleum and Other Thick Pollutant Films

Most of the sampling concepts discussed here have also been considered for the collection of petroleum and other pollutant films from a water surface. Other techniques have been described as well for these substances (Chang and Saner, 1974); a complete review of these methods is not within the scope of this text. The present consensus is that the most important purpose for sampling an oil spill is for source identification to aid in the determination of liability. In an application of this kind the sampler must not modify the collected petroleum or contribute any extraneous chemicals to it. Consequently, the use of hydrophilic teflon has been recommended as a sampling medium (Fortier and Jadamec, 1974). The wettable teflon is perforated with holes and attached to a rake-like sampler which is swept through an oil spill. The collection, although not quantitative, has been shown to be a valid qualitative representation of the spilled oil. Methods for the determination of oil-layer thickness are available, but because it is difficult to apply them to oil-spill situations under dynamic wind-wave conditions, they are seldom utilized. There is considerable evidence that oil spills in open water degrade rapidly to tar-like residues which may be collected by towed neuston nets. Thus a fresh, unweathered petroleum film is relatively rare in open water. Some of the hydrocarbons from these spills will be distributed throughout the water column, as well as in the surface microlayer, in dissolved form. Dissolved and dispersed petroleum constituents in surface layers may be sampled by most of the described microlayer samplers as long as they are constructed from materials which do not modify or contaminate the collected sample.

5. CONCLUSIONS

A number of surface microlayer samplers have been developed over the last decade to collect thin slices of surface water for chemical analyses, physical-chemical measurements, and biological studies. These samplers vary in their complexity, in the thickness of the surface layer which they remove, in the degree of contact with subsurface water, and in other features. As no single type is adaptable to all sampling requirements, selection must be based on sampler characteristics and on the nature of the research to be performed. While additional microlayer collection techniques may be developed, the existing samplers are adequate for a broad variety of studies. Additional intercomparison studies are recommended to broaden knowledge of the suitability of the various samplers for specific research goals.

REFERENCES

ADAM, N.K. 1937. A rapid method for determining the lowering of tension of exposed water surfaces with some observations on the surface tension of the sea and of inland waters. *Proceedings of the Royal Society of London,* B-122: 134-139.

BAIER, R.E. 1970. Surface quality assessment of natural bodies of water. *Proceedings of the 13th Conference on Great Lakes Research:* 114-127.

BAIER, R.E. 1972. Organic films on natural waters: their retrieval, identification and modes of elimination. *Journal of Geophysical Research,* 77: 5062-5075.

BAIER, R.E., D.W. GOUPIL, S. PERLUMUTTER and R. KING. 1974. Dominant chemical composition of sea-surface films, natural slicks and foams. *Journal de Recherches Atmospheriques,* 8: 571-600.

BALASHOV, A.I., Y.P. ZAYTSEV, G.M. KOGAN and V.J. MIKHAYLOV. 1974. A study of some components of the chemical composition of water at the ocean-atmosphere interface. *Oceanology,* (USSR) 14: 664-668 (American Geophysical Union translation).

BARGER, W.R., W.H. DANIEL and W.D. GARRETT. 1974. Surface chemical properties of banded sea slicks. *Deep-Sea Research,* 21: 83-89.

CHANG, W.J. and W.A. SANER. 1974. Evaluation of boat deployable thin film oil samplers. Preprint of paper for 6th Offshore Technology Conference, Houston, Texas, May 6-8: 467-486.

CROW, S.A., D.G. AHERN and W.L. COOK. 1975. Densities of bacteria and fungi in coastal surface films as determined by a membrane-adsorption procedure. *Limnology and Oceanography,* 20: 644-646.

DAUMAS, R.A., P.L. LABORDE, J.C. MARTY, and A. SALIOT. 1976. Influence of sampling method on the chemical composition of water surface film. *Limnology and Oceanography,* 21: 319-326.

DUCE, R.A., J.G. QUINN, C.E. OLNEY, S.R. PIOTROWICZ, B.J. RAY and T.L. WADE. 1972. Enrichment of heavy metals and organic compounds in the surface microlayer of Narragansett Bay, Rhode Island. *Science,* 176: 161-163.

FASCHING, J.L., R.A. COURANT, R.A. DUCE and S.R. PIOTROWICZ. 1974. A new surface microlayer sampler using the bubble microtome. *Journal de Recherches Atmospheriques,* 8: 649-652.

FORTIER, S.H. and J.R. JADAMEC. 1974. An oil slick sampling system. U.S. Coast Guard Report No. CG-D-71-75. National Technical Information Service, Springfield, Va. 22151, 18 pp.

GARRETT, W.D. 1962. Collection of slick-forming materials from the sea. Naval Research Laboratory Report #5761, Washington, D.C., 10 pp.

GARRETT, W.D. 1965. Collection of slick-forming materials from the sea surface. *Limnology and Oceanography,* 10: 602-605.

GARRETT, W.D. 1972. Impact of natural and man-made surface films on the properties of the air-sea interface. In *The Changing Chemistry of the Oceans,* Almqvist and Wiksell, Stockholm.

GARRETT, W.D. and W.R. BARGER. 1974. Sampling and determining the concentration of film-forming organic constituents of the air-water interface. Naval Research Laboratory Memorandum Report 2852, 13 pp.

GARRETT, W.D. and V.M. SMAGIN. 1976. Determination of the atmospheric contribution of petroleum hydrocarbons to the oceans. World Meteorological Organization, Report No. 440, Geneva, Switzerland, 27 pp.

HARVEY, G.W. 1966. Microlayer collection from the sea surface: a new method and initial results. *Limnology and Oceanography,* 11: 608-613.

HARVEY, G.W. and L.A. BURZELL. 1972. A simple microlayer method for small samplers. *Limnology and Oceanography,* 17: 156-157.

JGR. 1972. *Journal of Geophysical Research,* 77: 5059-5350. Papers from Working Symposium on Sea-Air Chemistry, Jan. 31-Feb. 3, 1972, Fort Lauderdale, Florida.

JRA. 1974. *Journal de Recherches Atmospheriques,* 8: 509-997. Papers from the International Symposium on the Chemistry of Sea/Air Particulate Exchange Processes, 4-10 October 1973, Nice, France.

LARSSON, K., G. ODHAM and A. SODERGREN. 1974. On lipid surface films on the sea. I. A simple method for sampling and studies of composition. *Marine Chemistry,* 2: 49-57.

LEDET, E.J. and J.L. LASETER. 1974. A comparison of two sampling devices for the recovery of organics from aqueous surface films. *Analytical Letters,* 7: 553-563.

MacINTYRE, F. 1968. Bubbles: a boundary layer "microtome" for micron thick samples of a liquid surface. *Journal of Physical Chemistry,* 72: 589-592.

MacINTYRE, F. 1972. Flow patterns in breaking bubbles. *Journal of Geophysical Research,* 77: 5211-5228.

MIGET, R., H. KATOR, C. OPPENHEIMER, J.L. LASETER and E.J. LEDET. 1974. New sampling device for the recovery of petroleum hydrocarabons and fatty acids from aqueous surface films. *Analytical Chemistry,* 46: 1154-1157.

MORRIS, R.J. 1974. Lipid composition of surface films and zooplankton from the Eastern Mediterranean. *Marine Pollution Bulletin,* 5: 105-109.

ROY, V.M., J.L. DUPUY, W.G. MacINTYRE and W. HARRISON. 1970. Abundance of marine phytoplankton in surface films: a method of sampling. *Proceedings of the Symposium on Hydrobiology,* Proceedings Series No. 8, Miami Beach, Florida, June 1970: 371-380.

SIEBURTH, J. McN. 1965. Bacteriological samplers for air-water and water-sediment interfaces. *Ocean Science and Ocean Engineering: Transactions of the Joint Conference and Exhibit,* 14-17 June 1965, Washington, D.C.: 1064-1068.

SIEBURTH, J. McN. 1971. Distribution and activity of oceanic bacteria. *Deep-Sea Research,* 18: 1111-1121.

TYSBAN, A.V. 1971. Marine bacterioneuston. *Journal of Oceanographical Society of Japan,* 27: 56-66.

UNESCO. 1976. Report of the International Co-ordination Group (ad hoc task team) for the Global Investigation of Pollution in the Marine Environment, Intergovernmental Oceanographic Commission; IOC/GIPME-III/8 rev., Paris, Feb. 16, 25 pp.

ZISMAN, W.A. 1941. The spreading of oils on water. Part III. Spreading pressures and the Gibbs adsorption relation. *Journal of Chemical Physics,* 9: 798-793.

26

Atmospheric Radiation Instruments

H. Hinzpeter

1. INTRODUCTION

Radiation measurements near the ocean-atmosphere interface are primarily made to determine the net radiation flux and the sea surface temperature for studies of the oceanic energy budget. For that budget, three energy fluxes are usually considered as important: the sensible heat flux, the flux of latent heat (evaporation), and the net radiation flux. The last is sometimes discussed separately as the short wave flux (coming from the sun) and the long wave flux (emitted by the atmosphere and the earth's surface). The radiation fluxes depend on different parameters, but mainly upon the height of the sun in the sky, the surface temperature, and the cloudiness. As an example, we find in the tropical ocean at noon about 800 W m^{-2} for the short wave net flux, 60 W m^{-2} for the long wave net flux, 60 W m^{-2} for the latent heat flux, and 10 W m^{-2} for the sensible heat flux. The last two also depend on the wind speed. The short wave radiation flux decreases with decreasing sun height, while the other three fluxes are nearly independent of the time of day.

Since the divergence of radiation fluxes is small in the lowest layer above the ocean, measurements at a distance of 10 to 20 m are representative of the fluxes at the surface. Since meteorologists are interested in the same radiation quantities near the continental surfaces, the instruments used over land and over sea are similar. But the instruments used on a moving vessel need special mountings and protection from sea spray. Only calorimetric instruments are considered (radiation temperature measurements are discussed by Katsaros [see Chapter 16] and will not be treated here).

The determination of the energy budget requires instruments sensitive to the total net flux (about $0.3 < \lambda < 50\ \mu m$). Since the long wave radiation ($3 < \lambda < 50\ \mu m$) does not enter the water, it is particularly desirable to measure the short wave flux (about $0.3 < \lambda < 3\ \mu m$) coming from above separately in order to determine the flux entering the ocean. For determination of the radiation fluxes within the water the upward and downward short wave fluxes at the surface form the boundary conditions. For optimal accuracy the total net flux components should be measured separately for each hemisphere.

Since the shape of the spectrum of the incoming radiation varies with turbidity, cloudiness and sun elevation, the total spectrum must be sampled and therefore grey or black receivers are needed. The quality of the black lacquers used in radiation instruments will not be discussed here.

2. CALORIC RADIATION FLUX MEASUREMENTS

Nearly all operational radiation instruments use the temperature difference between the receiver surface and a reference body as a measure of the radiation flux density. The following analysis is presented to illustrate the dependence of that difference on various factors and to show how various instrument designs attempt to minimize undesirable influences of some of these factors. For simplicity, we consider instruments with steady state conditions so that the thermal time constants of the instruments do not enter the analysis.

The variables used have the following definitions:

a the absorptivity of the receiver for short wave radiation;

b the absorptivity of the receiver for long wave radiation;

α the effective heat transfer coefficient (conduction and convection by the surrounding gas);

β the effective thermal conductivity (by metal and plastics) between the receiver and its environment;

T_i the absolute temperature of the receiver;

T the absolute temperature of the reference body;

S the hemispheric short wave flux density;

L the hemispheric long wave flux density;

σ the Stefan-Boltzmann constant;

F any other energy flux to the receiver;

I the electrical current; and

R the electrical resistance.

For simplicity it is assumed that the reference body and the air have the same temperature.

With these definitions the energy balance of the receiver is:

$$aS + bL - b\sigma T_i^4 + F = \alpha(T_i - T) + \beta(T_i - T) \qquad (1)$$

In practice, only ΔT, which is $T_i - T$, and T, the temperature of the reference body, are measured. For $\Delta T/T \ll 1$ we obtain

$$\sigma T_i^4 = \sigma T^4 + 4\sigma T^3 \Delta T \qquad (2)$$

A more convenient expression for the energy balance is then

$$aS + bL - b\sigma T^4 + F = (\alpha + \beta + 4\sigma T^3 b)\Delta T \qquad (3)$$

and

$$\Delta T = \frac{aS + bL - b\sigma T^4 + F}{\alpha + \beta + 4\sigma T^3 b} \qquad (4)$$

As mentioned above, nearly all operational instruments use the temperature difference, ΔT, to determine the radiation flux. This is only possible if a = b, if T is measured separately, and if F is known. If S or L is measured separately, the condition a = b is unnecessary.

In the denominator of Equation 4, α and β are weak functions of temperature but α is a strong function of ventilation. The temperature dependence can be empirically calibrated, but the dependence on wind speed must be reduced so calibration is possible.

Total net flux radiometers are normally constructed as two receivers mounted on opposite sides of a metallic, glass or plastic body. Using Equation 4 for both receivers gives

$$\Delta T_i = \frac{a_i S_i + b_i L_i - b_i \sigma T^4 + F_i}{\alpha_i + \beta_i + 4\sigma T^3 b_i} \qquad (5)$$

with i = 1,2, where index 1 is for the upper and index 2 for the lower receiver. If the absorptivities and the heat transfer coefficients are the same for both receivers, subtracting ΔT_2 from ΔT_1 gives

$$\Delta T = \Delta T_1 - \Delta T_2 = \frac{a(S_1 - S_2) + b(L_1 - L_2) + F_1 - F_2}{\alpha + \beta + 4\sigma T^3 b} \tag{6}$$

If $a = b$ and $F_1 = F_2$ or $F_1 = F_2 = 0$ the total net flux, $(S_1 - S_2 + L_1 - L_2)$, is proportional to the difference in temperature between the receivers.

The main problem with a total net flux meter is the dependence of the calibration factor on wind speed. Since the temperature dependence of the calibration factor is small it can be compensated for internally, over a wide range of temperature, by incorporating a resistor-thermistor-resistor network in the electrical thermometer circuit within the instrument. With this problem 'solved' it remains to design instruments in such a way that the variability of α due to ventilation is reduced to acceptable levels and F_i is sufficiently controlled. A brief discussion of different designs is presented below. For a more detailed discussion the reader is referred to Coulson (1975).

3. TOTAL RADIATION INSTRUMENTS

3.1 Net Flux Meters or Net Pyrradiometers

These instruments are designed to measure the total net flux through a horizontal plane. Their response is governed by Equation 6, which we rewrite with $a = b$

$$\Delta T = \frac{a(S_1 - S_2 + L_1 - L_2) + F_1 - F_2}{\alpha + \beta + 4\sigma T^3 a} \tag{7}$$

with $F_1 = F_2 = 0$ and full symmetry between the upper and lower receivers, the major design problem is to make $\alpha \approx$ constant.

3.2 Shielded Instruments

The condition $\alpha \approx$ constant can be nearly fulfilled if the receivers can be shielded from the wind. The material used must be transparent between 0.3 and 30 μm. A useful material for such a filter is polyethylene, first introduced by Schulze (1953). Later Bolle and Schulze (Schulze, 1962) were able to produce hemispheres of polyethylene which do not influence the geometry of the radiation field. Independently Suomi and Kuhn (1958) used polyethylene, but only in the form of plane filters extended parallel to the receiver plate.

Though often used, polyethylene has some disadvantages. It has strong absorption bands located in the 13 to 15 μm and the 6 to

8 μm region. It is, however, assumed that the influence is small since the atmosphere itself is nearly opaque in these regions due to the strong bands of CO_2 and H_2O respectively. Polyethylene transmittance for foils of different thicknesses is shown in Figure 1.

The molecular structure of polyethylene leads to a scattering of radiation which increases with decreasing wavelength and which is strong in the 0.3 to 0.4 μm region (Schulze, 1962). To avoid these disadvantages, foils of only 10 to 50 μm thickness are used. This is not difficult for instruments with plane stretched foils (Suomi-type) but instruments with hemispheric films of such thickness need a minimal overpressure below the hemisphere to maintain their shape. An obvious advantage is that polyethylene is not soluble in water.

Since some absorption (and emission) occurs in the foils themselves and since dirt accumulates, the temperature of the foils often deviates from air temperature. As a result of radiation exchange or heat transfer between the shield and the receiver an additional energy flux F occurs and ΔT is no longer a direct measure of the total net flux (this is also true for the black surface pyranometer, which is described later). Georgi and Haase (1951) were the first to obtain a thermal separation for these instruments by ventilation of the space between two shielding hemispheres. Beier (personal communication) used the same principle for his net flux meter, which also has two polyethylene hemispheres for each of the two receivers. For these designs the hemispheres have diameters of 0.03 and 0.06 m respectively and the airflow for both is about 5×10^{-4} m^3 s^{-1}. With these designs the total net flux can be determined with sufficient accuracy. Balance meters which have only one

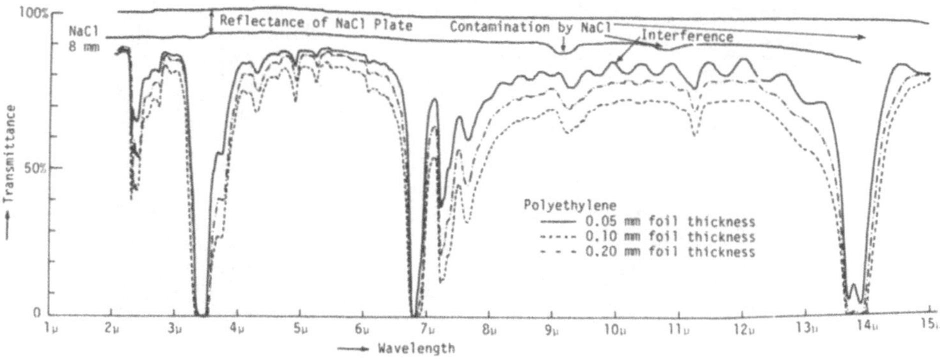

Fig. 1 Transmittance of polyethylene (Schultze, 1962).

hemisphere at each receiver are less convenient and are ventilated from the outside to maintain equilibrium with air temperature (Däke, 1972).

All net flux meters have black receiver plates. In the short wave region the pigment and in the long wave region the lacquer are responsible for absorption. Therefore it is difficult to get an absorptivity which is sufficiently independent of the wavelength. Generally, in the long wave region the flux meter has a somewhat lower absorptivity than in the short wave region due to both the different absorptivity of the receiver and the different transmissivity of the filter material. To compensate for this some polyethylene-shielded instruments (Funk, 1959) have a thin white or grey line or circle marked on the receiver. The absorptivities of the white line and of the black surface are equal in the long wave region but the white line reflects more in the short wave region. Using a white line of known reflectivity and geometry the absorptivity of the total receiver plate can be made equal for the short and the long wave regions respectively. For those instruments Equation 7 can be used to determine the total net flux. Instruments with differential absorptivities can measure the total net flux only during the night ($S_1 = S_2 = 0$). The determination of the total net flux during the day requires a separate measurement of the short wave net flux (Hinzpeter, 1967).

An example of a shielded instrument, the Funk polyethylene net flux meter, is shown in Figure 2.

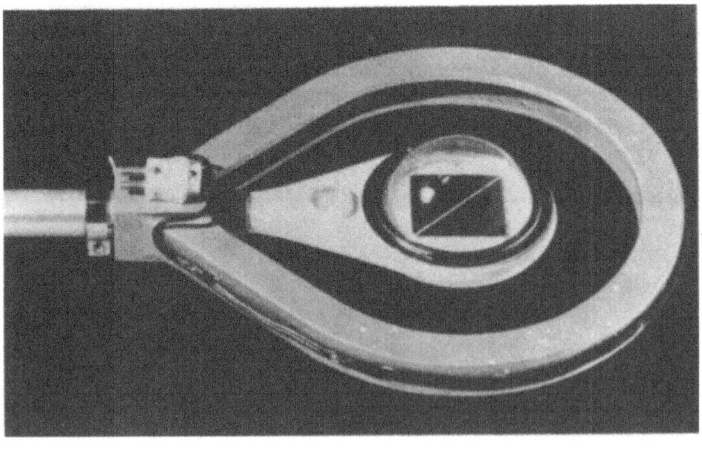

Fig. 2 Polyethylene-shielded net flux meter designed by Funk (Coulson, 1975).

3.3 Nonshielded Ventilated Instruments

The influence of varying wind velocity can be at least partly eliminated by artificial ventilation of the receiver plates. This has been done for a number of different instruments (Suomi et al., 1954; MacDowall, 1955; Gier and Dunkle, 1951; Courvoisier, 1950) and can be described simply by splitting the conductivity into two parts:

$$\alpha = \alpha_c + \alpha_w$$

where α_w is the varying conductivity due to the natural wind and α_c that due to constant artificial ventilation. If $\alpha_c \gg \alpha_w$, the influence of the wind speed is greatly reduced, but depends upon the direction of the wind relative to the receiver plate. With increasing wind speed the sensitivity of the instruments decreases, but over a speed range of 0 to 10 m s^{-1} the decrease is only about 3 to 5%. A disadvantage of some of these is that the ventilator box shields too great a space angle. All of these instruments have black receiver plates without any white line.

3.4 Nonshielded Nonventilated Instruments

The influence of the wind speed v can also be reduced if $\beta \gg \alpha$. Implicitly this technique employs long time constants so that the instruments can be calibrated with respect to the mean wind speed. With $F_1 = F_2 = 0$ and a = b we find

$$S_1 - S_2 + L_1 - L_2 = \Delta T\ f(v,T)$$

The principle was first introduced by Albrecht (1933). The Janishewski instrument (Janishewski, 1957) also works in a similar way. Without artificial ventilation the sensitivity of the instrument decreases by 30% over a wind speed range of 0 to 15 m s^{-1}.

3.5 Compensating Net Pyrradiometers

In these instruments ΔT is kept equal to zero by electrically heating the receiver plates. These heat fluxes F_1 and F_2 enter from Equation 6:

$$a(S_1 - S_2 + L_1 - L_2) = F_1 - F_2$$

During daytime we have $S_1 - S_2 + L_1 - L_2 > 0$, thus $F_2 - F_1 > 0$, and generally we expect $F_1 = 0$, $F_2 > 0$. During the night the total net flux is negative and generally we expect that $F_2 = 0$, $F_1 > 0$. The electric heating is I^2R and for R = constant we have

$$S_1 - S_2 + L_1 - L_2 \propto I^2 \qquad (8)$$

For a = b the calibration constant is determined only by the absorptivity, a, and the instrument can be considered as an absolute one (see Frankenburger, 1954).

In order to measure the net flux using compensation by heating, the effect of wind fluctuations should be eliminated by smoothing. Therefore, the instruments should have sufficiently large time constants. Courvoisier combined ventilation with the compensation method to obtain an absolute instrument.

3.6 A Compensating Net Pyrradiometer with Constant Heating

To avoid the servo-system used to maintain $\Delta T = 0$ in the radiometers described above, Albrecht (1933) constructed a balance meter which consists of two equal balance meters placed near each other, with $a_i = b_i = a$ for i = 1, 2, 3, 4 (i.e. short and long wave absorptivities should be equal within each receiver and among all four receivers). For system I, $F_2 = 0$ and $F_1 = F$ = constant, and for system II, $F_1 = 0$ and $F_2 = F$ = constant (F is a constant due to electric heating). The temperature differences are then

$$\Delta T_i = \frac{a(S_1 - S_2) + b(L_1 - L_2) \pm F}{\alpha + \beta + 4\sigma T^3 b}$$

with i = I: +F, and i = II: -F

Then

$$\Delta T_I + \Delta T_{II} = \frac{2[a(S_1 - S_2) + b(L_1 - L_2)]}{\alpha + \beta + 4\sigma T^3 b}$$

and

$$\Delta T_I - \Delta T_{II} = \frac{2F}{\alpha + \beta + 4\sigma T^3 b}$$

Hence

$$\frac{\Delta T_I + \Delta T_{II}}{\Delta T_I - \Delta T_{II}} = \frac{a(S_1 - S_2) + b(L_1 - L_2)}{F} = Q$$

Thus we have the total net flux if a = b

$$S_1 - S_2 + L_1 - L_2 = \frac{F}{a} Q \qquad (9)$$

Using electronic techniques to measure the temperature differences, an electrical analogue to the total net flux, Q, can be recorded. In practice it is nearly impossible to guarantee $\alpha_I = \alpha_{II}$ at low wind speeds, because one of the upward-facing receiver plates is always warmer, which, at least during the night, produces convection. Therefore, additional ventilation is required to ensure $\alpha_I = \alpha_{II}$.

4. INSTRUMENTS FOR SHORT WAVE FLUXES (PYRANOMETERS)

At least at present, instruments for short wave fluxes have a greater accuracy than the net flux meters described above. Very often the short wave flux is interesting in itself, especially over the sea, since it is that part of the total net flux that penetrates the sea to some depth. All short wave flux meters in current use are protected against changing natural ventilation by glass hemispheres. These are nearly transparent for all wave lengths between 0.3 and 3 μm (no absorption, only reflection).

In the long wave region ($\lambda \geqslant 3$ μm) the transmittance of glass is zero. Therefore, if the glass domes have the temperature of the reference body, the hemispheric long wave flux density is given by the glass emissivity and the black body radiation. Equation 4 is then reduced to

$$\Delta T = \frac{a\,S + F}{\alpha + \beta + 4\sigma T^3 b} \tag{10}$$

(any deviation of the glass temperature from the reference temperature can be taken into account in F).

4.1 The Black and White Pyranometer

This instrument has temperature sensors - generally junctions of a thermocouple - which are covered alternately by white and black plates. The system of white and black receivers is shielded by a glass hemisphere. In the long wave region glass radiates like a black body. Thus, as long as there is no absorption of solar radiation, the glass hemisphere has a lower temperature than the air and additional energy exchanges occur at the receiver due to both radiation and conduction.

The temperature difference between the white or black plates and the reference is then

$$T_i - T = \frac{a_i S_i + F_i}{\alpha_i + \beta_i + 4\sigma T^3\, b}$$

with $i = 1$ for the white plates,
and $i = 2$ for the black plates.

Since both white and black plates are located under the same dome, we have

$$F_1 = F_2 = F; \qquad S_1 = S_2 = S$$

$$\alpha_1 = \alpha_2 = \alpha; \qquad \beta_1 = \beta_2 = \beta$$

We find for the temperature difference

$$T_2 - T_1 = \Delta T = \frac{(a_2 - a_1)S}{\alpha + \beta + 4\sigma T^3 b} \; ; \tag{11}$$

that is, the temperature difference is a measure of S and does not depend upon the temperature of the glass dome. While this is a great advantage, these instruments have not been used as widely as the black pyranometers. One disadvantage is that the white surfaces age in such a way that $a_2 - a_1$ probably changes more quickly than a_2.

4.2 Black Pyranometers

Black surface pyranometers have only a black receiver plate exposed to the radiative flux. Generally, thermopiles are used to measure the temperature difference analogue to the short wave radiative flux. We may distinguish between two different types of thermocouples. For the first type the so-called active junctions are connected with the receiver plate and the passive junctions with the instrument body of reference temperature T. The difference is

$$T_i - T = \frac{a\,S + F}{\alpha + \beta + 4\sigma T^3 b} \tag{12}$$

In the case of the widely-used 'Moll' thermopile, active and passive junctions are exposed to the radiation flux but have very different thermal conductivities to the instrument body. Again, the difference in temperature between the junctions and the reference body is

$$T_i - T = \frac{a_i S_i + F_i}{\alpha_i + \beta_i + 4\sigma T^3 b_i}$$

with $i = 1,2$ for the low and high conductivity, respectively.

With

$$F_1 = F_2 = F; \qquad S_1 = S_2 = S$$

$$\alpha_1 = \alpha_2 = \alpha; \qquad a_1 = a_2 = a$$

we get

$$T_2 - T_1 = \Delta T \simeq \frac{(aS + F)(1 - \beta_1/\beta_2)}{\alpha + \beta_1 + 4\sigma T^3 b}$$

Thus in the case of a black surface pyranometer the temperature difference depends not only upon the radiation flux but also upon any additional energy flux between the glass dome and the receiver plate. To reduce the influence of the dome temperature, black surface pyranometers use two concentric domes (about 0.05 and 0.03 m diameters). The inner hemisphere exchanges heat by radiation and convection with the receiver and with the outer hemisphere, but reduces the influence of the temperature of the outer hemisphere. Even with the two domes the short wave flux can be in error by about 5 to 10% for low wind speeds and low solar elevations. This error can be greatly reduced by ventilation of the space between both hemispheric domes (Georgi and Haase, 1951). A diagram of a typical pyranometer is shown in Figure 3.

5. INSTRUMENTS FOR LONG WAVE FLUXES, NET PYRGEOMETERS

An instrument for long wave flux measurement must use a filter dome which is transparent only for wavelengths > 3 μm. No filter material fulfills all these conditions, but black polyethylene foil and crystals have been used. Different methods are applied to avoid

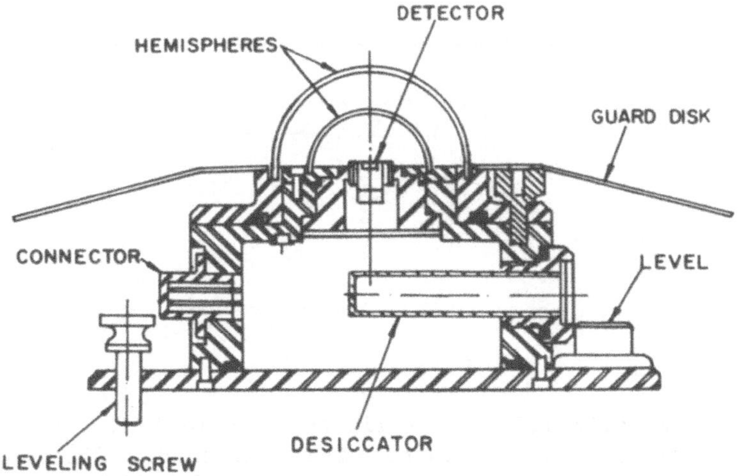

Fig. 3 Diagram of a pyranometer of solarimeter type (Coulson, 1975).

solar heating of the filters or the influence of heating on measurements. Crystals are covered with additional coatings which raise the reflectance in the region < 3 μm. Often the crystal is a binary thallium iodide-thallium bromide KRS 5 (Kristallschmelze 5) with a special coating. Recently a special coating of silicone has been tried which has a great natural reflectance in the short wave region. Since the long wave net flux density is only about a tenth of the short wave flux for cloudless days and sun elevations > 30°, the errors due to filter temperature deviations from the reference temperature are about ten times greater than for black pyranometers with a glass dome.

Experience with KRS 5 domes and theory both indicate that it is nearly impossible to measure the long wave flux with such filters during daytime. Also, during the night the errors with these instruments are great. One disadvantage of the crystal is its solubility; even with special coating KRS 5 ages. Silicone is not soluble but we have little experience with it as a filter; the transmissivity of silicone is shown in Figure 4.

Probably it is much more convenient to use black polyethylene as an additional filter. Some experience exists with a Funk net radiometer which is surrounded by an additional black polyethylene sphere. The space between the inner polyethylene sphere and the black sphere is ventilated. To avoid different energy fluxes from the black sphere to both receiver plates, the black filter is electrically rotated about the fixed radiometer. A rotation rate of once per second is sufficient to avoid differential heating of the plates. A diagram of the instrument is shown in Figure 5 (Paltridge, 1969). The black polyethylene filter used with the Funk instrument is composed of soot and polyethylene and its transmittance is shown in Figure 6. We have little experience with this

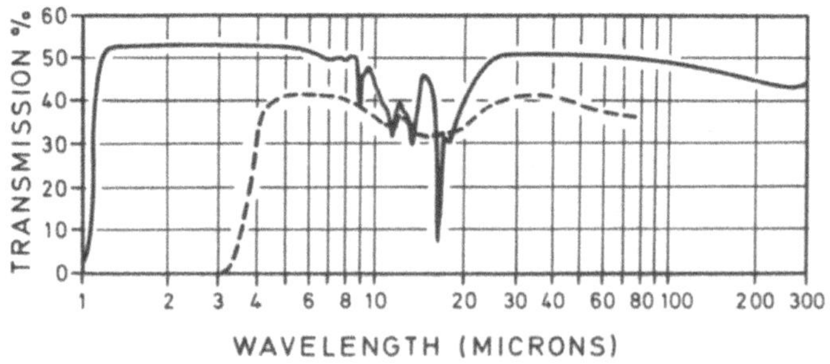

Fig. 4 Transmissivity of silicone: solid line - uncoated; dashed line - coated with increased reflectance in the short wave region (courtesy of Eppley Laboratories).

instrument, but the accuracy is estimated to be about 8% (Simpson, 1979) (more information on crystals and soot suspended in polyethylene is given in Nayer and So, 1977).

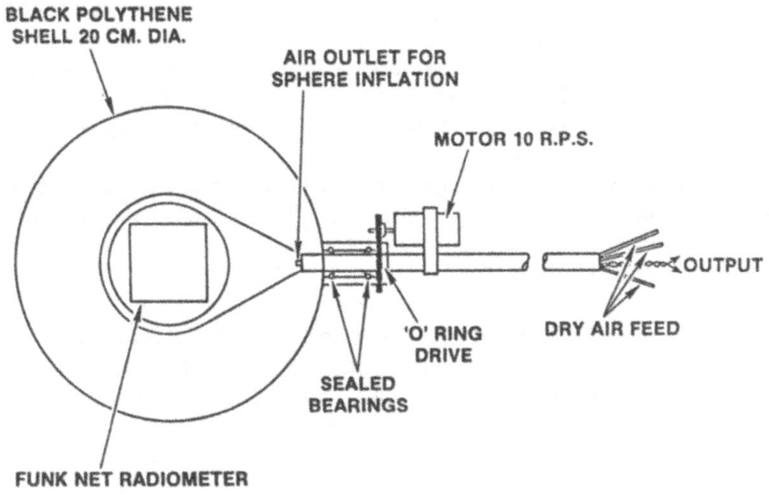

Fig. 5 Diagram of a net pyrgeometer with black polyethylene (Funk type) (Paltridge, 1969).

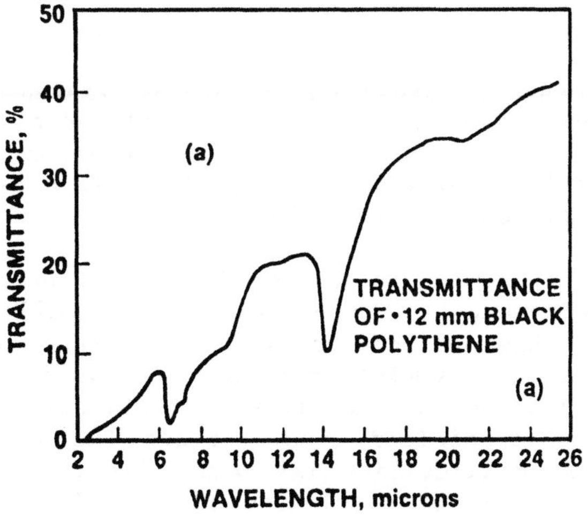

Fig. 6 Transmittance of black polyethylene (Paltridge, 1969).

6. PRACTICAL CONSIDERATIONS

Most total net flux meters (balance meters) are constructed in such a way that only the temperature difference between the upper and lower receiver plate is measured as an analogue of the total net flux density ($S_1 - S_2 + L_1 - L_2$). This is correct for the non-shielded instruments. The polyethylene shielded instrument designed by Schulze can separately measure the temperature difference between the upper plate and the body and the lower plate and the body. This provides for error control, but to satisfy Equation 4 the reference temperature must be recorded separately.

Short wave flux meters are generally constructed to measure the flux from one hemisphere only but can be simply combined to form short wave net flux meters; for reasons of error control they should be used separately.

All instruments should maintain a cosine dependence for radiation arriving from angles off the vertical. Generally, all instruments show deviations from the cosine law for large angles of incidence. These errors are small, except for the Suomi-Kuhn net flux meter with plane polyethylene foils, for which there can be an error of 10% for an angle of incidence of 65° (measured from the vertical). The cosine error increases with condensation at the top of the domes (inside or outside) and is especially high for rime. This can be avoided by drying the inner space with silica gel or by filling it with dry nitrogen and artificially ventilating the outer dome.

The influence of rain on shielded instruments is small and not important in practice since the fluxes themselves are very small for rainy conditions.

Instrumental accuracy should not be overestimated. The error of long term records, even for well calibrated and carefully attended short wave flux meters, is at least 3%. The error in the long wave flux during daytime, determined by the difference of measured total and short wave fluxes, can be 50% or more.

Instruments for measuring radiation fluxes must be mounted horizontally. On board ships or buoys gimballed installations should be used; often this has not been done. While errors may be acceptable for daily sums in the tropics, this is not true elsewhere. Shielded instruments are much simpler to use with a gimballed arrangement than ventilated nonshielded ones. Since the instruments should move freely some difficulties may occur due to forces exerted by the wires. These can be eliminated if the wires are led through the pivot joints. A small mean misalignment due to wind forces is also possible. To avoid large resonant amplitudes the free movement of the instrument should be damped.

In the fifties British weather ships used instruments of the MacDowall type, but more recently polyethylene shielded instruments were usually used. Even in calm conditions the domes of the instruments are soon covered with a thin film of sea spray. However, even for the long wave region the film seems to have very little influence upon the measured long wave transmission.

In general, it is highly desirable to use separate instruments for the measurement of upward and downward fluxes. The downward flux instruments can then be mounted near the top of a mast, where the influence of shadows from the ship's superstructure are considerably reduced. Instruments for the fluxes from below require a boom at the bow or other such good exposure. Of the two upward fluxes only the short wave one can be measured with sufficient accuracy, but precautions are necessary in order to minimize errors characteristic of those measurements. The glass domes reflect and scatter the sun's radiation. Especially for small sun elevations, the scattered light creates very large errors, making albedo determination at a small sun elevation very difficult. This problem can be reduced by using a circular screen which shields the glass dome against the sun's radiation but covers only nadir angles $>80°$. In addition, the screen has to cover the space angle occupied by the ship in order to reduce the ship's reflection (Heinrich and Hinzpeter, 1975). Instruments for measuring the total flux from below do not have sufficient accuracy. Therefore it is more reasonable to use a radiation thermometer for determining the surface temperature (see Katsaros, Chapter 16). Using this temperature, the emissivity of the sea, and Boltzmann's law, the long wave component of the total flux from the surface can be determined with sufficient accuracy.

During long sea cruises periodic calibration is necessary. If a calibrated pyrheliometer is available, a recalibration of pyranometers and pyrradiometers with the sun as a source is possible. Two methods can be used, each assuming the absence of clouds. In the first method, a circular disc (~ 0.05 m diameter) is used to shade the direct beam from the horizontal radiometer while it is being measured by the reference pyrheliometer. The change in flux measured between the shaded and unshaded condition is then equal to the vertical component of the beam. The shading disc should be located at a distance from the receiver such that its shadow is exactly equal to or slighty larger than the receiver area, in order to ensure that all of the direct beam is blocked but that the blocked amount of the diffuse radiation is equal to the amount of diffuse radiation measured by the pyrheliometer.

The second method temporarily converts the subject radiometer into a pyrheliometer by mounting a collimating tube over the instrument, such that the tube's long axis is normal to the receiver, and covering the instrument so that no radiation can reach the receiver except through the tube. The modified instrument is then directed at the sun and the measured flux compared with that of the reference pyrheliometer. Again the main concern is that the field of view of the tube-receiver system and the reference instrument be equal. It is not difficult to fulfill this condition, but the convection which generally exists below the glass dome may change with the modified receiver system. Then the calibration would be systematically in error, but little is known of the magnitude of this error.

REFERENCES

ALBRECHT, F. 1933. Ein Strahlungsbilanzmesser zur Messung des Strahlungshaushaltes von Oberflächen. *Meteorologische Zeitschrift*, 50: 55-62.

COULSON, K.L. 1975. *Solar and Terrestrial Radiation*. Academic Press.

COURVOISIER, P. 1959. Über einen neuen Strahlungsbilanzmesser. *Verhandlung der Schweizerischen Naturforschenden Gesellschaft*, 130: 155.

DAKE, C.U. 1972. Über ein neues Modell des Strahlungsbilanzmessers nach Schulze. *Berichte des Deutschen Wetterdienstes*, 16: 1-22.

FALCKENBERG, A. 1947. Ein Vibrationspyrgeometer. *Meteorologische Zeitschrift*, 1: 372.

FRANKENBERGER, E. 1954. Ergebnisse von Wärmehaushaltsmessungen. *Meteorologische Rundschau*, 7: 81-85.

FUNK, J.P. 1959. Improved polyethylene-shielded net radiometer. *Journal of Scientific Instruments*, 36: 267-270.

GEORGI, J. and L. HAASE. 1951. Ein belüftetes Pyranometer. *Meteorologische Rundschau*, 4: 188-192.

GIER, J.T. and R.V. DUNKLE. 1951. Total hemispherical radiometers. *Transactions of the American Institute of Electrical Engineers*, 70: 339-343.

HEINRICH, H. and H. HINZPETER. 1975. Radiation balance and albedo in the tropical Atlantic during ATEX 1969. *"Meteor" Forschungsergebnisse Reihe B*, 10: 56-64.

HINZPETER, H. 1967. Ergebnisse der Messungen zur Strahlungsbilanz während der Fahrtabschnitte zwischen Suez und Aden II der Indischen-Ozean-Expedition 1964-1965. *"Meteor" Forschungsergebnisse Reihe B*, 1: 1-13.

JANISHEWSKI, YU.D. 1957. *Aktinometric Instruments and Methods of Observation*. Hydrometeorological Publishing House, Leningrad, USSR (in Russian).

MacDOWALL, J. 1955. Total-radiation fluxmeter. *Meteorological Magazine*, 84: 65-71.

NAYER, P.S. and C.K. SO. 1977. Properties of some infrared filters. *Applied Optics*, 16: 289-290.

PALTRIDGE, A.H. 1969. A net long wave radiometer. *Quarterly Journal of the Royal Meteorological Society*, 95: 635-638.

SCHULZE, R. 1953. Uber ein Strahlungsmessgerät mit ultrarotdurchlässiger Windschutzhaube am Observatorium Hamburg. *Geofisica Pura e Applicata*, 24: 107-114.

SCHULZE, R. 1962. Uber die Verwendung von Polyäthylen für Strahlungsmessungen. *Archiv für Meteorologie, Geophysik und Bioklimatologie, Series B*, 11: 211-223.

SIMPSON, J.J. and C.A. PAULSON. 1979. Mid-ocean observations of atmospheric radiation. *Quarterly Journal of the Royal Meteorological Society*, 105: 487-502.

SUOMI, V.E., M. FRANSILLA, and N.F. ISLITZER. 1954. An improved net radiation instrument. *Journal of Meteorology*, 11: 276-282.

SUOMI, V.E. and P.M. KUHN. 1958. An economical net radiometer. *Tellus*, 10: 160-163.

27

Oceanic Radiation Instruments

C.A. Paulson

1. INTRODUCTION

The ability to measure the distribution of light and optical properties in the upper ocean is useful for a number of applications. Among these are: (1) the estimate of solar heating as a function of depth; (2) the study of biological processes; and (3) the investigation of the motion of surface waters using optical properties as tracers. The purpose of this chapter is to describe relatively simple instruments for measuring light and optical properties which would yield results useful for air-sea interaction investigations. The description of the instruments is prefaced by an elementary description of the distribution of light and optical properties in the upper ocean.

It is important to emphasize that the scope of this chapter is limited to instruments most appropriate for air-sea interaction studies. Optical oceanography has a long history and there has grown a large body of theoretical and observational literature which has been excellently summarized by Jerlov (1976). The theoretical aspects of radiative transfer in water have also been comprehensively treated by Preisendorfer (1976), and Ivanoff (1977) has recently reviewed radiative processes in the lower atmosphere and upper ocean with a view toward modelling the upper ocean. Tyler and Preisendorfer (1962) and Jerlov (1976) have described instruments used in optical oceanography.

Instrument developments have contributed greatly to advances made in optical oceanography. The design, construction, calibration, and use of oceanic instruments have been undertaken primarily by laboratories with extensive and specialized capabilities in optical oceanography. However, the simplest of existing optical instruments are the most suitable for air-sea interaction investigations and require only a modest amount of resources and experimental expertise for their successful use. The discussion below is limited to these instruments.

2. LIGHT IN THE UPPER OCEAN

A comprehensive description of the distribution of light and optical properties in the upper ocean is given by Jerlov (1976). Of the solar radiation striking the sea surface, about 7% on average is reflected and backscattered upward into the atmosphere (see, e.g., Payne, 1972) and the remainder is absorbed in the sea. The upward solar radiative flux just above the surface is mainly caused by reflection at the surface and secondarily by backscattering from below the surface. The contribution from backscattering is only about 0.5% of the radiative flux striking the surface (Payne, 1972; Ivanoff, 1977). Solar radiation is attenuated in sea water by both absorption and scattering with absorption dominating. The downward irradiance, defined as the flux of radiant energy per unit area arriving at a horizontal surface that is facing upward, falls to one-half the surface value within the uppermost metre of the sea. As

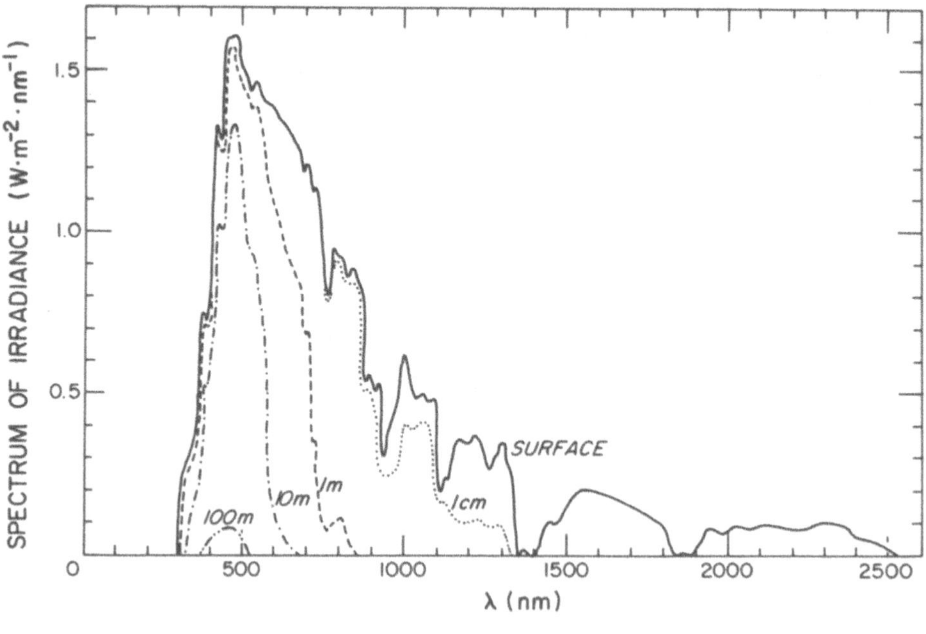

Fig. 1 The spectrum of downward irradiance at various depths in the ocean (after Jerlov, 1976)

shown in Figure 1, the attenuation is strongly dependent on wavelength; water acting as a monochrometer leaves only blue-green light below a depth of 10 m. The distribution of radiant energy as a function of direction depends on the location of the sun, but, with increasing depth, tends toward a maximum at the zenith, becomes symmetric about the zenith, and falls off rapidly with increasing zenith angle. The symmetric distribution about the zenith results because light rays following vertical paths are least attenuated.

Because of backscattering, there is also an upward irradiance: the flux density of radiant energy striking a horizontal surface that is facing downward. The ratio of upward to downward irradiance is less than 0.1 and is typically about 0.05 below a depth of 2 m (Jerlov, 1976, Table XXXIII, p. 148).

As shown in Figure 2, the downward irradiance falls nearly exponentially with depth below 10 m. Above 10 m, the rate of decrease with depth is much faster than exponential because of the dependence of attenuation on wavelength as shown in Figure 1. The curves drawn to the observations are the sum of two exponentials, as suggested by Kraus (1972):

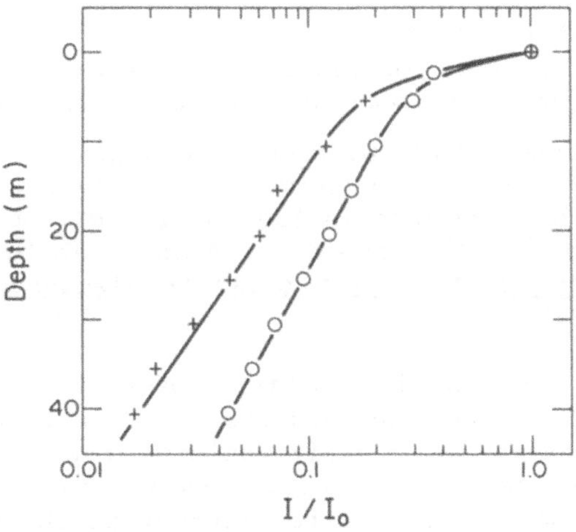

Fig. 2 Observations in the North Pacific (35°N, 155°W) reported by Paulson and Simpson (1977) of normalized downward irradiance as a function of depth. The circles are an average of five sets of observations under overcast skies with solar altitudes ranging from 30 to 38 degrees. The crosses are a single set of observations under clear skies with a solar altitude of 16 degrees.

$$I_d(z)/I_d(0) = R \exp(\gamma_1 z) + (1 - R) \exp(\gamma_2 z) \tag{1}$$

where I_d is the downward irradiance, z is the vertical space coordinate which is positive upwards with z = 0 at mean sea level, $I_d(0)$ is the downward irradiance just below the surface, R is a constant depending on water clarity and solar altitude, and γ_1 and γ_2 are attenuation coefficients for downward irradiance. The first term dominates near the surface and represents the attenuation of the long-wave components of the solar radiation; the second term represents the attenuation of approximately monochromatic blue-green light below a depth of about 10 m.

Heating of the upper ocean occurs because of the absorption of radiative energy. The heating rate, dQ/dt, is equal to the divergence of the net downward radiative flux:

$$\frac{dQ}{dt} = \frac{\partial}{\partial z}(I_d - I_u) ,$$

which may be rewritten:

$$\frac{dQ}{dt} = \frac{\partial}{\partial z}\left[I_d\left(1 - \frac{I_u}{I_d}\right)\right]$$

As was stated above, I_u/I_d is nearly independent of z below a depth of 2 m; thus the terms in the above equation containing I_u/I_d can be neglected. Above a depth of 2 m, I_u/I_d decreases with decreasing depth from about 0.05 to about 0.005 just below the interface. It can easily be shown that terms containing I_u/I_d contribute less than 5% to the heating rate in the upper 2 m; thus for many purposes the divergence of the upward irradiance can be safely neglected. Radiative heating of the upper ocean is discussed more extensively by Ivanoff (1977).

Oceanic water has variable optical properties depending on the amount and kind of suspended and dissolved material. Jerlov (1976) has devised a system of classifications of surface water based on spectral transmittance of downward radiation at high solar altitude. Surface water tends to be clearest in the open ocean at mid-latitudes, while coastal areas and high latitudes often have reduced clarity because of high biological productivity. Paulson and Simpson (1977) have computed values of R, γ_1, and γ_2 (Eq. 1) for each water type.

An optical property which is useful for characterizing sea water and which is relatively easy to measure is the beam transmittance, τ, defined as the radiant flux, F_t, transmitted over a distance r by a beam of infinitesimal width, divided by the incident flux:

$$\tau = \frac{F_t}{F_0}$$

The beam attenuation coefficient, c, is the proportional change in flux, $\Delta F/F$, across an infinitesimally thin layer of the medium normal to the beam divided by the thickness, Δr, of the layer:

$$c = \frac{\Delta F}{F \Delta r}$$

The beam attenuation coefficient and transmittance depend on wavelength. For a homogeneous medium

$$\tau = \exp(-c\, r). \qquad (2)$$

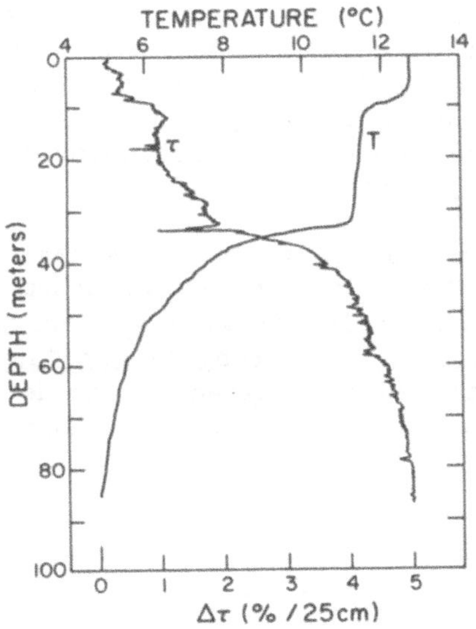

Fig. 3 Observations of the change of beam transmittance with respect to the value near the surface (path-length of 25 cm) and temperature versus depth in the North Pacific (50°N, 145°W), August 1977 (observations supplied courtesy of D.R. Caldwell and T.M. Dillon, School of Oceanography, Oregon State University, Corvallis, OR).

Vertical profiles of transmittance show whether the water column has uniform optical properties. An assumption underlying Equation 1 is that the attenuation coefficients γ_1 and γ_2 are independent of depth. This assumption can be tested by measurements of transmittance.

An example of measurements of transmittance and temperature as a function of depth is shown in Figure 3. The general increase of transmittance with depth is consistent with the expectation of a higher concentration of biological material near the surface. The small-scale variability in transmittance is caused by variability in the concentration of particles and may be useful for drawing inferences about mixing processes. The small-scale variability may also be of biological interest.

3. BEAM TRANSMITTANCE METER

As the name implies, the beam transmittance meter or transmissometer, as it is often called, measures beam transmittance, τ, from which one may obtain the attenuation coefficient, c, by use of Equation 2. A schematic diagram of a representative instrument is shown in Figure 4. The essential elements are a light source of constant intensity, optics to produce a parallel beam with a large length-to-width ratio, windows which protect sensitive elements and determine the fixed path length, a filter which transmits light in a particular wavelength band, and a cell which measures light intensity. Mirrors or internally reflecting prisms are sometimes used to minimize the size of the instrument for a given path length (e.g., Tyler et al., 1974).

An instrument for measuring beam transmittance has been described by Tyler et al. (1974), and is shown schematically in Figure 5. The light beam is 0.2 cm in diameter and traverses 1 m through the water. The source is a 20 W tungsten-iodide lamp. The intensity of the lamp is monitored on command by the detector through a light pipe and shutter as shown in Figure 5. The detector is a silicon

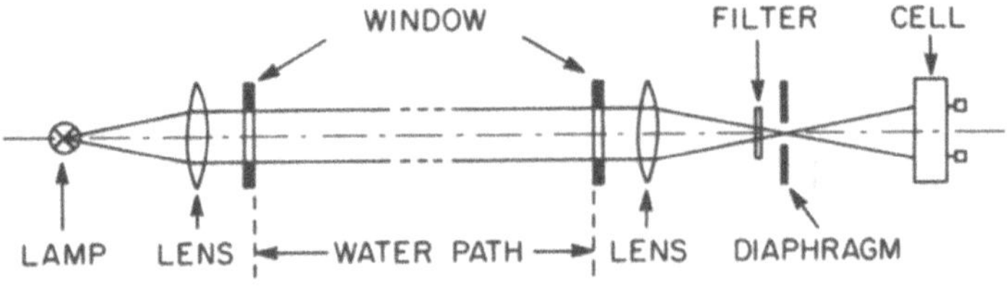

Fig. 4 Schematic diagram of a beam transmittance meter (after Joseph, 1949)

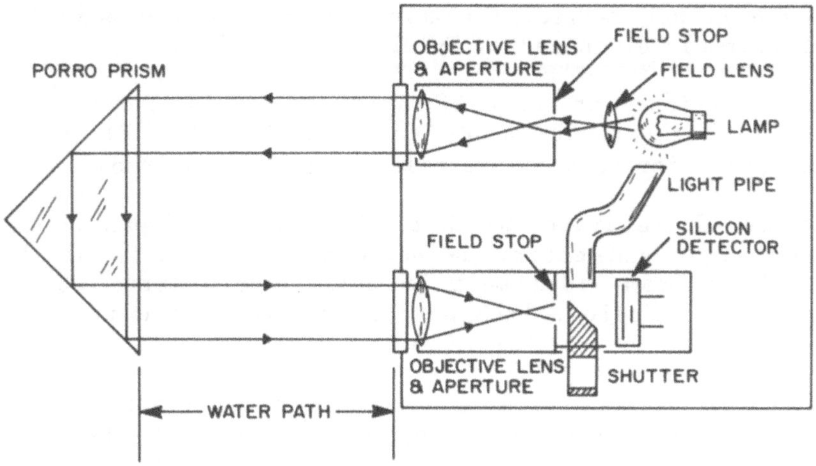

Fig. 5 Schematic diagram of a beam transmittance meter (after Tyler et al., 1974)

photovoltaic cell. Although not shown in Figure 5, there is placed in front of the detector a filter to restrict the sensitivity of the instrument to a band centered at a wavelength of about 530 nm. The spectral characteristics of the source and detector have to be considered together with the filter in obtaining the desired spectral sensitivity.

Jerlov (1976, p. 104) has reviewed the merits of different types of collectors. In addition to the silicon photovoltaic cells, possible choices include selenium photovoltaic cells and photomultiplier tubes. Photomultiplier tubes are more sensitive than photodiodes but are much larger in size, require a well stabilized high voltage, and are subject to changes in calibration; they are sensitive to light intensity over a range of approximately ten orders of magnitude compared to about seven for a photodiode.

The following is a list of precautionary measures to be taken in the design and use of a beam transmittance meter:

Care must be taken to minimize errors due to fluctuations in intensity of the light source. The intensity of the source may be monitored as in Figure 5.

Daylight must be excluded from the detector or its effects may be excluded electronically by blinking the source and measuring the amplitude of the signal. Care must be taken so that shielding does not reflect scattered light from the beam into the detector (as discussed by Tyler et al., 1974).

The sensitivity of the calibration of the instrument to variations in temperature needs to be minimized. Electronics to compensate for temperature effects may be required.

Because of the finite width of the beam, some extraneous scattered light can enter the detector. The ratio of the path length to beam width must be sufficiently large to minimize this error (Jerlov, 1976). The instrument shown in Figure 5 has an error due to this effect of less than 1%.

To obtain reliable absolute measurements, readings should be taken in air prior to a profile. Because the beam diverges less in water than in air, care must be taken to ensure that the beam is not blocked in air. The index of refraction of the windows must be known to obtain absolute values of τ by use of measurements in air.

There may be a false minimum in τ if the beam is tilted from the horizontal while passing through an interface separating fluid of two densities. The error occurs because of refraction at the interface.

Transmissometers are usually lowered from a ship, but may be towed (Joseph, 1957) or moored (Zaneveld et al., 1978). The measurements shown in Figure 3 were taken by a transmissometer, thermistor, and pressure sensor mounted on a microstructure profiler (see Chapter 37 by Lange) which has wings to cause rotation and slow the descent rate to several centimetres per second. The profiler is mechanically decoupled from the ship during descent and data is transmitted through a very light multiconductor cable.

4. IRRADIANCE METER

Irradiance meters measure the total amount of radiant flux striking a plane collector of known area. They are used most often with the collector horizontal and facing upward to measure downward irradiance but they may be inverted to obtain upward irradiance. A drawing of an irradiance meter is shown in Figure 6. The collector is diffusive, typically opal glass or opal plastic, and the dimensions of the materials surrounding the collector are adjusted so that the response nearly obeys a cosine law (Smith, 1969). There may be a filter between the collector and the quartz window depending on whether a particular wavelength band is of interest as in some biological investigations. A filter may also be required to compensate

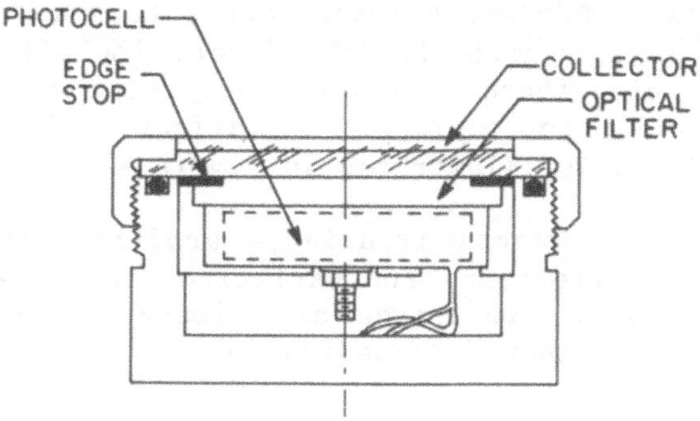

Fig. 6 Schematic diagram of an irradiance meter (after Tyler and Preisendorfer, 1962)

for spectral characteristics of the collector and detector. The detector is usually a photomultiplier tube or a photodiode. The signal from the detector is ordinarily amplified nonlinearly to obtain better signal-to-noise ratios for low light levels.

Tyler and Preisendorfer (1962) quote an uncertainty in irradiance measurements of 2% for the meter shown schematically in Figure 6.

Some precautionary measures to be taken in the design and use of an irradiance meter follow:

Small departures from a cosine response are not serious when measuring downward irradiance because the downward radiance field approaches a maximum near the vertical. However, errors may be introduced when measuring upward irradiance because the upward radiance is more nearly isotropic.

The spectral response of the collector is an important concern when the total irradiance is desired. Many collectors attenuate strongly in the short-wave end of the solar spectrum. Calibrations to determine the spectral response are required. The spectral response of the instrument used to obtain the measurements shown in Figure 2 was uniform within $\pm 5\%$ for wavelengths from 400 to 1000 nm.

When absolute values of irradiance are desired, one needs to make routine comparisons of the irrandiance meter in air with a well calibrated pyranometer. Comparisons are necessary because of possible sensitivity of the electronics to variations in temperature and because of changes in calibration of the detector.

When comparing underwater measurements to measurements in air, correction must be made for the 'immersion effect' which is a combination of effects caused by reflection at the sea surface and at the collector surface. The effect is discussed by Westlake (1965) and is illustrated in Figure 7.

When measuring vertical irradiance profiles, it may be necessary to make corrections for variations in surface irradiance, particularly when conditions are cloudy. Continuous surface measurements are therefore desirable.

The irradiance meter is usually lowered over the side of the ship. Care must be taken to minimize effects of shadowing or reflection caused by the supporting cable and the ship. Ship motion and waves may make it difficult to keep the collector horizontal, particularly near the surface. Accurate measurements can only be made in calm or moderate sea states. Weighting the base of the irradiance meter and gimballing it may help ensure that the collector is horizontal. It is probably possible to make accurate measurements in the upper few metres, only from stable platforms; such observations have been made from FLIP by Paulson and Simpson (1977) and from the Bouée Laboratoire by J.P. Bertoux (Ivanoff, 1977). Even from stable platforms, waves are a problem because of the variable depth of the water over the instrument. Linear averaging of the irradiance signal may not be appropriate unless the depth of the instrument below mean sea level is much greater than the amplitude of the waves. This problem arises because of the highly nonlinear dependence of I_d on z when the magnitude of z is small.

5. THE SECCHI DISC

One of the oldest oceanographic instruments is the 'Secchi' disc, named after Professor P.A. Secchi who, in the 1860s, investigated the effect of many variables on the visibility of a white disc suspended from a ship. The Secchi disc has a diameter of 30 cm and is painted white. The 'Secchi depth' is the depth at which the disc ceases to be visible by an observer above the surface. A description of the Secchi disc and instructions in its use are given in the Instruction Manual for Obtaining Oceanographic Data (U.S. Naval Oceanographic Office, 1968).

The Secchi disc is widely used because of its simplicity, but the interpretation of the observation, except as a rough measure of transparency, is difficult. The Secchi depth may depend on solar elevation, cloud type, cloud amount, and sea state. In spite of these difficulties, the numerous observations of Secchi depth may

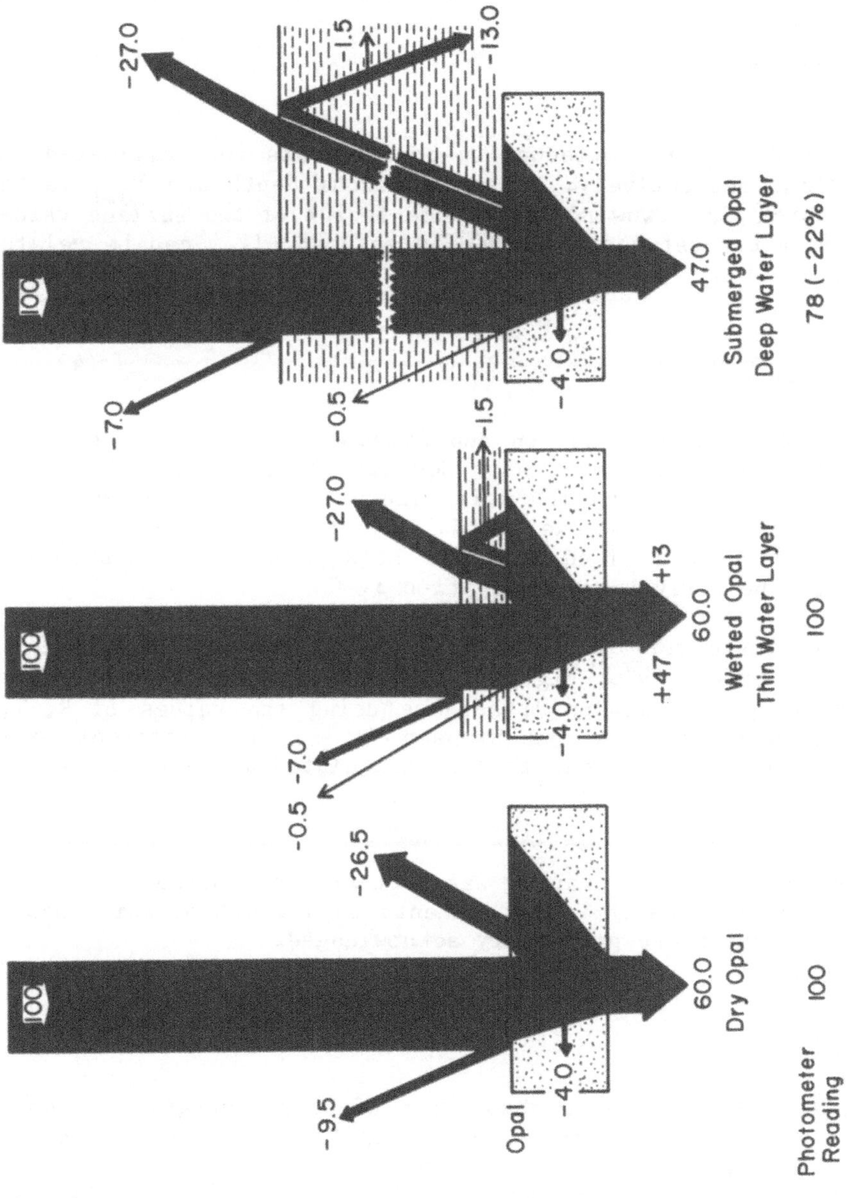

Fig. 7 A diagram illustrating the immersion effect (after Westlake, 1965).

be useful in estimating the heating rate in the upper ocean as a function of depth.

Tyler (1968) has described the theory of the Secchi disc observation and has suggested that

$$\frac{3.78\ K_d}{(c + K_d)} = \frac{D}{I_{d.1}} \tag{3}$$

where c and K_d are the attenuation coefficients for collimated and diffuse light respectively, D is the Secchi depth and $I_{d.1}$ is the depth at which the downward irradiance is 10% of the surface value. If a fixed ratio between c and K_d is assumed, $I_{d.1}$ can be related to D. Tyler (1968) plots observations which are in rough agreement with

$$I_{d.1} = D \tag{4}$$

for D less than 30 m. Paulson and Simpson (1977) also report observations in fair agreement with Equation 4 if cases of high wind speed (rough surface) and low solar elevation are neglected. Caution, however, should be exercised. As stated by Tyler (1968), c is not necessarily a fixed multiple of K_d, and other assumptions are made in the derivation of Equation 3.

Given an estimate for $I_{d.1}$ from Equation 4, one may determine the water type (Jerlov, 1976, Fig. 71, p. 136). One may then use Equation 1 to compute downward irradiance using the values of R, γ_1, and γ_2 given by Paulson and Simpson (1977) for different water types. The heating rate can then be calculated as shown above.

ACKNOWLEDGEMENTS

The preparation of this chapter was partially supported by the U.S. Office of Naval Research. The comments on a draft of this chapter by J.R.V. Zaneveld are gratefully acknowledged.

REFERENCES

IVANOFF, A. 1977. Oceanic absorption of solar energy. In *Modeling and Prediction of the Upper Layers of the Ocean,* edited by E.B. Kraus, Pergamon, Oxford: 47-71.

JERLOV, N.G. 1976. *Marine Optics.* Elsevier, Amsterdam, 231 pp.

JERLOV, N.G. and E.S. NIELSEN, editors. 1974. *Optical Aspects of Oceanography.* Academic Press, London, 494 pp.

JOSEPH, J. 1949. Durchsichtigkeitsmessungen im Meere im ultrabioletten Spektralbereich. *Deutschen Hydrographischen Zeitschrift.,* 2: 212-218.

JOSEPH, J. 1957. Extinction measurements to indicate distribution and transport of water masses. In *Proceedings of the UNESCO Symposium on Physical Oceanography,* 1955, Tokyo: 59-75.

KRAUS, E.B. 1972. *Atmosphere-Ocean Interaction.* Clarendon, Oxford, 275 pp.

PAULSON, C.A. and J.J. SIMPSON. 1977. Irradiance measurements in the upper ocean. *Journal of Physical Oceanography,* 7: 952-956.

PAYNE, R.E. 1972. Albedo of the sea surface. *Journal of Atmospheric Sciences,* 29: 959-970.

PREISENDORFER, R.W. 1976. *Hydrologic Optics.* Volumes I-IV, U.S. Department of Commerce, National Oceanic and Atmospheric Administration, Environmental Research Laboratories, Honolulu, Hawaii.

SMITH, R.C. 1969. An underwater spectral radiance collector. *Journal of Marine Research,* 27: 341-351.

TYLER, J.E. and R.W. PREISENDORFER. 1962. Transmission of energy within the sea: Light. In *The Sea,* edited by M.N. Hill, Interscience, New York: 397-451.

TYLER, J.E. 1968. The Secchi disc. *Limnology and Oceanography,* 13: 1-16.

TYLER, J.E., R.W. AUSTIN and T.J. PETZOLD. 1974. Beam transmissometers for oceanographic measurements. In *Suspended Solids in Water,* edited by R.J. Gibbs, Plenum, New York: 51-59.

U.S. NAVAL OCEANOGRAPHIC OFFICE. 1968. *Instruction Manual for Obtaining Oceanographic Data, 1968.* Publication No. 607, third edition, Washington, D.C.: 1314-1315.

WESTLAKE, D.F. 1965. Some problems in the measurement of radiation under water: A review. *Photochemistry and Photobiology,* 4: 849-868.

ZANEVELD, J.R.V., R. BARTZ and H. PAK. 1978. A transmissometer for profiling and moored observations. In: *Ocean Optics V.* American Society of Photogrammetry, *Proceedings of the Society of Photo-Optical Instrumentation Engineers,* Volume 160, Rochester, New York.

28

Precipitation Measurements over the Ocean

P.M. Austin and S.G. Geotis

1. INTRODUCTION

Two general types of rainfall measurements are considered in this chapter: (1) essentially local measurements where observations are restricted to the vicinity of an instrument, which may be mounted on a ship, aircraft, buoy, island, or coast; and (2) widespread measurements made with satellite-borne instruments. The techniques which will be discussed have all been tested, but applications to oceanic regions have been mostly in research or experimental context. Measurements of rain over the oceans are not generally made operationally and knowledge of oceanic precipitation regimes is still meager.

Traditionally, rainfall measurements have been made with gauges, which indicate the amounts of water falling into a collector in known intervals of time. A gauge is a device of type 1 above; it provides a direct measurement of rainfall but has the disadvantage of being a point observation which may not be representative of surrounding areas. Rain-gauge measurements over oceanic regions have been very limited, largely because of the obvious impracticality of installing any substantial network over such vast expanses of water. Another factor which has discouraged even isolated measurements at a few points is a general mistrust of measurements from shipborne gauges, which arises primarily from concern about the effects on collection efficiency of air flow around the ship.

Over land, radars are used to depict the small-scale time and space variations of precipitation and to extend the coverage provided by networks of gauges. Although its coverage is considerably greater than that of a gauge, the radar is also an instrument of type 1 above, and networks of radars would be required for widespread observations. Such networks already exist over some land areas but their extension over the ocean would obviously be impractical.

Although the available information is extremely limited and indirect, some estimates have been made of the annual and seasonal distribution of precipitation over the oceans. Tucker (1961) gives a brief survey of some of the early attempts. The first maps were simply based on extrapolations from continental and island stations. Early estimates based on oceanic data are those of Wüst who, in 1936, obtained mean latitudinal values of precipitation from an analysis of surface salinity by using the assumption that the mean salinity at any latitude reflects the difference in evaporation and precipitation. His results implied that the average rainfall is significantly less over the open ocean than over adjacent land areas.

In the late 1940s a number of weather ships were stationed in the North Atlantic and Pacific Oceans. These ships do not make routine measurements of precipitation amounts but regularly issue 'present weather' reports which include some precipitation information. Tucker (1961) devised a technique for obtaining quantitative estimates from these reports. In essence, he correlated the frequency of occurrence of the various weather code numbers with monthly precipitation amounts at a number of land stations where both types of observations were recorded. He then applied the resulting relationships to the 'present weather' reports of the ocean stations to deduce precipitation amounts. Tucker's analysis for the North Atlantic Ocean showed significantly smaller values for the ship stations than for most of the island and coastal ones. Reed and Elliott (1973), using Tucker's method to make a similar study for the North Pacific Ocean, also found less rainfall over the ocean than was indicated in previous estimates or was observed at coastal and island stations. Both the salinity technique and Tucker's method are semiquantitative at best so that there is considerable uncertainty in the estimated rainfall amounts. It should be noted, however, that the various studies all led to the same conclusion, that rainfall over the oceans is considerably less than over land.

The dearth of actual measurements of rain over the oceans and the evidence that the amounts may differ significantly from those at coastal or island stations point up the need for development and implementation of more adequate methods for measuring oceanic precipitation than have heretofore been available.

2. RAIN GAUGES AT SEA

2.1 Rain Gauges on Ships

The problems associated with measuring precipitation with rain gauges aboard ship have received considerable attention, and a summary of results prior to 1962 may be found in World Meteorological Organization Technical Note No. 47 (WMO, 1962). The main contention of the writers of that report is that air motions around and over a ship significantly affect collection by rain gauges and cause a great underestimate of the true rainfall. They cited the results of an experiment performed in Germany wherein rainfall measured with three rain gauges on a tower 34 m above the surface was compared with that obtained with a properly-installed gauge on the ground nearby. The results, summarized in Table 1, illustrate the decrease in collection efficiency with increasing wind speed.

On the basis of these results the authors of the WMO Technical Note concluded that gauges on a ship should be placed as high as possible to minimize the effects of air currents flowing around and over the ship. However, the substantial differences between the different gauges in the tower experiment suggest that the most important factor is the wind currents around the gauges themselves rather than those around the tower. Thus, the conclusion that rain gauges on ships should be placed as high as possible in order to minimize wind effects may well not be valid.

In the recent GATE experiment (Kuettner et al., 1974) a considerable amount of data pertinent to this question was obtained with rain gauges placed at a number of locations on some of the ships.

Table 1

Precipitation amount measured on a tower in percent of the total measured on the ground nearby, for different wind speeds (from WMO, 1962)

Type of rain-gauge used on the tower	Wind speed range in m s^{-1} 0-4.9	5.0-9.9	10.0-14.9
Ordinary rain-gauge	87%	79%	17%
Marine rain-gauge	117%	85%	50%
Streamlined rain-gauge	81%	64%	25%

Table 2

Comparison of rainfall amounts measured by five rain gauges on R/V Gilliss during GATE, 28 June-18 September 1974. Positions for each number are shown in Figure 1.

6-hr Rain gauge totals (mm)					
Stern (1)	Fantail (2)	Flying Bridge (3)	Mast (4)	Port Bow (5)	Mean
313.3	336.7	267.4	277.2	311.5	302.0

Estimated standard deviation = 17.2% for 6-hr samples.

On the R/V Gilliss there were four plastic wedge gauges and two recording siphon gauges distributed as shown in Figure 1. The wedge gauges were used to measure rainfall accumulations and the siphon ones for rain rate measurements. In addition, the siphon gauge on the mast emptied into a collector where accumulations could be measured and compared to values obtained with the wedge gauges.

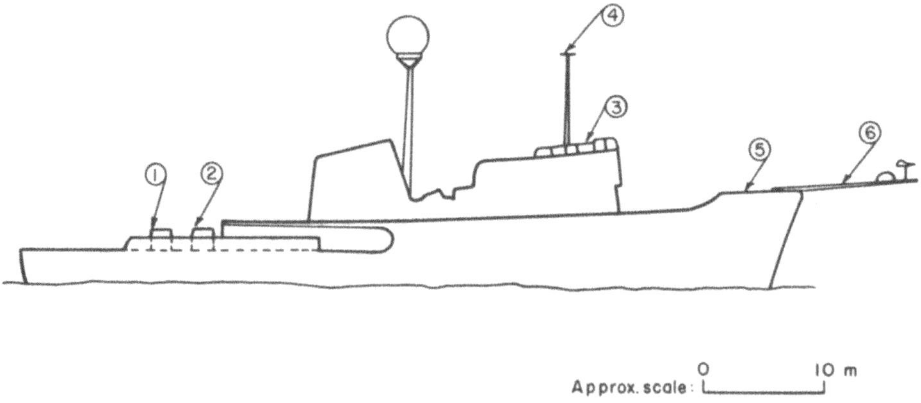

Fig. 1 Locations of rain gauges on R/V Gilliss during GATE. Identified in text and tables are wedge gauges (1) stern, (2) fantail, (3) flying bridge, (5) port bow; and siphon gauges (4) mast and (6) boom.

A comparison has been made between 6-hr rainfall amounts measured by the accumulation gauges for 35 periods when all 5 were operating. The total amounts are listed in Table 2.

The five gauges agreed to within 12% for the total, and the scatter of the individual 6-hr samples was only 17%. It should be noted that the majority of measurements in GATE were made with the ships nearly stationary and with ambient winds usually less than 10 m s^{-1} or so. Normally an attempt was made to keep the bow headed into the wind, but the onset of rain was often accompanied by wind shifts and, as a result, wind direction with respect to the ship was not consistent during rain periods. The smallness of the variability between gauges in such diverse locations suggests that shipboard measurements of rain, at least on more-or-less stationary ships and in the absence of strong winds, can be made with a fair degree of confidence. Corrections for effects of the ship's pitch and roll on the rain catch are suggested in WMO (1962). However, they were considered to be small compared with the other uncertainties involved and therefore unlikely to be particularly meaningful. Also, during GATE, sea spray was never serious enough to present a problem, as might well be the case in areas with more severe wind regimes.

It is interesting to note that the mast gauge, the highest and therefore generally considered to be least likely to be affected by air motions around the ship, collected a smaller amount of rain than most of the other gauges. Only the one on the flying bridge collected slightly less and this gauge, too, was located fairly high on the ship (see Fig. 1). Thus, on the basis of these measurements, it appears that rain gauges on ships are best placed at locations which are more-or-less protected, but not susceptible to sea spray or dripping from above. In view of the known problem of flow distortion around ships, unbiased results cannot be ensured by any special positioning of the gauges. Using more than one gauge and location would appear to be a good idea.

Finally, judging from previous results (WMO, 1962; Larson, 1971), gauge shapes and sizes should be chosen with the aim of minimizing unwanted wind effects. The plastic wedge gauge which was used in GATE appears to be a good example of a desirable type.

2.2 Measurements of Rainfall Rate

Most devices for measuring rain rate, such as the tipping bucket, are quite sensitive to changes in attitude and therefore not suitable for shipboard measurements. A siphon-type gauge, which does not have this difficulty, was designed by W. Everard (Seguin and Crayton, 1977) to measure rain rates on ships in GATE. Four U.S. ships had pairs of siphon gauges aboard, one on the mainmast and one on the forward boom, as shown in Figure 1 for the R/V Gilliss.

Table 3

Comparison of 3-min rainfall rates measured with pairs of Everard Siphon gauges on American ships in GATE

Ship	No. of pairs of 3-min samples	Gauge Totals (mm) Mast	Boom	Estimated Standard Deviation
Researcher	325	69.4	82.6	31%
Gilliss	101	28.0	31.7	59%
Dallas	242	18.9	23.1	38%
Oceanographer	18	1.4	1.4	16%

Comparisons of measurements with the pairs of gauges, for all periods when data from both were available, are shown in Table 3. The differences in the totals are undoubtedly a result of location, with the mast gauge collecting less in virtually every case. This result is consistent with the findings of the previous section which also suggested that a mast may not be a very desirable location for a rain gauge.

The deviations of the individual measurements, which result from both location differences and statistical scatter, are not much greater than those introduced by location difference alone. Thus, the siphon gauge appears to provide a promising means for measuring rainfall rate at sea, at least under conditions where the catch is not seriously affected by strong winds or sea spray.

Details of the time distribution of rainfall rate can also be derived from measurements of drop-size distributions. Such measurements are difficult to make, however, especially on ships, and would not be recommended simply as a method of measuring rainfall rate. They are generally undertaken for research purposes such as studies in cloud physics or for use in interpreting radar measurements.

2.3 Rain Gauges on Buoys

Rain gauges on buoys, with the information recorded automatically or telemetered, offer another possible way to measure rain at sea. Elliott and Reed (1973) report a fair amount of success in measuring rain on a spar buoy moored off the Oregon coast. On the buoy were two nonrecording gauges and one tipping-bucket gauge. The latter was "suspended in a barrel partially filled with water to provide damping and eliminate premature tips from accelerations of the buoy." Movements of the buoy were slow and the maximum tilt was 10°. The gauges were mounted at a height of approximately 11 m above the waterline, and wind effects, a troublesome unknown, were apparently taken care of by taking the maximum of the three gauge readings. Measurements on buoys are too few to make a determination of their accuracy, but it seems likely that they are comparable to those on a stationary ship.

2.4 Representativeness of Measurements

The chief limitation of rain-gauge measurements at sea is the lack of representativeness, a result of their extremely limited coverage. The same limitation, while not nearly so severe, also applies to land-based gauges. And land networks of gauges, at densities far greater than would ever be possible over the oceans, are able to delineate only the grossest features of rainfall patterns.

Thus, the main advantage of shipboard (or buoy) gauges is to provide point measurements for specific experiments which require only limited coverage or for serving as calibration points for methods with larger-scale coverage. The most notable among the latter are techniques involving satellite-borne instruments, which will be discussed in a later section.

3. RADAR MEASUREMENTS

In view of the significant small-scale variability of precipitation and the difficulty of obtaining representative samples with networks of gauges, the use of radar to measure rainfall amounts over an area has long been considered an attractive possibility. A single radar can cover an area of some tens of thousands of square kilometres with resolution on the order of a square kilometre, while a network of radars could measure the rain over a much larger area. Over ocean areas it would not be possible, of course, to achieve continuous coverage. Nevertheless, quantitative radar records from widely separated installations on islands, coastal stations and ships could provide valuable information on the local climatology and could be used for comparison with various types of satellite observations. Oceanic storms of particular interest, such as hurricanes and typhoons, are often monitored by radars on reconnaissance aircraft. At the present time only rough estimates of storm intensity are obtained, but the potential exists for more quantitative rainfall measurements with the airborne radars.

The need to use empirical relations and various assumptions to deduce rainfall rates from measured radar reflectivities introduces considerable uncertainty into the results. To date, operational use of radars for rainfall measurements has been very limited, but efforts are being made to develop techniques for reducing the uncertainty by coordinating radar and gauge observations (e.g. Wilson, 1976). In essence, the measurement of surface rainfall with a radar involves four steps:

(1) measurement of the average signal intensity for fluctuating weather echoes;

(2) calculation of the reflectivity of the storm from the average signal intensity;

Fig. 2 Instantaneous rainfall rates, in mm hr^{-1}, deduced from R/V Gilliss radar data with the relation $Z = 230\ R^{1.25}$. Format reversed at 3 mm hr^{-1}. Grid squares are 4 km by 4 km. Ship location, depicted by symbol, 9.3°N, 24.8°W. Data taken on 5 September 1974 during GATE.

(3) deduction of the precipitation rate from the radar reflectivity; and

(4) consideration of possible differences between the precipitation observed by the radar and that reaching the surface.

With the present state of technology for radar calibration and for recording and processing digital data, radar reflectivity can be measured (steps 1 and 2 above) with an acuracy of approximately 1 dB. Uncertainties associated with steps 3 and 4 above can amount to 2 or 3 dB, sometimes even more, and generally represent the limiting factors on the accuracy which can be attained.

To deduce the rainfall rate, R, from the radar reflectivity factor, Z (step 3 above), a relation between the two is needed. Since individual raindrops contribute to the radar reflectivity of a storm according to the sixth power of their diameters while their contributions to the rainfall are roughly proportional to the 3.5 power of the diameters (the mass times the fall velocity), the relationship is not unique but depends on the sizes of the raindrops in any particular situation. Empirical Z-R relations are obtained from measurements of raindrop-size spectra in actual storms.

Battan (1973) has listed a large number of such empirical Z-R relations for different geographical locations and storm types. The standard deviation for samples contributing to a given Z-R relation is generally in the vicinity of 0.17 in $\log_{10}R$. Differences between the various Z-R relations are of similar magnitude. Thus the uncertainty is somewhat greater if an empirical Z-R relation is applied to places other than the one where the drop-size measurements were made.

Very few drop-size measurements have been taken directly over the ocean. Some were obtained during GATE and yielded the relation

$$Z = 180\ R^{1.35}$$

where Z is expressed in $mm^6\ m^{-3}$ and R in $mm\ hr^{-1}$. This relation is quite similar to the one obtained by Stout and Mueller (1968)

$$Z = 221\ R^{1.32}$$

for the Marshall Islands.

Figure 2 shows an example of the type of rainfall data which can be obtained with a radar: in this case a map of instantaneous rainfall rates in a grid of 4x4 km squares. Time integration of such data would provide hourly or daily amounts. Similar data could be obtained for other oceanic areas with radars installed on islands, ships or shores.

Important features for a radar which is to be used to measure rain are:

(a) a wavelength in the 5 to 10 cm range,

(b) a narrow beam, preferably 1 to 2 degrees between half-power points, and

(c) capability for digital recording and processing of the data.

For shipborne and airborne radars a further requirement is some form of stabilization to compensate for the motion of the craft.

Use of shorter wavelengths is not recommended because the radiation is severely attenuated by heavy rain. At 5 cm the attenuation is often significant, but in the interest of attaining higher resolution this wavelength is often chosen, especially for airborne or shipborne installations where the antenna size must be limited.

Because of the complexity of the measurements and the spatial and temporal variability of the precipitation characteristics, it is not possible to make definitive assessments of the expected accuracy of radar measurements of rainfall rate. A rough estimate is that with a well calibrated radar, an instantaneous observation covering an area of a few square kilometres can yield a rainfall rate which is accurate within a factor of two. When integrated

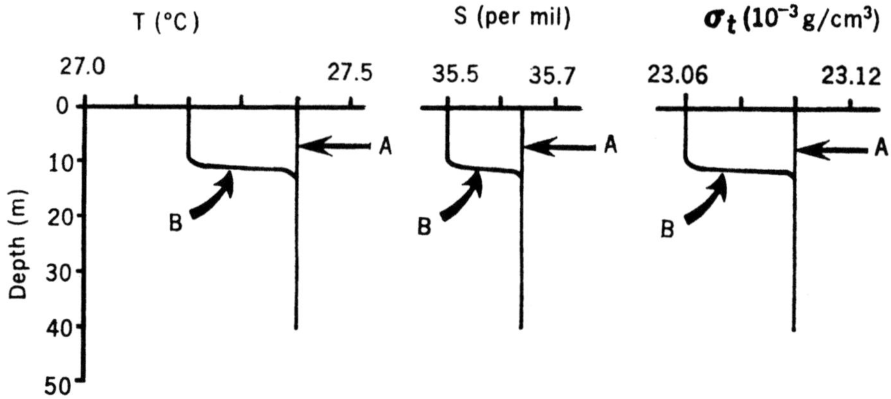

Fig. 3 Temperature (T), salinity (S), and density (σ_t) profiles before and after precipitation: (A) profiles at 0830, 19 July 1972; (B) profiles at 0230, 20 July 1972 (from Ostapoff et al., 1973).

over space and time the random errors tend to cancel so that for daily rainfall amounts over areas of 10^4 to 10^5 km^2 an accuracy of 25% may be expected.

4. RAINFALL DEDUCED FROM SALINITY PROFILES

On several occasions localized surface layers of low salinity have been observed during oceanographic experiments in tropical areas (Ostapoff et al., 1973). These pools have horizontal dimensions of a few tens of kilometres and depths of 10 or 20 m; they are produced by the fresh water from rain and are quite stable, persisting for a period of days or even weeks. The amount of fresh water required to produce an observed mass of low-salinity water can be computed with relative ease, and provides an estimate of the total rain which fell from a shower.

Figure 3 shows the modification in salinity, temperature, and density profiles observed by Ostapoff et al. (1973). On that occasion the computed rain amount was 46 mm as compared with 17.5 mm recorded by a gauge on the ship. Too few measurements of this type have been made to assess the accuracy or to determine whether the discrepancy in this case should be attributed to a high rain estimate from the salinity measurements, an underestimate by the rain gauge, or to horizontal variations in the precipitation intensity.

At the present time the salinity profile technique is not suitable for operational measurements of rainfall over the ocean. Any measurements made in a research context, however, can be very useful for comparing or coordinating with information obtained by other methods.

5. MEASUREMENTS FROM SATELLITES

5.1 General Considerations

Instruments carried on meteorological satellites are able to observe clouds and storms over the vast oceanic areas but are not able to measure precipitation per se. The sensors on operational satellites provide data at visual and infrared wavelengths while the experimental Nimbus-5 satellite carries a microwave radiometer as well. Maps from the visual data indicate the locations of cloud systems and their general configurations; the intensity of the image at any point depends on the cloud depth, the cloud water content, and the angle of solar illumination. Infrared images indicate the temperature at cloud top, and thus provide estimates of the heights of clouds. These types of data do not appear promising for the estimation of rainfall because the cloud characteristics they depict are not closely related to the intensity of rainfall. Although heavy rain usually falls from cloud with considerable depth (and therefore with bright images and cold temperatures), broad cirrus shields often show up strongly and yet deposit no significant rain.

The microwave radiometer measures brightness temperature, which is a function of the emissivity of the earth's surface modified by the intervening atmosphere. Since liquid water in the atmosphere is an important modifying factor, measurement of the brightness temperature can provide an indirect indication of the rainfall rate. The fact that it is an indirect measurement, requiring the use of empirical relations and models to deduce the rainfall, leads to rather large uncertainties in the estimated values.

For a number of years the possibility of placing a radar on a spaceship has been discussed, but has generally been dismissed as impractical. Recent technological developments, however, appear to be bringing this possibility into the realm of the feasible. Rainfall measurements with a spaceborne radar should be more accurate than those obtained with the microwave radiometer and should be comparable to ones from a ground-based radar of similar resolution.

Another serious limitation to estimates of total precipitation based on satellite data arises from the fact that a polar-orbiting satellite views any given area on the earth only once a day visually and only twice a day with infrared or microwave radiometers. Precipitation patterns and intensities sometimes change significantly in periods as short as an hour. Moreover, in many regions the precipitation regime tends to have a diurnal cycle so that the cloud patterns which prevail at the time of the satellite observation may not be representative of the day as a whole.

5.2 Rainfall Deductions from Visible and Infrared Satellite Imagery

In early attempts to estimate rainfall from visual satellite data, rain amounts measured at surface stations were related empirically to the percentage of area covered by clouds of different types as shown by the satellite observations. Barrett (1970) used this method for estimating monthly rainfall amounts over Australia. Martin and Scherer (1973) describe several experiments wherein a refined version of Barrett's technique was used to estimate daily precipitation amounts. This method is crude at best and is even less feasible over ocean than over land because of the lack of sufficient surface data to determine the needed empirical relations.

By the early 1970s methods for precise measurement of cloud brightness were developed and it was confirmed that bright clouds (in either visual or IR imagery) are often associated with rain areas, as was indicated by gauges or radar echoes. Some empirical expressions have been developed which relate the rain flux from convective cloud systems to the areas of clouds at a selected brightness level and the rate of growth (or diminution) of such areas (Griffith et al., 1978). Use of these empirical relations to estimate surface rainfall from satellite observations requires time histories of the cloud systems, data available only from geosynchronous satellites. This method is being used to determine rainfall amounts for portions of the GATE area which are beyond the range of the quantitative radars. Comparison with the radar data in the region where they are available provides an estimate of the accuracy. An accuracy of approximately 1.5 is claimed for daily amounts over areas ~10^5 km^2, and it deteriorates for smaller areas and time periods. For a single convective storm over a period of a half hour, the rms error is approximately equal to the average rain of the storm (NCAR, 1977).

5.3 Microwave Radiometer

The Electronic Scanning Microwave Radiometer (ESMR) system, which has been carried on the Nimbus-5 experimental satellite since 1972, responds selectively to liquid water in the atmosphere, and of the currently available instruments it appears to offer the greatest

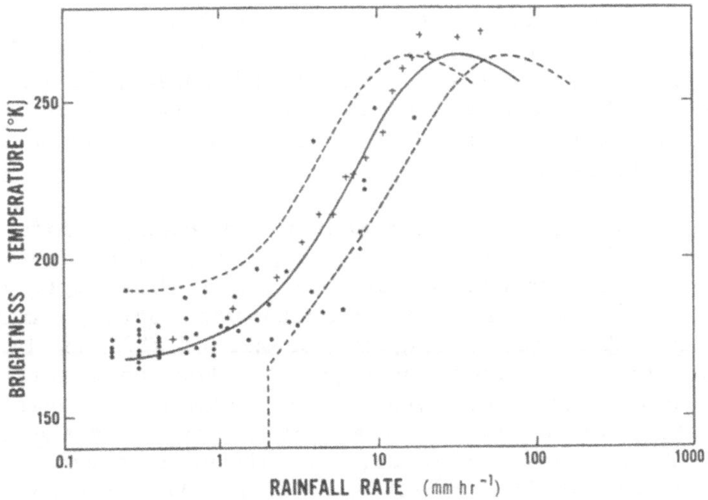

Fig. 4 Brightness temperature as a function of rain rate. Solid line is the calculated brightness temperature for a 4-km freezing level. Dashed lines represent departure of 2 mm hr^{-1} or a factor of 2 in rain (whichever is greater) from the calculated curve.
- ● Nimbus-5 ESMR versus WSR-57 radar
- \+ Inferred from ground-based measurements of brightness temperature and direct measurements of rain rate

(from Wilheit et al., 1977).

promise for measuring surface rainfall over broad oceanic areas. The ESMR system receives radiation emitted by the earth and the atmosphere at a frequency of 19.35 GHz. The antenna beam scans perpendicularly to the direction of satellite motion through an arc of 100° every 4 s. Spatial resolution of the image is a 25-km circle at nadir which degrades to an oval 45x160 km at the ends of the scan. Because the emissivity of land surfaces is large and highly variable, the effect of water in the atmosphere cannot be separated from other factors which contribute to variations in radiation received from land areas. The ocean emissivity, however, is relatively low and quite uniform so that changes in the radiation received from oceanic regions can be attributed to liquid water in the atmosphere. The theory relating brightness temperature to rainfall rate is presented by Allison et al. (1974). A height for the 0°C isotherm is assumed and ice particles above this level are presumed to have no effect on the brightness temperature. Below this level an exponential size distribution of raindrops is assumed and Mie's (1908) scattering theory is applied. Figure 4 shows the computed relation compared with some experimental results obtained by Wilheit et al. (1977) in the Miami, Florida, area. Two types of comparison are included. In one, ESMR brightness temperatures are plotted against rainfall rates for the area as deduced from a WSR-57 radar. In the other, brightness temperatures from ground-based radiometers are plotted against rainfall rates measured by surface gauges. For rainfall rates between 1 and roughly 15 mm hr^{-1} most of the observations are within a factor of two of the calculated curve. Higher rainfall rates are not measureable because the system becomes saturated.

Rao et al. (1976) have computed yearly and weekly rainfall amounts for 1973 and 1974 over all the oceans, except for arctic regions, in area blocks of four degrees of latitude by five degrees of longitude. A comparison of their results with data from shipborne gauges in the GATE area during the summer of 1974 is in Table 4; the ship positions are shown in Figure 5. For the overall average, agreement between the two types of measurement is fair with the ESMR indicating 30 to 40% less than the gauges. For the individual periods of 2 to 7 days, however, the differences are quite large and may be in either direction. Similar results were obtained from a comparison of instantaneous ESMR measurements with hourly amounts derived from the radar data for an area two degrees of latitude by two degrees of longitude, centered over the ship array. For 31 pairs of observations the ESMR indicated 38% less rain than the radars on the average, and the variance was large with a standard deviation of 80%. The fact that a significant fraction of tropical rain occurs at rates above the saturation level of the ESMR may partially explain the average underestimate. The pronounced small-scale variability in rainfall intensity doubtless contributes to the large variances, but sufficient comparisons have not yet been made to identify all of the sources of variance and to assess their relative contributions.

Table 4

Rainfall amounts collected on ships during GATE compared with those indicated by ESMR data from Rao et al. (1976). Ship positions and the 4°x5° grid rectangles, A and B, are shown in Figure 5. Rainfall amounts in mm.

Dates	Ships in Area A						ESMR	Ships in Area B				ESMR
(all 1974)	1	2	7	8	13	Aver.		5	11	12	Aver.	
29-30 June	28	0	0	0	0	6	0	106	21	90	72	62
1-7 July	182	7	65	0	2	51	17	90	1	10	34	17
8-14 July	74	17	61	0	10	32	17	132	79	175	129	34
29-31 July	0	1	M	0	0	0	17	10	0	2	4	17
8-14 Aug.	19	139	47	2	78	57	34	40	0	69	36	0
30-31 Aug.	6	0	19	3	3	6	17	9	0	21	10	34
1-7 Sept.	35	8	85	40	23	38	34	85	53	70	69	84
8-14 Sept.	96	192	77	111	25	100	50	43	47	96	62	17
15-19 Sept.	36	25	51	5	10	25	50	72	0	2	25	0
TOTALS	476	389	405	161	151	316	236	587	201	535	441	265

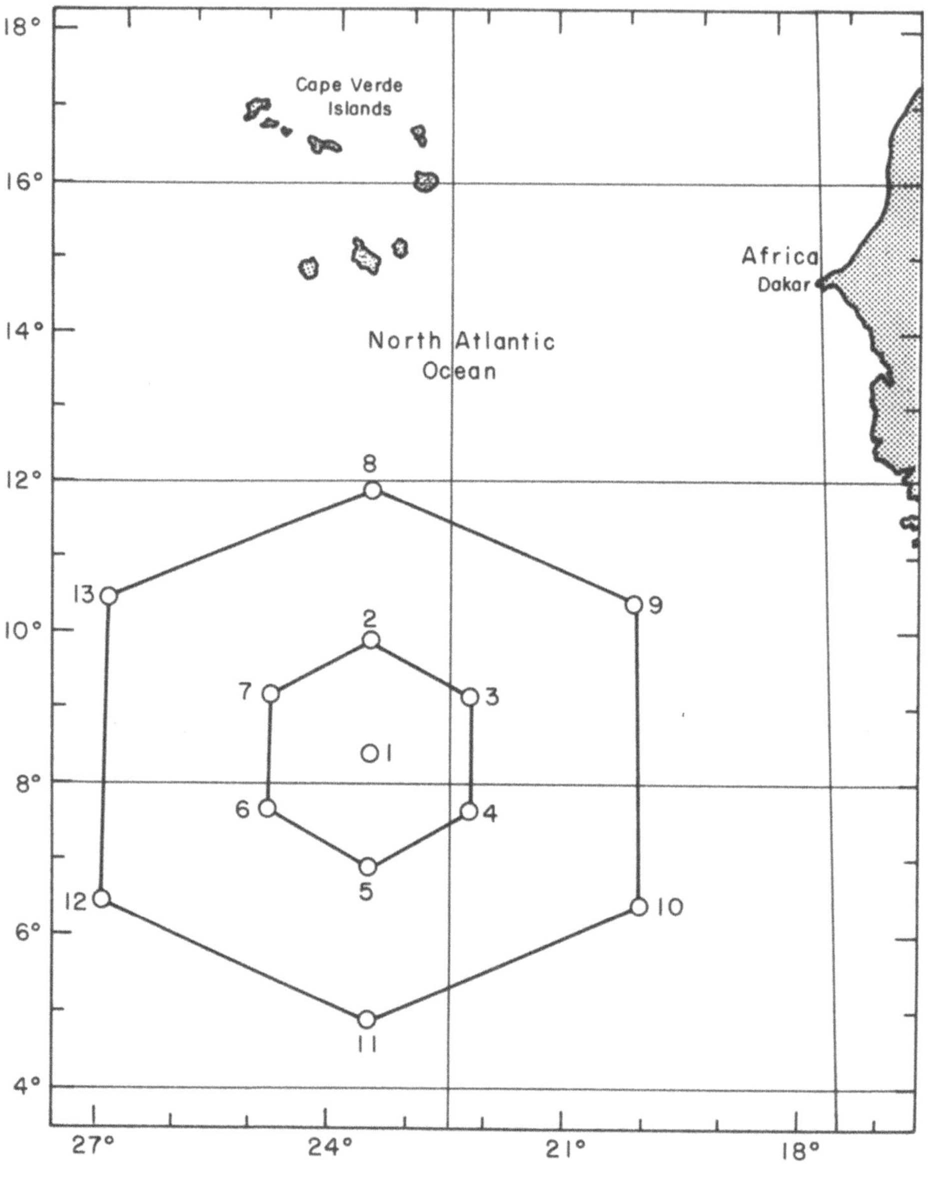

Fig. 5 Ship locations and ESMR quadrangles in GATE. 'A' quadrangle 8° to 12°N, 22.5° to 27.5°W; 'B' quadrangle 4° to 8°N, 22.5° to 27.5°W.

The differences in the rainfall amounts recorded at the different ship stations, shown in Table 4, point up the rather sharp gradients that can occur, even for seasonal amounts, and illustrate the need for complete coverage in order to obtain realistic distributions of rainfall over the oceans.

6. SUMMARY AND CONCLUSIONS

Various techniques have been described for measuring rainfall over the oceans. Each of them has certain advantages and limitations, and it would appear that satisfactory depiction of the distribution of precipitation over the oceans can best be obtained by coordinating several types of observations. Spatial coverage of the vast ocean regions can be achieved only with satellite-borne instruments. Of those, the electronically scanning microwave radiometer (ESMR) appears to offer the greatest promise as a device for measuring precipitation, even though the measurements have relatively coarse resolution in time and space and a relatively high level of uncertainty. Rain-gauge and radar measurements from installations at selected points can supply information on the smaller-scale details in time and space and, through comparisons, can be used to narrow the limits of uncertainty in the radiometer indications.

In the GATE experiment rainfall observations over the open ocean were, for the first time, made simultaneously with all of the available techniques. Data were recorded over a period of several months from satellite radiometers (visual, infrared and microwave), from shipborne gauges and radars, and from salinity probes. Comparison of measurements made with the different instruments in this experiment is yielding much useful information on their respective capabilities. When analysis of this comprehensive data set is complete, a more adequate evaluaton of techniques for measuring rainfall over the oceans will be possible.

For several decades, efforts to map rainfall over the oceans have depended on crude methods which sampled only miniscule portions of the total ocean areas. Currently emerging techniques, especially the use of satellite-borne sensors, presage much more comprehensive and reliable measurements in the future.

REFERENCES

ALLISON, J.A., E.B. RODGERS, T.T. WILHEIT, and R.W. FETT. 1974. Tropical cyclone rainfall as measured by the Nimbus 5 electrically scanning microwave radiometer. *Bulletin of the American Meteorological Society*, 55: 1074-1089.

BARRETT, E.C. 1970. The estimation of monthly rainfall from satellite data. *Monthly Weather Review*, 98: 322-327.

BATTAN, L.J. 1973. *Radar Observations of the Atmosphere*. University of Chicago Press, Chicago, 324 pp (see pp. 90-92).

ELLIOTT, W.P. and R.K. REED. 1973. Oceanic rainfall off the Pacific Northwest coast. *Journal of Geophysical Research*, 78: 941-948.

GRIFFITH, C.A., W.L. WOODLEY, P.G. GRUBE, D.W. MARTIN, J. STOUT and D.N. SIKDAR. 1978. Rain estimation from geosynchronous satellite imagery -- visible and infrared studies. *Monthly Weather Review*, 106: 1153-1171.

KUETTNER, J.P., D.E. PARKER, D.R. RODENHUIS, H. HOEBER, H. KRAUS and G. PHILANDER. 1974. GATE final international scientific plans. *Bulletin of the American Meteorological Society*, 55: 711-744.

LARSON, L.W. 1971. Precipitation and its measurement, a state of the art. Water Resources Series No. 24, Department of Hydrology, University of Wyoming, Laramie, 74 pp.

MARTIN, D.W. and W.D. SCHERER. 1973. Review of satellite rainfall estimation methods. *Bulletin of the American Meteorological Society*, 54: 661-674.

MIE, G. 1908. Beiträge zur Optik trüber Medien, speziell kolloidaler Metallösungen. *Annalen der Physik*, 25: 377-445.

NCAR. 1977. *Report of the U.S. GATE Central Program Workshop*. NCAR, Boulder, CO. 723 pp.

OSTAPOFF, F., Y. TARBEYEV and S. WORTHEM. 1973. Heat flux and precipitation estimates from oceanographic observations. *Science*, 180: 960-962.

RAO, M.S.V., W.V. ABBOTT, III, and J.S. THEON. 1976. Satellite-derived Global Oceanic Rainfall Atlas (1973 and 1974). NASA SP-410, Goddard Space Flight Center, Greenbelt, Md. Available from Superintendent of Documents, U.S. Government Printing Office, Washington, D.C. 20402. 186 pp.

REED, R.K. and W.P. ELLIOTT. 1973. Precipitation at ocean stations in the North Pacific. *Journal of Geophysical Research*, 78: 7087-7091.

SEGUIN, W.R. and R.B. CRAYTON. 1977. U.S. GATE B-scale Ship Precipitation Data. NOAA Technical Report EDS 23, 402 pp. CEDDA, Environmental Data Service, NOAA, Washington, D.C.

STOUT, G.E. and E.A. MUELLER. 1968. Survey of relationships between rainfall rate and radar reflectivity in the measurement of precipitation. *Journal of Applied Meteorology*, 7: 465-474.

TUCKER, G.B. 1961. Precipitation over the North Atlantic Ocean. *Quarterly Journal of the Royal Meteorological Society*, 87: 147-158.

WILHEIT, T.T., A.T.C. CHANG, M.S.V. RAO, E.B. RODGERS and J.S. THEON. 1977. A satellite technique for quantitatively mapping rainfall rates over the oceans. *Journal of Applied Meteorology*, 16: 551-560.

WILSON, J.W. 1976. Radar-rain gauge precipitation measurements. *Preprints of Conference on Hydro-Meteorology*, American Meteorological Society, Boston: 72-75.

WORLD METEOROLOGICAL ORGANIZATION. 1962. Precipitation measurements at sea. Technical Note No. 47, WMO - No. 124.TP.55, Secretariat of the WMO, Geneva, 18 pp.

29

Sodar and Lidar Measurements in the Atmospheric Boundary Layer at Sea

H. Ottersten and A. Hågård

1. INTRODUCTION

Echo-sounding instruments and methods that use beams of audible SOund or coherent LIght are well suited for probing the atmospheric boundary layer at sea. These instruments are often referred to as SODAR and LIDAR, respectively, since at their heart they are Detection And Ranging devices. They operate in a radar-like fashion and may conveniently be discussed with the aid of theory and terminology developed for their cousin. Sodar and lidar methods are attractive because they combine unique sensitivity to certain boundary-layer properties with the ability to interrogate the medium with good spatial and temporal resolution. Sodar is essentially a device for boundary-layer probing; lidar has a wider range of applications but some of its features are best exploited in a boundary-layer probe.

We will discuss mainly proven measurement concepts that appear feasible for use on buoys and ships. We will focus on monostatic systems, which have transmitter and receiver co-located. Bistatic systems offer additional measurement potentials but are less feasible for use from a ship than from shore or ice. We will exclude methods for the sensing of gaseous pollutants in order to concentrate on techniques for the measurement of meteorological properties.

Sodar can make visible the boundary-layer structure in clear air by mapping the small-scale variability in air density. With lidar, boundary-layer cross-sections can be obtained by mapping the aerosol density distribution. With lidar, cloud height and slant visibility can also be measured and it is possible to derive vertical profiles of air temperature and humidity using sophisticated molecular scattering or absorption techniques. Both lidar and sodar offer means for remote determination of wind, generally from the Doppler effect. All of these methods have been proven experimentally and appear feasible to adapt for shipboard use. Some of them have already been used for exploration of the boundary layer at sea.

Echo-sounding methods offer a number of notable advantages compared to in situ measurements. Volume averages are obtained, often with excellent space and time resolution, and these are generally more representative than point values. The risk of inadvertent modification of the medium is avoided and nearly continuous measurements can be obtained simultaneously in space and time along the beam. With steerable beams, probing in two or three spatial dimensions is also possible. In addition, lidar and sodar instrumentation is easily adaptable to automatic operation and data processing.

Beams of non-coherent light were employed for atmospheric sounding 40 years ago, but it was the advent of the pulsed laser, the heart of current lidar systems, that sparked the rapid development of atmospheric lidar sensing about 15 years ago. In 1963, lidar observations of the upper atmosphere were reported (Fiocco and Smullin, 1963) and simultaneously the first lidar returns from the boundary layer were being investigated at the Stanford Research Institute in the U.S.A. (Ligda, 1964). Brief historical background is included in the review by Collis (1970). The potential of utilizing advanced lidar concepts based on Raman shift backscattering, resonant absorption, and Doppler effect were quickly recognized and demonstrated by Leonard (1967), Schotland et al. (1966), and Huffaker et al. (1970), respectively. Today, atmospheric measurements by lidar span a wide range of applications as evidenced by the series of eight conferences on Laser Atmospheric Studies (1969-1977). A thorough review of the field has been edited by Hinkley (1976).

Acoustic echo sounding of atmospheric structure dates back more than 100 years, but the quick exploitation of sodar as we know it today started no more than 10 years ago and built upon experiments at the Weapons Research Establishment in Australia (McAllister, 1968). Much of the impetus for further development was provided by work at the NOAA Wave Propagation Laboratory in the U.S.A. with Little's (1969) analysis of the potentials of the technique and with the development of Doppler wind sensing methods (Beran et al., 1971; Beran and Willmarth, 1971). A summary of the early work has been provided by Hall (1972). The quickly expanding interest in sodar techniques and the rapid development toward operational applications have been reflected in the series of five workshops on Atmospheric Acoustics (1972-1978). Advances in the field were recently reviewed in detail by Brown and Hall (1978).

2. SODAR AND LIDAR PRINCIPLES

Monostatic sodar and lidar systems utilize the same basic principle. Pulses of energy are emitted within a beam into the atmosphere and backscattered energy is detected in a receiver. The range r to the scattering volume is obtained from $r = ct/2$ where t is the time delay between pulse emission and detection and c is the propagation velocity. The receiver signal versus time contains information about the distribution of scattering elements along the beam of radiation.

The instantaneous received power P(r) is given by

$$P(r) = P_t(ct_p/2)\ A\ r^{-2}\sigma(r)\tau(r) \qquad (1)$$

where P_t is the emitted pulse power, t_p the pulse duration, A the effective receiver aperture area, $\sigma(r)$ the backscattering coefficient, and $\tau(r)$ the two-way radiation transmittance which accounts for the round-trip atmospheric extinction.

The received power P(r) is the sum of signal contributions from the elementary scattering volume defined by the beam of radiation and the resolution length $(ct_p/2)$. The term Ar^{-2} is the solid angle of the effective receiver aperture area seen from the scattering volume, and σ is a measure of the backscattering cross section of the medium per unit volume per unit solid angle ($m^{-1}\ sr^{-1}$, sr = steradian). The transmittance $\tau(r)$, which accounts for the round-trip attenuation of radiation returning from range r, often decreases rather slowly or predictably with increasing range and Equation 1 may then be used to derive from P(r) the relative variation of σ with range along the beam. This is the most commonly used principle in sodar and lidar probing and is the basis for mapping of boundary-layer structure; e.g., by producing height versus time displays of returns from the air mass drifting through a vertical beam, or by scanning a lidar in elevation to obtain vertical cross sections of atmospheric structure. With calibrated systems, Equation 1 is also often used to measure σ as a function of range, although this requires that $\tau(r)$ be predicted. Measurements of σ are then used to derive values of various atmospheric parameters, since σ can be related to atmospheric properties as discussed in the following.

2.1 Acoustic Backscattering and Extinction

Acoustic waves are scattered in the atmosphere due to the variability in air density and wind velocity. For isotropic homogeneous turbulence within the inertial subrange (Tatarski, 1961), the backscattering coefficient σ_a for acoustic waves may be related to the acoustic refractive-index structure parameter (structure function coefficient) C_n^2 ($m^{-2/3}$).

$$\sigma_a = 0.030\lambda^{-1/3} C_n^2 \quad (2)$$

Only refractive-index inhomogeneities of scale sizes 0.5 λ contribute to the backscattering (Ottersten et al., 1973), and for commonly used acoustic wavelengths (e.g. $\lambda \sim 0.2$ m) the relevant scale sizes can be expected to fall within the inertial subrange. C_n^2 depends on the structure parameters C_T^2 ($K^2m^{-2/3}$) for temperature variations and, to a lesser extent, C_e^2 ($mb^2m^{-2/3}$) for water vapor pressure fluctuations. The influence of wind variability on C_n^2 depends on the scattering geometry and vanishes for pure backscattering (see Ottersten et al., 1973, Eq. 10).

If temperature and humidity fluctuations in the inertial subrange are assumed to be uncorrelated, the refractive-index variability for acoustic backscattering may be expressed (Wesely, 1976).

$$C_n^2 = (2T)^{-2}C_T^2 + 0.094(2p)^{-2} C_e^2 \quad (3)$$

where T (K) is the average ambient temperature and p (mb) is the average total pressure.

Over dry land, the contribution of water vapor fluctuations C_e^2 in Equation 3 may be neglected. According to Equation 2, σ_a then provides a direct measure of the small-scale temperature variability C_T^2; this has been verified by Neff (1975) over a considerable range of meteorological conditions. In the moist marine boundary layer, however, C_e^2 may produce a significant contribution in Equation 3 and the correlation between temperature and humidity fluctuations may also need to be accounted for, as explained in detail by Wesely (1976) and discussed by Gaynor et al. (1976). Over land, σ_a may reach values in the order of 10^{-8} m^{-1} ($\lambda \sim 0.2$ m) in regions of enhanced variability, while the maximum values over sea generally will be one order of magnitude lower.

The atmospheric extinction of acoustic waves at audio frequencies is essentially the sum of three different terms. Classical absorption, which includes viscous and heat conduction effects, is small and can generally be ignored in comparison with molecular absorption and the so-called excess attenuation of sound propagating in turbulent air. As demonstrated recently by Brown and Clifford (1976), this excess attenuation is due to the beam broadening produced by turbulence-induced phase fluctuations. This implies that the excess attenuation is not an intrinsic property of turbulent air but is also dependent upon path and beam geometry. It is not simply related to path length but is most sensitive to the turbulence closest to the sound source. Molecular absorption, which depends primarily on the influence of water vapor on the sound excitation of oxygen molecules, varies strongly with the acoustic frequency (Harris, 1966) and limits the useful range of sodar systems, particularly at higher frequencies. Although water vapor content is the important factor, the molecular absorption is also a function of atmospheric temperature and pressure. In conditions of strong variability of wind and temperature, the excess attenuation will further limit the sodar range. Quantitative evaluation according to Equation 1 of the backscattering coefficient σ_a from sodar measurements is often of limited value because of uncertainties in the prediction of the acoustic extinction. With increasing frequency, acoustic extinction increases rapidly while ambient acoustic noise generally decreases, and as a result the optimum sodar frequency for boundary-layer probing will generally fall in the 1000 to 2000 Hz interval, although lower frequencies may be preferable if ranges exceeding 1 km are desired.

2.2 Optical Backscattering and Extinction

Optical radiation is scattered by gas molecules and aerosol particles. The scattering is called elastic when the wavelength λ is unchanged by the scattering process. Inelastic scattering implies interaction with rotational-vibrational states of molecules and reradiation of wavelengths that differ from the incident wavelength.

Elastic scattering by molecules (Rayleigh scattering) generates dipole fields with backscattering cross sections proportional to λ^{-4} and typically in the order of 10^{-31} m^2 sr^{-1} for single molecules and visible radiation. The backscattering coefficient σ for pure molecular air at sea level is approximately 1.5×10^{-6} m^{-1} sr^{-1} at $\lambda = 0.55$ μm.

Elastic scattering by spherical particles depends in a complex way on particle size and refractive index as described by the classic Mie theory. For given particle size distributions and refractive index, Mie theory may be used to obtain a relation between the aerosol backscattering coefficient σ and the density of atmospheric aerosols. Since aerosol particles are often non-spherical and their size distribution and refractive index vary with altitude and meteorological conditions, it is desirable to calibrate the lidar return by comparison with a direct measure of the aerosol density in a representative aerosol sample. Aerosol density profiles may then be inferred from lidar returns provided that particle properties do not vary drastically along the lidar beam. Typical σ-values for atmospheric aerosols and visible radiation range from 10^{-6} to 10^{-2} m^{-1} sr^{-1}. Normally σ decreases with increasing wavelength except for large particles, such as in fog, rain and snow, for which σ is nearly independent of wavelength. For non-spherical particles and multiple scattering the state of polarization is changed by the scattering process. The subject of elastic scattering is discussed further by Collis and Russell (1976).

Inelastic scattering is potentially of great interest since the spectrum of re-emitted radiation may be used to identify specific molecules. Raman scattering is the inelastic scattering process which holds greatest promise for boundary-layer probing. Although Raman scattering is a very weak interaction, every irradiated molecule emits Raman scattering in narrow sidebands shifted from the incident wavelength with amounts specific for the molecule. The incident radiation need not be tuned to a specific wavelength, as for resonant elastic and inelastic scattering processes, but the cross section for a given molecule is proportional to λ^{-4} and a short irradiating wavelength is preferable. Figure 1 from Inaba and Kobayasi (1972) shows the Raman spectrum in an atmospheric return obtained with an N_2 laser operating at 0.337 μm. The Raman backscattering cross section for a single N_2 molecule is 10^{-34} m^2 sr^{-1} and the N_2 Raman backscattering coefficient for air at sea level is 7.4×10^{-9} m^{-1} sr^{-1}. Cross sections for other molecules are given in the excellent review by Inaba (1976) on Raman scattering. With the use of Raman lidar, humidity and temperature profiles can be measured.

Optical extinction is caused by scattering and absorption in aerosol particles and gas molecules. For single scattering, the two-way transmittance of monochromatic optical radiation can be written

$$\tau(r) = \exp[-2 \int_0^r \gamma(x)\, dx] \qquad (4)$$

where $\gamma(x)$ is the extinction coefficient. The aerosol extinction is proportional to the aerosol number density but also depends on particle shape, size distribution, and refractive index. The coefficient γ may be calculated from standard Mie theory programs for spherical particles, although assumptions of size distribution and refractive index introduce uncertainties. The aerosol extinction coefficient varies slowly with wavelength and decreases typically a factor of 10 from 0.5 μm to 10 μm. Extinction due to fog, rain, and snow is rather insensitive to wavelength, however, since particles are large. The extinction coefficient at λ = 0.55 μm is related to the visibility V_M by the so-called Koschmieder formula (Middleton, 1958)

$$\gamma = 3.91\ V_M^{-1} \qquad (5)$$

obtained from $\exp(-\gamma\ V_M) = 0.02$ where V_M is defined as the distance of 2% contrast transmission in a homogeneous atmosphere.

The absorption in gas molecules varies rapidly with wavelength and displays a line structure. For visible radiation the gaseous absorption can often be ignored but in the infrared (IR) region the

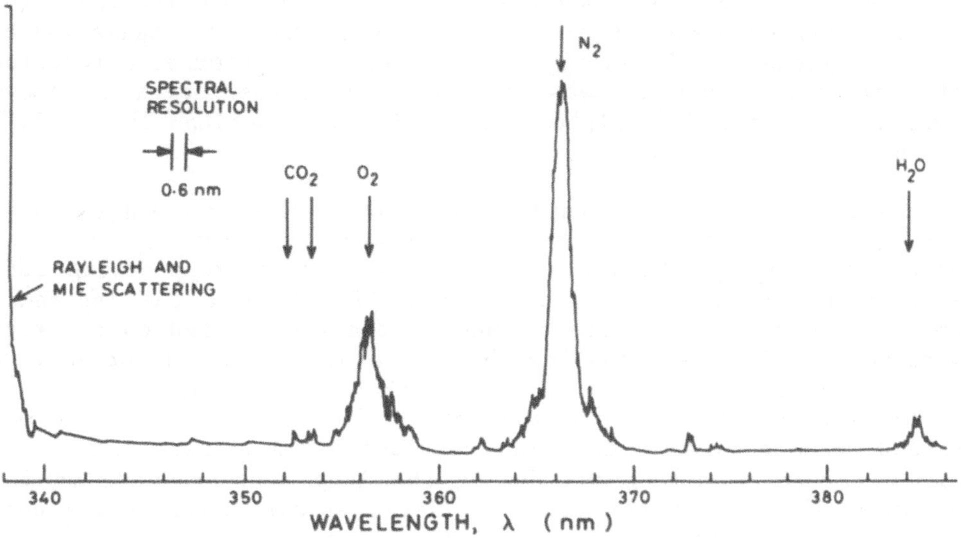

Fig. 1 Measured spectrum of Raman-shifted and unshifted backscatter from the ordinary atmosphere using N^2 laser illumination (after Inaba and Kobayasi, 1972).

density of absorption lines for atmospheric molecules is high. Molecular absorption can be utilized in the so-called DASE-method to measure concentration profiles of gaseous constituents, such as water vapor (see Section 4). More information on extinction and molecular absorption and an extensive bibliograpy is given by Derr et al. (1974).

2.3 System Configurations

Monostatic sodar systems transmit and receive through a common horn antenna, parabolic dish, or loudspeaker array with apertures typically in the order of 1 m^2. An acoustic shield normally surrounds the antenna in order to suppress man-made noise and wind noise which generally set the limit for reception of the weakest echoes. Typically, acoustic power of 10 to 100 W is transmitted with pulse durations of 10 to 100 ms and repetition rates of 0.1 to 1 s^{-1}. Antenna beam widths are often in the order of 10° and frequencies ordinarily range from 1000 to 5000 Hz. Typically, ranges from 500 to 1000 m are obtained. Ordinarily, a vertically pointing beam is used together with a facsimile recorder synchronized to the sound pulses in order to generate a continuous height versus time diagram where intensity modulation by backscattered returns displays the structure of the air mass being advected overhead. Provisions are often made to determine vertical velocities as a function of height by detecting Doppler frequency shifts in the returned echoes. Similarly, Doppler measurements in two tilted, orthogonal beams may be used to synthesize the average horizontal wind vector as a function of altitude. Sodar systems of general characteristics as above have been adapted for shipboard operation as described in the following.

Figure 2 shows the sodar antenna used by Mandics and Owens (1975) during sea trials of a monostatic vertically pointing system on board the Oceanographer. The 1.8 m tall, noise-suppressing cuff surrounding the antenna is not shown. To minimize the incident, ambient acoustic noise, the antenna system was located on the forward weather deck near the bow about 10 m above the sea surface. A vibration-isolated, pitch- and roll-stabilized platform was used to avoid structure-borne noise and echo signal loss due to ship motions, which otherwise may cause considerable deviations in transmit and receive directions during the long turn-around time for acoustic signals. With this installation sodar sounding was possible even while the ship was steaming at speeds up to 8 m s^{-1}. As

Fig. 2 Ship-mounted, stabilized sodar antenna with noise-suppressing cuff removed to show parabolic dish, transducer and feed horn (after Mandics and Owens, 1975).

long as the relative wind speed did not greatly exceed 12 m s^{-1}, acoustic returns could be received from altitudes up to 300 m and Doppler frequency shifts could be extracted in order to indicate ascending and descending air motion.

Ottersten et al. (1974), anticipating problems with ship motions and ambient noise during tests of a monostatic, vertically pointing sodar system on board the Planet, placed their parabolic dish in a vibration-free mount on the gyro-stabilized platform located on the upper deck, in front of the mast and funnel. A 400 kg, 4 m high, noise-suppressing shroud (Fig. 3) surrounded the antenna and rested unstabilized on the fixed walkway around the platform. With the ship at anchor, difficulties due to vibrations and ship motions did not arise and good results were obtained both with the parabolic antenna and a second, non-stabilized horn antenna. Useful sodar records up to altitudes of 500 to 800 m were obtained in wind speeds less than 10 m s^{-1}. Higher winds generated noise in various structures surrounding the antenna and almost prohibited sounding, but it became obvious that much of this problem can be avoided by a careful choice of antenna location. Other noise sources also impede sodar reception. In particular, with the ship machinery working, sounding was almost impossible due to interference from the nearby fans.

Fig. 3 The 4 m high noise-suppressing shroud surrounding sodar antenna mounted on the stabilized platform at the Planet upper deck (from Ottersten et al., 1974). A cut along the upper rim allowed the ship's 10-cm radar antenna to sweep above part of the shroud.

Lidar systems usually consist of a pulsed laser with beam expansion optics co-located with a receiver telescope such that the beams overlap and can be scanned in vertical and horizontal direction. Laser and receiver beam divergences are usually of the order of 1 mr. In the receiver, a narrow-band interference filter for the laser wavelength suppresses background light and a photodetector converts the backscattered energy to electrical signals, which may be displayed on an oscilloscope as a function of range, and photographed. Digital data processing now often replaces photography for storage and display of lidar signals (Uthe and Allen, 1975). The Stanford Research Institute Mark IX mobile ruby lidar system, shown in Figure 4, is built on these principles. It is housed in a 6 m long van and is designed to withstand hostile environments characterized by vibrations, high humidity, and marine air. The system can be operated while the van is moving and has recently been used on board a cruising ship (R.T. Collis, personal communication). A CO_2 lidar system has also been used successfully on board a ship for monitoring air flow at very short range (Lockheed Report, 1975).

When lidar systems are used at sea, operational problems can arise due to beam deflection caused by ship movement and due to contamination of optical surfaces by sea spray. Careful mounting, if necessary with a motion compensating mirror system, and ordinary housekeeping services are normally sufficient remedies, at least in

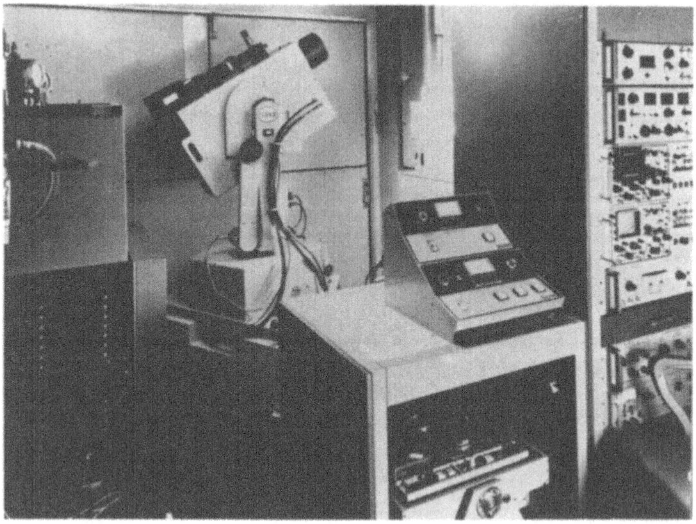

Fig. 4 The SRI Mark IX pulsed ruby lidar (from Collis and Russell, 1976). The lidar unit, mounted on the pedestal, may be automatically scanned in asimuth and/ or elevation. Data are recorded magnetically in both analogue and digital form.

research operations. One of the authors (Hågård, unpublished work) has successfully operated a lidar on board a ship without taking special precautions.

With lidar systems ranges of the order of several kilometres are commonly obtained. Laser pulse lengths may range from 10 to 1000 ns and peak powers from 10 W to 100 MW. Ruby and neodymium lasers at 0.69 and 1.06 μm, respectively, are common and typically generate 20 ns long pulses of 50 MW peak power at pulse rates of 10 s^{-1} or less. GaAs lasers giving 100 W pulses with 1 kHz rate or modulated gas lasers generating continuous power from 1 mW to 1 W are also used. Wavelengths vary from ultraviolet (UV) to 10.6 μm (CO_2 laser). Tunable lasers exist for the visible region and some regions in the IR. In systems designed for Doppler measurements and in order to achieve optimum receiver sensitivity, coherent detection is used, which implies mixing of the backscattered radiation with a local oscillator beam on the detector.

Commercial sodar and lidar instruments for the simplest applications have been available for several years. For the more sophisticated uses, instruments for simple field operation are not yet available. A lidar user, in particular, may have to deal with complex and sensitive equipment and the novice will be well advised to consult or cooperate with an experienced lidar group.

In particular, a lidar operator should observe the potential hazards of laser radiation (ANSI, 1976). Visible or near IR high power laser radiation can be damaging to the eye retina at ranges of tens of kilometres and radiation reflected out of the beam can also reach dangerous intensity. Other wavelengths are less hazardous and the high intensities that can cause damage to the eye cornea and the skin are usually not obtained with lidars.

3. BOUNDARY-LAYER STRUCTURE AND AEROSOL DENSITY DISTRIBUTION

Sodar and lidar mapping of air density variability and aerosol density, respectively, are often used to obtain time-height cross sections (and range-height in the lidar case), which may be interpreted in terms of boundary-layer structure and processes.

A typical sodar facsimile or strip-chart record (Fig. 5) gives a height-time image of the structure of the air mass being advected by the wind over the antenna. Dark areas correspond to strong sodar echoes. Two major types of echoes, typical in sodar records, are seen in Figure 5. The grass-like, or spiky, vertically oriented echoes at low altitudes represent convective plumes or 'thermals' that occur over surfaces warmer (or much more humid) than the air aloft. Echoes are obtained from intermixed warm (humid) and cool (dry) air in regions of turbulent entrainment associated with warm (humid) and generally rising currents. Quasi-horizontal echo bands, generally associated with temperature inversions, originate from statically stable zones where turbulence induced by vertical shear in horizontal wind generates enhanced small-scale temperature variability. At times, sodar records also reveal strong wind shear, which is evidenced by perturbed or braided echo bands indicative of breaking wave structures.

The sodar record in Figure 5 was obtained by Ottersten et al. (1974) during JONSWAP 2 (Joint North Sea Wave Project) and illustrates the boundary-layer structure over the North Sea in the autumn period when the ocean surface generally is warmer than the air throughout the day and night. Shipborne sodar has also been used for studies of the tropical marine boundary layer (Mandics et al., 1975; Gaynor and Mandics, 1978) during GATE (Global Atmospheric Research Program Atlantic Tropical Experiment). Recently, Schacher et al. (1977) investigated the influence of the marine inversion on atmospheric turbidity from the Acania cruising in the southern California coastal waters, using a commerical sodar in a simple antenna mounting within a standard enclosure placed on the fantail of the ship about 2 m above the water. Sodar measurements have also been performed in the North Polar Sea from drifting multi-year ice during AIDJEX (Arctic Ice Dynamics Joint Experiment; Carsey, 1976). Marine boundary-layer behavior has also been explored with land-based sodars, and extensive lists of references (e.g. Thomson, 1975) cover the use of sodar sounding over land. Demonstrated operational applications include the monitoring of convective or statically stable mixing layers and the delineation of the upper boundary of stratus and fog, which generally coincides with a temperature inversion.

Lidar systems are used not only to detect cloud, haze and other visible aerosols, but also to outline the distribution of aerosols in the visually clear atmosphere. In the case of cloud and haze particles, the backscattering cross sections are so large that they can be detected by eye-safe lidars of modest power, particularly in applications related to the lower troposphere. The determination of cloud ceilings, and of the morphology and transformation of cloud and haze layers were among the first applications of lidar in atmospheric research. Boundary-layer applications have mostly concerned mapping of the aerosol density distribution on height versus time diagrams or in vertical cross sections which may be analyzed in a semi-quantitative way. Monitoring of mixing heights and of the dispersion of smoke plumes are among the practical uses of these techniques.

Quantitative lidar measurements of aerosol extinction profiles may be of great value in such applications as radiative transfer studies and slant visibility determination, although some difficulties have to be considered. As follows from Equations 1 and 4 the lidar signal is influenced by two aerosol parameters: the backscattering

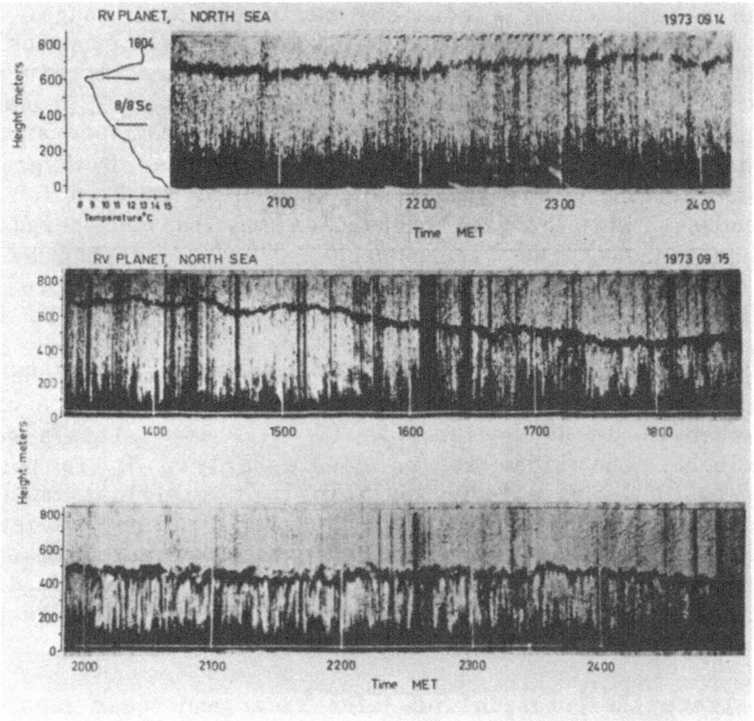

Fig. 5 Sodar records of convective plumes beneath a temperature inversion over the sea (after Ottersten et al., 1974). Echoes are displayed as dark areas in height-time diagrams visualizing the advecting air mass structure. In the rising plumes, turbulent temperature and humidity fluctuations produce echoes, whereas slowly subsiding undisturbed air in between is echo-free. The quasi-horizontal dark band outlines a temperature inversion where echoes originate from small-scale temperature contrasts created by mechanical mixing of the temperature gradient. Vertical bright stripes every hour show acoustic background noise during 60 s sodar transmitter shutdown. Intermittent vertical dark stripes across the entire diagram are caused by various interfering sounds and by occasional wind noise. A radiosonde ascent from the ship 2 hr before the start of the upper record showed a strong temperature inversion capping the stratocumulus deck (upper left). The inversion persists throughout the following day, as inferred from the echo band in the lower records.

coefficient σ and the extinction coefficient γ. Evaluation of one of these requires information about the ratio σ/γ; for example, a measure of γ from an in situ instrument at the lidar site together with the assumption of a constant ratio σ/γ with range. This assumption applies if the aerosol particle size distribution and composition may be considered independent of range. If, in addition, the aerosol density can be considered homogeneous, γ can be obtained directly from the slope of the lidar return on a power versus range display. This latter method was used by Viezee et al. (1973) to determine visibility V_M, according to Equation 5, along slant paths. When aerosol properties may be considered horizontally homogeneous, the "slant path method" described by Hamilton (1969) can be used, which implies extracting vertical profiles of γ from lidar measurements along two paths of different elevation angles. The methods and considerations described above also hold for quantitative determination of aerosol density although in this case calibration of the lidar return for the actual type of aerosol is necessary. In dense aerosols ($V_M < 3$ km) multiple scattering may influence the lidar return considerably. It is possible to account for this effect by analytical or empirical methods. By measurement of the depolarization of the scattered radiation, information on particle shape or indication of multiple scattering can be obtained (McNeil and Carswell, 1975). A more detailed discussion of quantitative lidar aerosol measurements is given by Collis and Russell (1976).

To illustrate the potential of joint lidar and sodar remote sensing in boundary-layer studies, some experimental results obtained over land by Uthe and Russell (1974) are shown in Figure 6. A surface layer of high aerosol density rises gradually from about 300 m to about 1000 m shortly before noon, when an abrupt rise to about 2000 m takes place. In the afternoon, the layer of high aerosol density fluctuates strongly with sharp peaks indicative of convective activity. In the evening the lidar record displays strong variations in echo intensity at elevations below about 800 m, and a layer of more continuous backscattering above. The distribution of aerosols in the first part of the lidar record may readily be interpreted as a mechanically forced mixing layer within which pollutants released near the surface are contained and dispersed. Under the influence of solar heating the forced mixing zone transforms into a convective layer which expands by entrainment of the much clearer air from above a capping inversion. During the evening, the convection dies out and a new, mechanically forced mixing layer develops at lower levels. Using the lidar data alone, however, the limits of the latter mixing layer are not so easily interpreted. The sodar record during the morning hours shows a surface layer of enhanced backscatter, the upper boundary of which indeed shows an almost perfect agreement with the mixing height inferred from the aerosol density distribution. During the period of high-reaching convection in the middle of the day the sodar record shows ground-based convective cells. In conditions with strong and high convection, the effective probing range of the sodar is generally too low to record any echoes at the top of the convective layer. During the evening, the sodar well depicts the new, mechanically forced mixing layer.

Joint use of sodar and lidar data allows a clearer interpretation and better understanding of boundary-layer processes. Moreover, qualitative sodar and lidar records such as in Figure 6 are particularly valuable in combination with quantitative in situ measurements, for example with radiosondes, of the vertical distribution of wind, temperature and humidity. The remote-sensing devices may then monitor any structural changes of the boundary layer in between the intermittent quantitative measurements. Joint sodar and lidar probing is also of value for short-term predictions of fog

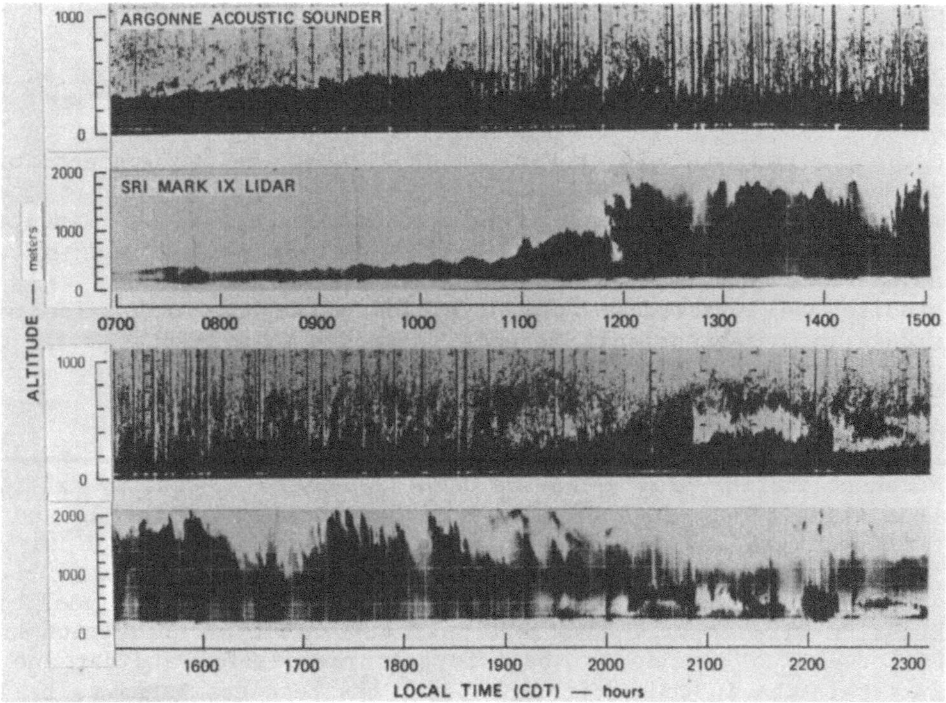

Fig. 6 Atmospheric boundary-layer records obtained over land by joint sensing with vertically pointing lidar and sodar (acoustic sounder) in St. Louis, Mo., on 14 August 1972 (after Uthe and Russell, 1974). The vertical scale for the lidar data is twice that for the sodar data.

and stratus dissipation, since their combined use allows delineation of upper and lower boundaries of stratus and fog and identification of structural changes; for example, the transformation from stratiform to cumuloform cloud type.

4. LIDAR MEASUREMENT OF VERTICAL PROFILES OF TEMPERATURE AND HUMIDITY

For lidar measurement of atmospheric temperature or humidity, their influence on a molecular interaction such as backscattering or extinction may be utilized. Profiles along the beam may be retrieved from range-resolved measurements, generally using pulsed lidars. Potentially useful optical interactions are, for temperature: Raman scattering from N_2 and O_2, Doppler frequency broadening of scattering from air molecules, and temperature-dependent resonant gas absorption. For humidity measurements, Raman scattering from water vapor or absorption by water vapor can be utilized. DIAL methods (DIfferential Absorption Lidar, also called DASE, Differential Absorption of Scattered Energy) imply measurement of gas absorption with a lidar operating at two wavelengths and detecting elastic scattering from particles. With one wavelength on an absorption line of the analyzed gas and the other wavelength just outside the line, the range derivative of the signal difference between the two channels provides a profile of the gas concentration along the beam. A double-wavelength or a rapidly tunable laser source is needed in order to obtain the required almost simultaneous measurements at two wavelengths. A DASE method using a tunable laser and heterodyne detection (Inaba and Kobayasi, 1975) promises an improvement in sensitivity of several orders of magnitude over conventional methods, although experimental verification has not yet been achieved.

4.1 Lidar Measurements of Air Temperature

Raman scattering from N_2 may be used to measure N_2 density profiles from which temperature profiles can be calculated with the use of the ideal gas law and the hydrostatic equation. Strauch et al. (1971) used this method at night to measure temperature fluctuations at a height of 30.5 m with an N_2 laser and obtained good correlation with in situ measurements. Obtainable accuracy with this method, which depends on absolute calibration of the lidar and on uncertainties in the extinction term, has been estimated to be $\pm 2K$ for temperature profiles in the clear troposphere.

Cooney (1972) has proposed a technique using the Raman scattering due to the interaction with pure rotational states of N_2 and O_2 molecules. The temperature dependence of the shape of the spectrum for pure rotational Raman scattering (Fig. 7) is utilized. By simultaneous measurement of the scattering at two wavelengths in the spectrum and determination of the difference as a function of height, a temperature profile can be obtained. Cooney (1974) calculated the uncertainty of temperature measurements to be ±0.5 K at a height of 2000 m at night for a ruby lidar of 6 J pulse energy and a receiver aperture of 0.4 m^2. Integration of 10 pulses was used. In a field test, temperature profiles to a height of 1100 m were measured and compared to a radiosonde profile with encouraging results. This technique shows promise of being applicable also in daylight with high background noise levels. Haze, fog, rain, and cloud would reduce the range but since relative measurements are used some attenuation can be tolerated.

Air temperature may also be derived from the Doppler broadening of the frequency spectrum of Rayleigh scattering from air molecules, which has been measured by Fiocco et al. (1971). With the recent development of high spectral resolution lidar the significance of this method has increased.

A DASE method for air temperature measurement proposed by Mason (1975) utilizes the temperature dependence of the distribution of rotational states in an absorption band for an atmospheric gas molecule. The development of tunable laser sources such as parametric oscillators for the IR region makes this a very promising method, although no experimental verification is known by the authors.

4.2 Lidar Measurement of Atmospheric Humidity

Raman scattering from water vapor can be utilized to measure

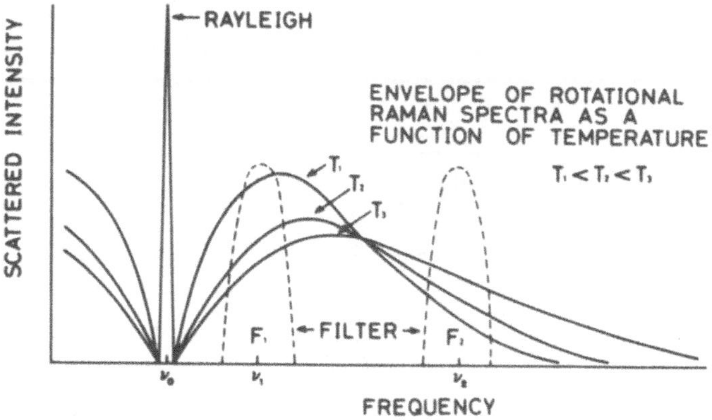

Fig. 7 Envelopes of rotational Raman spectra for three different temperatures. Filter positions F_1 and F_2 for temperature sounding are indicated with their spectral transmission curves (from Inaba, 1976).

humidity profiles. Melfi et al. (1969) and Melfi (1972) measured water vapor profiles with frequency doubled ruby lasers emitting high power pulses at 342.7 nm. The Raman scattering from N_2 was measured simultaneously and thus the water vapor mixing ratio was obtained. With this method of comparing two signals, problems with lidar calibration and compensation for attentuation are eliminated. Mixing ratios to an altitude of 2 km have been obtained during night time. Figure 8 shows a lidar mixing ratio profile compared with a radiosonde profile. Strauch et al. (1972) measured Raman scattering from water vapor using an N_2 laser emitting at 337.1 nm and obtained good correlation with direct measurements. From these results they estimated that water vapor mixing ratios can be measured to an altitude of 4 km at night. Accuracies in the order of ±10% or better should be obtainable with Raman methods. Daylight operation of Raman lidars will be possible especially when lasers emitting in the UV at less than 290 nm become available. In this region ozone attenuation reduces the sky light.

The DASE method was used by Schotland et al. (1966, 1971) to measure water vapor profiles up to 2 km with a ruby laser thermally tuned on and off the absorption line at 694.38 nm. Recently Murray et al. (1976) published results from DASE (or DIAL) measurements of water vapor, utilizing a CO_2 laser emitting 1 J pulses at 10.3 μm. Water vapor concentration to a horizontal range of 1.2 km was

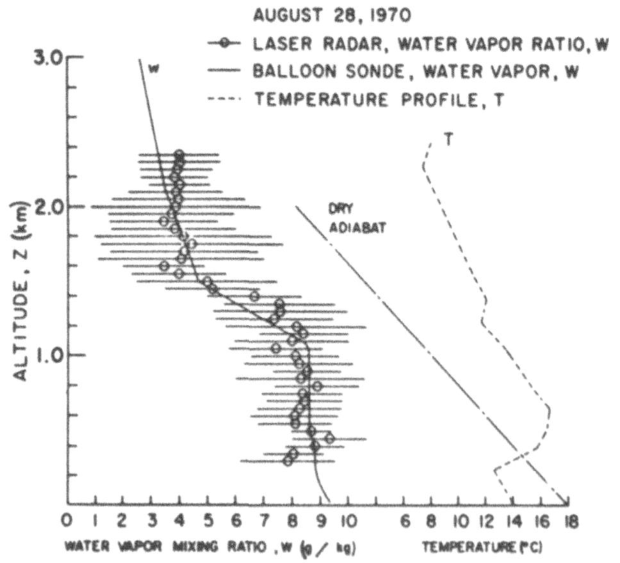

Fig. 8 Profiles of water vapor mixing ratio from Raman lidar and balloon sonde (after Melfi, 1972).

measured. With increased pulse energy the maximum range is estimated to be more than 5 km. This method seems very promising, especially since measurements can be performed in full daylight.

5. MEASUREMENT OF WIND VELOCITY

Lidar returns from aerosol particles and sodar scattering from turbulent inhomogeneities may also be used for the determination of air motion. The velocity can be obtained either from the Doppler effect or by correlation methods applied to returns from separate volumes.

The Doppler effect gives the scattered wave a frequency shift f_D which for monostatic systems is proportional to the radial component U_r of the wind vector

$$f_D = 2\lambda^{-1}U_r = 2\ fc^{-1}U_r \tag{6}$$

where λ is the transmitted wavelength, f is the transmitted frequency and c is the propagation velocity. The Doppler shift is positive (increasing frequency) when the scatterer is moving toward the instrument and negative for the opposite direction. When turbulent and thermal motion is superposed on the mean wind, the Doppler shift is distributed around a mean value representing the mean wind. Doppler techniques are used in both sodar and lidar wind sensing.

5.1 Lidar Measurement of Wind Velocity

A straightforward method to measure the Doppler shift of a lidar return is to use a scanning interferometer in the receiver to analyze the spectral distribution of the backscattered radiation (Benedetti-Michelangeli et al., 1972). A more sensitive technique to exploit the high coherence of laser radiation is to use 'heterodyne' detection for Doppler wind measurement. Here the received backscattered radiation is mixed with a local oscillator (L.O.) beam which is frequency shifted relative to the transmitted radiation. (If the L.O. shift is zero we have a 'homodyne' system.) In the detector the mixed radiation generates a signal at the difference frequency. With a spectrum analyzer following the detector, the Doppler shift and thereby the wind speed is obtained.

With heterodyne detection the highest attainable sensitivity is approached. The sensitivity increases with increasing wavelength, which is one reason why CO_2 lasers at 10.6 μm are suitable for Doppler measurements. Focused continuous wave lasers have generally been used. These systems give poor range resolution at long ranges although resolution lengths less than 100 m can be obtained out to ranges of 1 km with CO_2 laser systems having apertures of about 0.5 m diameter. When pulsed Doppler heterodyne systems become available better range resolution will be obtained.

The theory for Doppler heterodyne detection and an experimental feasibility demonstration have been described by Huffaker (1970). Figure 9 shows lidar-measured wind velocity data compared with simultaneous conventionally measured data. In such comparisons, the error in the lidar Doppler measurement is generally small compared to deviations that occur because the two techniques cannot sample the same air volume.

A Laser Doppler Velocimeter (LDV) system with a CO_2 laser and heterodyne detection has been developed (Lockheed Report, 1975). A conical scan called VAD from Velocity Azimuth Display is used to obtain wind speed and direction from altitudes up to 1 km. The system, which is van mounted, is operable day and night and has also been used on board a ship. A CO_2 Doppler lidar has also been used by Schwiesow and Cupp (1976) for measurements of atmospheric vortices. With Doppler lidar and VAD scan, mean wind profiles in the boundary layer should be obtainable with accuracies better than ± 1 m s^{-1}, although as long as continuous wave lasers are used a range scan of the focal point will also be required.

In Dual-Beam Laser Doppler Velocimeters (DBLDV) two beams are transmitted and directed so that they cross each other within a volume where interference fringes of radiation are formed. Particles that move through the fringes scatter light, and a modulated intensity is detected in the lidar receiver. The frequency is a measure of the motion component in the plane of the beams and perpendicular to their bisector. Alternatively, the signal can be

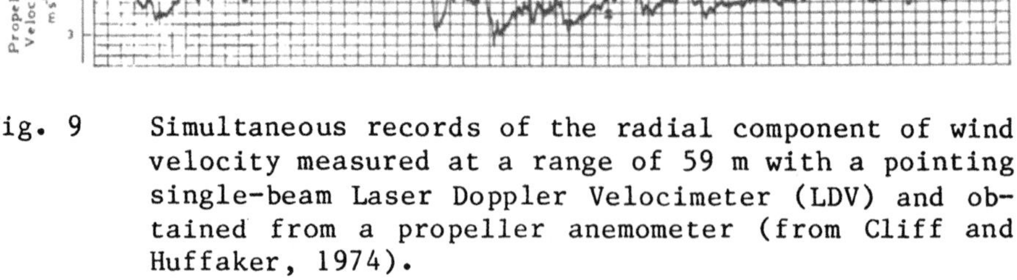

Fig. 9 Simultaneous records of the radial component of wind velocity measured at a range of 59 m with a pointing single-beam Laser Doppler Velocimeter (LDV) and obtained from a propeller anemometer (from Cliff and Huffaker, 1974).

described as the result of interference between the Doppler-shifted scattered waves due to the two beams. The range of DBLDV systems is limited by coherence degradation due to atmospheric turbulence and, possibly, by mechanical vibrations. Bartlett and She (1976) reported measurements with an argon laser system with a power of 0.35 W. They were able to measure wind speeds at 60 m distance and predicted a range of 300 m with realistic system improvements.

The correlation method determines the drift velocity of wind tracers, such as aerosols, by observing and correlating two returns from separate volumes. Spatial inhomogeneities in aerosol density drift through the volumes and produce time variations in the two returns. The wind component parallel to the line connecting one pair of volumes is obtained from the time lag which gives maximum correlation between the two signals. The method may be used with two or more spaced beams or a conically scanned beam (Derr and Little, 1970). The vertical profile of horizontal wind may be obtained by observation of volume pairs at various altitudes and along two different azimuths. The method has been used successfully with monostatic lidar for the determination of wind speeds in the boundary layer by aligning the lidar beam with the mean wind direction, pointing the beam at 10° elevation, and determining drift velocities along the beam (Eloranta et al., 1975). Results were obtained at distances up to 1 km. The reliability of the correlation method depends on the scale structure and persistence of aerosol inhomogeneities. Investigations by Zuev et al. (1976) show that the aerosol inhomogeneities are often stable enough to allow reliable wind measurements with the correlation method.

5.2 Sodar Measurement of Wind Velocity

For sodar sensing of winds, the Doppler technique has received widespread interest. Bistatic configurations of acoustic echo sounding are superior for Doppler measurement since velocity fluctuations contribute to the forward scattering, which is typically much stronger than monostatic sodar returns. Various bistatic system configurations have been used for the determination of Doppler shifts in three different directions as a function of altitude, from which the mean profile of the total wind vector may be synthesized. If vertical motions can be neglected or accounted for, measurements in two directions suffice. Winds in agreement with values from conventional methods can be measured regularly with bistatic Doppler sodar up to at least 500 m (Beran, 1974; Balser et al., 1976). No attempts to operate bistatic sodar systems at sea have been reported, however. Bistatic systems should be of particular interest for use on small islands where the required 200 to 300 m base lines in two or three directions can easily be obtained.

Monostatic Doppler sodar systems have also been used successfully for determining the vector wind profile (Beran et al., 1973) and these systems may become feasible for use on board ships. A three-axis monostatic sodar was gyro-mounted in the bow of the Oceanographer during GATE, to obtain data on the structure of the marine boundary layer and to measure wind velocity in the subcloud layer (Mandics et al., 1975). To resolve the vector wind velocity from Doppler frequency shifts in the backscattered signals along three axes, one antenna was pointed vertically while two were elevated 45° from the horizontal and separated by 90° in azimuth. An accelerometer was used to eliminate the ship's up and down motions from the Doppler-derived vertical wind velocity. Attempts to evaluate horizontal wind components have not been successful, since in the tilted antennas sea scatter and noisy ship environment produced markedly inferior signal-to-noise ratios compared to the vertical channel, from which good quality, quantitative backscattered data were obtained. The facsimile-recorded echo intensity provided valuable information on the structure and dynamics of the tropical marine boundary layer up to 800 m in height and the vertical velocity field could be obtained from the Doppler shift in the acoustic returns (Fig. 10 and 11).

6. CONCLUSIONS

To summarize, we conclude that valuable information on the marine atmospheric boundary layer may be obtained from ship-mounted, monostatic sodar and lidar systems. Their greatest value to air-sea interaction studies appears, at present, to be the capability of

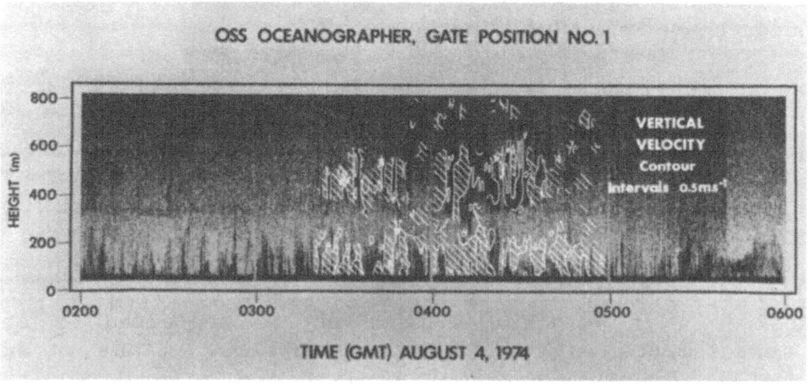

Fig. 10 Contours of Doppler-derived vertical velocities overlaid on sodar record showing convective plumes extending from the sea surface and hummock-shaped echoes associated with low-level tropical cumuli above (from Gaynor et al., 1976). Hatched areas indicate downward motion and contour intervals are 0.5 m s^{-1}.

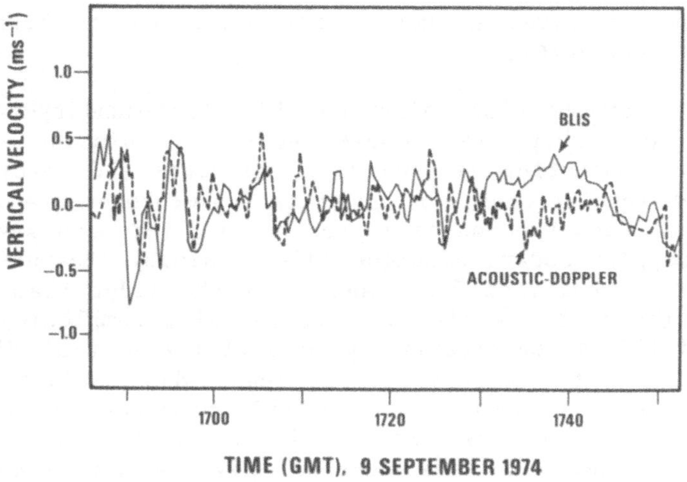

Fig. 11 Comparison of acoustic Doppler-derived vertical velocities at 138 m above the sea surface to values from balloon-borne anemometer (BLIS) obtained from the Oceanographer during GATE (after Mandics and Hall, 1976).

providing nearly continuous time-height and, in the lidar case, range-height displays of atmospheric structure, which support and complement data obtained by other techniques. These displays vividly depict boundary-layer processes and structural changes and allow monitoring of thermal plumes, temperature inversions and atmospheric waves, and delineation of cloud and haze layers, especially when sodar and lidar are used in concert.

The quantitative remote sensing from a ship of meteorological properties in the boundary layer holds great promise, although few systems have been used at sea. Experiments over land have demonstrated that sophisticated lidar techniques, in particular, offer means of obtaining vertical profiles of temperature, humidity and wind in the boundary layer with precision and resolution comparable to that of radiosondes and often more representative of mean conditions, if averaging is applied to remote measurements of high temporal resolution. Although we expect that further development of lidar techniques will result in systems of excellent accuracy and resolution even during daylight operation, clouds and fog will always limit their range. Adaptation to shipboard use of proven lidar techniques requires care but offers no insurmountable obstacles. For wind measurement, the Doppler sodar method will be a good complement since the acoustic waves easily penetrate cloud and fog.

Shipboard Doppler sodar measurement of vertical air motion is feasible, but available sodar systems for vector wind measurement have not proven themselves at sea, where noise due to high winds is their main limitation.

For a first time user of sodar and lidar for boundary-layer observations from a ship, we propose vertically pointing systems for qualitative monitoring of turbulent structure and aerosol density from time-height records. If quantitative measurements are attempted, digital data processing is recommended for both systems. The vertical Doppler sodar technique adds a valuable feature of modest complexity. Provisions for scanning of the lidar beam in azimuth and elevation increase flexibility, but also complexity, and with calibrated lidars quantitative mapping of the spatial distribution of aerosol density is possible, at least when extinction is low. In high extinction conditions, available concepts for quantitative aerosol measurements by lidar and inferences about visibility should be applied with care because they are highly dependent on assumptions about the aerosol properties.

For wind measurement, the best system currently available is the Laser Doppler Velocimeter using a CO_2 laser and heterodyne detection, but initially the complexity and cost of such systems may prove prohibitive. With suitable scanning systems and data processing units, simpler lidars employing the correlation method may be used to obtain mean winds from drifting aerosol inhomogeneities.

Raman lidar, at present, appears to be the most promising method for obtaining profiles of both temperature and humidity although in the future both of these quantities may be obtained more reliably by DASE (or DIAL) methods, which measure differential absorption of lidar returns. Experimenters familiar with Raman and DASE techniques should accompany their instruments, however, on the first cruise.

ACKNOWLEDGEMENT

The authors gratefully acknowledge discussions with Björn Holmgren of Uppsala University and his contributions to Section 3 in particular.

REFERENCES

ANSI. 1976. American National Standard for the Safe Use of Lasers. Z 136.1 - 1976, American National Standards Institute, 1430 Broadway, New York.

BALSER, M., C.A. McNARY and A.E. NAGY. 1976. Remote wind sensing by acoustic radar. *Journal of Applied Meteorology*, 15: 50-58.

BARTLETT, K.G. and C.Y. SHE. 1976. Remote measurement of wind speed using a dual beam backscatter laser Doppler velocimeter. *Applied Optics,* 15: 1980-1983.

BENEDETTI-MICHELANGELI, G., F. CONGEDUTI and G. FIOCCO. 1972. Measurement of aerosol motion and wind velocity in the lower troposphere by Doppler optical radar. *Journal of Atmospheric Sciences,* 29: 906-910.

BERAN, D.W. 1974. Remote sensing wind and wind shear system. Interim Report FAA-RD-74-3, National Technical Information Service, Springfield, Va. 22151, 115 pp.

BERAN, D.W. and B.C. WILLMARTH. 1971. Doppler winds from a bistatic acoustic sounder. *Proceedings of the 7th International Symposium on Remote Sensing of Environment,* University of Michigan: 1699-1714.

BERAN, D.W., C.G. LITTLE and B.C. WILLMARTH. 1971. Acoustic Doppler measurements of vertical velocities in the atmosphere. *Nature,* 230: 160-162.

BERAN, D.W., W.H. HOOKE and S.F. CLIFFORD. 1973. Acoustic echosounding techniques and their application to gravitywave, turbulence, and stability studies. *Boundary-Layer Meteorology,* 4: 133-153.

BROWN, E.H. and S.F. CLIFFORD. 1976. On the attenuation of sound by turbulence. *Journal of the Acoustical Society of America,* 60: 788-794.

BROWN, E.H. and F.F. HALL, JR. 1978. Advances in atmospheric acoustics. *Review of Geophysics and Space Physics,* 16: 47-110.

CARSEY, F.D. 1976. The AIDJEX acoustic radar and some preliminary results. AIDJEX Bulletin No. 31, University of Washington, Seattle, Wash. 98105: 1-19.

CLIFF, W.C. and R.M. HUFFAKER. 1974. Application of a single laser Doppler system to the measurement of atmospheric winds. NASA TM X-64891, NASA, Marshall Space Flight Center, Alabama, 18 pp.

COLLIS, R.T.H. 1970. Lidar. *Applied Optics,* 9: 1782-1788.

COLLIS, R.T.H. and P.B. RUSSELL. 1976. Lidar measurement of particles and gases by elastic backscattering and differential absorption. In: *Laser Monitoring of the Atmosphere,* edited by E.D. Hinkley, Springer-Verlag, Berlin: 71-151.

COONEY, J.A. 1972. Measurement of atmospheric temperature profiles by Raman backscatter. *Journal of Applied Meteorology,* 11: 108-112.

COONEY, J.A. 1974. Measurement of atmospheric temperature profiles. AD 785741, National Technical Information Service, Springfield, Va. 22151, 20 pp.

DERR, V.E., M.J. POST, R.L. SCHWIESOW, R.F. CALFEE and G.T. McNICE. 1974. A theoretical analysis of the information content of lidar atmospheric returns. COM-75-10393, National Technical Information Service, Springfield, Va. 22151, 284 pp.

DERR, V.E. and C.G. LITTLE. 1970. A comparison of remote sensing of the clear atmosphere by optical, radio, and acoustic radar techniques. *Applied Optics,* 9: 1976-1992.

ELORANTA, E.W., J.M. KING and J.A. WEINMAN. 1975. The determination of wind speeds in the boundary layer by monostatic lidar. *Journal of Applied Meteorology,* 14: 1485-1489.

FIOCCO, G., G. BENEDETTI-MICHELANGELI, K. MAISCHBERGER and E. MADONNA. 1971. Measurement of temperature and aerosol to molecule ratio in the troposphere by optical radar. *Nature (Physical Sciences),* 229: 78-79.

FIOCCO, G. and L.D. SMULLIN. 1963. Detection of scattering layers in the upper atmosphere. *Nature,* 199: 1275.

GAYNOR, J.E. and P.A. MANDICS. 1978. Analysis of the tropical marine boundary layer during GATE using acoustic sounder data. *Monthly Weather Review,* 106: 223-232.

GAYNOR, J.E., P.A. MANDICS, A.B. WAHR and F.F. HALL, JR. 1976. Studies of the tropical marine boundary layer using acoustic backscattering during GATE. Preprint Volume 17th Conference on Radar Meteorology, American Meteorological Society, Boston, Mass.: 303-306.

HALL, F.F., JR. 1972. Temperature and wind structure studies by acoustic echo-sounding. In: *Remote Sensing of the Troposphere,* edited by V.E. Derr, U.S. Government Printing Office, Washington, D.C. 20402, 26 pp.

HAMILTON, P.M. 1969. Lidar measurement of backscatter and attenuation of atmospheric aerosol. *Atmospheric Environment,* 3: 221-223.

HARRIS, C.M. 1966. Absorption of sound waves in air versus humidity and temperature. *Journal of the Acoustical Society of America,* 40: 148-159.

HINKLEY, E.D. 1976. *Laser Monitoring of the Atmosphere.* Springer-Verlag, Berlin, 380 pp.

HUFFAKER, R.M. 1970. Laser Doppler detection systems for gas velocity measurement. *Applied Optics,* 9: 1026-1039.

HUFFAKER, R.M., A.V. JELALIAN and J.A.L. THOMSON. 1970. Laser-Doppler system for detection of aircraft trailing vortices. *Proceedings of the IEEE,* 58: 322-326.

INABA, H. 1976. Detection of atoms and molecules by Raman scattering and resonance fluorescence. In: *Laser Monitoring of the Atmosphere,* edited by E.D. Hinkley, Springer-Verlag, Berlin: 153-236.

INABA, H. and T. KOBAYASI. 1972. Laser-Raman radar. *Opto-electronics,* 4: 101-123.

INABA, H. and T. KOBAYASI. 1975. Infrared laser radar technique using heterodyne detection for range-resolved sensing of air pollutants. *Optics Communications,* 14: 119-122.

LEONARD, D.A. 1967. Observation of Raman scattering from the atmosphere using a pulsed nitrogen ultraviolet laser. *Nature,* 216: 142-143.

LIGDA, M.G.H. 1964. Meteorological observations with lidar. *Proceedings 11th Weather Radar Conference,* World Conference on Radio Meteorology, University of Colorado: 482-489.

LITTLE, C.G. 1969. Acoustic methods for the remote probing of the lower atmosphere. *Proceedings of the IEEE,* 57: 571-578.

LOCKHEED REPORT. 1975. Laser remote atmospheric monitoring system. LMSC-HREC SD D 496584, Lockheed Huntsville Research and Engineering Center, Huntsville, Al. 35807, 28 pp.

MANDICS, P.A. and F.F. HALL, JR. 1976. Preliminary results from the GATE acoustic echo sounder. *Bulletin of the American Meteorological Society,* 57: 1142-1147.

MANDICS, P.A., F.F. HALL, JR., E.J. OWENS and D. WYLIE. 1975. Observations of the tropical marine atmosphere using an acoustic echo sounder during GATE. Preprint volume 16th Radar Meteorological Conference, American Meteorological Society, Boston, Mass.: 257-259.

MANDICS, P.A. and E.J. OWENS. 1975. Observations of the marine atmosphere using a ship-mounted acoustic echo sounder. *Journal of Applied Meteorology,* 14: 1110-1117.

MASON, J.B. 1975. Lidar measurement of temperature: a new approach. *Applied Optics,* 14: 76-78.

McALLISTER, L.G. 1968. Acoustic sounding of the lower troposphere. *Journal of Atmospheric and Terrestrial Physics,* 30: 1439-1440.

MELFI, S.H. 1972. Remote measurements of the atmosphere using Raman scattering. *Applied Optics,* 11: 1605-1610.

MELFI, S.H., J.D. LAWRENCE, JR. and M.P. McCORMICK. 1969. Observation of Raman scattering by water vapor in the atmosphere. *Applied Physics Letters,* 15: 295-297.

McNEIL, W.R. and A.I. CARSWELL. 1975. Lidar polarization studies of the troposphere. *Applied Optics,* 14: 2158-2168.

MIDDLETON, W.E.K. 1958. *Vision through the atmosphere.* University of Toronto Press, 250 pp.

MURRAY, E.R., R.D. HAKE, JR., J.E. VAN DER LAAN and J.G. HAWLEY. 1976. Atmospheric water vapor measurements with an infrared (10-μm) differential-absorption lidar system. *Applied Physics Letters,* 28: 542-543.

NEFF, W.D. 1975. Quantitative evaluation of acoustic echoes from the planetary boundary layer. NOAA TR ERL 322-WPL 38, U.S. Government Printing Office, Washington, D.C. 20402, 34 pp.

OTTERSTEN, H., K.R. HARDY and C.G. LITTLE. 1973. Radar and sodar probing of waves and turbulence in statically stable clear-air layers. *Boundary-Layer Meteorology,* 4: 47-89.

OTTERSTEN, H., M. HURTIG, G. STILKE, B. BRÜMMER and G. PETERS. 1974. Shipborne sodar measurements during JONSWAP II. *Journal of Geophysical Research,* 79: 5573-5584.

SCHACHER, G., C. FAIRALL, T. HOULIHAN and K. DAVIDSON. 1977. Observations of the marine inversion with a shipboard acoustic sounder. Technical Digest Topical Meeting on Optical Propagation through Turbulence, Rain, and Fog. Optical Society of America, Washington, D.C. 20036, 3 pp.

SCHOTLAND, R.M., J.T. BRADLEY and A.M. NATHAN. 1966. Water vapor studies utilizing a thermally tuned ruby laser radar. Optical Sounding II, Final Report, Part 1, GSL-TR-66-9, New York University, 16 pp.

SCHOTLAND, R.M., A. PACE, K. SASSEN and R. STONE. 1971. Studies of atmospheric water vapor and hydrometeors by laser radar. PB 107 917, National Technical Information Service, Springfield, Va. 22151, 52 pp.

SCHWIESOW, R.L. and R.E. CUPP. 1976. Remote Doppler velocity measurements of atmospheric dust devil vortices. *Applied Optics,* 15: 1-2.

STRAUCH, R.G., V.E. DERR and R.E. CUPP. 1971. Atmospheric temperature measurement using Raman backscatter. *Applied Optics,* 10: 2665-2669.

STRAUCH, R.G., V.E. DERR and R.E. CUPP. 1972. Atmospheric water vapor measurement by Raman lidar. In: *Remote Sensing of Environment,* 2: 101-108.

TATARSKI, V.I. 1961. *Wave propagation in a Turbulent Medium.* McGraw-Hill, New York, 285 pp.

THOMSON, D.W. 1975. Acdar meteorology: the application and interpretation of atmospheric acoustic sounding measurements. Preprints 3rd Symposium on Meteorology Observations and Instrumentation, American Meteorological Society, Boston, Mass.: 144-150.

UTHE, E.E. and P.B. RUSSELL. 1974. Experimental study of the urban aerosol structure and its relation to urban climate modification. *Bulletin of the American Meteorological Society,* 55: 115-121.

UTHE, E.E. and R.J. ALLEN. 1975. A digital real-time lidar data recording, processing and display system. *Optical and Quantum Electronics,* 7: 121-129.

VIEZEE, W., J. OBLANAS and R.T.H. COLLIS. 1973. Evaluation of the lidar technique of determining slant range visibility for aircraft landing operations. AFCRL-TR-73-0708, Air Force Geophysics Laboratory, Bedford, Mass. 01730, 132 pp.

WESELY, M.L. 1976. The combined effect of temperature and humidity fluctuations on refractive index. *Journal of Applied Meteorology,* 15: 43-49.

ZUEV, V.E., Y.M. VOREVODIN, G.G. MATVIENKO and I.V. SAMOKHVALOV. 1976. Investigation of structure and dynamics of aerosol inhomogeneities in the ground layer of the atmosphere. *Applied Optics,* 16: 2231-2235.

30

Aircraft

B.R. Bean and C.B. Emmanuel

1. INTRODUCTION

The past decade has seen aircraft come of age as research platforms for air-sea interaction studies. As a platform they have the advantage of being independent of sea motion; this allows sensors to be operated with much more confidence than on a ship or buoy. Airflow about an aircraft is relatively well understood compared to a surface platform, which must contend with airflow from any direction. Surface platforms measure atmospheric properties transported by the prevailing wind while the aircraft may measure along- and cross-wind properties such as those present with roll-vortices (Bean et al., 1972; Le Mone, 1976). Although the basic instrumentation for temperature, humidity, wind, and solar radiation have remained essentially unchanged, in recent years methods of recording and analysis have advanced dramatically. At the start of the decade recording was most commonly done in analogue fashion via either strip charts or FM modulated magnetic tape. In either case, data reduction was long and tedious. Digital recording was common by the middle of the decade but still frequently involved difficult interfacing with large central computing centres. By the end of the decade inexpensive minicomputers could be dedicated to data acquisition and onboard processing with a resultant decrease in analysis time. For example, the airborne turbulence data for GATE (1974) and NORPAX (December 1977-January 1978) resulted in roughly an equal number of magnetic tapes per aircraft but the time for data reduction was reduced from one year to less than two months. We are on the verge of yet another quantum step in our abilities for aircraft experimentation. The growing availability of microprocessors holds out the promise of running real-time, onboard quality control and general consistency checks on the output of individual sensors, as well as actual data reduction.

The following sections will be concerned with general aircraft capabilities and problems as well as inertial navigation systems, turbulence instrumentation, and data reduction.

2. AIRCRAFT CAPABILITIES

A modern turboprop research aircraft can be utilized for investigations at heights as low as 15 m over distances ranging from a few hundred metres to several thousand kilometres. The height of an aircraft above local terrain when flying at 150 m or less may be determined to a fraction of a metre by utilizing either Doppler or CW-FM radio altimetry (McBean and Paterson, 1975). Aircraft, for example the Lockheed Orion WP-3D (NOAA's long-range 'hurricane hunter'), can operate effectively to about 9000 m, loiter at speeds between 90 and 110 m s^{-1}, cruise at 165 m s^{-1}, and attain 'dash' speeds in excess of 200 m s^{-1}. Aircraft performance data for several aircraft are presented by Aanensen and Zipser (1974) and NCAR (1973), where it is seen that the aircraft have the range and endurance to study large-area phenomena or, once on station, to study the time evolution of mesoscale phenomena.

Most of the normal variables of interest to air-sea interaction studies are measured with off-the-shelf instrumentation and sampling techniques such as vortex thermometers, differential pitot tubes, etc., as listed in Table 1.

The WP-3D aircraft digitized meteorological radar system (two C-band and one X-band) was designed primarily as a meteorological research tool. As such, reflectivity information provided by the radar is based on the moisture content of the cloud system and on the gradient of moisture content. At ranges of 100 km or less the radar offers 'normalized' target capability (i.e., if a 10 mm hr^{-1} precipitation rate is detected at 100 km the return will not grow in amplitude as the range to the return decreases). The system features range coverage to 400 km and accurate contoured display of precipitation to 100 km. The nose radar (C-band, PPI) has an area of coverage of $\pm$120° in azimuth and $\pm$20° in elevation (referenced to the aircraft's longitudinal axis). The lower fuselage radar (C-band PPI) has an area of coverage of 360° perpendicular to the longitudinal axis and $\pm$25° fore and aft of the normal to the longitudinal axis of the aircraft.

One may also measure temperature versus depth in the sea with aircraft expendable bathythermographs (AXBT) which are described by Barnett and Bernstein (see Chapter 20). A typical oceanic cross-section is given in Figure 1. These observations were taken every degree of latitude and every one-half degree in the interval 10°N to 1°S. This involved alternating the three radio frequencies available to avoid interference from signals of AXBTs released earlier. The end result was that the aircraft AXBT sections so determined permitted studies of the equatorial currents to a detail not otherwise available. The aircraft completed this cross-section (Hawaii to Tahiti) in 12 hours compared to 21 days for an oceanographic research ship.

TABLE 1

WP-3D Aircraft Instrumentation

Meteorological Sensors:
- Temperature (total)
- Pressure (static-dynamic)
- Winds
- Dewpoint

Cloud Physics:
- Cloud droplet spectrum
- Hydrometeor size spectrum
- Cloud liquid water
- Total water content
- Total liquid water content
- Ice-water discriminator
- Cloud condensation nuclei
- Bulk water sampling system
- Nuclei (total dust, ice, condensation, millipore filter)

Radiation:
- Sea surface temperature
- CO_2 air temperature
- Microwave radiometer

Radar:
- Nose C-band PPI, 240° scan
- Lower fuselage C-band PPI, 360° scan
- Tail X-band RHI, 360° scan

Miscellaneous:
- Gust probe (vane and hot film)
- AXBT (external/internal systems)
- Flare seeding system
- Photography
- Radar altimeter
- Laser altimeter
- Omega dropwindsonde

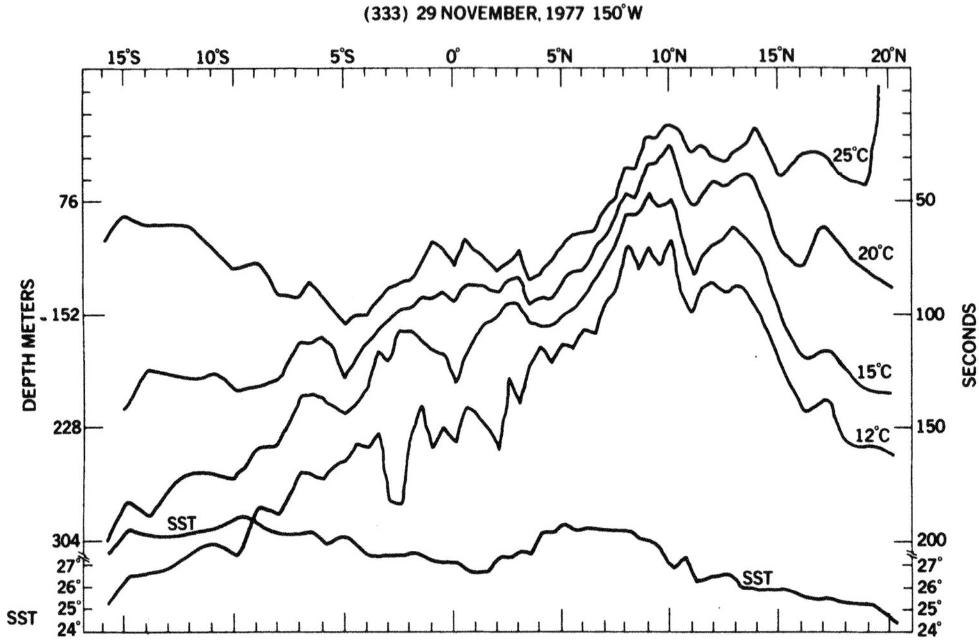

Fig. 1 AXBT cross-section for 29 November 1977, along 150°W longitude.

3. PROBLEMS

Aircraft make excellent platforms for over-water research in that they are unaffected by sea motion and are relatively unaffected by salt and particle contamination (sensors may be cleansed with distilled water before and after each flight, as well as during the flight by flying through rain or clouds), and may still operate under high wind conditions such as hurricanes.

One disadvantage is that the instrumentation required for resolving atmospheric scales of 100 m or less must have a time constant considerably less than 1 s; thus instrumentation is required with faster response than that commonly used on towers and buoys.

The airflow characteristics of an aircraft must be known for proper placement of the sensors. For example, errors in the measured static pressure are induced by pressure disturbances caused by the aircraft fuselage. The errors are a function of air speed, Mach number, angle of attack, and the variations in flow geometry caused by flaps, spoilers, gear and control surfaces. From information available from the aircraft manufacturer an optimum mounting location and aerodynamic compensation are selected to minimize these influences. Figure 2 shows a pressure pattern for a particular aircraft, where the distribution is a function of $\Delta P/q_c$ (ΔP is the local static pressure minus true static pressure; q_c is the pitot pressure minus static pressure). Points 1 through 6 in this figure are sites of minimum static pressure error. Generally, best accuracy is obtained by locating the pitot-static tube on a nose boom extended well ahead of the fuselage (Point 1 in Fig. 2). Maintenance, radar installations, and storage considerations often make the forward fuselage or a leading edge more desirable locations. The distribution in Figure 2 is typical for a single flight condition and configuration, and the influence of other variables is not included. Further analysis of the influence of these variables under all conditions within the flight capabilities for a particular vehicle should be made before selecting a location. The extension of the 'pressure-dome' in front of the aircraft is well defined. An excellent discussion of the problems of measurement of air motions and of atmospheric fluxes can be found in articles by MacPherson (1973), McBean and Paterson (1975), and, especially, Telford et al. (1977). Additional references are included at the end of this chapter.

One of the most vexing and time-consuming problems in aircraft measurements is maintaining the soundness of the electrical ground system. Electrical ground loops are very common and difficult to

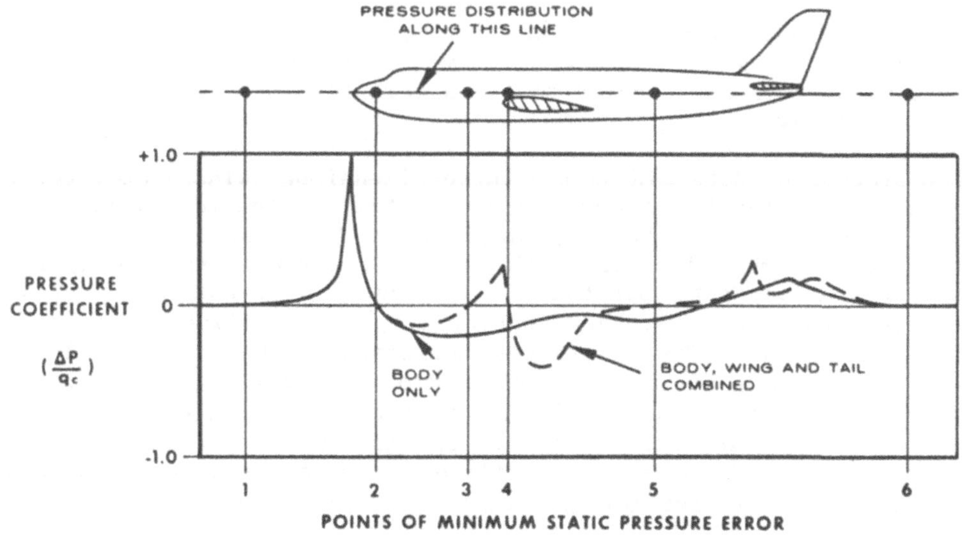

Fig. 2 Pressure coefficient versus location for typical subsonic static pressure distribution pattern (compliments of Rosemont Inc., Bulletin 1013).

isolate and eliminate. Since an aircraft is usually made of many pieces of aluminum welded or riveted together, corrosion, even if slight, at the juncture of sections can create impedance variations and, hence, problems in interpretation of electric potentials. It is difficult to find two separated points on an aircraft where a potential difference does not exist. Great care must be exercised to establish a common ground for all equipment working together. Large copper conductor ground straps or bars should be used rather than the skin of the aircraft. This is especially important where the aircraft operates a considerable amount of electronic equipment such as radars, inverters, 400 cycle motors, radios, etc. These devices produce noise which is easily picked up by a questionable ground system and can appear in the recordings of relatively weak signals.

A great deal of information may be obtained from the suppliers of off-the-shelf atmospheric sensors to be used on aircraft. This information involves sensor design, performance specifications, and recommended calibration procedures. For this reason the remainder of this chapter will emphasize some of the less common techniques, not covered in the literature, for measurement of turbulent properties of the atmosphere and navigation or position determination. The results presented below, although based on data most familiar to the authors, would be expected to be essentially the same for other research aircraft.

4. THE INERTIAL NAVIGATION SYSTEM (INS)

Highly detailed mapping of meteorological variables in both time and space can be achieved when atmospheric and oceanographic measurement systems are coupled to the aircraft's inertial navigational system (INS).

The process of determining the three-dimensional wind vector referenced to the local earth coordinate system via the INS involves the vector sum of angles (pitch, yaw, roll). To maintain the inertial reference frame coincident with the local earth coordinate system, the proper aircraft altitude, earth curvature, and rotation are combined to form a feedback control loop - accomplished by a dedicated computer. The aircraft's horizontal and vertical velocity components and position are derived from integration and double integration of the horizontal and vertical accelerometer outputs. The altitude calculation requires the internal vertical accelerometer output, and an external altitude reference such as a barometric altimeter or, for low levels, an FM-CW radar altimeter.

Typical INS manufacturer's published error values are:

(1) average horizontal velocity error of ~1 m s^{-1},

(2) average vertical velocity error of ~10 cm s^{-1},

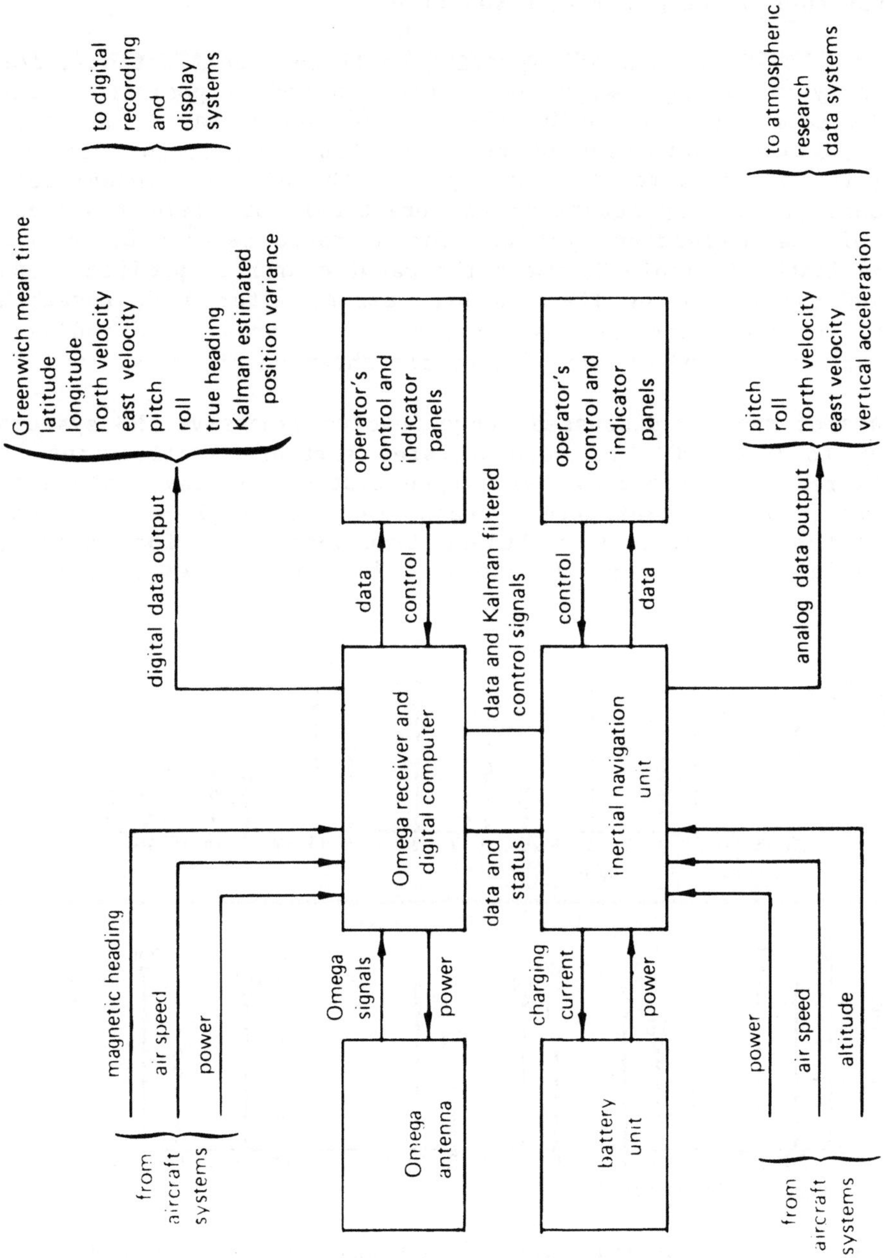

Fig. 3 Typical Omega inertial system.

(3) aircraft attitude angle error of ~1 arc min, and

(4) radial position error to be within 2.5 km at the end of the first hour of flight 50% of the time.

There are, however, ways of improving on these specifications, for instance by utilizing 'Omega' navigation to update position (Fig. 3). The Kalman filter (NATO, 1970) yields an unbiased, minimum-variance, linear estimation of the navigation parameters. By returning the aircraft to the same spot on the airfield immediately upon landing, one may determine the practical, or field residual, errors of the navigation system. The averaged results of such a test are listed in Table 2, where the manufacturer's specifications are seen to be met by the residual errors after long research flights involving intricate patterns and many turns. The individual errors at the end of each flight are shown on Figure 4.

The class of systems discussed above involve expensive INS systems ($75,000 to $250,000) and sophisticated software applications in the data reduction process. With such systems one may quite successfully adapt angle-of-attack measurements to replace the gust-probe system for vigorous turbulent transport (see, for example, the articles by McBean and Paterson, 1975, and Bean et al., 1975,

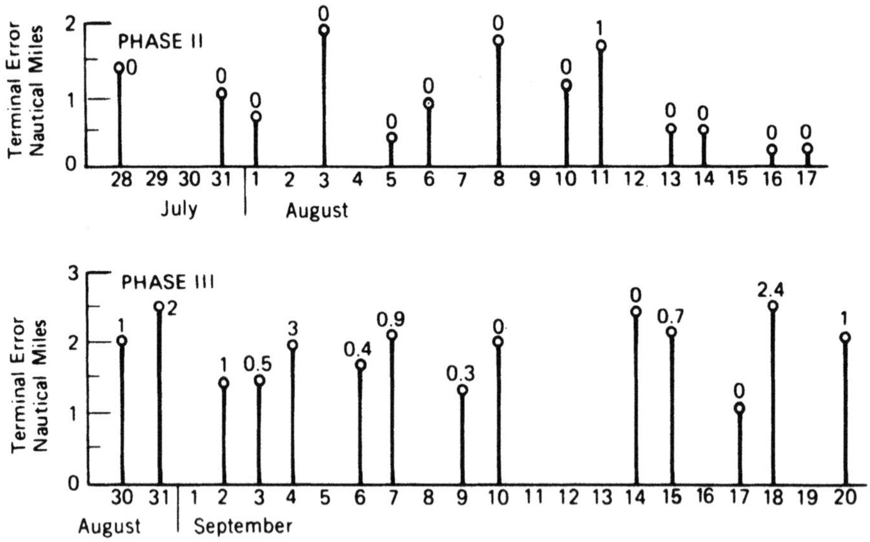

Fig. 4 Omega-Inertial Terminal Errors. Numbers at end of bars are velocity errors in knots. Data were obtained during GATE. The results for Phases II and III are representative of a well functioning system.

TABLE 2

Typical Omega-Inertial Performance

Total flights:	27
Total flight time:	281 hr
Average flight duration:	8.1 hr
Average position error:	2.6 km
Average velocity error:	0.26 m s^{-1}
Position error rate	0.10 m s^{-1}

wherein quite different equipment was used to study airmass modification on the same day over Lake Ontario).

One may also utilize a relatively inexpensive small aircraft system, wherein the nose radar of a two-engined aircraft is replaced by an aluminum nose cone with the gust probe system modified by replacing the INS with rate gyroscopes (approximate cost $4000 for the rate gyroscope and $6000 for the custom nose cone). Such a system has been successfully flown for calibration against a 444 m instrumented tower at Norman, Oklahoma. The aircraft was flown along the wind, as close to the tower as was practical (200 m), to permit data comparison. The samples were normalized to the aircraft's true air speed and the wind speed to ensure that the two systems were sampling the same air mass. Table 3 shows the results. The averages indicate that the two systems compare well considering their horizontal separation. Vertical separation (tower 444 m, aircraft 481 m) could account for some of the bias in the temperature and pressure results; for example, the aircraft temperatures could be increased by 0.2°C assuming a standard atmosphere or 0.3°C assuming an adiabatic temperature distribution. Similarly, the average pressure at the aircraft could be increased by 3.7 mbar to correct for height differences. Both of these corrections would bring the two sets of measurements into closer agreement. The gust probe vanes were expected to be more sensitive to turbulence than the bivanes on the tower, with the result that the range of aircraft-determined vertical velocities exceeded that measured on the tower. The agreement obtained in this intercomparison is unusually good, perhaps because care was taken to obtain the measurements in the early part of the afternoon when the lower atmosphere is relatively well mixed.

TABLE 3

Aircraft-Tower Flyby
Comparison of Aircraft Data and Tower Data,
Norman, Oklahoma, 1977

Run		Temperature (°C)	Water Vapour Density (gm m^{-3})	Pressure (mbar)	Range Vertical Wind (m s^{-1})
1. Upwind:	Tower	21.9	3.9	939.5	-4.0 +4.4
	Aircraft	21.3	3.6	935.1	-6.3 +5.4
2. Downwind:	Tower	21.9	3.7	939.2	-2.1 +3.5
	Aircraft	21.3	3.7	936.8	-2.7 +4.5
3. Upwind:	Tower	22.0	3.5	939.2	-2.1 +5.5
	Aircraft	21.5	3.6	936.8	-6.3 +5.4
4. Downwind:	Tower	21.9	3.6	939.1	-2.1 +5.5
	Aircraft	21.7	3.6	937.0	-5.4 +4.9
5. Upwind:	Tower	21.8	3.6	939.0	-1.9 +5.7
	Aircraft	21.6	3.5	937.8	-8.1 +5.9
Average:					
	Tower	21.9	3.7	939.2	-2.4 +4.8
	Aircraft	21.5	3.6	936.5	-5.7 +5.2

5. TURBULENCE INSTRUMENTATION

The sensor package for the turbulence measurements is normally mounted on a boom 3 to 5 m in front of the nose of the aircraft. Great care must be taken when mounting a boom on an aircraft to keep unwanted vibrations out of the measured signals. Both static and dynamic vibration tests have to be carefully made. Output of accelerometers at various points on the boom are measured for the static tests (the boom is struck sharply with a mallet) with and without the engines running. The dynamic tests consist of recording data obtained during sinusoidal or 'roller-coaster' flights of various amplitudes, wavelengths, and indicated air speeds in 'quiet air', normally clear conditions at 5 to 7 km above sea level. The data are then reduced to remove the deliberate signals introduced by the roller-coaster flight patterns. The high-frequency vibrations of the probe mounts and the natural resonance of each individual sensor require all signals to be low-passed with identical filters prior to being digitized, to avoid aliasing errors. To reduce aliasing errors (i.e., to less than 1%) active filters are used in conjunction with a sampling rate well in excess of the desired upper frequency of interest. All signals must be filtered identically due to the phase lag introduced by all analogue filters. The results of tests normally encourage one to introduce ahead of the recording of the signals from the sensors a rather sharp low-pass filter with a decrease of about 24 dB per octave. (The authors usually employ four-pole Butterworth filters with a cutoff frequency of 11.5 Hz, well below the resonant frequency of the boom; a sampling rate of 80 Hz is used; the desired upper frequency limit is 10 Hz). Additional filtering can be done during the numerical data reduction.

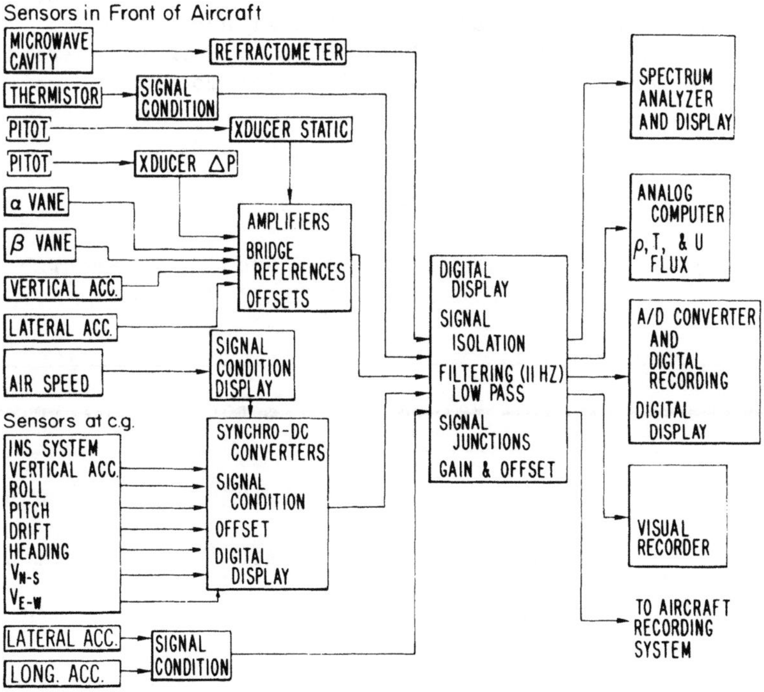

Fig. 5 Configuration of a typical airborne turbulence system. (Note: A Lyman-α humidiometer could equally well be utilized in place of the microwave refractometer.)

The sensor package may include:

(1) a small bead thermistor for temperature measurement,

(2) a Lyman-α humidiometer or a microwave refractometer cavity for water vapour measurements, and

(3) a gust probe or a hot film sensor for air velocity measurements.

The sensors are located at a distance of about one metre or less from each other to minimize the time of transport of air between sensors to about one-tenth that corresponding to the highest practical frequency that can be recorded.

A gust probe measures fluctuations in the components of the wind. It can consist of two vanes placed 3 to 5 m ahead of the aircraft on a boom (one horizontal vane that measures the vertical force of the wind and, hence, the vertical wind, w; and a vertical vane that measures the horizontal crosswind, v) and a centrally-located pitot tube to measure the fluctuations in the along-axis wind. All three of these components are corrected for motion of the aircraft (primarily pitch, roll, and yaw) as well as for motion due to the boom,

TABLE 4

Typical Errors (rms) due to Response of Sensors of an Airborne Gust Probe System

Parameter	Typical Values	Error	Units
u'	1.0	2×10^{-2}	$m\ s^{-1}$
w'	0.3	6×10^{-2}	$m\ s^{-1}$
T'	0.4	0.5×10^{-2}	°C
ρ_v'	0.6	1×10^{-2}	$g\ m^{-3}$
$u'w'$	0.4	1.2×10^{-2}	$m^2\ s^{-2}$
$T'w'$	0.4	1.2×10^{-2}	$°C\ m\ s^{-1}$
$\rho_w'w'$	0.2	1.2×10^{-1}	$g\ m^{-2}\ s^{-1}$

as measured by accelerometers located on the gust probe and at the INS. A schematic of an overall system is shown on Figure 5, while typical measurement errors are given in Table 4.

A wind tunnel calibration of the system is required. The expression used to determine the attack angle of the airflow is

$$\Delta\alpha = 2 \frac{\Delta F_\alpha + m \, \Delta a_N}{C_N \, \rho \, V_T^2 \, A}$$

where $\Delta\alpha$ is the angle of attack in radians, ΔF_α is the force on the horizontal (α) vane, m is the mass of the vane, Δa_N is the normal acceleration to the vane, C_N is the vane constant (i.e., the angle of attack as a function of the normal force), ρ is the air density, V_T is the true air speed, and A is the area of the vane. It is necessary to determine whether C_N is constant, independent of air velocity and angle of attack. In a typical wind tunnel test the vane is mounted on a special jig in the centre of the test section of the tunnel. The angle of attack is set from outside the tunnel. A static calibration is first performed to check the electronic gain factor by setting the vane at 0° angle of attack before loading it with a range of weights corresponding to expected aerodynamic forces. The tunnel is then closed and the air velocity is raised to, say, 50 m s^{-1}. The attack angle of the vane is then varied (both positive and negative) and the output voltage noted for each setting. The same procedure is followed with a tunnel speed of 70 m s^{-1} and 110 m s^{-1}. C_N is then computed from these data. A typical value of C_N is 2.8 rad^{-1}.

Temperature is measured with a small bead thermistor chosen for excellent stability, low noise, and uniform characteristics for interchangeability. Each thermistor should be carefully calibrated prior to use. To minimize nonlinearity, the system is constrained to operate over a 10°C range centered on the ambient temperature.

Since the thermistor is mounted on a relatively fast-moving aircraft, a correction is made for dynamic heating. This is done both in the wind tunnel and in flight in clear, 'quiet' air via the expression

$$T_a = T_i \, (1 - 0.20 \, K \, M^2)$$

where T_a is the ambient temperature (°C), T_i is the indicated temperature (°C), M is the Mach number, and K is the form factor. The form factor for a bead thermistor varies from about 0.5 to 1.0. Typically, research aircraft travel at about 100 m s^{-1} (M = 0.294) with a resultant dynamic heating of 5 to 6°C. T_a can be determined from a vortex thermometer which decelerates the ambient air; its sensing element is a bead thermistor. The form factor can then be determined from the above equation.

Conventional methods of measuring humidity do not have adequate responses for flux measurements. The microwave refractometer and the Lyman-α humidiometer have been suggested as instruments that could overcome these difficulties (see Chapter 22 by Hay).

Using a microwave refractometer in conjunction with other sensors permits the refractivity, temperature, and pressure of the atmosphere to be measured accurately to at least 10 Hz, which permits determination of fluctuations in water vapour density to 10 Hz. The absolute accuracy of the microwave hygrometer is ± 0.2 g m^{-3} and the resolution is 0.02 g m^{-3}. In practice, one must be careful to check for plating erosion of the cavity due to salt spray, impact of ice, etc. It is good practice to rinse all sensors in distilled water before and after each flight.

The Lyman-α humidiometer is also in wide use. In this system the attenuation of a Lyman-α beam is recorded to give a direct measure of the water vapour content of the intervening atmosphere (Buck, 1976). This system also had some difficulties with window corrosion, constant transmitter output, etc., but these have been largely overcome. It is best for the prospective user of the Lyman-α or the radio refractometer or other sophisticated instrumentation to visit, or better yet, work with a team familiar with the day-to-day vagaries of these systems.

6. DATA REDUCTION

The data obtained by systems such as those described above are subject to all of the usual pitfalls of experimental investigation plus the added complexity arising from the intermittent nature of atmospheric processes. 'Intermittent' means that a significant portion of the desired signal is contained in a small portion of the record; that is, the signal tends to come in random bursts. It is difficult to devise an automatic data-editing procedure to remove blanks or spurious spikes in the data that will not also edit the bursts. Consequently, such data require a great deal of visual examination to distinguish signal from noise. Care must also be taken when examining spectral properties. It is prudent to low-pass filter the data by averaging successive measurements such that the Nyquist frequency reflects the highest frequency of interest. In addition, any linear trend should be removed (the linear correlation coefficients between the variables and time are normally quite small; thus there is usually very little trend). The data are then recalculated relative to the regression line. If the resultant spectra display significant spurious high energy at low or high frequency, one returns to a visual examination of the data to determine if higher-order filtering is required.

An example of the above procedure is illustrated on the 'flow-chart' shown on Figure 6. Note that the basic flight level

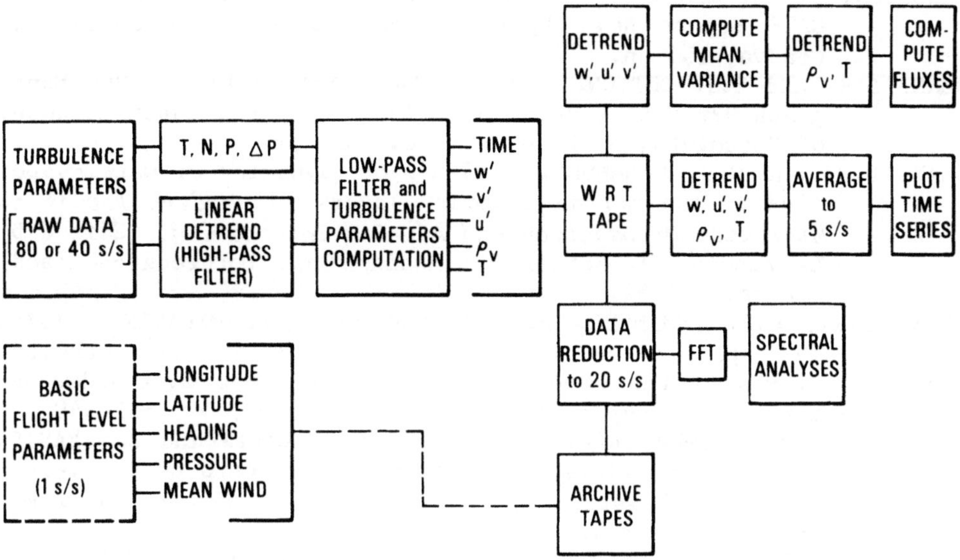

Fig. 6 Flow diagram for reduction of aircraft data.

parameters are recorded at 1 Hz while the turbulence parameters are recorded at either 80 or 40 Hz. The basic operating tape is the WRT (Wind, Rho-v, and Temperature). Note that detrending occurs at almost every step of the way for the turbulence data.

A final basic note: all of the above takes a great deal of time. When the data have to be edited, processed, re-edited and reprocessed, one finds that a well documented log book maintained by an experienced onboard observer is invaluable. The Engineer's Log will normally keep track of centre values, dynamic range, reference voltages, etc., and record all in-flight changes; it serves as the primary reference for reducing the data. The Observer's or Chief Scientist's Log serves to interpret the experiment since it will give sea and atmospheric state with reference to how the experiment progressed and, in particular, why in-flight changes were made. For example, if cloud streets are observed during a pre-planned free convection experiment, the flight patterns might be changed to give longer downwind than crosswind runs in the subcloud layer.

This chapter is necessarily a brief summary of some of the advantages and problems of working with aircraft as a research platform. Additional references are given to provide further insight into these matters as well as types of studies particularly amenable to study by aircraft.

REFERENCES

AANENSEN, C.J.M. and E.J. ZIPSER. 1974. The GATE aircraft plan. GATE Report No. 11, World Meteorological Organisation, Geneva, 154 pp.

BEAN, B.R. and E.J. DUTTON. 1966. Radio Meteorology. NBS Monograph 92, U.S. Government Printing Office, Superintendent of Documents, Washington, D.C., 435 pp.

BEAN, B.R. and C.B. EMMANUEL. 1973. The dynamics of water-vapor flux in the marine boundary layer. In *Modern Topics in Microwave Propagation and Air-Sea Interaction.* Edited by A. Zancla, D. Reidel Publishing Co., Dordrecht-Holland/Boston: 51-64.

BEAN, B.R., C.B. EMMANUEL, R.O. GILMER and R.E. McGAVIN. 1975. The spatial and temporal variations of the turbulent fluxes of heat, momentum and water vapor over Lake Ontario. *Journal of Physical Oceanography,* 5: 532-540.

BEAN, B.R., R. GILMER, R.L. GROSSMAN, R.E. McGAVIN and C. TRAVIS. 1972. An analysis of airborne measurements of the vertical flux of water vapor during BOMEX. *Journal of Atmospheric Sciences,* 29: 860-869.

BEAN, B.R., R.O. GILMER, R.F. HARTMANN, R.E. McGAVIN and R.F. REINKING. 1976. Airborne measurement of vertical boundary layer fluxes of water vapor, sensible heat and momentum during GATE. NOAA Technical Memo, ERL WMPO-36, U.S. Department of Commerce, NOAA/ERL, Boulder, Colorado, 83 pp. plus 436 microfiche.

BUCK, A.L. 1976. The variable-path Lyman-alpha hygrometer and its operating characteristics. *Bulletin of the American Meteorological Society,* 57: 1113-1118.

GROSSMAN, R.L. and B.R. BEAN. 1973. An aircraft investigation of turbulence in the lower layers of a marine boundary layer. NOAA Technical Report ERL 291-WMPO 4, U.S. Government Printing Office, Superintendent of Documents, Washington, D.C. 20402, 166 pp.

HSUEH, Y. 1968. Mesoscale turbulence spectra over the Indian Ocean. *Journal of Atmospheric Sciences,* 25: 1052-1057.

LEMONE, M.A. 1976. Modulation of turbulent energy by longitudinal rolls in an unstable planetary boundary layer. *Journal of Atmospheric Sciences,* 33: 1308-1320.

LENSCHOW, D.H. 1970. Airplane measurements of planetary boundary structure. *Journal of Applied Meteorology,* 9: 874-884.

LENSCHOW, D.H. 1974. Model of the height variation of the turbulent kinetic energy budget in the unstable planetary boundary layer. *Journal of Atmospheric Sciences,* 31: 465-474.

MacPHERSON, J.I. 1973. A description of the NAET-33 turbulence research aircraft, instrumentation and data analysis. National Research Council of Canada, Division of Mechanical Engineering, Quarterly Bulletin of the Division of Mechanical Engineering and the National Aeronautical Establishment, No. 4, Ottawa.

MALKUS, J.S. 1962. Large-scale interactions. In *The Sea,* Vol. 1, *Physical Oceanography,* edited by M.N. Hill, Interscience Publishers, New York.

McBEAN, G.A. and R.D. PATERSON. 1975. Variations of the turbulent fluxes of momentum, heat and moisture over Lake Ontario. *Journal of Physical Oceanography,* 5: 523-531.

MERCERET, F.J. 1976. Measuring atmospheric turbulence with airborne hot film anemometers. *Journal of Applied Meteorology,* 15: 482-490.

NATO AGARDOGRAPH 139. 1970. Theory and applications of Kalman filtering. Edited by C.T. Leondes, North Atlantic Treaty Organization Advisory Group for Aerospace Research and Development, 7 Rue Ancelle 92200, Neuilly sur Seine, France.

NCAR. 1973. Aircraft research systems, inertial navigation systems, etc. *Atmospheric Technology,* 1, National Center for Atmospheric Research, Boulder, CO, 76 pp.

NCAR. 1975. Instruments and techniques for probing the atmospheric boundary layer. *Atmospheric Technology,* 7, National Center for Atmospheric Research, Boulder, CO, 91 pp.

NCAR. 1975. Instrumentation in cloud physics. *Atmospheric Technology,* 8, National Center for Atmospheric Research, Boulder, CO, 65 pp.

PENNELL, W.T. and M.A. LEMONE. 1974. An experimental study of turbulence structure in the fair-weather trade wind boundary layer. *Journal of Atmospheric Sciences,* 31: 1308-1323.

TELFORD, J.W., P.B. WAGNER and A. VAZIRI. 1977. The measurement of air motion from aircraft. *Journal of Applied Meteorology,* 16 (2): 156-166.

WORLD METEOROLOGICAL ORGANIZATION (WMO). 1974. GATE: final international scientific plans. *Bulletin of the American Meteorological Society,* 55 (7): 711-744.

31

Tethered Balloons

N. Thompson

1. INTRODUCTION

Tethered balloon systems have played an important role in many detailed studies of the structure of the planetary boundary layer. They provide a convenient method for obtaining vertical profiles of wind, temperature and humidity up to a kilometre or so above the surface, or alternatively time series of these (and similar quantities) at a number of heights simultaneously. They have been used for at least 30 years in studies of the boundary layer over the sea: among the early measurements were series of observations in the summers of 1946-1948 off the north coast of Russia (Vorontsov and Selitskaya, 1955). Their potential has been exploited much more fully in the last decade during which they have been used increasingly to support sensors sampling the turbulence structure as well as mean properties of the atmosphere.

The majority of recent data obtained by tethered balloons has been from within the tropical boundary layer during the Barbados Meteorological Experiment (BOMEX) (Garstang et al., 1971; Ropelewski, 1975), and the Atlantic Tropical Experiment (GATE) preceded by its Intercomparison Sea Trials (Andreev et al., 1974; Ropelewski, 1976; Thompson, 1976a). More limited observations in middle latitudes have been obtained over the East China Sea by Yokoyama et al. (1969) and in the same area during the Air Mass Transformation Experiment (AMTEX) by Ootsuka et al. (1975); Thompson (1972a) discussed data obtained over the Northeast Atlantic in the Joint Air-Sea Interaction Experiment (JASIN).

The problems associated with operating from ships have led to the adoption of much smaller balloons than those used in some of the successful studies of the boundary-layer over land (Jones and Butler, 1958; Caughey and Readings, 1975). This in turn has stimulated the development of sophisticated lightweight sensor systems such as the Boundary Layer Instrument Package (BLIP) (Burns, 1974) and has encouraged the use of lightweight dielectric tethering cables. In spite of these advances it is probable that, up to at least 1976, no shipboard balloon system was able to probe the convective boundary layer up to a satisfactory height (at least 1500 m) with sufficient vertical resolution in fixed level operation. On the other hand it was demonstrated during GATE that balloons could be inflated under optimum conditions in harbour and then raised well clear of the ship and towed to the experimental area 1000 km away. Thus there is now no difficulty in principle in using larger balloons with ample net lift which would allow measurements to greater heights than hitherto.

A balloon and its tethering line provide an inherently unstable platform from which to suspend instruments. The variations in height and horizontal position of the balloon produced by atmospheric turbulence and the consequent motion of sensors attached to the line result in spurious contributions to measured velocities which may bias the measured momentum fluxes (Thompson, 1969) even when tethered to a fixed point on land. The wave-induced movement of a ship produces additional line and sensor motion which may seriously affect the measured velocities (Thompson, 1972a).

In severe weather (strong winds, heavy rain, or lightning) the tethered balloon technique has proved less successful up to the present. Ship motion in strong winds produces violent movements of the tether line which are likely to preclude useful velocity data being obtained by the suspended sensors (the high drag on the balloon may even break the line). Heavy rain may prevent proper functioning of the sensors. Heavily-laden balloons often become unstable in the strong downdrafts which are sometimes associated with the rain and then undergo large changes of altitude. Electrical discharges down the wet line may destroy it. Despite these limitations tethered-balloon systems used in fixed-level or profiling modes have amply proved their worth in investigations of the convective boundary layer over the sea.

2. TYPES OF BALLOON AND METHODS OF INFLATION

2.1 Types of Balloon

The chief components of a typical balloon are shown in Figure 1. The main envelope has a ratio of length to diameter ('fineness ratio') of about 2.5. The envelope fabric is usually rip-stop nylon which has been sealed with several applications of polyurethane to reduce the gas loss to less than about 1% per day. Balloons in recent use have also been constructed from thin plastic sheet but

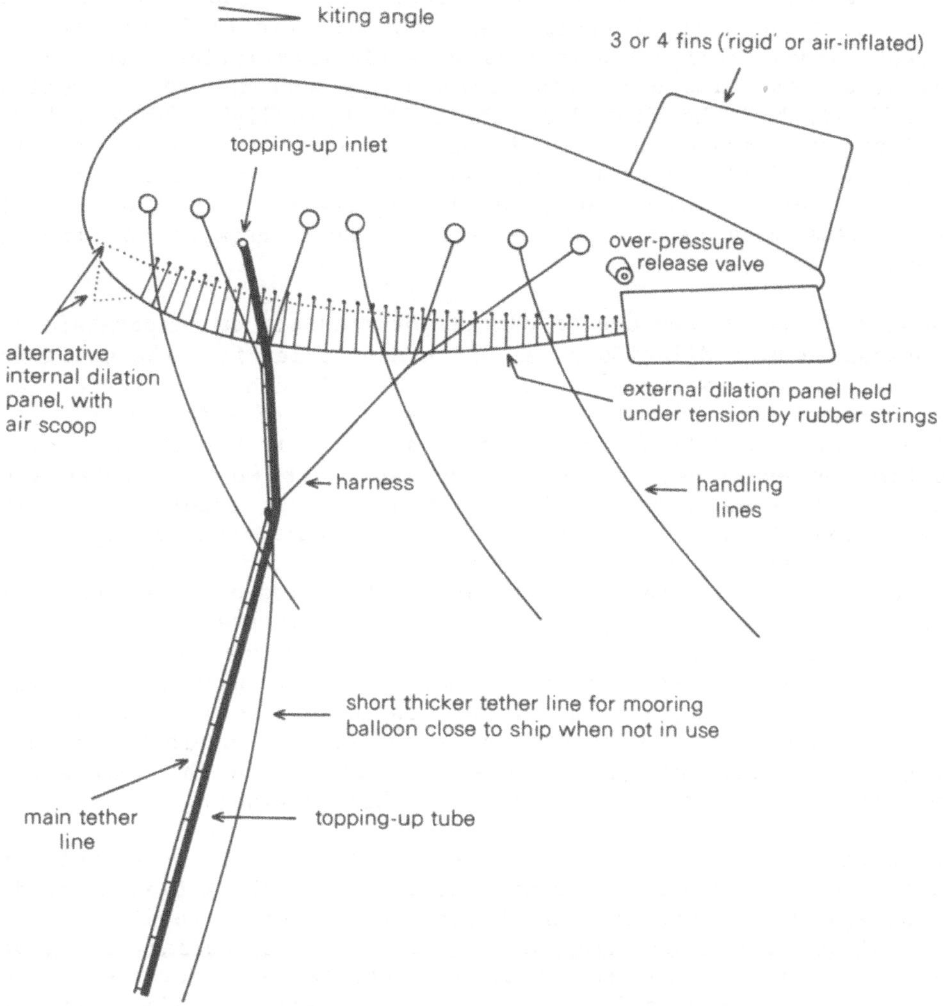

Fig. 1 Typical tethered balloon

they are probably less durable. The rubberized cotton used in early balloons has been superseded entirely because of its high weight, but neoprene-coated nylon has been employed in the manufacture of very large balloons flown over land and might have applications at sea whenever it is possible to use larger balloons there. This last fabric has the advantage of high resistance to abrasion and high tolerance to folding, unlike that coated with polyurethane whose porosity appears to increase rapidly with repeated handling. The fins are usually of fabric stretched over a rigid tubular metal frame. Some early balloons flown over land with rigid fins were unsuccessful due to the fragile nature of the fins (Harrington et al., 1972) but careful design has overcome this difficulty in the latest types. Balloons with hollow fabric fins inflated through an air-scoop have also been used successfully over the sea as well as over land, but the fins lose their shape in light winds and the balloon may become unstable if then struck by a sudden squall. Hollow fins inflated by the balloon gas itself have been tried but have the serious disadvantage of losing their shape when the balloon loses much gas: the balloon is then unstable in stronger winds.

Changes of gas volume due to changes of pressure or temperature are compensated by a dilation panel which may be let into the underside of the envelope, whose shape is then maintained by a series of lateral rubber strings under tension across the dilation panel. Alternatively the balloon incorporates a small air-filled compartment within the envelope, underneath the gas cavity and separated from it by a dilation panel: air is drawn into or expelled from the compartment through a scoop as the gas volume changes and so the balloon's shape is preserved. Gas may valve off through a safety valve if the dilation panel becomes fully distended due to overfilling, increase of temperature or decrease in pressure. The system of volume compensation which uses the external dilation panel appears to be the better since it usually copes fully with all volume changes without valving off and consequent frequent topping up. On the other hand it distorts the shape of the envelope when contracted to compensate for volume contraction and this may affect the balloon's aerodynamic stability in higher winds.

The balloon is attached to the tethering line by a harness which is rigged to produce a significant kiting angle. The kiting is needed to obtain the dynamic lift which compensates for increasing drag on the balloon and tether line as the wind speed increases. In strong winds the angle of the tether line at the balloon increases to 20° or so from the vertical. A wide range of kiting angles has been used, ranging from (in zero wind) at least 20° for balloons flown from U.S. ships in GATE to 10° or less for balloons flown in JASIN and GATE from British ships. The lower angle produces a more stable flying behaviour but with reduced dynamic lift.

2.2 Methods of Inflation

Helium should be used exclusively for filling balloons flown at sea: the risks involved in the use of hydrogen are unacceptable. Inflation in harbour or at sea is best carried out on a flat area, well sheltered from the wind and at least the size of the balloon. A suitable area on a ship is a helicopter flight deck astern of the helicopter's hangar. Except in very light winds the ship has to head into wind to shelter the balloon, but turbulent eddies shed from the superstructure will still buffet the balloon and it is probably best to inflate it under a net in these circumstances rather than relying on the handling lines to hold it down. The balloon is most at risk when being raised off the deck or recovered since it then experiences the greatest turbulence. It is likely to yaw violently during these operations in winds above about 7 m s^{-1}, and above 10 m s^{-1} the recovery, in particular, of the balloon to deck level is likely to be hazardous. For this reason the balloon should not be brought down to the ship after initial inflation until it is finally deflated, and any interim topping-up which might be required is performed through an 'umbilical cord' long enough for the balloon to fly at least 25 m above the ship during the operation. The author has used rigid-walled plastic tube of less than 0.01 m bore for this purpose but flexible-walled tubing of larger bore has been found more suitable (M. Garstang, pers. comm.).

Gas losses higher than the nominal 1% per day mentioned above will occur as the porosity of the envelope fabric increases, especially in stronger winds, but may still be acceptable in terms of the cost of helium involved. However, there is a higher rate of diffusion of air into the gas cavity associated with the increased loss of helium and hence a gradual dilution of helium within the cavity. It may eventually become necessary to deflate completely and re-inflate the balloon to maintain lift.

Typical balloons of volume about 70 m^3 have a gross lift near 70 kg and a net lift, after deducting the weight of balloon and harness, of about 35 kg.

3. TETHERING LINES

Details of the main types of tethering line are summarised briefly in Table 1. Steel rope has been largely superseded because of its low strength-to-weight ratio. Its advantages are high resistance to abrasion, low susceptibility to wind-induced oscillations and the fact that it can be wound directly onto a winch drum without de-tensioning. The dielectric lines are all significantly lighter than steel and for this reason they are now used almost exclusively. Their disadvantages include the need for expensive winches with traction drive to de-tension the line before winding on the drum, thus preventing crushing of the inner layers on the drum (Almazan, 1972), and their susceptibility to wind-induced oscillations (see Section 5, below) which may be severe enough to damage instruments suspended from the line. The nylon and terylene lines are also rather bulky, with the penalty of larger wind drag and the need for larger winches.

Table 1

Balloon Tether Lines
(for nominal breaking load 500 kg)

Material	Type of Construction	Mass/Length $kg\ m^{-1}$	Approximate Diameter (mm)
Steel	Cable-laid	0.025	2.5
Nylon	Cable-laid, braided or 'Kernmantel' (straight fibres jacketed by woven sheath)	0.015	5.0
Terylene	Kernmantel (jacketed with polyethylene)	0.015	5.0
Aromatic Polyamide (Aramid)	Kernmantel (jacketed with polyurethane or braided)	0.006	3.0

Many of the studies carried out recently with tethered-balloon systems have involved measurements in convective conditions, sometimes with large precipitating cumulus or cumulonimbus and their attendant electric fields; in these circumstances the non-conducting lines may be safer than steel. After use for some time at sea, however, they are likely to become partially impregnated with salt and therefore partially conducting at high relative humidities: this has led to line failure near large cumulus clouds, caused by currents flowing along and fusing the line (Almazan, 1972). The jacketed lines are probably less prone to destruction in this way because the salt has more difficulty in penetrating to the main load-bearing core of the line.

The small net lift of typical balloons necessitates the use of lines which are as light as possible in order to make measurements at heights of at least 1000 m. This had led in the past to occasional breakages in winds stronger than about 20 m s^{-1}. The net lift of a balloon in zero wind is given approximately by

$$N = G\,(1 - 0.0001Z) - M \qquad (1)$$

where G is the gross lift at surface and is approximately equal to V kg where V is volume in cubic metres, Z is height above surface in metres, and M is mass of balloon. The ratio M/G decreases with increasing volume, for example from 0.6 for 40 m^3 to 0.4 for 100 m^3 in the case of one type of balloon used in GATE. Small balloons have therefore a disproportionately small net lift.

The tether-line load (T) at the point where the line is attached to the balloon is

$$T \simeq G\,(1 - 0.0001Z) - M + aU^2 \qquad (2)$$

where U is the wind speed, a is a coefficient determined by air density, size of balloon, and its lift and drag, and given approximately by 0.017 $V^{2/3}$ where V is the volume in cubic metres (a varies slightly with balloon shape and kiting angle). For balloons used at wind speeds up to 20 m s^{-1} at heights of 1000 m the maximum cable load for 40 and 100 m^3 balloons are then approximately 90 and 200 kg. The corresponding ratio of diameters of tether lines giving the same percentage margin of safety for each balloon is 1:1.5, showing that the relative drag on the line decreases with increasing balloon size. At a given wind speed the angle of the tether line from the vertical is therefore larger for the smaller balloons and this discourages their use in stronger winds.

In order to assure stability in lighter winds, when the aerodynamic lift of the balloon is small, the gross lift of the balloon should exceed by at least 15% its mass and the masses of the line and suspended instruments. On this basis the maximum payload (tether line and instruments) of a 100 m^3 balloon flying at 1000 m is 37 kg. A line with a breaking load of 500 kg will give an adequate margin of safety for wind speeds up to 20 m s^{-1}, and if this line is of nylon or terylene then the maximum instrumental load is 22 kg in light winds. The corresponding estimates for a 40 m^3 balloon with the same 2.5 to 1 safety margin for the tether line in 20 m s^{-1} winds is 1 kg. Payloads increase to 31 kg and 5 kg if aramid line is used. It is clear that for multilevel observations up to 1500 m with a number of instrument packages each weighing 1 kg or more the selected size of balloon will be substantially larger than 40 m^3.

The balloon is normally tethered at a height of about 100 m when not in use. At heights much less than this it is likely to yaw considerably because of the short length of tether and therefore be subjected to unnecessary strain. A few tens of metres of strong line (≃ 1000 kg breaking load) are usually attached to the balloon harness and tied off to the main tether-line: when not required, or in periods of strong winds the balloon is pulled down to near the ship, the strong line is clipped to a similar one 50 m or so long attached to the ship and then the balloon is raised until the strong line takes over as the tether.

4. LINE WINCHES AND METHODS OF TETHERING

The main functions which the winch has to perform are:

(i) storage of the line,

(ii) payment or recovery of the line (usually at a measured speed),

(iii) de-tensioning of the line before winding on the storage drum,

(iv) spooling the line onto the drum to prevent bunching, and

(v) measuring the amount of line paid out.

Desirable supplementary functions are:

(vi) measurement of line tension,

(vii) measurement of the angle of elevation of the line, and

(viii) measurement of the direction (azimuth) of the line.

Winches with direct drive are not suitable for use with the dielectric lines since in any but light winds the inner layers on the drum will become crushed as further line is wound on. Aramid line with an elongation at failure of only 3% may suffer less from crushing but the sound practice is to use it also with indirect drive, especially in strong winds.

Winches have forward and reverse drive and variable speed to provide complete control over line payout and recovery. Typical line speeds are about 1 m s^{-1}: the maximum need not exceed 3 m s^{-1}. Hydraulic rather than electric drive probably gives more flexibility in operation.

The line can run directly from a suitably designed winch to the balloon but often it is not possible to find a position where this can be done without the risk of the tether fouling the superstructure; it is then necessary to pass the line from the winch through a snatch block placed at a more convenient point. The optimum siting of winch or snatch block is roughly amidships to minimize the effects of ship motion (see Section 5, below) but more often the position has to be nearer the stern where the effects of pitching are larger. Wherever the location is chosen it should allow the winch operator a clear view of the balloon at all times when near the ship and especially when very close to deck level.

5. DYNAMIC STABILITY OF TETHERED BALLOONS AND THEIR SENSOR SYSTEMS

5.1 Turbulence-induced Balloon Motion

The characteristic motion of a tethered balloon is slow periodic migration acrosswind. On this is superimposed additional lateral motion on shorter time scales (and also longitudinal and vertical movements) caused by changes of wind speed and direction, and vertical gusts. There are few published data on the magnitude of the resulting effects on measurements being made by the balloon-borne sensors. Thompson (1969) made limited theodolite measurements on large balloons over land from which it appeared that these motions may cause significant errors in measured variances of horizontal and vertical wind speed, and in the vertical flux of momentum. Direct comparisons of data from sensors at heights of 150 m and 300 m on a balloon cable and on an adjacent tower (Haugen et al., 1975) showed significant differences between measured speeds but only small differences between momentum fluxes from tower and balloon data. Balloons used over the sea are about one-tenth the volume of those in the studies just described and are relatively heavily laden and so would be expected to be more sensitive to turbulence. On the other hand turbulence over the sea is usually less than over land. Some evidence for the effects of balloon motion on data obtained over the sea is provided by cospectra of the momentum flux obtained by Thompson (1972a, 1972b) which showed unexpectedly small contributions at frequencies around 0.03 Hz, possibly attributable to balloon motion. Clearly there is a need now for direct measurements of the sensor velocities caused by balloon motion [Andreev et al. (1974) reported that they applied corrections for balloon motion to vertical velocity data obtained during GATE but neither details of the technique used nor magnitudes of the spurious velocities have been published].

5.2 Wave-induced Motion of the Tether Line

Ship motion produced by surface waves is imparted to the sensors suspended from the tether line and may result in large spurious contributions to the velocities measured by the sensors. The magnitude of the introduced errors depends on sea state, ship's response to waves, position of the point of tether on the ship, and the exact shape of the line's catenary. The motion of the ship includes both pitch and roll, with periods between about 6 and 15 s for typical ships. Surface waves usually have similar periods and the motion may be therefore amplified by resonance every few periods to produce much larger displacements than on average. The most important result of the motion is the production of an alternate slackening and tightening of the line catenary which causes mainly horizontal displacements of the suspended sensors. Sensors near the middle of the tether line will experience the largest horizontal displacements. Thompson (1972a, 1972b) measured typical spurious horizontal velocities produced in this way of around 1 m s^{-1}, with a predominant frequency near 0.1 Hz and a corresponding maximum acceleration of about 0.1 g. These observations were obtained with the tether line's fairlead close to the stern of the ship, in moderate seas; location of the fairlead near the centre of the ship where heave is substantially less than at the stern results in appreciably lower noise, in the same sea state. Tethered balloon measurements of vertical velocity have been calculated in most cases from the measured inclination of the airflow to a local horizontal which has been defined using a damped pendulum: since this pendulum responds to horizontal accelerations the result has been large spurious contributions (up to ±1 m s^{-1}) to the derived vertical velocities. Examples of velocity spectra affected in this way have been given by Thompson (1972a, 1972b), Berman (1976), and Ropelewski (1975, 1976).

Temperature and humidity measurements are affected much less by ship motion because the lapse rate for each of these quantities is usually small, so only small amounts of noise are produced by vertical displacements of the sensors. In the case of temperature, for example, for which the lapse rate is close to dry adiabatic (-0.01°C m^{-1}) in the convective boundary layer, typical periodic vertical displacements of the order of ±1 m caused by ship motion produce spurious rms fluctuations of less than 0.01°C, usually at least an order of magnitude less than the actual temperature fluctuations. Eddy fluxes of heat and water vapour deduced using concurrent measurements of vertical velocity will include some contribution from ship motion, but this is usually negligible. For example, temperature fluctuations produced by periodic vertical displacements of ±1 m, if in exact phase or antiphase with spurious vertical velocities of ±1 m s^{-1}, result in contributions to the heat flux of around 5 W m^{-2}. On the other hand the measured momentum fluxes may be severely contaminated with signal-to-noise ratios as low as 0.1 (Thompson, 1972a, 1972b).

The spurious contributions may be reduced by tethering the balloon to a point near the centre of the ship (to reduce the effects of heave) and by making measurements with the ship alongwind (minimizing the effects of roll) but complete removal of the contributions requires measurements of the sensor motion, or servostabilization of the tethering point on the ship (Thompson, 1976b). Fortunately, the main contributions to the eddy fluxes of momentum, heat and water vapour at heights a few tens of metres above the surface are at frequencies lower than those associated with ship motion and therefore the noise may be filtered out without significant loss of the genuine contributions to the cospectra. However, the inertial subrange of spectra spans the frequency range of ship motion and it may be difficult therefore to deduce from tethered balloon data the spectral densities in the subrange in order to estimate dissipation rates of the turbulent fluctuations unless high-response sensors and high sampling rates are used.

5.3 Tether-line Vibrations

Oscillations of the tether line at relatively high frequencies may occur, particularly for the dielectric lines, with associated accelerations sometimes large enough to damage the attached sensors. Steel lines appear much less likely to experience significant vibrations because of their high density and low extensibility. Vortex-shedding from the line produces vibrations at a frequency

$$n \simeq 0.2\ U/D \tag{3}$$

where D is the line diameter. n is thus a few hundred hertz for typical values of U and D.

Oscillations with frequencies of a few tens of hertz and amplitudes of a few centimetres have also been found. The oscillations at higher frequencies may be damped out by relatively heavy instrument packages attached to the line, or alternatively by attaching damper weights to the line. Those at lower frequencies are harder to remove but their effects may be minimized by attaching the instrument packages to appropriate shockmounts, at the expense of extra weight and complexity.

The vibrations may introduce special difficulties in the case of measurements made by mechanical sensors such as cup anemometers or wind vanes. If, for example, the tilt of the sensor package is measured by a damped pendulum then any nonlinearities in the damping combined with vibrations are likely to produce significant mean displacements of the pendulum from the vertical.

5.4 Reduction of the Effects of Sensor Motion

The characteristic motions of tethered balloons (predominantly crosswind migrations) are unavoidable, although their magnitude may be minimized by ensuring that the balloon always has an adequate excess of lift over payload. The effects of the motions on velocities measured by the sensors attached to the tether line can be eliminated, in principle, by suitable monitoring. There is no opportunity for theodolite tracking of the motion and so the measurements have to be made directly, which involves the use of three orthogonal accelerometers and three rate gyroscopes. As yet measurements of this complexity have not been made for balloon systems. Since these sensors would also have to monitor the much larger accelerations produced by ship motion <u>and</u> withstand line vibrations, it is unlikely that their drift and hysteresis and other nonlinearities would then make them suitable for complete elimination of the effects of the characteristic balloon motions.

The effects of wave-induced ship motion on measured velocities may be reduced along the lines suggested above, or by decoupling as far as possible the ship from the tether. A simple though only partially effective way is to lead the tether line from the winch through a pulley attached to elastic cords; this method is successful only if the cords are very long and the cable tension is constant. A better alternative is to use a servo-winch or similar system to adjust continuously the amount of line paid out to compensate for ship motion (Thompson, 1976b). This latter procedure is probably essential if useful measurements of turbulent fluctuations are to be made in rough seas.

Cable oscillations occur at frequencies which are usually higher than those required to be resolved and are to that extent unimportant but their side effects may be serious. Ways of minimizing the effects have been described in subsection 5.3.

5.5 Mounting of Instruments on the Tether Line

A variety of methods has been used to attach instruments to the tether line and it is beyond the scope of this paper to describe them in detail. The design requirement is usually for a mounting on the line which allows the vane to which the sensors are attached to pivot freely into wind for all angles and orientations which the line itself may take up. Gimbal mountings are sometimes employed but then damping is usually necessary to avoid 'corkscrew' motions of the vanes. Descriptions of different kinds of mountings have been given by Garstang et al. (1971), Almazan (1972), Burns (1974), and Thompson (1976b).

6. OPERATIONAL MODES AND CORRESPONDING RELIABILITY OF RESULTS

6.1 Fixed-level Operation

The effects of sensor motion have been discussed already. The problem of the optimum vertical spacing of sensors in boundary-layer investigations is outside the scope of this chapter. However, it is appropriate to consider the duration of sampling required to describe adequately the turbulence structure of the atmosphere. The low-frequency contributions to the spectra or cospectra are determined by the spacing of cumulus clouds and if these clouds are small and spaced about 2 km apart then in winds of 5 m s^{-1} a sampling duration of about 2 hours will ensure passage of sufficient clouds to achieve reasonable statistical reliability. Convective rolls (aligned nearly alongwind) move rather slowly acrosswind, and in their case and also in the case of large cumulus clouds spaced several kilometres apart it is then necessary to measure over periods perhaps as long as 10 hours; it may be difficult to obtain this data sample before synoptic variability causes a change in the turbulence structure.

6.2 Profile Operation

During profiling the balloon is usually winched up or down at 1 to 2 m s^{-1}, with a transit time of about 10 minutes from surface to 1000 m. In this time different levels within a slice of the atmosphere several kilometres long are sampled by the suspended sensors. Clearly a single profile will not adequately represent the mean vertical structure, especially if horizontal inhomogeneity is accentuated by cumulus convection.

Significant distortions in the measured profiles may be caused by finite sensor time constants, especially near regions with large vertical gradients in the parameters being measured (Ropelewski, 1976). Averaging of up and down profiles will remove most of the distortions except near large vertical gradients where it may be necessary to carry out an objective correction procedure (Sanders et al., 1975).

Horizontal winds are exaggerated during descent and reduced in ascent by alongwind displacements of the sensors produced by profiling. Simply averaging the ascent and descent data to obtain the mean speed is not wholly satisfactory because it assumes that drag on the balloon, cable angle, and cable tension are the same at any one point in ascent or descent (so that the balloon follows the same curve in the x-z plane in each case). If the balloon is in nearly static equilibrium then its path during profiling may be obtained from a knowledge of cable tension, cable angle, and drag on balloon and cable (Gathman, 1975; Wright, 1976) but because of gusts and, in particular, just after change-over from ascent to descent, the balloon system is not in equilibrium. A more elaborate treatment is needed then, requiring comprehensive data on the balloon's dynamic behaviour, to eliminate the effects of profiling from the data; the necessary information on which to base such a treatment does not yet appear to be available.

7. OPERATIONAL LIMITATIONS

Most tethered balloons are robust enough to allow them to be used in winds of at least 25 m s^{-1} with a sufficiently strong tether line. In practice the ship motion may then be violent enough to prevent useful data being obtained unless some form of servo-winch is used. The large amount of salt spray in these conditions will rapidly contaminate the sensors.

The effects of rain on the electrical circuits of sensor systems may be minimized by good design, but sensor performance may be degraded, for example, by water on (and evaporating from) thermometers, or by raindrops altering the speed of rotation of cup anemometers.

Lightning must be assumed a major hazard whether dielectric or steel tether lines are used. The risk to balloons is reduced by tethering them near the surface but may be minimized only by bringing them down to deck level. Smaller static discharges may fuse the dielectric line and release the balloon and it must be remembered that this is a hazard not necessarily limited to occasions with nearby cumulonimbus.

An unnatural hazard is that from research aircraft in the vicinity of the ship flying the balloon. Aircraft operators are naturally wary of approaching close to balloons, especially during conditions with restricted visibility. In practice the solution is that aircraft only approach within a few kilometres if the ship has a navigation aid such as a DME beacon (Distance Measuring Equipment); if not the balloon is hauled down since the aircraft's measurement program usually has the greater priority.

Manoeuvers of the ship while the balloon system is operating will make the interpretation of the velocity measurements from the balloon very difficult unless a continuous record of the ship's speed and direction is maintained. Best of all is to have the ship stationary while making measurements.

Radio-frequency interference from high-power HF ship transmissions has been experienced with most tethered-balloon systems, and also some interference from radar. The only reliable solution up to now to this very serious problem has been scheduling of the transmissions and data-recording periods so that they do not overlap.

8. ACKNOWLEDGEMENTS

Informaton supplied by E. Augstein, G.D. Emmitt, M. Garstang, and O. Yokoyama has been of great help in preparing this chapter.

REFERENCES

ALMAZAN, J.A. 1972. The BOMEX boundary layer instrument package. Second Symposium on Meteorological Observations and Instrumentation, March 27-30, 1972, San Diego, California, American Meteorological Society.

ANDREEV, V.D., V.N. IVANOV, V.C. KOROLEV, V.M. LINKIN, T.F. MASAGUTOV, A.V. SMIRNOV and Y. HOLTZ. 1974. The results of direct measurements of turbulent characteristics for planetary boundary layer. Preliminary scientific results (Vol. 1) of the GARP Atlantic Tropical Experiment, GATE Report No. 14 (ICSU, WMO): 233-244.

BERMAN, E.A. 1976. Measurements of temperature and downwind spectra in the buoyant subrange. *Journal of the Atmospheric Sciences,* 33: 495-498.

BURNS, S.G. 1974. Boundary-layer instrumentation system. *Atmospheric Technology,* 6 (Winter 1974-1975, NCAR): 123-128.

CAUGHEY, S.J. and C.J. READINGS. 1975. An observation of waves and turbulence in the earth's boundary layer. *Boundary-Layer Meteorology,* 9 (3): 279-296.

GARSTANG, M., M. MUNDAY, W.R. SEGUIN, J.D. BROWN and N.E. LaSEUR. 1971. Fluctuations in humidity, temperature and horizontal wind as measured by a subcloud tethered-balloon system. *IEEE Transactions on Geoscience Electronics,* GE-9, 4: 199-208.

GATHMAN, S.G. 1975. Characteristics of the tether line for airborne vehicles at sea. NRL Report No. 7919, Naval Research Laboratory, Washington, D.C. 20375.

HARRINGTON, E.V., A.S. CARTEN and C.D. CORBIN. 1972. The Hugo II tethered balloon system, Bedford, Mass. Air Force Cambridge Research Laboratory, Special Report No. 152: 71-91.

HAUGEN, D.A., J.C. KAIMAL, C.J. READINGS and R. RAYMENT. 1975. A comparison of balloon-borne and tower-mounted instrumentation for probing the atmospheric boundary layer. *Journal of Applied Meteorology,* 14 (4): 540-545.

JONES, J.I.P. and H.E. BUTLER. 1958. The measurement of gustiness in the first few thousand feet of the atmosphere. *Quarterly Journal of the Royal Meteorological Society,* 84 (359): 17-24.

OOTSUKA, S., N. SHISHIDO, N. HONDA, S. NEMOTO, S. KOINUMA, M. HAYASHI and M. MIYAKE. 1975. Observations of planetary boundary layer by tethered balloons and lower tropospheric radiosonde. Scientific Report of the Fourth AMTEX Study Conference, *AMTEX Report No.* 8, Management Committee for AMTEX: 70-73.

ROPELEWSKI, C.F. 1975. Bomex wind spectra derived from the boundary layer instrument package (BLIP). NOAA Technical Memorandum EDS BOMAP-17, Centre for Experiment Design and Data Analysis, Washington, D.C.

ROPELEWSKI, C.F. 1976. An evaluation of the meteorological data from the GATE Boundary Layer Instrument System (BLIS). NOAA Technical Memorandum EDS CEDDA-9, Centre for Experiment Design and Data Analysis, Washington, D.C.

SANDERS, L.D., J.T. SULLIVAN and P.J. PYTLOWANY. 1975. Correction of Bomex radiosonde humidity errors. NOAA Technical Memorandum EDS BOMAP-16, Centre for Experiment Design and Data Analysis, Washington, D.C., May 1975.

THOMPSON, N. 1969. The effect of sensor motion on data collected by captive balloon-borne turbulence instrumentation. Meteorological Office, Met O 14, Turbulence and Diffusion Note No. 5 (Unpublished, available from Meteorological Office, Bracknell, England).

THOMPSON, N. 1972a. Turbulence measurements over the sea by a tethered-balloon technique. *Quarterly Journal of the Royal Meteorological Society,* 98 (418): 745-762.

THOMPSON, N. 1972b. An investigation of turbulence structure and fluxes over the sea by a tethered-balloon technique. Ph.D. Thesis, University of London (available from Meteorological Office, Library, Bracknell, England).

THOMPSON, N. 1976a. Observations of the planetary boundary layer during GATE. Seminar on the Treatment of the Boundary Layer in Numerical Weather Prediction, Sept. 6-10, 1976, Shinfield Park, England, European Centre for Medium Range Weather Forecasting: 183-204.

THOMPSON, N. 1976b. Meteorological observations from HMS HECLA during the final phase of GATE. *Meteorological Magazine,* 105 (1250): 272-282.

VORONTSOV, P.A. and V.I. SELITSKAYA. 1955. Methods of sounding the atmosphere by aerostat. Leningrad, Glavnoe Geofizicheskoe Observatoriya, Trudy No. 51 (113): 3-16 (in Russian).

WRIGHT, J.B. 1976. Computer programs for tethered balloon system design and performance evaluation. Air Force Geophysics Laboratory, Hanscomb AFB, Massachusetts, Report No. AFGL-TR-76-0195.

YOKOYAMA, O., M. HAYASHI and Y. OGURA. 1969. Measurements of vertical profiles in the atmospheric layer, up to 1000 m high above the sea surface. Preliminary Report of the Hakuho Maru Cruise KH-69-3, Ocean Research Institute, University of Tokyo: 24-25.

32

Flow Distortion by Supporting Structures

J. Wucknitz

1. INTRODUCTION

Distortions of the wind field by structures supporting anemometers are not unique to air-sea interaction studies. Indeed, most of our knowledge about this subject comes from investigations over land or from wind tunnel experiments. However, these results are, in principle, transferable to marine conditions. Generally, the unavoidable effects of wind field distortion will be greater at sea because more mechanical stability of masts, spars, and booms is required.

In air-sea interaction studies of the wind field we want to measure mean values, mean vertical gradients, and fluctuations. Of these the determination of the gradient is the most demanding; admissible errors ought to be not more than about 1% of the mean velocity.

There are two regions of disturbed flow. Close to the obstacle (at distances smaller than the length scale of the obstacle, which in most cases is its diameter) the flow distortion is rather large and complicated, because it is partly determined by the boundary layer developing at the surface of the obstacle and therefore is rotational. This type of flow distortion is met if ships or large platforms are used as instrumental supports. For these cases it is generally impossible to describe the disturbed flow by a simple theory; empirical investigations are needed (some of these are presented in Section 4). In contrast, at larger distances from the structure and its wake (say at more than two diameters), the flow can be considered as potential flow, and for long masts and booms can be approximated by a two-dimensional flow. The simplest approach is to consider it as flow around a circular cylinder (see

Section 2). The flow distortion is determined not only by the mast diameter, but also by the wake forming in the wind shadow region. Deterministic models of the wake structure are available only for small 'mast Reynolds number' ($<10^2$), where the flow in the wake is stationary. For higher Reynolds numbers, which is the case for most of the present investigation, the wake becomes nonstationary or even turbulent and the behavior of the wake must be inferred from experiments. For our purpose it is not necessary to investigate the details of the wake structure but only the effect of the wake on the ambient flow.

Unfortunately, field investigations are rather sparse and the relevance of wind tunnel investigations is restricted because the range of mast Reynolds number attainable in wind tunnels is limited to Re $< 5x10^5$ (in a few cases up to $2x10^6$). Masts under atmospheric conditions correspond to Reynolds numbers between 10^4 to 10^7. Also, the results from wind tunnel experiments are different from those of natural flows, even for the same mast Reynolds number, due to different turbulence intensities and different surface roughness effects.

A semi-theoretical approach for two-dimensional flow around a circular cylinder is applied with some success to masts of lattice-type (see Section 3).

2. FLOW AROUND A CIRCULAR CYLINDER

2.1 Basic Problems

Two-dimensional flow around a circular cylinder is approximately realized in the atmospheric flow around a vertical pipe or smoke stack, provided that at the height under consideration the flow is influenced neither by the top of the mast nor by the boundary below. Furthermore, the vertical wind shear due to the logarithmic profile as well as the vertical flux of momentum are assumed to have no effect on the distortion of the wind field by the obstacle. The flow around a circular cylinder and its drag force have been studied many times, both experimentally and theoretically. Nevertheless, the problem is not completely solved, especially for higher Reynolds' numbers (between 10^4 and 10^7). The phenomena associated with a wake behind an obstacle in parallel flow can be understood from the flow around this simply-shaped body.

For the circular cylinder a simple semi-theoretical flow model has been formulated below, which describes the flow around the mast as it is known from observations. Fortunately, we need no details of the flow in the wake. Within the wake the wind field is disturbed too much for it to be acceptable for wind measurements. Hence we only have to look for the expected extension of the wake in order to predict the region which is to be excluded for wind measurements. On the other hand, we need some measure of the total intensity of the wake in order to model its effect on the environmental flow field, which we approximate by a potential flow. The potential flow around a circular cylinder combined with an (idealized) wake differs considerably from the potential flow around a circular cylinder without a wake (see Fig. 1). The lines on which the flow velocity u is equal to the undisturbed velocity U_∞ appear at $|\phi|$ = 135° for a circular cylinder without a wake; they tend to approach $|\phi|$ = 90° with increasing wake intensity. The line of maximum positive velocity distortion, which is found at $|\phi|$ = 90° for a cylinder without a wake, is shifted downwind towards the edge of the wake. Even opposite the wake, on the negative axis, significant flow distortion occurs. Accordingly, a flow model has to include not only the length scale of the obstacle (i.e. the diameter of the circular cylinder) but also suitable quantities characterizing the wake intensity.

Unfortunately, the characteristics of the wake, such as the position of separation points or the wake width (or intensity), are highly variable with varying Reynolds number in the range of interest. The wake is also sensitive to roughness elements on the cylinder surface and to the intensity of the turbulence in the

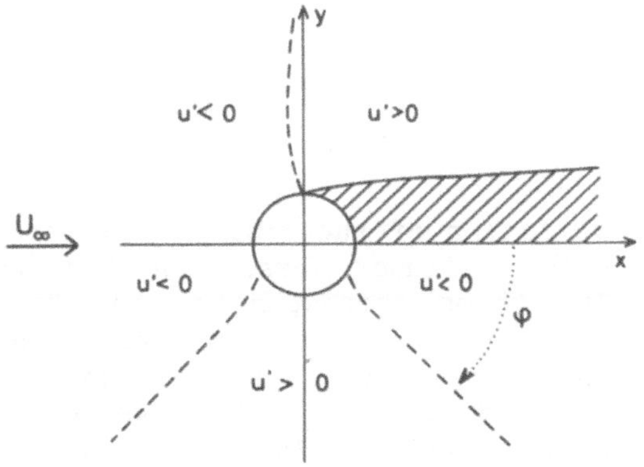

Fig. 1 Characteristics of the flow around a circular cylinder with a wake (upper part), and without a wake (lower part). The hatched area indicates the region of the wake. On the dashed lines the wind velocity is equal to the undisturbed velocity U_∞. The maximum negative wind field distortion is found on the negative X axis in both cases, the maximum positive wind field distortion is found near the wake if there is one, and on the Y axis if there is no wake. u' is the flow distortion.

undisturbed flow. The relationship between the wake intensity and the drag coefficient C_D of a circular cylinder is

$$C_D = \frac{2}{D} \int_{-\infty}^{+\infty} [u'(y)/U_\infty] dy \qquad (1)$$

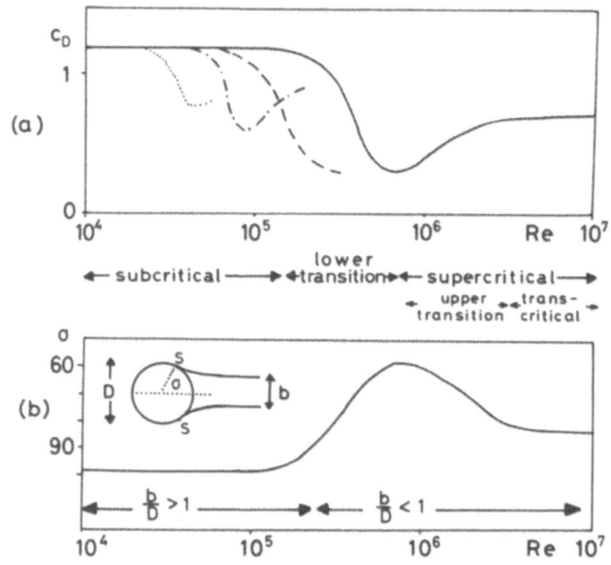

Fig. 2 (a) Variation of the drag coefficient C_D of a circular cylinder versus the Reynolds number Re. The terms corresponding to the different Re intervals refer to the full line, i.e. for a smooth cylinder in a wind tunnel of low turbulence intensity. The dashed line is for a smooth cylinder in a wind tunnel with higher turbulence intensity (Wieghardt, 1969). For a wind tunnel of medium turbulence intensity the effect of the surface roughness of the cylinder is demonstrated by the dashed-dotted curve ($\delta/D = 4x10^{-3}$) and the dotted curve ($\delta/D = 2x10^{-2}$), where δ is the diameter of the roughness elements (after Fage and Warsad, 1930).

(b) Variation of separation points S and the wake width b with varying Reynolds number. This corresponds to the full line in (a).

(see e.g. Schlichting, 1968). D is the cylinder diameter, y is the cross-wind coordinate, and u' is the flow distortion.

For a finite wake width b and an undisturbed flow outside the wake, the infinite limits of the integral in Equation 1 can be replaced by -b/2 and b/2;

$$W_b = \int_{-b/2}^{+b/2} (u'/U_\infty)dy. \qquad (2)$$

where W_b is the 'wake intensity.'

The idea of varying wake intensity with Reynolds number, Re, is also supported by experimental results on separation points. In Figure 2 the wake width b is sketched qualitatively. As pointed out by Roshko (1961) in the subcritical regime ($Re < 2x10^5$ for a smooth surface) the wake width is greater than the diameter D of the cylinder. In the lower transition regime ($2x10^5 < Re < 6x10^5$) the wake width decreases and becomes distinctly smaller than D. For $Re > 6x10^5$, in the so-called supercritical regime, the wake width increases again, but remains smaller than D. For $Re > 3x10^6$ the flow may become independent of Re with a constant wake width.

Observations of the flow field around circular cylindrical masts used in this paper are represented in Figure 3. These investigations cover the range of mast diameters useful as structures on which to mount meteorological instruments. We are concerned with flows between the subcritical and the transcritical regime (primarily with flows within the lower transition) where the variation of C_D is rather large. Roughness elements on the mast surface shift the limits of the critical regime (Szechenyi, 1975) and we are not able to predict the wake intensity from the mast diameter alone. It is therefore necessary to analyze observations of the distorted wind field.

2.2 Model of Flow Around a Circular Cylinder

None of the investigations presented in Figure 3 yields a complete picture of the obstacle-disturbed wind field between about r/R = 2 and r/R = 30, where r is the distance from the mast centre and R is the mast radius. In some cases the observed area is restricted to a circle around the mast with constant radius r. Furthermore, in most cases no measured value of the 'undisturbed' wind velocity U_∞ exists as a reference value. Instead measurements are compared from two anemometers at different positions, which are differently disturbed. This rather inconsistent set of experimental results will be summarized and discussed by comparing it with a simple flow model. The empirical parameters of the model have to be chosen to fit the observations. Parkinson and Jandali (1970) have developed a model of the flow around different cylindrical obstacles, including the circular cylinder. They simulate the obstacle in the usual manner by adding a doublet to the mean flow. The wake is modelled by placing two additional sources and a sink at suitable positions. The final body shape is obtained by conformal mapping of the basic model. The parameters to be specified empirically are the angle of points of separation and the 'base pressure coefficient' at the back of the cylinder within the wake. In the experimental results neither of these two quantities is given. For our purpose, we combine the wake characteristics into a single

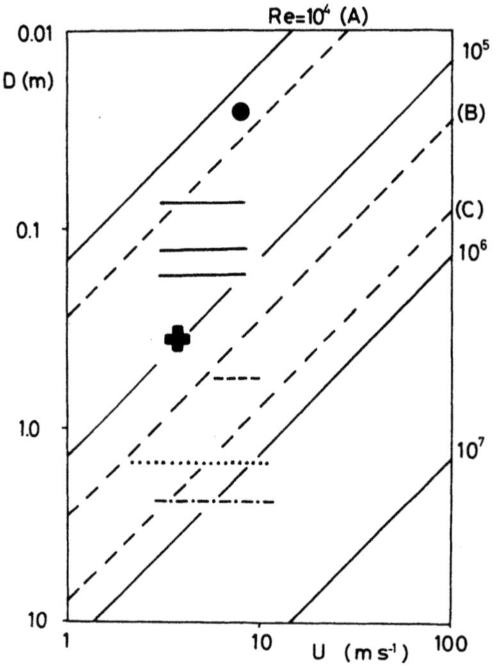

Fig. 3 Experimental investigations of the flow around a circular cylinder, characterized by the mast diameter D and the wind velocity U. Rider (1960), full circle; Wucknitz (1977), full bar; Gill et al. (1967), cross; Dabberdt (1968b), dashed bar; Link (1966), dotted bar; Borovenko et al. (1963), dashed-dotted bar. Lines of constant Reynolds number are plotted including: A, the upper limit of the subcritical regime for a rough cylinder ($\delta/D = 2 \times 10^{-2}$); B, the upper limit of the subcritical regime for a smooth cylinder in a wind tunnel of low turbulence intensity; and C, the minimum drag coefficient for the same conditions as B.

parameter, which can be varied to give a best fit to the observations.

The following simple model, of course, shows some deviations from Parkinson and Jandali's model. The upwind contour of the present model is only approximately a circular arc. However, the wind field distortion at a distance greater than about 1 D from the body and the wake (which is the region of interest) is not very sensitive to details of the contour of the body and idealized wake.

For our model we add to the undisturbed parallel flow, $u = U_\infty$, a source Q of strength $2U_\infty y_1$ and, at a distance 'a' downstream, a sink S of smaller strength $-2U_\infty y_2$ (see Fig. 4). The complex potential is given by

$$F(z) = U_\infty\left(z + \frac{y_1}{\pi} \ln z - \frac{y_2}{\pi} \ln (z - a)\right), \tag{3}$$

where $z = x - iy$.

The corresponding contour of the body simulating mast and wake is given by

$$y + \frac{y_1}{\pi} \arctan \frac{y}{x} - \frac{y_2}{\pi} \arctan \frac{y}{x-a} + y_2 - y_1 = 0 \tag{4}$$

(see e.g. Tietjens, 1960) where $a = R/2$, and y_1 and y_2 are chosen such that the upstream part of the contour approaches a circular arc with the given radius R. The free parameter is the wake intensity; $W_b = 2(y_1 - y_2)$. To estimate W_b consider the momentum loss (compared to the undisturbed velocity U_∞) well downstream from the cylinder along a line x = constant. The momentum loss within the modelled wake is $2(y_1 - y_2)$; the momentum gain outside the wake where the flow is accelerated is $(y_1 - y_2)$. The total momentum deficit is then $y_1 - y_2 = W_b/2$, and it follows from Equation 1 that

$$C_D = W_b/D. \tag{5}$$

Hence we expect relative wake intensities W_b/D of the order of 1.2 to 0.3. Some examples of corresponding model contours are shown in Figure 4.

Experimental observations of wake intensity were used to verify Equation 5 and to obtain W_b. In order to compare the model wind field distortions outside the wake to those observed, the modulus of the wind velocity was computed from $|F'(z)|$, where $F'(z)$ is the derivative of $F(z)$ in Equation 3. The coordinates r, ϕ are used with origin C as indicated in Figure 4.

2.3 Observations

There are only a few observations available for the disturbed wind field around solid masts with circular cross-section. Link (1966) and Borovenko et al. (1963) use large masts (some hundred metres high) having a diameter of some metres; Wucknitz (1977) considers a slim mast of only 8 m height. Special studies of structure-induced wind field distortion were done by Gill et al. (1967) in a wind tunnel, and by Dabberdt (1968b) in the atmospheric surface boundary layer.

Figure 5 shows quantitative wake observations for circular cylindrical masts in the atmosphere and includes the wind tunnel results from Gill et al. (1967). The wake occupies a wedge-shaped area extending from about $\phi = +30°$ to $-30°$ behind the obstacle. This area has to be excluded from wind velocity measurements. From Dabbert's results, the wake width increases approximately proportional to $x^{\frac{1}{2}}$; this was predicted by Schlichting (1968). Simultaneously the velocity deficit decreases as $x^{-\frac{1}{2}}$ so the wake intensity is nearly

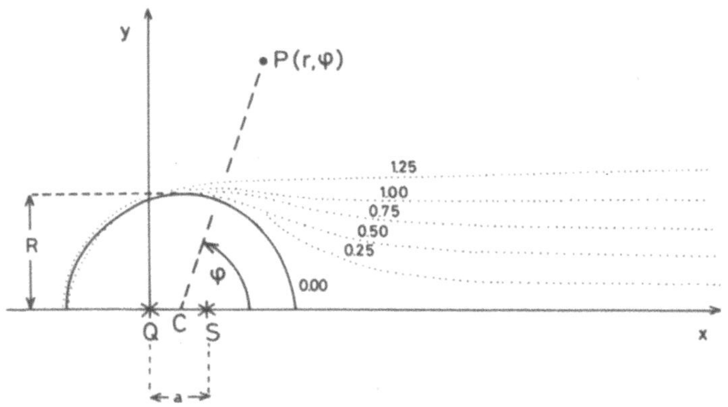

Fig. 4 Contours of bodies corresponding to Equation 4, simulating a circular cylinder combined with a wake. Q and S are the source and the sink with intensities $2\ U_{\infty} y_1$ and $-2\ U_{\infty} y_2$, respectively, where $y_1 > y_2$. Q is the origin of the coordinate system used in Equations 3 and 4, while C is the centre of the approximated circular arc on the upwind side of the contour. C is the origin of the polar coordinate system used to describe the disturbed wind field, indicated by P. The numbers on the individual curves give the values of $(y_1 - y_2)/R$, which is equal to the relative wake intensity W_b/D.

independent of the downstream distance for $x/r \geq 5$.

Although the data are rather fragmentary, the resulting wake intensity W_b/D seems to depend on the Reynolds number (as expected from the consideration of the drag coefficient). The greatest values of W_b/D appear in the presumably subcritical regime $Re \leq 10^5$ [from Gill et al. (1967) $Re \simeq 10^5$, $W_b/D \simeq 1.1$; and Wucknitz (1977) $Re \leq 10^5$, $W_b/D \simeq 1.0$]. For $Re \geq 2x10^5$ the wake intensity becomes smaller [Link (1966), $Re \geq 3x10^5$, $W_b/D \simeq 0.75$; and Dabberdt (1968b), $Re \geq 2.5x10^5$, W_b/D between 0.3 and 0.65]. These results are rather inconclusive since: for Gill et al. the limited width of the wind tunnel ($\simeq$7D) may influence the wake formation; for Dabberdt the low height seems to affect the wake as indicated by the differences in the wake between z = 1.6 m and 3.2 m and by the nonsymmetrical shape of the wake; for Wucknitz the mast radius R is not well defined because of an irregular thickening of the cylindrical mast at

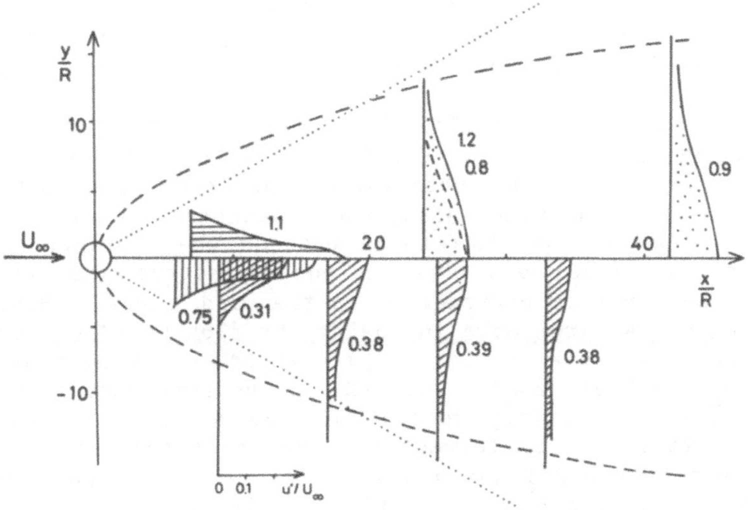

Fig. 5 Observed shape of wakes at different distances behind a circular cylindrical mast: Dabberdt (1968b), diagonally hatched area; Gill et al. (1967), horizontally hatched area; Wucknitz (1977), dotted areas; Link (1966), vertically hatched area. The numbers are the relative wake intensities W_b/D. The wake region can be enclosed between the dashed lines $y \sim \pm x^{\frac{1}{2}}$, or approximately by the dotted lines $\phi = \pm 30°$.

some heights; and for Link the wake determination only holds for a rather short distance behind the mast, r/R = 6 (the same restriction applies to data from Gill et al.).

In considering the wind field distortion outside the wake (which is the area of interest for wind measurements) several difficulties arise in determining the 'undisturbed' wind velocity U_∞. Wucknitz (1977) and Link (1966) did not measure any reference velocity. Instead, the relationship between two (or three) disturbed wind values is given, measured at a constant distance r/R from the mast but separated by an angle $\Delta\phi$ of 146° or 120°. Hence, these experiments give relative information only. Borovenko et al. (1963) used the measured wind velocity on a long boom (r/R = 10 to 20) as the 'undisturbed' reference velocity. Even these values may be affected by the tower by up to 3%. Dabberdt (1968b) obtains the reference wind from a separated thin mast. Unfortunately, these observations suffer from the low measuring heights used: 1.6 and 3.2 m. The influence of the ground is indicated by differences in the wind field distortion between the two heights. Also, the distorted wind field is nonsymmetric with respect to the mean wind direction. Hence the application of a simple two-dimensional model is rather doubtful. In Gill et al.'s (1967) wind tunnel observations, the wind field is influenced by the limited cross section of the wind tunnel; hence the reference wind velocity, measured with the obstacle removed, is not very useful.

In Figure 6 the observed wind field distortions $u^+ = (u - U_\infty)/U_\infty$ are compared to the model. The data of Gill et al. are shifted by adding a constant so that observation and model agree with respect to the average value taken along the respective line r/R = constant outside the wake. The difference between the averaged wind field distortion for model and observations is assumed to be at least partly caused by flow contraction in the wind tunnel. Since observations showed a large wake intensity the flow model was calculated using a large value $W_b = 2(y_1 - y_2)$. In the lower part of Figure 6 a value $W_b/D = 2$ was used. For $r/R > 5$ the agreement between model and experiment is fairly good. For $r/R < 3$, the observed wind field distortion varies between -25% and -32% for the interval ϕ = 40° to 180°; the model gives values between -16% and -22%. This discrepancy may be explained partly by wall effects.

All the other observations are represented in the upper part of Figure 6 and are compared to the model using $W_b/D = 1$. For Dabberdt (1968b) the average value from the two measuring heights (1.6 m and 3.2 m) is used, and the (nonsymmetrical) wind field was averaged with respect to the apparent axis of symmetry. Nevertheless the resulting functions $u^+(r/R,\phi)$ show a rather irregular shape, deviating from the smooth curves of the model. On the whole the observations indicate a smaller wind field distortion than the model; this may be due to the wake intensity being smaller than D, as indicated by the direct wake observation. However, the large differences in u^+ between the two heights (up to 2% for $\phi = 180°$ and up to 5% for $\phi = 60°$) remind one to be cautious when explaining the observations with the model. The low heights used and the small mast length to diameter ratio (5 m/0.56 m) prevent a description using a two-dimensional model.

The results of Borovenko et al. (1963) also show some nonsymmetrical behavior, and some scatter (see Fig. 6). Considering this and

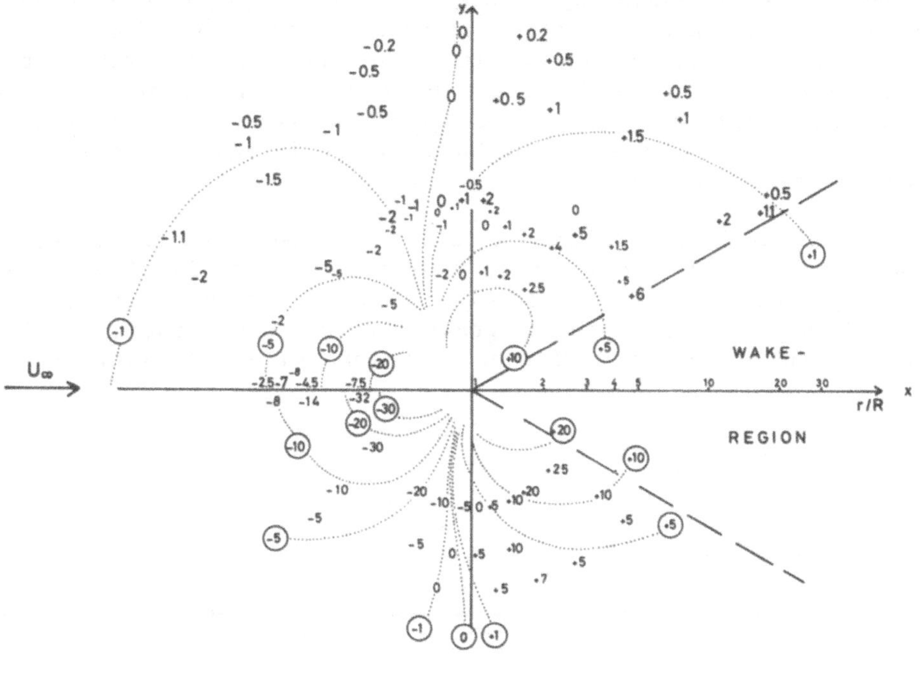

Fig. 6 Comparison between observed wind field distortion and calculations from the source-sink model (Eq. 3). The normalized wind field distortion $u^+ = (u-U_\infty)/U_\infty$ is given in percents; $u = |F'(z)|$.
Upper part: dotted lines with encircled figures are isotachs from the model with $W_b/D = 1$. Experimental values are from: Wucknitz (1977), large numbers at $r/R \geq 10$; Borovenko et al. (1963), large numbers at $r/R \leq 10$; Dabberdt (1968b), medium sized numbers; and Linke (1966), small.
Lower part: dotted lines: isotachs from the model with $W_b/D = 2$. The numbers are from observations by Gill et al. (1967).

instrumental errors, as well as the uncertainty of the reference value (mentioned earlier), the mean wind field distortion is explained rather well by the model.

Link's (1966) observations, where a reference value is missing, are approximated by the model using $W_b/D = 0.75$. The corresponding values $u^+(r/R,\phi)$ are given in Figure 6. The agreement with the model using $W_b/D = 1$ is very close, within a range of 1%. The results of Wucknitz (1977) are presented in a similar way. The amplitude of the wind field distortion at the height of 1.5 m is only half as large as at 2 m, although the ratio r/R is nearly the same. This may be partly due to some irregular thickening of the mast at 2 m and hence the corresponding values are given in Figure 6 at a somewhat smaller distance r/R from the mast centre. This does not totally remove the discrepancy. It is assumed that the wake intensities may actually be different between 1.5 and 2 m. Near the critical Reynolds number the wake width may be determined by small differences in the surface roughness or by irregular surface elements of the cylindrical mast.

The experimental results are, within the scope of experimental accuracy, predictable by a simple source-sink potential flow model, given by Equation 3. This model is able to explain the observed wind field distortions for a greater area than a potential flow model around a circular cylinder, say for $|\phi| > 30°$ (i.e. outside

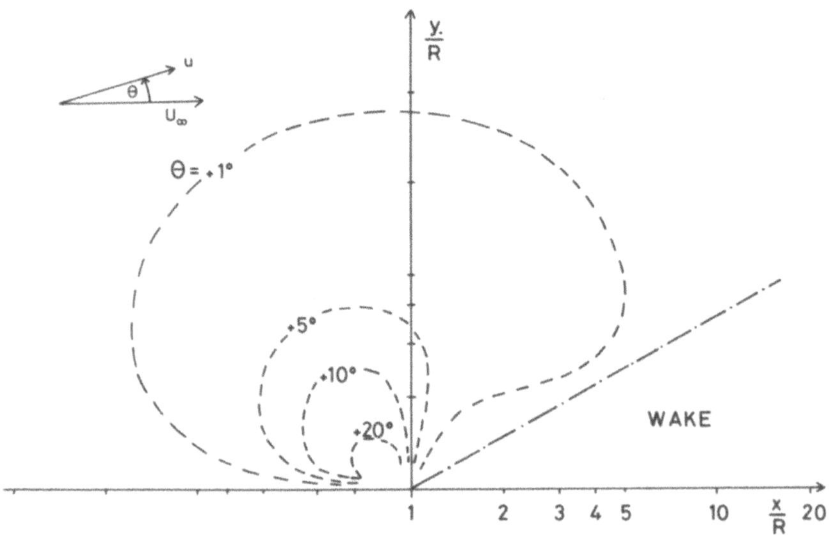

Fig. 7 Isolines of wind direction deflection for the model flow using $W_b/D = 1$.

the wake). For mast diameters used in the atmosphere the relative wake intensity W_b/D is of the order of 1, varying between 0.4 and 1.2. For a given mast the wake intensity seems to be determined essentially by roughness elements on the mast surface. The dependence on the mast Reynolds number is not so clear; in no case does the wind field distortion and/or the wake intensity seem to depend on the wind velocity for a given mast. This may be due to the relatively small range of wind speed used. The usefulness of wind tunnel experiments for investigating the wind field distortions by an obstacle is limited because of the limited range of Reynolds numbers obtainable in a laboratory, wall effects, and the difference in turbulence intensities between the wind tunnel and natural flows.

2.4 Distortion of the Wind Direction

Observations of the wind direction in the environment of a mast are rather few and the available values are inconclusive because the small observed distortions in the wind direction are of the same order as experimental errors. Thus, for estimating distortion of the wind direction, it seems better to derive the expected values from the flow model used in the previous section. In Figure 7 isolines of wind direction deflection are plotted for the model using $W_b/D = 1$. It can be seen that for distances $r/R \geq 5$ the influence of the mast on the wind direction is smaller than 5° (which is a reasonable accuracy in some cases but not good enough, for example, for determination of large scale divergences).

3. FLOW AROUND LATTICE-TYPE TOWERS

3.1 Basic Problems

The description and prediction of wind field distortion induced by lattice-type towers are more complicated than those for a cylindrical mast. Typical tower cross sections are quadratic or triangular but not symmetrical with respect to the centre, and their construction is not purely two-dimensional but consists of a number of bracings and additional structural members in several directions. Each member generates its own wake. The formation of an individual wake involves the problems discussed in the previous section, insofar as the individual members can be approximated using circular cross-sections. The wake formation will be changed if the structural members have sharp edges, and streamline separation will be rather independent of Reynolds number. The individual wind field distortions by the structural members are superimposed with some interaction between them. At some distance behind the structure the wakes will coalesce forming one single wake (see Fig. 8). For masts in natural flows the combining of the individual wakes into a single one will appear at a distance of 2 to 3 diameters from the tower (this is influenced by the variance of wind direction). The wind field distortion in the area used for wind measurements (say 2 to 3 diameters from the tower and its wake) will be determined essentially by the integral wake intensity and will be nearly independent of the internal structure of the total wake regime.

It seems to be possible to simulate the wind field distortion produced by a lattice-type tower with the wind field distortion from a suitable circular cylindrical tower, provided that the distance from the tower is sufficiently large, and that the tower is quasi-two-dimensional (i.e. essentially independent of height). An 'effective tower radius', R_{eff}, can be found, which is the radius of a circular cylindrical tower which would produce the same wind field distortion as the given lattice-type tower. R_{eff} will be determined by the length scale of the tower cross section and by the ratio of obstructed to unobstructed areas. The corresponding 'effective wake intensity', W_{eff}, will also depend on the cross sections and the surface conditions of the individual structural members as well as on the 'effective Reynolds number', Re_{eff}, which is based on the average diameters of the individual structural members. It is important that Re_{eff} is generally smaller than the mast Reynolds number for the circular cylindrical masts used earlier. With the aid of Figure 3 it can be estimated that in some cases Re_{eff} is of the order of 10^4 or even smaller; that is, within the lower transition or even in the subcritical flow regime. This means that the 'effective relative wake intensity' W_{eff}/D_{eff} may vary widely, taking on large values in the subcritical regime and lower values in the lower transition regime. Observations by Moses and Daubek (1961) include the transition from the subcritical to the critical regime; these are the only observations at present which show a dependency of the wind field distortion on the (effective) Reynolds number. Re_{eff} is rather small due to the low

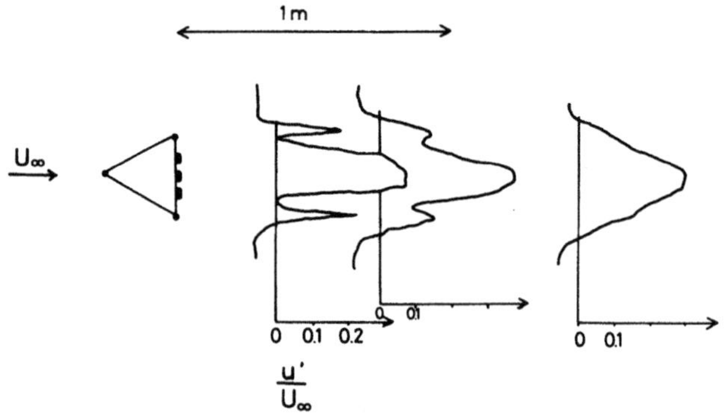

Fig. 8 Confluence of individual wakes into a single wake behind a triangular mast of lattice-type in a wind tunnel (after Cermak and Horn, 1968).

wind velocity and the small diameter of the structural members of about 0.25 m. In Figure 9 the directly observed relative wake intensity $W_b/(qD_0)$ is plotted against $Re_{eff} = 0.25\ U_\infty/\nu$ for various wind speed intervals U_∞; $D_0 = 2R_0$ is the mean diameter of the tower cross section (defined later), and q is the 'structural member density', i.e. the ratio of obstructed to unobstructed area of the projection of the tower on a vertical plane. We set q = 0.21. The decrease of $W_b/(qD_0)$ is very similar to the decrease of C_D with increasing Reynolds number; this confirms Equation 5.

3.2 Observations and Description from the Model

Although lattice-type masts are used frequently for wind measurements there are only a few systematic investigations of the disturbed wind field available; Moses and Daubek (1961) as well as Izumi and Barad (1970) consider towers of square cross section, while Dabberdt (1968a) and Cermak and Horn (1968) use towers of triangular cross section. In the last paper the field investigation is supplemented with an extensive wind tunnel study. A similar laboratory study is published by Gill et al. (1967).

These observations have been considered in terms of our circular cylindrical tower approximation; the results are listed in the table. In order to compare masts of different cross-sectional areas, a length R_0 is defined as the radius of a circle having the area of the tower cross section. The distance between an anemometer and the centre of the respective tower is denoted by r. In

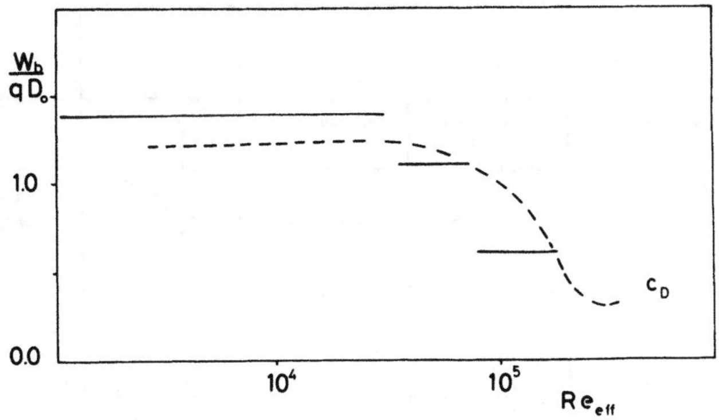

Fig. 9 Solid lines: dependency of the relative wake intensity on the "effective Reynolds number" after Moses and Daubek (1961).
Dotted line: variation of C_D for a circular cylinder of 0.25 m diameter, which is approximately the diameter of the structural members of the tower.

Authors (anemometer height above ground)	Cross-section of the tower (anemometer positions are indicated by crosses)	q "structure member density"	R_o (m) Radius of a circle of same area as the tower cross section	r(m) Distance between anemometer and center of the tower cross section	R_{eff} (m)	R_{eff}/R_o
MOSES and DAUBEK (1961) (5.7 m)	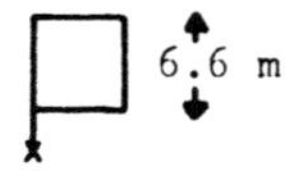	0.21	3.7	6.85	1.3	0.35
DABBERDT (1968a) (30.5 m)	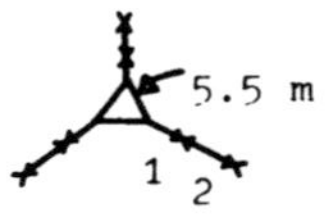	0.3	2.0	r_1=5.9 r_2=8.7	0.8	0.40
IZUMI and BARAD (1970) (5.7 and 22.6 m)	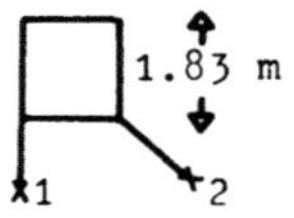	0.2	1.0	r_1=3.66 r_2=3.12	0.56	0.56
GILL et al. (1967) (Wind tunnel)	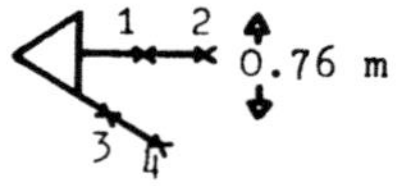	0.26	0.28	r_1=0.60 r_2=0.98 r_3=0.79 r_4=1.16	0.13	0.46
ÇERMAK and HORN (1968) (Wind tunnel)	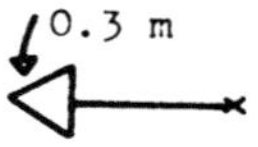	0.3	0.11	0.89	0.12	1.1

Table 1.

the following columns the observed wind field distortions are compared with model calculations in order to find the best fitting model. In contrast to flow around a solid cylinder, where the radius R is given, the effective radius R_{eff} is determined from the wind field distortion. It can be shown that the available observations can be approximated sufficiently well by a set of models where W_b/D is set equal to 1. Hence 2 $R_{eff} = W_{eff}$, which is the only parameter to be determined for the model of a given tower. For the individual investigations the ratio r/R_{eff} from the respective best fitting model was computed; the corresponding R_{eff} is given in the table. The individual observed curves $u^+(\phi)$ show, in part, considerable scatter around the model curves. However, this scatter is well within the experimental accuracy.

It is important to relate R_{eff} to the mean radius R_0 of the tower and to q, the 'structural member density'. Unfortunately, most of the authors give no definite value for q. In some cases estimates of q can be taken from photographs and sketches. Generally, R_{eff}/R_0 is between 0.35 and 1.1, and is greater than q, which is of the order of 0.2 to 0.4 for typical towers. R_{eff}/R_0 seems to increase for decreasing tower cross section. Small towers with thin structural members generate a greater relative wake intensity. Presumably the wake broadens with decreasing Reynolds number Re_{eff}, which is expected as the subcritical regime is approached.

The ratio (R_{eff}/R_0) to q, which equals $W_b/D = C_D$ for a solid mast, is distinctly greater (about 1 to 3) than for a solid circular cylinder (0.3 to 1.2). This can be explained by the greater drag coefficient of noncircular cylindrical obstacles. Presumably the structural members of complicated lattice-type structures have sharp edges and irregularities, which tend to enhance the drag force.

4. THREE-DIMENSIONAL FLOW DISTORTIONS

In this section distortions of the wind field are considered that cannot be treated two-dimensionally. Such complications become important near the top and base of masts and towers, as well as in the environment of irregular-shaped instrumental supports, such as buoys, ships, and light-towers. In some of these cases the wind field at distances smaller than the length scale of the obstacle is relevant for wind measurements. The corresponding flow field cannot generally be described by a simple semi-empirical potential flow model such as the flow around a quasi-two-dimensional mast.

Only the effect of the top of a mast can be estimated in a general way. Above the end of a mast, say higher than about 3 diameters, the wind field distortion can be approximated by potential flow around a sphere, located at the top of the real mast. The flow around a sphere, including the wake, can be treated analogously to the flow around a circular cylinder, because the problem is two-dimensional when referred to polar coordinates. The flow distortion u', which is $Rx/(\pi r^2)$ at a distance r from the centre of a circular cylinder of radius R, becomes $R_0^2 x/(4r_0^3)$ at a distance r_0 from the centre of a sphere of radius R_0, where x is the downwind coordinate (Tietjens, 1960). Using these formulae the effect of a sphere on the wind field can be determined with the aid of Figure 6. It is important to know that for $r_0/R_0 > 1$ the flow distortion decreases faster above the top of the mast than near the two-dimensional mast farther below. The sphere approximation becomes more valid with increasing distance above the top of the mast.

Another type of complication appears near the base of a tower, where the shear of the boundary layer flow becomes important. The height variation of the stagnation pressure on the forward side of an obstacle generates a detached vortex near the ground and upstream of the obstacle (Mollo-Christensen, 1979). The vortex separation distance can be scaled using the crosswind dimension of the obstacle and the boundary layer thickness. The position of the vortex may be unstable, even in a steady flow.

Both the above kinds of flow distortion can aid in the understanding of the flow around large instrumental supports as ships, light-towers, etc. For an indication of the effects of obstacles on pressure measurements, see Dobson, Chapter 13. Wind measurements with an accuracy of 5 to 10% can be expected at distances of at least one scale length from the hull of the obstacle and from the wake; the latter can be considered as a downwind extension of the obstacle, with increasing cross-section (cf. Fig. 5). The wind field near the obstacle, where most wind measurements are made, can be distorted by more than 10%. No simple formula can be given for the estimation because of the great variations in geometry of obstacles. In order to find special positions within this strongly disturbed region where the deviation from the undisturbed flow velocity could be small, the specific structure must be tested experimentally in the field, or in a wind tunnel using a small scale model. The Reynolds number, which is of the order of 10^6 to 10^7 for the real obstacle in natural flow, should be at least $3x10^5$ for the model experiment. For $Re > 3x10^5$ the drag coefficient of a circular cylinder, for example, varies only slowly (see Fig. 2), which implies flow similarity between model and real obstacle.

Several experimental investigations are available that discuss the flow near special structures. An extensive field investigation for a research platform is published by Thornthwaite et al. (1965); a wind tunnel investigation of a model of a similar light-tower has been published by Mollo-Christensen and Seesholtz (1967). The wind field distortions are roughly as follows: upwind the region of decelerated flow extends nearly one length scale, sideways the flow is accelerated by about 10 to 20% within a region of one length scale, and downwind a wake is formed in which a velocity reduction is observed. The region of 50% reduction in speed extends to half a length scale. Augstein et al. (1974) compared wind values measured at a ship's mast at 25 m height to nearly undisturbed values on a buoy. The ship's values were systematically too small by about 5 to 10% when the ship was heading into the wind, and by 10 to 20% when the ship was lying broadside to the wind. Ching (1976) compares values from a mast at about 25 m height with values from a boom extending 10 m forward from the bow and 10 m above the sea surface, for different ships. For the ship heading into the wind, the wind velocities agree within a few percent. When the ship is oriented broadside to the wind, the mast wind speeds are lower than the boom values by about 10%. The considerable variability of wind field distortions due to different ships is well demonstrated by Kidwell and Seguin (1978), who analyze wind measurements on the forward mast and boom at the bow for four ships during GATE. They have available no undisturbed wind values. The dependency of mean measured wind velocities on the relative wind direction, which cannot be accidental due to the large number of observations, must be caused by the ship's influence. A careful model study was conducted for RV Flip (Mollo-Christensen, 1968), in order to find special positions within her environment where the flow velocity approaches the undisturbed value. Such positions will exist between regions of decelerated and accelerated flow regimes, at least for a certain range of flow directions and flow velocities. Large platforms of irregular shape cause considerable wind field distortions in the environment. The best positions for mounting anemometers may be found by means of special studies, preferably by model tests in a wind tunnel. Nevertheless, wind measurements on such platforms are not usable for gradient measurements but only for mean wind velocity and, to some extent, for bulk fluxes.

Information on wind direction measurements near large supporting structures are sparse and do not lend themselves to generalization. For example, the results of Kidwell and Sequin (1978) demonstrate the quantity and variability of wind direction distortions for different ships. Careful field and model investigations are generally necessary.

5. EDDY DEFORMATION

The influence of flow distortion on the eddy correlation technique has not been treated in the preceding sections. No experimental investigation concerning this problem is known to us, except the statement that there are such influences (see, for instance, Pond et al., 1971). It should be noted that negligible wind speed error in certain positions near an obstacle is not indicative of negligible error in cross correlation. Also, for example in the simple case of two-dimensional flow around a circular cylinder, the area of minimum wind speed deviation shows the largest deviation in wind direction (compare Fig. 6 and 7).

From the theoretical point of view the turbulent flow around two-dimensional cylindrical bodies has been extensively investigated by Hunt (1973). The modification of turbulent wind components in the environment of the obstacle (well outside the wake) can be calculated only in the limiting situations where the turbulence scale is very much larger or smaller than the length scale of the obstacle. In the atmospheric surface layer the scale of the momentum transporting eddies is, however, of the order of the measuring height or larger. Hence only for slim towers and masts can the theoretical results help. In this case the effect can be considerd as vortex stretching in the deformed mean two-dimensional flow. But the vortices of turbulent motion are three-dimensional, and the theoretical treatment requires homogeneity or isotropy assumptions for the spatial structure of the undisturbed turbulence field. In order to avoid these complications for eddy correlation measurements the distance from the tower should be large enough, say, in a range where the deflections of mean wind speed and direction are negligible.

In the case of less two-dimensional structures, and in close proximity to a structure (compared to its length scale), the distortions of the turbulent field are still more complicated. There is not only the vortex stretching and shrinking due to the mean field but also the blocking effect and the concentration of vortex lines (Hunt, 1973). In a shear flow the generation of vortices upwind of the obstacle has been described by Mollo-Christensen (1979), as mentioned in Section 4. Since the boundary layer flow is already rotational, variation of vorticity by the tilting term may occur. Therefore, if eddy correlation measurements are to be made near a supporting structure, detailed tests both in the wind tunnel and in the field will be necessary.

The same words of caution are applicable to measurements of stress via the so-called dissipation technique. Although in this case it is even less obvious what influences to expect, a first guess is that they will be of the same order as for the eddy correlation technique or even greater.

REFERENCES

AUGSTEIN, E., H.HOEBER, and L. KRÜGERMEYER. 1974. Fehler bei Temperatur-, Feuchte- und Windmessungen auf Schiffen in tropischen Breiten. *"Meteor" Forschungsergebnisse, Reihe B*, 9: 1-10.

BOROVENKO, E.V., O.A. VOLOVOVITSKII, L.M. ZOLOTAREV, and S.A. ISAEVA. 1963. Estimation of the effects of the 300-meter meteorological mast structure on the wind-gauge readings. In: *Investigation of the bottom 300-meter layer of the atmosphere*. Edited by N.L. Byzova, Israel Program Scientific Translation, Jerusalem, 1965: 83-92.

CERMAK, J.E., and J.D. HORN. 1968. Tower shadow effect. *Journal of Geophysical Research,* 15: 1869-1876.

CHING, J.K.S. 1976. Ship's influence on wind measurements determined from BOMEX mast and boom data. *Journal of Applied Meteorology,* 15: 102-106.

DABBERDT, W.F. 1968a. Tower-induced errors in wind profile measurement. *Journal of Applied Meteorology,* 7: 359-366.

DABBERDT, W.F. 1968b. Wind disturbance by a vertical cylinder in the atmospheric surface layer. *Journal of Applied Meteorology,* 7: 367-371.

FAGE, A., and J.E. WARSAD. 1930. Aeronautical Research Council R&M No. 1283. cited by Szechenyi (1975).

GILL, G.C., L.E. OLSSON, J. SELA, and M. SUDA. 1967. Accuracy of wind measurements on towers or stacks. *Bulletin of the American Meteorological Society,* 48: 665-674.

HUNT, J.C.R. 1973. A theory of turbulent flow round two-dimensional bluff bodies. *Journal of Fluid Mechanics,* 61: 625-706.

IZUMI, Y., and M.L. BARAD. 1970. Wind speeds measured by cup anemometers and influenced by tower structure. *Journal of Applied Meteorology,* 9: 851-856.

KIDWELL, K.B., and W.R. SEGUIN. 1978. Comparison of mast and boom wind speed and direction measurements on U.S. GATE B-scale ships. NOAA Technical Report EDS 28. Department of Commerce, Washington, D.C., 44 pp.

LINK, A. 1966. Uber den Einfluss eines Rohrmastes auf Windgeschwindigkeitsmessungen an demselben. Berichte Institut für Meteorologie, Technische Hochschule Darmstadt (unpublished report), 74 pp.

MOLLO-CHRISTENSEN, E. 1968. Wind tunnel test of the superstructure of the R/V FLIP for assessment of wind field distortion. Report 68-2, Fluid Dynamics Laboratory, M.I.T., Cambridge, U.S.A., 29 pp.

MOLLO-CHRISTENSEN, E. 1979. Upwind distortion due to probe supports in boundary layer measurements. *Journal of Applied Meteorology,* in press.

MOLLO-CHRISTENSEN, E., and J.R. SEESHOLTZ. 1967. Wind tunnel measurements of the wind disturbance field of a model of the Buzzards Bay Entrance Light Tower. *Journal of Geophysical Research,* 72: 3549-3556.

MOSES, H., and H.G. DAUBEK. 1961. Errors in wind measurements associated with tower-mounted anemometers. *Bulletin of the American Meteorological Society,* 42: 190-194.

PARKINSON, G.V., and T. JANDALI. 1970. A wake source model for bluff body potential flow. *Journal of Fluid Mechanics,* 40: 577-594.

POND, S., G.T. PHELPS, J.E. PAQUIN, G. McBEAN, and R.W. STEWART. 1971. Measurements of the turbulent fluxes of momentum, moisture, and sensible heat over the ocean. *Journal of Atmospheric Sciences,* 28: 91.

RIDER, N.E. 1960. On the performance of sensitive cup anemometers. *Meteorological Magazine,* 89: 209-215.

ROSHKO, A. 1961. Experiments on the flow past a circular cylinder at very high Reynolds number. *Journal of Fluid Mechanics,* 10: 345-356.

SCHLICHTING, H. 1968. *Boundary Layer Theory.* McGraw-Hill, New York, 6th edition, 747 pp.

SZECHENYI, E. 1975. Supercritical Reynolds number simulation for two-dimensional flow over circular cylinders. *Journal of Fluid Mechanics,* 70: 529-542.

THORNTHWAITE, C.W., W.J. SUPERIOR, and R.T. FIELD. 1965. Disturbance of airflow around Argus Island Tower near Bermuda. *Journal of Geophysical Research,* 70: 6047-6052.

TIETJENS, O. 1960. *Strömungslehre I.* Springer-Verlag, Berlin-Göttingen-Heidelberg, 536 pp.

WIEGHARDT, K. 1969. *Theoretische Strömungslehre.* Teubner-Verlag, Stuttgart, 226 pp.

WUCKNITZ, J. 1977. Disturbance of wind profile measurements by a slim mast. *Boundary-Layer Meteorology,* 11: 155-169.

33

Surface Followers

O.H. Shemdin and G. Tober

1. INTRODUCTION

'Wave follower' is a name ascribed to mechanisms which maintain instruments in the close proximity of undulating water surfaces. The sensors that are used to measure dynamical properties near the air-water surface are normally designed to operate either in air or in water. Only a few sensors have been shown to operate in both fluids sequentially (Wills, 1976). The importance of studying the near surface air layer above undulating surfaces came into focus following the theoretical developments by Phillips (1957) and Miles (1957) on the generation of surface waves by wind. The momentum transfer from air to water was considered by Miles to occur in a thin layer above the interface denoted by the 'critical layer' and defined by $U(z_c) = C$, where $U(z)$ is the wind velocity at elevation z above the mean water level and C is the phase speed of the waves.

In considering the nature of wind boundary layers in laboratory facilities and the associated wind generated waves it becomes evident that critical layer heights fall in the range 0.1 mm to 10 cm. This can be increased to 1.0 m if mechanically generated waves are used. Typical critical layer heights in the laboratory fall in the range 1 to 5 cm with typical wave heights of 5 to 15 cm.

The role and existence of the critical layer was seriously questioned in theoretical developments by Davis (1972) and Townsend (1972) who emphasized the role of turbulence in the wind boundary layers over waves. This, however, did not detract from the need to measure the aerodynamic pressure above surfaces perturbed by water waves in order to assess the magnitude of the momentum transfer, $\vec{\tau}_w$, given in the most simple case by:

$$\underset{\to}{\tau}_w = \overline{P_s \frac{\partial \zeta}{\partial x}} , \qquad (1)$$

where P_s is the surface pressure and ζ is the surface displacement of waves travelling in the x-direction. The latter was considered to be central to the verification of the various theoretical models proposed and referred to above.

Under typical <u>ocean</u> wave generation conditions the phase speed of the dominant waves, C, is equal to about $29u_*$, where u_* is the shear velocity as defined in a logarithmic boundary layer U(z),

$$U(z) = \frac{u_*}{\kappa} \ln \frac{z}{z0} , \qquad (2)$$

where κ is the von Karman constant and $z0$ is the roughness height which is of order 0.5 mm for winds in the range 0 to 20 knots. The critical layer height for a logarithmic boundary layer is computed from $U(z_c) = C$, or

$$z_c = z0 \exp \left[\kappa \frac{C}{u_*}\right] . \qquad (3)$$

Since $C = g/\omega$ for deep water gravity waves, where ω is radian frequency, the thickness of the critical layer in the field varies from 50 m for the dominant waves to a few centimetres for the slower, high-frequency waves at three times the dominant wave frequency. It is the high-frequency waves which gain energy and momentum from the wind field, and it is this fact which has led to the use of wave followers for field studies of wave generation, mostly in short fetch conditions in the wave frequency range where $30 > C/u_* > 5$, and where the water depth is sufficiently small that the instruments can be bottom-mounted (Dobson and Elliott, 1978).

The height of the constant stress layer is another important parameter that must be considered in determining the need for use of wave followers. The constant layer height in a turbulent boundary layer lies in the range 0.1 to 0.2 δ, where δ is the boundary layer height. In laboratory facilities δ is typically 0.5 m so that the constant stress layer height is typically 5 cm. Over the ocean δ is of order 500 m so that the constant stress layer height is of order 50 m.

One of the difficulties in comparing laboratory and field studies is that in the laboratory, particularly for generation studies on mechanically-generated waves, the ratio of constant stress layer thickness to critical layer thickness is typically about 1, while in the field the ratio is more like 10 to 100 in the range of active wave growth. This means that the interactions between the air-turbulence and the wave-induced secondary flows in the air are not modelled well in the wind tunnel, unless special precautions are taken to thicken the boundary layer in the tunnel while working with relatively small critical heights, as was done by Shemdin and Lai (1973).

Consideration of both critical and constant stress layers suggests that under laboratory conditions the sensors must be allowed to follow the wavy surface if they are to provide representative measurements within these layers. Under field conditions Eulerian measurements above the highest waves are adequate for studying the air flow above and the growth rate of the dominant waves within both the constant stress and critical layers. Field wave follower studies are necessary, however, to measure the growth rates of the slower, high frequency waves, where, in fact, most of the energy and momentum transfer occurs. Eulerian measurements are also needed to define the vertical structure of the wave-induced air velocity and pressure fields, and to keep the wave follower sensors 'honest'; the Eulerian instruments do not typically remain within the critical layer of the growing waves, and coherences with the waves fall off as e^{-kz}, where k is wave number and z is measurement height.

Wave followers may also be used to investigate the dynamics of short wind-generated waves and their modulation by long ocean waves. For such measurements high response laser-optical sensors are used which give the time history of the surface normal vector. The sensor's optical-path constraints are such that the mean-water-level variations must not exceed ± 30 cm. This requirement can only be met in the field if the laser-optical sensor unit is mounted on a wave follower. Then the water surface displacement associated with long gravity waves is automatically subtracted from water level changes in the optical path. The slopes of short waves can then be detected as they ride over the long waves (see Chapter 24 by Stewart).

In summary, the use of wave followers is mandatory for experimental investigations of certain air-sea interaction processes. The scientific objectives for their use in the field differ considerably from those in the laboratory. This paper discusses the various environmental, structural, and electro-mechanical factors that affect the design of wave followers intended for use under laboratory and field conditions.

2. GENERAL REQUIREMENTS

An ideal wave follower is one that is able to maintain sensors at a fixed, predetermined elevation above, or below, a wavy water surface without disturbing the surrounding fluids. This is not possible in the real world and a designer seeks to develop wave followers that follow the water surface with minimum error in the frequency range of most interest to the scientific objectives being investigated. The design must minimize interference with the flow processes being investigated. In this context some general requirements are listed below to highlight areas of concern when a wave follower design is being contemplated.

(1) System response - A wave follower must follow waves of interest with errors not exceeding a small fraction of the wave height (since most are servo systems, which measure tracking error anyway, this constraint may be relaxed if the variable being measured has a well understood height variation near the sea surface). The frequency response of a wave follower system is determined by the available driving power, sensor payload, mass of the moving parts, and system friction.

(2) Power requirements - The required power is calculated as the product of the force applied to the moving parts of the system and the velocity of the moving parts. The force is dependent on the mass of the moving parts, including the sensor payload, and their acceleration. Friction, at sliding surfaces or roller bearing points, increases this force. The dead weight of the moving parts, if not neutralized by buoyancy or a spring, will impose an asymmetric force. More power will then be required to move the sensors vertically upward than downward. Required power must be reduced to a minimum to maximize ease of operation (especially in the ocean).

(3) Interference with surrounding flow - This requirement governs the physical size of the wave follower and the size of elements supporting the sensors. To minimize flow obstruction the structural elements must be kept slender. The sensors must be placed at least 10 structural-element diameters away in the upstream flow direction (see Chapter 32 by Wucknitz).

(4) Sensor platform stability - Sensors used in air-sea interaction studies are sensitive to structural vibrations. Sufficient structural integrity must be incorporated to reduce vibration levels to below those tolerated by the sensors. This consideration is important in the design of slender tubes that support instrument packages. The reader is referred to Timoshenko (1964) for detailed insight into structural vibration problems.

(5) Sensor orientation - Precise orientation of sensors is usually required in investigations of pressure, velocity, and turbulence above the air-sea interface. This requirement imposes provision for rigid sensor platforms even in the wave-following mode. Floating buoys usually violate this requirement and re-

quire corrections for buoy motion at the data analysis stage.

(6) Environmental forces - The wave follower must be constructed to withstand environmental forces in the ocean under extreme conditions. Such conditions are normally beyond the scope of experimentation but are imposed because of operational constraints.

(7) Operational considerations - Ease of installation, testing, and data collection are important aspects of wave follower design, especially in field operations. Rapidly changing weather imposes the necessity of planning for swift discontinuation of testing and often removal of the support vessel from the test site.

(8) Water depth - The above requirements also determine the water depth in which a wave follower can be expected to operate satisfactorily. In shallow water, rigid structures can be supported on the bottom to satisfy most of the requirements listed above. In deep water, free-floating or moored buoys can be used provided that buoy motion effects can be tolerated. Use can also be made of deep ocean platforms such as FLIP (see Chapter 34 by Blendermann) from which wave followers can be deployed. The influence of the platform on surrounding air or water motions is of primary concern in the latter case.

The requirements for laboratory wave followers differ considerably from those in the field. The physical variables and scale of apparatus for the latter are one order of magnitude greater than in the laboratory. This in itself imposes vastly different design requirements.

3. LABORATORY WAVE FOLLOWERS

Laboratory wave followers are designed with small moving parts to follow waves in the frequency ranges 0.5 to 10 Hz and wave heights of order 10 cm. The wave follower must track waves with errors not to exceed 10% of the wave height. Smaller errors (of order 3%) are achievable with careful designs. A design that can accomplish these objectives is made of a freely moving rod (or tube) that can travel vertically inside linear bearings that are supported inside a fixed hollow shaft which in turn is attached to the structural frame of the wind and wave facility. The drive is provided by a high-response electric motor which is linked to the moving rod via a rack and gear assembly or via a cable and pulley assembly. The motor control is provided by a wave sensing device. The wave height sensor is mounted on the lower end of the vertically moving rod, then immersed in water to its midpoint and the motor voltage regulated to indicate no motion at that immersion level. A simple

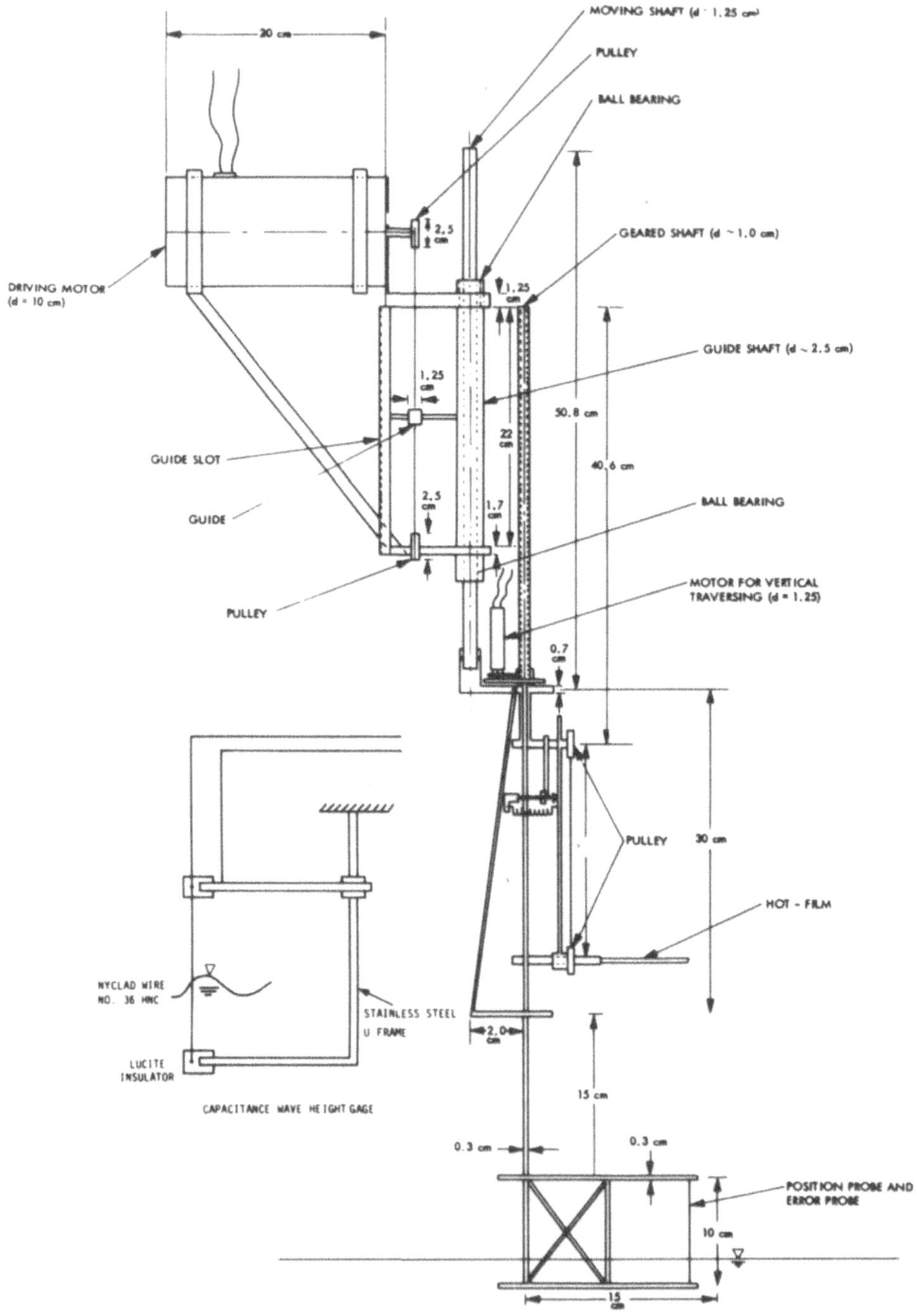

Fig. 1a Side view of a laboratory wave follower.

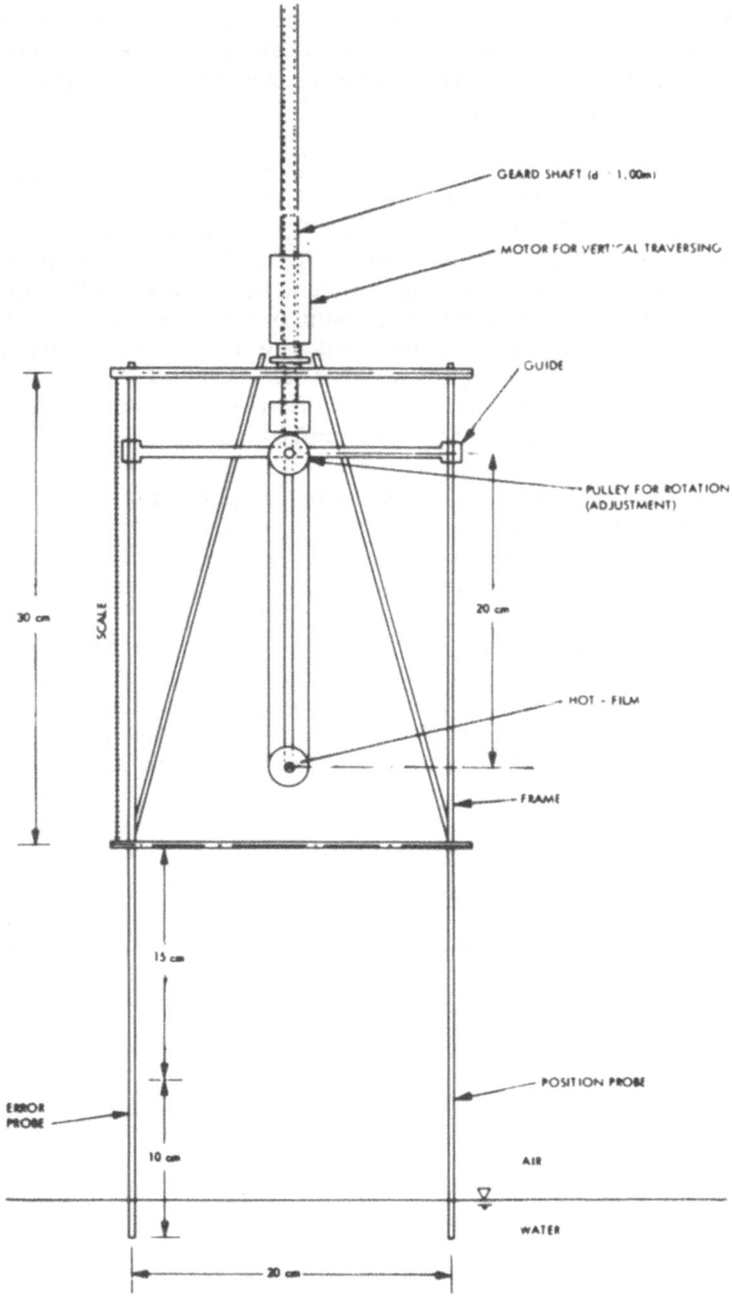

Fig. 1b Front view of instrument traverse mechanism mounted on a laboratory wave follower.

feedback control loop maintains the wave height sensor at a constant immersion level even when the water surface is displaced vertically by wave action. The difference between the instantaneous water surface elevation and the position of the vertically moving rod is a voltage error which drives the motor in the direction needed to achieve zero error.

The above basic design was first developed by Shemdin and Hsu (1967) and later in modified forms by Chang et al. (1970), Shemdin and Lai (1973), and Baldy et al. (1978). The mechanical design of Shemdin and Lai (Fig. 1a and 1b) is typical of laboratory type mechanisms. Considerable thought and effort goes into the design of frames and mechanisms needed to support the sensors. The latter is beyond the scope of this paper and the reader is referred to the above references for insight on interactions between sensor requirements and platform motion. The references also contain circuit diagrams for the control electronics.

Control may be achieved with one wave height gauge simply by comparing the signal from the gauge with a fixed reference voltage and amplifying the difference to drive the gauge vertically and reduce the difference (a simple servo system). Improvement in response can be achieved by inserting lead-lag electronics or by using two wave gauges to provide the control system with a slope (D in Fig. 2), which is used to anticipate surface elevation events. The schematic diagram for a two-gauge control system is shown in Figure 1, after Shemdin and Lai (1973). The modelling of such control systems has been the subject of considerable work and documentation

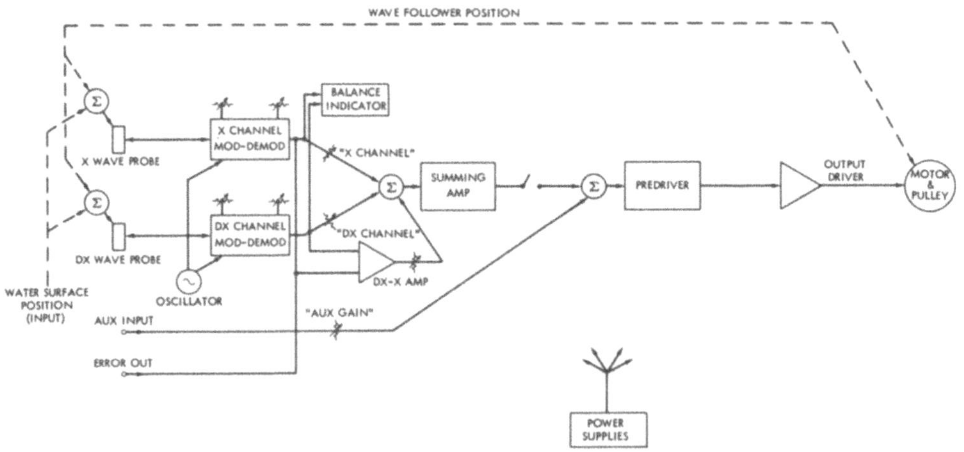

Fig. 2 Schematic of a laboratory wave follower control with two wave sensors.

by electrical engineers. The reader is referred to Sinha et al. (1974) for a more detailed discussion of this subject.

The performance of a wave follower system is determined from the input-output relationship in the standard manner (Bendat and Piersol, 1971). The input and output spectra are defined as $G_{xx}(f)$ and $G_{yy}(f)$, respectively, where f is the frequency in hertz. The cross spectrum, $G_{xy}(f)$, is defined in terms of its real and imaginary components:

$$G_{xy}(f) = C_{xy}(f) + i\,Q_{xy}(f). \tag{4}$$

The transfer function, T(f), is defined as:

$$T(f) = \frac{G_{xy}(f)}{G_{xx}(f)} = |T(f)|\exp\{i\theta(f)\}. \tag{5}$$

The system coherence function $\gamma(f)$ is defined as:

$$\gamma^2(f) = \frac{|G_{xy}(f)|^2}{G_{xy}(f)\;G_{yy}(f)}\;. \tag{6}$$

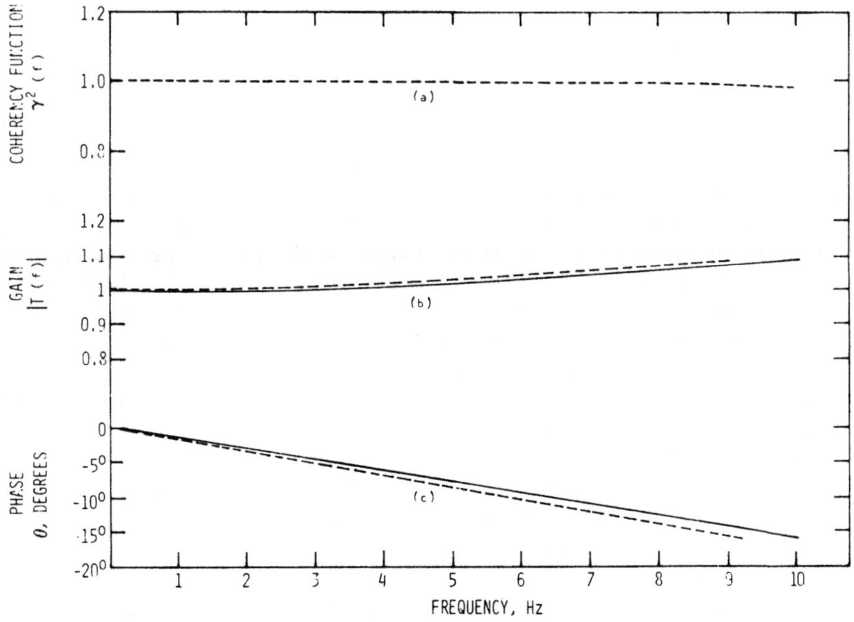

Fig. 3 IMST laboratory wave follower frequency response performance results (after Baldy et al., 1978): (a) coherency, (b) gain, and (c) phase lag.

Linear systems have ideal coherencies equal to unity (see Bendat and Piersol), and the system linearity is often verified by this technique. The transfer function, T(f), provides a complete description for a linear system.

The laboratory wave followers described above are all linear systems. A careful performance test for such a system was given by Baldy et al. (1978). Their tracking performance results are displayed in Figure 3. The unity in coherence verifies the system linearity. The gain, $|T(f)|$, and phase lag, $\Theta(\omega)$, are also shown. Subject to lag corrections the system is found to follow waves to within 3% error bounds up to 9 Hz.

4. OCEAN WAVE FOLLOWERS

Ocean wave followers are designed to operate in water depths in excess of several metres and under the following desirable environmental conditions:

Wind speed = 10 m s^{-1}

Significant wave height = 1.0 to 2.0 m

Maximum wave height = 2.0 to 4.0 m

Tidal range = 1.0 to 2.0 m

Tidal current = 0.2 to 0.5 m s^{-1}

The requirements for mass in the moving part, and the response prohibit the use of electric motors with direct mechanical links. Alternatively, response can be achieved by employing hydraulic systems in which energy is stored under high pressure and controlled by servo-mechanisms. Peep and Flower (1969) at the Chesapeake Bay Institute successfully implemented such a design concept for a field wave follower. Harvey and Dobson (1976) produced an improved version of the same design to investigate the near-surface aerodynamic pressure above waves. Both systems used a piston and connecting rod assembly to provide linear motion for tracking and water surface displacement. Control was achieved through a feedback system that employed a capacitance wave-height gauge for a sensor. The stroke was limited to 90 cm in order to limit the required oil to a manageable volume.

The Harvey-Dobson wave follower was operated successfully in the JONSWAP and Bight of Abaco experiments. Its primary limitation lies in its 90 cm stroke. The likelihood of ocean wave heights exceeding 90 cm is high, and consequently this wave follower is more useful for low wind speeds and in sheltered waters (see Dobson and Elliott, 1978).

A 3.8 m stroke wave follower was developed at the University of Florida by the authors. The system was built to operate in 10 m water depth when placed on the bottom and in deeper water when placed on a pedestal. The maximum stroke was governed by an anticipated 1.8 m tidal range and 2.0 m maximum wave height. It was required that a flat frequency response be achieved from dc to 1 Hz

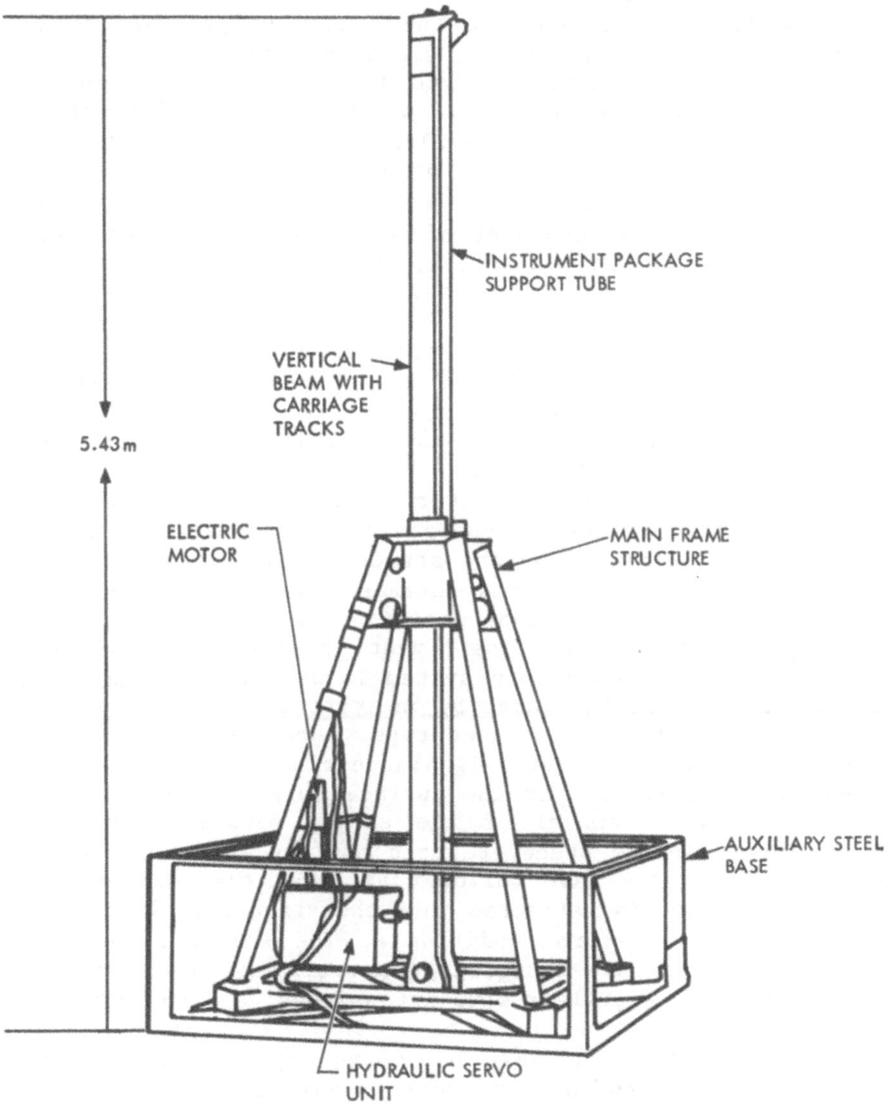

Fig. 4 University of Florida wave follower underwater structure.

when tracking waves having spectral characteristics of the JONSWAP type (Hasselmann et al., 1973) with a peak at 5.0 s period and a significant wave height of 1.2 m.

The rod and piston design concept used by Peep and Flower requires a large volume of oil. Excessive oil is difficult to handle in the ocean and poses a serious surface contamination threat if leakage occurs; an alternate design was therefore considered. Unlike the Peep and Flower device, the new design incorporated an intermediate stage before converting hydraulic pressure into linear motion. A servo-valve was used to convert the hydraulic pressure into controlled rotary motion through the use of a hydraulic pump. The rotary motion was then converted into linear motion through a chain and sprocket mechanical assembly. A carriage linked to the chain was allowed to travel vertically along linear tracks. A neutrally buoyant surface-piercing tube was attached to the carriage to carry the instrument payload in a wave following mode. In such a design the required oil is independent of stroke and can be reduced to a manageable volume. The mechanical, hydraulic and electronic control subsystems are described below.

The underwater support structure of the University of Florida's wave follower is shown in Figure 4. The structure supports and powers a vertically restrained, neutrally buoyant tube assembly that in turn supports instruments near the ocean surface. Hydraulic power source, servo amplifiers, control, and follower data electronics are mounted at the structure's base. Electrical power and a serial multiplex of digital command and status signals are routed over two cables leading from the structure to a nearby control and data collection point (a moored surface vessel). The system's operational mode is a closed-loop position feedback system that maintains alignment between the instrument package and local vertical water motion. The closed loop system is depicted schematically in Figure 5. Loop error is sensed by an instrument-mounted error wave gauge which is of the capacitance type, shown in Figure 1. Error, together with instrument data signals pertinent to a particular experiment, are telemetered to the surface ship via an RF link, thus allowing for quick removal of the sensor package from the wave follower tube assembly whenever rapid buildup in sea state occurs. In severe weather the wave follower vertical tube assembly, minus the sensor package, is retracted into the structural frame which is designed to survive storm conditions. The sensors, however, are fragile, since they are designed to measure extremely small pressure fluctuations, capillary wave intensities, and turbulence.

In favourable weather the ship is moored approximately 500 m from the wave follower structure in order to minimize flow disturbances (including backscattered waves) caused by the ship. The terminals at the mooring site are connected with those on board the vessel. All sensor signals including the error signals are transmitted only

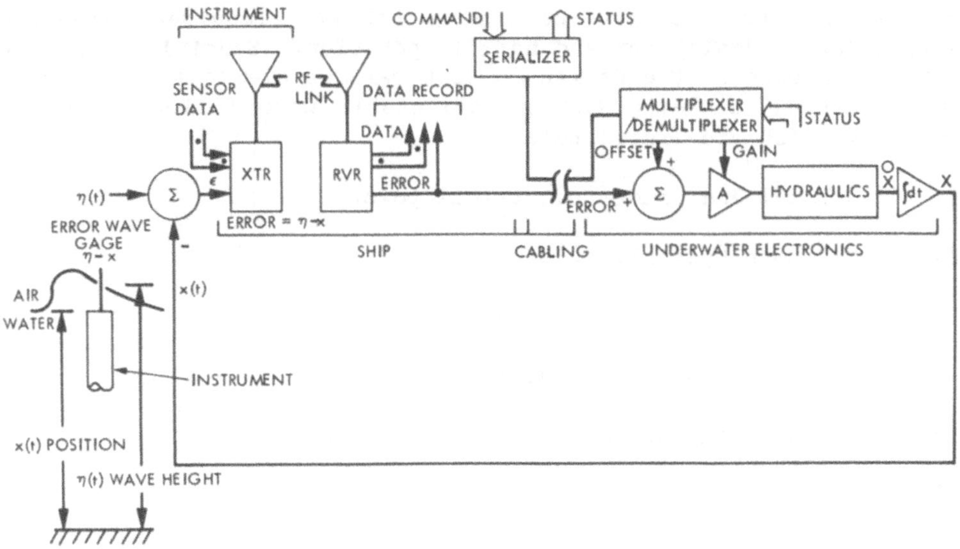

Fig. 5 Schematic representation of University of Florida ocean wave follower closed loop data paths.

through the RF link. Error is separated from data at the surface ship and routed to the underwater structure electronics where the feedback loop is closed via the hydraulic servo system. Operational mode, and loop gain and offset values are remotely controllable from the surface vessel.

Fail-safe features of the site electronics include:

(1) Hydraulic pump motor quick trip breakers and branch circuit fusing.

(2) Pump motor shut-down for excessive oil temperature.

(3) Appropriate limit switch interrupt of open loop slew commands.

(4) Limit switch interrupt of closed loop via:

  (a) Slew-up fast and stop on upper limit-switch for a lower limit-switch contact.
  (b) Stop on upper limit-switch. for an upper limit-switch contact.

Closed loop exit initiated from the control point can either simulate the lower limit-switch contact or exit to any open loop function.

The complex nature of an ocean wave follower requires extensive testing during deployment and normal operation. Special provisions must be included in the design to test various elements of the system. This is achieved through selectable wave follower command functions which are listed below:

(1) Local and remote electrical power

(2) Remote hydraulic power

(3) Open loop control of vertical velocity

(4) Closed loop mode control

(5) Monitoring of several selected functions

(6) Loop error source selection

(7) Digital link testing

A command for the closed loop mode (with the instrument at least partially out of the water) results in a site-initiated 'slew-down-slow' command and loop closure on the first zero crossing from the error gauge. Loop error offset adds a variable reference to the error signal and hence adjusts the zero error (wave following point) on the wave gauge. The 'monitor select' capability allows monitoring of one of four site status flags. Because the loop error signal is routed through the control point, an alternate (test) error signal may be injected in place of the wave gauge.

The performance of the wave follower was determined by the mass of the moving components, available power in the hydraulic drive, and friction in the bearing surfaces. The response of the servo-valve was not limiting. Preliminary estimates of these factors were used in the design stage to define the power requirement. The reader is referred to Merrit (1967) for an extensive treatment of hydraulic control systems and their modelling.

The wave follower system has been tested at various field sites for performance evaluation. An open-loop forward velocity transfer function was derived from recordings made of vertical velocity output versus a band-limited Gaussian excitation input. Coherency, $\gamma^2(f)$, as a measure of system linearity, is shown in Figure 6a. The velocity transfer function, as derived from auto- and cross-spectral analysis according to Equations 4 and 5, is shown in Figure 6b. From Figure 6b it was concluded that flat response was obtainable to 1.0 Hz; this is consistent with the imposed design requirements. From Figure 6c the phase lag at 1.0 Hz is 15° and less than 5° at 0.5 Hz. The shape of the coherency function in Figure 6a is attributed to friction and other nonlinearities in the wave follower system. At low frequencies the initiation friction at the point of velocity reversal in the wave following cycle generates considerable nonlinearities in the shape of the position signals. Such nonlinearities show as reduced energy at low frequencies and increased energy at their higher harmonics.

An actual performance record in closed loop (wave following) mode is shown in Figure 7. The tracking error is ±20 cm and is somewhat higher than was predicted from the open loop tests; this is attributed to the increase in instrument payload weight (a factor of 2)

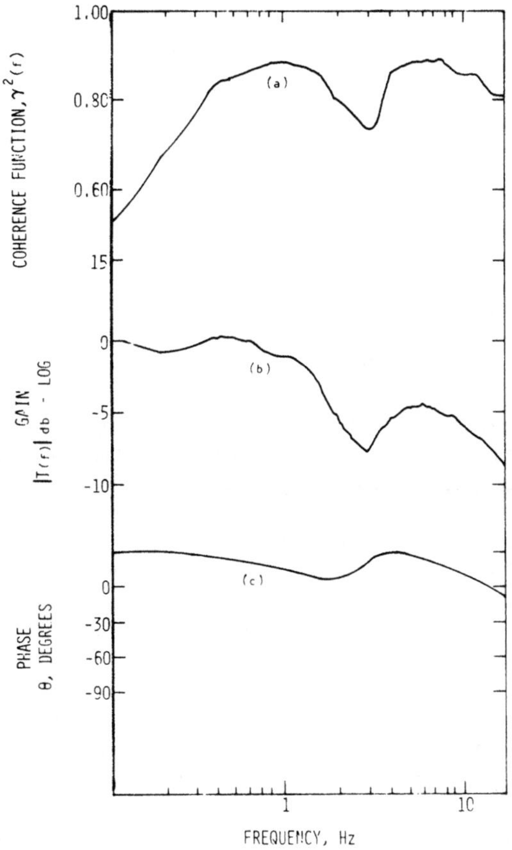

Fig. 6 University of Florida ocean wave follower frequency response performance results: (a) coherency, (b) gain, and (c) phase lag.

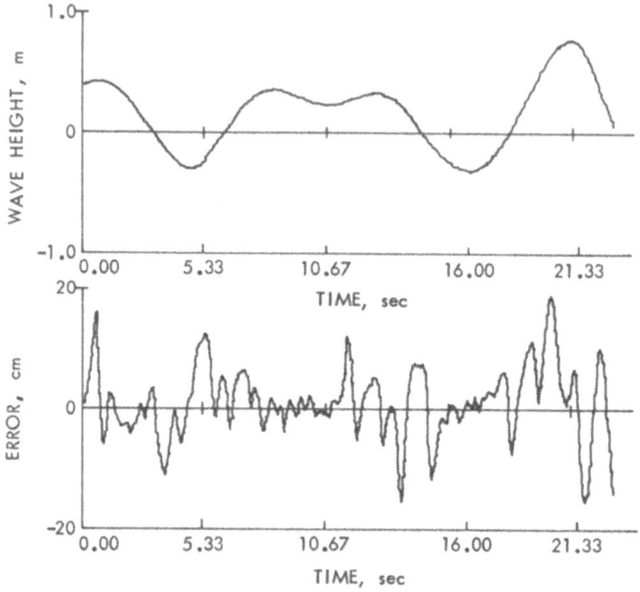

Fig. 7 University of Florida wave follower closed loop operation record.

used in the closed loop runs.

5. SUMMARY AND CONCLUSIONS

The advent of wave followers was prompted by the need for experimental verification of the mechanisms of atmospheric transfer from air to water. In the laboratory the use of wave followers is mandatory since the critical layer and constant stress layer are both small compared to the heights of wind-generated waves. Laboratory followers are relatively simple in design because they can be supported above the water by the frames of the wind and wave facilities. The small moving parts permit attainment of high-response linear performance (up to 10 Hz).

Field wave followers are supported by structures placed underwater to minimize interference with the air flow above. The large structural frame required combined with salt-water operation results in a problem of greater dimensions. Unpredictable and changing weather combined with survival requirements in severe storms make field operations very difficult. The rigid sensor platforms, required to provide accurate sensor orientations, presently restrict use of wave followers to shallow waters. The technology for shallow-water wave followers is developed. Deployment from research towers would simplify logistics problems during extreme weather conditions.

The need exists for obtaining near surface data in deep water. A floating buoy was used by Longuet-Higgins, Cartwright and Smith (1963) to measure the near-surface pressure. Adequate resolution of pressure phase shifts with respect to surface displacement could not be achieved, presumably because of buoy motion and the aerodynamic influence of buoy shape on near-surface pressure. The technology for making sufficiently accurate near-surface measurements in deep water is not yet in hand.

In the future, wave followers will probably continue to be used in both field and laboratory, to further understanding of the wave generation process. In the field, the most promising use to which a wave follower might be put is to measure wave-induced momentum fluxes from air pressure-wave correlations at short fetches and high wind speeds, in the range $1 < C/u_* < 30$. Since the devices measure their own tracking error, the waves would not have to be followed perfectly; rather the wave follower would serve as a platform to keep the sensors as close as possible to the water surface. An understanding of the hydrodynamics of modulation of short waves by long waves is another critical problem that will require future use of wave followers to measure (a) the modulation of short waves by long waves directly, and (b) to measure the modulating near-surface shear stress which acts as a forcing function on the short waves. Such new insights are required to understand the atmospheric stress balance at the air-sea interface and to understand how long waves are imaged by the newly-advanced radar techniques.

REFERENCES

BALDY, S., A. RAMAMONJIARISOA AND M. COANTIC. 1978. Description and characteristics of a wave follower system for energy exchange studies in the vicinity of an air-water interface. *Review of Scientific Instruments*, 49: 1077-1082.

BENDAT, J.S. and A.C. PIERSOL. 1971. *Random Data; Analysis and Measurement Procedures*. John Wiley and Sons, Interscience, 407 pp.

BONNETT, J.W. 1975. Analysis of the Lindquist wave follower. Report No. AD-A019282, Naval Postgraduate School, Monterey, California: 100 pp.

CHANG, P.C., A. GOROVE, R.L. ATCHLEY and E.J. PLATE. 1970. A self-adjusting probe positioner for measuring flow fields in the vicinity of wind generated water surface waves. *Review of Scientific Instruments*, 41: 1544-1549.

DAVIS, R.E. 1972. On prediction of the turbulent flow over a wavy boundary. *Journal of Fluid Mechanics,* 52: 287-306.

DOBSON, F.W. and J.A. ELLIOTT. 1978. Wave-pressure correlation measurements over growing sea waves with a wave follower and fixed-height pressure sensors. In *Turbulent Fluxes Through the Sea Surface, Wave Dynamics and Prediction,* edited by A. Favre and K. Hasselmann, Plenum, New York: 421-432.

ELLIOTT, J.A. 1972. Microscale pressure fluctuations near waves being generated by the wind. *Journal of Fluid Mechanics,* 54: 427-448.

HARVEY, D.R. and F.W. DOBSON. 1976. The Bedford Institute Wave Follower. Report Series BI-R-76-13, Bedford Institute of Oceanography, Dartmouth, N.S., Canada, 67 pp.

HASSELMANN, K., T.P. BARNETT, E. BOUWS, H. CARLSON, D.E. HASSELMANN, P. KRUSEMAN, A. MEERBURG, P. MÜLLER, D.J. OLBERS, K. RICHTER, W. SELL and H. WALDEN. 1973. Measurements of wind-wave growth and swell decay during the Joint North Sea WAve Project (JONSWAP). *Deutsches Hydrographisches Zeitschrift, Suppl. A.,* 8: 95 pp.

LONGUET-HIGGINS, M.S., D.E. CARTWRIGHT and N.D. SMITH. 1963. Observations of the directional spectrum of sea waves using the motions of a floating buoy. In *Ocean Wave Spectra,* Prentice Hall, Englewood Cliffs, N.J.: 111-132.

MERRITT, H.E. 1967. *Hydraulic Control Systems.* John Wiley and Sons, 358 pp.

MILES, J.W. 1957. On the generation of surface waves by shear flows. *Journal of Fluid Mechanics,* 3: 185-204.

PEEP, M. and R.J. FLOWER. 1969. The Chesapeake Bay Institute wave follower. Technical Report No. 58, Chesapeake Bay Institute, The Johns Hopkins University, 76 pp.

PHILLIPS, O.M. 1957. On the generation of waves by turbulent wind. *Journal of Fluid Mechanics,* 2: 417-445.

SHEMDIN, O.H. and E.Y. HSU. 1967. Direct measurement of aerodynamic pressure above a simple progressive gravity wave. *Journal of Fluid Mechanics,* 30: 403-417.

SHEMDIN, O.H. and R.J. LAI. 1973. Investigation of the velocity field over waves using a wave follower. Technical Report No. 18, Coastal and Oceanographic Engineering Laboratory, University of Florida, Gainesville, Florida, U.S.A.

SINHA, N.K., C.D. DICENZO and B. SZABADOS. 1974. Modeling of DC motors for control applications. *IEEE Transactions on Industrial Electronics and Control Instrumentation,* IECI-21: 84-88.

TIMOSHENKO, S. 1964. *Vibration Problems in Engineering.* D. Van Nostrand Co., Inc., New York, 468 pp.

TOWNSEND, A.A. 1972. Flow in a deep turbulent boundary-layer over a surface distorted by water waves. *Journal of Fluid Mechanics,* 55: 719-735.

WILLS, J.A.B. 1976. A submerging hot wire for flow measurements over waves. *DISA Information,* 20: 31-34.

34

Buoys

W. Blendermann

1. INTRODUCTION

In this chapter emphasis is laid on design aspects of buoy and buoy-cable systems rather than on the practical aspects of system employment. A survey of the pertinent problems is given, and references are made for further studies. Details which are not treated in the literature are presented more thoroughly. Mooring dynamics are covered by Berteaux, Chapter 35; drifting buoys (Lagrangian drifters) by Vachon, Chapter 11.

Not all possible buoy motions can be treated in a general way. For instance, little can be said on the rotational motion of a buoy about its vertical axis (yawing) since this motion depends on the combined effects of forces from the tether line, the wind, the waves, and the currents. A practical expedient is to avoid yawing motions by making the buoy rotationally symmetric with as few appendages as possible above and below the water surface and, especially, no asymmetrical ones.

Another aspect of buoy motion that is not dealt with in this chapter is its influence on scientific measurements taken using the buoy as a platform; this depends very much on the specific scientific problem. Of course, it is the primary objective of every buoy design to keep the effects of buoy motion sufficiently small. But, since oceanographers are more and more interested in small-scale processes, sophisticated solutions are necessary in order to avoid such effects; in practice it is often not possible to keep them negligibly small. In such cases it may be necessary to measure buoy motion with the help of accelerometers, gyro horizon, compass azimuth, etc.

2. GENERAL

2.1 Buoy Classification

Buoys play an important role in the study of oceanographic phenomena. They are used for short-range measurements or are required to operate for perhaps a year or more, often under severe weather conditions. They are equipped with instruments to gather, convert, and transmit oceanographic and meteorological data. There is no activity at sea which does not profit somehow from data gained using oceanographic buoys. Though measurements by airplane and satellite certainly will replace some measurements hitherto performed using buoys, buoys will increasingly be employed, especially networks of long-range telemetering buoys which are needed for long-term series measurements and the synoptic coverage of certain parts of the oceans.

Historically, the term 'buoy' refers to an anchored float that aids navigation by showing navigable courses, reefs, etc. In ocean engineering it is sometimes extended to manned platforms for oceanographic research, large storage and tanker loading units, and even subsurface floats, but in this chapter 'buoy' is restricted to an unmanned surface-penetrating float. Dynamic response to waves is similar for buoys and manned platforms, except for the influence of scale (there is one order of magnitude difference in size). Buoys, however, have already reached remarkable dimensions: the largest realized buoy is the so-called SPAR (Seagoing Platform for Acoustic Research, Spiess, 1968). This 4.8 m diameter cylindrical buoy is 106.5 m long and reaches 90 m below the ocean surface. In contrast, one of the smallest buoys used in oceanography is the so-called 'Waverider,' a spherical buoy of only 0.7 m diameter (Draper, 1975).

Among the great variety of existing buoys five generic forms can be distinguished (Fig. 1) which, at the same time, imply certain features with respect to their behavior in waves:

- Disc (any circular flat float)
- Torus (any annular flat float)
- Sphere
- Boat (any horizontally elongated float)
- Spar (any vertically elongated float)

All buoys can be classified into these basic forms. For example, a rectangular barge can be regarded as a buoy of the boat type. Buoys may be combinations of two or more basic float forms of the same or different kind to profit from the dynamic behavior of the specific basic form or to get properties of the compound float which the single component does not possess. If, for instance, two slender boat hulls are connected (catamaran), high lateral stability (strong righting moment) can be achieved, while preserving the low resistance of the slender hulls in a stream. All semisubmersibles follow similar principles.

2.2 Dynamic Response of Buoys

Buoys of forms (a) to (d) of Figure 1 are usually called surface-following and those of elongated shape, form (e), surface-decoupled. Surface-following means the buoy undergoes strong heaving motions, generally associated with strong pitching motions, whereas surface-decoupled means reduced heaving and pitching motions. The amplitude of buoy oscillations depends on the ratio of the frequency of the exciting wave to the natural frequency of the float. If the ratio is much less than one, any float will follow the waves (elevation, slope). However, if the ratio is much greater than one, the float is hardly affected by the waves. In between, strong responses can occur depending on buoy properties.

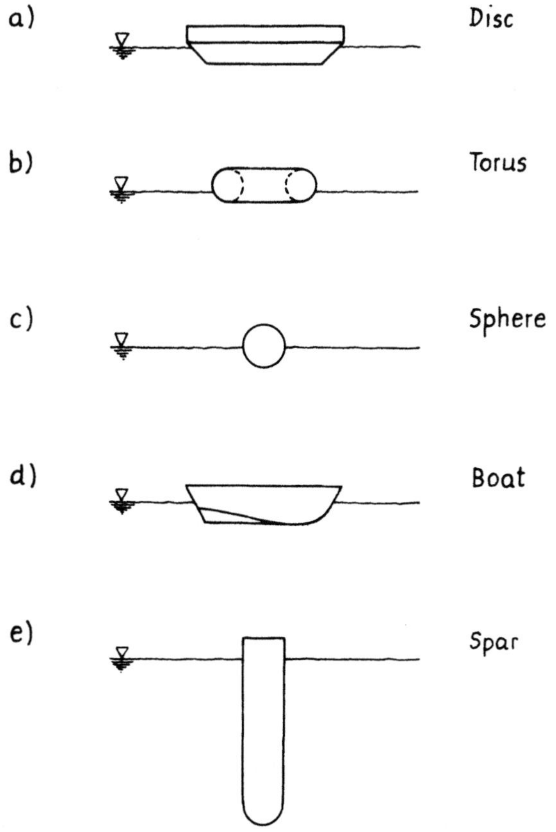

Fig. 1 Generic buoy forms.

The ratio of the amplitude of buoy motion to the amplitude (or, alternatively, slope in the case of pitching and rolling) of the exciting wave is the frequency response or transfer function. When nothing else is assigned, the translational oscillations may conveniently be referred to the buoy's centre of gravity. As an example of a buoy's response, Figure 2 shows experimental transfer functions of pitching (pitch amplitude over wave slope), heaving, and surging motions for an unrestrained disc buoy used for meteorological data acquisition. Also plotted is the transfer function of combined heave, pitch, and surge oscillations; that is, the transfer function of ship motions in the direction of a towed rope at the ship's stern. This computed prediction is for the German research vessel Meteor (length 72.8 m, displacement 29,500 kN).

The disc buoy shows the typical moderate heave amplification of a surface follower, whereas its pitch amplification is more pronounced, although already effectively diminished by the shaft. The rate of motion amplification for the ship is rather small. But this is

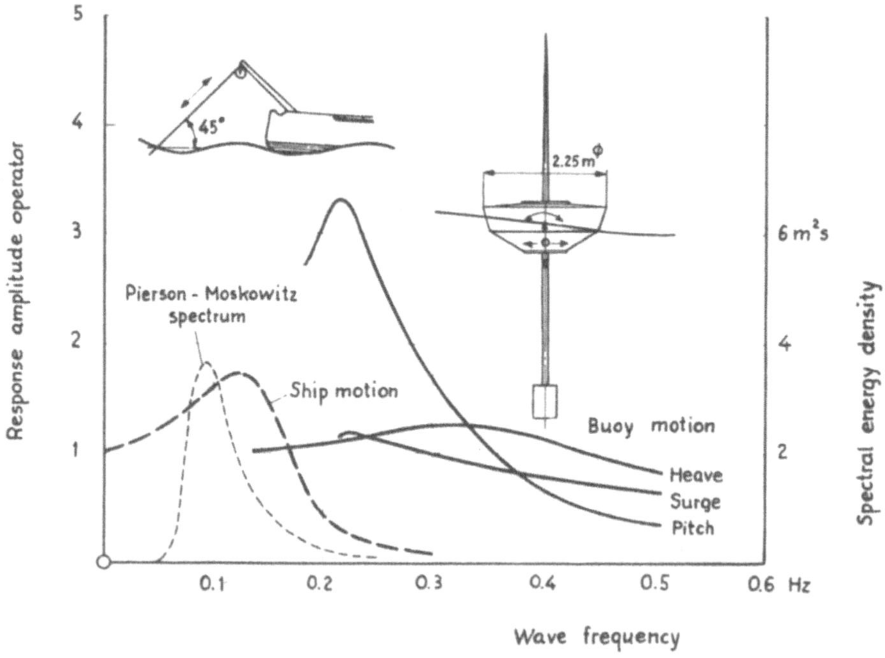

Fig. 2 Transfer functions of oscillations for an unrestrained buoy and a transfer function of ship oscillations in the direction of a towed cable. The ship moves with 2 knots in head-on seas. Also plotted is the energy spectrum of a fully developed sea.

of less importance, for the point is whether a resonance of the specific float motion (or other physical quantities) occurs in the zone of wave frequencies where the energy is high. As Figure 2 shows, this proves true for the ship. As a result, strong dynamic loads may stress the towing arrangement. On the contrary, resonance of the disc's pitching motion, which is generally rather troublesome, occurs beyond the zone of high wave energy.

Large spar buoys, for which heave transfer functions with high and sharp peaks at resonance are typical, avoid the frequency zones with expected high ocean wave energy. Their resonant frequencies are chosen lower than the frequencies of significant wave energy (e.g. the natural heave frequency of the SPAR is 0.05 Hz, Spiess, 1968), so that they virtually do not respond in the range of wave excitation frequencies. This gives rise to their property of being stable platforms in rough seas. On the other hand, the natural frequencies of vertical cylinders with low ratio of length to diameter, which also exhibit marked amplification at resonance, are well beyond the range of the predominant wave energy; that is, they do follow the waves, though they may be driven to resonance by wind-generated chop.

In most cases buoys are moored or horizontally tethered. This can alter their response to waves considerably depending on the mooring arrangement (i.e. position of attachment point, relation of the oscillating mooring line forces to the hydrodynamic forces on the buoy hull, etc.). A taut mooring, for instance, obstructs the buoy's heaving motion thus increasing pitching if the buoy is moored eccentrically; for example, a boat type buoy moored at its bow. Similar effects can be observed at horizontally tethered buoys, where, although heaving is hardly affected, the pitching motion may be increased due to surging.

2.3 Evaluation of a Buoy's Dynamic Performance

A first approximation of a buoy's response to waves can be found from the natural frequencies of free oscillations. The natural frequency of heaving motion (without damping) is given by the square root of the ratio of the restoring force to the effective mass, in analogy to the single degree of freedom mass-spring system,

$$f_z = \frac{1}{2\pi}\sqrt{\frac{\rho g A_w}{m + m'}} \qquad (1)$$

where ρ is the water density, g the acceleration of gravity, and A_w the buoy's waterplane area. The effective buoy mass, $m + m'$, is given by the buoy's own mass, m, increased by the effects of the surrounding water on the buoy, the so-called added mass or hydrodynamic mass, m'. If a body immersed in a fluid is accelerated, the fluid exerts, besides drag, a force on it which depends essentially on acceleration; m' is the factor of proportionality. The added mass of a sphere, for instance, is half the mass of its displaced water volume. A compilation of hydrodynamic masses for various body forms is given by Patton (1965).

The natural pitch frequency is expressed by

$$f_\psi = \frac{1}{2\pi}\sqrt{\frac{\overline{MG}mg}{\Theta + \Theta'}}, \tag{2}$$

where $\overline{MG}$ is the metacentric height, Θ the moment of inertia, and Θ' the added moment of inertia. $\overline{MG}$ (see Fig. 3b) is the distance between the centre of gravity, G, and the intersection , M, of the buoyancy vector with the buoy axis for small inclinations. The metacentre M is given by

$$\overline{MF_0} = \frac{I}{\nabla} , \tag{3}$$

where I is the moment of inertia of the waterplane area about its axis, ∇ the displaced volume of the freely floating buoy, and F_0 the centre of buoyancy. The metacentre must, of course, lie above G for stable floatation. A discussion of float stability is found, for example, in Comstock (1970).

For floats at the water surface the hydrodynamic masses and moments are complicated by their dependence on the frequency of oscillations because of free-surface effects (wave radiation). They cor-

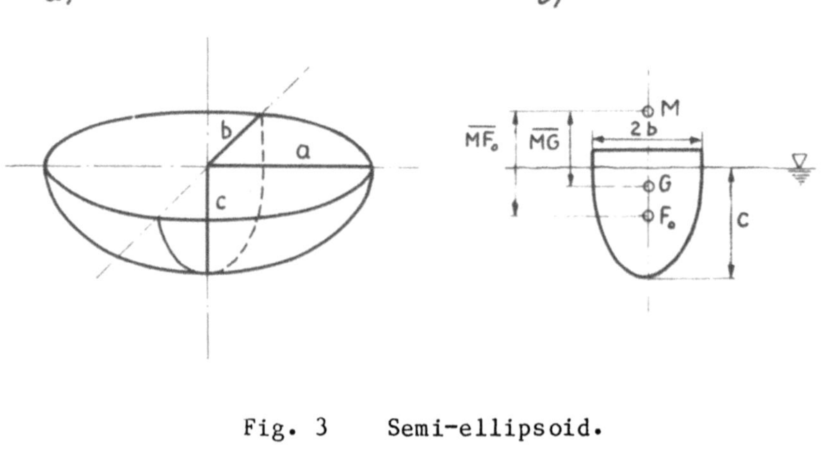

Fig. 3 Semi-ellipsoid.

respond with the results for deeply submerged bodies for high frequencies only (i.e. vibrations, strictly speaking $f = \infty$), which, in this connection, are beyond the range of interest. Nevertheless, by neglecting free-surface effects, Equations 1 and 2 are approximations of a buoy's natural frequencies. But even these can cause difficulties, for estimations of the added mass and, especially, the added moment of inertia of an arbitrary buoy are generally rather uncertain. One method is to convert the buoy into a semi-ellipsoid with the same waterplane area and displacement (Fig. 3a), for which the added masses and moments of inertia are well known (Kotchin, 1954). This, of course, will lead to reasonable values only if the buoy does not differ too much from its equivalent semi-ellipsoid. The horizontal axes of the semi-ellipsoid, a and b, are chosen such that it has, in addition to the waterplane area, A_w, the buoy's aspect ratio of length to breadth, L/B:

$$\frac{a}{b} = \frac{L}{B} . \tag{4}$$

Thus,

$$a = \left(\frac{1}{\pi} A_w \frac{L}{B}\right)^{\frac{1}{2}} \tag{5}$$

$$b = \left(\frac{1}{\pi} A_w \frac{B}{L}\right)^{\frac{1}{2}} \tag{6}$$

The vertical axis follows from

$$c = \frac{3}{2} \frac{\nabla}{A_w} . \tag{7}$$

The equivalent semi-ellipsoid has still another useful application. It can be used to classify the wide variety of realized buoys in a simple manner. If the buoy's waterplane area, whatever shape it has, is transformed into a circle, one arrives at a semispheroid with radius $R = \sqrt{A_w/\pi}$ and vertical axis c, from Equation 7. Then the ratio R/c represents the degree of virtual vertical slenderness of the buoy. It characterizes, to some extent, the buoy type, taking into account integral geometric features of the buoy only. The ratio becomes 1 for a hemisphere, and is greater or smaller than 1 for increasing or decreasing waterplane area with decreasing or increasing displacement, respectively.

A total number of 22 realized buoys have been mapped in Figure 4, a plot of the equivalent cube length, $\nabla^{1/3}$, representing the buoy size, against R/c. Two large manned research platforms have been included (FLIP, region II, 13, and Bouée Laboratoire, region II, 19

[Spiess, 1968]). The buoys fall clearly into two categories, surface-following buoys (I), and surface-decoupled buoys (II). For comparison, the same formal procedure has been applied to seagoing ships (region III, 1 is the research vessel Meteor), semisubmersibles (region IV), and storing and loading units for oil and gas fields (region V). All occupy distinct regions of the diagram.

The modes of motion of an elongated float (e.g. a ship) are the three translational components, surging along the longitudinal axes, swaying along the lateral axis, and heaving along the vertical axis, and the three rotational components about these axes, rolling, pitching, and yawing, respectively. For buoys of rotational symmetry, only three modes of motion can be discerned: heaving, and, say, surging and pitching.

Describing the motion of a buoy in a particular sea state is a complex process. Not only do the hydrodynamic effects of the surrounding water on the buoy depend on buoy geometry, mode of motion, and frequency of oscillations, but also a fairly strong coupling exists between different modes. For ordinary buoy hull forms there is

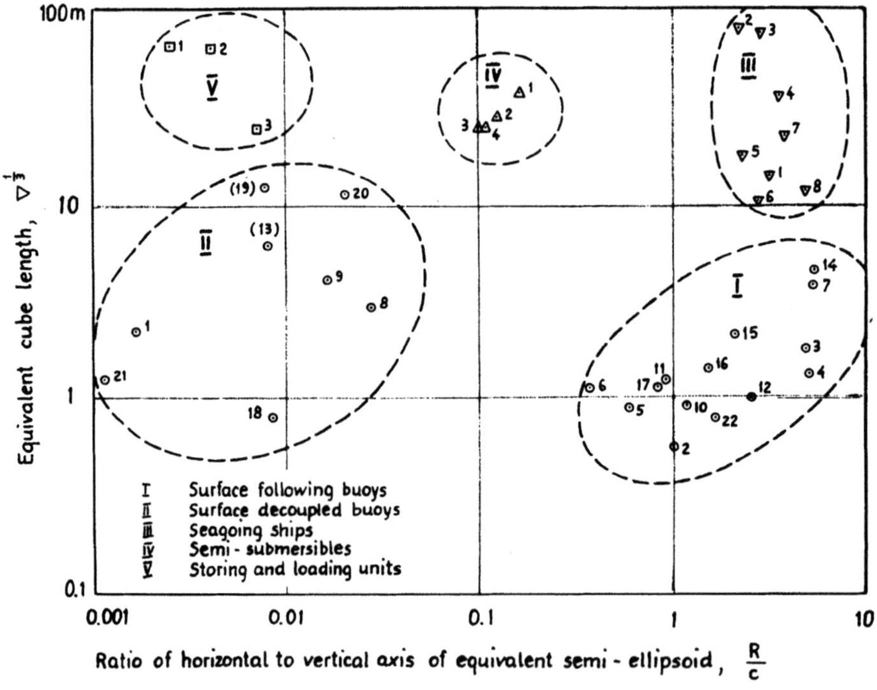

Fig. 4 A comparison of various types of buoys and other large structures shown in terms of their equivalent cube length and equivalent semi-ellipsoid.

linear coupling in the longitudinal plane between pitch and surge, and, analogously, coupling between roll and sway in the lateral plane. Nonlinear coupling is found, especially at higher amplitudes. Thus tilting moments may be exerted on a spar-like buoy by its heave motion, but these are of little influence if the buoy's natural frequencies of heaving and pitching are quite different, which usually is the case. Or a rotationally symmetric buoy with low damping may show pronounced lurching motions in the resonant zone resulting from pitching and pitch-induced rolling (i.e. coupling between modes of motion in different planes, but with equal natural frequencies).

The mathematical treatment of buoy motions in regular waves is conveniently simplified by separating motions in the plane of symmetry from motions in other planes (see, for example, Comstock, 1970). From the three possible cases two are of importance for buoys: float perpendicular to the wave crests with heaving, pitching, and surging motions; and float parallel to the wave crests with heaving, rolling, and swaying motions.

(19)	Scripps Institution of Oceanography FLIP	Spar
20	U.S. Naval Ordnance Laboratory SPAR	Spar
21	Woods Hole Oceanographic Institution (WHOI) Spar buoy	Spar
22	WHOI Torus	Torus

( ) • Manned research platform

<u>Seagoing ships</u>
Region III

1 Research vessel "Meteor"
2 Very large crude carrier (VLCC)
3 VLCC
4 Container ship
5 Cargo ship
6 Ocean tug
7 Cruiser
8 Destroyer

<u>Semi-submersibles</u> (at operating draught)
Region IV

1 Natural gas liquefaction platform
2 Drilling platform "Mark 2"
3 Drilling platform "SedcoH"
4 Drilling platform "AkerH-3"

<u>Storing and loading units</u>
Region V

1 Liquefied natural gas (LNG) storing and loading unit
2 LNG Storing and loading unit
3 Tanker loading unit

<u>Buoys</u>
Regions I and II

1	Comex Equipment Borem buoy	Spar
2	Datawell Waverider	Sphere
3	General Dynamics SODS/16	Disc
4	General Dynamics SODS/12	Disc
5	General Electric DLCB	Sphere with cylinder
6	General Electric Sea Robin	Cone with cylinder
7	Gesellschaft für Kernenergieverwertung in Schiffbau und Schiffahrt (GKSS) Disc buoy	Disc
8	Hagenuk AMOB	Cylinder with torus
9	Hagenuk STAMOB	Cylinder with torus
10	Meteorologisches Institut der Universität Hamburg (MI Hamburg) Measuring buoy 3	Disc with shaft
11	MI Hamburg Measuring buoy 4	Cone with shaft
12	MI Hamburg Measuring buoy 5	Disc with shaft
(13)	Musée Oceanographic de Monaco Bouée Laboratoire	Spar
14	National Oceanic and Atmospheric Administration (NOAA) Data Buoy Office MONSTER buoy	Disc
15	NOAA NOMAD buoy	Boat
16	NOAA Horizontal Cylinder	Cylinder, horizontal
17	NOAA Vertical Cylinder	Cylinder
18	NOAA MCP	Cylinder

Legend for Figure 4.

There is a further simplification for freely-floating/drifting or centrally moored, rotationally symmetric buoys. In the range of linear response heaving can be treated separately from pitching and surging.

2.4 Buoy and Mooring as a System

Generally, the performance of a moored buoy cannot be treated sufficiently without consideration of the mooring line. Therefore a brief description of the diverse mooring configurations and their main characteristics shall be given here. For details refer to Berteaux, Chapter 35.

A buoy-cable system held by one anchoring point only is called a single-point mooring. The mooring line can be taut or slack (Fig. 5a and b). The advantage of a taut mooring is small deflections of the buoy under shifting current conditions and hence less motion of instruments and sensor packages attached to the mooring line beneath the buoy. Thus signal errors are diminished. On the other hand, a taut mooring means high static loads due to ocean currents and high dynamic stresses due to wave-induced buoy motions. The mooring line stresses can be reduced considerably by using a slack mooring. But buoy and instruments will undergo larger movements, and it must be decided whether the influence on the oceanographic data of system oscillations and movements is tolerable, and, moreover, whether inherent signal errors can be appropriately corrected. For deep-sea moorings one often makes use of both systems by

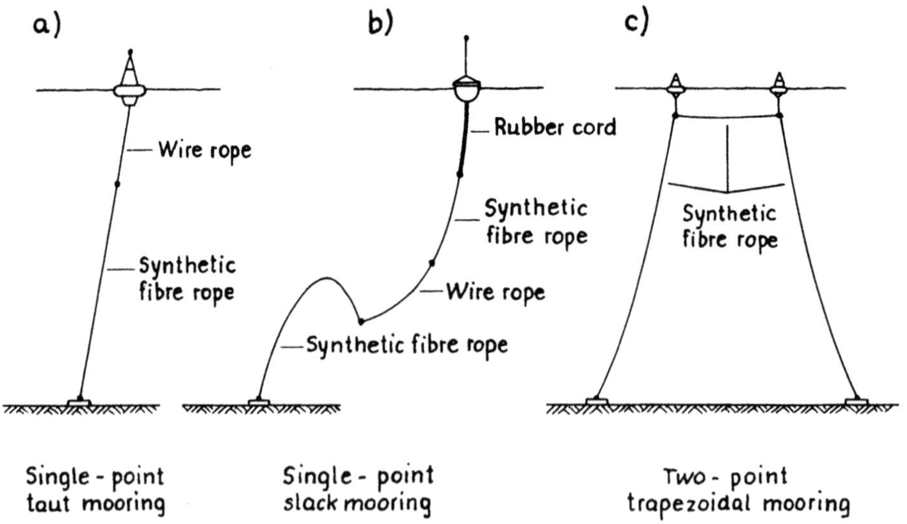

Fig. 5 Mooring systems

supporting the mooring line tautly with buoyancy elements at a certain depth below the ocean surface, where the waves have died out. The remaining mooring, with the buoy, is then made slack. In special cases, such as wave measurements, extremely slack moorings are indispensable. As an example, Figure 5b shows a mooring arrangement over continental shelves (with a section of highly extensible rubber cord nearest to the buoy) for the Waverider spherical buoy (Draper, 1975).

Multipoint moorings, which are more expensive due to material consumption and time for deployment and retrieval, are usually employed if station keeping and/or low signal errors are important; for example, in studying small scale phenomena in the ocean. Figure 5c shows a trapezoidal mooring (two-point mooring), which is used when the zone of interest is extended horizontally, as for internal-wave measurements. By increasing the number of anchor points the variety of multipoint moorings is practically unlimited. They can be given any desired property, but at the cost of an increasing amount of time and money. Recently this mooring system has been developed to its extreme in the so-called tension-leg moorings of platforms for oil and gas production; these moorings are fixed to the ocean bottom by a multiplicity of steel hawsers and chains to minimize the platform's horizontal and rotational motions.

3. ANALYSIS

3.1 Optimization of Buoy Dynamic Response

Buoys are subjected to wind, wave, and current forces, all of which can drive the system into oscillations. For the theoretical hydrodynamic fundamentals see, for instance, Newman (1977). A mean current may seriously affect the performance of a moored buoy by periodically shedding vortices from its hull, thus causing the buoy to oscillate in all possible modes of motion. Periodic gusts may shift a moored buoy from its mean position. Nevertheless, a buoy system is expected to have certain dynamic properties; the buoy pitch angle may be limited for data transmission, or the vertical movement of an instrument line suspended from the buoy must be kept within certain limits.

In practice, optimization of buoy system response is a difficult task. Besides the desired dynamic behavior, many different and sometimes opposing requirements must be considered. Handling, attendance, and control usually are important in assigning the principal dimensions of the system. Buoy size and weight may be limited so that it can be shipped. Servicing activities may require entering a buoy on position, so spare buoyancy must be provided. Moreover, the requirements on buoys for long-term measurements, which might include operating under severe weather conditions, are more restrictive than those on buoys with a ship standing by.

In buoy dynamics, wind and current are regarded as steady environmental conditions. They are responsible for the buoy's deflection and mean draught if it is moored, and for its mean inclination. The wave-excited buoy motions are superimposed on these effects. For the wave-excited motions, statistical analysis is generally employed as developed for ship theory (Price and Bishop, 1974; Newman, 1978; Wehausen, 1971). Linear theory is used, and the transfer functions of buoy motions, $Y_{B/S}$, are solutions in the frequency domain of the coupled equations of motion (e.g. Goodman et al., 1972; Kaplan et al., 1972). If the sea state is given by its power spectrum $S_S(\omega,\theta)$, the energy of the specific physical quantity of the buoy system in the frequency band $\Delta\omega$ and the sector of wave directions $\Delta\theta$ is equal to the transfer function squared multiplied by the wave power spectrum. Thus

$$R_B(\omega,\theta) = Y_{B/S}^{\,2}(\omega,\theta) \cdot S_S(\omega,\theta) \tag{8}$$

is the response energy spectrum, for instance, of a buoy's vertical oscillations, if $Y_{B/S}$ is the transfer function of heaving motion. In actual buoy applications linear conditions are often a rather gross approximation of reality. For spectral response in the case of nonlinear oscillations and seaways see e.g. St. Denis (1975) for a survey.

If the random mooring line stresses are evaluated by assuming a linear response of the mooring system to wave excitation, the above equation yields the response power spectrum of the dynamic mooring line load. On this basis the energy density of the fluctuating stresses can be determined at any desired point on the mooring line. But having accomplished this, the real problem, which as yet has not been solved sufficiently even for on-land application, is to relate fatigue to the stresses (i.e. will the system stand the high number of load cycles expected during long-term measurements?). A pragmatic method seems to be Miner's hypothesis of cumulative damage (Jasper, 1957): the material fatigue of an alternately loaded structure is estimated by relating the number of stress cycles suffered to the total number of stress cycles endured under the respective load, and summing them up. Thus one arrives at a cumulative damage rate $0 < D < 1$ per unit of time, for example per year, where $D = 1$ stands for material breakdown.

3.2 Hydrodynamic Forces on Buoys

In the range of linear response the frequency-dependent hydrodynamic (pressure) forces on an unrestrained buoy in waves can be resolved into the forces exerted on a buoy performing forced oscillations in still water and the forces exerted on a stationary buoy by passing waves (the same is true for the moments),

$$F_{hydrodynamic}(t) = F_{oscillation}(t) + F_{excitation}(t) \qquad (9)$$

Theoretical solutions are available for typical buoy forms (boat, multihull, disc, spar). The evaluation of the hydrodynamic forces on boat-shaped buoys follows the analytical method used for conventional ship forms, i.e. the so-called strip method (Grim, 1960; Wang, 1976), which assumes two-dimensional flow about each cross-section of the hull. The forces are then integrated along the hull length. Since boat hull buoys are not slender as assumed by the strip method, corrections for three-dimensional effects are necessary. A pragmatic approach is available from submerged spheroids (Havelock, 1956). The single hulls of a multihull buoy, for instance a catamaran, are generally quite slender, so that the strip method applies directly, but interaction effects between the separate hulls must be accounted for (Ohkusu, 1970; Ohkusu and Takaki, 1971). The forces on disc hulls can be deduced from Kim's (1962) results for shallow draught ships. Newman (1963) investigated the forces on spar hulls. Kaplan et al. (1972) used the above buoy hull forms in their computer simulations and compared computed buoy motions with experimental results. The hydrodynamic forces may also be determined through representing the buoy hull by a suitable distribution of sources (see e.g. Newman, 1978; Hooft, 1975; Faltinsen and Michelsen, 1975). This method should yield reliable values at least for simple buoy forms.

Model basin experiments are used to generate data and to provide controls for computational values of hydrodynamic forces on buoys. As in the resolution of the forces in Equation 9, buoy models are oscillated in still water and held fixed in passing waves and the forces and moments exerted on them are recorded. The buoy models are made extremely light, for instance out of styrofoam, in order not to contaminate the results with the buoy's own inertia in the case of forced oscillations.

Figure 6 shows experimental data from Mercier (1971) for the hydrodynamic forces on buoys of various forms performing forced heave oscillations. The reference parameter, $\rho\nabla\omega^2\bar{z}$, is the water mass displaced by the buoys times the amplitude of the acceleration, where $\bar{z}$ is the amplitude of heaving motion. The hydrodynamic forces are generally out of phase with the excitation. For convenience they are resolved into their in-phase and out-of-phase components. If the in-phase force components are diminished by the hydrostatic part, the remainder, which is proportional to buoy acceleration, gives the added mass (k_{zz} is the coefficient of added mass in the vertical direction due to heaving motion). The out-of-phase component is also called the damping force, because it is in phase with buoy velocity (h_{zz} is the coefficient of damping in the vertical direction due to heaving motion). The work done by the oscillating buoy results in the generation of waves (and possibly vortices, by flow separation and friction).

Such model tests on a few specific buoy hull forms can give

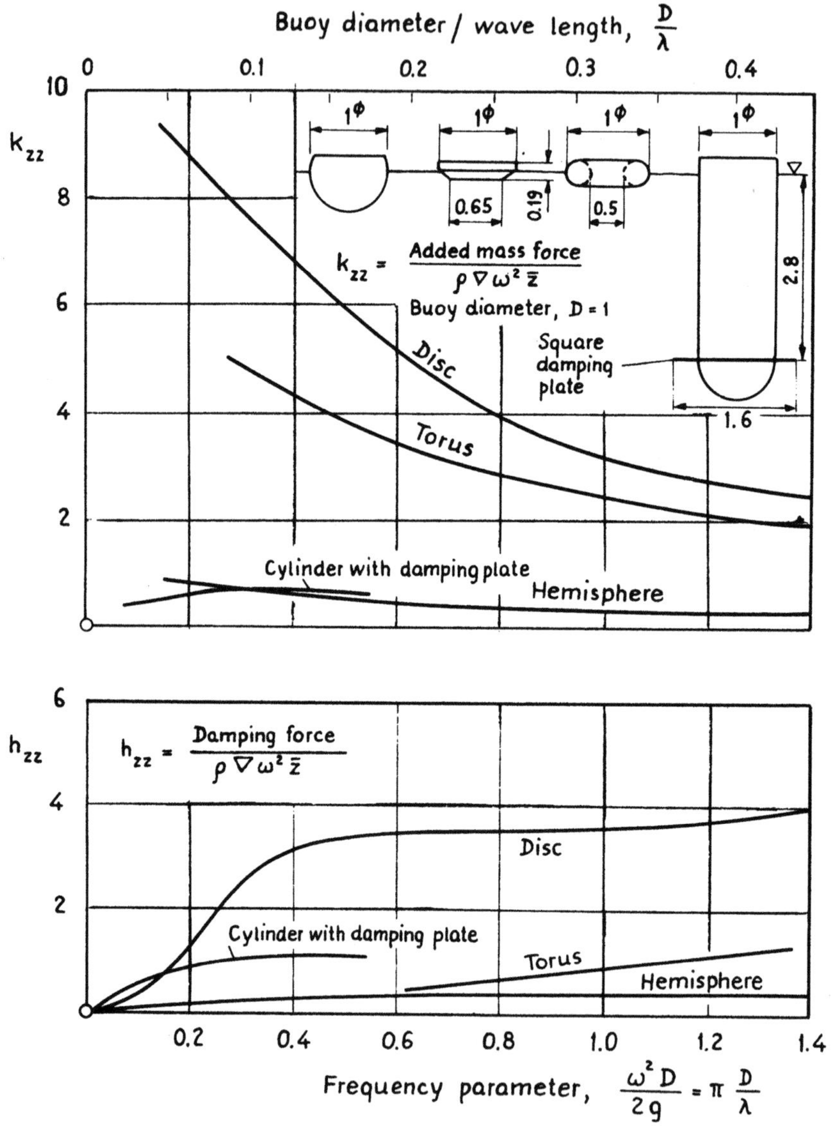

Fig. 6 Nondimensional coefficient of added mass (in-phase) force, k_{zz}, and of damping (out-of-phase) force, h_{zz}, in vertical direction due to forced heaving oscillations for various buoy forms (after Mercier, 1971).

reliable data even for buoys differing from the tested forms provided that the appropriate physical reference parameters are used. This is illustrated in Figure 7, where Mercier's reference parameter, the buoy mass (see Fig. 6), has been replaced by the mass displaced by a hemisphere with the cross-section of the buoy, A_M. Obviously, the buoy's cross-sectional area is of more physical importance than its displaced volume (essentially $\rho A_M^{3/2}$), since the new coefficients differ less than the former for rather different buoy forms. This appears strikingly at frequency parameter values of about 0.4. Here the coefficients of added mass, c_{zz}, are almost the same for such differing buoy forms as a cylinder with a damping plate on one hand, and a disc and a torus on the other hand, whereas the k_{zz} values of Figure 6 differ nearly by an order of magnitude. In the case of a cylinder with damping plate there is no doubt that the cross-sectional area of the reference hemisphere must be the same as the area of the damping plate; the graph in Figure 7 confirms this assumption. A reference sphere with the smaller cylinder waterplane area would yield six times the values.

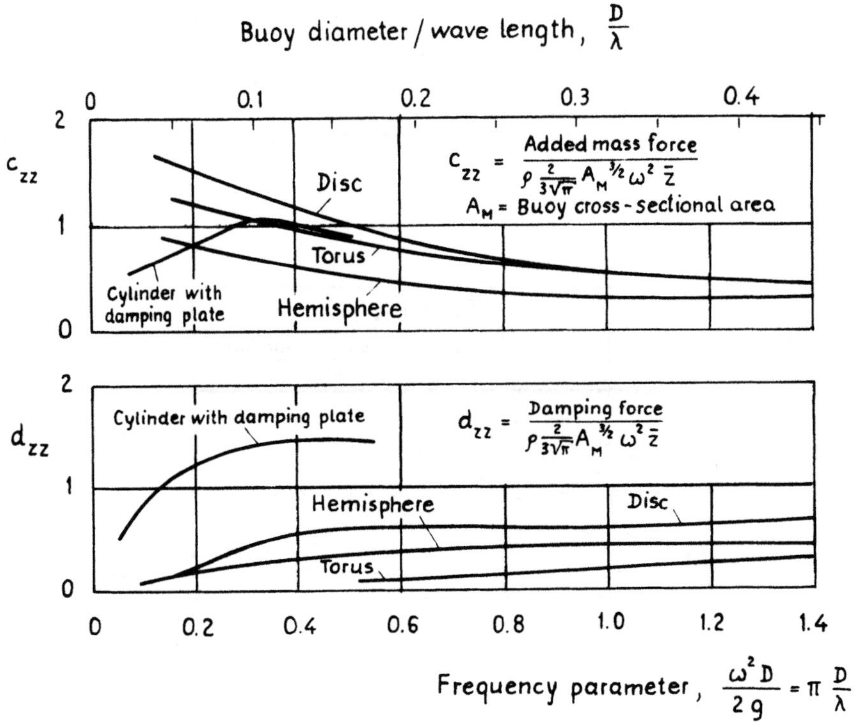

Fig. 7 Nondimensional coefficient of added mass (in-phase) force, c_{zz}, and of damping (out-of-phase) force, d_{zz}, in vertical direction due to forced heaving oscillations for the buoy forms shown in Figure 6.

For a torus there is some ambiguity as to which area is more suitable, the real waterplane area or an area with the outer diameter at the waterline. Comparison with the other buoy forms shows that the latter fits better, at least for the added mass.

The second component of the hydrodynamic effects on buoys in waves (Eq. 9) is the wave-excited force. As an example, Figure 8 shows Mercier's (1971) results for the vertical wave force (amplitude, and phase related to the wave crest) on a semisubmerged sphere. The reference parameter $\rho g \pi\ (D^2/4)\ \bar{\zeta}$ is the weight of water in a cylinder formed by the buoy waterplane area and the wave amplitude. The force leads the waves by an amount which increases with decreasing wavelength; for long waves it is hydrostatic ($D/\lambda \to 0$: $E_{z,r} = 1$, $E_{z,i} = 0$).

3.3 Buoy Survivability

Linear theory is used to design buoy systems with predetermined dynamic properties. But because of the generally small size of buoys in contrast to ships and other large oceanic floats it is realistic for moderate sea states only. For the situation where the sea shows white caps and begins to spill, the assumption of linear response is unrealistic. Nevertheless, for long-term measurements, severe sea conditions are of major importance. For instance, whether a buoy maintains its upright position and whether instruments

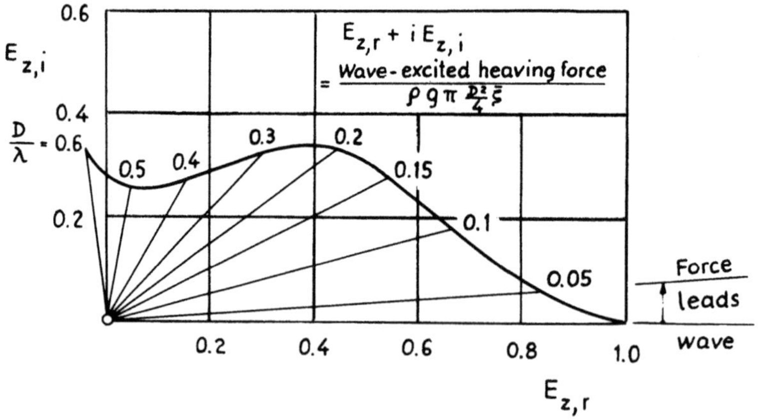

Fig. 8 Nondimensional coefficients of wave-excited heaving force on a restrained hemisphere as a function of wavelength. $E_{z,r}$ is the in-phase component and $E_{z,i}$ the out-of-phase component with reference to the wave (after Mercier, 1971).

and sensors will stay intact become important design considerations. In the extremes of storms and hurricanes the question will be whether the system is likely to survive.

System survivability is not only affected by the ocean environment; man can be even more destructive. Without doubt, research buoys are a hazard to navigation, and, by the same token, ships are a hazard to buoys. To make a buoy less sensitive to damage of its shell, subdivision of its hull and/or filling the unemployed volume with synthetic foam is recommended.

Vandalism presents serious problems and prosecution is difficult because no maritime law at present applies to the protection of buoys; moreover, the originator of the damage is generally unknown. Once a buoy has been boarded, protection against vandalism is minimal. Thus boarding should be prevented, provided that it is consistent with maintenance demands. As experience shows, a sheer side without any handles gives effective protection.

Some controversy exists as to the buoy forms best suited to severe sea conditions. In fact, the properties of a moored buoy in a storm sea are difficult to quantify. A useful criterion might be the extremes of the buoy's response in a train of breaking waves; for instance, the maximum angle of inclination (will the buoy capsize?) or the maximum tension in the mooring line (will it break?). Certainly, such values are not realistic because of the randomness of the sea. Nevertheless they could be used for a comparison of different buoy forms and mooring configurations, which would diminish the present uncertainties in evaluating nonlinear buoy system response.

If, for instance, two buoys of contrary shape, a disc and a spar, which are moored in the same way, are exposed to a train of breaking waves, the pitch angle of the disc will increase more with increasing wave height than that of the spar. As the wave height increases, the spar inclines progressively, thus diminishing its resistance; the disc also inclines progressively, but here the resistance increases with inclination, which makes disc buoys sensitive to capsizing. This statement is rather general, since the performance of a moored buoy in a rough sea cannot be properly treated without quoting the actual mooring configuration.

Under certain circumstances capsizing may be tolerable as long as the buoy returns to its upright position and its instrumentation continues to function. As a rule, under extreme sea conditions buoys should be employed that have a single preferred position and are not subject to catastrophic inclinations. For freely floating/drifting buoys, rotationally symmetric or even polygonal forms are superior to horizontally elongated forms because of the latter's tendency to broach. However, a slender boat hull might be preferred where a buoy is to be moored in strong ocean currents.

3.4 Towing Tank Experiments

The reliability of computational results for the dynamics of buoys depends on how much is known of the effects of environment on the system and of the validity of wind, wave, and ocean current data. Reliable environmental data are becoming more important for the design of ocean platforms. Oceanographers are hardly interested in the sea-keeping qualities of a buoy system on the basis of linear response if such conditions are not realistic. They ask for the actual performance at sea, which imposes considerable problems on the ocean engineer (see e.g. St. Denis, 1975). But he can easily generate a controlled environment in special test facilities and subject small-scale models to it. For the theoretical basis of model testing see, for example, Newman (1977).

The purposes of model tests are twofold: to provide the ocean engineer with the fundamental data necessary for his mathematical simulations of buoy and buoy-cable dynamics, and to determine or verify the dynamic behavior of a specific float design. Often the actual buoy deviates considerably from the theoretical buoy hull form and incorporates appendages such as damping plates, keels, and shafts, and other details, which cannot be accounted for analytically. Here model tests are necessary.

Buoy dynamics can be studied in towing tanks. These tanks are equipped with instrumented carriages capable of travelling at various speeds, and wave makers capable of generating regular waves of different heights and periods, and irregular waves by superimposing regular waves. Test results are scaled by the Froude frequency parameter

$$\frac{\omega^2 D}{2g}, \qquad (10)$$

where ω is the wave circular frequency and D is the buoy diameter (or another characteristic length). If the float moves with constant forward speed U relative to the surrounding water, or if the current velocity is -U, in the case of the moored buoy at an angle μ_e to the wave direction, ω is replaced by the encounter frequency

$$\omega_e = \omega - \frac{\omega^2 U}{g} \cos\mu_e . \qquad (11)$$

The hydrodynamic effects of an additional current on a buoy's response to waves are shown in Figure 9, where the pitch transfer function of a horizontally weakly-tethered buoy has been plotted on the basis of the encounter frequency in nondimensional form.

The characteristic parameter of a float in a stream at the water surface is the Froude number

$$Fr = \frac{U}{\sqrt{gD}} . \tag{12}$$

Small currents running with or opposite to the waves (low Froude number) can be neglected in practice, whereas for a strong stream (increased Froude number) the effects of wave encounter should be considered. For the condition of no relative-flow (Fr = 0) the frequency parameter corresponds to the ratio of buoy diameter to wavelength, D/λ:

$$\frac{\omega^2 D}{2g} = \pi \frac{D}{\lambda} . \tag{13}$$

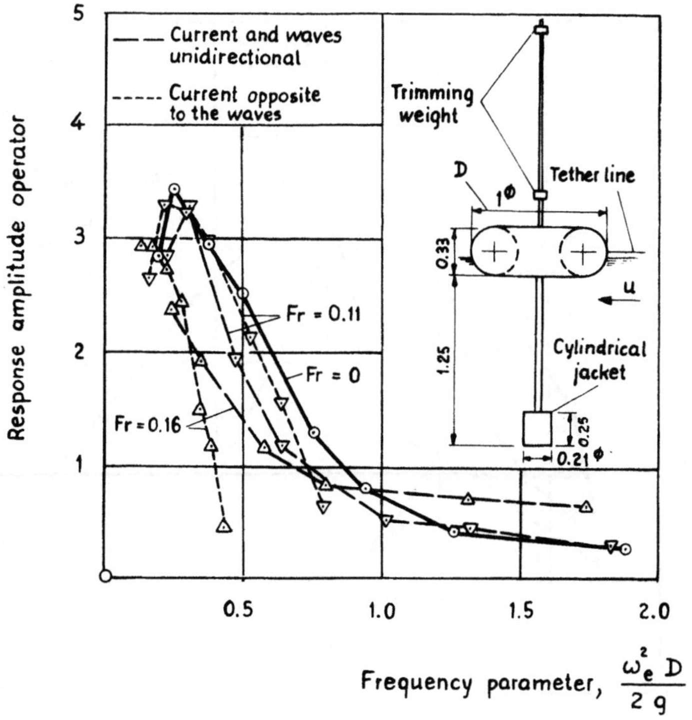

Fig. 9 Transfer function of pitching motion (pitch amplitude over wave slope) for a horizontally tethered toroidal buoy with shaft, in a stream unidirectional with and opposed to the waves and without a stream [from towing tank tests at Institut für Schiffbau der Universität Hamburg (ISUH)].

3.5 Fully-scaled Model Tests

If a model is allowed to oscillate freely, it must be dimensionally as well as dynamically similar to its full-scale counterpart (prescribed mass, centre of gravity, and radius of gyration). Tests using such models are the most versatile method for studying a specific buoy design; in towing tank facilities practically any desired buoy arrangement and sea condition can be modelled. Care must be taken especially with small models, that an extremely light motion sensing apparatus is used in order not to alter the model's motion. Recently, opto-electronic means have been developed.

Some test results with a fully-scaled dodecagonally annular buoy (such buoys may be more easily manufactured than circular ones) are

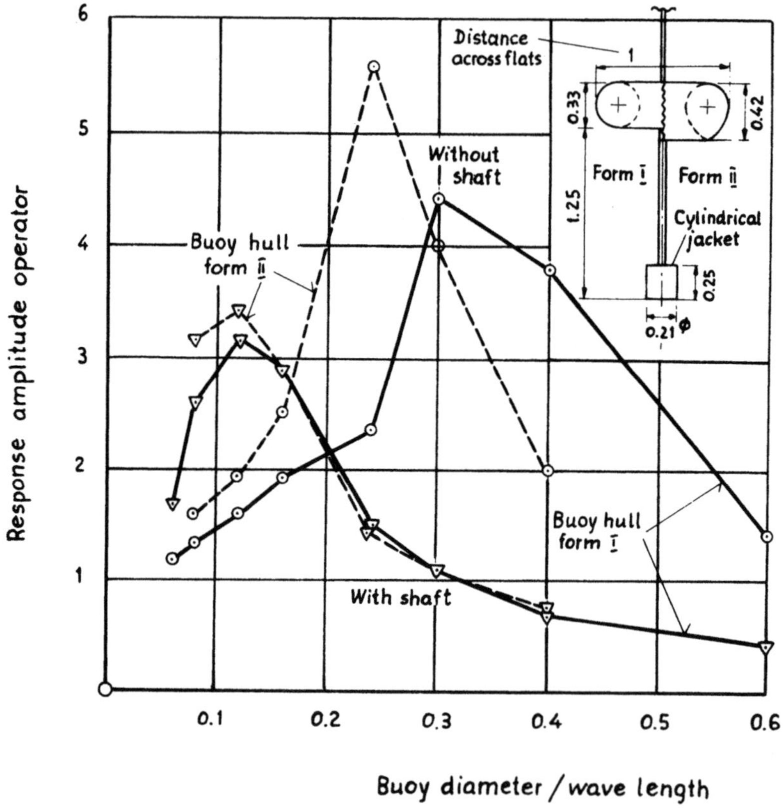

Fig. 10 Transfer function of free pitching motion (pitch amplitude over wave slope) for a dodecagonally annular buoy, with and without a shaft (from ISUH towing tank tests).

plotted in Figures 10 and 11. Figure 10 shows the transfer function of free-pitching motion for the buoy's possible operational version with a shaft, and for the buoy hull alone; the transfer functions are not significantly different from those obtained with circular symmetry. Of course, no oceanographer would employ the buoy without a shaft or other equivalent appendage. Hulls were given the cross-section of a circle (Form I) and of a circle-arc triangle (Form II, i.e. a more ship-like cross-section). As might be expected, the pitch amplitudes in the zone of resonance are greater for hull form II than for hull form I because the former is less damped. If the hulls are fitted with shafts, the transfer functions nearly coincide. This might suggest that the buoy hull form is not important to the buoy's sea-keeping performance. But what cannot be read directly from the transfer functions is that hull form II shows little slamming (impact of water surface on buoy hull in relative motion) even in high waves, whereas noticeable slamming occurs on hull form I for moderate wave heights, due to a

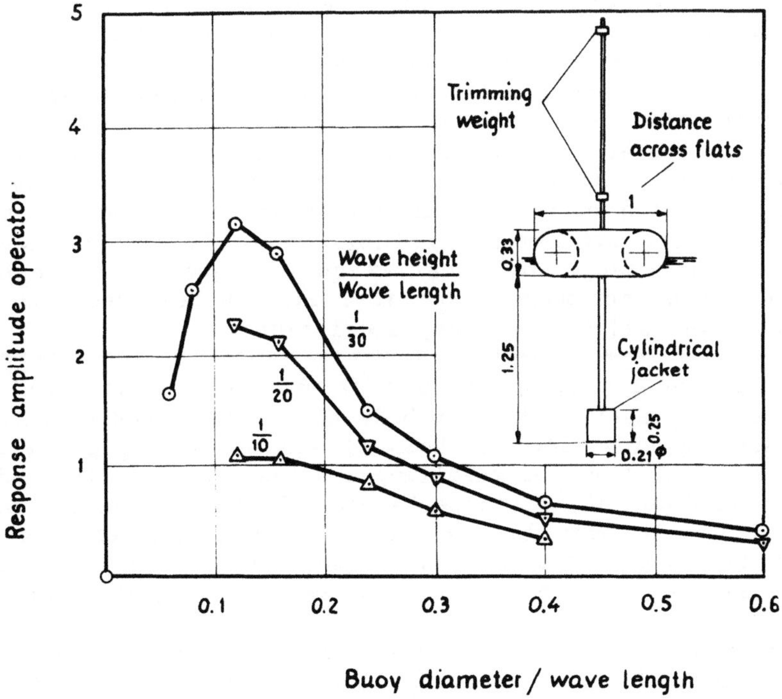

Fig. 11 Effect of wave height on pitch amplitude. Transfer function of free pitching motion (pitch amplitude over wave slope) for a dodecagonally annular buoy with shaft (from ISUH towing tank tests).

partially flat bottom associated with its small draught. Slamming may cause vibrations of instruments, for instance on a slim mast.

Figure 11 illustrates the effects of wave height on the pitching motion of a buoy. Here nonlinearities become clearly visible at wave heights of one-twentieth the wave length, where Kaplan et al. (1972) in their tests with various buoy hull forms still observed linear behavior. Linear response not only demands moderate waves, but also implies that no serious nonlinearities are induced by the buoy hull geometry or by damping. The latter is true for the case in point.

Experimental work with fully-scaled models of various hull forms (e.g. Hoffman et al., 1973; Kaplan et al., 1972) can be used to verify computational results in the range of linear response. The reliability of the theoretical methods depends on the equations and the coefficients of the hydrodynamic effects. Computations for a specific buoy using the method of coupled equations are very accurate if the coefficients are determined experimentally. Using theoretically determined coefficients the computations are less perfect, since the real buoy deviates from the basic theoretical buoy form and three-dimensional flow effects on non-slender hull forms can be accounted for only approximately. Sloping sides at the waterline are a further problem. Also, computations show a tendency to overestimate the actual buoy response, especially in the vicinity of resonance. On the whole, it can be said that numerical results are sufficient if the actual buoy does not deviate too much from its mathematical generic form. The principle of linear superposition of motions in irregular waves, which applies to ships, has been confirmed for buoys (e.g. Kaplan et al., 1972). But beyond the range of linear response, especially for buoys in breaking waves, not much theoretical aid can be expected. This is the domain of fully-scaled buoy model tests.

The effects of a mooring on buoy dynamics can be modelled to some extent in the restricted water body of a towing tank by turning the model mooring line with the help of sheaves and by using suitable elements in the line for modelling its mass, elasticity, and hydrodynamic damping. The distance between the buoy model and the guide pulley below it must be such that buoy-induced oscillations of mooring line angles greater than those found at the prototype mooring are avoided; that is, there must be a sufficient water depth.

Results from respective model tests with a toroidal buoy are plotted in Figure 12. The prototype buoy-cable system consisted of a 2.4 m diameter buoy and a single-point deep-sea mooring (water depth 4000 m) supported by buoyancy balls 800 m below the water surface. Certainly, the transfer functions of the moored buoy must be interpreted qualitatively, but nevertheless it can be concluded that a mast on the prototype buoy will keep rather quiet.

Full-scale tests, finally, are required if the dynamics of the mooring are the prime requisite, since it is, as a rule, not possible to simultaneously scale all the significant properties of a buoy-cable system. Generally the mooring line length is such that, if the mooring is scaled, the model buoy would become inadmissibly small. A further problem is the elasticity of the mooring line.

3.6 Wind Tunnel Experiments

Wind tunnel tests are used to investigate the fluid dynamic forces and moments on buoys and buoy system components above or below the water surface in steady state flow. Small-scale models are mounted on balances in the wind tunnel air stream. Test results are scaled by the Reynolds number

$$Re = \frac{UD}{\nu} , \qquad (14)$$

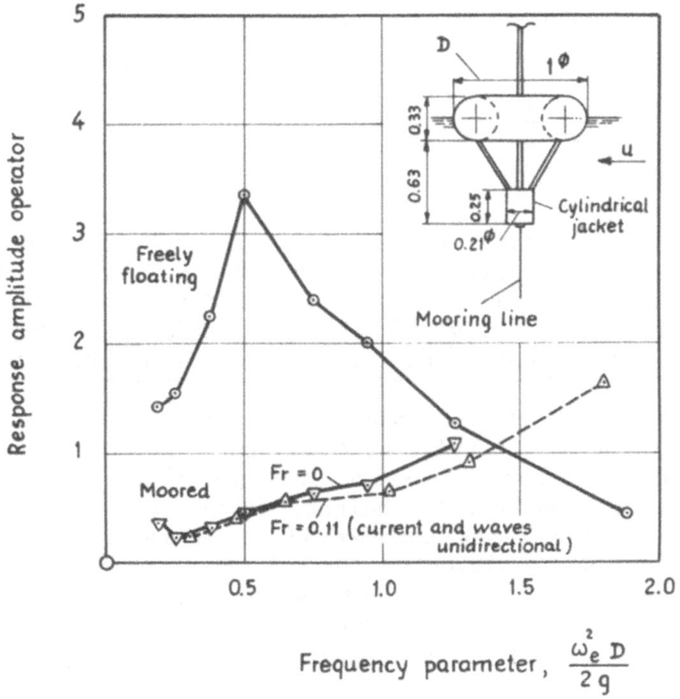

Fig. 12 Effect of a single point taut mooring on buoy motion. Transfer function of pitching motion (pitch amplitude over wave slope) for a toroidal buoy (from ISUH towing tank tests).

where U is a characteristic velocity, D a characteristic length, and ν the kinematic viscosity. Tests with models of small scale may not reach the full-scale Reynolds number. But this can be neglected provided that the object has relatively sharp edges so that zones of flow separation are fixed. A water surface is simulated either by a flat plate or by reflecting the object at the water surface (double model). A wind profile (without achieving similarity of turbulence) can be produced by inserting a grid upstream. Tests for which both a wind profile and turbulence similarity are indispensable are executed in special wind tunnels with long entrance sections fitted with roughness elements. Wind tunnel tests are usually performed in a uniform flow even though in nature this may not be true. Compilations of fluid dynamic coefficients are available in the literature (Hoerner, 1965); these are also based on uniform flow conditions. If the buoy system includes various components which have little influence on each other, for instance a meteorological mast fitted with instruments, the total resistance is determined by summing up the resistances of the component parts to the local wind velocities.

For ocean engineering purposes the logarithmic law of the natural wind may be replaced by the potential formula (e.g. Wieghardt, 1972)

$$\frac{U}{U_A} = \left(\frac{z}{z_A}\right)^{\frac{1}{n}} , \qquad (15)$$

where U is the velocity at the height z, and U_A the velocity at the reference height z_A (e.g. the anemometer height). The exponent becomes one-tenth for near-neutral stratification of the atmosphere (wind velocities $\gtrsim \sim 12$ m s^{-1} at 10 m height) and about one-sixth for unstable atmospheric conditions (e.g. light breezes).

If the resistance of a compact structure is known for a uniform flow, the value for field conditions can easily be estimated with the help of the dynamic pressure averages of the wind (see Chapter 14) over either the object's height, or the 'wetted' circumference, or the shaded area (see Chapter 33). Of these, the shaded area value is defined consistently for arbitrary objects. In order to arrive at the force on an object in a natural wind, its resistance area (resistance divided by dynamic pressure) under uniform flow conditions is multiplied by the respective dynamic pressure average of the wind. Which of the three possible averages is to be preferred depends on the geometry of the object. For instance, the circumferential pressure average yields good results for plates in cross-flow. On the other hand, the shaded area value seems to be more suitable for three-dimensional, multiform objects like ships, and clearly is the only possible reference parameter for structures with perforated shaded areas as is shown in Figure 13, which compares test results of resistance area for a specialized buoy (for plankton investigations in foil tanks) with results for the research vessel Meteor. The method yields good results if the main flow characteristics in the wind are not distinct from those in a uniform flow. For instance, on plates and sails flow separation for natural wind occurs at a greater angle of attack than for uniform flow. Thus, at certain angles of attack the flow at a

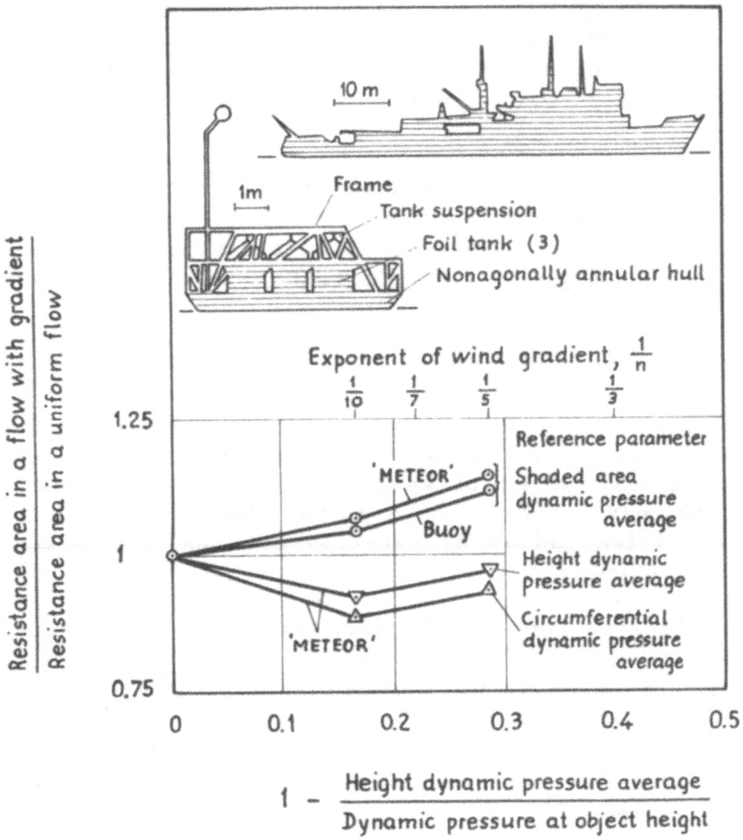

Fig. 13 Resistance area in a flow with gradient, related to the resistance area in a uniform flow, for a specialized buoy (for plankton investigations) and for the research vessel Meteor (from ISUH wind tunnel tests).

plate in natural wind can still be attached (high lateral force), whereas it would already have separated for a uniform flow (low lateral force). Here the method fails, but such conditions are encountered primarily on simple geometric forms.

4. OCEANOGRAPHIC BUOY SYSTEM DESIGN

4.1 System Design Process

A good buoy system is, inevitably, a compromise. For instance, a buoy whose dynamic response characteristics demand a low reserve buoyancy may have to be entered for servicing. Often initial requirements concerning dynamic behavior must be restricted in the interest of overall performance (performance restrictions). The system response strived for in a moderate sea may be unfavourable under extreme sea conditions. As a rule, the available transport facilities and handling requirements limit a buoy's overall dimensions (handling bounds and logistic constraints), and thus influence its dynamics. Finally, the funds may be limited (budget constraints).

Experience gained with other systems is an integral part of every buoy and mooring design. It does not replace computation, but complements it. Computation reaches its limit sometimes, for instance in regard to mooring-line fatigue where experience proves essential.

4.2 Surface-following Buoys

Buoys with relatively large waterplane areas but small displacements tend to follow the waves. Actually, their heave response is affected by a wide range of wave frequencies, but shows moderate amplification only. For pitch response, the position of the centre of gravity and the radius of gyration are needed, in addition to buoy shape and mass.

A surface-following buoy can be made wave slope-following or wave slope-decoupled. Accordingly, surface-following buoys may be split into two groups, namely, wave elevation/slope-following buoys and wave elevation-following/slope-decoupled buoys. The latter shall be called intermediate. As an example of a wave elevation/slope-following buoy, Figure 14 shows the General Dynamics SODS/16 buoy (region I,3 of Fig. 4) which is used for environmental data acquisition and telemetry.

4.3 Surface-decoupled Buoys

For surface-decoupled buoys (which are described more precisely as wave elevation/slope-decoupled), small waterplane areas together with relatively large displacement volumes are typical. Often

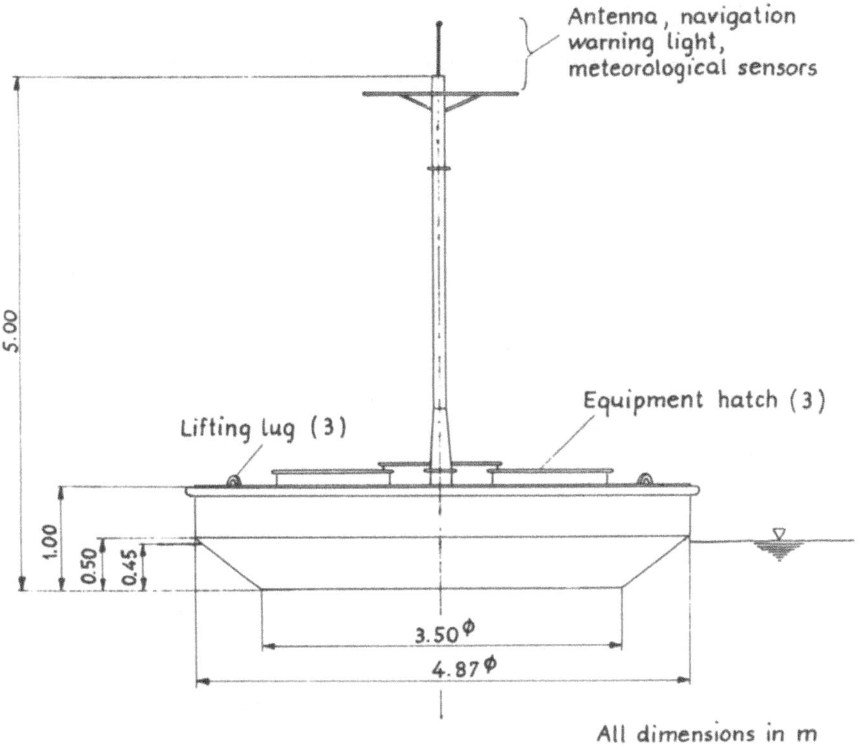

Fig. 14 General Dynamics SODS/16-buoy (Small Ocean Data Station).

these buoys are the very slender 'spar' buoys. As a rule, surface-decoupled buoys exhibit a marked resonance, and generally high amplification, in the case of the slender forms, but at frequencies other than the buoys' natural frequencies there is not much motion. For a spar of constant cross-section to be effective even at large wave periods, it must be of considerable length. Therefore, the largest buoys ever built for research purposes belong to this category (e.g. SPAR, FLIP, Bouée Laboratoire; Spiess, 1968).

v

If a spar is to be moored in deep water, it must be remembered that such buoys have a relatively large drag but not much reserve buoyancy to compensate vertical mooring line forces. Problems can be avoided by horizontally tethering the buoy with the help of auxiliary buoys or subsurface floats. Sometimes freely-floating spars are employed which use dynamic positioning.

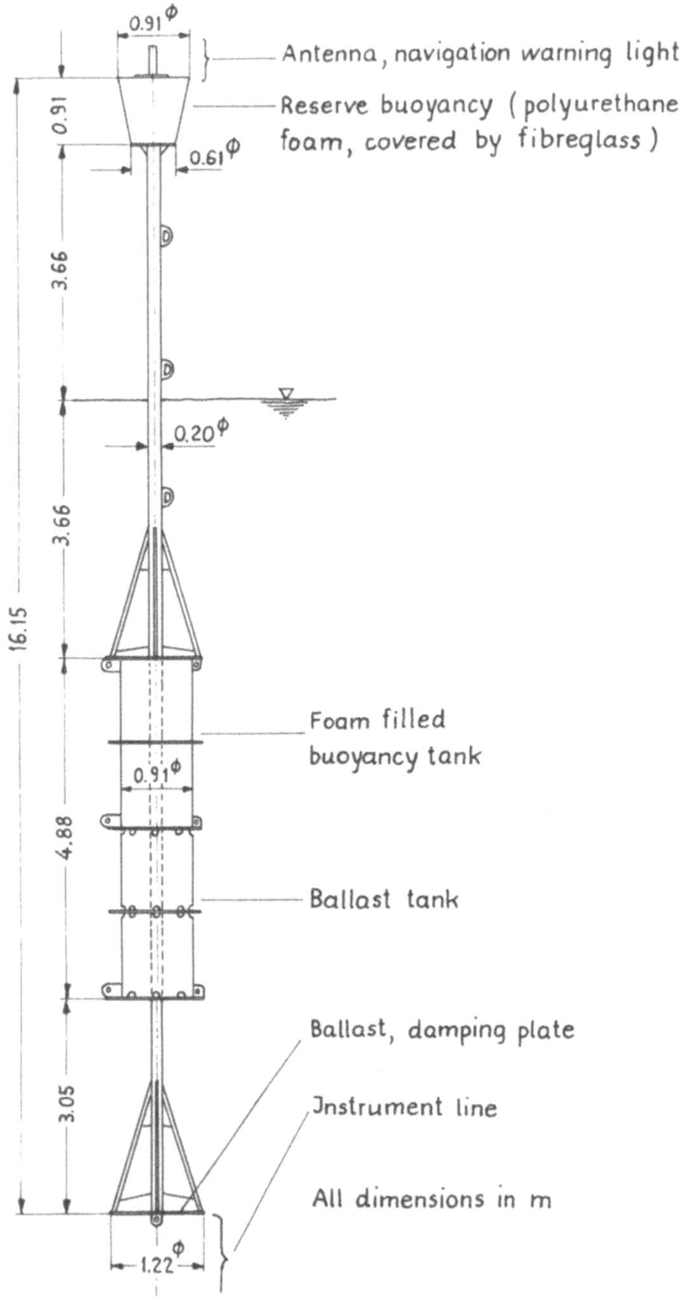

Fig. 15 Woods Hole Oceanographic Institution (WHOI) - spar buoy.

As a typical example of a wave elevation/slope-decoupled buoy for oceanographic application, Figure 15 shows the Woods Hole Oceanographic Institution (WHOI) Spar (region II,21 of Fig. 4). This has an R/c ratio of 1.14×10^{-3} (i.e., effectively a very slender buoy).

4.4 Intermediate Forms

The intermediate type includes buoys which, with some simplification, follow wave elevation but not wave slope. The most important representative of this group is the spherical buoy. Since passing waves cannot exert any exciting moments on a semi-submerged sphere, the spherical buoy is slope-decoupled. The favourable hydrodynamic properties of a sphere are sometimes utilized by giving the buoy a convex hull form at its floatation line.

Every wave elevation-following buoy can be made slope-decoupled with respect to certain ranges of wave frequencies, for instance wind-generated chop, by reducing its metacentric height and thus reducing the natural pitch frequency (Eq. 2). But as a consequence, static inclinations due to wind and current forces or icing will increase, and capsizing tendencies will be more pronounced.

Such effects on buoy stability are, as a rule, not acceptable. Another way of reducing the natural pitch frequency is to augment the buoy's moment of inertia (Eq. 2). A shaft with a lumped mass or a floodable tank below a disc will increase the buoy's own moment of inertia considerably, but at the same time increase its mass, thus reducing even the natural heave frequency. Such unintended results can be avoided by light plates and other appendages that alter the buoy's response with respect to specific modes of motion only. Vertical plates on or a cylindrical jacket around the lower end of a shaft below the buoy hull have little influence on heaving but reduce pitching by adding a restraining moment and increasing the moment of inertia. At the same time, strong pitching in the resonant range is diminished by the separation of vortices at such appendages. Similar means are applied to other buoys, especially spars.

A buoy of the intermediate type (region I,10 of Fig. 4) is shown in Figure 16. It is employed by the Meteorologisches Institut der Universität Hamburg for meteorological data acquisition, especially wind fluctuations, for which the support of sensing elements in the true upright position and at a constant height above the instantaneous water surface is essential. Residual tilting of the buoy mast is eliminated by a mechanical pitch and roll compensator.

4.5 Material Selection

Materials for use in ocean engineering applications must be selected for their resistance to deterioration in the oceanic environment. The ocean has its own environmental categories (salinity,

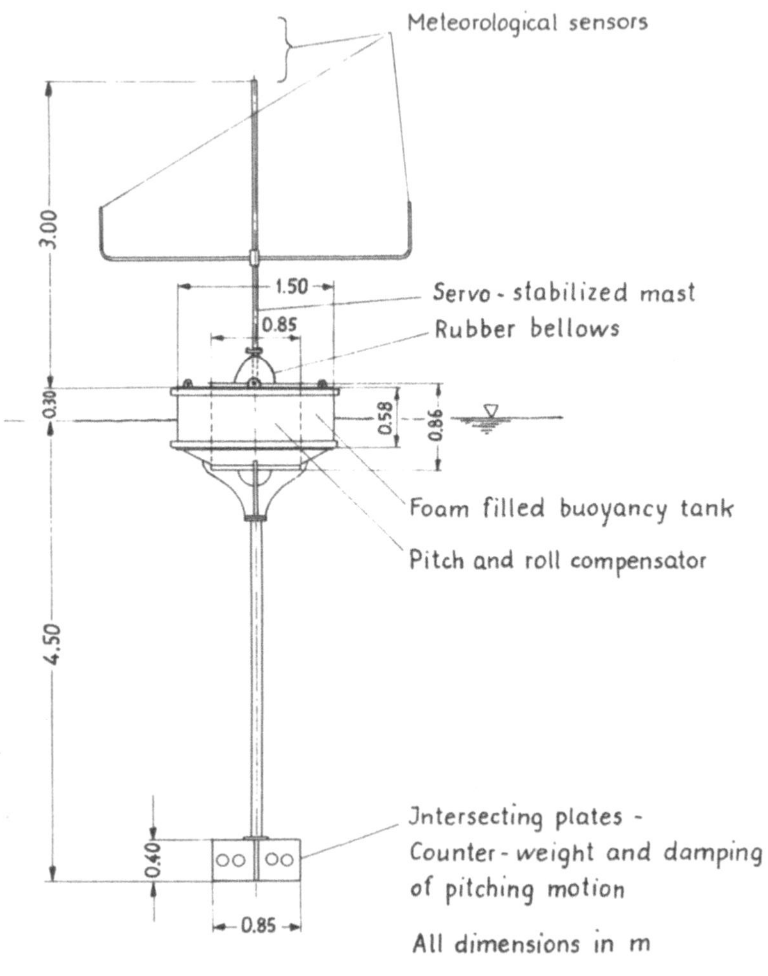

Fig. 16 Meteorologisches Institut der Universität Hamburg - disc buoy with shaft.

marine life, and pressure) that impose difficult problems. The usual properties of fabrication and strength are essentially the same as applications in other engineering lines. A reference source for materials suited to oceanic use is Masubuchi (1970).

One of the most difficult environmental problems is the corrosive action of sea water on the metallic components of buoy systems. Steel wire ropes, which are difficult to protect, are seriously affected. Marine corrosion can be defined as the electro-chemical reaction of metals with the sea water/saline atmosphere. For corrosion causes and protection see Uhlig (1971). Related to corrosion are material wear and abrasion, that is, the effects on interacting surfaces in relative motion, called tribologic effects. Abrasion increases in the presence of sea water. Great care should be taken to avoid the creation of unintended galvanic cells by connecting unprotected metals of different electromotive potentials. Rapid and severe material deterioration and hence failure of the structure or the mooring may result. Special galvanic cell elements are sometimes inserted in moorings in order to free components of the mooring after a certain time for subsequent retrieval.

Fouling is another problem which does not exist on land. The standard protection is to use antifouling paints (Berendsen, 1975). A certain relation exists between fouling and fish attack on a buoy-cable system; the former is often the start of a food chain at the mooring site. An effective protection against fish bite is the use of steel wire ropes down to about 1000 m.

The third characteristic property of the ocean environment is pressure. Many new materials and techniques have been developed in connection with deep-sea moorings to meet the enormous pressures at great depths. Certainly, metallic shells can be made to resist the highest occurring pressures, but they would be useless as buoyancy elements because of their weight. Instead, thin shells can be filled with fluids lighter than water, or with gas, and the pressure differential across the shell can be kept small by a convenient method, for example a diaphragm. Further nonmetallic materials are in use. Of these, glass and syntactic foam, a composite material of small glass balls and plastics, is most effective and suited for the greatest depths. A single element is restricted to small dimensions by manufacturing techniques, but they can be clustered or attached to frames, thus producing floats of desired boyancy. Many materials, especially plastics, change their properties with increasing depth, which may influence the materials selected for deep-sea moorings. For mooring components see Chapter 35.

4.6 Deployment and Retrieval Requirements

Two methods are used for the deployment of one-point buoy moorings (Berteaux, 1976). Either the buoy or the anchor is launched first. In the former method, after buoy launch, the mooring line, probably equipped with instruments and sensors, is paid out while the ship is slowly moving ahead. Finally, the anchor is lowered to the ocean bottom or allowed to fall freely. The second method (anchor-first) is, in principle, the opposite. The anchor is dropped overboard first and lowered, or fixed to an auxiliary buoy and dropped when the whole line has been paid out. Which method is preferred depends on diverse conditions, for example precision of reaching a predetermined location, weather and ship traffic at the site, or instrumentation requirements of the mooring line. If an auxiliary buoy is used in the anchor-first case, there is essentially no difference between the two methods. The first method is generally favoured, since instrument lines loaded with the anchor weight are difficult to handle.

In contrast to one-point moorings, complex multipoint moorings need much more planning, including computer simulations. The deployment techniques are essentially identical, with the distinction that complex moorings demand more precision.

The retrieval of buoy-cable systems is straightforward and the same for one-point and multipoint moorings. In case a system fails to be retrieved as a whole, the major objective is to gain the experimental data and to recover the instruments and the most expensive parts of the mooring arrangement. Therefore, buoyancy elements and release facilities are appropriately inserted directly in the mooring line so that as much as possible of the failed end of the mooring comes to the surface. The buoyant elements can be lumped in one package immediately above the anchor release or discretely distributed along the mooring line. Berteaux (1971) has reported on experience gained with the deployment of buoys.

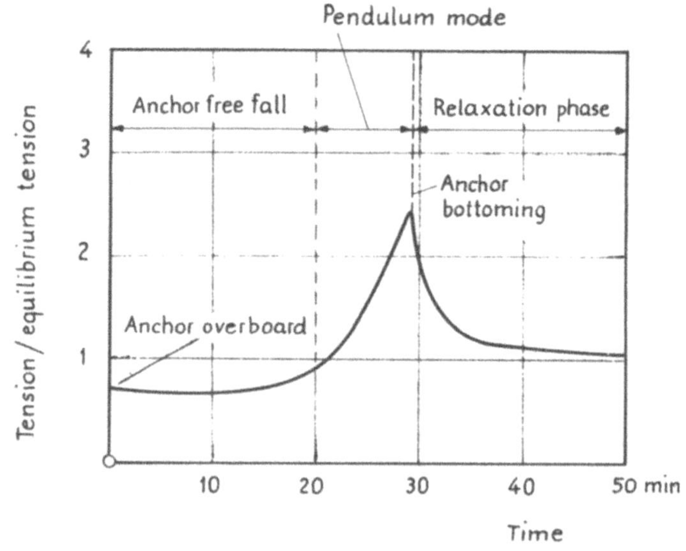

Fig. 17 History of tension at the buoy during launching of a taut mooring; buoy-first deployment, freely falling anchor (after Berteaux, 1971).

With regard to system design, the deployment, mission phase, and retrieval of a buoy-cable array must be regarded as a whole. The array should be examined carefully as to when and where peak loads are likely to occur. Often the highest mooring line tensions are found during deployment. In the anchor-first procedure the mooring line is loaded with the anchor weight until the anchor reaches the bottom. Moreover, dynamic forces due to ship motion stress the mooring line. High peaks of mooring line stress can occur in the procedure of buoy-first deployment with the anchor falling freely (Fig. 17). This must be accounted for in taut moorings, for which highly extensible fibre ropes are used, because the permanent mooring line strain is generated by the peak loads. A summary of mooring line material properties was given by Wilson (1969).

If a mooring fails and is lost, not only has the experiment been in vain, but no information is gained on the cause of failure, which might initiate successful corrective measures. To achieve progress oceanographers should not forget that as much information as possible on the behavior of successful buoy systems should be delivered to the ocean engineer, who often is only engaged in the design and perhaps, trial phase but not in the field application of the system.

REFERENCES

BERENDSEN, A.M. 1975. *Ship painting manual.* Deboer Maritiem-Verfinstituut, Delft, Netherlands, 197 pp.

BERTEAUX, H.O. 1971. An engineering review of the Woods Hole Oceanographic Institution buoy program. *Proceedings, Colloque International sur l'Exploitation des Oceans,* Bordeaux, France, March 1971, Vol. 1, Thème V-G1-09. 21 pp. Centre National pour l'Exploitation des Oceans (CNEXO), Paris France.

BERTEAUX, H.O. 1976. *Buoy engineering.* John Wiley and Sons, New York, U.S.A., 314 pp.

COMSTOCK, J.P. (ed.) 1970. *Principles of naval architecture.* The Society of Naval Architects and Marine Engineers, New York, U.S.A., 827 pp.

DRAPER, L. (ed.) 1975. Waverider discussion. *Proceedings, Conference, National Institute of Oceanography, Wormley, U.K.,* 1972, 148 pp.

FALTINSEN, O.M., and F.C. MICHELSEN. 1975. Motions of large structures in waves at zero Froude number. The dynamics of marine vehicles and structures in waves, edited by R.E.D. Bishop and W.G. Price, pp. 91-106. The Institution of Mechanical Engineers, London, U.K., 1975.

GOODMAN, T.R., P. KAPLAN, T.P. SARGENT, and J. BENTSON. 1972. Static and dynamic analysis of a moored buoy system. Oceanics, Inc., Plainview, N.Y., U.S.A. Prepared for National Data Buoy Center, Contract NAS8-26879, 62 pp.

GRIM, O. 1960. A method for more precise computation of heaving and pitching motions both in smooth water and in waves. *Third Symposium on Naval Hydrodynamics,* Scheveningen, Netherlands, September 1960, pp. 483-523, edited by S.W. Doroff, Office of Naval Research, Department of the Navy, Washington, D.C., U.S.A.

HAVELOCK, T.H. 1956. The damping of heave and pitch: A comparison of two-dimensional and three-dimensional calculations. *Transactions, Royal Institute of Naval Architects,* 98: 464-468.

HOERNER. S.F. 1965. *Fluid-dynamic drag.* Published by the author, Midland Park, New Jersey, U.S.A., 454 pp.

HOFFMAN, D., E.S. GELLER, and C.S. NIEDERMAN. 1973. Mathematical simulation and model tests in the design of data buoys. *Transactions of the Society of Naval Architects and Marine Engineers (SNAME),* 81: 243-273.

HOOFT, J.P. 1975. Motions of stationary structures. The dynamics of marine vehicles and structures in waves, edited by R.E.D. Bishop and W.G. Price, pp. 68-79. The Institution of Mechanical Engineers, London, U.K., 1975.

JASPER, N.H. 1957. Statistical distribution patterns of ocean waves - induced ship stresses and motions, with engineering applications. David Taylor Model Basin (DTMB) Report 921, Washington, D.C., U.S.A., 59 pp.

KAPLAN, P., A.I. RAFF, and T.P. SARGENT. 1972. Experimental and analytical studies of buoy hull motions in waves. Report 72-89, Oceanics, Inc., Plainview, N.Y., U.S.A. Prepared for National Data Buoy Center, Contract NAS8-26879, 161 pp.

KIM, W.D. 1963. On the forced oscillations of shallow-draught ships. *Journal of Ship Research,* 7 (2): 7-18.

KOTCHIN, N.J., I.A. KIBEL, and N.W. ROSE. 1954. *Theoretical Hydrodynamics.* (Theoretische Hydromechanik, in German, translated from Russian), Vol. 1, Akademie-Verlag, Berlin, 507 pp.

MASUBUCHI, K. 1970. *Materials for ocean engineering.* The MIT Press, Massachusetts Institute of Technology, Cambridge, Mass., U.S.A., 542 pp.

MERCIER, J.A. 1971. Hydrodynamic forces on some float forms. *Journal of Hydronautics,* 5 (4): 109-117.

NEWMAN, J.N. 1963. The motions of a spar buoy in regular waves. David Taylor Model Basin (DTMB) Report 1499, Washington, D.C., U.S.A., 27 pp.

NEWMAN, J.N. 1977. *Marine hydrodynamics.* The MIT Press, Massachusetts Institute of Technology, Cambridge, Mass., U.S.A., 402 pp.

NEWMAN, J.N. 1978. The theory of ship motions. *Advances in Applied Mechanics,* edited by Chia-Shun Yih, Vol. 18: 221-283. Academic Press, New York, U.S.A.

OHKUSU, M. 1970. On the motion of multihull ships in waves (I). Reports of the Research Institute for Applied Mechanics, 18 (60), 33-60. Kyushu University, Japan.

OHKUSU, M., and M. TAKAKI. 1971. On the motion of multihull ships in waves (II). Reports of the Research Institute for Applied Mechanics, 19 (62): 75-94. Kyushu University, Japan.

PATTON, K.T. 1965. Tables of hydrodynamic mass factors for translational motion. Technical Paper No. 65-WA/UNT-2, American Society of Mechanical Engineers (ASME), New York, U.S.A., 7 pp.

PRICE, W.G. and R.E.D. BISHOP. 1974. *Probabilistic theory of ship dynamics.* John Wiley and Sons, New York, U.S.A., 311 pp.

SPIESS, F.N. 1968. Oceanographic and experimental platforms. *Ocean Engineering,* edited by J.F. Brahtz: 553-587, John Wiley and Sons, New York, U.S.A.

ST. DENIS, M. 1975. On the motion of oceanic platforms. The dynamics of marine vehicles and structures in waves, edited by R.E.D. Bishop and W.G. Price, pp. 113-134. The Institution of Mechanical Engineers, London, U.K., 1975.

UHLIG, H.H. 1971. *Corrosion and corrosion control.* John Wiley and Sons, New York, U.S.A., 381 pp.

WANG, S. 1976. Dynamical theory of potential flows with a free surface: A classical approach to strip theory of ship motions. *Journal of Ship Research,* 20: 137-144.

WEHAUSEN, J.V. 1971. The motion of floating bodies. *Annual Review of Fluid Mechanics,* edited by M. Van Dyke et al., 3: 237-268, Annual Reviews, Inc., Palo Alto, California, U.S.A.

WIEGHARDT, K. 1972. On the wind profile at sea (In German; original title: Zum Windprofil über See). *Schiffstechnik,* 19: 35-37.

WILSON, B.W. 1969. Elastic characteristics of moorings. *Topics in Ocean Engineering,* edited by C.I. Bretschneider, 1: 45-81, Gulf Publishing Company, Houston, Texas, U.S.

35

Mooring Dynamics

H.O. Berteaux

1. INTRODUCTION

Because of their inherent capacity of providing long-term series measurements of meteorological and oceanographic variables, a relatively large number of moored buoy systems are each year deployed in the world's oceans.

The simplest and most common buoy systems consist of a surface or subsurface float, a single mooring line which supports the sensors, and one anchoring point (Fig. 1 and 2). More complex systems may have several floats and several anchoring points. Buoy systems, their advantages and disadvantages, as well as many practical details on mooring deployment and retrieval, are reviewed in detail in this text by Blendermann (see Chapter 34).

Moorings are constantly subjected to the action of ocean waves and currents. As a result, the sensors these moorings support are constantly moving in the environment which they try to monitor. This motion is a source of errors, or noise, that both scientists and instrument designers must and do recognize. The severity of this problem is discussed in this text by several authors (see Halpern, Chapter 7, and McCullough, Chapter 6).

To evaluate and possibly correct these errors, two types of models are needed: one which describes the dynamic response of specific moorings to the environmental excitation, and one which describes the response of specific instruments to mooring motion. A brief review of the techniques for predicting, measuring, and eventually reducing mooring motion is hereafter presented.

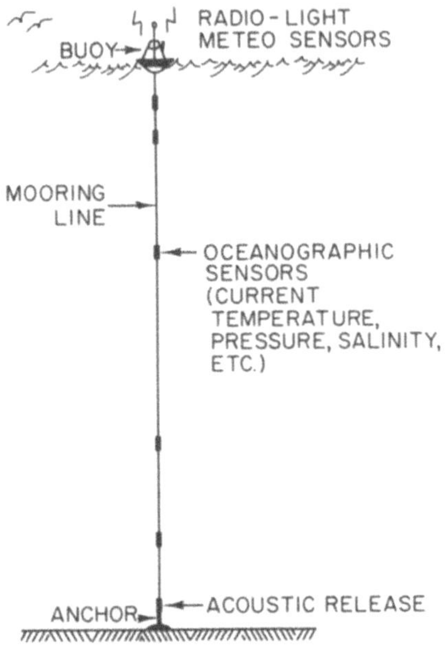

Fig. 1 Single point moored surface buoy system.

2. PREDICTION OF MOORING MOTION

Methods of mooring dynamic analysis vary with the types of buoy systems to be studied and the types of excitation considered.

The low frequency mooring motion resulting from slowly varying current is often investigated by comparing the successive equilibrium configurations that moorings will assume when subjected to a sequence of steady state current profiles. This approach is based on mooring static analysis, a subject reviewed in detail by Berteaux (1976).

When studying the high frequency response of buoy systems to wave excitation, it is often practical to separately investigate the motions of the buoy from the motions of the anchoring lines.

2.1 Buoy Dynamics

Like other free-floating bodies, buoys can undergo three

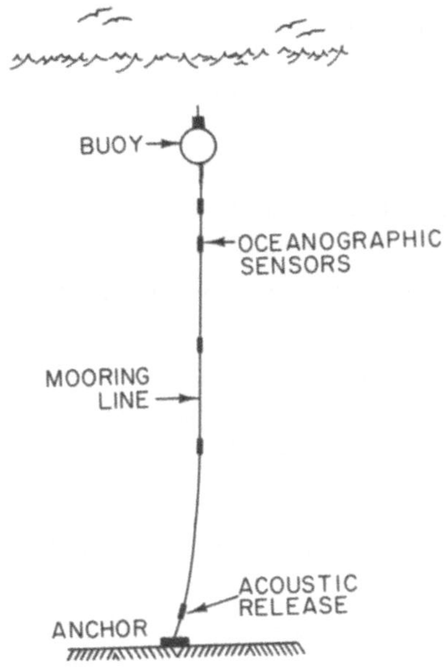

Fig. 2 Single point moored subsurface buoy system.

translational motions (surge, sway, and heave) and three rotation motions (pitch, roll, and yaw). In the case of axisymmetric buoys, only four degrees of freedom need to be considered, heave and roll often being of predominant concern.

The approach normally followed to investigate the dynamic behavior of a given buoy is to first consider how this buoy responds to simple harmonic waves. To this end, the linear differential equations expressing the various types of buoy response (heave, roll, etc.) to a simple harmonic wave of unit amplitude and frequency ω are formulated and integrated. The functions $H_i(\omega)$ obtained from this integration process are called frequency response functions. They describe, in terms of the physical parameters at play and of the wave frequency ω, how the input (wave) and the output (buoy motion) are related both in amplitude and phase. The amplitude of a transfer function $|H_i(\omega)|$ is called the gain, or magnification factor, or the response amplitude operator (RAO). The immediate usefulness of RAOs is that they permit easy prediction of the response of a certain buoy to a wave of known amplitude and frequency.

Suppose, for example, that one is interested in comparing the heave response of two cylindrical buoys of different aspect ratios. One, a flat disk, should be a perfect wave follower over most wave frequencies. The other, a long and slender cylinder, properly ballasted so as to float upright, will have a behavior which is difficult to quantitatively predict.

The equation of heave motion of a cylindrical buoy of cross section A, mass m, and draft D is given by:

$$cx + b\dot{x} + m_v\ddot{x} = F_0 \cos(\omega t + \sigma) \qquad (1)$$

where $c = \rho g A$ is the buoy restoring constant, b is the buoy linearized damping coefficient, $m_v = m + m'$ is the virtual mass of the buoy, m' being the added mass, $F_0 = re^{-kD}[(c - m'\omega\delta)^2 + b^2\omega^2]^{\frac{1}{2}}$ is the exciting force due to a wave of amplitude r, wavenumber k, and frequency ω, and $\sigma = \tan^{-1} [b\omega/(c - m'\omega^2)]$ is the phase angle between the force and the wave.

The corresponding heave RAO is in turn given by:

$$RAO = \frac{[(c-m'\omega^2)^2 + (b\omega)^2]^{\frac{1}{2}}}{m_v[(p^2-\omega^2)^2 + 4n^2\omega^2]^{\frac{1}{2}}} e^{-kD} \qquad (2)$$

where $2n = \frac{b}{m_v}$ and $p^2 = \frac{c}{m_v}$.

Graphs of the heave RAOs of the two buoys considered are shown in Figure 3. With the help of these graphs, one can quantitatively predict the heave amplitude of either buoy when excited by waves of given periods and amplitudes. For example, the heave amplitude of the flat disk and of the spar will be 1.0 m and 3.3 m respectively when both buoys are excited by a wave with a 5 s period and with an amplitude of one metre.

As expected, the RAO of the flat disk has a value of one over most wave periods. On the other hand, the spar buoy shows that it will magnify the amplitude of the waves with periods between 4 s and 10 s, at which time it becomes itself a wave follower. The first buoy could therefore constitute a good platform from which to make wave ampitude measurements by simply recording heave motion. Obviously, making the same measurements with the second buoy would be problematic.

Regular swells are exceptional sea states, and indefiniteness of the sea surface prevails most of the time. Confronted with this irregularity, one must resort to statistical analysis to describe

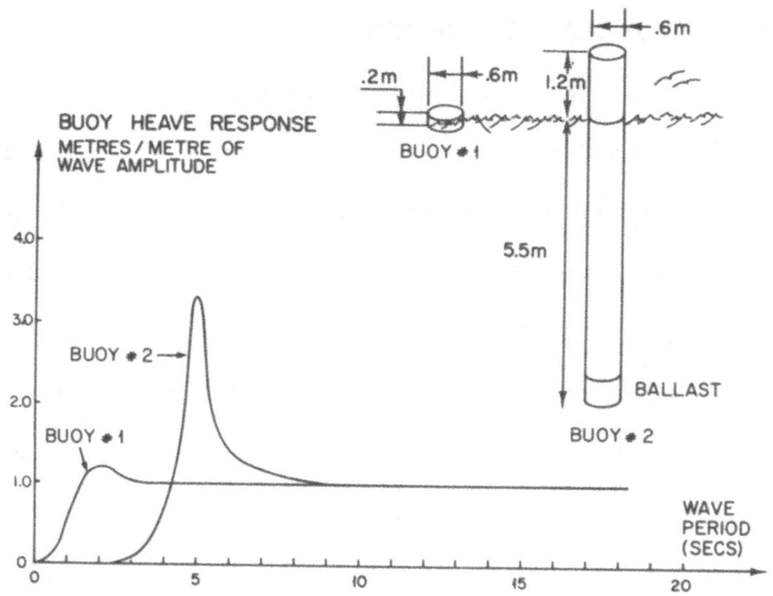

Fig. 3 Heave response amplitude operators of two cylindrical buoys.

buoy dynamic response in terms of means and maxima which are likely to occur. Examples of these could be the significant heave, that is, the mean of the one-third highest heave amplitudes, or the largest roll angle that the buoy could see in a thousand waves.

Experience has shown that the amplitudes of heave, roll, vertical acceleration, and so forth, of ships often do follow the Rayleigh distribution function

$$p(r) = \frac{2r}{\overline{r^2}} e^{-r^2/\overline{r^2}} \tag{3}$$

where $\overline{r^2}$ is the mean square of the response considered (Comstock, 1967). When this probability function is used to compute the expectation of response means and maxima, the results are found to be proportional to the root mean square (rms) value of the response amplitudes (see table).

Table 1

Buoy Responses for Rayleigh Statistics

1. AVERAGE BUOY RESPONSE AMPLITUDE	
The most frequent amplitude will be	$0.707 \sqrt{R}$
The significant amplitude	$1.416 \sqrt{R}$
The average of the 1/10 highest	$1.800 \sqrt{R}$
The mean amplitude	$0.886 \sqrt{R}$
2. EXPECTED BUOY MAXIMUM RESPONSE	
In 50 waves the response may be as large as	$2.12 \sqrt{R}$
In 100 waves it may be	$2.28 \sqrt{R}$
In 1000 waves it may be	$2.78 \sqrt{R}$
In 10,000 waves it may be	$3.13 \sqrt{R}$
In 100,000 waves it may be	$3.47 \sqrt{R}$

When response measurements are available, the rms can be directly obtained from the record. On the other hand, if no measurements are available, and if the response is to be predicted for a certain sea state, then the rms must be derived by computation. To this end, the assumption is made that the buoy response to a sum of sinusoids describing the sea state equals the sum of the responses to the individual sinusoids. If $S(\omega)$ is the spectral density of the sea state, the response of the buoy to a component wave of the spectrum is given by

$$\lim_{d\omega \to 0} H_i(\omega)[S(\omega)d\omega]^{\frac{1}{2}} \tag{4}$$

where $H_i(\omega)$ is the RAO previously defined. The quantity

$$H_i^2(\omega)S(\omega)d\delta \tag{5}$$

thus represents the mean square value of the response in the frequency band $d\omega$.

The mean square value of the response to all component waves will therefore be

$$\overline{r^2} = \int_0^\infty H_i^{\,2}(\omega)S(\omega)d\omega = \int_0^\infty R(\omega)d\omega = R \qquad (6)$$

from which the rms can be simply computed, i.e.

$$rms = \sqrt{R} \ .$$

To illustrate the use of this technique, let R be the area under the heave response spectrum of a particular buoy. Then, using the results of the table, the average heave will be given by 0.886 $\sqrt{R}$, whereas the largest heave in a thousand waves would be 2.78 $\sqrt{R}$.

That linear Response Amplitude Operators can be used to provide meaningful predictions of ship and buoy dynamics has been proposed and verified by a number of authors (Marks 1963; Kaplan et al., 1972; Price, 1976). The application of spectral analysis and statistics to investigate the stochastic behavior of ships (and thus buoys) is particularly well reviewed by Price and Bishop (1974).

2.2 Mooring Line Dynamics

The dynamic response of buoy anchoring lines to the excitation of the waves has been investigated with the help of two different models.

In the first, the mooring line is treated as a continuous medium with an elastic behavior characterized by a known equation of state. In the second, the mooring line is represented by a finite number of discrete, or lumped, masses connected to each other by linear or nonlinear restoring and damping elements. Deterministic and - under linearized conditions - stochastic solutions have been obtained with either type of model.

Continuous elastic medium. The simplest model in this class of solutions is one where the vertical mooring line is treated as continuous, linearly elastic, undamped cable which is fixed at one end and excited by wave action at the other (see Fig. 4). This model is particularly useful when studying single point taut and homogeneous moors (Nath, 1970; Yamamoto et al., 1974).

In this solution the longitudinal displacement of a point at a distance 's' from the origin $\xi(s,t)$ is governed by the one-dimensional elastic wave equation

$$\frac{\partial^2\xi}{\partial s^2} = \frac{1}{a^2}\frac{\partial^2\xi}{\partial t^2} \qquad (7)$$

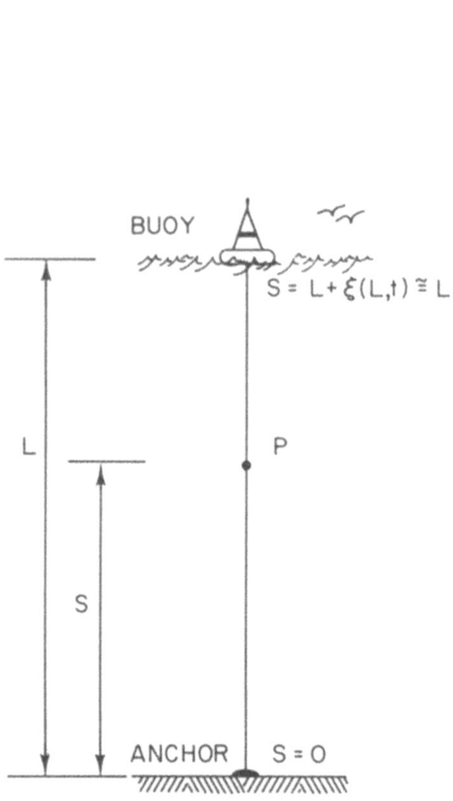

Fig. 4 Mooring line continuous elastic medium model.

where $a = (AE/\mu)^{\frac{1}{2}}$ is the speed of propagation of the longitudinal displacement along the line, A being the line cross section, E the line modulus of elasticity, and μ the virtual mass per unit of length of the immersed cable.

The integration of the wave equation is straightforward: a general solution is of the form

$$\xi(s,t) = \sum_{i=1}^{\infty} (A_i \cos q_i t + B_i \sin q_i t)(C_i \cos \frac{q_i}{a} s + D_i \sin \frac{q_i}{a} s) \quad (8)$$

The values of the four constants can be derived from specified initial and boundary conditions. The tension T(s,t) in the line is then given by

$$T(s,t) = ps + To + EA \frac{\partial \xi}{\partial s} \quad (9)$$

where p is the immersed weight per unit of length, and T the original static tension.

The following example illustrates the usefulness of this type of solution. Let us consider a single point taut moor with a surface buoy of physical characteristics such that inertial forces are predominant. Let $s = 0$ be at the sea bottom, and $s = L$ be at the surface. Then the boundary conditions of the system are:

- at the bottom, $\xi(0,t) = 0$ (no motion)

- at the surface, the sum of the forces on the buoy including the line tension yields the equation of heave motion which must be satisfied when $s = L$, namely

$$m'(\ddot{\xi}\big|_L - \ddot{y}) + C(\xi\big|_L - y) + EA\,\frac{\partial\xi}{\partial s}\Big|_L = m_v\ddot{\xi}\big|_L \qquad (10)$$

where $m_v = m + m'$ is the buoy virtual mass, C is the restoring constant, and y is the water surface, i.e. $y = r \sin \omega t$, r being the wave amplitude.

When the initial condition $\xi(s,0) = 0$ and the two boundary conditions above are introduced in the particular solution of the wave equation given by

$$\xi = (A\cos qt + B\sin qt)(C\cos \frac{q}{a}\, s + D\sin \frac{q}{a}\, s), \qquad (11)$$

the result is

$$\xi(s,t) = \left(\frac{r(c - m'\omega^2)}{\sin(\omega L/a)(c - m_v\omega^2) + EA\,\frac{\omega}{a}\cos\frac{\omega L}{a}}\right) \sin \frac{\omega}{a}\, s \,\sin \omega t. \qquad (12)$$

This result expresses the steady state dynamic response of the mooring line to a wave of amplitude r and frequency ω. Resonant frequencies of the system are given by the roots of the denominator. At $s = L$, $\xi(L,t)$ gives the heave motion of the moored buoy. When divided by r, it becomes the heave RAO of the buoy.

The vertical displacement of a point 'i' in the mooring line at a distance $s = s_i$ is, of course, $\xi(s_i,t)$. The cyclic stresses in the line, at any point 'i' can be obtained from

$$\sigma(s_i,t) = \frac{T}{A} = E\,\frac{\partial\xi}{\partial s}\Big|_{s = s_i} \qquad (13)$$

Closed loop solutions of this type apply only to a small number of mooring situations. They can be used in first approximation studies and as limiting cases for comparison with results obtained from more complex models.

In the majority of moorings, the cable path is three-dimensional. Furthermore, in addition to the buoy, at least part of the cable and attached instrumentation is subjected to time-varying hydrodynamic forces. To investigate the cable response to this excitation, the differential equation of motion of a cable element is first established and then integrated along the cable path.

As usual, the equation of motion of the cable element of length ds and mass μds is obtained from Newton's law

$$\sum_i \vec{F}_i = \mu ds\, \vec{a}$$

where the forces $\vec{F}_i$ to be considered must include the resultant of the gravity forces on the cable element (difference between weight and buoyancy), the normal and tangential hydrodynamic drag forces resulting from the respective relative velocity components, the inertial force due to the water added mass effect produced by the relative acceleration, and the tension vectors at both ends of the cable element. Summing these forces and accounting for the material elasticity result in a complex vectorial nonlinear differential equation which is difficult to manipulate. The interested reader can find excellent descriptions of integration techniques and related computer programs by Reid (1968), Nath (1969), Patton (1972), and Breslin (1974).

Lumped mass. The dynamic response of buoy systems where many mass discontinuities exist along the cable is often investigated using a lumped mass model. The value of the restoring and damping elements connecting the 'n' masses of the system is determined by the physical characteristics of the moored array components such as line elasticity and weight, drag and buoyancy of inserted floats, instruments, etc. The multiple degrees of freedom system thus obtained is then studied under different forcing conditions. Such a model is depicted in Figure 5.

The matrix equation to be solved in this approach is:

$$[M]\{\ddot{x}\} + [C]\{\dot{x}\} + [K]\{x\} = [F(x,\dot{x},\ddot{x},t)] \tag{14}$$

where [M] is the inertial coefficients matrix, [C] is the damping coefficients matrix, and [K] is the stiffness coefficients matrix. $\{x\}$, $\{\dot{x}\}$ and $\{\ddot{x}\}$ are the column matrices of nodes' displacement amplitudes, speeds, and acceleration, and $[F(x,\dot{x},\ddot{x},t)]$ is the matrix of the external forces applied at the nodes.

Most of the time wave action is considered to act exclusively on the float at the beginning of the mooring line. With this assumption, the force matrix is reduced to one term and the analysis is considerably simpler (Polachek et al., 1963; Brainard, 1971).

Valuable information can be obtained from the free response of the system. The solution of

$$[M]\{\ddot{x}\} + [K]\{x\} = 0 \tag{15}$$

will yield the spectrum of the system's natural frequencies and the principal modes of vibration. These results can then be used to investigate the system response when excited at frequencies close to resonance and also to evaluate the degree of coupling between nodes. The merit of placing damping elements in the line or changing line elastic characteristics can thus be established.

Stochastic solutions. When the equations of motion are linear (or linearized), as for example in the first model described (elastic wave equation), then the dynamic response of moorings to the random excitation of the sea is best investigated in the frequency domain. In this approach the first step is again to solve the linear differential equations describing the motion of selected material points along the mooring line, assuming that its upper end is subjected to the action of a wave which has an amplitude of one unit and a frequency ω. The result of this integration process is a

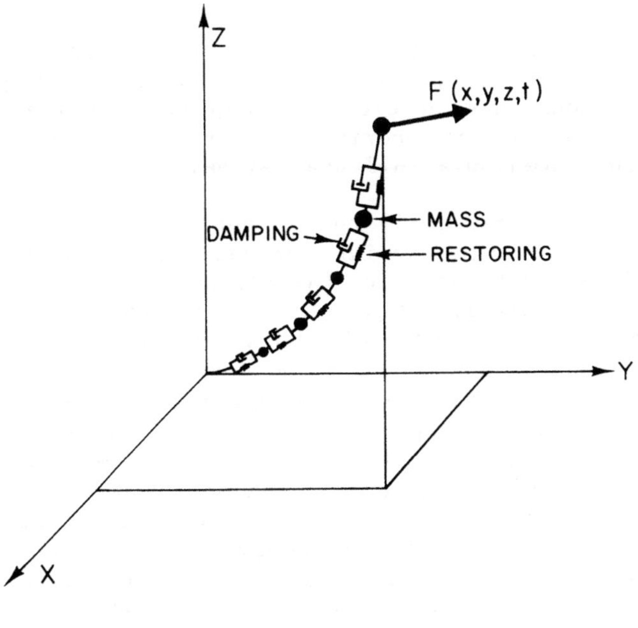

Fig. 5 Mooring line lumped mass model.

set of RAOs $\{H_i(\omega)\}$ which express as a function of ω the response of the 'i' material points to this particular wave. The RAOs thus obtained can be used in the manner previously described to investigate the response of the line to regular waves of given amplitude and frequency or in conjunction with a known wave spectral density to predict certain expectations of response means and maxima. The possibility of using this powerful method to study mooring and cable dynamics deserves more attention than it has received (Hong, 1972; Firebaugh, 1972).

The numerous analytical methods for dynamic simulation of cable moored buoy systems have been reviewed and classified by Casarella and Choo (1973), Dillon (1973), and Albertsen (1974).

3. MEASUREMENTS OF MOORING MOTION

The validation of the mathematical models describing the motions of buoys moored in deep water rely on the acquisition of experimental data. These data are obtained from scale or full size tests. Scale models of various buoy hulls have been extensively tested in wave tanks, and test results have been published by Newman (1963), Devereux and Jennings (1966), DeSaix (1968), Adee and Bai (1970), Kaplan et al. (1972), and Hoffman et al. (1973). The response of full size buoys to random seas has also been investigated, but published results are few (Price, 1976). Scale models of moored buoy systems have been tested with a limited degree of success (Hoffman et al., 1973; Pattison, 1977).

Experiments to measure simultaneously the forcing functions of the environment and the response of moorings are relatively few. Because of the difficulty of scaling down the large lengths of mooring lines commonly encountered in deep sea applications and the elastic behavior of the mooring line material, such experiments have to be conducted on a full scale level.

A first attempt at measuring the response of a subsurface mooring when acted on by slowly varying currents was conducted jointly by the Woods Hole Oceanographic Institution and the Massachusetts Institute of Technology Draper Laboratory in 1972. The mooring extended down from a depth of 500 m to the anchor at 5460 m. Measurements of relative current speed were obtained with the help of current meters inserted in the mooring line. An acoustic transceiver placed near the top of the mooring provided range measurements from three bottom-mounted transponders. Motion of the top of the line could thus be monitored. The motion was found to be periodic (~12 hr period) and rotary in a clockwise direction. Maximum horizontal excursion of the transceiver was approximately 100 m. Space configuration of the mooring from top to bottom was inferred from measurements of pressure and line inclination made along the line. Data thus obtained were used to validate a mooring dynamics computer program (Chhabra et al., 1974). Velocity records from meters placed on moorings for which this validated model apply could thus be corrected to account for mooring motion (Chhabra, 1977).

In a more recent and somewhat similar experiment, the motion of acoustic transceivers located close to the top of two identical subsurface moorings was monitored with great accuracy for a period of nine days. The moorings, each 3850 m long, were deployed in an area southwest of Bermuda in waters 5200 m deep. The observed mooring motion was found to be quasi-circular with a diameter of approximately 100 m and a period close to the local semidiurnal tide period (Spindel et al., 1977).

Recent and major experiments to measure the response of moorings include the Sea Lane Buoy Project (NOAA, 1970), the shallow water mooring dynamics experiment conducted by the Research and Development Center of the U.S. Coast Guard (Bitting and Lincoln, 1978), and the Hawaii Mooring Dynamics Experiment (Walden et al., 1977). Mooring configurations included in this experiment are shown in Figure 6. These configurations were:

- Two surface moorings, one taut, one slack, with a discus buoy at the surface.

- A surface spar buoy attached to a moored subsurface buoy by buoyant tether cable.

- A deep ocean current measuring system (DOCMS) with a spherical subsurface buoy, a number of instruments, and a wire rope mooring supported by distributed clusters of glass balls.

- A simpler subsurface mooring whose buoy was displaced and released for transient relaxation measurements.

The experiment took place in October 1976 at the Pacific Missile Range Facility, Hawaii. The instrumentation required to monitor the environment (currents, waves) and the mooring dynamic responses (heave, roll and sway of surface buoys, three-dimensional displacement of surface and subsurface buoys, tension in the mooring lines, etc.) are fully described by Walden et al. (1977).

This experiment has made available a considerable amount of excellent mooring dynamics data. Full interpretation of these data has yet to be completed.

The motion of the DOCMS has been found to be within theoretical predictions. The mooring being quite stiff and the currents at the time of the experiment being relatively weak, the motions experienced were reported to be small (5 m horizontal excursion, and 1 m

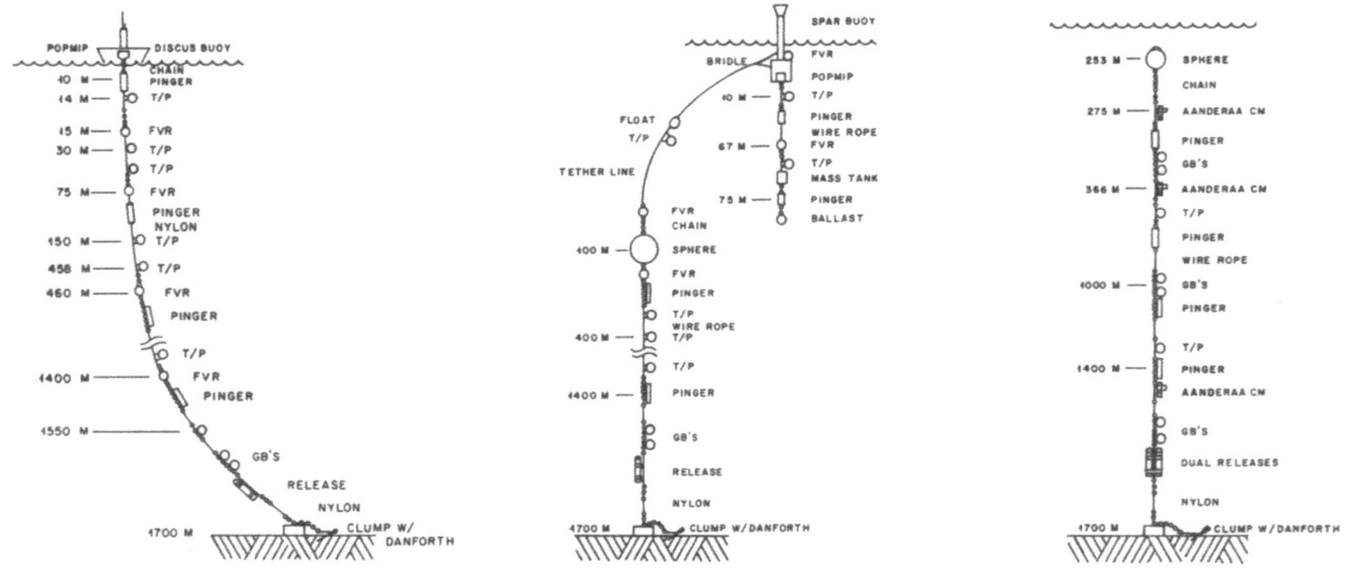

SURFACE MOORING W/DISCUS BUOY SPAR/SPHERE BUOY SYSTEM DOCMS SUBSURFACE MOORING

Fig. 6 Hawaii mooring dynamics experiment - three mooring configurations. Notation: GB = glass balls; T/P = temperature pressure sensors; VACM = vector averaging current meter; CM = current meter; FVR = force vector recorder.

vertical dip; Walden et al., 1977). The data of the relaxation experiments have been partly processed and some results reported (Meggitt and Dillon, 1978). Data from the spar/sphere buoy system are presently being analyzed by the M.I.T. Draper Laboratory.

4. REDUCTION OF MOORING MOTION

The speed and the acceleration of sensors mounted on surface following buoys which heave, roll, and surge as waves pass by can severely contaminate the measurements of barometric pressure and temperature, and of the air and water velocity fields made from these buoys. One alternative to palliate this problem is to reduce sensor motion to a level at which errors become negligible (Pollard, 1973).
Reducing mooring and sensor motion is a difficult proposition to implement in deep ocean locations. Over the years a number of schemes have been devised and sometimes tried with various degrees of success. The 'watch circle' of a single point moored buoy, and therefore the ensuing motion of the attached instruments, is reduced if the mooring is made taut. Unfortunately, however, the tauter the line, the worse its high frequency response to wave action.

A buoy system made of a surface spar buoy tethered to a subsurface buoy, moored itself to the bottom, seems to offer a working compromise for making both deep and near-surface measurements (see Fig. 6). Being decoupled from wave action, the subsurface buoy and its mooring line can be made fairly stable. Sensors on this buoy and the mooring line can monitor the water column from the bottom up to the depth of the subsurface buoy. Due to its property of filtering heave and roll over a wide frequency band, a well designed spar buoy can in turn be a relatively stable platform from which to make atmospheric and sea surface measurements (Berteaux et al., 1977). Furthermore, sensors attached to a line hanging from the spar could monitor the water column from the spar down to the depth of the subsurface buoy with a reduced noise level. This system, which has been used in shallow waters (Martinais, 1971), has yet to be fully evaluated in the deep ocean.

Large moored spar buoy systems, such as TOTEM, FLIP, and the French Bouée Laboratoire BOHRA, have been successfully deployed as deep sea stable platforms. Use of these buoys, however, is limited because of their size and correspondingly large operational cost.

The motion of single-point-moor subsurface buoys can be reduced by increasing the buoyancy-to-drag ratio of the buoy and the mooring line. The tauter the line, the smaller the dip. Multimoors, such as the IWEX trimoor (see Fig. 7), if properly designed, offer much greater stability than single-point moors (Walden and Berteaux, 1974).

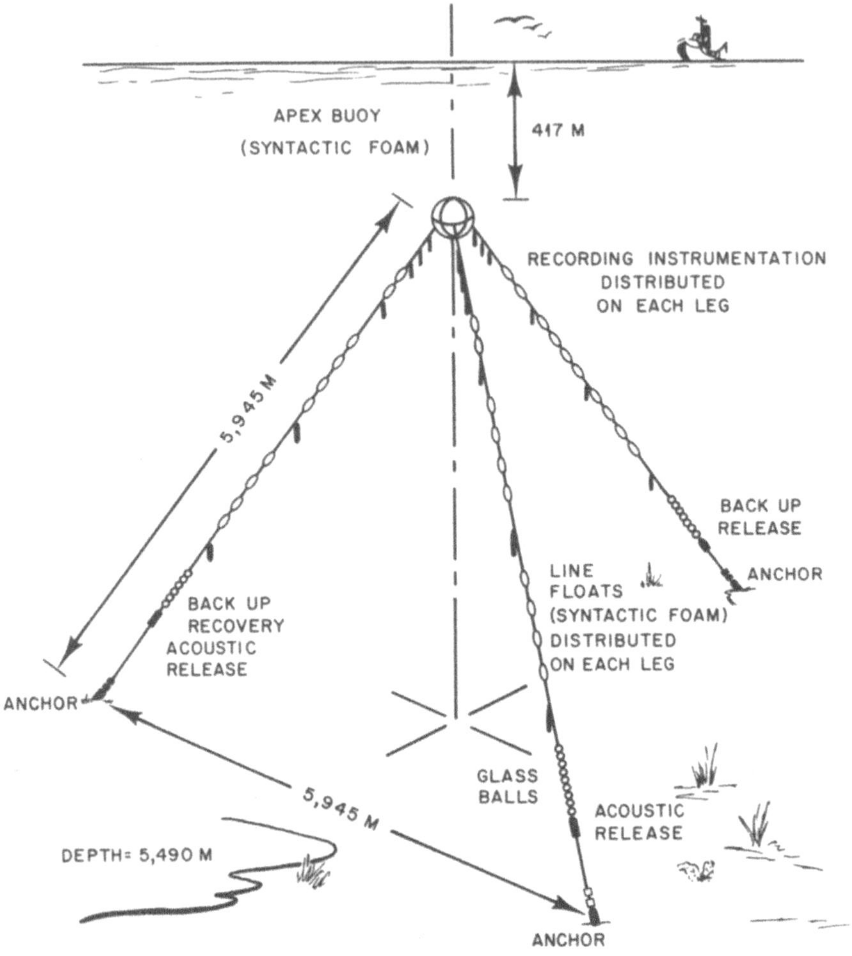

Fig. 7 IWEX Trimoor

Finally, the high frequency vibrations induced by vortex shedding, although perhaps marginal to the general field of mooring mechanics should be mentioned because of their practical importance. When the frequency of vortex formation (Strouhal frequency) is close to the natural vibration frequency of the taut mooring line, strumming will occur. Instruments, if free to oscillate about their point of mooring attachment, will do so if the Strouhal frequency is close to the natural pendulum frequency of the instrument. These vibrations and oscillations will naturally be sources of noise in the recorded data. The problem can be particularly severe when making acoustic measurements. Means of preventing cohesive vortex formation include fairings attached to the mooring (Hays et al, 1975), splitter plates behind instrument casings, and herring bone patterns of ropes wrapped around the casings (Walden and Berteaux, 1974).

ACKNOWLEDGEMENTS

The assistance received from Mr. Dahlen and Dr. R. Spindel in reviewing recorded mooring motion data, and the constructive comments of Dr. E. Hays and Dr. C. Maillard who proofread the manuscript, are gratefully acknowledged. The writing of this chapter was partly supported by the Woods Hole Oceanographic Institution Sea Grant Program.

REFERENCES

ADEE, B.H. and K.J. BAI. 1970. Experimental studies of the behavior of spar type stable platforms in waves. Report No. NA-70-4, College of Engineering, University of California.

ALBERTSEN, N.D. 1974. A survey of techniques for the analysis and design of submerged mooring systems. U.S. Navy Civil Engineering Laboratory, Technical Report R-815, 34 pp.

BERTEAUX, H.O. 1976. *Buoy Engineering*. Wiley Interscience Publication, John Wiley and Sons, New York, 314 pp.

BERTEAUX, N.O., R.A. GOLDSMITH and W.E. SCHOTT, III. 1977. Heave and roll response of free floating bodies of cylindrical shape. Woods Hole Oceanographic Institution, Technical Report 77-12, 112 pp.

BITTING, K.R. and W.B. LINCOLN. 1978. Mooring configuration analysis. Interim Reports 782702.4.2, R&D Center, U.S. Coast Guard.

BRAINARD, J.P. 1971. Dynamic analysis of a single point taut compound mooring. Woods Hole Oceanographic Institution, Technical Report 71-42, 59 pp.

BRESLIN, J.P. 1974. Dynamic forces exerted by oscillating cables. *Journal of Hydronautics*, 8: 19-31.

CASARELLA, M.J. and Y.-I. CHOO. 1973. A survey of analytical methods for dynamic simulation of cable-body systems. *Journal of Hydronautics*, 7: 137-143.

CHHABRA, N. 1977. Correction of vector-averaging current meter records from the MODE-1 central mooring for the effects of low-frequency mooring line motion. *Deep-Sea Research*, 24: 279-287.

CHHABRA, N., J.M. DAHLEN and M.R. FROIDEVAUX. 1974. Mooring dynamics experiment - determination of a verified dynamic model of the WHOI intermediate mooring. The Charles Stark Draper Laboraotry, R-823, 305 pp.

COMSTOCK, J.P. 1967. *Principles of Naval Architecture.* The Society of Naval Architects and Marine Engineers, New York, N.Y., 827 pp.

DeSAIX, P. 1968. Model tests of three dynamic scale model buoys. Letter Report 1340, Davidson Laboratory, Stevens Institute of Technology.

DEVEREUX, R.S. and F.D. JENNINGS. 1966. A special report: ONR ocean data systems. *Geo-Marine Technology,* 2.

DILLON, D.B. 1973. An inventory of current mathematical models of scientific data-gathering moors. Hydrospace-Challenger, Inc., Technical Report 4450 0001, 57 pp.

FIREBAUGH, M.S. 1972. An analysis of the dynamics of towing cables. M.I.T. Department of Ocean Engineering, Doctor of Science thesis, 121 pp.

HAYS, E.R., R. NOWAK and P. BOUTIN. 1975. Strumming tests in two faired cables. Woods Hole Oceanographic Institution, Technical Report 75-47, 10 pp.

HOFFMAN, D., E.S. GELLER and C.S. NIEDERMAN. 1973. *Mathematical simulation and model tests in the design of data buoys.* The Society of Naval Architects and Marine Engineers, pp. 243-269.

HONG, S.T. 1972. Frequency domain analysis for the tension in a taut mooring line. University of Washington, Department of Civil Engineering, Technical Report No. SM 72-1, 77 pp.

KAPLAN, P., A. RAFF and T.P. SARGENT. 1972. Experimental studies of buoy hull motions in waves. NOAA Data Buoy Center, 58 pp.

MARKS, W. 1963. The application of spectral analysis and statistics to seakeeping. The Society of Naval Architects and Marine Engineers, *Technical and Research Bulletin* No. 1-24, 94 pp.

MARTINAIS, J. 1971. A lightweight automatic relay buoy. Colloque International sur l'Exploitation des Oceans, Theme V., Tome 1, 12 pp.

MEGGITT, D. and D. DILLON. 1978. At sea measurements of the dynamic response of a single point mooring during an anchor last deployment. Civil Engineering Laboratory Technical Memo M-44-78-9, 115 pp.

NATH, J.H. 1970. Analysis of deep water single point moorings. Colorado State University Technical Report (unpublished manuscript), 109 pp.

NEWMAN, J.N. 1963. The motions of a spar buoy in regular waves. DTMB Report 1499.

NEWMAN, J.N. 1977. *Marine Hydrodynamics.* The MIT Press, Cambridge, MA, and London, England. 402 pp.

NOAA Data Buoy Systems. Sea lanes test-operations report. Technical Report, NDBCM W 6222-1.

PATTISON, J.H. 1977. Components of force generated by harmonic oscillations of small-scale mooring lines in water. David W. Taylor Naval Ship Research and Development Center, Report No. SPD 589-01.

PATTON, K.T. 1972. The response of cable-moored axisymmetric buoys to ocean wave excitation. NUSC Technical Report 4331, 379 pp.

POLACHEK, H., T.S. WALTON, R. MEJIA and C. DAWSON. 1963. Transient motion of an elastic cable immersed in a fluid. *Mathematics of Computation,* 17: 60-63.

POLLARD, R. 1973. Interpretation of near-surface current meter observations. *Deep-Sea Research,* 20: 261-268.

PRICE, D. 1976. Buoy response amplitude operators obtained from step response tests. Offshore Technology Conference No. 2467, 20 pp.

PRICE, W.G. and R.E.D. BISHOP. 1974. *Probabilistic Theory of Ship Dynamics.* John Wiley and Sons, New York, 352 pp.

REID, R.O. 1968. Dynamics of deep-sea mooring lines. Texas A&M Technical Report 68-11F, 215 pp.

SPINDEL, R.C., R.P. PORTER and J.A. SCHWOERER. 1978. Acoustic phase tracking of ocean moorings. *IEEE Journal of Oceanic Engineering*, 3: 27-32.

WALDEN, R.G. and H.O. BERTEAUX. 1974. Design and performance of a deep-sea tri-moor. Nation Needs and Ocean Solutions. *Proceedings of the 10th Annual Conference of the Marine Technology Society*: 81-96

WALDEN, R.G., C.W. COLLINS, JR., P.R. CLAY and P. O'MALLEY. 1977. The mooring dynamics experiment - a major study of the dynamics of buoys in the deep ocean. *Proceedings of the 9th Annual Offshore Technology Conference,* 10 pp.

WALDEN, R.G., C.W. COLLINS, JR., P.R. CLAY and P. O'MALLEY. 1977. Validation testing of the DOCMS intermediate mooring. Woods Hole Oceanographic Institution Technical Report 77-53, 96 pp.

YAMAMOTA, T., J.H. NATH and C.E. SMITH. 1974. Longitudinal motions of taut moorings. *Journal of the Waterways Harbors and Coastal Engineering Division of the American Society of Civil Engineers*: 35-50.

36

Profiling Devices

J.C. Van Leer

1. INTRODUCTION AND SCOPE

This chapter will cover a brief background of profiler development and sampling problems in the upper ocean. Secondly, the advantages and disadvantages of profile measurements compared to other means of determining the mean properties of the surface layers of the ocean will be discussed. Profiling devices known to be in active use in the upper ocean as of July 1977 will be covered, including depth duration and accuracy limitations. Profile techniques are of broad interest and undergoing rapid development at a number of laboratories, so it would be difficult to trace all such instruments, many of which are not yet tested or described in the general oceanographic literature. For example, the work of Dahlen et al. (1977) or the profiling current meters used in Kiel, Germany, by Müller et al. (1974) are not described in detail in the open literature. Finally, conclusions on the present state and future promise of profile measurement techniques will be presented.

The techniques discussed here are aimed at measuring the mean and slowly varying properties (especially velocity) of the upper ocean (taken to be the upper 300 m) which are the most important to the general topic of air-sea interaction. Vertical resolution of metres and time resolution of hours, rather than the high frequency in time or microstructure in space (which are adequately covered elsewhere: see Chapters 17 to 20) are the primary topics of discussion. See the article by Cox and Lange (Chapter 37) in this volume for free-fall vehicles. Also see the articles by Nasmyth (Chapter 38) on towed vehicles and Halpern (Chapter 7) on instruments moored at fixed depths. Sensors will not be covered here, since they too are adequately addressed elsewhere in this volume.

The comprehensive nature of this book precludes total lack of overlap; for example, the nature of profiling devices discussed here will cover some free-fall acoustically-tracked instruments which measure profiles.

2. HISTORICAL BACKGROUND

Meteorologists have long recognized the value of synoptic profile measurements in understanding and predicting future states of the atmosphere. Radiosondes and rawinsondes have been used to probe the lower atmosphere, while sounding rockets and satellite techniques have been used in sounding the upper atmosphere. High resolution profile methods, such as sensor masts and pibal balloons (see Hasse and Schriever, Chapter 5), have also been used to study the structure of the boundary layer near the ground or sea surface.

In the upper ocean, early profiling was carried out by mechanical bathythermograph (BT) or lowered current meters of Ekman or other types. Mechanical BT measurements developed by Spilhaus (1938) and others quickly showed how much spatial resolution was missing in discretely sampled water bottle casts. It became apparent that accurate estimation of mixed layer heat or salt storage must depend on high resolution vertical profile measurements. The subsequent development of conductivity-temperature-depth (CTD) profile devices by Brown (1974) and others made it abundantly clear that temperature and salinity profiles with complex vertical structure existed within the seasonal thermocline, the main thermocline, and below.

Investigators such as Stommel and Fedorov (1967) and Cox et al. (1969) began to explore this small-scale T-S microstructure and began to notice the troublesome noise induced by ship heave transmitted down a taut wire to a rigidly tethered instrument. Thus, Cox et al. (1969) at Scripps started using free-drop instruments. While the scalar properties are hard enough to measure with a rigidly tethered instrument, the accurate estimation of vector velocity becomes nearly impossible. Woods (1968) and Van Leer (1971) established the existence of velocity structure on the same small scale as the temperature and salinity microstructure near the seasonal and in the permanent thermoclines, respectively, with dye streak techniques. Van Leer (1971) successfully isolated ship heave from tethered current meters with a long flexible spar buoy. Düing and Johnson (1972) isolated vertical heave with a roller coupled instrument descending down a taut wire. Free-fall profilers of the Electro-Magnetic Velocity Profiler (EMVP) type were developed by Sanford (1975) and several acoustically-tracked sinking probes were used by Rossby (1969) and more recently by Luyten and Swallow (1976). Van Leer et al. (1974) developed a moored autonomous velocity profiler for the upper ocean. Winch lowered sensors and acoustic Doppler methods have been used on motion-isolated platforms such as ice islands by Neshyba et al. (1972) and

FLIP by Pinkel (1975) (see also Chapter 10).

3. SAMPLING THE UPPER OCEAN IN SPACE AND TIME

Cost and time considerations prevent oceanographers from making high temporal and spatial frequency measurements for extended times over large oceanic regions. One is forced to make sampling choices, depending on the nature of the phenomenon being sampled. If one chooses to make microstructure measurements, for example, short periods of intense observations in space and time are dictated. For large-scale and mesoscale studies, other lower frequency sampling programs are possible with conscious trade-offs being made to extend record lengths or increase the size of the observation area.

The first question to ask is: What are the temporal sampling demands? We have numerous examples of recordings from fixed-level current and temperature sensors to guide us in our choices. First, we can consider the shape of typical oceanic auto spectra observed below the zone of surface wave interference. Briscoe (1975) reported data from the IWEX mooring in the main thermocline on the Hatteras Abyssal Plain. While these data are from a greater depth than our zone of interest, they are uncontaminated by surface wave-induced mooring motion. The observations show that the great bulk of horizontal and vertical kinematic energy is contained in the mean, tidal, and inertial currents. At frequencies above the semi-diurnal tidal frequency, the energy spectrum falls with a slope of about -2 with increasing frequency. Above the local buoyancy frequency, the energy spectrum falls even more steeply (at a slope of about -4). Neshyba et al. (1972) and Pinkel (1975) have both shown that the use of profile measurements to infer the vertical displacement spectrum shows that energy levels above N, the local Brunt-Väisälä frequency cutoff, fall with a slope of about -8 with increasing frequency. When they compute the same spectrum with data from fixed depths, they find the energy levels fall with the -2 slope above N due to spatial microstructure aliasing, as predicted by the theories of Phillips (1971) and Garrett and Munk (1971). One must conclude from these observations that away from the sea surface, where surface wave orbital velocities are unimportant, very little energy exists at temporal frequencies above local N. From fixed-level current meters Van Leer (1974a) has estimated the energy content above 12 cycles per day to be less than 5% of the total horizontal kinetic energy at 100 m depth on the continental shelf off Oregon. Figure 1 shows as a percentage the integral of horizontal kinetic energy above a given frequency plotted against the frequency of sampling; these data were gathered by R. Smith and the Current Meter Group at Oregon State University, using a five-minute sampling rate and a subsurface mooring. Had this curve been extended to lower frequency, the energy would rise to about 35% at 2 samples per day and 55% at 1 sample per day, with abrupt increases at the inertial period and the daily and semi-daily tides.

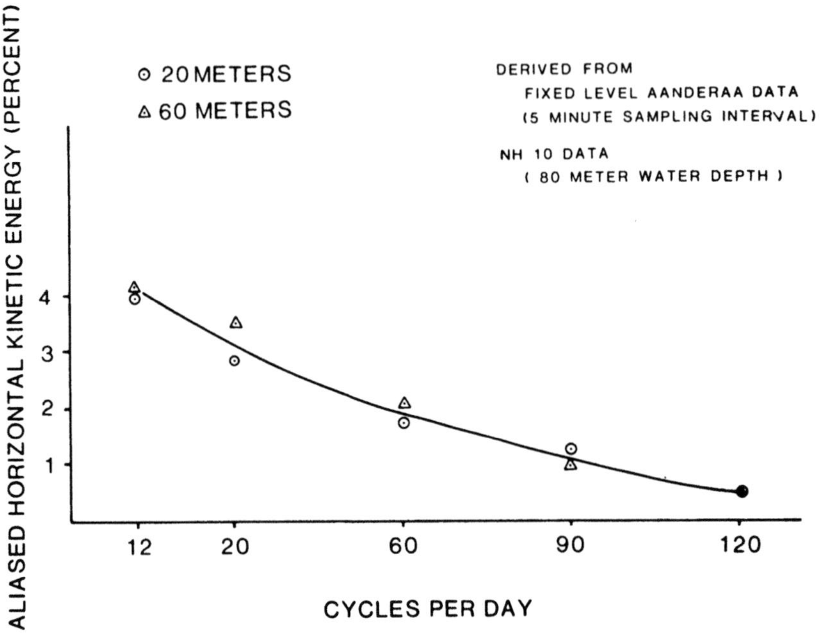

Fig. 1 Aliasing errors as a function of sampling rate shown as the cumulative horizontal kinetic energy above a given frequency plotted versus frequency.

In view of the findings of Neshyba et al. (1972) and Pinkel (1975) above, the high frequencies are clearly an overestimate. The IWEX data and most other uncontaminated current meter data show a clear inertial period peak (often containing the greatest energy levels in the spectrum). Webster (1968) has noted the lack of coherence in the vertical at the inertial period at separations of 100 m or more. It appears that typical current and temperature records from a fixed depth are strongly contaminated by spatial aliasing of unresolved inertial waves. In fact, as one approaches the inertial frequency, there is no limit to the vertical wave number except dynamic instability.

At the high frequency end of the internal wave spectrum, the reverse seems to be true. Pinkel (1975) has shown that in the upper 220 m of the ocean near Hawaii, above 3 to 4 cycles per hour, the coherence squared is consistent with first- or very low-mode internal waves, while from 2 cycles per hour to tidal frequency, higher modes (4-6) predominate. Johnson (1976) has also shown that modes from 1 to 7 are required to represent 95% of the energy in 100 m water depths off Oregon. Figure 2 shows the cumulative variance from the first eight Empirical Orthogonal Eigenfunctions given by Johnson (1976) in his Table 5 averaged over three Cyclesonde moorings in 100 to 200 m water depth.

3.1 Examples of Profile Data

Perhaps the best way to illustrate the complex vertical structure of the upper ocean is to give several examples (Fig. 3-5) from existing profile data. These data show the considerable vertical structure expected not only in the inertial wave and tidal frequency bands, but also in the several-day mean of profiles. Figure 3 is a most striking example of a month-long series of profile records from the equatorial Indian Ocean (Luyten and Swallow, 1976). These data show striking coherence in eastward velocity from profile to profile over the entire month, with what appears to be a series of equatorial currents and counter currents. Johnson et al. (1976) observed, off Oregon, several-day mean currents with

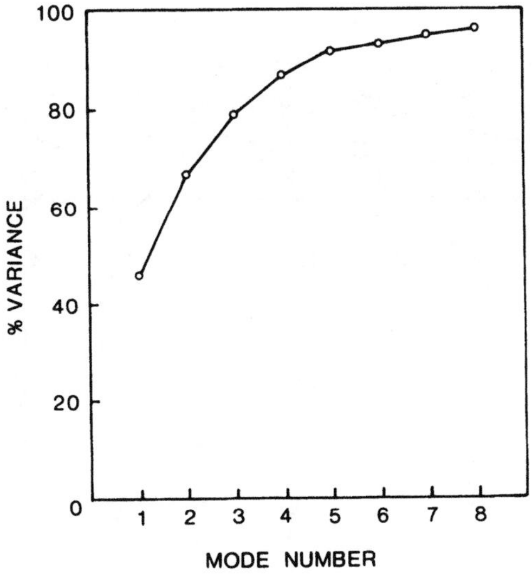

Fig. 2 Cumulative percentage variance in Empirical Orthogonal Eigenfunctions starting with the lowest mode plotted against mode number (from Johnson, 1975).

several zero crossings (see Fig. 4). Examples of the energetic tidal and near-inertial wave motion (Fig. 5a) have been given by Van Leer and Leaman (1978) by contouring a time series of vertical profiles.

Each of these samples of mean and near inertial wave motion would have been difficult to sample without profile techniques. It is abundantly clear that in the upper ocean, sampling resolution of 10 m or better is required to resolve the near-inertial motion and in many cases the mean motion itself, particularly in a direction normal to a front or shoreline where the mean velocity must be nearly zero and frictional effects may be important (as in Fig. 4).

4. ADVANTAGES OF PROFILERS OVER CONVENTIONAL INSTRUMENTS

4.1 Sensor Stability and Accuracy

One of the principal advantages of profile techniques is ease of calibration. If, instead of many, one set of sensors is used, higher quality, more stable sensors can be used, such as platinum thermometers instead of thermistors. Fewer units needing to be calibrated results in time and cost savings. In any one of the profile data sets presented in Figures 3 to 5, 10 to 50 fixed level sensors would have been required to resolve most of the spatial

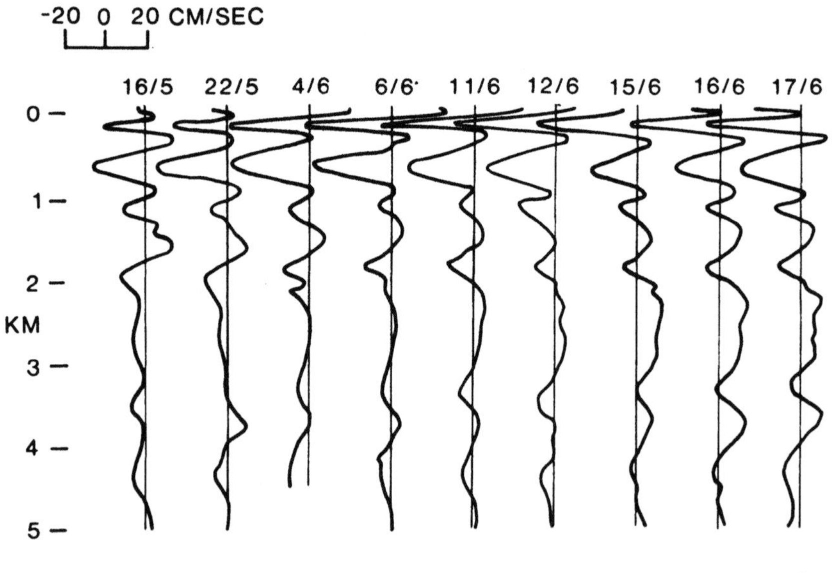

Fig. 3 Eastward components of velocity in vertical profiles at 0°, 53°E: INDEX Equatorial Profiles from the Indian Ocean (Luyten and Swallow, 1976) over a one-month period.

structure observed. Johnson (1976) showed (Fig. 2) that six or more empirical orthogonal vertical modes were required to represent 95% of the variance in his Oregon Cyclesonde analysis. This is consistent with Garrett and Munk (1971) and Sanford (1975) for data from the deep sea. A profiling sensor with 10% or 20% accuracy might well provide a more representative picture of each data set than six ideal error-free fixed-level sensors.

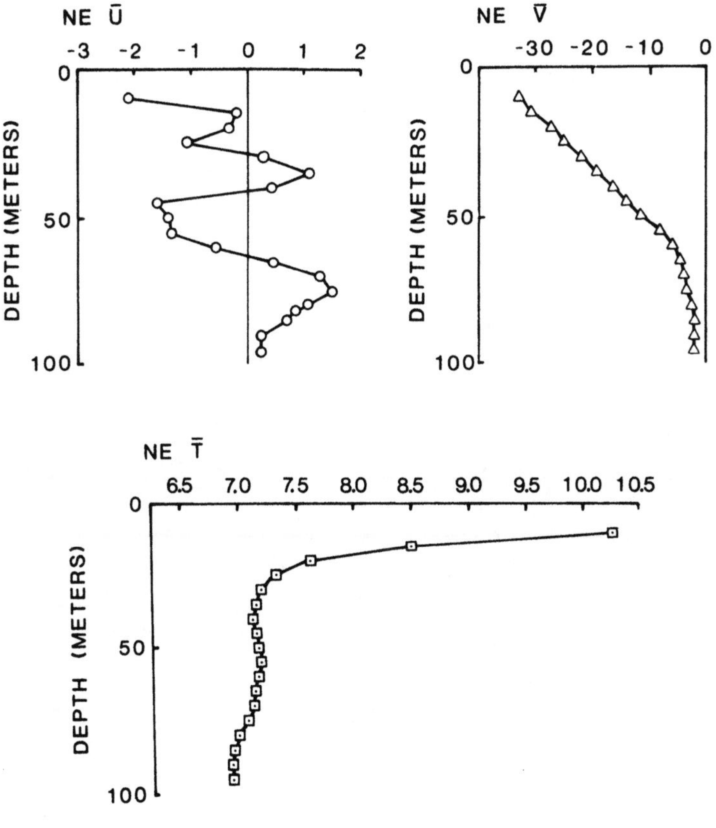

Fig. 4 CUE-II Cyclesonde Station (after Johnson et al., 1976). Lat.: 45°18.2'N, Long.: 124°07.5'W. Start date: 27 Aug. 1973; Stop date: 30 Aug. 1973; Start time: 2300 GMT; Stop time: 1300 GMT. Speeds are given in centimetres per second and temperature in degrees Celsius.

Derivative quantities such as the Brunt-Väisälä frequency, shear or Richardson number can be computed with a greater accuracy because the differences will no longer depend on the long-term stability of a large number of individual sensors. A drift-prone sensor, such as for dissolved oxygen, can be self-calibrating in some instances. For example, off the coast of Peru, an anaerobic near-bottom layer and an oxygen-saturated surface layer permit the end points of a calibration curve to be established on each round trip profile. Other examples exist where deep temperature or salinity structure could be used to monitor sensor drift, such as the 4°C water in the bottom of deep lakes. Lag and other errors can be detected in profile data by comparing up and down going profiles. Since these

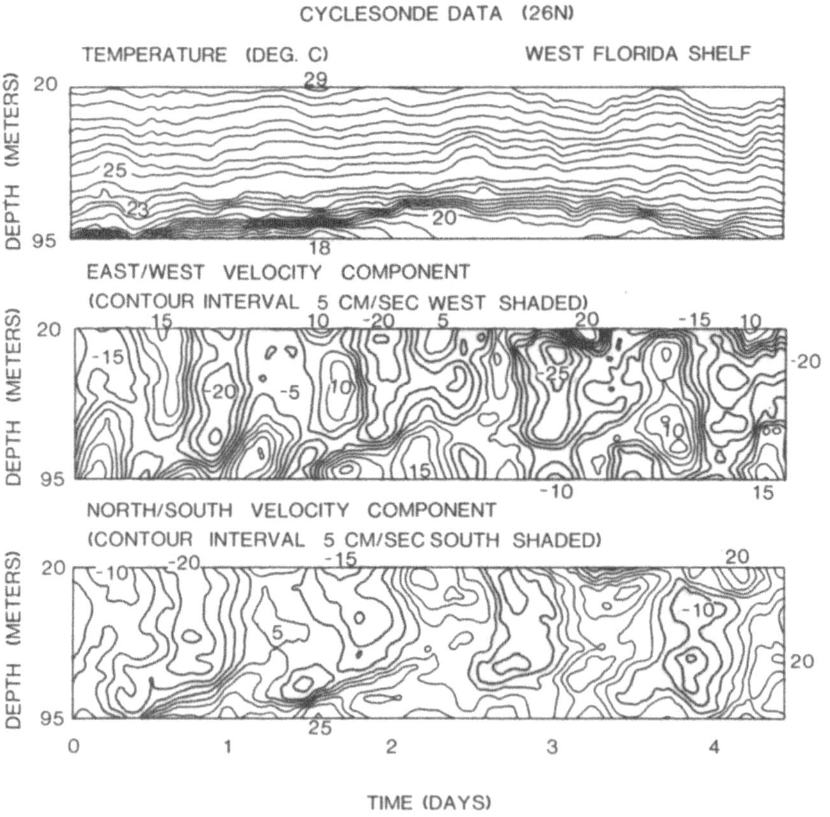

Fig. 5a Mark II Cyclesonde data from the West Florida Shelf at 26°N in a water depth of 100 m. These contours are based on 210 profiles and show a period of rapid bottom boundary layer evolution rich in inertial and tidal energy (Van Leer and Leaman, 1978).

errors usually have the opposite sign on the respective profiles, hysteresis-like curves result when lagging quantities are plotted against depth [especially if the mean of many up profiles is plotted in the same scale with the mean of the corresponding down profiles (see Fig. 5b)]. Other errors, such as sensor flushing, shading, tilt or lift, and velocity-induced errors can be similarly detected.

4.2 Near-surface Fouling

Sensors held near the sea surface for extended periods of time tend to foul more rapidly than those held at several hundred metres depth. This is due to their exposure to the euphotic zone with its greater abundance of primary producers and pelagic eggs and larvae. In tropical oceans, algae and gooseneck barnacles commonly grow on current meters of all types, while in mid-latitudes, mussels, oysters, barnacles, worms and soft fouling organisms are common.

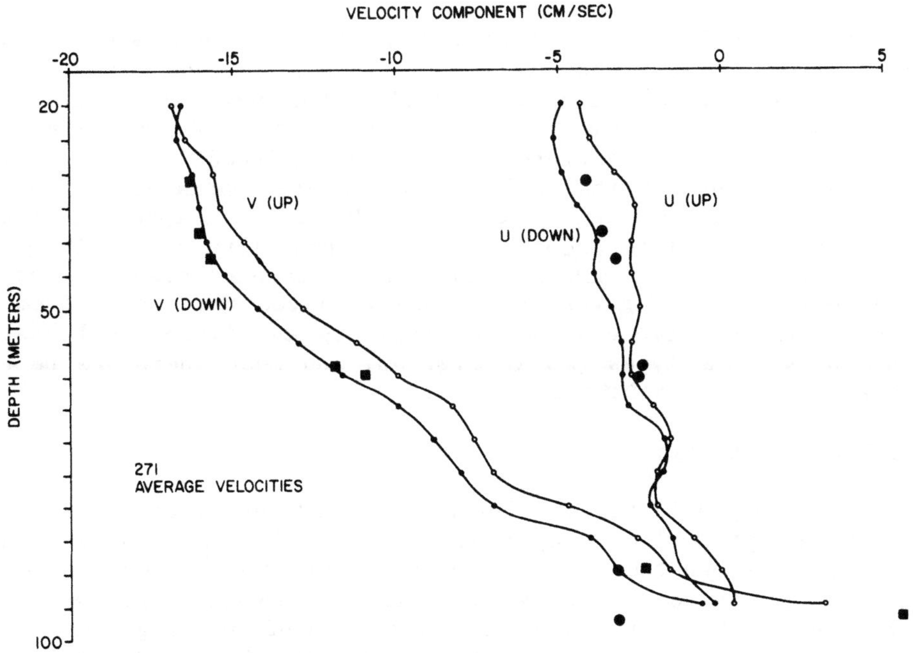

Fig. 5b Mean U and V components for 105 up going and down going Cyclesonde profiles contoured in Figure 5a. Large square and round symbols are mean V and U components observed by Aanderaa current meters during the same time period 0.5 km away. Both Aanderaa and Cyclesonde moorings were subsurface.

Profilers are either kept dry between profiles or, if moored, can be scheduled to spend the minimum time in the euphotic zone. Minimum exposure to the near surface wave zone also minimizes the chance of instrument breakage due to wave abuse.

4.3 Errors Due to Poor Synchronization

If currents are measured by one instrument and temperature and salinity by another instrument with a separate data recording system, there may be large errors in computed, dynamically important quantities like the Richardson number.

The labour in merging diverse data sets can also be time-consuming and error-prone. Profilers are ideal for combined instruments because only one recording system and sensor cluster is required per instrument (or ship or mooring). The use of inexpensive temperature-current profilers by Düing et al. (1975) during the GATE equatorial experiment is a good example of uniform synoptic profile data acquisition from ships of diverse type and nationality.

5. TYPES AND EXAMPLES OF PROFILERS PRESENTLY IN USE

The table in this section gives representative examples of the profilers most widely used; they are grouped by type depending on their method of profiling or velocity sensing. Other profilers still undergoing testing or about to be tested, which are not described in current literature, are not included. One promising type not included in the table is the Doppler backscatter acoustic profiler used on FLIP and described by Pinkel in Chapter 10 of this volume. The absolute accuracies of all profilers should be judged against a standard. Since no generally accepted standard current meter exists, the figures for accuracy in the table come from intercomparisons between existing profilers and fixed-level instruments of various types (as in Fig. 5b) and must therefore be considered only as consistency checks.

5.1 Electromagnetic

The Electromagnetic Velocity Profiler (EMVP) has recently been described in great detail by Sanford et al. (1978). The motion of sea water through the earth's magnetic field induces a weak voltage field which can be sensed with electrodes immersed in the moving sea water. These voltages have been sensed at large horizontal separations using the GEK techniques developed at WHOI by von Arx (1962). However, electrode and amplifier drift precluded the use of small electrode spacing. The EMVP solves this problem by rotating the entire free-fall instrument; its electrodes act as a mechanical 'chopper' so that slow electronic drift is unimportant and only signals with the basic revolution frequency are saved. These ac signals are then referred to the earth's magnetic field by means

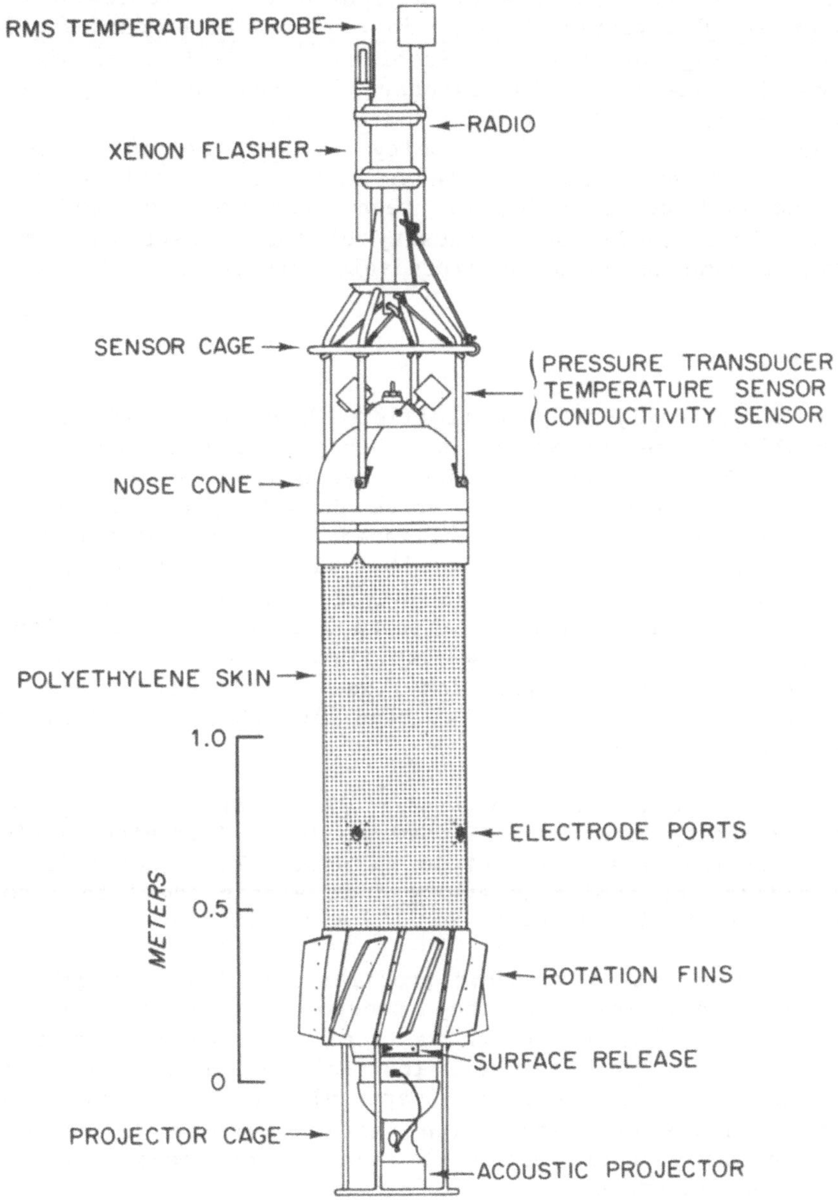

Fig. 6 Free-fall electromagnetic velocity profiler (EMVP) developed at WHOI (after Sanford et al., 1978).

of a detecting coil. The entire instrument is shown in Figure 6.

The chief virtues of the system are its portability without moorings or bottom transponders or navigational references and high precision at 10 m vertical resolution. Such instruments are ideal for wide-ranging survey work. Problems include loss of signal at the magnetic equator and the expense of a dedicated ship for each profile, making time-series measurements expensive. Also, a constant and unknown offset in velocity must be subtracted from the EMVP data to recover the absolute velocity profile (which could be determined with acoustic Doppler techniques from the EMVP). The high cost (~$100,000) and complexity of these instruments may be overcome to some extent by an expendable version now under development.

5.2 Acoustic

Several acoustically-tracked profilers have been made and used successfully in recent years, by Pochapsky and Malone (1972) at Lamont, Rossby and Webb at WHOI (personal communication), and Schmitz and Koehler at WHOI ('White Horse'; personal communication). These instruments are ballasted with expendable weights to sink at rates of 0.2 to 0.4 m s^{-1}. While they sink, the range between the profiler and acoustic transponders on the bottom is measured to typical accuracies of 1 to 3 m of horizontal displacement. In the case of acoustical profilers such as White Horse (see table), the time between position fixes in mid-ocean water depths is a minimum of 10 to 15 s travel time plus about 5 s delay, while all reverberations die out.

Since the difference between a pair of positions is required to find the velocity, the total $\Delta T \simeq 40$ to 50 s. This gives a highest possible vertical resolution of 16 to 50 m. To reject noises, adjacent points are typically averaged to attain the 1 to 3 cm s^{-1} accuracy estimate in the table for 100 m resolution.

The chief advantage of acoustic profilers is that they give absolute velocity profiles since they incorporate their own navigational reference. However, their vertical resolution is a bit too coarse to resolve the energetic inertial period motions or near-surface mixed layer processes (particularly in the presence of entrained bubbles or rain). They also require a dedicated ship plus a bottom acoustic array at each desired measurement site, which can be expensive.

5.3 Winched

When a very stable platform is available, such as FLIP, an ice island or semi-submersible oil rig, profile measurements can be made with a direct winch-lowered system without introducing excessive measurement noise (for a description of such platforms see Blendermann, Chapter 34). Such techniques have been used by Pinkel (1975) from FLIP and Neshyba et al. (1972). These techniques have the advantages that rather conventional commercial instruments can be used with minimum modification and that high rates of vertical profiling are possible, since power is plentiful. They are ideal for studying high frequency, internal waves and near-surface mixed layer processes. They have the disadvantage that very expensive special platforms and/or logistics are required for each measurement point desired. Array data are likely to be prohibitively expensive, except on very small horizontal scales (similar to the platform size). Platform interference with the measurements is also likely to be a problem (see Wucknitz, Chapter 32).

5.4 Wire-guided

Since most of the noise in measurements made with ship-suspended instruments is caused by tension variation in the supporting wire due to pitch and roll, it makes sense to decouple the instrument from the wire with a roller and allow it to sink with its own ballast. This approach was first employed by Düing and Johnson (1972) at Miami, and later at Kiel, Germany, by Meincke (personal communication). All that are required are a simple but rugged roller, a hull which acts as a vane, and a buoyancy element to support a small current meter. Vertical resolution of 3 to 5 m is then set by the sampling rate of the current meter, and the sinking rate of 10 to 15 cm s^{-1} (which is in turn determined by the overweight, wire angle, and drag of the instrument).

The main advantage of the technique is that it is simple and inexpensive to operate, so that many ships can be equipped with profilers, as in GATE (see Düing et al., 1975). At present, the main disadvantages are the lack of accurate, short-term navigation of the ship so that profiles of velocity measured relative to the ship can be converted to absolute profiles, and the use of a dedicated ship. Anchoring the ship, radar reference buoys and land-based references have been used with reasonable success. Major improvements to provide all-weather 24-hour precision satellite navigation, which should be forthcoming in the early 1980s, will make wire-guided profilers even more attractive for depths from 0 to 750 m.

5.5 Moored Automatic

If the buoyancy of a wire-guided profiler can be controlled to be alternately positive and negative in water, then an automatically cycling profiler can be produced. The Cyclesonde is such an instrument; it was first developed at Miami by Van Leer et al. (1974) with subsequent improvements described in Van Leer (1976). A

Table I: Some Profilers in Use as of July, 1977

	TYPE OF PROFILER	MEANS OF PROFILING	METHOD OF VELOCITY SENSING AND ACCURACY (OUT OF WAVE NOISE)
Electro-magnetic Profiler	Electromagnetic Velocity Profiler (EMVP) (Electromagnetic)	Free-fall ~ 1 m s^{-1} with expendable weight release and excess buoyancy	Senses with electrode pairs sea water potential caused sea water motion in the earth's magnetic field-velocity ± 1 to 3 cm s^{-1} @ 10 m resolution
	Telemetering Position and Sensed Data Float (Acoustic)	Free-fall $\sim .2$ m s^{-1} with expendable weight	Acoustic range on bottom-mounted transponders Range accurate to ± 1 m.
Acoustic Profilers	Rossby/Acoustic (Acoustic)	Free-fall $\sim .3$ m s^{-1} expendable weight and excess buoyancy	Acoustic range on bottom transponders ± 1 or 2 cm s^{-1}
	White Horse (Acoustic)	Free-fall $\sim .4$ m s^{-1} expendable weight and excess buoyancy	Acoustically self-navigating dropsonde ± 2-3 m; gives ± 2-3 cm s^{-1} accuracy with 100 m resolution
Winched Profilers	FLIP Profiler (Winched)	Winch-lowered and raised at 4 m s^{-1}	Doppler acoustic (accuracy unknown)
Wire-Guided Winch-Recovered Profilers	Profiling Current Meter (PCM) (Free-falling roller coupled & winched recovered)	Descends at .1-.3 m s^{-1} with 1 kg overweight; returned to surface by ship's winch	Aanderaa rotor vertical oriented ± 5 cm s^{-1}; $\pm$ship navigational errors during profile at 5 m vertical resolution
Moored Automatic Profiler	Cyclesonde (autonomous, moored profiling & telemetry package)	Compressed helium inflated bladder periodically giving 100-1000 profiles/tank full (depending on depth) .05-.15 m s^{-1}	Twin Aanderaa rotors horizontally oriented ± 2 cm s^{-1} at 1-5 m vertical resolution. Speed estimated by pressure change between samples (see Fig. 3b)
	(Moored Automatic Profiler)	Number of profiles possible per tank $N = \frac{2}{\Delta B}\left(\frac{2000}{P + 1.5} - 12\right)$ where Δ B=bladder displacement, P=pressure at bottom of profile in atmospheres.	(.6 liters typical Δ B)

SENSORS AND RESOLUTION	LIMITATIONS	COMMENTS ON USE	CONTACT POINT FOR INFORMATION
Temperature ±.01°C Conductivity ±.02 millimho cm.$^{-1}$ Pressure ±3 dbar Direction ±3°	Must be retrieved by ship after each profile pair; velocity measured includes an unknown constant (4 cm s^{-1} in MODE)	Makes 6000 m profile in ~1½ hrs; vertical resolution of velocity 10 dbar has recorder profiles in time series or survey mode. Use near magnetic equator or anomalies may be restricted.	Robert Drever and Thomas Sanford WHOI
Temperature ±.001°C Pressure ±1 dbar	Must be used within miles of bottom transponders; must be retrieved by ship after each profile	Can be used to 10,000 psi. Has been used rather infrequently	T.E. Pochapsky Lamont
Temperature ±.02°C Pressure ±5 dbar Webb	Must be used near hydrophone array and retrieved by ship after each profile	Can be used over entire ocean depth for surveys near hydrophone array as in Bermuda.	Tom Rossby URI Doug Webb WHOI
Temperature ±.001°C Conductivity ±.001 millimho cm^{-1} Pressure ±3 dbar	Must be used with bottom transponders and retrieved by ship after each profile. Can't be used during rainstorms due to acoustic noise	High quality combined sensor profiler for survey of time series use. Coarse vertical - resolution	W.J. Schmitz & Richard Koehler WHOI
3 temperatures ±.01°C Pressure ±1 dbar	Can be used from FLIP only to depths of 440 m	Extremely fast profiling rate - 30 profiles/hr with 3 independent temperature/pressure probes	Robert Pinkel SIO
Temperature ±.025°C Conductivity ±.05 millimho cm^{-1} Pressure ±.5% full scale Direction ±3°	Must be lowered from ship with good navigation or have radar reference buoys installed	Has been widely used to survey strong currents on the equator & in Gulf Stream; about 1000 profiles to 750 m	Walter Duing RSMAS Univ. of Miami Jens Meincke Kiel, Germany
Temperature ±.005°C Conductivity ±.01 millimho cm^{-1} Pressure ±.1 dbar Direction ±0.5°	Depth limitation 200-300 m; current speed limitation: MKI & MKII 60 cm s^{-1}; MKIII 200 cm s^{-1} (est.); 0.25-2 profiles/hr typical. Profile schedule & rest periods & telemetry transmissions are controlled by a programmed clock.	Used on taut wire morings to gather time series profiles lasting typically 4 days to 1 month; ~15,000 profiles	John Van Leer RSMAS Univ. of Miami

typical shallow water mooring is shown in Figure 7 with the Cyclesonde free to ride vertically between the upper and lower ends of the taut wire mooring. It can cycle automatically between 100 and 1000 profiles depending on the water depth and bladder displacement. The valves can be cycled according to a preset time schedule or on radio command and telemetry can be received up to 30 km away without resorting to directional antennae. The instrument (Fig. 8) measures and internally records conductivity, temperature, depth, and current data at intervals from 2 s to 1 min. It is coupled by a low-friction roller block to a smooth plastic-coated wire and moves vertically at rates of 5 to 10 cm s^{-1}. This corresponds to a vertical resolution of 0.5 to 5 m. Its principal advantage is its operation without a ship standing by, so it is ideal for recording long time series of profiles. It also has no navigational problems and can remain on station from 4 days to 1 month. With telemetry,

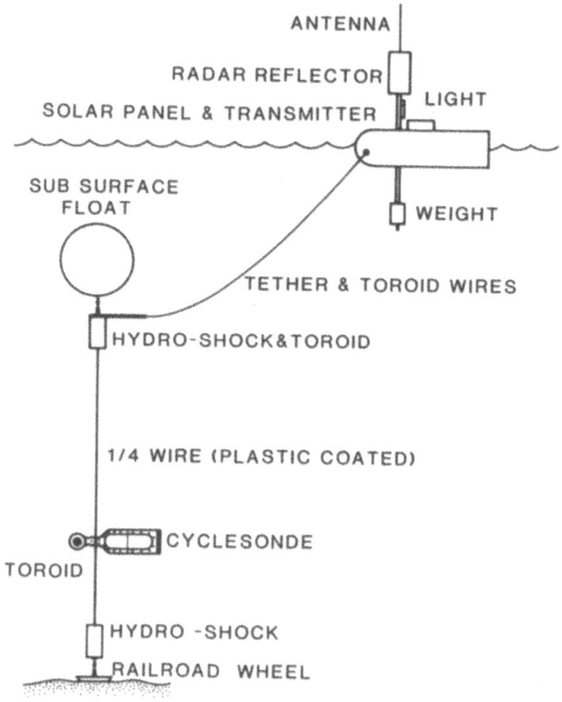

Fig. 7 Subsurface Cyclesonde mooring for shallow water (0-300 m).

Fig. 8 Mark II Cyclesonde lifts off bottom stop during a test near Miami. Note the horizontal rotor orientation.

a synoptic array can be maintained with a single ship. Its principal disadvantages are the added complexity of a cycling mechanism, precise ballasting, and its maximum current limitations of 60 cm s^{-1} for the Mark I and Mark II Cyclesondes and about 200 cm s^{-1} (estimated) for the Mark III.

6. SOURCES OF ERROR IN PROFILERS

6.1 Temporal Sampling Errors

To estimate the low frequency motion and tidal and inertial motion within 5% of the total kinetic energy at each depth usually requires at least 12 profiles per day (Fig. 1). At greater time intervals, the tidal and inertial energy present in the system start to leak into lower frequency current estimates. If one samples twice per inertial period, as Sanford (1975) has done, most of the inertial signal can be rejected but some leakage from tidal period motion is likely. A greater number of samples per inertial period will reduce the probability of aliasing from high frequency internal waves or surface gravity waves at the expense of increased ship time for attended profilers or gas used for Cyclesondes.

Within each profile, averaging should be done over a time interval equivalent to the desired vertical resolution. For near-surface current profile measurements, this average should be a true vector average over several gravity wave periods; for example, 30 s to a minute or longer in extreme storm conditions. This, in turn, will limit the profiler's ascent or descent rate. For example, a one-minute averaging period with a 3 m vertical resolution requirement limits the vertical profiling speed to 3 m min^{-1}. Thus spin-stabilized free-fall profilers will be of little use in the depths where wave orbital motion is significant.

For air-sea interaction studies, it is highly desirable to average temperature and salinity over these same depth ranges so that accurate integrals of heat and salt content can be made over desired depth ranges without the spatial contamination associated with microstructure.

6.2 Reduction of Surface Wave-Induced Errors

Practically all vertical profilers can be decoupled totally or partly from the effects of wire-transmitted wave heave motions. Many profilers are free-fall and not coupled to either ship or moorings. Others are coupled by roller to a ship or mooring-supported wire. Since the primary transmission of wave heave motion to substantial depths is along the axis of the wire, a low-friction roller can be quite effective at decoupling wave heave. Transverse wire motion is greatly reduced by lateral drag forces on the wire and thus exists only near the sea surface. The degree of transmission of lateral motion increases with increasing tension. These motions seem to transmit to about one wavelength depth for 135 kg of tension. The rms difference between up and down going Cyclesonde velocity measurements gathered from a mooring with a surface buoy is shown in Figure 9. The dominant waves during this observation period had a U-direction wavelength of the order of 100 m. In regions with strong surface currents, such as the GATE C scale experiment, transverse wire-transmitted motions tend to die out more rapidly with depth, owing to a large mean drag force. The use of subsurface flotation or lateral drag elements at depths of 10 to 40 m, depending on sea state, can reduce these effects dramatically.

The vertical velocity of profilers is usually determined by an excess of weight or buoyancy balanced by friction drag or lift forces. Motion-related problems like rotor pump-up and salinity spiking can be reduced or eliminated by the profiler's slow, monotonic rising or sinking relative to the water. For example, a temperature sensor on a CTD (time constant $\simeq$ 0.3 s) rigidly fixed to the end of a ship-lowered wire can easily attain peak vertical speeds exceeding 1 m s^{-1} relative to the water. In a near-surface thermocline with temperature gradient of 0.3°C m^{-1} such a vertical speed gives an error in temperature estimate of about 0.1°C. The same temperature sensor mounted on a profiler moving vertically with a speed of 5 to 10 cm s^{-1} would give corresponding lag errors in

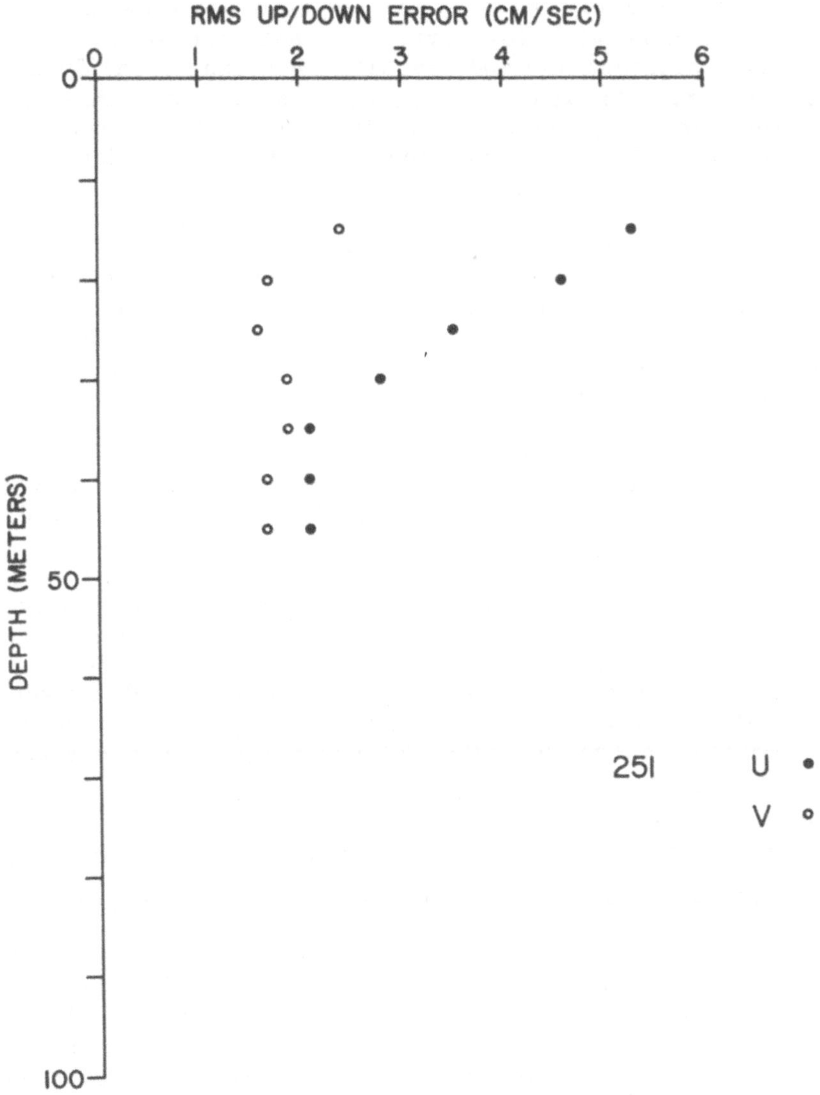

Fig. 9 Uncertainty in horizontal velocity V_h estimate for a Mark I Cyclesonde as a function of pressure and speed uncertainty (from Van Leer et al., 1974).

temperature of 0.005 to 0.01°C. In fact, sensors can be biased away from low speed problems like rotor stall (note the horizontal rotor orientation in Fig. 8), self heating or flushing.

6.3 Spatial Sampling Errors

From the data shown above, one can readily see that inertial and internal tidal motions often contain the major portion of the horizontal kinetic energy. It is poor sampling practice to spatially alias these signals with low vertical resolution so that they are included as a spurious low mode result. Between 10 and 50 discrete vertical measurements would be required to get an unaliased estimate of the current profiles shown in Figures 3, 4, and 5. Most profile methods offer high vertical resolution, as determined by the interval between samples (ΔT) multiplied by the instrument's vertical velocity (V_v). Spin-stabilized free-fall instruments have typical vertical velocities of $0.2 \leq V_v \leq 1.0$ m s^{-1}. In the case of acoustical profilers given earlier, the present vertical resolution is 16 to 50 m. To attain the 2 to 3 cm s^{-1} accuracy estimate in the table, they must be averaged over about 100 m. The EMVP can average 5 to 10 samples to attain a vertical resolution of about 10 m with a noise level of about 1 cm s^{-1}.

Wire-guided profilers such as the Cyclesonde and PCM can attain vertical resolutions of from 1 to 10 m depending on the chosen sampling rate, current resolution and the full scale of the depth sensor.

Figure 10, from Van Leer et al. (1974), shows how the precision of estimated vertical velocity V_v and uncertainty in resultant speed S influence the estimate of the horizontal velocity V_H for a Cyclesonde with horizontal rotors. The winter data were taken with coarse resolution and the summer data with relatively fine resolution. Clearly if the velocity ratio of V_v/V_H is minimized, the error in estimate of V_H is also minimized. Conversely, if a velocity profiler has a large vertical velocity V_v, which is being measured, large percentage errors may result in the estimate V_H when the vertical speed is removed from the signal by data processing.

Time lag errors in a moving profiler, such as angular response, translate into spatial errors in the vertical over a distance proportional to the velocity ratio V_v/V_H. These errors show up as hysteresis-like curves when up and down going profiles are plotted on the same axes. For this reason, profile measurements should be made in both directions as a quality control step. Errors induced by lift and drag forces such as the gliding of a free-fall profiler tilted at an angle to the vertical and poorly spin-stabilized or angular offset errors in a wire-guided profiler will increase with

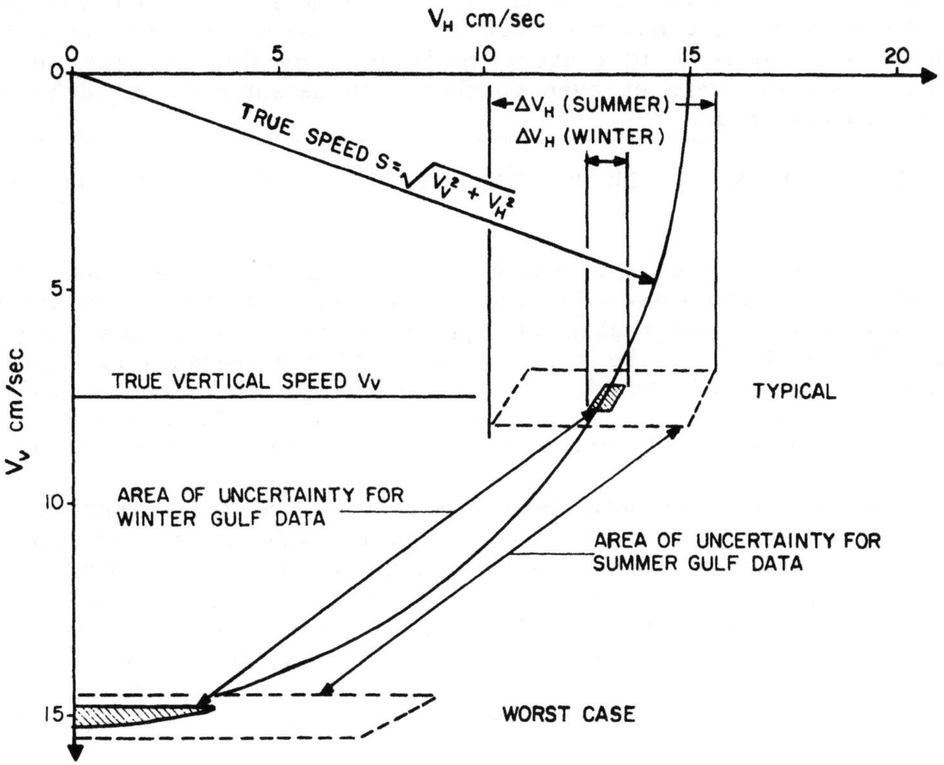

Fig. 10 RMS difference between up and down going profiles for a Cyclesonde suspended beneath a surface float. Note the decreasing error with depth.

the square of the velocity ratio $(V_v/V_H)^2$ (see Van Leer, 1974b).*

Therefore, the desire for rapid profiles to reduce the 1 to 4% temporal aliasing at any given depth must be carefully traded off against measurement errors induced by a rapid vertical velocity. In regions of surface wave contamination sufficient averaging time must be allowed.

To estimate accurately the heat or salt content of the upper mixed layer also requires high resolution profile measurement. If, for

---

* Note: The EMVP of Sanford and Drever does not sense the relative velocity between the profiler and the water and is thus not sensitive to this type of error.

example, the base of the mixed layer were measured within typical 20 to 50 m current meter or bottle cast resolution, the error in the estimation of heat content could be equivalent to the total heat flux over weeks or even months; such measurement would be of very limited value.

7. CONCLUSIONS ON THE PRESENT STATE AND FUTURE PROMISE OF PROFILERS

Profile data (particularly velocity) are relatively new, but have slowly become accepted. In terms of absolute accuracy, the three acoustically-tracked probes are probably best for scales larger than 100 m, but require that an array of transponders or hydrophones be in place and a ship to find and retrieve the profiler after each deployment. They are also subject to substantial errors of spatial omission of small but energetic vertical scale motion.

The Sanford or Düing profilers are best suited for surveys over large areas and in strong currents where moorings are difficult. The FLIP acoustic Doppler profilers are best suited to 40 m horizontal scale high frequency (30 sample hr^{-1}) profiles. Arrays of 4 to 5 Cyclesondes, equipped with radio telemetry (Fig. 7) have been used successfully in synoptic scale arrays of 10 to 50 km. Profile rates from 2 per hour to one profile every 4 hours either in the deep sea or on continental shelves (either moored or drifting) have been used.

While several complex and relatively accurate profilers exist today, the greatest potential will come from simple, less expensive profilers which can be used by a greater number of investigators. Such cheap, expendable, velocity profilers would find wide application. Extending the time duration and depth range of moored profilers could also reduce the cost per profile by reducing the use of expensive ship time. The higher cost of long duration profilers must be carefully weighed against the increasing chance of surface mooring loss with long duration moorings, particularly in deep water.

The cost of the measurements at each desired geographic point can be reduced by making one vertically profiling sensor package do the job of many instruments at fixed depths. This would permit synoptic sampling with greater lateral coverage. If multiple profilers, either moored or operated from ships, are available and if radio telemetry is also available, then synoptic data suitable for forecasting are possible. Meteorologists have used radiosonde data in this way for years.

Advances in near-surface velocity measurements will come rapidly once improved low-power cosine response sensors become available. Once these improved sensors are routinely in use, it will be necessary to use averaging techniques in time or depth to compact the data. Profile data will get great impetus as economic or military use of continental shelves, such as for oil production and mining, require more accurate forecasting.

REFERENCES

BROWN, N. 1974. A precision CTD microprofiler. *Proceedings of Ocean 74 IEEE Conference on Engineering in the Ocean Environment,* Halifax, Nova Scotia, vol. 2: 271-278.

BRISCOE, M.G. 1975. Preliminary results from the Trimoored Internal Wave Experiment (IWEX). *Journal of Geophysical Research,* 80: 3872-3884.

COX, C., Y. NAGATA, and T. OSBORN. 1969. Oceanic fine structure and internal waves. *Bulletin of the Japanese Society of Fisheries Oceanography,* Special Issue (Professor Uda's Commemorative Papers): 67-71.

DAHLEN, J.M., N.K. CHHABRA, J.F. McKENNA, J.R. SCHOLTEN, J.T. SHILLINGFORD, F.J. SIRACO, and W.E. TOTH. 1977. Draper laboratory profiling current and CTD meter (prototype design description). Report R-1095, Charles Stark Draper Laboratory, Cambridge, MA. 122 pp.

DREVER, R.G. and T.B. SANFORD. 1970. A free-fall electromagnetic current meter instrumentation. *Proceedings of the IEEE Conference on Electronic Engineering in Ocean Technology,* IEEE, London: 353-370.

DÜING, W. and D. JOHNSON. 1972. High resolution current profiling in the Straits of Florida. *Deep-Sea Research,* 19: 259-274.

DÜING, W., P. HISARD, E. KATZ, J. MEINKE, L. MILLER, K.V. MOROSHKIN, G. PHILANDER, A.A. RIBNIKOV, K. VOIGT, and R. WEISBERG. 1975. Meanders and long waves in the equatorial Atlantic. *Nature,* 257: 280-284.

GARRETT, C. and W. MUNK. 1971. Internal wave spectra in the presence of fine-structure. *Journal of Physical Oceanography,* 1: 196-202.

JOHNSON, W.R. 1976. Cyclesonde measurements in the upwelling region off Oregon. Technical Report, University of Miami, Miami, Florida, U.S.A.

JOHNSON, W.R., J. VAN LEER, and C.N.K. MOOERS. 1976. A Cyclesonde view of coastal upwelling. *Journal of Physical Oceanography,* 6: 556-574.

LUYTEN, J.R. and J.C. SWALLOW. 1976. Equatorial undercurrents. *Deep-Sea Research,* 23: 999-1001.

MÜLLER, T.J., F.A. SCHOTT, G. DIEDLER, and K.P. KOLTERMANN. 1974. Observations of overflow on the Iceland Faeroe ridge. *'Meteor' Forschungsergebnisse, Reihe B,* 15: 49-55.

NESHYBA, S., V.T. NEAL, and W.W. DENNER. 1972. Spectra of internal waves: In situ measurements in a multiple-layered structure. *Journal of Physical Oceanography,* 2: 91-95.

PERKINS, H. and J.C. VAN LEER. 1976. Simultaneous current-temperature profiles in the equatorial counter current. *Journal of Physical Oceanography,* 7: 264-271.

PHILLIPS, O.M. 1971. On spectra measured in an undulating layered medium. *Journal of Physical Oceanography,* 1: 1-6.

PINKEL, R. 1975. Upper ocean internal wave observations from FLIP. *Journal of Geophysical Research,* 80: 3892-3910.

POCHAPSKY, T.E. and F.D. MALONE. 1972. A vertical profile of deep horizontal current near Cape Lookout, North Carolina. *Journal of Marine Research,* 30: 163-167.

ROSSBY, H.T,. 1969. A vertical profile of currents near Plantagenet Bank. *Deep-Sea Research,* 16: 337-385.

SANFORD, T.B. 1975. Observations of the vertical structure of internal waves. *Journal of Geophysical Research,* 80: 3861-3871.

SANFORD, T.B., R.G. DREVER, and J.H. DUNLAP. 1978. A velocity profiler based on the principles of geomagnetic induction. *Deep-Sea Research,* 25: 183-210.

SPILHAUS, A.F. 1938. A bathythermograph. *Journal of Marine Research,* 1: 95-100.

STOMMEL, H. and K.N. FEDOROV. 1967. Small scale structure in temperature and salinity near Timor and Mindanao. *Tellus,* 19: 306-325.

VAN LEER, J.C. 1971. Shear of small vertical scale observed in the permanent oceanic thermocline. Sc.D. Thesis, MIT and WHOI, Woods Hole, MA., U.S.A. 209 pp.

VAN LEER, J.C. 1974a. A note on expected temporal aliasing errors in the CUE-I Cyclesonde frontal station. *Tethys,* 6: 433.

VAN LEER, J.C. 1974b. Progress report on Cyclesonde development and use. RSMAS Technical Report, #74029, University of Miami, Miami, Florida, U.S.A. 77 pp.

VAN LEER, J.C. 1976. An automatic oceanographic profiling instrument. Instrument Society of America, *Instrumentation in the Aerospace Industry,* 23 (7626B): 489-500.

VAN LEER, J.C. and K.D. LEAMAN. 1978. Physical oceanographic research using the attended profiling current meter (APCM) and the Cyclesonde. *Proceedings of a Working Conference on Current Measurement,* Technical Report DEL-SG-3-78, College of Marine Studies, University of Delaware, Newark, DE: 71-93.

VAN LEER, J.C., W. DUING, R. ERATH, E. KENNELLY, and A. SPEIDEL. 1974. The Cyclesonde: An unattended vertical profiler for scalar and vector quantities in the upper ocean. *Deep-Sea Research,* 21: 385-400.

VON ARX, W.S. 1962. *An Introduction to Physical Oceanography.* Addison-Wesley, Reading, Mass., 422 pp.

WEBSTER, F. 1968. Observations of inertial-period motions in the deep sea. *Review of Geophysics,* 6: 473-492.

WOODS, J.D. 1968. Wave-induced shear instability in the summer thermocline. *Journal of Fluid Mechanics,* 32: 791-800.

37

Free Fall Vehicles

R.E. Lange

1. INTRODUCTION

Since the early sixties, a class of oceanographic observations have been undertaken that required a unique platform. Some of these observations were directed at understanding vertical shear structure; this required an instrument to track the small horizontal speed fluctuation with depth in a known way. Other observations attempted to measure the smallest scales of the heat, salt, and shear structures found in the ocean, and required slow, vibrationless fall rates to accomodate sensitive but slow response probes, or to be free from vibration-induced signals.

These measurements precluded the use of cable-lowered devices, in order to remove ship-induced motions of the vehicle, and generated a class of instruments that record internally or telemeter data through a small wire or acoustical link. All of these devices descend to a preset depth, change buoyancy, and ascend to the surface, without on-board ship control.

The advantages of such devices include relatively vibrationless descent, a tendency to be Lagrangian in character, and a controlled and repeatable descent rate; also, they generally leave the tending vessel free to conduct other operations. The disadvantages include the necessity for self-contained data acquisition systems, the need for a highly mobile ship or a skiff and recovery crew, generally a long time interval between successive profiling (making a synoptic survey awkward at best), and the inherent danger of catastrophic loss.

This article will concern itself with some generalities about fall characteristics applicable to many free-fall vehicles and a brief survey of some vehicles presently in use.

There are appended here two sets of references for the reader interested in further information: a scientifically oriented set of journal articles taken to be representative of the use of free-fall vehicles in ocean exploration, and a set of references in which some vehicles are described in engineering terms, i.e. their construction, evaluation, and testing.

Two classes of instruments can be defined: those relying upon Lagrangian behavior in tracking shear flows and those that attempt to profile small-scale features where the Lagrangian character of the instrument is not of importance. In both of these classes of instrument some understanding of fall dynamics is necessary. The basic problem of all the free-fall vehicles is that to obtain predictable fall characteristics either low overmass (weight excess in water) and low-turbulent form drag devices or large overmass and auto-rotating air foils utilizing high lift/drag ratios to slow descent must be used and understood well.

Few instrument systems measure fall behavior directly, either because it is not deemed necessary for the understanding of the bandwidth of phenomena investigated, or because the vehicle's behavior is believed to be well understood from numerical models, by intercomparison drops with another instrument, or from a tracking system.

Some representative types of free-fall devices are described by Sanford et al. (1978), Osborn (1974), Richardson and Schmitz (1965), and Mortensen and Lange (1976).

Richardson's 'dropsonde' (Richardson and Schmitz, 1965) is a long, cylindrical tube whose only active element is a hydrostatic release pin that drops ballast weights at a predetermined depth; the dropsonde then rises to the surface. By deploying several such instruments simultaneously, each set for a different depth, an integrated measurement of shear and mean transport can be made by measuring their return positions precisely. Knowledge of the Lagrangian behavior of this device is essential for a correct interpretation of the data. Another similar device, using tri-moored acoustic transmitters for positioning by triangulation, is the 'Whitehorse' (Luyten and Swallow, 1976). It records arrival times of acoustic pulses as it descends, and, since it is assumed to track shear, a profile of shear can be inferred.

These two devices depend upon knowing the response of the instrument to horizontal motion. A shear-measuring device that does not is Sanford's 'electromagnetic velocity profiler' (Sanford et al., 1978). Unique among free-fall instruments, this device measures the electric field induced in ocean flows within the earth's magnetic field, and is not dependent upon the instrument's ability to track flows. The flow around the instrument appears as a second-order noise source. This device achieves slow descent by a low overmass (weight in water), falls to the sea-floor, and releases ballast. The changed relationship of centre of mass to centre of buoyancy then causes the instrument to flip upside down, and a profile is made vertically upwards on its return to the sea surface.

Osborn, who attempts to measure micro-scales of shear (and whose sensing element is described elsewhere in this book, see Chapter 19), achieves a stable and slow fall rate by using a collar of filamentous material, providing a high drag coefficient, and yet inhibiting the shedding of vortices, a source of tilting and oscillating behavior of the instrument (Osborn, 1974). This instrument was not intended to measure low-wavenumber shear, focusing attention rather on the highest wavenumbers measurable. However, the understanding of vehicle fall characteristics is still important for understanding aliasing from low-wavenumber shear and cross-flow variations due to tilting and fishtailing through the water.

Mortensen and Lange (1976) describe a vehicle that employs auto-rotating hydrofoils to achieve slow descent rates and some degree of spin-stabilization. The high lift-to-drag ratio permits large overmasses to be used, which lowers the centre of gravity with respect to the centres of buoyancy and of drag. This apparently reduces off-axis tilts and axial fishtailing considerably (as will be shown later) and permits very slow descent rates to be used.

2. THE DYNAMICS OF FALL CHARACTERISTICS

The large differences in geometries and bandwidths of phenomena measured with free-fall instruments make generalizations about them difficult. Auto-rotating devices have a nontrivial angular inertia and momentum, while drag-limited cylinders may not.

To analytically model the ability of free-fall instruments to track lateral motions we will assume that the device is well represented by a point drag force, the shear scales of interest are large compared to the device, and the instrument does not suffer from tilts or does not fishtail through the water.

Given these assumptions, the drag force vector can be written as

$$\underset{\sim}{D} = \tfrac{1}{2}\, \rho C A_D \underset{\sim}{V} |V| \qquad (1)$$

where ρ is the density of sea water, A_D is the projected area of the device along the flow vector $\underset{\sim}{V}$, and $\underset{\sim}{V}$ is the relative flow velocity directed at an angle ϕ from the vertical. C is the drag coefficient, which is assumed for this simplified case to be scalar and constant. In reality, because of the lift characteristics of some irregularly-shaped objects on the vehicle, C is in general a tensor.

The momentum equations are written as

$$m\dot{u} = D \sin(\phi)$$
$$m\dot{w} = g - D \cos(\phi) \tag{2}$$

where m is the mass (actual plus virtual mass), $\dot{u}$ is the time rate of change of horizontal velocity anomaly, $\dot{w}$ is the time rate of change of vertical velocity anomaly, D is the drag force magnitude, g is the overweight, and ϕ is the angle from vertical.

These can be rewritten as

$$m\dot{u} = \tfrac{1}{2} \rho C A_D |V|^2 \sin(\phi)$$
$$m\dot{w} = g - \tfrac{1}{2} \rho C A_D |V|^2 \cos(\phi) \tag{3}$$

where $A_D = A_x \sin(\phi) + A_z \cos(\phi)$, the area projected along the drag force vector.

This then resolves itself into

$$m\dot{u} = \tfrac{1}{2} \rho C A_x (U-u)^2 + \tfrac{1}{2} \rho C A_z (U-u)(W-w)$$
$$m\dot{w} = g - \tfrac{1}{2} \rho C A_z (U-u)(W-w) - \tfrac{1}{2} \rho C A_z (W-w)^2 \tag{4}$$

where U and W are respectively the horizontal fluid velocity and instrument drop speed, both relative to fixed coordinates, and u and w are the horizontal and vertical components of velocity relative to the instrument. The cross terms arise because the instrument is assumed to stay vertical and not align itself with the flow vector. This description, so far, follows that of Hendricks and Rodenbusch (1977). We now will depart from their simplified analysis and assume only that $w/W \sim 0$, to get the two equations

$$m\dot{u} = \tfrac{1}{2} \rho C A_x (U-u)^2 + \tfrac{1}{2} \rho C A_z (U-u) W$$
$$m\dot{w} = g - \tfrac{1}{2} \rho C A_x (U-u) w - \tfrac{1}{2} \rho C A_z w^2 \tag{5}$$

The horizontal equation, assuming the device enters a step change in horizontal velocity, has the solution

$$U - u = \frac{U e^{-z/\ell}}{1 + (U/W_s)(1 - e^{-z/\ell})} \tag{6}$$

where z is the vertical distance into the layer, $s = A_z/A_x$, and ℓ is equal to $m/(\tfrac{1}{2}\rho C A_z)$.

This solution exhibits an important, and often overlooked, aspect of the tendency of free-fall devices to follow horizontal flows: the stronger the vertical shear, the more quickly the instrument will accelerate to the speed of the water. This follows because of the nonlinear nature of Equations 5. There is thus a tendency for the instrument to track stronger vertical shears more quickly than weaker ones.

This special case (assuming the instrument is a point-drag source) is relatively easy to handle analytically. However, in many such instruments, an extension to higher wavenumbers is sought. The assumption that the instrument is a point-drag source is no longer appropriate, and fall rate variability must be considered. In this case, the momentum equations become

$$m'u' = \tfrac{1}{2}\rho \frac{1}{h}\int_0^h C(z')A_x(z')[U(z') - u]^2 dz'$$

$$+ \tfrac{1}{2}\rho \frac{1}{h}\int_0^h C(z')A_z(z')[U(z') - u][W(z') - w]dz'$$

(with a similar equation for y) and (7)

$$m'w' = g - \tfrac{1}{2}\rho \frac{1}{h}\int_0^h C(z')A_x(z')[U(z') - u][W(z') - w]dz'$$

$$- \tfrac{1}{2}\rho \frac{1}{h}\int_0^h C(z')A_z(z')[W(z') - w]^2 dz'$$

This analysis considers nonrotating instruments that experience a cross-flow distributed along the axial length of the vehicle. The vehicle is assumed to remain vertical and not to experince tilts resulting from the cross-flow forces. Despite these assumptions, the equations are still analytically intractable. A few simplifying assumptions are made by Hendricks and Rodenbusch (1977), but those assumptions seem restricted to the case of $A_x << A_z$, which is not typical of most free-fall devices.

The opposite case, of an auto-rotating, spin-stabilized vehicle, must be handled differently. Angular momentum, variable form drag, and attitude (tilt) changes must all be considered. The problem can be cast in the following form: assuming cylindrical symmetry about the vertical or axial coordinate,

$$\frac{d}{dt}\left(\frac{\partial L}{\partial q_i}\right) - \frac{\partial L}{\partial q_i} = Q_i \qquad (8)$$

where q_i is a set of generalized coordinates, and Q_i are forces and torques. The Lagrangian L may be constructed using symmetry considerations of the body in question. The Q_i forces and torques can be resolved into orthogonal components and may be quite complicated for autorotating, winged instruments. Mortensen and Lange (1976) have implemented this for a single instrument by using a computer-driven model.

As in the simplified case above, an instrument is assumed to enter a step-like shear layer, and the response time of the instrument is calculated. The result for the type of instrument under consideration is highly nonlinear. The advantage of this approach is that some information on tilt behavior is available from the analysis; the reader is referred to Mortensen and Lange (1976) for further discussion.

3. IN SITU MEASUREMENTS

Aside from a few acoustically-tracked instruments (tested at the tracking range in St. Croix, Virgin Islands), the author is aware of only two instruments that actually measure vertical acceleration and tilts in normal data acquisition - Elliott and Oakey's (1976) 'Octuprobe', and Lange's velocity microstructure recorder (VMSR).

As the data of the VMSR are readily available, and because it represents a class of instruments for which vibrations, tilts, nutations and pendular motions are important, while Lagrangian behavior is not, those data will be presented here.

The general geometry of the VMSR is shown in Figure 1. It is a 2 m long by 25 cm diameter aluminum pressure casing with four 2 m long aerodynamic foils. With upper and lower end-caps, the instrument is a total of 3.2 m long (excluding antennas). It falls at approximately 80 cm s^{-1} to a preset depth where the wings are deployed. The descent rate then slows to 8 to 10 cm s^{-1}, and the instrument rotates (about once in 20 s) as it falls. Upon termination of the data acquisition, the excess ballast weight is released, the wings fold down, and the instrument ascends at about 60 cm s^{-1}.

The wings and other parts that are irregularly shaped have been matched in weight so the centre of mass of the instrument lies along the axial line of the pressure casing. The centre of mass of the instrument lies at point A, and the centre of buoyancy at point B in Figure 1. The centre of drag forces is ill defined in a system such as this, but a detailed numerical model (Mortensen and Lange, 1976) predicts it to be about 10 cm below the plane of the wings, for a nearly vertical flow. For a horizontal flow not reaching the wings, the drag approaches the uniform distribution of a cylinder. However, when the wings (which are rotating) enter a cross-flow, a highly increased drag (over that calculated by projected area) is experienced, and the instrument has a tendency to

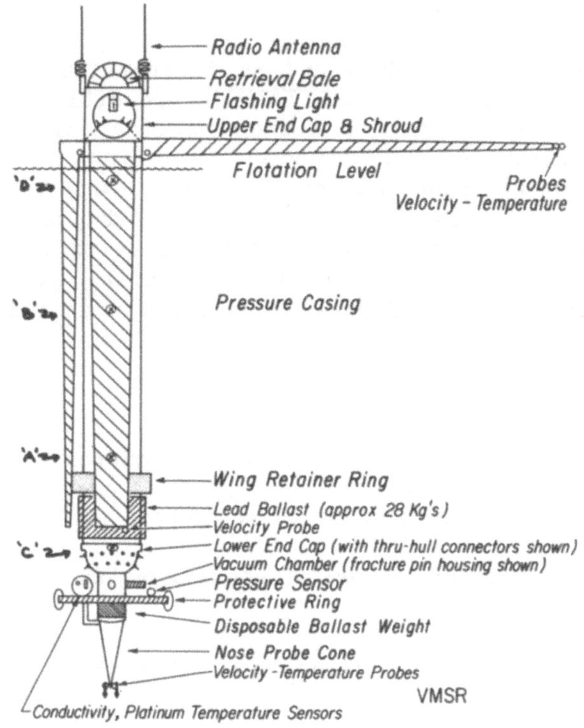

Fig. 1 Velocity microstructure recorder (VMSR).

tilt, as the drag is not symmetric.

Three orthogonal accelerometers, whose responses are recorded digitally for later analyses, are located at point C on Figure 1. The resolution of the accelerometers, limited by least-count noise of digitization, is 0.2 mm s^{-2}. Their long term accuracy is estimated at 1 cm s^{-2} and is due to temperature-induced drifts. At point D on Figure 1 is located a flux gate magnetometer sensing the lateral component of the geomagnetic field perpendicular to the instrumental axial length to within one part in 3000.

The three accelerometer records, the magnetometer record, and the pressure transducer data yield a set of measurements of fall characteristics that, although not complete, are sufficient to infer probable variations in fall characteristics. Ideally, one would like to have an inertial response system for pitch, roll and yaw, and sets of three-axis accelerometers distributed in the instrument (since the instrument does not act as a point mass in response to distributed flows). This seems prohibitively expensive and presently difficult, and may not be necessary. Many interesting questions are addressable with the suite of sensors presently on the VMSR.

Before proceeding to the data and an interpretation, some ambiguities should be pointed out. It is impossible to discriminate lateral accelerations (due to sudden vertical shear) from tilts, using the accelerometer data alone. Two points per rotation from the magnetometer are representative of tilts, and this added information helps in some cases to remove that ambiguity. Centrifugal acceleration of the instrument about the axes of the accelerometer package will be seen only as an apparent tilt. This ambiguity is resolved by portraying horizontal acceleration as tilt only.

Displayed in Figures 2 through 5 are data from four different cases:

(1) winged instrument at 400 m (below wave action)

(2) winged instrument descending from surface

(3) nonwinged instrument at 200 m, showing ballast release and rise characteristics

(4) nonwinged cylinder descending from surface.

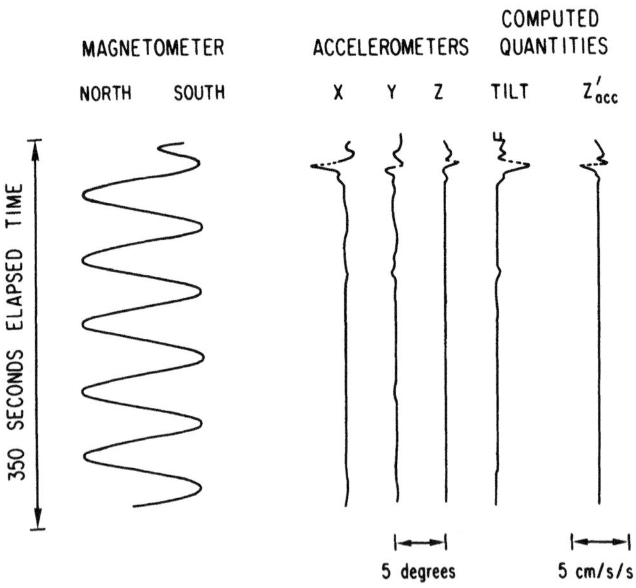

Fig. 2 VMSR data: 4 wings deployed - 400 to 415 m.

The measured data are the orientation of the instrument in the geomagnetic field, and three orthogonal measures of acceleration. Computed data are absolute tilts (always positive), and axial acceleration (down the length of the cylinder) with acceleration anomaly (due to tilt in the gravitational field) removed.

The fall rate of the winged instrument is 8 cm s^{-1}, while that of the nonwinged instrument is approximately 50 cm s^{-1}. Each plot shows 150 s worth of data. In Figure 4 where the instrument drops ballast and ascends, the rise rate is approximately 85 cm s^{-1}.

The two horizontal accelerometers are assumed to be responsive to tilts only, and not to horizontal accelerations of the body, or to centripetal accelerations arising from off-axis rotation. This assumption can be shown to be true for this particular instrument for tilts less than 5°.

The following relation was used to compute absolute vertical tilt:

$$\theta_z = \text{arc cos } [1 - \sin^2(\theta_x) - \sin^2(\theta_y)]^{\frac{1}{2}} \qquad (10)$$

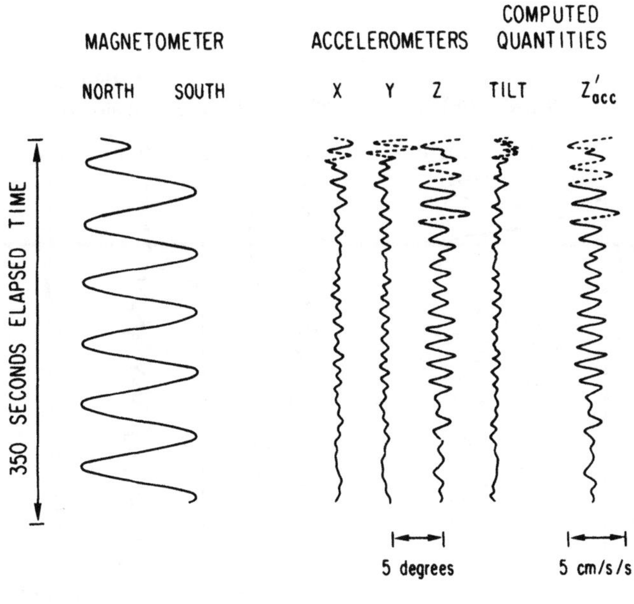

Fig. 3 VMSR data: 4 wings deployed - 10 to 30 m (surface data).

Given the tilt, the apparent acceleration anomaly due to tilt was computed as

$$\delta = 984\ [1 - \cos(\theta_z)] \qquad (11)$$

in cm s^{-2}. The true axial acceleration down the length of the instrument (z-axis) with tilt removed is

$$Z_{acc}' = Z - \delta \qquad (12)$$

The plotted data are then x and y components of tilt, measured z-acceleration (Z), absolute tilt, and axial acceleration (Z_{acc}').

Figures 2 through 5 speak for themselves. It is clear that a winged instrument is most stable well below the surf zone. Characteristic tilts are 0.1° ±0.03°, and the axial acceleration is below least count, or less than 0.2 mm s^{-2}. Below the surf zone the cylindrical body has a wobble associated with it of up to 1°, with a period of about 5 s. This is interestingly the frequency of eddies shed from the instrument as calculated by the Strouhal number. The instrument in this configuration is not streamlined, and is subject to vortex shedding. In shallow water (50 m depth), the instrument

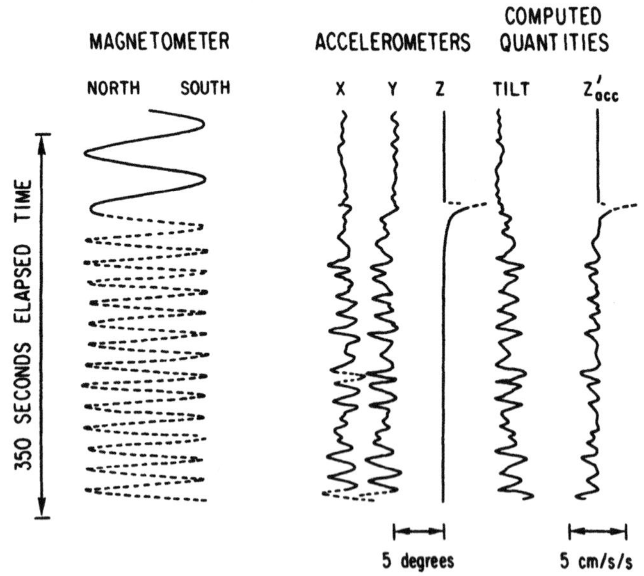

Fig. 4 VMSR data: no wings, turn around at 200 m (200 to 100 m).

is subject to wave action, and this is especially apparent in the z-accelerometer data of Figure 3. The winged instrument has greater vertical acceleration anomalies than the nonwinged instrument, but apparently smaller off-axis tilts. It should be emphasized, however, that the winged instrument is highly damped, strongly coupled to vertical flow, and may be tracking the vertical component of the wave field quite well.

A curiosity appears in Figure 4. Here the instrument is ascending rapidly (85 cm s^{-1}). There are substantial tilts; however, the z-accelerometer is near least-count. Physically this means that the measured axial acceleration anomalies are nearly perfectly cancelled by tilt-induced anomalies. The instrument may be accelerating when it is tilted and decelerating when it is righting itself, and/or experiencing a nontrivial centripetal acceleration.

4. GENERAL REMARKS

Free-fall instruments have demonstrated their unique capabilities (the reader is referred to the list of journal articles on scientific results in the list of references). As observations of heat and salt fluctuations in the ocean are extended to smaller and smaller scales, and as shear-measuring devices look to higher frequencies, some care must be exercised. Free-fall instruments do vibrate; they tend to wobble, no matter how good the stabilization

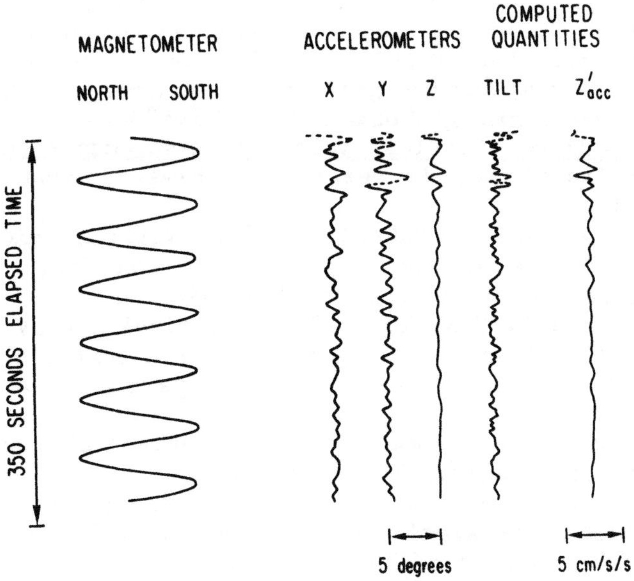

Fig. 5 VMSR data: no wings, 10 to 80 m depth.

is; and errors can creep into the data. It is perhaps worth noting that free-fall instruments tend to follow high shears more quickly than low shears; thus there is a tendency for them to follow convergence fields. How this affects the statistics of heat, salt and velocity microstructure, and the measurement of high modal number shears is not known, but this property should be considered in evaluating descriptions of oceanic properties. Equally important, as scales of variability are pushed to ever higher wavenumbers, is the apparent signal introduced by low amplitude, high frequency oscillations in a smooth gradient field; this must be explored thoroughly. Winged instruments seem to exhibit a vertical oscillation of sporadic occurrence at from 5 to 15 Hz; this is probably due to wing flexure. Although the amplitude may be small, so are the variations in the physical properties of the ocean which are being measured.

ACKNOWLEDGEMENT

The author wishes to acknowledge the helpful and consistent support of the U.S. Office of Naval Research, without whose support this instrument (and many others of its kind) could not have been developed.

SCIENTIFIC REFERENCES

CRAWFORD, W.R. 1976. Turbulent energy dissipation in the Atlantic equatorial undercurrent. Ph.D. thesis, University of British Columbia, Vancouver, British Columbia, 150 pp.

ELLIOTT, J.A. and N.S. OAKEY. 1976. Spectrum of small-scale oceanic temperature gradients. *Journal of the Fisheries Research Board of Canada,* 33: 2296-2306.

GALLAGHER, B. 1976. Vertical velocity measurements in the ocean. Unpublished MS, University of Hawaii, Manao, Honolulu, Hawaii.

GREGG, M., C. COX and P. HACKER. 1973. Vertical microstructure measurements in the Central North Pacific. *Journal of Physical Oceanography,* 3: 458-469.

LEAMAN, K.D. and T.B. SANFORD. 1975. Vertical energy propagation of inertial waves: a vector spectral analysis of velocity profiles. *Journal of Geophysical Research,* 80: 1975-1978.

LUYTEN, J.R. and J.C. SWALLOW. 1976. Equatorial undercurrents. *Deep-Sea Research,* 23: 999-1001.

OSBORN, T.R. 1974. Vertical profiling of velocity microstructure. *Journal of Physical Oceanography,* 4: 109-115.

RICHARDSON, W.S., W.J. SCHMITZ, JR. and P.P. NIILER. 1969. The velocity structure of the Florida Current from the Straits of Florida to Cape Fear. *Deep-Sea Research,* 16: 225.

ROSSBY, H.T. 1969. A vertical profile of currents near Plantagenet Bank. *Deep-Sea Research,* 16: 337-385.

ROSSBY, H.T. and T.B. SANFORD. 1976. A study of velocity profiles through the main thermocline. *Journal of Physical Oceanography,* 6: 766-774.

SANFORD, T.B. 1975. Observations of the vertical structure of internal waves. *Journal of Geophysical Research,* 80: 3861-3871.

SCHMITZ, W.J., JR. 1976. Equatorial undercurrents. *Deep-Sea Research,* 23: 1005-1007.

SIMPSON, J.H. 1975. Observations of small-scale shear in the ocean. *Deep-Sea Research,* 22: 619-627.

WILLIAMS, A.J., III. 1974. Salt fingers observed in the Mediterranean Outflow. *Science,* 185: 941-943.

ENGINEERING REFERENCES

BROWN, J.F. and C.S. COX. 1973. Design of light cylindrical pressure cases. *Engineering Journal*: 35-37.

BURT, K.H. 1974. Autoprobe: an autonomous observational platform for microstructure studies. *Ocean 74; IEEE International Conference on Engineering in the Ocean Environment,* Halifax, Nova Scotia, 1974. New York, IEEE 1974. 1: 171-176.

CALDWELL, D.R., S.D. WILCOX and M. MATSLER. 1975. A relatively simple freely falling probe for small-scale temperature gradients. *Limnology and Oceanography,* 20: 1035-1042.

HENDRICKS, P.J. and G. RODENBUSCH. 1977. Interpretation of velocity profiles measured by freely sinking probes. Woods Hole Oceanographic Contribution #4006, Woods Hole, Massachusetts, 15 pp.

JOHNSON, B.P. 1974. Free-fall microstructure recorder for temperature studies in lakes and oceans. National Needs and Ocean Solutions; *Proceedings of the 10th Annual Conference of the Marine Technology Society,* Washington, D.C., 1974: 651-662.

MORTENSEN, A.C. and R.E. LANGE. 1976. Design criteria for winged-stabilized free-fall vehicles. *Deep-Sea Research,* 23: 1231-1240.

OAKEY, N.S. 1977. An instrument to measure oceanic turbulence and microstructure. Bedford Institute of Oceanography Report Series, BI-R-77-3, May 1977, Dartmouth, Nova Scotia, 52 pp.

RICHARDSON, W.S. and W.J. SCHMITZ, JR. 1965. A technique for the direct measurement of transport with application to the Straits of Florida. *Journal of Marine Research,* 23 (2): 172-185.

SANFORD, T.B. and N.G. HOGG. 1977. The North Atlantic fine microstructure cruise Knorr 52 and Eastward 75-12. Woods Hole Oceanographic Institution Reference #77-11, Woods Hole, Massachusetts (NTIS AD-A038622), 88 pp.

SANFORD, T.B., R.G. DREVER and J.H. DUNLAP. 1978. A velocity profiler based on the principles of geomagnetic induction. *Deep-Sea Research,* 25: 183-210.

WILLIAMS, A.J. 1974. Free sinking temperature and salinity profiler for ocean microstructure studies. *Ocean 74; IEEE International Conference on Engineering in the Ocean Environment,* Halifax, Nova Scotia, 1974. New York, IEEE 1974. 1: 279-283.

38

Towed Vehicles and Submersibles

P.W. Nasmyth

1. INTRODUCTION

In studies on air-sea interaction, a knowledge is required of many parameters, relating to both ocean and atmosphere. The discussion in this chapter is limited to some of the system design characteristics of importance in measurement of subsurface parameters leading to determination of the heat budget and vertical heat flux in the upper layers of the ocean, and the exchange of heat between the ocean and the atmosphere. Other air-sea exchanges - water vapour and momentum, for example - are of perhaps equal importance, but the techniques of measurement are different and are beyond the scope of this section.

To understand and predict the exchange of energy between the ocean and the atmosphere, one would like to be able to determine the total energy available for interaction in a volume of the ocean, and the transport mechanisms, both vertical and horizontal, whereby energy may be carried from depth within the volume to the surface layer (or away from the surface layer) where the interaction takes place. To map completely an ocean volume to the detail required is impractical with present technology and is likely to remain so for the foreseeable future. Techniques have been evolved, however, by which reasonable estimates can be derived. The measurements involved are relatively straightforward in principle and can be performed without much difficulty in the laboratory. In actual measurements at sea, however, signals are often just on the threshold of detection, and interference, frequently very difficult to distinguish from real signal, may arise from many sources. A good deal of engineering effort has gone into system design to minimize interference and to identify that which cannot be eliminated. A great deal of care is still required in calibration of instruments and in interpretation of results, in order to avoid seriously misleading conclusions.

Several classes of systems have been developed, each complementary to the others and capable of obtaining some portion of the information required. Probably the simplest and most economical are the 'free-fall' instruments (treated separately in another chapter) which are dropped from a ship, either with internal recording equipment, or with a slack 'umbilical' by which signals are transmitted to the surface for recording. These systems produce vertical profiles, repeated as frequently as desired (or as practical) in an ocean area. The instrument has to be brought back to the surface, retrieved and recycled between 'drops', and area coverage tends to be slow. Thermistor chains, with thermistors spaced along a vertical or near-vertical cable measure temperature at a number of depths, and when towed by a surface ship give continuous or periodic temperature information along a number of horizontal lines in a vertical plane behind the ship. Temperature contours in the vertical plane can be constructed. This chapter deals with some of the characteristics, capabilities, and limitations of two other classes of systems in which the sensors are (i) towed by a surface vessel, or (ii) mounted on a manned submersible. These two yield essentially the same kinds of information along horizontal or near-horizontal paths. Each has its own strengths and weaknesses, but both are potentially more powerful than the free-fall technique and more versatile than thermistor chains. Because the ocean is usually horizontally stratified, spatial correlations in the horizontal plane tend to be much stronger than in the vertical, and data taken along a path of only slowly varying depth will yield much of the same information as a vertical profile. By varying depth in a cyclic manner about a horizontal mean path it is therefore possible to obtain a mix of vertical and horizontal information at the same time.

Several towed systems have been developed in the U.S.A., Canada, the USSR, and elsewhere. So far as the author is aware, only one system for use on a manned submersible has been developed to the operational stage (in Canada). A second is at an early planning stage in the USSR. The following discussion will be primarily qualitative because in most cases systems have been developed and techniques evolved on an empirical, trial-and-error basis, and only limited quantitative design data exist. There are a few exceptions. 'Batfish', which will be described later, was designed in a systematic way, and a new system just now reaching an operational stage at the University of California, San Diego, is a product of modern aeronautical and hydrodynamic design techniques. Coverage cannot be exhaustive but mention will be made of the more important systems known to the author. In developing a plan for the chapter it has become apparent that it will be difficult to achieve a uniform coverage. Much of the information is of a type not usually found in scientific publications, but in the notebooks and minds of the scientists and engineers involved. Some has been obtained by personal communication with the designers of other systems but, because more detail is available, coverage will be more complete on those systems with which the author has been directly involved.

The fundamental parameters required to be measured are: temperature and temperature gradient, salinity (usually by measuring electrical conductivity), and turbulent fluctuations in water velocity - all as functions of time and space. Measurement of temperature and conductivity with something approaching the necessary precision and resolution is within the capability of present technology and can, with sufficient care, be accomplished even under the sometimes difficult conditions encountered at sea. From temperature and salinity, density may be derived. From temperature and at least one of the space derivatives of temperature, available heat and vertical heat flux may be estimated. Assumptions must be made, however, which are sometimes difficult to substantiate, and confidence in the resulting values cannot be high. By other methods, turbulent energy dissipation and an eddy coefficient of diffusivity may be calculated from actual measurements of water velocity and turbulent fluctuations in velocity. Assumptions must still be made, but these assumptions tend to be more believable and confidence in the result is higher. It is the measurement of velocity which proves most difficult but, because of its importance in deriving a reliable vertical heat flux, a considerable amount of effort has been devoted by several groups to the development of techniques of measurement.

The most commonly used sensor for velocity measurement, and one of the few which offer the frequency response and spatial resolution that one would like to achieve, is a heated platinum film used in the manner of a hot wire anemometer. There are at least three difficulties of some magnitude in using such a device at sea. One arises from the fact that the ocean consists of a very impure mixture of water with varying concentrations of solid or semi-solid particles, mainly of biological origin. For lack of a more precise descriptive term, we refer to them as plankton. The response of the hot film is dependent primarily on the form and thickness of the boundary layer over the film, and when a plankton particle contacts or passes close to the tip of the probe the boundary layer is momentarily disturbed and a sharp transient occurs in the output signal. These 'plankton spikes' are readily identified and, if they do not occur too frequently, can be deleted in the analysis process without seriously affecting the result. The geometry of the probe determines the thickness of the boundary layer over the sensitive element and thereby the frequency response. A heated cylinder (or wire) offers the best high frequency resonse, but a conical probe tip is used by most researchers because it has been found to shed plankton particles more efficiently than any other practical shape. Nevertheless, a particle occasionally sticks on the tip with an effect equivalent to a thickening of the boundary layer, and a sudden and persistent decrease in indicated mean velocity is observed. There are places and depths where at times the concentration of plankton is so high that the data cannot be interpreted in a meaningful way. When the plankton population is reasonably low, however, there are corrective measures which can be taken.

The second and most severe difficulty in successfully operating a hot film velocity sensor at sea results from the effects of mechanical vibration. The probe is sensitive to the relative motion of water over the film, and cannot distinguish between real variations in water velocity in a turbulent field and apparent variations in velocity due to vibratory motions superimposed on its mean forward motion. Some quite elaborate measures have been taken to minimize the effects of vibration, but no complete solution has been found.

The other unresolved difficulty in operating velocity probes of this sort at sea is that the output will often be contaminated to some extent by fluctuations in temperature which almost always accompany (and are frequently generated by) the turbulent activity which we are attempting to measure. Since the operation of the probe depends upon the rate of transfer of heat from the platinum film, it will exhibit some sensitivity to fluctuations in the temperature of the water. The response to a change in mean temperature is easily measured, but the dynamic response to rapid fluctuations in temperature is much more difficult to determine. An estimate of this factor has been made for hot wires or films in air by operating two probes side-by-side at different temperatures and therefore different ratios of temperature sensitivity to velocity sensitivity. The same technique should be possible in water but, so far as the author is aware, it has not been attempted. No definitive measurements have yet been made and we have no reliable quantitative measure of the magnitude of the error.

The velocity sensor is the only component of the system which is critically sensitive to angle of attack. With special care in design and manufacture, it has been possible to produce hot film probes which vary in sensitivity by only about 1% for angles of attack in the range $\pm 10°$. Somewhat larger changes in sensitivity - up to 4 or 5% perhaps - can be tolerated if the associated problems are recognized. Pitching motions of the body, for example, will generate two kinds of interference (and other second order effects): one, a periodically changing sensitivity, which will be interpreted by the system as a fluctuating velocity, and the other a periodic motion of the probe through the water, which will come through also as a fluctuating velocity. The problem is complex and a full discussion would require a chapter in itself. Perhaps it is sufficient to say here that in the design of a platform for this kind of measurement, secondary motions of pitch, roll and yaw must be kept to a minimum and, for reliable interpretation of the output, there must be some means of measuring attitude and residual motions at the position of the velocity sensor(s).

2. TOWED SYSTEMS

The several towed systems which have been developed are essentially similar in their major components, each consisting of a towing ship on the surface, and a towed body carrying a cluster of sensors and containing electronic instrumentation for initial signal conditioning. The interconnecting cable combines a strength element with electrical conductors as necessary to transmit power and command information from the ship to the towed body, and information from the various sensors back up to the ship for secondary conditioning and recording. All systems include a subsystem for body depth control, which must do two things. It must cause the body to follow some preselected pattern in depth, and it must decouple the body from vertical motions of the towing ship which may become both violent and irregular in high sea states. Three techniques have been used. In the simplest case, there is a small group of systems designed to operate at fixed (though adjustable) depths with no control other than passive devices such as mechanical accumulators to decouple the body from the vertical motions of the towing ship. In the second case a steep cable angle is achieved by using a depressor or simply a heavy ballasted body. Body depth is maintained or controlled by continuous adjustment of cable length. A special winch is required. Thirdly, there is a growing family of systems in which depth control is accomplished by means of active control surfaces on the body. The body is 'flown' on the end of a tether. Representative systems will be described in some detail, with briefer mention of a limited number of others, including one or two which combine two or more depth control techniques.

In all cases, the towing cable and depth control system tend to be the factors limiting performance. Because of hydrodynamic drag on the cable, large depressing forces are required on the body in most cases, to achieve the desired operating depths. Cable tension is high. Vibrations are caused by 'eddy shedding' from the cable, and the cable vibrates as a violin string in one or several modes simultaneously. Vibrations are transmitted through the cable to the towed body and in some cases, even with the most elaborate precautions, remain the major source of 'noise' against which some of the measurements must be made. Hydrodynamic fairings of several kinds have been used with considerable success to reduce cable drag and vibration levels, but the problem remains.

2.1 Batfish

The most widely used system of the class using active surfaces for depth control, and the only one, so far as the author is aware, which has been engineered to the production stage, is the one known as 'Batfish', developed by the Bedford Institute of Oceanography (Dessureault, 1976).

The Batfish concept employs a light body with low-drag streamline contour. A set of horizontal wings with cambered airfoil section and variable angle of attack is used to develop downward thrust to achieve operating depths at high speed, or lift as necessary to carry some of the weight of the towing cable when the body is 'flown' close to the surface. The 'standard' model will operate to depths of 200 m at towing speeds of 2.5 to 7 m s^{-1}. An optional 'wide wing' model will reach 400 m at 2.5 to 4.5 m s^{-1}. The standard wing span is 0.75 m (wide wing, 1.25 m); overall length is 1.30 m; and total weight is 70 kg in air, 20 kg in water (wide wing slightly heavier). Two horizontal tails provide lift and stability in pitch. All horizontal surfaces are fitted with end plates to increase the effective aspect ratio and keep the span to a minimum. The major characteristics of the body are shown in Figure 1 and the handling arrangement for launching and retrieval in Figure 2.

The body itself acts as a vertical tail surface providing damping in roll and yaw, while a unique, gravity-operated vertical trim tab on the back of the body provides stability in roll and yaw as well as improved tracking characteristics.

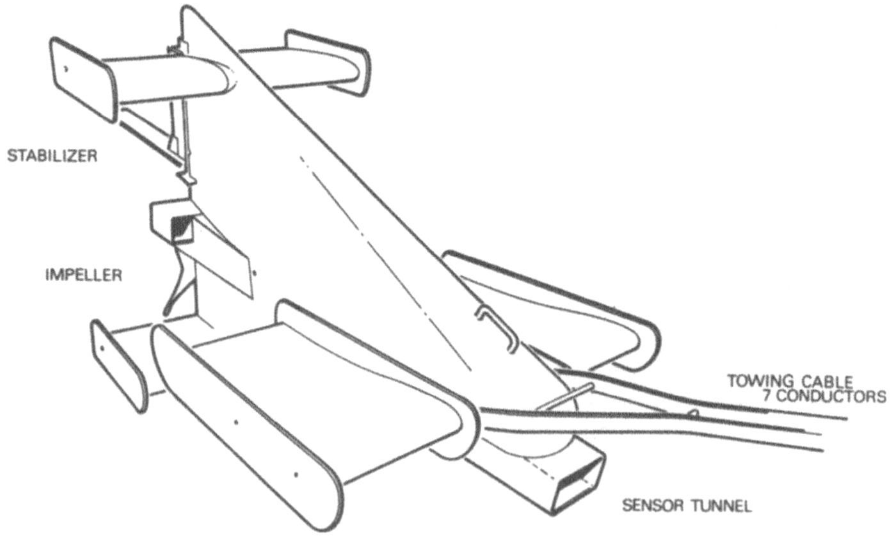

Fig. 1 Batfish body (courtesy Guildline Instruments).

The main wings are operated through an electro-hydraulic servo mechanism by electrical command from the towing ship, with hydraulic actuating power produced by an impeller in the slipstream behind the body. It is stated by Dessureault (1976) that a "constant depth command is maintained within 1 m in most cases" but no reference to sea state is made in association with that statement. In other depth-keeping modes the depth error will increase with ship's speed and with the rate of climb or dive required. For best results and minimum vibration a sectional plastic fairing is used on the towing cable. Under these conditions a maximum depth of 400 m can be attained with 500 m of towing cable, at towing speeds up to 4.5 m s^{-1} (wide wing model). In this operating mode a maximum rate of dive or climb of about 1.5 m s^{-1} (or angle of climb of approximately ±20°) can be achieved. Within these limitations, any desired pattern can be traced in the vertical plane behind the ship.

Batfish was not designed specifically for air-sea interaction studies, but is readily adapted and has been used successfully for some of the measurements required. Sensors may be mounted externally or in an instrument tunnel through the centre of the body. Most commonly, standard sensors of the type used in CTD

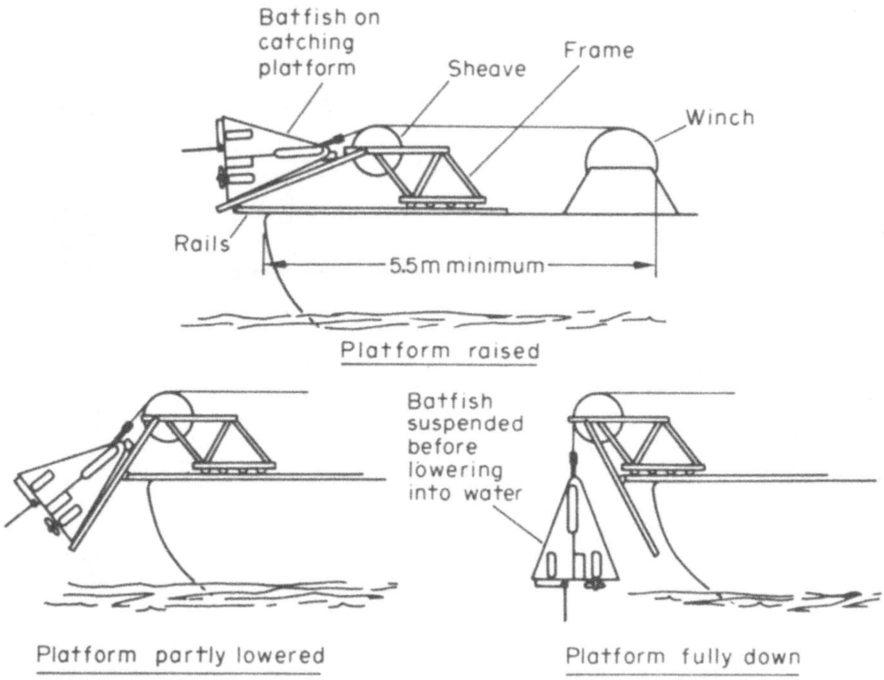

Fig. 2 Batfish handling gear (after Dessureault, 1976).

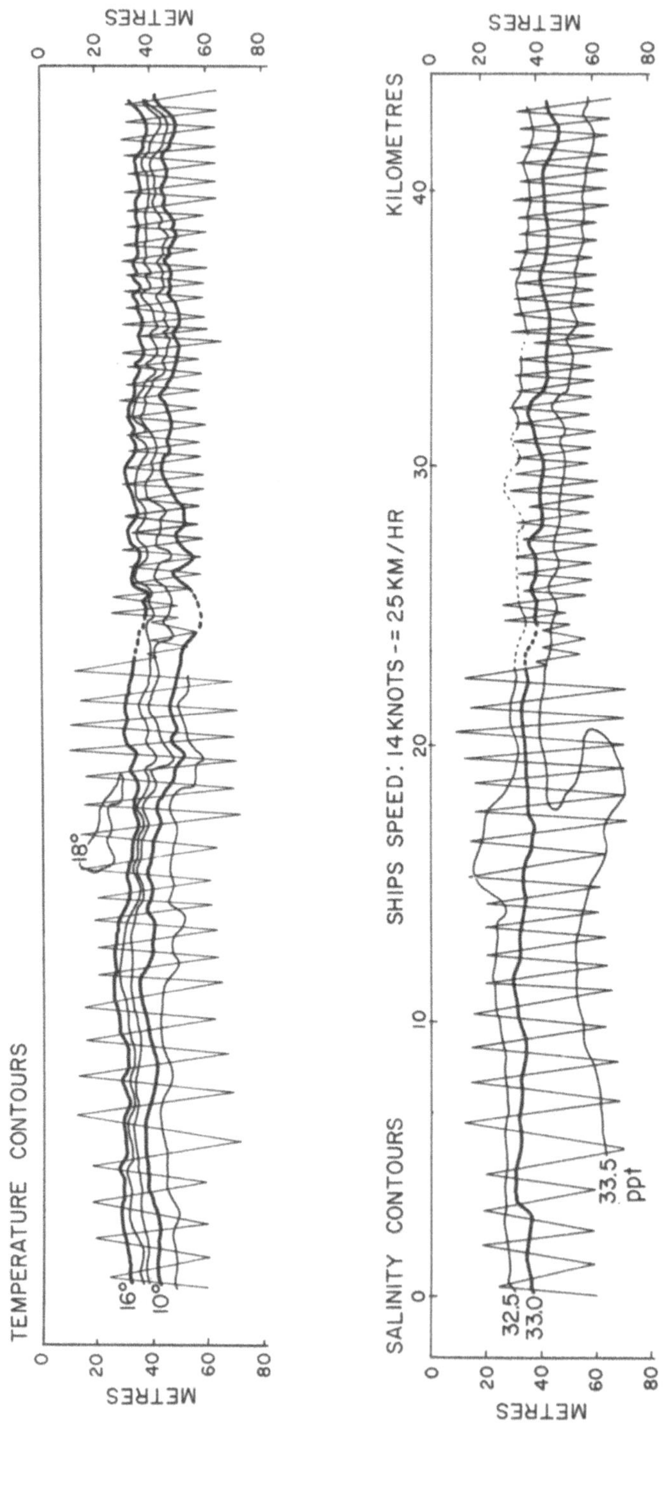

Fig. 3 Vertical contours of temperature and salinity drawn from Batfish data (courtesy Guildline Instruments).

(conductivity/temperature/depth) instruments have been fitted in the instrument tunnel. By generating an undulating or 'saw-tooth' body path, then, contours of temperature and conductivity can be developed in the vertical plane intersecting the ship's path. Similar contours of salinity and density or σ_t may, of course, be derived by computation. A sample of the way the output data may be treated is shown in Figure 3. Batfish was used extensively in this way during the GARP Atlantic Tropical Experiment (GATE) of the Global Atmospheric Research Program (GARP).

Because relatively high towing speeds are required for the wings to develop sufficient lift or depressing force for effective depth control, spatial resolution is limited and the smaller features of ocean microstructure cannot easily be resolved. It is probably not possible for example, to observe the viscous subrange of the turbulence spectrum, and, so far as the author is aware, this has not been attempted.

Dessureault (1976) states that "vibrations in any of the three modes were less than 0.1 g rms for any frequencies between 4 and 50 Hz" at towing speeds up to about 4.5 m s^{-1}, and that "at 20 km hr^{-1} (~5.5 m s^{-1}) the peak-to-peak amplitude was about 0.7 g at 40 Hz," most of which he attributes to hydrodynamic flow around the wings. On the basis of experience at the Institute of Ocean Sciences, these figures would seem to preclude the use of either hot films or 'shear probes' (Osborn, 1974) of which further mention will be made later, for velocity measurement.

2.2 WHOI Towfish

A similar system of the same class as Batfish, but with certain fundamental differences, has been developed at the Woods Hole Oceanographic Institution. The system employs two bodies -- an upper one primarily for depth control, and the other carrying the principal sensors and electronics, on a short faired cable below. Depth penetration to 1000 m is achieved by the weight of the tow cable and the two bodies, and depth variation is induced by hydrodynamic lift forces. A shallow cable angle and low cable tension result in low vibration and good decoupling from the heaving motions of the ship (Katz et al., 1977) but no values are available to the author at this time.

The two bodies are shown in Figure 4a (depth control body) and 4b (instrument body). The depth control body is about 2.3 m long and weighs 160 kg (45 kg in water). Its ring tail encloses an impeller, driven (as in Batfish) by the slipstream, and providing hydraulic power by command from the surface to rotate the entire tail assembly about a transverse axis to achieve hydrodynamic lift. Maximum depth of tow is dependent on ship's speed and cable length.

Fig. 4a WHOI depth control body shown upside down with bottom covering removed (courtesy E.J. Katz).

Fig. 4b WHOI instrument body (courtesy E.J. Katz). The Doppler scattering flowmeter on top near front of body has been used successfully to measure horizontal currents at depth.

With 6000 m of unfaired cable out at a towing speed of 6 knots, a depth of 800 m can be reached and enough lift can be developed to raise the bodies 200 m above that level.

The dihedral wings of the depth control body combined with the weight (68 kg in water) of the instrument body suspended below it provide good stability in roll, and the two-body configuration should result in very good isolation from vibration. Several instrument bodies have been developed, and a wide range of sensors can be used. Most commonly, conductivity-temperature-depth instrumentation (a CTD) has been carried and the depth control system has been programmed to follow isobaric, isopycnal, or isothermal surfaces. A saw-tooth path is also possible, yielding a series of quasi-vertical profiles. The limiting rate of climb or dive is something like 25 cm s^{-1}, for a relatively flat climb angle of ± 4 to 5°.

Using a two-component Doppler scattering current meter (top front of body in Figure 4b) for total horizontal current a stepping cycle has been used to sample current sequentially at a series of depths. By comparison between depths, vertical shear can be estimated. Auxiliary sensors in the depth control body can be used also for determination of mean vertical gradients over the range of body separation (10 to 70 m has been used). For more details see Hess and Nowak (1974) and Katz and Nowak (1973).

2.3 MIT Hydroglider

Another similar system has been designed in the Department of Meteorology of the Massachusetts Institute of Technology -- again with certain fundamental differences. A light body some 2 m long and weighing about 80 kg is slightly buoyant in water. Positive or negative lift can be generated by control surfaces on a horizontal tail plane with electric power from the surface. Roll stability is achieved by ailerons on a main forward wing, driven by mechanical linkages from pendulum weights in wing-tip 'tanks'. Towing speed is in the range 4 to 10 knots. With almost neutral buoyancy the 'glider' has been designed specifically for near-surface measurements, and has been used successfully in the depth range from zero to 30 m. For further information see Morey and Mollo-Christensen (1975).

2.4 University of California Towed Profiling System

A new system which combines the significant feature of this category (i.e. active control surfaces) with the use of passive depressors is under development in the Department of Applied Mechanics and Engineering Sciences of the University of California, San Diego, in cooperation with the Applied Physics Laboratory of The Johns Hopkins University, the U.S. Naval Coastal Systems Laboratory, and a commercial research firm. Two depressors, separated by 3 to 30 m at the lower end of the towing cable, maintain a taut and relatively straight section of cable between them. A catamaran instrument body rides freely up and down this section of cable on a sheave near its centre of mass (and lift) (Fig. 5).

The body, resembling a twin fuselage aircraft, has a main wing and both horizontal and vertical tail fins. The trailing section of the main horizontal tail fin connecting the two halves of the body is controllable by command from the surface, and with this the body can be flown in a 'cycling' or 'profiling' mode between the two depressors. A pair of slack, neutrally buoyant, trailing tethers attach the body to the two depressors, and carry power, command information, and data. These tethers also serve to limit the vertical excursions of the body during controlled profiling, and act as hydroelastic springs, allowing the body, when uncontrolled, to 'float' in a neutral position with minimum accelerations while the main cable and depressors may be pulled up and down by the heave of the towing ship.

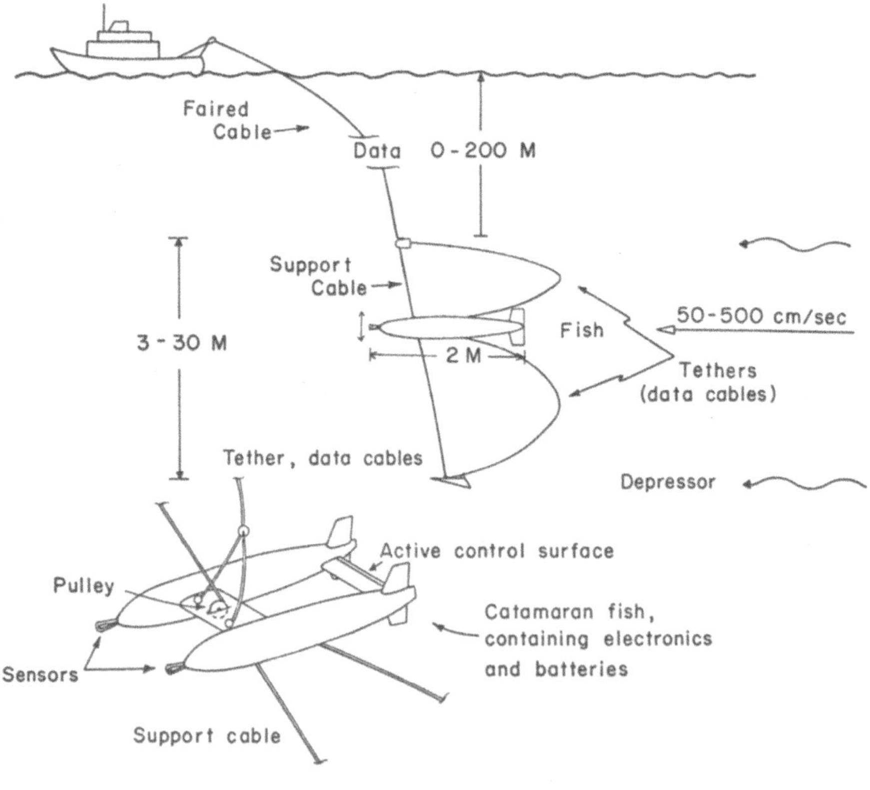

Fig. 5 University of California catamaran system (courtesy C.H. Gibson).

Overall length of the body is 1.88 m, width 1.17 m, and height 0.46 m. It weighs approximately 180 kg in air and is neutrally buoyant in water. The main towing cable and tethers are fitted with a newly designed fairing consisting of a nylon fabric sheath, fastened with a Velcro strip just behind the wire, and easily attached and removed by hand as the wire is paid out or hauled in.

An initial sensor package will consist of a thermistor micro-bead for temperature measurement and a heated thermistor for velocity. A very small conductivity cell (described elsewhere in this volume) will measure conductivity with a spatial resolution of 3 to 4 mm but with relatively poor long term stability. The absolute value of conductivity and slow variations will be determined with a larger cell with better stability but poorer resolution. There will, in addition, be a depth sensor, a propeller current meter, and accelerometers to sense body attitude, motions, and vibrations.

The system has recently undergone initial trials at sea. Few performance data are yet available, but linear accelerations have been quoted as less than 0.01 g rms over the frequency range from 0 to 500 Hz (Gibson, 1977). This is significantly better than has been achieved with any other towed system known to the author.

2.5 USSR Systems

Over the past 10 years a group at the P.P. Shirshov Institute of Oceanology in Moscow has developed a series of towed systems, and has probably covered more miles at sea and collected more microstructure/turbulence data than any other agency. Early systems in the series utilized only passive decoupling between ship and towed body and had no means of depth control other than the length of towing cable put out. The velocity data that we have seen from these early configurations are heavily contaminated by vibration.

More recently the group has developed a more sophisticated configuration, similar in some of its characteristics to the University of California system just described but preceding it in time. A catamaran instrument body (Fig. 6) is used with two auxiliary bodies -- one generating lift above the catamaran, and the other a hydrodynamic depressor at the extreme lower end of the cable as in Figure 7. Decoupling of the catamaran from the higher frequencies of ship motion is good because of the shallow angle of the upper section of cable between the ship and the lift body.

Several configurations have been described in the USSR literature (Ozmidov, 1973, 1975; Paka, 1974) with variations in body design and sensor arrangement. The information available is incomplete, but in most cases the catamaran carries a heated platinum film and either a cold film or small thermistor bead for high resolution velocity and temperature measurement, respectively. One reference

Fig. 6 USSR catamaran (after Ozmidov, 1975). Several sensor configurations have been used. A typical arrangement is shown. Individual sensors are not identified.

mentions drawing some of the glass coating off a commercially available thermistor bead to improve its high frequency response - but no figures are given. A conductivity probe is carried, for which a spatial resolution of 1 cm and noise level equivalent to 10^{-4}°C at constant salinity are quoted. In some cases additional thermistors are spaced above and below the high resolution probes for a measure of local mean temperature gradient, and to "compare the intensity and structure of turbulence with the characteristics of internal waves" (Paka, 1974).

In at least one system configuration the wings of the depressor body are controllable so that the depth of the catamaran can be varied over a 20 m range (Fig. 7) to a maximum depth of 100 m. Without the depth control feature the system has been used to a depth of 200 m.

Vibration and fouling by plankton are both noted as problems relating to the velocity sensor. Using a sectional plastic fairing to reduce cable vibration, a noise level of 0.5 cm s^{-1} is quoted (this is assumed to be a broad band, rms level measured over the range of sensitivity of the instrument from 1 to 200 Hz). Towing speed is in the range from 150 to 600 cm s^{-1}.

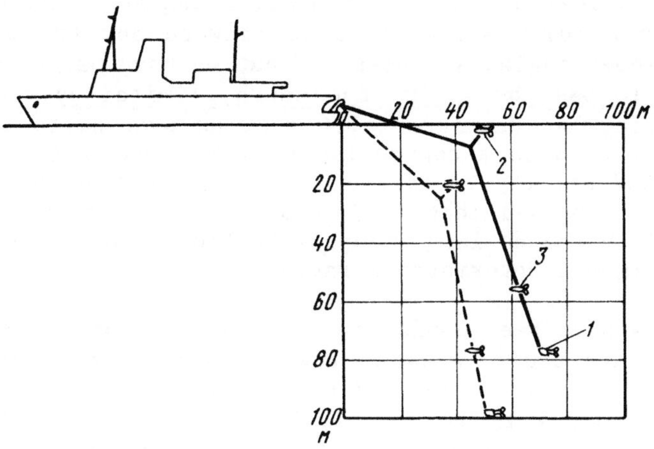

Fig. 7 USSR catamaran towing system, showing depth control (after Paka, 1974).

2.6 Charlie

Probably the most highly developed of the systems employing a servo winch for depth control is the one of which the towed body is affectionately known to his masters as 'Charlie.' It was developed initially by the Pacific Naval Laboratory (now the Defence Research Establishment Pacific) at Esquimalt, British Columbia, in the mid-1960s, and later further developed and used by the Institute of Ocean Sciences, Patricia Bay. Charlie has a ballasted body some 0.80 m in diameter, 3.70 m long, and weighing approximately 1000 kg. The body design (and some of the hardware) is that of one of the early variable depth sonars. It carries a set of fixed crossed stabilizing fins on its tail and a cluster of sensors in front. Figure 8 is a partial view of the body, showing a recent stage in an evolving series of sensor arrangements. The launching arrangement is shown schematically in Figure 9.

A double armoured cable about 2.85 cm in diameter with a multiconductor electrical core connects the body to a hydraulic powered servo winch on the towing ship. This cable acts as strength member for lifting and towing, conveys power and command information down to the body, and brings signals from the sensors back up for recording on the ship. With inputs from a depth sensor on the body and an accelerometer on the stern of the ship, the winch compensates for ship motion and maintains body depth at any desired value; or, in a 'cycling mode', is programmed to repeatedly vary depth at a constant rate over any desired depth interval so that, at constant towing speed, the body follows a saw-tooth path through the water. Depth can be maintained within ±50 cm of the prescribed path in sea states up to 5. Climb/dive angles of 20 to 40° have been used in a fully automatic mode between any two preset depth limits. Steeper angles are possible but we have had no operational requirement to use them. Maximum depth, limited by the length of wire available, has been about 350 m at an average towing speed of 1.50 m s^{-1}, chosen as a compromise between sensitivity of the velocity film and ability to control the ship in winds up to 15 m s^{-1} (55 km hr^{-1}). Beyond that limit the sea state becomes such that the servo winch can no longer cope with the motion of the ship, and system performance deteriorates rapidly.

No flow tests have been conducted, but the assumption has been made that through the usual range of towing speeds from 1.25 to 1.75 m s^{-1} the water at a point three diameters ahead of the maximum diameter of the body will be relatively undisturbed by the pressure field of the advancing body. Putting it another way, the sensor array has been designed so that no component of the body or support structure protrudes outside a conical surface with an apex angle of 20°, centered on the more critical sensors - except thin support struts and the tow-bar.

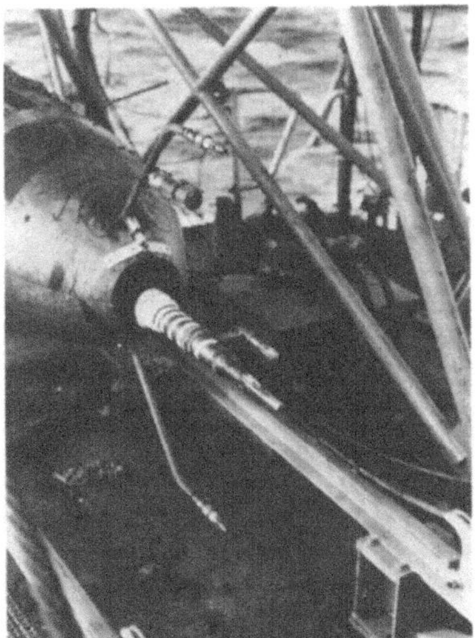

Fig. 8 Forward portion of Charlie's body, showing a recent sensor arrangement. Hot and cold platinum film probes (with protective covering in place) on extreme forward end of vibration isolated nose spar. Flow-through conductivity cell offset to side of nose spar. Two high response thermistors 1 m above and below centre-line. Propeller current meter on upper thermistor mount.

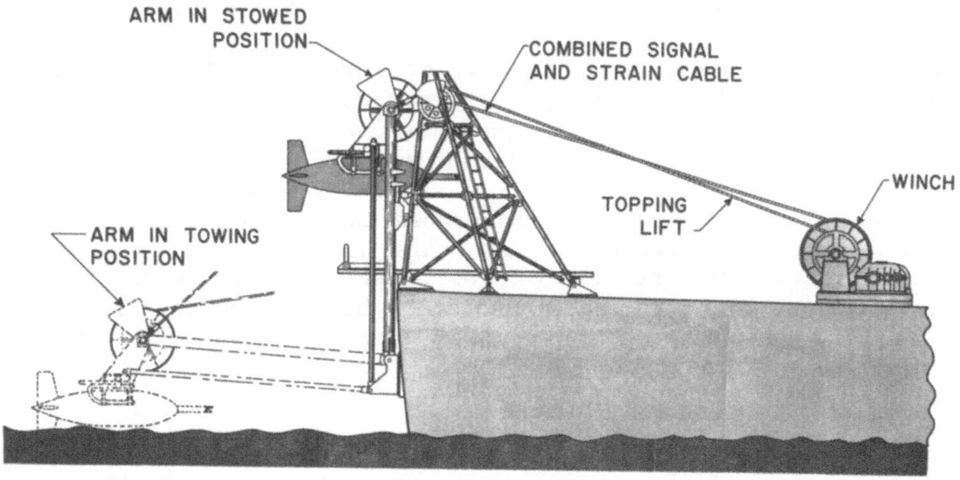

Fig. 9 Handling gear for launching and retrieving Charlie at sea (courtesy Defence Research Establishment Pacific).

The sensor array has evolved through a series of configurations, most recently consisting of two platinum film probes - one a heated film to measure velocity and the other a cold film for temperature - a conductivity cell containing a thermistor for temperature in the computation of salinity, and one or more additional thermistors spaced vertically for a measure of mean local temperature gradient. Auxiliary sensors include a propeller current meter to measure mean speed, a depth gauge, and two accelerometers to measure body attitude and vibration.

The two platinum films, forming the heart of the systems, are placed about 4 mm apart near the centreline of the body. The unheated film measures temperature fluctuations with a frequency response extending to 1 kHz (equivalent to a spatial resolution better than 2 mm) with a broad band rms noise level equivalent to approximately 10^{-4}°C. The heated film measures total velocity and fluctuations in velocity with similar response and spatial resolution, and with a broad band noise level equivalent to about 3×10^{-3} cm s^{-1} while towing in quiet water. The response of both probes is such that essentially the whole of the turbulent energy dissipation spectrum $k^2\phi(k)$ can be observed, where k is a wavenumber defining scales of turbulent motion and $\phi(k)$ is a one-dimensional energy density function. The proximity of the two probes is such that the phase of temperature and velocity can be compared within individual 'eddies' for which k is less than about 3 cm^{-1}. By Taylor's

Hypothesis, time derivatives from both of these probes have been taken as space derivatives in interpretation of the data.

The conductivity cell is of the direct measurement type, using six platinum electrodes in sets of three facing each other across a flow-through channel. The design combines sensitivity and stability with the best spatial resolution possible. One would like to be able to achieve the same sort of spatial resolution that is possible with the two platinum film probes for temperature and velocity (i.e. a few millimetres), but this has not yet been achieved without sacrificing stability. A temperature measurement taken with a thermistor micro-bead immediately behind the electrode array is used in conjunction with conductivity to compute salinity and density with a resolution of about 5 cm. A short section of

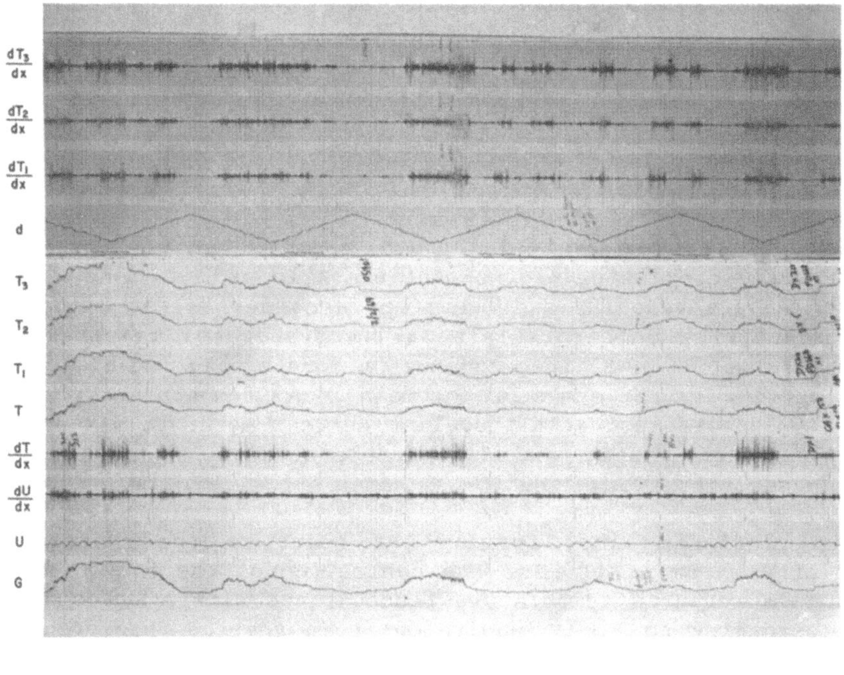

Fig. 10 Sample of raw data from Charlie. From the top: output of three thermistors at top, centreline (in conductivity cell) and bottom, differentiated to emphasize high frequencies; depth showing cycling operation over 10 m range; undifferentiated output from the three thermistors; platinum film temperature probe; differentiated output from temperature probe; differentiated output from hot film velocity probe; undifferentiated velocity; conductivity.

analogue output, recorded in parallel with digital tapes, is shown in Figure 10.

It has been pointed out that there are difficulties associated with the effective operation of a hot film anemometer at sea. There are difficulties also, though less obvious, in very high frequency measurements of temperature. Firstly, the accuracy of both measurements suffers from fouling of the probes by plankton. There are times when reliable measurements are completely impossible. When the plankton population is low, however, we have solved the problem reasonably satisfactorily in the following way. When a plankton particle becomes lodged on the tip of the velocity probe, it causes a sudden and permanent decrease in indicated mean velocity. Such events are identified by a side-by-side visual display of the output of the hot film and mean forward velocity from a propeller current meter. The particle may 'come unstuck' after a few seconds, but it is sometimes necessary to remove it by 'washing' the probe. For this purpose a pump and system of piping have been developed, capable of generating a high velocity reverse flow over the probe tip. A 'wash' is carried out by manual control from the towing ship with a break of only a few seconds in the continuity of data, and without reducing towing speed or otherwise interrupting the recording sequence.

It is reasonable to assume that the temperature probe, being of similar size and shape, becomes fouled as frequently, on the average, as the velocity probe. Again, the effective thickness of the boundary layer is increased. One would expect a drop in sensitivity in the upper part of the frequency spectrum, but there is no recognizable change in visual output. The practice adopted is to wash both probes whenever the velocity probe becomes fouled. By this procedure, if it is assumed that for each probe the time interval between contact with particles that stick is governed by a Poisson distribution and that such events at the two probes are statistically independent, then the temperature probe should be expected to be clean for 70% of the time (Grant et al., 1968). In the most recent arrangement, with only about 4 mm separation, the events are probably not completely independent, and the temperature probe should be clean something more than 70% of the time.

Vibration is the most serious and persistent problem with all towed systems; the towing cable itself is one of the major sources of interference through the generation of eddy-shed frequencies given by

$$n = \frac{0.2\ [\text{velocity (cm s}^{-1})]}{\text{cable diameter (cm)}}$$

and vibration in one or several 'violin' vibrating string modes. In our case we have experienced a band of interference from about 7 to 24 Hz, which includes the fundamental eddy-shed frequency of the cable as well as others from unidentified sources. Vibrating string modes tend to fall in another band below about 2 Hz.

To minimize vibration, it is important, of course, that the body itself should be of a reasonably streamlined form and of sufficiently rigid construction that no parts of it can vibrate appreciably in turbulent flow. The next obvious and probably most effective measure is to apply a hydrodynamic fairing to the towing cable. Sectional plastic fairings, continuous rubber fairings, and 'hairy' fairings have been used with varying degrees of success. The sectional plastic type is probably best and is available commercially now from a number of suppliers. On Charlie's cable we have used a continuous rubber type because it was available. Although it has inherent deficiencies and is not highly recommended for the purpose, it does substantially reduce cable-induced vibration. Both types increase the complexity and cost of the system because they require special winches and special care in handling. In our case we have faired only the lower 100 m of cable so that the faired portion can be laid in a single final layer on the winch drum as the body is being retrieved. Hairy fairings are less costly and easier to manipulate, but are also less effective.

Further improvement in performance may be accomplished by mounting the velocity probe on a vibration isolating device, of which several designs have been tried with varying degrees of success. The higher frequencies can be attenuated quite successfully. To be effective down to, say, 5 Hz, however, requires a mounting which is flexible in three dimensions, with a natural period of 1 to 2 Hz. This in itself is possible, but even with very good control on the towing winch or isolation between ship and body, some components of ship motion will come down the cable in the frequency range from 0.2 to 1 Hz; this requires that the mounting device have a travel of at least a few centimetres, even though the frequencies we wish to eliminate may have a maximum amplitude only in the order of 10^{-2} mm. If, by a large excursion at very low frequency, the device is caused to 'hit its stops' then, of course, many higher frequencies will be generated, with performance deteriorating as sea state increases.

The most recent device used on Charlie has a peak-to-peak free travel of something over 10 cm on all three axes, with a natural frequency of about 1 Hz. In calm water, vibration peaks are barely distinguishable in the spectrum above noise from other sources. Performance is acceptable up to sea state 5 or 6, but beyond that the servo system cannot keep up with the heave of the ship (usually running downwind for minimum pitch and heave). The antivibration mount hits its stops with increasing frequency and the ability to reliably interpret the hot film data is substantially curtailed.

The sensitivity of velocity probes currently in use to changes in mean temperature amounts to something like 10 cm s^{-1} $°C^{-1}$. Correction for changes in mean temperature has been made manually in the past but a new bridge has now been designed to compensate automatically.

To compensate or correct for high frequency fluctuations in temperature is quite another matter. It seems unlikely that the response of a hot film to temperature at higher frequencies would be greater than it is at zero frequency, and one would expect it to be lower due to the effect of the boundary layer. It would not be surprising if the form of the response were similar to that derived for the temperature probe in which the amplitude, related to the zero frequency response, is given by

$$\frac{A}{A_0} = \exp[-\text{const}(\text{frequency/Hz})^{\frac{1}{2}}]$$

for a given mean temperature and mean velocity (Fabula, 1968). The best estimates we have been able to make indicate that the constant in this expression must be greater than 0.25 (if, indeed, the expression is valid at all for the velocity probes), noticeably larger than typical values for the temperature probes we have used on Charlie, which range around 0.1. This is not a surprising conclusion, because the velocity film is farther away from the tip of the cone and the boundary layer will be thicker than in the case of the temperature probe. Neither is it a fortuitous result, of course. It was a primary consideration in the original design of the two probe types -- to minimize the response of each to the unwanted signal of the other kind.

Rigid criteria are difficult to apply, but in practice we place limited confidence in velocity results obtained in the presence of strong temperature microstructure. On the other hand, velocity contamination of the temperature signal should be very small indeed, and we have never had either theoretical or experimental reason to believe that it is ever a matter of any concern.

2.7 Other Systems

Other systems known to the author have been developed in Spain (Cruzado et al., 1970), the USSR (Zhuravle, 1969), the UK (Glover, 1967), and the Federal Republic of Germany (Joseph, 1962). Biological sampling has been one of the major interests behind several of these developments (including Batfish) but each represents a possible approach, adaptable to air-sea interaction measurements. There are still others, but no attempt will be made here at a complete listing.

3. SUBMERSIBLE SYSTEMS

Much has been said about decoupling a towed instrumented body from the motions of a towing ship on the surface, and about vibration of the towing cable and its attendant problems. The ultimate solution would appear to be to have no coupling link with the surface and no cable to vibrate. Some of the free-fall instruments satisfy these criteria, but are limited to vertical profiling. To obtain horizontal information or the sort of horizontal/vertical combination that has proven so useful from towed systems, one can think of several possibilities. A free gliding body, for example, descending along a straight sloping path or wide spiral could produce useful information. There is, in fact, reference to such a vehicle in the USSR literature (Ozmidov, 1975), but little information is available other than a photograph (Fig. 11) and accompanying notation that it carries instrumentation for measuring temperature, velocity and turbulent fluctations in velocity (presumably a high resolution, hot film instrument of the type known to be used by the P.P. Shirshov Institute in Moscow), as well as other instruments for measuring 'fine structure' and 'background' information. With internal recording equipment, it glides from the surface in free flight to some preset depth, drops a ballast weight and glides back to the surface. An almost vibration-free measurement should be possible in this way, but repetitive cycling about a mean depth, as has been done with towed bodies, would require a more sophisticated approach.

An unmanned, self-propelled body, preprogrammed or under remote command from the surface, is another possibility. A body of this type, to be equipped with a variety of sensors including a hot film

Fig. 11 USSR glider (after Ozmidov, 1975).

is under development at the University of Washington, but no test results are yet available.

The closest approach to an ideal platform, particularly for velocity measurements, would seem to be a true submarine and, in fact, a few very successful measurements have been made in this way by the Defence Research Establishment Pacific in 1962 (Grant et al., 1968). Instruments, including a hot film, were mounted through one of the forward torpedo tubes. At low speed (1.50 m s^{-1}) and on battery power, vibration is almost unmeasurable and the submarine motion is not affected appreciably except by very large turbulent eddies in which the period is so long that it can be ignored for most purposes. Full size military submarines, however, are expensive and not readily available, and cannot be seriously considered as an alternative.

One of the small, deep-diving, research submersibles may be the next best choice. They offer complete freedom from the surface and from the problems of towing cables, greater depths than any towed system so far, and great flexibility in instrumentation and mode of operation. They are not, however, without problems of their own.

3.1 Pisces IV

A system of instrumentation has been developed at the Institute of Ocean Sciences for use on Pisces IV, one of the series of Pisces submersibles designed and manufactured in Canada. Pisces IV is designed for a maximum depth of 2000 m but so far has been tested and certified only to 800 m. It carries a maximum crew of three -- usually, for this purpose, a pilot and two scientist/operators.

The first requirement to be met was for a minimum operating speed close to 1.0 m s^{-1} because at lower velocities the sensitivity curve of the hot film probes becomes very nonlinear. Because of the rather blunt hull form, characteristic of most of the small submersibles, it was found in early trials that Pisces IV developed a slow but persistent pitching motion of considerable amplitude at that speed and could not be operated stably.

A series of wind tunnel tests on a model hull revealed that the damping factor in pitch was about 0.008 - clearly too low for stable flight. A set of stabilizing fins was designed, raising the damping factor to about 0.17, with fully satisfactory results both on the model and in full scale. The fins were designed to serve two other purposes at the same time. Twin vertical surfaces provide damping and improved stability in both roll and yaw, while an adjustable vertical trim-tab is used to remove any bias in yaw so that it is not necessary to make course corrections while recording. The vertical angle of attack of the entire fin structure is adjustable to facilitate fore-and-aft trim in level flight and the

Fig. 12 Pisces IV on launching frame with stabilizing fins attached. Instrument spar at lower right.

adjustment of trim in a climbing-diving mode of operation, similar to the cycling mode of some of the towed systems. A view of Pisces IV with fins attached is shown in Figure 12.

The system as presently designed includes hot and cold films, a conductivity cell and propeller current meter for mean forward velocity, all as on Charlie, and several additional sensors. A 'shear probe' as developed at the University of British Columbia (Osborn, 1974) and described elsewhere in this volume is mounted near the hot film. This probe is sensitive to the two cross-stream components of velocity fluctuation, while the hot film is sensitive primarily to the axial component. Three acoustic flowmeters, mounted near the forward end of the instrument spar, measure mean flow past the sensors in three orthogonal directions. Five accelerometers - two in the submersible and three in a pressure housing on the instrument spar - measure vibration and attitude. A recent sensor arrangement is shown in Figure 13.

Vibration again is a serious problem, but of lesser magnitude than any of the towed systems with which the author is familiar. The shear probe is even more subject to interference than the hot film, at least partly because the mass of the probe tip must be accelerated by forces transmitted through the piezoelectric element. Amplitudes, however, are lower and independent of sea state, and the spectral distribution of vibration frequencies is such that they can be more easily dealt with. The main source of vibration is the two geared propulsion motors, and interference appears as a series of very sharp harmonically related peaks in the spectrum below about 60 Hz. No serious attempt has yet been made to design a vibration isolation mounting for the probes, but the design of the instrument spar is such that none of its members has a mechanical resonance near the dominant vibration frequencies, and it does not, therefore, amplify the vibration levels in the submarine hull.

At a normal operating speed of just under 1.0 m s^{-1} the rms vibration level over the frequency range 0 to 150 Hz (which includes all the vibration peaks) has been measured at less than 0.002 g; this corresponds to a peak-to-peak displacement of less than 10^{-2} mm. This is better by at least a factor of 2 than Charlie can do in quiet water, and it does not increase with sea state.

The first comprehensive trials were carried out in November 1976 and certain system deficiencies are now being corrected; a new digital data acquisition subsystem is being constructed which will accept and record up to 32 channels of analogue and 32 channels of digital information in multiplexed form. It will incorporate an automatic change-over between two digital cassette tape units, so that a continuous record can be obtained for as long as the supply

Fig. 13 Pisces IV sensor system. Three orthogonal acoustic flowmeters at top, left, and bottom; conductivity cell, right; shear probe and two platinum film probes at extreme forward end (left, foreground) of array.

of tapes lasts (without this feature it seems that an 'end-of-tape' always occurs during the most interesting portion of a recording sequence). On the basis of further wind-tunnel tests on the model a new instrument spar is being built, which will place the sensors some 50 cm farther ahead than their previous location; tests indicate that, at the new position, cross flow caused by the pressure field of the submersible is no more than 5% of the forward velocity. In full scale operation in an effectively infinite sea, as compared with the restricted flow in the wind tunnel, the distortion of flow should be even lower by perhaps a factor of 2.

With side-by-side operation of a hot film and a shear probe, it becomes possible for the first time at sea to measure all three components of velocity in a turbulent field. In the past most researchers have assumed isotropy because there has been no basis for any other assumption. Turbulence in the ocean, however, usually occurs in thin elongated patches. The Reynolds number cannot be high and there is no reason to believe that the field should be truly isotropic. A measure of isotropy will add significantly to our understanding of the processes involved.

Another major advantage expected of the submersible system is that it will permit operation in both surface and bottom boundary layers of the ocean, where it is difficult to reach with a towed system without very precise depth control and some means of obstacle avoidance. Preliminary trials in both these modes were carried out successfully in 1976.

In 1977-78 the Pisces system has been used successfully in surface and bottom boundary layers, regions of entrainment at the boundaries of advective flow, breaking internal waves, and in regions of flow separation around a submerged obstacle - the underwater equivalent of a mountain range. Data are on hand from which a quantitative measure of isotropy can be derived.

REFERENCES

CRUZADO, A., A. JULIA and A. BALLESTER. 1970. A remotely commanded depressor for continuous physical and chemical analysis. NATO Subcommittee on Oceanographic Research, Technical Report No. 52, Brussels, 30 pp.

DESSUREAULT, J.-G. 1976. 'Batfish' A depth controllable towed body for collecting oceanographic data. *Ocean Engineering*, 3: 99-111.

FABULA, A.G. 1968. The dynamics of towed thermometers. *Journal of Fluid Mechanics*, 34: 449-464.

GIBSON, C.H. 1977. Personal communication. Department of Applied Mechanics and Engineering Science, University of California, San Diego.

GLOVER, R.S. 1967. The continuous plankton recorder survey of the North Atlantic. Symposia Zoological Society of London, 19: 189-210.

GRANT, H.L., A. MOILLIET and W.M. VOGEL. 1968. Some observations of the occurrence of turblence in and above the thermocline. *Journal of Fluid Mechanics,* 34: 443-448.

GRANT, H.L., B.A. HUGHES, W.M. VOGEL and A. MOILLIET. 1968. The spectrum of temperature fluctuations in turbulent flow. *Journal of Fluid Mechanics,* 34: 423-442.

HESS, F.R. and R.T. NOWAK. 1974. A medium-depth controllable towed instrument platform. IEEE International Conference on Engineering in the Ocean Environment Record, 1: 187-191.

JOSEPH, J. 1962. Der Dolphin, ein Messgerät zur Untersuchung von Oberflächennaben. Temperature Schichtungen im Meere. *Deutsches Hydrographisches Zeitschrift,* 15: 16-23.

KATZ, E.J. AND R.T. NOWAK. 1973. A towing system for a sensing package: experiences and plans. *Journal of Marine Research,* 31: 63-76.

KATZ, E.J., W.E. WITZELL, JR., F. HESS and R.T. NOWAK. 1977. Personal communication. Woods Hole Oceanographic Institution.

MOREY, K.A. and E.L. MOLLO-CHRISTENSEN. 1975. Design, development and field trials of a towed instrument glider. Department of Meteorology, Massachusetts Institute of Technology, Unpublished MS, 63 pp.

OSBORN, T.R. 1974. Vertical profiling of velocity microstructure. *Journal of Physical Oceanography,* 4: 109-115.

OZMIDOV, R.V. 1973. Research on oceanic turbulence. P.P. Shirshov Institute of Oceanology, Academy of Sciences, USSR, 16 pp.

OZMIDOV, R.V. 1975. From the Tropic of Cancer to the Antarctic Circle (in Earth and Universe), Academy of Sciences, USSR, 8 pp.

PAKA, V.T. 1974. Hydrophysical and hydrooptical research in the Atlantic and Pacific Oceans. P.P. Shirshov Institute of Oceanology, Academy of Sciences, USSR, 11 pp.

ZHURAVLE, V.F. 1969. A depressor for towed oceanographic instruments. *Oceanology, USSR,* 9: 138-141.

Authors and Addresses

P.M. Austin
MIT Weather Radar Laboratory
Massachusetts Institute of Technology
Cambridge, MA 02139
U.S.A.

T.P. Barnett
Scripps Institution of Oceanography
La Jolla, CA 92093
U.S.A.

B. Bean
Weather Modification Program Office
NOAA Environment Research Laboratory
U.S. Department of Commerce
Boulder, CO 80302
U.S.A.

R.L. Bernstein
Scripps Institution of Oceanography
La Jolla, CA 92093
U.S.A.

H.O. Berteaux
Woods Hole Oceanographic Institution
Woods Hole, MA 02543
U.S.A.

W. Blendermann
Institut für Schiffbau
Universität Hamburg
Lammersieth 90
D-2000 Hamburg 60
West Germany

N.E. Busch
Risø National Laboratory
DK-4000 Roskilde
Denmark

O. Christensen
Risø National Laboratory
DK-4000 Roskilde
Denmark

M.F. Coantic
Institut de Mécanique Statistique
de la Turbulence
12 Avenue du Général Leclerc
13003 Marseille, France

W.R. Crawford
Institute of Oceanography
University of British Columbia
2075 Wesbrook Place
Vancouver, B.C. V6T 1W5
Canada

R.E. Davis
A-030
Scripps Institution of Oceanography
La Jolla, CA 92093
U.S.A.

E.L. Deacon
4 Haldane St.
Beaumaris, BIC 3193
Australia

T.K. Deaton
Scripps Institution of Oceanography
La Jolla, CA 92093
U.S.A.

F.W. Dobson
Air/Sea Interaction
Ocean Circulation Division
Bedford Institute of Oceanography
P.O. Box 1006
Dartmouth, N.S. B2Y 4A2
Canada

R.A. Duce
Graduate School of Oceanography
Narragansett Bay Campus
University of Rhode Island
Kingston, RI 02881
U.S.A.

M. Dunckel
Max-Planck Institut für Meteorologie
Bundesstrasse 55
D-2000 Hamburg 13
West Germany

C.A. Friehe
Department of AMES
Mail Code B-010
University of California
La Jolla, CA 92093
U.S.A.

W.D. Garrett
Interface Chemistry Section
Ocean Sciences Division
Naval Research Laboratory
Washington, DC 20375
U.S.A.

S.G. Geotis
MIT Weather Radar Laboratory
Massachusetts Institute of Technology
Cambridge, MA 02139
U.S.A.

C.H. Gibson
University of California
Department of AMES
La Jolla, CA 92093
U.S.A.

M.C. Gregg
Ocean Physics Department
Applied Physics Laboratory
1013 Northeast Fortieth St.
Seattle, WA 98195
U.S.A.

T. Gytre
Marine Research Laboratory
Nordnesparken 1
N-5000 Bergen
Norway

A. Hågård
Försvarets Forskningsanstalt
P.O. Box 1165
S-581 11 Linköping
Sweden

D. Halpern
U.S. Department of Commerce
NOAA Pacific Marine Environmental
Laboratory (RF 28)
3711 15th Ave. N.E.
Seattle, WA 98195
U.S.A.

L. Hasse
Meteorologisches Institut
Universität Hamburg
Bundesstrasse 55
D-2000 Hamburg 13
West Germany

D.R. Hay
Department of Physics
University of Western Ontario
London, Ont. N6A 3K7
Canada

H. Hinzpeter
Meteorologisches Institut
Universität Hamburg
Bundesstrasse 55
D-2000 Hamburg 13
West Germany

J. Højstrup
Meteorology Section
Risø National Laboratory
DK-4000 Roskilde
Denmark

E.P. Jones
Chemical Oceanography Division
Bedford Institute of Oceanography
P.O. Box 1006
Dartmouth, N.S. B2Y 4A2
Canada

I.S.F. Jones
RAN Research Laboratory
P.O. Box 706
Darlinghurst, NSW 2010
Australia

J.C. Kaimal
NOAA Wave Propagation Laboratory
Boulder, CO 80302
U.S.A.

K.B. Katsaros
Department of Atmospheric Sciences
AK-40
University of Washington
Seattle, WA 98195
U.S.A.

L. Kristensen
Risø National Laboratory
DK-4000 Roskilde
Denmark

L. Lading
Risø National Laboratory
DK-4000 Roskilde
Denmark

R.E. Lange
Scripps Institution of Oceanography
La Jolla, CA 92093
U.S.A.

S.E. Larsen
Meteorology Section
Risø National Laboratory
DK-4000 Roskilde
Denmark

J.R. McCullough
Woods Hole Oceanographic Institution
Woods Hole, MA 02543
U.S.A.

P.W. Nasmyth
Institute of Ocean Sciences
P.O. Box 6000
9860 West Saanich Road
Sidney, B.C. V8L 4B2
Canada

T.R. Osborn
Institute of Oceanography
University of British Columbia
2075 Wesbrook Place
Vancouver, B.C. V6T 1W5
Canada

H. Ottersten
Försvarets Forskningsanstalt
P.O. Box 1165
S-581 11 Linköping
Sweden

C.A. Paulson
School of Oceanography
Oregon State University
Corvallis, OR 97331
U.S.A.

A.M. Pederson
General Manager
Ocean Physics Department
Applied Physics Laboratory,
1013 Northeast Fortieth St.
Seattle, WA 98195
U.S.A.

R. Pinkel
Scripps Institution of Oceanography
Ocean Research Division
La Jolla, CA 92093
U.S.A.

D. Schriever
Max-Planck Institut für Meteorologie
Bundesstrasse 55
D-2000 Hamburg 13
West Germany

O.H. Shemdin
Jet Propulsion Laboratory
Mail Station 183-501
4800 Oak Grove Drive
Pasadena, CA 91103
U.S.A.

S.D. Smith
Air/Sea Interaction
Ocean Circulation Division
Bedford Institute of Oceanography
P.O. Box 1006
Dartmouth, N.S. B2Y 4A2
Canada

R.H. Stewart
Institute of Geophysics & Planetary
Physics A-025
Scripps Institution of Oceanography
La Jolla, CA 92093
U.S.A.

N. Thompson
Meteorological Office
(Met O 8a)
London Road
Bracknell, Berkshire
England RG12 2SZ

G. Tober
DBA Systems, Inc.,
P.O. Drawer 550
Melbourne, FL 32901
U.S.A.

W.A. Vachon
Arthur D. Little, Inc.
Room 20-511
Acorn Park
Cambridge, MA 02140
U.S.A.

J.C. Van Leer
University of Miami
4600 Rickenbacker Causeway
Miami, FL 33149
U.S.A.

R.A. Weller
Woods Hole Oceanographic Institution
Woods Hole, MA 02543
U.S.A.

J. Wucknitz
Meteorologisches Institut
Universität Hamburg
Bundesstrasse 55
D-2000 Hamburg 13
West Germany

Index

N.B.: Types of sensors are listed under the entry "Sensors".

GPSR Compliance

The European Union's (EU) General Product Safety Regulation (GPSR) is a set of rules that requires consumer products to be safe and our obligations to ensure this.

If you have any concerns about our products, you can contact us on ProductSafety@springernature.com

In case Publisher is established outside the EU, the EU authorized representative is:

Springer Nature Customer Service Center GmbH
Europaplatz 3
69115 Heidelberg, Germany

Batch number: 10838053

Printed by Printforce, the Netherlands